MW00761162

Highlights

Derived from Extension of Lorentz Group Covariance to the Complex Lorentz subgroup of GL(4) with Superluminal Transformations:

- Four species of fermions: charged leptons, neutrinos, up-type quarks, down-type quarks.
- Parity violation.
- SU(2)⊗U(1)⊕U(1) ElectroWeak Sector with two WIMP species: leptonic WIMPs and Quark WIMPs. (Dark Matter?).
- Color SU(3).

Derived from an additional GL(4) space-time:

- Four Generations of each Species.
- Fermion and Vector Boson Masses from a non-Higgsian Mechanism analogous to the appearance of a mass constant in the separation of equations for the Robertson-Walker metric.
- Generation Mixing.

∴ The Known Form of The Standard Model plus some additional features such as WIMPs. Its basis is GL(4)⊗GL(4) Space-time, beneath which is matrix Asynchronous Logic.

Philosophy & Logic – The only realistic basis for The Standard Model:

- Resolution of Logical Paradoxes including Gödel's Theorem establishing validity of Logic based on the concept that statements are function calls.
- New Spin ½ Matrix Formulation of Logic with 4-valued and 9-valued IEEE 1164 Asynchronous Logic for Asynchronous Circuits.
- Spin ½ Matrix temporal and spatial Asynchronous Logic leads to The Standard Model fulfilling the Platonic Paradigm of Ideals & Reality.

Interstellar Starship Appication

Our formulation of the Standard Model makes faster than light starships possible. Parts of *All the Universe* whose previous edition led to NASA Starship Program.

Tachyon Quantum Field Theory

Excerpts from *The Origin of The Standard Model* – the book that correctly describes tachyon field second quantization.

Renormalization:

This volume also includes Blaha's Book, *Quantum Theory of the Third Kind*, on Two Tier Renormalization that eliminates all infinities in Perturbation Theory for the Standard Model and Quantum Gravity.

From Asynchronous Logic to The Standard Model to Superflight to the Stars

Third Edition of *Relativistic Quantum Metaphysics*

Stephen Blaha, Ph.D.

Pingree-Hill Publishing

ISBN: 978-0-9845530-3-7

Cover Credits
Cover by Stephen Blaha © 2011. Photographs of the earth and stars courtesy of NASA.

rev. 00/00/05 August 20, 2011

To Margaret With Love

Some Other Books by Stephen Blaha

The Algebra of Thought & Reality: The Mathematical Basis for Plato's Theory of Ideas, and Reality Extended to Include A Priori Observers and Space-Time; Second Edition (ISBN: 9780981904931, Pingree-Hill Publishing, Auburn, NH, 2009)

All the Universe! Faster Than Light Tachyon Quark Starships & Particle Accelerators with the LHC as a Prototype Starship Drive (ISBN: 978-0-9845530-1-3, Pingree-Hill Publishing, Auburn, NH, 2011)

The Metatheory of Physics Theories, and the Theory of Everything as a Quantum Computer Language (ISBN: 097469584X, Pingree-Hill Publishing, Auburn, NH, 2005)

A Complete Derivation of the Form of the Standard Model With a New Method to Generate Particle Masses SECOND EDITION (ISBN: 9780981904900, Pingree-Hill Publishing, Auburn, NH, 2008)

The Origin of the Standard Model: The Genesis of Four Quark and Lepton Species, Parity Violation, the ElectroWeak Sector, Color SU(3), Three Visible Generations of Fermions, and One Generation of Dark Matter with Dark Energy (ISBN: 0974695882, Pingree-Hill Publishing, Auburn, NH, 2007)

Physics Beyond the Light Barrier: The Source of Parity Violation, Tachyons, and A Derivation of Standard Model Features (ISBN: 0974695874, Pingree-Hill Publishing, Auburn, NH, 2007)

A Derivation of ElectroWeak Theory based on an Extension of Special Relativity; Black Hole Tachyons; & Tachyons of Any Spin (ISBN: 0974695866, Pingree-Hill Publishing, Auburn, NH, 2006)

Quantum Theory of the Third Kind: A New Type of Divergence-free Quantum Field Theory Supporting a Unified Standard Model of Elementary Particles and Quantum Gravity based on a New Method in the Calculus of Variations (ISBN: 0974695831, Pingree-Hill Publishing, Auburn, NH, 2005)

Quantum Big Bang Cosmology: Complex Space-time General Relativity, Quantum Coordinates,™ Dodecahedral Universe, Inflation, and New Spin 0, ½, 1 & 2 Tachyons & Imagyons™ (ISBN: 0974695815, Pingree-Hill Publishing, Auburn, NH, 2004)

A Unified Quantitative Theory Of Civilizations and Societies: 9600 BC - 2100 AD (ISBN: 0974685858, Pingree-Hill Publishing, Auburn, NH, 2006)

SuperCivilizations: Civilizations as Superorganisms (ISBN: 978-0-9819049-8-6, McMann-Fisher Publishing, Auburn, NH, 2010)

Available on bn.com, Amazon.com, and other web sites as well as at better bookstores (through Ingram Distributors).

ABSTRACT/PREFACE of the Third Edition

This edition is the latest in a series of books by this author that have appeared in the past ten years that seek to make sense of the form of The Standard Model. Previously The Standard Model was viewed as a hodgepodge of particles symmetries and features that worked experimentally but was only an approximation to a "true" fundamental theory. The overall purpose of this series of books was to show that the form of The Standard Model is based on certain fundamental principles that ultimately emanate from Logic. We show that parity violation, particle symmetries $SU(3)\otimes SU(2)\otimes U(1)\oplus U(1)$, and the spin ½ fermion spectrum including generations are based on an Asynchronicity Principle from 4-valued Logic, the complex Lorentz group, and the (Yang-Mills) locality of the interaction symmetries.

If one considers the various other possible fundamental theories of physics they are all based on assumptions of a fundamental physical construct such as strings or some set of particles that satisfy certain assumed principles that ultimately lead to the form of The Standard Model of Elementary Particles. However, there is no apparent selection principle that would "choose" any fundamental theory as "The" fundamental theory of Nature out of the universe of possible fundamental physical theories.

In the absence of such a selection principle and eschewing fundamental constructs that have no inherent basis for belief, we have proposed Logic, suitably formulated, as the fundamental ground upon which the theory of Elementary Particles should be built.

If this theory is correct, as we believe it is since its many derived features correspond to experiment: parity violation, color SU(3), ElectroWeak $SU(2)\otimes U(1)\oplus U(1)$, four fermion species, four fermion generations (three are known and experiment is currently suggesting a fourth generation), and so on, then it is possible that faster than light starships can be built to travel to the stars. (This is the only mechanism for faster than light travel that does not have a "showstopper.") The key property for faster than light travel is complex quark 3-momenta. Complex 3-momenta enable the singularity at $v = c$ to be avoided. Quark thrust can power a starship faster than the speed of light – a primary practical consequence of this theory.

Returning to the Logical basis of the theory we note that since every physical theory must necessarily be created using Logic it is apropos that Logic also be the foundation of the fundamental theory. In proposing Logic as the basis of the fundamental theory of physics we are aware that various unsettled issues exist in Logic. These issues, logical paradoxes and Gödel's Undecidability

Theorem, have been much discussed with much handwringing as if the cornerstone of reason was in serious trouble. Realizing this problem area we begin with a lengthy discussion of logical paradoxes and Gödel's Undecidability Theorem. We show these "paradoxes" and the Undecidability Theorem are resolved if we understand logic statements to be a form of function whose function arguments have a domain of validity. The verbal and symbolic statements that are at the heart of paradoxes use function arguments that are outside the domain of the functions embodied in their statements. Invalid arguments are the source of these seeming paradoxes and Gödel's Undecidability Theorem.

We do not address other issues relevant to Logic such as the Continuum Hypothesis, Gödel's L construct, Woodin's "Ultimate L" and related matters in set theory and levels of infinity. These issues are not relevant for contemporary Physics. Physicists assume that mathematicians will ultimately resolve these issues and are "content" to use mathematical techniques such as the path integral formulation of quantum field theory which cannot be rigorously established just as Newton and others used calculus although its rigorous development (still in question) did not take place until the 19th century. Thus our construction/derivation of The Standard Model is physically rigorous but not mathematically rigorous.

Most scientists, Logicians and Mathematicians are familiar with mathematical functions that have numerical arguments and calculate a numerical value or set of numerical values. However, those familiar with Computer Science know that a function, in general, has two types of output: its numeric value(s), and a status value indicating whether the computation that the function performed was successfully done or failed. Typically the status value is a zero or one,[1] but it is understood as true or false also. (True – computation successful; false – computation failed for some reason) Thus we categorize functions as being in one of three broad categories:

Types of Functions

A Mathematical Function Producing A Value(s) Only Example: sin function	A Function Producing a Value(s) and a Status Value Example: a Programming Function	A Function Producing a Status Value Only Example: a logical statement

Having reestablished the integrity of Logic we can now use Logic as the basis of the construction/derivation of The Standard Model from Logic and physical postulates. Since the truth of many Logic statements is coordinate position

[1] In many cases the status value could be one of a set of values indicating various stages of success and/or how the computation failed.

dependent we see that coordinates are needed and that Locality is necessarily a part of Logic as it is of Physics.

If we wish to avoid the introduction of secondary time constraints in Asynchronous Logic circuits, then we must use multivalued Logic. Thus we construct multivalued matrix representations of Logic. We find that the four-valued Logic of Fant (for VLSI asynchronous computer chip design) maps to 4 component spinors that are naturally identified as Dirac-like spinors upon the introduction of coordinates and generalized Lorentz boosts. The asynchronous computer circuits that are based on four-valued Logic correspond to the asynchronous time behavior in Feynman diagrams containing particles and anti-particles. (Feynman's portrayal of antiparticles as negative energy particles moving backwards in time is a pertinent symptom of the time asynchronicity of states of Dirac-like particles and antiparticles.) So the choice of four component spinors to represent four Logic states maps to particle-antiparticle time asynchronicity.

For particle dynamics at or below the speed of light the Lorentz group is the appropriate coordinate group. For particle dynamics above the speed of light the complex Lorentz group[2] supplemented by an $SU(3)\otimes SU(2)\otimes U(1)\oplus U(1)$ symmetry group is required. The $SU(3)\otimes SU(2)\otimes U(1)\oplus U(1)$ symmetry group leads to the ElectroWeak and Strong interactions by requiring locality and implementing it with Yang-Mills gauge fields. It also leads to WIMPs – Weakly Interacting Massive Particles – a candidate for Dark Matter.

Another type of asynchronicity is spatial asynchronicity in particle states. For example neutrinos are now known to oscillate between various types as they travel from the sun to the earth. Other types of particles also may be expected to oscillate with other particles of their species. This type of mixing introduces a *spatial* asynchronicity between particles of a generation as the particles move through space.

Thus we are led to introduce another four-valued Logic to handle this asynchronicity. We identify these four values as corresponding to four generations of particle species – three of which are presently known.

Time asynchronicity gives us particles and their antiparticles through Dirac-like equations. Spatial asynchronicity gives us the generations of particles. Thus we are lead to the form of the fermion spectrum through these considerations, and other considerations, that lead to $SU(3)\otimes SU(2)\otimes U(1)\oplus U(1)$ symmetry.

Thus we see a logical development from Logic to the form of The Standard Model. This new edition provides more detail on the construction of The Standard Model particularly with respect to the role of Asynchronous Logic in leading to Dirac-like equations of motion and particle generations.

The absence of Higgs particles in LHC experiments at the current time leads us to propose an alternate mechanism for the generation of fermion and vector boson masses. It has more detail than the discussion in the previous edition.

---

[2] Streater and Wightman pointed out that the complex Lorentz group was required for the proof of the CPT Theorem in axiomatic Quantum Field Theory.

Lastly, we describe the unique possibility of faster than light travel by taking advantage of the complex spatial 3-momenta of quarks (in quark-gluon plasmas) to evade the conventional limit of the speed of light on velocities. This approach to superlight travel requires only incremental advances in particle beam and magnet technology. Thus it has no "showstoppers" unlike all other approaches to superlight travel. Consequently, our Standard Model opens the door to the universe for Mankind.

New and significantly revised sections and chapters are marked with an asterisk. The previous edition was entitled "The Standard Model's Form Derived from Operator Logic, Superluminal Transformations and GL(16)."

PREFACE to the Second Edition

Charles Darwin once remarked that he was tired of continually creating revisions and new editions of *The Origin of the Species*. In the present case this author regrets to present a new edition of a work that has evolved over the past seven years or so to a complete derivation of the form of The Standard Model from quantum theory and the extension of the Theory of Relativity to superluminal transformations. The much derided form of The Standard Model can be established from a consideration of Lorentz and superluminal relativistic space-time transformations. So much so that other approaches pale in comparison.

This edition contains a major extension of the derivation of the Standard Model presented in the most recent edition, *Relativistic Quantum Metaphysics*. We now can show that the SU(2)⊗U(1) Weak Interaction sector follows directly from the extension to superluminal (faster-than-light) Lorentz transformations. ElectroWeak symmetry actually has an SU(2)⊗U(1)⊕U(1) symmetry that naturally include WIMPs (Weakly Interacting Massive Particles), a candidate for Dark Matter. Together with the derivation of color SU(3) from superluminal Lorentz transformations presented in previous editions we see that most of the form of the Standard Model is dictated by an extension of Special Relativity to include Superluminal transformations. (The word "metaphysics" appears in the title of the prior edition because we show that a new formulation of Logic – Operator Logic – has a matrix formulation that can be viewed as a precursor to the Dirac equation. If we assume the truth or falsity of a statement is local – dependent on the spatial location and time – as it is in general, then combining the new matrix form of Logic with space-time coordinates, and generalized Lorentz transformations, leads to the Dirac equation and other spin ½ dynamic equations. This sequential development implements Plato's view of Ideas and Reality.)

This edition expands on the presentation of the prior edition in Part 1 to include a derivation of ElectroWeak SU(2)⊗U(1)⊕U(1) symmetry with WIMPs and a derivation of broken SU(4) four fermion generations as well as a non-Higgsian derivation of fermion and ElectroWeak gauge boson masses.

In addition, it also has *The Quantum Theory of the Third Kind* in part 3. This book, and earlier books, describes our renormalization theory in detail. This renormalization theory yields finite results (No Feynman diagram calculations diverge!) in The Standard Model and Quantum Gravity calculations to all orders in perturbation theory.

The logical order of the chapters was changed to place the particle physics part at the beginning. The discussion of Logic (which leads to particle theory) is in Part 2. Thus Part 1 of this book consciously begins with chapter 15.

Part 2 contains the description of Operator Logic, Probabilistic Operator Logic, Quantum Operator Logic (for q-number statements). It is a revision of chapters 1 – 14 of the first edition with new material.

PREFACE to the First Edition
Entitled
Relativistic Quantum Metaphysics

A reading of a goodly number of the many books on the various branches of Philosophy and Metaphysics leads this author, a physicist by way of education and an investigator in many areas of science, economics and history, to conclude that Philosophy and Metaphysics have yet to come to terms with modern scientific research in physics, mathematics, logic, biology, computer theory, and cognition. Part of the problem – particularly as regards studies of Reality – is a lack of understanding, or incomplete understanding, of what has become the language of these areas of endeavor – mathematics. Mathematics is a requirement to appreciate what we have learned of Reality not only in physics but increasingly in biological areas related to understanding the human Mind and its relation to Reality. We see this, for example, in computer networks developed to simulate neural processing, learning and other aspects of processes taking place in the human mind.

Failing to understand mathematics and not completely realizing the implications of modern science particularly quantum theory and the theory of relativity, philosophers and metaphysicians revert to verbal descriptions that answer some questions but do not truly give a correct view of Reality as it is known today from scientific investigations. These efforts make sense for everyday phenomena but at the quantum level and at high velocities these efforts fail to come to grips with Reality. *And so the claims of universality of existing philosophy and metaphysics are simply not true.* The discussions of Reality are simplistic (although sometimes subtle) in view of our current knowledge of the universe, use terminology that is not well-defined and often vague, and often engage in verbal gymnastics that do not grasp the essence of aspects of Reality. Despite these shortcomings there are genuine grains of wisdom in philosophical and metaphysical studies.

Our plan in this book is to examine the true core of philosophy and metaphysics, taking account of quantum and relativity theory as it applies to physical Reality, and to develop a line of reasoning that ultimately leads us to Reality as it is currently understood at the most fundamental level – the Standard Model of Elementary Particles and General Relativity. *We will derive the form of The Standard Model in detail showing the origin of parity violation, the Strong interaction and the origin of its peculiar symmetry.*

This book develops new formalisms for Logic that are of interest in themselves and also provide the Platonic bridge to Reality. The bridge to Reality will then be explored in detail. We anticipate that the current "fundamental" level of physical Reality may be based on a still lower level and/or may have additional aspects remaining to be found. However the effects of certain core features such as quantum theory and relativity theory will persist even if a lower level of Reality is found, and these core features suggest the form of a new natural Philosophy and a new Metaphysics of physical Reality. We have coined the phrase "Relativistic Quantum Metaphysics" for this new metaphysics of physical Reality. One might also describe our generalization of the encompassing Philosophy as Relativistic Quantum Philosophy.

It will be objected that vast areas of Philosophy and Metaphysics exist outside of the domain of physical Reality. To that objection we answer that most of those areas – the materialistic areas – emerge ultimately from physical Reality. As science evolves it is becoming increasingly clear that all events and phenomena in our universe ultimately follow from fundamental physical Reality. Some phenomena, such as the processing of abstract concepts in the human mind, are very distantly related to fundamental physical Reality, and it may be difficult (and perhaps presently impossible) to derive a complex phenomena from fundamental physical Reality. Nevertheless the connection exists despite our inability to derive it. So we regard all things (except the spiritual) as emanating from fundamental physical Reality. Our present inability to derive connections is a matter of happenstance; but the principle of a unitary Reality based on fundamental physical Reality has the force of historical trends behind it as the developments in the twentieth century clearly show.

This book starts by describing aspects of Philosophy and Metaphysics relevant to the study of current physical Reality introducing a new natural Philosophy and a new Metaphysics of physical Reality based on quantum theory: Quantum Metaphysics. Part of this development are new Logics, Operator Logic, Probabilistic Operator Logic, and Quantum Operator Logic, developed in earlier books by this author. Using them we are led to develop a connection to the beginnings of The Standard Model of Elementary Particles. Then we derive the form of The Standard Model in detail based on the new Operator Logics. This derivation is an expanded version of Blaha (2008) and (2009). Metaphysical implications of The Standard Model are then considered – Relativistic Quantum

Metaphysics and a Generalized Metaphysics for all possible types of universes. (The Operator Logics described here are revisions and expansions of the Operator Logic formalisms described in earlier books by this author. They are the same as Blaha (2010a) and the first part of this book through chapter 13 is Blaha (2010a) reproduced for the sake of completeness.) While mathematics is essential in the latter stages of the book we have tried to present it with sufficient text discussion to make what it is doing understandable to the non-mathematical reader. Generally we will avoid using the jargon of Philosophy, Logic and Physics as much as possible. But the second part of this book is necessarily mathematical. In it we describe the basis of the form of The Standard Model in a more fundamental set of principles.

CONTENTS

TABLES AND FIGURES

*0. The Form of this Book

This book consists of five parts. In this chapter we overview the various parts of the book to guide the reader as we anticipate a broadly based readership from disciplines ranging from Logic to Philosophy to Physics to persons interested in superlight travel to the stars (which is based on The Standard Model presented here.) The various parts are more or less independent. So a physicist can skip part 1 (Logic) and start reading in part 2 and parts 4 and 5. Similarly a Logician can focus on part 1. A space scientist might choose to read part 3. We hope the reader will be happy with this flexibility.

Part 1. Logic, A Resolution of its Paradoxes, A New Formalism, and The Mathematical Transition To The Standard Model Via Asynchronous Logic

This part begins with an epitome of the author's view of Epistemology and Metaphysics in the light of Physics advances in the past 100 years. It then proceeds to analyze the paradoxes of Logic and shows that they are all resolvable, including Gödel's Undecidability Theorem if one treats statements as functions with a specific domain for arguments (subjects). Having established the validity of Logic it proceeds to develop an operator and a matrix formulation of 2-valued Logic, and then of multi-valued Logic. Probabilistic and Quantum Logics are defined. Then we consider Logic in relation to physical Reality. In chapter 14 we discuss asynchronous Logic – particularly Fant's 4-valued Logic – and show how time asynchronicity leads from 4-valued matrix Logic to Dirac-like equations which have an inherent asynchronicity when interactions are introduced with perturbation theory expansions. This asynchronicity is associated conceptually with the picture of positrons as negative energy electrons going backwards in time. Another asynchronicity is associated with spatial displacements – the ability of particles such as neutrinos to oscillate between generations as they travel through space. This asynchronicity is the basis of the four generations of fundamental fermions. Thus we pass from Logic (Asynchronous) to the fundamental features of fermions.

Part 2. The Construction/Derivation of The Standard Model

This part describes the construction of The Standard Model of Elementary Particles from a set of postulates that begins with a complex space-time postulate and the complex Lorentz group. It develops four species of fermions, each having a Dirac-like equation, from four types of complex Lorentz group boosts that boost from a mass with zero 3-momentum (a rest frame) to a frame where the energy is either a real or imaginary number.[3] The four species of fermions are identified with neutrinos, charged leptons, up-type quarks and down-type quarks. The differentiating factor between leptons and quarks is that leptons have real 3-momenta and quarks have complex 3-momenta. These new features emanate from the extension of the Lorentz group to superluminal (faster than light) transformations. And they imply that neutrinos and one of the quark species – we assume the down quark species – are tachyonic. We define free field theories for each species with our light front formulation of tachyon quantum field theory for neutrinos and down-type quarks – the only valid formulation of which we are aware. Then we define lepton and quark doublets that are initially covariant under the four types of Lorentz boosts. Then seeing that any physical transformation must be between real coordinates and yet some complex Lorentz group transformations relate real coordinates to complex coordinates we introduce a new group to transform complex coordinates, so generated, to real coordinates and find the group has the form SU(2)⊗U(1)⊕U(1). The combination of complex Lorentz transformations and these new transformations leads us to redefine the doublets making them into triplets. The third component of the triplets we identify with WIMPS (Weakly Interacting Massive Particles). Thus we now have a leptonic WIMP and a quark WIMP (an SU(3) singlet) species raising the total number of species to six. Subsequently we show that color SU(3) naturally arises for quarks ultimately due to thei complex 3-momenta. Upon requiring local SU(3) and SU(2)⊗U(1)⊕U(1) symmetry implemented by Yang-Mills fields we obtain the *form* of The Standard Model when we introduce four generations of quarks from translational asynchronicity. Symmetry breaking and generational mixing can result from Higgs fields. If Higgs particles are not found then we propose an alternate mechanism suggested by the appearance of a separation constant with the dimension of mass squared in the Robertson-Walker metric of General Relativity. Thus through a sequence of postulates we obtain the form of The Standard Model within the framework of Quantum Field Theory. The values of interaction constants, masses, and mixing constants is not determined.

Part 3. SuperLight Travel To The Stars Based On The Standard Model

In 2008 and 2009 we wrote two books proposing the research and development of a starship capable of journeys to other stars and galaxies in short periods of time. In 2010 and 2011 we provided a more detailed description of superlight travel in two new books. In these books we considered particle accelerators to accelerate faster than light particles using electromagnetic fields. We also examined the possibility of colliding compacted macroscopic spherules (little spheres) of particles rather than individual particles to produce macroscopic

---

[3] An imaginary energy can be transformed into a real number as we will show.

globules of quark-gluon plasma. The spherules would each be compacted to almost nuclear density by laser or particle beams just prior to collision. After the quark-gluon globule is produced its expansion is confined by magnetic fields, lasers, or particle beams to "one" direction – out the rear of the starship. The result is a quark "ion" drive engine with an exhaust speed greater than the speed of light. This engine can take Man to the stars with short travel times. Speeds of 5,000c and beyond are feasible as well as travel to stars and galaxies in months or a few years.

The key factor that makes the superlight, quark drive possible is complex spatial momenta – the reason faster than light quarks are possible.[4] The Standard Model proposed here has quarks and gluons with complex spatial momenta. Thus it is the key ingredient needed for superlight starships. All other approaches to travel to the stars either are based on undoable gravity effects or require travel times of the order of generations.

This part of the book will present some of the details of superlight dynamics – perhaps the major practical benefit of our Standard Model.

Part 4. Tachyon Quantum Field Theory – Chapters 3 - 5 of *The Origin of the Standard Model*

This part of the book contains a detailed description of tachyon quantum field theory. Tachyon quantum field theory is quite different from conventional quantum field theory. Among other unusual features it requires the use of light front coordinates. Neutrinos and down-type quarks are assumed to be tachyons in our Standard Model.

Part 5. Two-Tier Renormalization – *Quantum Theory of the Third Kind*

The renormalization of quantum field theories has been a recurring issue. The peak of the success of the standard renormalization program was the successful proof of the renormalizability of ElectroWeak theory by t'Hooft and Veltmann. This success is now somewhat in jeopardy since Higgs bosons – an essential component of the proof – have not as yet been found. Many years ago this author developed a new form of quantum field theory that possessed many of the features of conventional quantum field theory but generated Feynman propagators for all particles with a gaussian exponential factor in addition to the usual factors. This factor serves to eliminate ultraviolet infinities in loop integrations. Consequently all Feynman diagrams to any order were finite in any theory with a polynomial lagrangian. Traditional power counting rules to identify divergent

---

[4] Current ion colliding beam experiments that produce quark-gluon plasma display a very rapid transition from the collision state to the plasma state that is not understandable according to theoretical analyses based on sublight quarks and gluons. A re-analysis assuming faster than light quarks and gluons as well as sublight particles might resolve this mystery.

diagrams were no longer relevant due to gaussian exponential damping. Thus with or without Higgs bosons ElectroWeak theory, and indeed The Standard Model in the usual formulation or in our formulation with complex quark spatial momenta, are all finite term by term in perturbation theory. *An important example of this finiteness is the finiteness of the axial anomaly triangle diagram in Two Tier Quantum Field Theory.* Also Quantum Gravity of the Einstein variety, and other similar forms, are also divergence free in a weak field expansion around a flat spacetime. Thus a unified theory of The Standard Model and Quantum Gravity can be obtained by simply "gluing" them together – perhaps in a vierbein formulation. Such a theory would be finite order by order in perturbation theory.

This part of the book reprints *Quantum Theory of the Third Kind* for the sake of completeness and for the convenience of the reader including a discussion of finite Quantum Gravity.

Part 1. Logic, A Resolution of its Paradoxes, A New Formalism, and The Mathematical Transition To The Standard Model Via Asynchronous Logic

1. Epistemology

This chapter is the beginning of a study that leads to a contemporary understanding of physical Reality at its most fundamental level – the Standard Model of Elementary Particles combined with the General Theory of Relativity.[5] Some primary issues that we will examine in the next few chapters with a view towards our goals are Epistemology – the nature, scope and limitations of human knowledge; Metaphysics – the nature of ultimate Reality; and Logic – the true relationships between the parts of ultimate Reality. The author believes that the impact of Quantum Theory and the Special Theory of Relativity on these areas of study has yet to be fully appreciated. One of our goals is to provoke a thorough reanalysis of these subjects from an Operator Logic viewpoint.

1.1 Epistemology

Epistemology is the study of human knowledge. Since our goal is to understand one specific area of knowledge – physical Reality – our approach to epistemology will be limited to the nature, scope, and limitations of human knowledge of physical Reality. Physical Reality can be viewed as consisting of three parts: cosmological physical Reality (the universe in the large), everyday earthly (planetary) physical Reality, and quantum physical Reality (Reality at the quantum level on the very smallest distance scales). These parts of Reality conveniently break up into distance scales: very large distances – cosmological distances; the distances and sizes of everyday experience – ordinary distances; and very small distances – quantum scale distances. Each of these areas of human knowledge has a different epistemology. Almost all work in the epistemology of knowledge of physical Reality has been at the level of everyday earthly Reality.

Before turning to a study of the epistemology of very large-scale and very small-scale knowledge we should understand the makeup of epistemological theory. One way to begin is to consider an isolated human mind with absolutely no sensory inputs, and then to consider what physical senses are available and how

[5] It is clear that the current "fundamental" level is either an interim level preliminary to a deeper level or is only part of a broader fundamental level. The major evidence for this view is Dark Matter and Dark Energy – phenomena that have appeared in Cosmological attempts to account for observed astronomical data. Nevertheless the Standard Model of Elementary Particles has repeatedly been proved to be an accurate description of known earthly experimental data and most cosmological data and so will survive.

their capabilities and limitations affect the ability of the human mind to acquire knowledge and process it.

So let us first consider an awake human mind (or any intelligent mind) in *total* isolation with no prior, or current, sensory input and ask what sort of knowledge it can obtain. First it is clear that any knowledge that it might have would be the result of imagination.[6] Conceivably the mind could form some concepts. Perhaps the simplest concept would be number. But not having anything to count, why should it conceive of number.[7] Similarly, all other concepts derive from sensory perceptions and so it is difficult to believe that any concepts would occur to the isolated mind except possibly concepts/ideas of a genetic origin – if such exist.

But there is a difficulty in the mind itself. Modern research has established that sensory inputs affect the wiring of the human mind as the mind responds by processing the inputs. So a totally isolated mind could be viewed as an "unwired" mind. The thought processes that would take place in such a mind would appear to be haphazard. Thus we are led to conclude that an isolated mind without sensory input and without inherent genetic information would in general have no knowledge and be incapable of organized thought. The young Helen Keller, who lacked hearing and sight but had a sense of touch, is an (imperfect) example of an isolated mind. Ms. Keller was not rational until after much schooling.

If we now consider the human mind with its usual physical senses: sight, hearing, smell and touch, then a totally different view emerges. For the senses, by providing input to the mind, give it items to consider, behavior to analyze, and motives (self-preservation and the acquisition of wealth and power) for survival.

1.2 Origin of Human Knowledge of Physical Reality

The nature of human knowledge of physical Reality at the everyday level of experience has been a subject of discussion for thousands of years. We will not enter to that still ongoing discussion other than to say that it is largely based on direct observation using the human senses of physical Reality.

Rather we will examine the nature of human knowledge at the quantum and cosmological levels. In both cases the nature of human knowledge is observation of physical Reality through instrumentation of varying degrees of complexity, which is interpreted using a theoretical analysis that can be quite complex and often uses probabilistic interpretations. As a result the nature of human knowledge in these areas of Reality is indirect. Theoretical analyses stand between our knowledge of Reality and experimental results. A change in the

---

[6] The possibility of inherent genetic "knowledge" present in the mind (such as Jungian archetypes) is not excluded.

[7] Even for people who have sensory input the concept of number is not easy. For example members of some tribes cannot count beyond 1 or 2.

analysis of an experiment usually leads to a different "knowledge" of the corresponding part of Reality.

1.3 Scope of Human Knowledge of Physical Reality

Thus the scope of knowledge of Reality is in a sense limited by our ability to analyze what we are observing. And so quantum and cosmological knowledge of Reality has a layer of theory between the mind and observational data. In some areas of science such as elementary particle physics the theoretical analysis of the data is extremely complex and problematic. As our ability to do experiments to determine features of physical Reality becomes greater the theoretical analysis will correspondingly become more critical.

1.4 Nature of Human Knowledge of Physical Reality

In the three areas of knowledge of physical Reality we see a pattern or structure of knowledge. First one has a set of facts relating to some aspects of physical Reality. Then these facts are usually organized into a theory based on principles or axioms that presumably are "deeper" then the amalgamation of facts – from which the facts may be derived. Lastly, in an increasingly larger number of theories of various aspects of physical Reality, the theory is put into a mathematical form. Euclidean geometry is considered the example par excellance of a mathematicized theory deduced from Reality. Many theoretical areas of physical Reality, emulating Euclid, espouse the goal of becoming a mathematical-deductive theory – a theory with a set of fundamental axioms from which theorems can be derived that describe all phenomena in the area.

Presumably all aspects of physical Reality follow from a basic Theory of Everything[8] but the connections between the various specialized theories is often hard to derive. The Theory of Everything is expected by many scientists to follow from a unified theory of fundamental particles and gravitation. But the derivation of psychological phenomena, for example, from the Theory of Everything is a long and tortuous process that has many gaps currently – primarily at the level of the

---

[8] Some thinkers have recently asserted that a theory of everything is not possible. This assertion is not true if one believes that ultimately the understanding of each area of physical Reality can be reduced to a physical theory (a set of mathematical or logical rules) that describe phenomena in that area correctly. If one does not accept that assertion then one denies the basis of scientific thought for the past 400 years. Accepting the existence of a correct theory of each area of physical Reality, one can construct a theory of everything by joining all the experimentally verified areas in one composite theory. The composite theory may be unesthetic and cumbersome but it would be a theory of everything. If the objection is raised that parts of the composite theory might conflict, then the objection is refuted by noting that any parts of the composite theory that overlap must be in agreement since an experiment in the overlap area can give only one result showing one or the other parts to be true and the other false – contrary to the assumption that all parts are correct. Thus a theory of everything must exist.

understanding of the chemistry of the human brain and its relation to psychological states.

1.5 Limitations on Human Knowledge of Physical Reality

The sources of human knowledge are observation, direct or indirect, and rational analysis of the facts acquired through observation. To obtain factual knowledge about Reality we must be able to obtain information by direct observation using our senses or through instruments that play the role of extending the senses. For example, we cannot see ultraviolet light with our eyes but we can devise an instrument that can "see" ultraviolet light.

When everyday phenomena were the stuff of knowledge then it appeared that we could have a total knowledge of physical reality. However the situation changed dramatically in the twentieth century. First quantum theory showed us that certain combinations of facts could not simultaneously be ascertained at the quantum level (small distances). For example the position and momentum of an electron cannot be simultaneously exactly determined due to the Uncertainty Principle.

In addition, more recently, it has become clear that the amount of matter and the amount of energy in the cosmos are different from that determined by astronomical observations. There is an "unseen" part of the universe. All of our senses and instruments for observation are based ultimately on the four known forces of nature: gravity, electromagnetism, the weak interactions and the strong interaction. The latter two forces are only detectable at high energies (short "quantum-scale" distances). The unseen matter in the universe, Dark Matter (about eighty-five per cent the total matter in the universe) is only known from its gravitational effects at the present time. It seems to be either undetectable through the stronger forces of nature or , perhaps, only very weakly detectable through the stronger interactions. That being the case, Dark Matter can "interpenetrate" (occupy the same space) as the ordinary matter that we see, and form its own unseen part of the universe. Similarly we see the effects of Dark Energy[9] in the cosmos – particularly in the rate of expansion of the universe. Here again we see the effects but do not see the entity.

We conclude that our senses, even our extended senses through instrumentation, cannot *currently* acquire significant knowledge of most of the universe.

Everyday epistemology and metaphysics did not anticipate, and has not explained, the unseen, and unseeable, aspects of the universe. Therefore any claims that contemporary metaphysics explains ultimate Reality are incorrect. The mission of this book is to begin the extension of metaphysics to the current empirical views of Reality.

---

[9] Energy of unknown origin constituting about seventy-five per cent of the energy in the universe.

1.6 Mind-Body "Problem", and Reality

A topic of interest for many years has been the "Mind-Body" problem. Many people see a discrepancy between the intellectual processes taking place in the mind and the qualities of the human body—particularly the human brain. However, recent research on the human brain using new techniques such as MRI have revealed that specific parts of the brain handle input from the senses, emotions, and even certain thought processes such as number processing (mathematics). We are therefore developing an understanding of the relation between the mind or consciousness and the physical-chemical structure of the human brain.[10] In a very real sense the human mind is now being seen as a very complex computer, much beyond our bigest and fastest electronic computers, yet still a computing apparatus. Mathematical networks have been developed that simulate aspects of learning, thinking and memory.[11]

If all the aspects of the mind can be reduced to mechanisms of the human brain then the major remaining issue is the nature of consciousness. Consciousness is a conglomeration of the features of the mind that give us a coherent "picture" from which we can observe, process, and respond to external and internal events. Consciousness also generally has a continuity of transitions from one state to another. Perhaps the best simple description of consciousness is a self-controlled computer or television set that undergoes transitions based on internal processing of internal states and external sensory input.

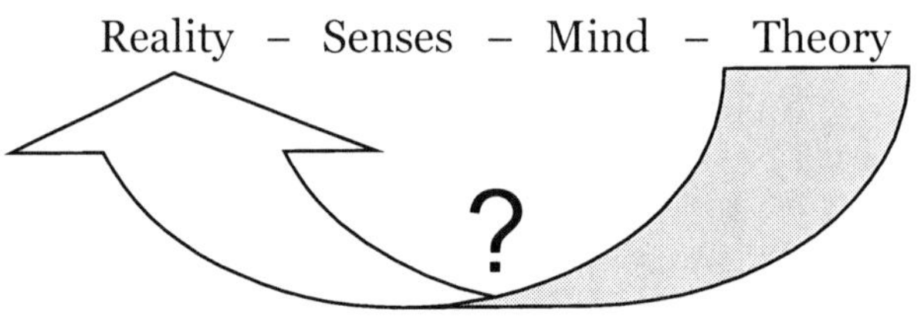

Figure 1.1. Why do theories created by the human mind so closely mirror Reality? The author believes it is due to Reality being "derived" from Logic, the mechanism of thought. This topic will be discussed in detail later.

---

[10] Scientists using fMRI (functional Magnetic Resonance Imaging) have recently found signatures of consciousness. They have detected coordinated activity across the entire brain. They have also detected conscious processing of images in the mind. See A. Schurger et al, Science, DOI: 10.1126/Science .1180029 (November, 2009).

[11] Cf. the papers and books of Professor Steven Grossberg of Boston University and collaborators, and colleagues in the field.

We now have a strong beginning to understanding the role of the various parts of the brain in processing sensory input, and in making the body perform mechanical activity, memory, creative thought and abstract thinking in general. Abstract thought takes one of two forms: coordinated brain activity and connectivity of thinking through logical progression. Creative thought arises from the association of ideas, the logical progression of ideas, changed views of concepts (a familiar word or idea that is evoked but viewed differently from its previous view in the mind), and serendipitous "mistakes."[12]

Based on this line of reasoning the phrase "Mind-Body Problem" is a misnomer at best. It would better described as the Mind-Reality Problem. We create theories with our minds that describe natural phenomena. Mathematical theories of physics have been extremely accurate in describing natural phenomena – physical Reality. On a number of occasions physicists have raised the question: how can our somewhat strange mathematical theories accurately describe (or perhaps better said – accurately mirror) physical Reality. Does Reality do the computations in real wall clock time (actually more seemingly instantaneously) to make events happen according to our calculations? That is a fundamental question which metaphysics should address.

Epistemology frames our view of knowledge of Reality and addresses the question of the theories that organize our knowledge of Reality. Thus it provides the backdrop for the study of metaphysics.[13]

---

[12] An interesting area that requires investigation is the role of writing as a memory aid. Before writing was developed, and up to the past few centuries, the human mind carried the burden of remembering a train of thought. Since writing became common there has been a remarkable increase in theoretical knowledge – particularly in the sciences – due to the interaction between thoughts in the mind and thoughts written on paper. A poignant case in point is that of Professor Stephen Hawking who has remarked that he regrets the impact on his work of not being able to work out thoughts on paper or a blackboard.

[13] The human mind is capable of simulating sensory input. Hypnotists do this when they cause a subject to think they see or feel something illusory. An individual can do this as well. It commonly happens in dreams. It can also happen through a form of self-hypnosis. We are familiar with this phenomena at the level of the senses. But it is possible that a mind can create a mental image of another universe and cause this universe to evolve according to some set of physical laws concocted by the mind. Then we would have a mental implementation of a form of the "many worlds" hypothesis that appears in metaphysics and modal logic. We might call this type of thought process creating "Universes of the Mind."

2. What is Metaphysics?

2.1 Definitions

Metaphysics means different things to different people. Some say it is the "Theory of Ultimate Reality."[14] Others say it is the "Theory of the Nature of Abstract Entities."[15] Yet others express alternate views.

We shall take Metaphysics to be the contemporary theory of ultimate Reality realizing that our views of Reality have changed drastically in the past hundred odd years and are likely to change again in the future. Rather than wait for "ultimate Reality" to surface[16] we will construct the Metaphysics of presently known Reality. In doing so we shall also explore the abstract entities that make up the world[17] as we know it focusing particularly on quantum Reality. Quantum Reality is quite different from everyday Reality – the primary focus of most metaphysical discussions. The theory of Relativity also dramatically alters our view of Reality.

The combination of the mathematics of quantum theory and relativity open a new world of metaphysics that we call *Relativistic Quantum Metaphysics*. This new metaphysics is simpler in some ways than traditional discussions of the metaphysics which take place at the level of everyday experience. It differs from everyday metaphysics in many significant ways. Since relativity and quantum theory are at a deeper level than everyday phenomena we can say with assurance that relativistic quantum metaphysics is closer (actually as close as we can currently get) to ultimate Reality.

We will thus view Metaphysics as composed of three parts (See Fig. 2.1.): the Metaphysics of everyday experience,[18] Quantum Metaphysics,[19] and Cosmic Metaphysics.[20] Because physics has shown us that the union of Relativity Theory and Quantum Theory is essential to the understanding of natural phenomena we consider a combined metaphysics based on the union of these physical theories.

---

[14] van Inwagen (2009).
[15] Lowe (2002).
[16] If perhaps it has not as yet surfaced.
[17] The word world represents our universe or the set of all universes if there is more than one.
[18] This is the metaphysics of the past 2500 years.
[19] The metaphysics of the ultra-small and of high energies whose empirical basis was only discovered in the twentieth century. Its nature is very different from "everyday" metaphysics as subsequent chapters show.
[20] The metaphysics of the cosmos based on the Special and General Theories of Relativity. This metaphysics also is vastly different from "everyday" metaphysics.

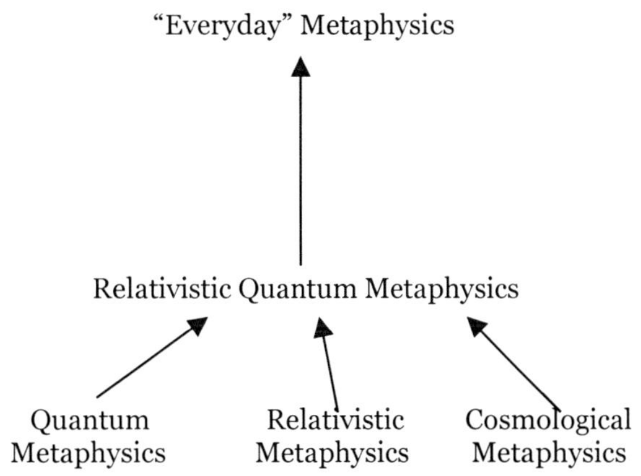

Figure 2.1. The three sources of metaphysics and their interrelations with "Everyday" Metaphysics.

"Everyday" Metaphysics has traditionally been applied to many areas of experience including physical phenomena, thought, the mind, ethics, and Theology. Excepting Theology we view all aspects of "Everyday" Metaphysics as ultimately based on Relativistic Quantum Metaphysics. The chain connecting Relativistic Quantum Metaphysics to aspects of "Everyday" Metaphysics may be quite long and involved—the links of the chain in many cases are not as yet forged. Yet we will take a solidly empiricist position[21] that this chain is fact and that progress on all fronts is being made. "Everyday" metaphysics is derivative from the much deeper Reality of Relativistic Quantum Metaphysics.

2.2 A Summary of Classical ("Everyday") Metaphysics

Classical Metaphysics[22] is the metaphysics everyday experience. Therefore it addresses issues relevant at that level of Reality.

Classical Reality consists of many different substances with many different properties. As a result Classical Metaphysics is in part concerned with categorizing the many substances and exploring the interrelationships between them and their properties. It considers substratum and bundle theories relating groups of properties and substances or entities. It also considers the nature and role of propositions (essentially statements), states of affairs (situations), events, and facts.

---

[21] In this choice we see our view as similar to that of Hume and other empiricists.

[22] It seems appropriate to call "everyday" metaphysics classical based on an analogy with the nomenclature of classical physics and quantum physics.

Classical Metaphysics has found it necessary to introduce the concept of a plurality of possible worlds.[23] The "existence" of this set of worlds enables the Modal concepts of necessity and possibility to have meaning. Is a property necessary to an entity or possibly part of an entity? – A modal question. With many worlds we can (in a simplified way) consider whether an entity in the many worlds always has a property (necessary) or sometimes has a property (contingent).

Another topic of interest in Classical Metaphysics is Causation: the study of cause and effect, their connection, and whether causes necessitate their effects. The study of classical causation is interesting because it attempts to describe these features in general terms that to this author does not reflect what actually happens in ultimate Reality according to scientific empirical data in the quantum and relativistic regime.

Actions and events are viewed as types of entities in Classical Metaphysics. Their nature is analyzed in a qualitative, abstract manner in contrast to what we encounter at the deeper levels of physical Reality.
A similar statement can be made about studies of the Nature of Time and Space, as well as some studies of space-time that seek to bring in a measure of the Reality of the Special Theory of Relativity. Attempts are made to abstract the nature of space and time that, in this author's view, do not correspond to Reality. As a statement becomes more abstract, it loses content. And the classical studies of space and time simply do not elucidate their nature.

Change takes place in time. The Classical Metaphysics of change is perhaps meaningful at the level of everyday change but does not correspond to what we have learned of change at the quantum level – the ultimate source of change.

We conclude – knowing that we have only briefly discussed Classical Metaphysics – that it is limited to the subject area from which it originated – everyday objects, properties, and change. It does not address the profound issues raised by Quantum Theory and the Theory of Relativity. Classical Metaphysics does not address ultimate physical Reality. In fact, it tends to mask it with a veneer of depth and generality that it truly does not have. Classical Metaphysics generally cannot be extended to quantum and relativistic entities.

2.3 Are There Any Truths in Classical Metaphysics?

Van Inwagen (2009) raised two questions that support the somewhat negative tone of the previous section:[24] "Why is there no such thing as metaphysical information? Why has the study of metaphysics yielded no

---

[23] The multiple worlds concept has no relation to many universes theories in physics since the multiple worlds exist at the conceptual level and are used to clarify the notions of modality. Does an entity have a necessary property (absolutely required) or a contingent property (a property that it might have in one world but not in another world).

[24] van Inwagen (2009) p. 11.

established facts?" To be fair there are certain principles of metaphysics that were proposed by Leibniz[25] that can be viewed as metaphysical information. Leibniz proposed three basic principles:[26]

1. The Principle of Sufficient Reason
2. The Principle of Identity
3. The Principle of Perfection

These principles, and their consequences, are to be understood in Leibniz's view as axioms within the framework of generalized subject-predicate logic. Leibniz viewed propositions (statements) about substances or entities (existents) as having a subject-predicate form[27] or as reducible to a combination of sub-statements having subject-predicate form. He used a form of analysis (*Leibnizean analysis*) of propositions that first determined the set of properties of a subject (substance) and then scanned the list to see if the predicate (a property) appeared in the list. Thus a proposition is true if the predicate is in the list of properties of a subject and false otherwise.

The Principle of Sufficient Reason states that every true proposition is Leibnitzean analytic. Thus it asserts every true proposition can be reduced to subject-predicate form or a combination thereof, and each predicate can be shown to be in the set of the subject's properties using Leibnitzean analysis.

The Principle of Sufficient Reason is illustrative of the abstract nature of the above three principles. Leibniz's four subsidiary principles are similarly abstract:[28]

1. The Identity of Indiscernibles
2. The Principle of Plenitude
3. The Law of Continuity
4. The Principle of Harmony

The reader is directed to Rescher (1967) and other sources for information on these principles since they will not be germane to Relativistic Quantum Metaphysics. They are too general to be of value in the study of ultimate physical Reality. However Leibniz's focus on subject-predicate Logic as the underlying framework of his metaphysics will be seen to be a precursor to part of our approach to Relativistic Quantum Metaphysics.

In addition to his principles Leibniz suggested that there are many possible worlds in a physical, tangible sense – not as a conceptual construct. And he gave a criterion for the selection of our world as the world of Reality:

---

[25] Rescher (1967).
[26] Rescher (1967) chapter 2.
[27] Example: "The house is green." has house as the subject and "is green" as the predicate.
[28] See Rescher (1967) chapter 4.

> The world which is most perfect … is the simplest
> in its axioms and the richest in phenomena.[29]

This principle is an extremal principle in the sense of extremal principles in mathematics and physics. Leibniz is famous for developing calculus (at the same time as Newton) and for his extremal principles which have been of great value in physics.

We will discuss the possibility of a selection principle for physical Reality with our universe as the solution (or a possible solution) at a subsequent point in our presentation.

We conclude our brief view of Classical Metaphysics with a figure, Fig. 2.2, that indicates where it applies in the range of Reality in this universe (world).

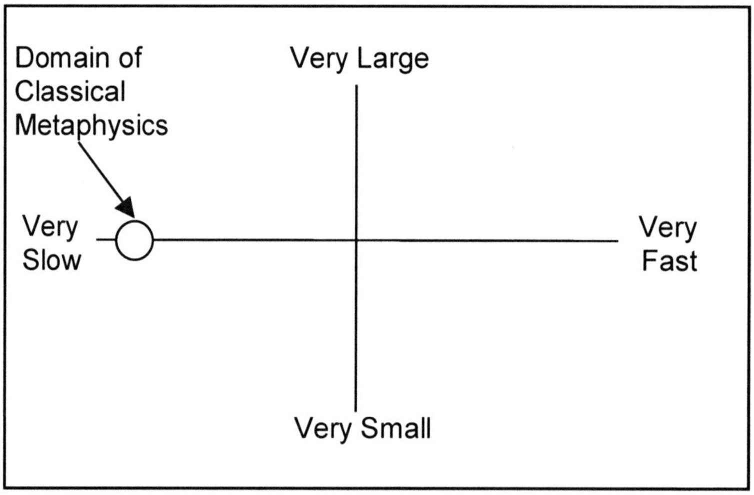

Figure 2.2. A visualization of the domain of applicability of Classical Metaphysics.

---

[29] Quoted in Rescher (1967) p. 19.

3. Relativistic Quantum Metaphysics

3.1 Reduction of Reality to the Presently Known Ultimate Reality

Classical Metaphysics was developed over the millenia based on concepts that arose in everyday experience with a small admixture from early science up to at most the nineteenth century. Since 1900 there has been an explosion of knowledge – primarily scientific knowledge – that calls for a new metaphysics of physical Reality.

First, we have come to the realization that everyday phenomena which appear complex in nature and involve a staggering variety of different substances, properties, and events at the physical level, the mental level and the social level are all ultimately derivative[30] from a vastly simpler underlying physics that is embodied in the Standard Model of Elementary Particles and the General Theory of Relativity.[31]

We will begin with a description of the overall nature of this "all encompassing" physics and then proceed to develop Relativistic Quantum Metaphysics. Fig. 2.1 shows our view of Classical Metaphysics in relation to Relativistic Quantum Metaphysics.

The progressive depth of view of physical Reality (omitting gravity for a while in our discussion) is:

All Everyday	→	Molecular	→	Atomic	→	Elementary Particles
Phenomena		Level		Level		Level

All substances at the everyday level with all their properties are consequences of the nature and behavior of elementary particles. The set of elementary particles, their nature and their behavior, are extremely well described by The Standard

[30] We will not consider Theological questions and the nature of God because we view this area as not part of physical Reality although the spiritual and the physical may perhaps interact with each other.

[31] We say that realizing that aspects of gravity and the cosmos (dark matter and dark energy for example) as well as features of elementary particles (including the issue of whether there is a deeper theory of elementary particles such as a Superstring theory) remain to be understood. These unknowns are not relevant to "everyday" phenomena and thus our development of Relativistic Quantum Metaphysics does provide a complete underpinning for "everyday" phenomena and thus the domain of Classical Metaphysics.

Model of Elementary Particles. We will examine features of The Standard Model in some detail in subsequent chapters.

3.2 A Metaphysical View of Elementary Particles

At this point we will consider general features of elementary particles and The Standard Model of interest to metaphysics.[32] We also will consider aspects of the theories of Special and General Relativity from a metaphysical perspective.

3.2.1 Particles and Substances

In everyday Reality we see a multitude of substances. In the case of elementary particles it appears that there is only one substance[33] but the substance can take different stable or semistable forms with different properties. Each form is a type of elementary particle.[34] Since the late 1960's it has become clear that there is a basic set of particles: eighteen known types of quarks and six types of leptons[35] that combine to make protons, neutrons, atoms, and other less familiar bound states (combinations). The forces of nature are gravity, electromagnetism, the Weak Interactions and the Strong Interactions. Quarks and leptons exert these forces on each other. Each force has an associated particle or set of particles that embody the force: gravity has gravitons, electromagnetism has photons, the Weak Interactions have vector bosons, and the Strong Interactions have eight gluons. In total, there are thirteen "force particles." Thus the number of known elementary particles is thirty-seven. Many physicists believe that there are more particles such as Higgs bosons, WIMPs, and a fourth generation of fermions remaining to be discovered. The known particles comprise The Standard Model.

If there are thirty-seven plus types of elementary particles how do we know they are composed of the same substance? The reason is quite simple. These particles can transform into each other through the decay of a particle into other particles, and can transform into each other when they interact (or exert a force) on each other. Although they differ in properties their ability to undergo transformations into each other strongly supports the idea that there is one substance that can take thirty-seven (or more) different forms with different

---

32 Again we note that The Standard Model, although it describes elementary particles and their behavior very well, is either an incomplete theory (Dark Energy and Dark Matter are not in The Standard Model.) or it is based on a deeper theory of particles such as Superstring theory. In later chapters we provide a line of reasoning (a derivation with some steps remaining to be justified) of the major features of The Standard Model.

33 The nature of Dark Matter and Dark Radiation is an open question. They may constitute another substance or possibly two substances. However experimental results reported recently suggest that Dark Matter can transform or decay into normal matter and energy. If these results are correct then Dark Matter is of the same substance as normal matter.

34 The vacuum is also a form of this substance.

35 An electron is a lepton. Neutrinos are also leptons. We note that a fourth fermion generation and WIMPs (hitherto not discovered) should be added to the list of particles when they are found experimentally.

internal properties. These properties include the spin of a particle and numbers associated with their form that we call internal quantum numbers.

Can we take the alternate view that there are thirty-seven different substances? Yes, but it does not change the fact of transformations between the particles which allows us to consider the universe as composed of one substance. Ockham's Razor – the simplest solution is usually the correct solution – supports the view that we have one substance capable of assuming thirty-seven different forms. This discussion shows that Relativity and Quantum Theory, upon which this Reality rests, gives us a far simpler metaphysics than Classical Metaphysics.

Having reduced the number of substances to one substance, the question naturally arises what is that substance? The answer is that it is unknowable. The various forms it can assume are known but the substance in itself is an unknowable. We can only specify the forms it takes and their properties. Taking a note from Logic we simply call it a *primitive term* – an undefinable term which we can use to build a mathematical theory. To the reader who might object to this answer we point to Euclid's geometry, which has several primitive terms in its axioms. Thus primitive terms are to be expected in a theory. Some day if a more fundamental theory than The Standard Model appears, then it also can be expected to have primitive terms as well.

Lastly, the reader might ask what of the vacuum? Is it nothing or a substance? The Standard Model, which is a Relativistic Quantum Field theory, has a vacuum which is composed of an infinite number of all the different particles. But these particles are uniformly distributed throughout space so the forces that they might have exerted on any individual particle of matter cancel each other[36] and thus we think of the vacuum as empty. The vacuum is thus also composed of the substance of particles in infinite, uniform, quantity.

Sometimes the vacuum undergoes a quantum fluctuation creating a particle-antiparticle pair for an instant. Thus the vacuum has a dynamic aspect. But the extremely short existence of the particle-antiparticle pair usually makes vacuum fluctuations unobservable. Yet we know that they happen because they exert effects on the observed properties of matter.

Our discussion has led us to the conclusion that there is only one kind of substance in the deepest known physical theory. And so we have our first Relativistic Quantum Metaphysics principle:[37]

I. *There is only one substance in the universe. It can assume a variety of forms as particles, which we call fundamental elementary particles, and as the vacuum.*

---

[36] This is a simplification in that a charged particle, for example, can distort the vacuum attracting particles of opposite charge to it which "mask" (renormalize) part of the charge.

[37] SuperString Theories, one set of possible theories of ultimate Reality, assume all elementary particles in the universe are made of extremely small mathematical strings. Thus these theories conform to principle I.

3.2.2 Forms and Properties

Classical Metaphysics is much concerned with properties. It considers them individually for entities, in bundles of properties that entities might possess, and as forming an entity in the form of a substratum.

We have learned from the study of the physical nature of matter that the thirty-seven known elementary particles (and any new ones that might be found) have bundles of properties consisting of energy, momentum, spin, and internal quantum numbers. We will call the bundle of properties of a particle its *form*. Thus we come to a second Relativistic Quantum Metaphysics principle:

II. Each particle's form consists of a finite number of properties: its energy, momentum, spin, and internal quantum numbers.

This extreme simplicity makes the theoretical developments in the study of properties in Classical Metaphysics not relevant to the ultimate Reality of the world (universe) that we inhabit.

However, Classical Metaphysical studies may be relevant to the study of other possible worlds if they are extended to a wider framework than the picture of ultimate Reality just discussed. In that wider framework Relativity and Quantum theory, may be supplemented, or replaced, by other theories. Thus a correspondingly wider metaphysics would be required.

3.2.3 Change, Interactions and Causation

Causation and change have been much studied in the abstract in Classical Metaphysics. The abstractness of the work in these areas and the variety of everyday phenomena have led to a plethora of ideas. But this plethora is little related to the provisional ultimate Reality embodied in Relativistic Quantum Field theory in general and The Standard Model of Elementary Particles in particular – the deepest verified theory of physical Reality – supported by a large amount of positive experimental data with no significant contradictory experimental data.[38] (The Standard Model is a specific Quantum Field Theory.)

In the Standard Model causation and change result from the interactions (forces) between elementary particles. The interactions can be at very small distances or large distances between individual elementary particles, or when particles combine (through forces such as electromagnetism or the strong interaction) the interactions can accumulate to cause tangible effects at everyday distances between everyday objects.

---

[38] We state again that it is known that there are areas that are not part of the Standard Model and there may be a deeper theory waiting in the wings. These difficulties are not roadblocks to viewing the Standard Model provisionally as ultimate physical Reality since The Standard Model will still hold for current experimental data within a larger theory or a deeper theory. In these cases it becomes a good approximation for its domain of experimental data.

At the most fundamental level the decay of an elementary particle and the interaction between two elementary particles take place through the thirteen fundamental interaction particles mentioned earlier: gravitons, photons, three vector bosons, and eight gluons. They can be visualized using Feynman diagrams. These diagrams, developed by Richard Feynman, associate a straight line with each of the eighteen "matter type" particles and a "wiggly" line with each of the thirteen "interaction particles." They are viewed from left to right. (Time can be considered as increasing from left to right in these diagrams although that is not strictly correct.) See Figs. 3.1 – 3.3 for simple Feynman diagrams illustrating decays and interactions.

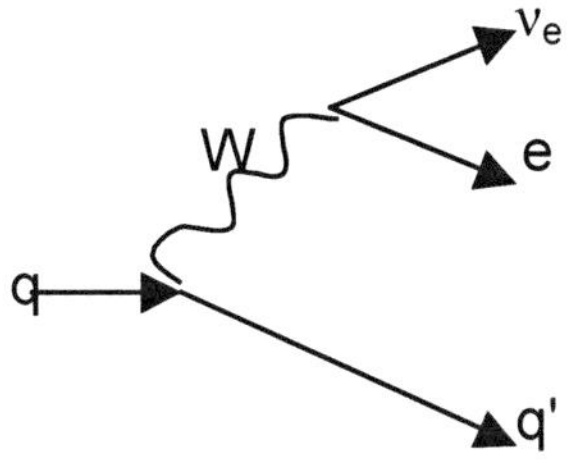

Figure 3.1. A Feynman Diagram for "Causation" and "Change" in The Standard Model. It depicts the decay of a heavy quark particle into three particles: a lighter quark q', an electron e and an electron-type neutrino ν_e. It illustrates one form of Change.

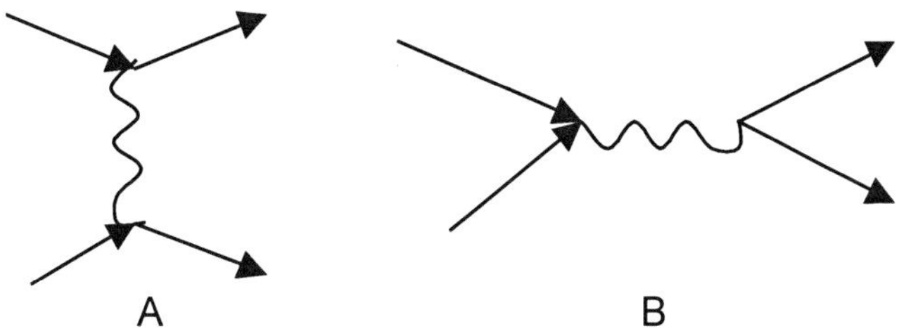

Figure 3.2. Additional Feynman Diagrams for "Causation" and "Change" in The Standard Model. It depicts the interaction (force exchange) between two particles which can lead them to change direction or to change the type of particles that they are. There are two simple diagrams shown. Actually these are the simplest diagrams of an infinite set of diagrams that contribute to the change from two initial particles to two final particles.

Thus we again see that the complex issues and ensuing discussions in Classical Metaphysics are reduced in Relativistic Quantum Metaphysics to interactions between particles or combinations of particles – much simpler, less abstract, and not subject to dispute. Causation and Change are reduced to single or multiple particle interactions.

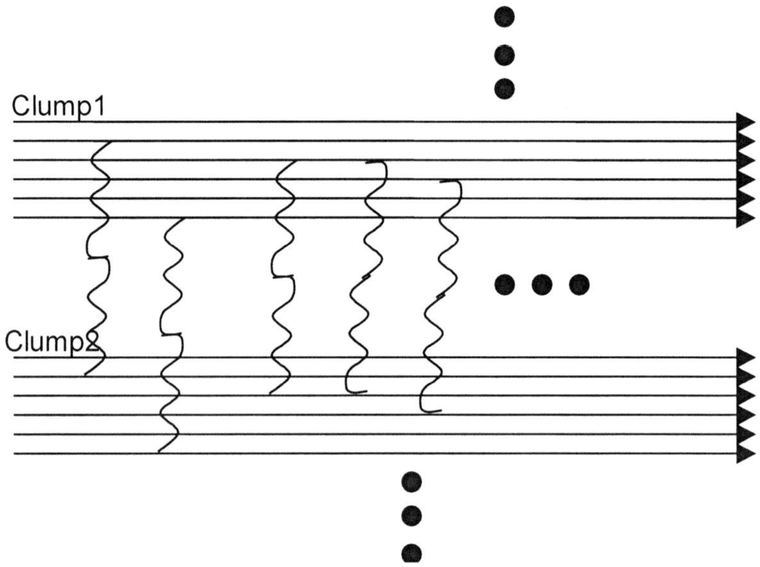

Figure 3.3. Yet another Feynman Diagram for "Causation" and "Change" in The Standard Model. It depicts the gravitational interaction (force exchange) between two clumps of matter via graviton particles on particles within the clumps, which lead the clumps to change direction. This illustrates the force of gravity at the quantum level. As yet the understanding of quantum gravity is subject to dispute. But most physicists would agree that gravitons exist and are the fundamental "carriers" of the force of gravity. Again these are the simplest diagrams of an infinite set of diagrams that contribute to the process of change. Clump1 and Clump2 each consist of a large number of particles in lumps of matter. The thick "dots" represent lots of particles and lots of gravitons being exchanged between the particles in the clumps. The result of the extremely large number of gravitons exchanged is the clumps change direction and approach each other since gravity is a strictly attractive force according to the General Theory of Relativity.

The general, abstract approach of Classical Metaphysics to Causation and Change is defective in not properly dealing with quantum phenomena.[39] However

[39] For example Leibniz and others suggest that Change must be continuous. It is a known fact that electrons make jumps ("quantum leaps") between orbits in atoms. They are never between atomic

if one wishes to study all possible universes then Classical Metaphysics suitably generalized to include Quantum and Cosmological Reality may support the study of alternate universes. This would open a new sphere of activity for a generalized metaphysics.

Considerations of the sort illustrated by by Figs. 3.1 – 3.3 lead us to the following additional principle for Relativistic Quantum Metaphysics:

III. At the level of ultimate Reality, Causation and Change take place through interactions (forces) mediated (conveyed) by particles.

3.2.4 Modality

Modality is concerned with questions of necessity and contigency. Is a property necessary to an entity or is a property merely a possible part of an entity? Modality appears in Classical Metaphysics in relation to questions about the nature of entities.

Part of the framework of Modality is the existence of multiple worlds in at least a conceptual sense. Thus if a certain entity existed in all worlds and always had a certain property then that would imply that the property was *necessary* to the entity. If, on the other hand, a certain entity existed in all worlds and had the property in some worlds but not in others then the property would be a *contingent* part of the entity.

It is perhaps appropriate to raise the question of Modality with respect to Relativistic Quantum Metaphysics. We were led to consider a generalization of Classical Metaphysics due to the Theory of Relativity and Quantum Theory. However, it is not clear that we should not consider an even more general metaphysics where Relativity and Quantum Theory are contingent "properties." This *Generalized Metaphysics* would consider worlds where Relativity and/or Quantum Theory did not necessarily hold and where alternatives to Relativity and Quantum Theory might exist. Generalized Metaphysics would then have the goal of determining why our world, which embodies ultimate physical Reality is based on the Theory of Relativity (Special and General) and on Quantum Theory.

3.2.5 From Generalized Metaphysics to The Standard Model

Generalized Metaphysics as we conceived it in the previous subsection assumes the three fundamental principles stated earlier but does not assume the Theory of Special or General Relativity, or Quantum Theory. It is consistent with the existence of an infinity of possible worlds (universes). It then would proceed to develop selection criteria that lead to the ultimate Reality with which we are familiar. Thus the space-time of Special and General Relativity must be selected

---

levels. Quantum Theory shows the inadaquacy of Classical Metaphysics view of Causation and Change.

somehow.[40] Secondly, the selection criteria must somehow choose Quantum Theory as we know it. Finally the detailed theory of elementary particles and their interactions must follow from the selection criteria.[41] The result would be a true ultimate theory of Reality. Fig. 3.4 symbolizes the steps just outlined.

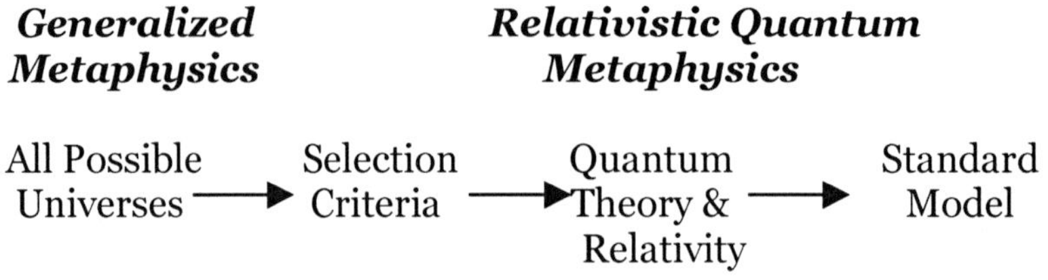

Figure 3.4. A visualization of the steps leading from a Generalized Metaphysics to Relativistic Quantum Metaphysics and thence to The Standard Model.

[40] There is an ongoing discussion of various possible forms of General Relativity. Einstein's formulation or something like it is the most favored form at present. Consequently the selection ctriteria, if found, could be of help in choosing the correct form.

[41] It is of course possible that Quantum Theory and the General Theory of Relativity emerged through mere chance or some arbitrary fluke. In these cases there would be no selection criteria. Ultimate Reality would then be an "accident."

4. Logic and Physics

4.1 Logic is the Key to Metaphysics

Classical Metaphysics is largely attempts to define fundamental principles, to clarify concepts and to categorize entities. The key basis of these attempts is Logic. This was pointed out many years ago by such philosophers as Leibniz.[42] Logic enables philosophers to use fundamental principles to derive new principles. Logic enables philosophers to clarify concepts and their differences. Logic is used to categorize and differentiate between entities.

Thus Logic is the engine of Philosophy and Metaphysics, as Reality is the destination.

In considering Metaphysics Leibniz felt that propositions of the subject-predicate form, or reducible to a combination of clauses of the subject-predicate form, sufficed to meet the needs of metaphysical discussion and analysis. This view has been disputed by some metaphysicians but from the viewpoint of physical Reality it appears the view of Leibniz is correct.

Consequently we will consider (physical?) Logic as applying to axioms and propositions that have subject-predicate form or are reducible to subject-predicate form.

We will relate logical propositions to experiments on physical Reality. We will then show that defects in Logic that lead to paradoxes can be remedied by a new form of logic called *Operator Logic* in chapter 5 or alternately by viewing propostions as function calls with arguments chosen from the domain of the function. Subsequently we will use Operator Logic to directly construct a basis for physical Reality (chapter 14) in a manner reminiscent of Plato's Theory of Ideas and Reality, and their connection.

4.2 A Logic View of Physical Experiments

At first glance experiments on physical Reality have no direct connection to Logic. However as Blaha (2009), and his earlier work, has shown a quantum experiment (specifying intermediate stages of the experiment and a specific result) can be stated as a proposition in subject-predicate format or a proposition consisting of a set of clauses in subject-predicate format. Thus quantum experiments can be mapped in a one-to-one fashion onto propositions. We describe the form of the map in the next section.

---

[42] Rescher (1967) pp. 22-23.

4.3 Filtration Stages in a Quantum Experiment and Their Mapping to the Form of a Proposition

If we consider a proposition such as

The car is a small, green, Ford.

we see that the proposition is equivalent to

The car is small, and the car is green, and the car is a Ford.

Now if we consider the set of all cars then we see that the first clause restricts the set of all cars to the subset of small cars. The second clause restricts the subset of small cars to green cars, and the clause then restricts the subset to the smaller subset of Ford cars. Each restriction to a yet smaller subset we can call a *filtration* in the sense that each restricting clause filters a set to a smaller subset.

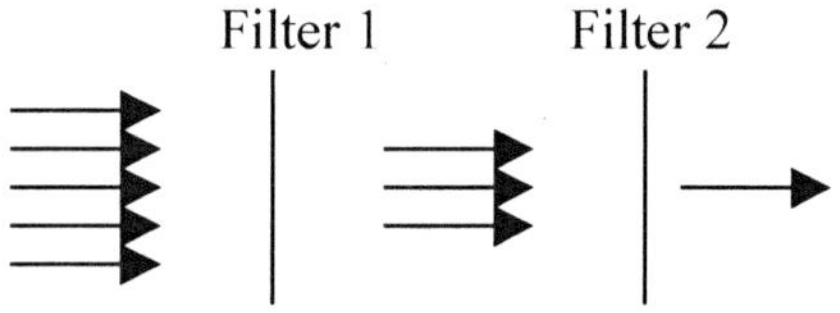

Figure 4.1. A two filter experiment that selects particles with the exact velocity of 100 miles per second from a stream of incoming particles with a variety of velocities.

Now let us consider a (marginally) quantum experiment[43] that progressively filters a stream of particles that have a range of velocities so that only particles with a velocity of "exactly" 100 miles per second emerge. Let us assume two filters: filter 1 only lets particles with velocities greater than or equal to 100 miles per second pass through, filter 2 only lets particles less than or equal to 100 miles per second pass through. As a result only particles with a velocity exactly equal to 100 miles per second pass through the two filters. Fig. 4.1 illustrates this simple experiment. This experiment can be expressed as a proposition:

"A stream of particles passes through a filter that allows only particles with a velocity greater than or equal to 100 miles per second through and then passes through a filter that only allows particles of velocity less than or equal to 100 miles

---

[43] The filters are assumed quantum in nature. The quantum aspect enters when an inquiry as to the position of the particles is made. A particle with an <u>exact</u> velocity (and thus momentum) has a indeterminate location. This situation is a consequence of the Heisenberg Uncertainty Principle.

per second through so that only particles with a velocity that is exactly 100 miles per second" emerge."

Or, in short,

"A stream of particles with a range of velocities" "is reduced to a stream of particles with a velocity of 100 miles per second."

We use quotes "..." to indicate the subject and predicate.

Clearly there is a conceptual correspondence between logical propositions and experimental propositions. In both cases predicates can be described as filters that perform a selection upon the set specified by the subject or by the implied set that the subject is generally a member of. In the above logic proposition, the subject begins as a member of the set of all cars and is then filtered to a very specific subset of the set of all cars.

Therefore it is clear that we can view subject-predicate logic propositions and their generalizations to subject-predicate clauses as a series of filtrations similar to the filtrations that take place in quantum physics experiments. We define this feature as the fourth principle:

IV. Subject-Predicate propositions in Logic can be viewed as filtrations analogous to filtrations that take place in quantum physics experiments.

With this concept as a beginning point we proceed to examine Logic, point out some paradoxes,[44] and develop a new form of Logic that we have called Operator Logic. This new logic resolves the issue of paradoxes including Gödel's Undecidability Theorem (An alternate approach is to view propostions as function calls with arguments chosen from the domain of the function. This approach will be considered later.) It also furnishes a starting point for a transition from the world of Ideas (Logic propositions) to the world of Reality. Pursuing this line of development we eventually reach the form of The Standard Model of Elementary Particles.

4.4 Logical/Mathematical Rigor in Physics

Physics since before Newton's time has used mathematical techniques that were not rigorously established. Currently, elementary particle physicists use the path integral formulation of quantum field theory in calculations. It works. It give the correct results. Yet it is not rigorously established – probably due to our incomplete understanding of integration over a continuum of fields – a mathematical issue related to our understanding of the levels of infinity and issues in set theory. These issues have not been resolved by mathematicians and

[44] There are an infinite number of paradoxes in traditional Logic.

logicians. Therefore we do not address these other issues, relevant as they are to Logic, such as the Continuum Hypothesis, Gödel's L construct, Woodin's "Ultimate L" and other related matters in set theory and the levels of infinity.
These issues are not critical for contemporary Physics.[45] Physicists assume that mathematicians will ultimately resolve these issues and are "content" to use mathematical techniques such as the path integral formulation of quantum field theory knowing that they cannot be rigorously established just as Newton and others used calculus although its rigorous development (still in question) did not take place until the 19th century. Thus our construction/derivation of The Standard Model is physically rigorous, but not mathematically rigorous, within the framework of the Logical developments in this book.

[45] They can be descibed as mathematicians' "universes of the mind" since they are constructs of the mind, at best, unrelated to contemporary physics to the extent of our knowledge.

5. Operator Logic

In this chapter we will establish a new formalism for Logic that we call Operator Logic. Operator Logic provides a mathematical method of eliminating paradoxes including Gödel's Undecidability Theorem which is not wrong but only illustrative of the error of choosing a subject not in the domain of a predicate function – an invalid function call.

5.1 Logic and its Paradoxes

Modern Logic is a very well developed field of studies with many notable achievements. A cloud hanging over Logic is the presence of a number of paradoxes. The number of paradoxes described in the literature is of the order of ten. But they represent an infinity of paradoxical propositions so the problem they represent is significant. Among the better known paradoxes is Gödel's Undecidability Theorem. This theorem asserts that for certain mathematical-deductive systems (and there are many) there exist propositions that can be proven to be true or proven to be false – thus the undecidability.

A number of philosopher-logicians[46] have suggested a mechanism to avoid paradoxes. This mechanism applies to propositions having a subject-predicate form or reducible to a set of clauses – each of which has a subject-predicate form. The mechanism is to assume that each predicate has a set of allowed subjects – a *domain* of allowed subjects. Subjects that are not in the set are excluded from a subject-predicate clause for that predicate. By limiting a domain of subjects, paradoxical propositions can be avoided. As Blaha (2009) points out the Gödel Undecidability Theorem is also seen to be of little consequence as it is an example of a function call with an argument outside the domain of the function.

The procedure of specifying a domain of allowed subjects for each possible predicate is a "manual" procedure in contemporary Logic. However it is possible to create a mathematical formalism that implements domains for predicates and enables propositions to be evaluated as inner products in a linear vector space. This formalism is called Operator Logic because it associates projection operators for subjects and predicates that lead to zero-valued inner products for undecidable or paradoxical propositions.[47]

While Operator Logic is of interest in its own right as a means of defining paradox free propositions and mathematical-deductive formalisms, it's matrix

[46] Leibniz may have been the first to suggest this mechanism but without thought of resolving paradoxes.
[47] Blaha (2009) and earlier work.

formulation is the starting point of the development of a metaphysics that leads to our current understanding of ultimate physical Reality. We shall pursue that possibility later in this book.[48]

5.2 Predicate Calculus

We start by considering the internal logical form of statements[49] and develop a *first order predicate calculus*. The first order predicate calculus deals with simple statements of the form (in English):

subject predicate

as well as more complex forms. Predicates typically contain a verb and often express a property. For example, a simple predicate is "is mortal". A statement generated from this predicate would have the form:

X is mortal.

where X is a subject such as Socrates. In more complex cases there can be multiple subjects and multiple clauses with differing predicates in a sentence.

A more general predicate calculus can be defined that enables more complex statements to be reduced to the form of subject-predicate clauses that are joined by connectives such as "and," "or," and "not." For example the complex statement

The sky and my car are the same color blue.

can be restated as

The sky is blue and and my car is the same color blue.

Each clause before and after "and" has a subject-predicate form. Statements about substances in Metaphysics all appear to fall within the framework of general subject-predicate statements.

5.3 Deductive Systems and Their Calculi

Theories usually consist of two types of statements: axioms (assumptions) and theorems. There are many theories of various aspects of mathematics, physics, and other sciences. Some examples include Euclidean geometry, quantum field theory, elementary number theory and so on. All or parts of these theories can be formalized with a set of axioms expressed in terms of primitive concepts, from which theorems are derived. The axioms plus the set of derivable theorems constitute an *axiomatic system*.[50]

---

[48] It is very similar to the development in Blaha (2009).
[49] We use statement and proposition interchangeably.
[50] Kleene (1967) pp. 198 – 201.

Axiomatic systems consist of two types: deductive systems[51] and calculi.[52] A *deductive system* has *semantic*[53] characteristics—it defines the primitive terms of the axioms (usually in terms of "real" things such as real lines, real physical particles and their properties, real properties of integers, and so on). Thus the implications of the axioms are "real" as well and describe the reality associated with the deductive system. In physics a deductive system describes an aspect of nature. Examples include the Landau-Ginzberg theory of superconductivity, and Newton's mechanics. In mathematics examples are Euclid's geometry and Gödel's theory of integer numbers. Statements derived in a deductive system are true – a semantic property.

A *calculus* is an axiomatic system without a definition of primitive terms. Theorems are derived following specified rules of derivation with no reliance on semantics. Thus they have syntax without semantics. Most calculi have one or more corresponding deductive systems. It is possible to have a calculus without any known corresponding deductive system. Despite that, a calculus is best viewed as the skeleton of a deductive system. The deductive system is the flesh.

The description of a *calculus*, its definition and its consequences, is the *metatheory* or *metalanguage* or *syntax language* of the calculus. The investigation of a calculus is done within the metalanguage and called *metamathematics* or *proof theory*. The calculus itself is often called the *object theory* or *object language*.

While there is much more to be said on this topic, the preceding discussion is all that we need to develop the features of Operator Logic. The reader is referred to the References at the end of this book, and other Logic books, for more details.

5.4 Universes of Discourse

A generalization of the constituent properties of a statement is needed to conclusively eliminate paradoxical statements from Logic. This generalization has an analogue in the measurement theory of Quantum Mechanics.[54] The generalization is based on the axioms:

0. Every statement appears in a universe of discourse. A *universe of discourse* is a set of statements and phrases. We denote a universe of discourse state in the form $|\Omega_u>$ and treat it as a state in a linear vector space using the notation and concept of quantum state introduced by Dirac (described later).

---

[51] Another name for some deductive systems is mathematical-deductive system.

[52] Other names used for calculi (singular calculus) are formal system, formalism, or logistic system.

[53] We use semantic in its usual sense as the "meaning" of something.

[54] See Blaha (2009). The reader who may be apprehensive of the use of linear vector spaces, which is implicit in this introductory discussion, may be reassured somewhat by noting that linear vector spaces together with their mathematical rules can be viewed as a language on a par with human languages and symbolic languages for use in Logic.

1. Every statement, and every part of a statement, in a universe of discourse has an associated logical status operator[55] S, which is a projection. These operators when applied to the universe of discourse state $|\Omega_u\rangle$ have two possible types of results. They can yield a zero value or a non-zero state. (A statement yielding a zero value does not have a truth value. A statement yielding a non-zero value does have a truth value – it is either true or false.) Thus $S|\Omega_u\rangle = 0$ indicates undecidable, while the expectation value[56] (inner product) $\langle\Omega_u|S|\Omega_u\rangle \neq 0$ indicates decidable – either true or false.

2. The status operator S of a statement is the product of the status operators of the phrases (terms) in the statement. A phrase can be one word or a finite set of words comprising a concept (an idea) or object. So symbolically we can write

$$S = S_1\ S_2\ S_3\ S_4\ S_5\ \ldots \tag{5.1}$$

   where S_1, S_2, S_3, S_4, S_5 and so on represent the status operators of the phrases in the statement S.

3. The status operator of a phrase S_i is determined by its use in the universe of discourse. It is a projection operator satisfying $S_i^2 = S_i$.

4. Projection operators are not necessarily commutative. (This will be discussed in more detail subsequently.)

If we now consider a paradoxical statement and form a trivial universe of discourse with

This sentence is true. (5.2)

and some related terms, we find a simple universe of discourse consists of the subject "This sentence" and the possible predicates "is true" and "is false." We now assign illustrative status operators:

Item	Status Operator
"This sentence"	P
"is true"	P
"is false"	1 – P

(5.3a)

with the result

[55] Also called a truth value operator.

[56] Inner products are often denoted in mathematics with parentheses such as (x, Ay) where x and y are vectors and A is an operator. Dirac's equivalent notation is $\langle x|A|y\rangle$ if A is Hermitean.

Statement	Status Operator Product
"This sentence is true"	PP = P
"This sentence is false"	$P(1 - P) \equiv 0$

(5.3b)

where P is a projection operator. Note $P(1 - P) = (P - P) \equiv 0$. Consequently we find "This sentence is false." does not have a truth value while eq. 5.2 does have a truth value. A paradoxical statement becomes a non-paradox with the introduction of a new formalism for statements, subjects, and predicates: namely status operators.

Another simple example is the Grelling paradox:

"Heterological is heterological."

If we define a simple universe of discourse with the status operators:

Item	Status Operator
"Heterological"	P
"is heterological"	1 – P

(5.4a)

then

Statement	Status Operator Product
"Heterological is heterological."	$P(1 - P) \equiv 0$

(5.4b)

we see the paradox is resolved: the statement eq. 5.4b does not have a truth value. A predicate containing a word, which is also a subject, may have a different status operator.[57]

In the case of more complex universes of discourse, one introduces as many status operators as needed: P_1, P_2, P_3, and so on in such a way as to exclude paradoxical statements. This can be systematically implemented by establishing all combinations of subjects and predicates that are paradoxical (function calls with bad arguments not in the domain of the function) and defining projection operators so as to exclude paradoxical statements from the set of valid statements.

The discussion in this section is meant to give the flavor of Operator Logic without obscuring it with numerous technical details. The details follow.

---

[57] The reader will note that the operator formalism implements the concept that the predicate of a statement is a function and the subject is the argument of a function call. The operator formalism is a mathematical mechanism for implementing the domain concept for function arguments.

5.5 Operator Representation of a Statement

In the previous section we introduced status operators, which are projection operators. We can extend the operator formalism to include interpreting statements as the "eigenvalues" of sequences of eigenvalue operators where the "eigenvalues" are words, phrases, or sequences of logical symbols.[58] We will call these strings of characters "eigenvalues" as well, although eigenvalues are normally numeric.[59] A *term* is one or more words or symbols. The terms that compose a statement can be viewed as occurring in consecutive "stages" (although the exact ordering is language dependent.)
We will develop an operator formalism that will yield statements as a sequence of eigenvalues of operators.

5.5.1 Dirac Notation vs. (x, y) Notation for Inner Products

A simple example that illustrates the use of eigenvalue operators and Dirac notation is to define a state vector with two "eigenvalues" such as

$$|\Omega_1> = |\text{"The car", "is red"}> \tag{5.5}$$

and two operators, Subject and Predicate, satisfying the "eigenvalue" equations:

$$\text{Subject}|\Omega_u> = \text{"The car"}\ |\Omega_u> \tag{5.6}$$
$$\text{Predicate}|\Omega_u> = \text{"is red"}\ |\Omega_u> \tag{5.7}$$

with the resulting inner product[60]

$$\begin{aligned}<\Omega_u|\text{Subject} + \text{Predicate}|\Omega_u> &\equiv (\Omega_u,(\text{Subject} + \text{Predicate})\Omega_u)\\ &= \text{"The car is red"}\end{aligned} \tag{5.8}$$

where a + sign is used as a separator between the operators.

In the present example the "eigenvalues" are strings of characters. Mathematically, we can relate each string of characters (symbols) in a one-to-one manner[61] to a real numeric value by mapping strings into Gödel numbers, denoted

---

[58] Thus terms can be primitive terms, defined terms, functions, variables, and logical and mathematical symbols.

[59] We convert terms to numeric values – Gödel numbers – at a later point in the discussion. Thus treating terms as eigenvalues is legitimatized by a mapping from terms to Gödel numbers.

[60] The Operator Logic eigenvalue operators and the projection operators are all hermitean in this book. Thus $(\alpha, O\beta) = (O\alpha, \beta)$ due to O's hermiticity and we can use Dirac notation $<\alpha|O|\beta>$ without ambiguity due to O's hermiticity.

[61] A consequence of the "Fundamental Theorem of Arithmetic", namely, that every integer number greater than one, which is not a prime itself, has a unique representation in terms of prime number factors.

gn(i). By mapping strings into Gödel numbers[62] we can establish numbers as eigenvalues for our eigenvalue operators. The set of operators can then be defined within the context of a linear vector space formalism. Thus we reduce the various eigenvalue operators of interest to self-adjoint linear vector space operators.

5.5.2 Gödel Numbers

Gödel numbers are real positive numbers. They can be viewed as the eigenvalues of self-adjoint operators A_i corresponding to a Subject and a Predicate. For example

$$A_{\text{Subject}}|\Omega_u> = \text{gn}(\text{“The car”})|\Omega_u> \qquad (5.9)$$
$$A_{\text{Predicate}}|\Omega_u> = \text{gn}(\text{“is red”})|\Omega_u>$$

with $|\Omega_u>$ the direct product of eigenstates of A_{Subject} and $A_{\text{Predicate}}$

$$|\Omega_u> = |\ \text{gn}(\text{“The car”})>|\ \text{gn}(\text{“is red”})> \qquad (5.10)$$

Using plus signs as *separators* we express the statement as

$$(A_{\text{Subject}} + A_{\text{Predicate}})|\Omega_u> = \text{“The car is red”}|\Omega_u> \qquad (5.11)$$

A string expression such as

AyyBx

is assigned a Gödel number by combining the assigned numerical values of each symbol in the string. The numerical values become exponents of a product of prime numbers greater than one. For example the symbols in the preceding string expression could have the symbol numbers: A ≡ 3, y ≡ 9, B ≡ 5 and x ≡ 7. Then the string's Gödel number is

$$\text{gn}(\text{“AyyBx”}) = 2^3 3^9 5^9 7^5 11^7 \qquad (5.12)$$

where the numbers being exponentiated are the ordered set of prime numbers.
In general, if E is an expression (string) of symbols, $\beta_1, \beta_2, \beta_3, \beta_4, \ldots \beta_n$ and $v_1, v_2, v_3, v_4, \ldots v_n$ is the sequence of integers corresponding to these symbols, then the Gödel number of E is the integer

$$\text{gn}(E) = \prod_{m=1}^{n} \text{Pr}\,(m)^{v_m} \qquad (5.13)$$

[62] It should be noted that Leibniz developed a procedure to map strings to integers. See Rescher (1967) p. 132 and cited references to Leibniz.

where $Pr(m)^{v_m}$ is the mth prime number raised to the power v_m with the 1st prime number taken to be 2. If E is an empty string, then gn(E) = 1.

The definition of Gödel numbers, and their consequent uniqueness, results in the following conclusions and definitions:

1. If g = gn(E), then the inverse relation is defined to be E = gi(g).

2. If E_1 and E_2 are expressions (terms) such that $gn(E_1) = gn(E_2)$, then $E_1 = E_2$.

3. If $E_1, E_2, E_3, \ldots, E_n$ are a finite sequence of expressions, then the Gödel number of the sequence is

$$gn(E_1, E_2, E_3, \ldots, E_n) = \prod_{m=1}^{n} Pr(m)^{gn(E_m)} \tag{5.14}$$

4. The Gödel number of an expression is never equal to the Gödel number of a sequence.

5. If $E_1, E_2, E_3, \ldots, E_n$ and $E'_1, E'_2, E'_3, \ldots, E'_n$ are both finite sequences of expressions and

$$gn(E_1, E_2, E_3, \ldots, E_n) = gn(E'_1, E'_2, E'_3, \ldots, E'_m) \tag{5.15}$$

then n = m, and for k = 1, 2, ..., n

$$E_k = E'_k$$

and the strings to which they correspond are identical.

5.5.3 Statements Expressed in Gödel Numbers

Statements can be expressed in terms of Gödel numbers. In this subsection we describe the general interpretation of expressions involving eigenvalue operators.

If eigenvalue operators are formally "added" as in eq. 5.11 then the sum of eigenvalue operators can be viewed as a symbolic list of Gödel numbers that is not arithmetically added or multiplied. As we read a statement from word to word the statement "unfolds" and becomes particularized, more precise and more specific. In this sense the symbolic sum of eigenvalue operators specifies the statement when applied to the state that represents the system.

Placing eigenvalue operators adjacent to each other[63] is interpreted as equivalent to the logical relation "or."

As an example, we add an additional term "is green" with operator $A_{\text{is green}}$ to the previous example and use the state

$$|\Omega_2\rangle = |\ gn(\text{"The car"})\rangle|\ gn(\text{"is red"})\rangle|\ gn(\text{"is green"})\rangle$$

Then the statement

$$(A_{\text{Subject}} + A_{\text{is red}}A_{\text{is green}})|\Omega_2\rangle \equiv \text{"The car is red or is green."}|\Omega_2\rangle$$

exemplifies the "or" case. Alternately, we could define "or" as a term with a corresponding Gödel number.

$$(A_{\text{Subject}} + A_{\text{is red}} + A_{\text{or}} + A_{\text{is green}})|\Omega_3\rangle \equiv \text{"The car is red or is green."}|\Omega_3\rangle$$

where

$$|\Omega_3\rangle = |\ gn(\text{"The car"})\rangle|\ gn(\text{"is red"})\rangle|\ gn(\text{"is green"})\rangle|\ gn(\text{"or"})\rangle$$

---

[63] This corresponds with the "or" notation of Hilbert (1950) p. 12 where XY = X v Y.

6. Linear Vector Spaces for Universes of Discourse

We will define two types of universes of discourse: a semantic universe of discourse and a calculus universe of discourse. A *calculus universe of discourse* consists of undefined primitive terms, axioms, and the theorems derived from these axioms. Statements in a calculus are provable, not provable or undecidable.

A *semantic universe of discourse* consists of primitive terms defined in some concrete fashion, axioms, and the theorems derived from the axioms. Statements in a semantic universe are true, false, or undecidable.

A calculus universe of discourse can be viewed as a semantic universe of discourse stripped of its definition of primitive terms and reduced to a "skeleton of algebra" with truth reduced to provability.

6.1 The Semantic Universe of Discourse

A *semantic universe of discourse* consists of a set of primitive and defined terms, related to some Reality,[64] that are used to formulate the axioms and theorems of a deductive system. The terms are mapped to Gödel numbers. Each term E has a unique Gödel number gn(E). Each Gödel number is assumed to be an eigenvalue of a unique, self-adjoint, eigenvalue operator that we denote A_E. Each A_E has two eigenvalues: E = gn(" ... "), the Gödel number of a non-empty string, or E = gn("") = 0. We denote an eigenstate as |E>. The eigenstate |0> corresponds to the empty string. Thus

$$A_E|s\rangle = gn(s)\,|s\rangle$$

$$A_E|0\rangle = gn(\text{``''})\,|0\rangle = 0$$

where s is a non-empty string.

If there are n primitive terms in a universe of discourse, then the *universe of discourse state* $|\Omega_u\rangle$ is a direct product of all term eigenstates with the form:

$$|\Omega_u\rangle = |E_1, E_2, E_3, \ldots, E_n\rangle = \prod_{m=1}^{n} |E_m\rangle \tag{6.1}$$

where each E_i can be a non-zero Gödel number, or zero. Note: since n is the total number of terms in the semantic universe of discourse it must be finite since the

[64] Related to Reality in the sense that each primitive term is "defined" in terms of real world features.

number of words in the English language, or any language expressing semantic meanings, is finite. Each eigenvalue operator A_{E_k} satisfies the equation[65]

$$A_{E_k}|E_1, E_2, E_3, \ldots , E_n> = gn(E_k)|E_1, E_2, E_3, \ldots , E_n> \qquad (6.2)$$

for k ε [1, n]. *All eigenvalue operators in a universe of discourse* in this chapter *commute and are thus compatible.*[66]

A semantic statement has the symbolic form

$$\sum_{n \in \{m\}} A_{E_n} \qquad (6.3)$$

for some subset {m} of the set of terms in the universe of discourse. If applied to the eigenstate eq. 6.1 we obtain the *symbolic* sum

$$\sum_{n \in \{m\}} A_{E_n}|E_1, E_2, E_3, \ldots , E_n> = \sum_{n \in \{m\}} gn(E_n)|E_1, E_2, E_3, \ldots , E_n> \qquad (6.4)$$

The Gödel numbers in eq. 6.4 are not literally added together. Rather the terms in the symbolic summation on the right side are individually rendered in a language[67] using the inverse relation of Gödel numbers gi(g), (which is property 1 in section 5.5.2) on each term of the symbolic sum. Additional words such as "a" and "the" are added to the statement implicitly to make it grammatically correct.

Thus we have a linear vector space formalism for the semantics of a deductive system. We thus "turn the tables" and use the very well known mathematics of linear vector spaces as the "language" of our semantic universe—rather than English or an artificial symbolic language, which both contain the possibilities of paradox and inconsistency.

One immediate benefit of a linear vector space formulation is that each self-adjoint eigenvalue operator A_E has a corresponding projection operator[68] P_E that projects a composite state. For example, suppose

---

[65] **Each eigenvalue operator and its corresponding projection operators operate on the corresponding eigenstate in the product of eigenstates comprising $|\Omega_u>$. Thus, although we do not make it explicit to avoid complicated expressions, one can consider each eigenvalue operator and its projection operator to be a direct product with the implicit factors being identity operators for the spaces spanned by the other eigenvectors in $|\Omega_u>$ of the other eigenvalue operators.** This also applies in the discussions in the following chapters.

[66] In Quantum Operator Logic, described in a later chapter, eigenvalue operators do not necessarily commute.

[67] Spaces and other required grammatical constructs such as "a" and "the" are assumed to be automatically inserted when translated into a language.

[68] Mackey (1963) pp. 64-65.

$$|E_{tot}> = a|E_1> + b|E_2> \quad (6.5)$$

Then projection operators project states into a specific eigenvalue state $|E_1>$ or $|E_2>$:[69]

$$P_{E_1}|E_{tot}> = a|E_1> \quad (6.6)$$

In addition, the operator $1 - P_E$ projects a state $|E>$ to 0. Thus

$$(1 - P_{E_1})|E_1> = 0 \quad (6.7)$$

We call the projection operators P_E *eigenvalue projection operators.*[70] The eigenvalue projection operators are combined with *semantic status projection operators*[71] denoted P'_E to determine whether a statement has a truth value, and, if it has a truth value, its trueness.[72] Combining these projections we have the *total projection*

$$P_{Etot} = P'_E P_E \quad (6.8)$$

The sets of total projector operators $\{P_{Etotk}, (1 - P_{Etotk})\}$, are the status operators discussed in section 5.4. Note: *All projection operators in a universe of discourse in this chapter commute.* The initial[73] general form of the projection operator expression corresponding to eq. 6.4 that determines whether a statement has a truth value is

$$F(P'_{E_1}, P'_{E_2}, \ldots) = \prod_{n \in \{m\}} P'_{E_n} \quad (6.9)$$

where {m} specifies the subset of terms in the statement.
If

$$<\Omega_u|F(P'_{E_1}, P'_{E_2}, \ldots)|\Omega_u> = 0 \quad (6.10)$$

then the corresponding statement has no truth value. If

$$<\Omega_u|F(P'_{E_1}, P'_{E_2}, \ldots)|\Omega_u> \neq 0 \quad (6.11)$$

then the corresponding statement has a truth value that may be true or false.

---

[69] The state $|0>$ corresponds to the empty string.
[70] We also call these operators *state projection operators.* These operators become significant when we discuss subuniverses in chapter 9. In the case of universe states such as those in this chapter eigenvalue projection operators always yield one when applied to a universe state: $P_E|\Omega>$.
[71] We also call these operators *truth value projection operators.*
[72] The trueness of statements are discussed in detail in chapter 7.
[73] The general form becomes a more complex expression in chapter 7.

The form of $F(P'_{E_1}, P'_{E_2}, \ldots)$ given above is for a simple subject-predicate statement. The form of $F(P'_{E_1}, P'_{E_2}, \ldots)$ is more complex for more complex statements that are reducible to a set of subject-predicate clauses (that depend on the logical constructs "and", "or", and so on). It will be discussed in chapter 7.

We can view a projection operator as projecting or "filtering" a state and producing a new state in general.[74] Each projection operator in a series of projection operators filters a state step by step to produce a final state. The most general form of a projection for a simple subject-predicate statement is:

$$F(P'_{E_1}, P'_{E_2}, \ldots) \prod_{n \in \{m\}} P_{E_n}|E_1, E_2, E_3, \ldots, E_n> \qquad (6.12)$$

where $\{m\}$ specifies the subset of terms in the statement. It determines whether a statement has a truth value (a non-zero value) or not (a zero value) when one takes its inner product with the initial universe state:

$$<E_1, E_2, E_3, \ldots, E_n|F(P'_{E_1}, P'_{E_2}, \ldots) \prod_{n \in \{m\}} P_{E_n}|E_1, E_2, E_3, \ldots, E_n> \qquad (6.13)$$

It yields either a zero (no truth value) or non-zero (has a truth value) value.

6.2 The Calculus Universe of Discourse

A *calculus universe of discourse* consists of a set of primitive terms, and terms defined in terms of them. They are unrelated to any reality and have only a symbolic significance. They are used to formulate the axioms and theorems of a deductive system. A calculus may correspond to one or more semantic deductive systems. The formulation of a calculus universe of discourse is analogous to that of a semantic universe of discourse.

The Operator Logic for a calculus universe of discourse is identical to that of the semantic universe of discourse as described in section 6.1.

6.3 Variables in Operator Logic

If a primitive term is in fact a function of one or more variables $x_1, x_2, \ldots$, then the corresponding eigenvalue operator and projection operator are also functions of these variables: $A_E(x_1, x_2, \ldots)$ and $P_{Etot}(x_1, x_2, \ldots)$ respectively for semantic and calculus universes of discourse.

If a free variable v appears in a statement, then it has an eigenvalue operator $A_v(v)$, and a projection operator, $P_{vtot}(v)$. A function f(x) has an

[74] This is similar to the role projection operators play in the analysis of quantum physics experiments. See Blaha (2009) for a more detailed discussion.

eigenvalue operator $A_f(x)$ with projection operators $P_{ftot}(x)$. The generalization to functions of several values is direct.

6.4 Linking Universes of Discourses Linear Vector Spaces

The calculus and semantic linear vector space statements are linked using a generalization of Tarski's method of combining a calculus sentence with its semantic meaning through eq. 6.14:

$$\text{SM is true if, and only if, S} \tag{6.14}$$

where S is a statement and SM is its semantic equivalent. If S is true, then SM is true. If S is false, then SM is false. If S is Gödel undecidable, then SM can be true or false.

Clearly, statement by statement, we can map semantic statements in a deductive system to syntactic statements in its corresponding calculus. We can always define a *perfect map* at the operator level in which eigenvalues and their corresponding eigenvalue operators have a one-to-one map as do the projection operators.

For example, consider a semantic statement with the form of eq. 6.4. Its corresponding product of projection operators is

$$P_S = F(P'_{E_1}, P'_{E_2}, \ldots) \prod_{n \in \{m\}} P_{E_n} \tag{6.15}$$

The product of projection operators for the corresponding calculus statement is

$$Q_S = F(Q'_{E_1}, Q'_{E_2}, \ldots) \prod_{n \in \{m\}} Q_{E_n} \tag{6.16}$$

where Q_{E_n} is the calculus projection operator corresponding to P_{E_n}, and Q'_{E_n} is the status operator corresponding to P'_{E_n}. The operator equivalents of eq. 6.14 and its associated comments are:

1. Q_S is non-zero if, and only if, P_S is non-zero (indicating that the statement can be proven true or false, or provable or not provable, in both universes of discourse.)

2. Q_S is zero if, and only if, P_S is zero (indicating that the statement is Gödel undecidable in both universes of discourse – it cannot be proved and cannot be disproved.) However, the statement may be shown true or false (or provable or not provable) by other means – for example, from experiments.

It is assumed that these projection operator expressions are evaluated between universe states.

The reader will note that these points differ from the Tarski formulation (eq. 6.14) in several ways. Tarski's formulation assumes that the statements, S and SM, are either true or false – contrary to the implications of Gödel's Undecidability Theorem. Points 1 and 2 above are consistent with the concept of undecidability. Item 2 provides a calculational approach to determine "decidability."

English Language Statement	**Semantic System Statement**	**Calculus Statement**
Eigenvalue Operator Sum	Semantic Eigenvalue Operator Sum	Syntax Eigenvalue Operator Sum
Status Projection Operator Product	Semantic Projection Operator Product	Syntax Projection Operator Product

Figure 6.1 A table of the syntactical and semantic operators associated with a statement.

Earlier we considered a simple example of a two statement semantic system, which is also its calculus:

(1) This sentence is false.
(2) This sentence is true.

Statement (1) is undecidable – being neither true nor false. Statement (2) is true. We do not differentiate between the semantic system and the calculus in this case since they are essentially identical.

The general case of the operator representation of an English (or other) language statement in a semantic system and its corresponding calculus statement is displayed in Fig. 6.1.

6.5 Predicate Calculus

Returning to an earlier example (subsections 5.5.1 – 5.5.3) we see that the predicate calculus can be implemented in our linear vector space representations. First we note the statement generated from the Subject and Predicate operators is

$$((\text{Subject}) + (\text{Predicate}))|\Omega_u> = \text{“The car is red”}\ |\Omega_u> \qquad (6.17)$$

which we again note is implemented via corresponding self-adjoint operators with Gödel number eigenvalues.

One can also define projection operators:

$$P_{\text{Subject(``The car'')}} = P_1$$

and

$$P_{\text{Predicate(``is red'')}} = P_2 \qquad (6.18)$$

The simplicity of the example allows us to treat semantic and calculus operators similarly. If the product of the projection operators for the statement, when evaluated for the universe state is not identically zero, then the statement has a non-zero truth value of true or false.

This simple example can be directly extended to more complex statements with multiple subjects and clauses.

6.6 General Approach to Subjects and Domains of Predicates

In a given universe of discourse we can introduce semantic status projection operators denoted P'_i for each predicate, or for primitive terms within a predicate as the case may be. Then, for each predicate, we can determine the class of acceptable subjects (subjects, which when combined with the predicate to form a statement, form a valid statement that is either true or false). For the jth subject in the class of allowed subjects, we can assign a semantic status projection operator P'_{ij}.

Unacceptable subjects for predicate i (subjects, which, when combined with a predicate form an invalid statement that has no truth value) are assigned a projection operator P' such that $P'_i P' = 0$.
Thus, in a universe of subjects and predicates, each predicate has a class of acceptable subjects (its *domain* D of allowed subjects) and a complementary class of unacceptable subjects.

We implement domains of predicates assigning a set of projections to each predicate—one semantic status projection operator (truth value projection) for each subject in the domain of subjects for the predicate. For predicate p we will denote its *predicate projection operator* as

$$P_{ptot} = P'_p P_p \qquad (6.19)$$

using the projection operators defined in the discussion of eqs. 6.5 – 6.8.

An eigenvalue *projection operator* P_p when applied to an eigenstate $|p\rangle$ of the predicate satisfies $P_p|p\rangle = |p\rangle$ if p is not a predicate consisting of an empty string (which has Gödel number zero). ($P_p|0\rangle = 0$ for an empty string.) The predicate semantic status projection operator (truth value projection operator) P'_p is defined by[75]

$$P'_p = \prod_{i \in D} P'_i \prod_{j \notin D} (1 - P'_j) \qquad (6.20)$$

[75] We have called these operators, and corresponding subject operators, status operators in earlier chapters.

where the combined set of projections denoted above as P'_i and P'_j are the truth value projection operators for all subjects in the universe of discourse for predicate p. D is the domain of subjects which can combine with predicate p to form statements with truth values.

For each subject s we associate a subject projection operator:

$$P_{stot} = P'_s P_s \tag{6.21}$$

where P_s is the subject eigenvalue (state) projection operator. When P_s is applied to the subject eigenstate |s> it satisfies $P_s|s> = |s>$ if the subject is not an empty string (which has Gödel number zero). ($P_s|0> = 0$ for an empty string.) P'_s is the subject truth value projection operator of the subject s.

P'_p satisfies

$$P'_p\, P'_s = P'_s\, P'_p = P'_p \tag{6.22}$$

by eq. 6.21 for predicate p with subject s contained in D, and

$$P'_p\, P'_s = P'_s\, P'_p = 0 \tag{6.23}$$

for predicate p with subject s *not* contained in D.

6.7 Formalism for Operator Logic

In this section we will look at the operator formalism for a semantic universe of discourse. A *statement* is represented by an inner product[76] containing a symbolic sum of the eigenvalue operators of the statement:

$$\text{statement} = \langle\Omega_u|A_1 + A_2 + A_3 + \ldots\ |\Omega_u\rangle \tag{6.24}$$

where A_i is the ith eigenvalue operator, and its truth value[77] is given by

$$TV = \text{Truth Value} = \langle\Omega_u|\ F(P'_{E_1}, P'_{E_2}, \ldots) \prod P_{E_n}\ |\Omega_u\rangle \tag{6.25}$$

TV = 0 indicates the statement has no truth value. TV $\neq$ 0 indicates the statement has a truth value: true or false. The projections appearing in F and the product are those corresponding to the eigenvalue operators in the statement.

---

[76] Expressions of the form <α|O|β> are inner products that are called *expectation values* in quantum mechanics and are often denoted as (α, Oβ) in mathematics texts where O is a hermitean operator. The Operator Logic eigenvalue operators and the projection operators are all hermitean. Thus (α, Oβ) = (Oα, β) due to O's hermiticity and we can use Dirac notation <α|O|β> without ambiguity due to O's hermiticity.

[77] In the case of a calculus universe of discourse truth value becomes provability.

We will define the *universe state* $|\Omega_u\rangle$ for a universe with q primitive terms as

$$|\Omega_u\rangle = |E_1, E_2, E_3, \ldots, E_q\rangle \tag{6.26}$$

where each eigenvalue E_i is a non-zero Gödel number and the set of integers 1, 2, …, q is denoted {m} and $\langle\Omega_u|\Omega_u\rangle = 1$. The hermitean conjugate of $|\Omega_u\rangle$ is denoted $\langle\Omega_u|$. Our equations use Dirac's bra-ket notation for states and inner products.

A statement is composed of adjoining (a symbolic sum of) Gödel numbers: gn(i). Using the inverse function gi(gn(i)) on each adjoining Gödel number (separated by symbolic plus signs) the statement can be "translated" into English or some other human or symbolic language. The plus signs act as separators and are subsequently discarded. Grammatical constructs such as "a" and "the" can be added as needed to make a grammatically correct statement.

6.8 A Simplified Example

One simple example of a semantic universe has the primitive terms (and corresponding eigenvalue operators):

Primitive Term	Eigenvalue Operator	Truth Value Projection Operator
"This sentence"	A_1	P'_1
"Some Sentence"	A_2	P'_2
"is true"	A_3	$P'_3 = P'_1$
"is false"	A_4	$P'_4 = 1 - P'_1$

where we assign projection operators to each term using an operator P' defined below. The primitive terms consist of two predicates, "is true" and "is false" and the subjects: "This sentence" and "Some sentence".

We can construct four simple statements from these primitive terms:

English Statement	Eigenvalue operator Sum	Projection Operator Inner Product[78] $\langle\Omega_1\|\text{product}\|\Omega_1\rangle$
"This sentence is true"	$A_1 + A_3$	$P'_1P'_3 = P'_1 \neq 0$
"This sentence is false"	$A_1 + A_4$	$P'_1P'_4 = P'_1(1 - P'_1) \equiv 0$
"Some sentence is true"	$A_2 + A_3$	$P'_2P'_3 = P'_2P'_1 \neq 0$
"Some sentence is false"	$A_2 + A_4$	$P'_2P'_4 = P'_2(1 - P'_1) \neq 0$

where the zero and non-zero values in the third column result from taking the inner product $\langle\Omega_1|\text{product}|\Omega_1\rangle$ using the universe state $|\Omega_1\rangle$.

[78] The universe states are not displayed in this column to save space.

From the preceding table we see that the statements represented by $A_1 + A_3$, $A_2 + A_3$ and $A_2 + A_4$ have truth values (either true or false). The statement represented by $A_1 + A_4$ does not have a truth value.

The statements represented by $A_1 + A_3$, $A_2 + A_3$ and $A_2 + A_4$ have the truth value true from semantic considerations. This result is not due to the projection operator results. Projection operators only determine whether a statement has a truth value, or does not have a truth value, *at this point in our discussion.*[79]

For each of the predicates A_3 and A_4 we identified a domain of subjects that lead to statements that have a truth value. For A_3 we find A_1 and A_2 are subjects that leads to statements with a truth value and thus constitute the domain of A_3. In the universe specified above we find A_4 has a domain consisting of A_2 only.

It is clear that the identification of the domain of subject terms for each predicate that leads to statements with a truth value is the key to excluding paradoxical statements. We will see this in more detail after discussing primitive terms in general in our formulation. This discussion can be directly generalized to more complex statements.

6.9 The Primitive Terms of Universes of Discourse

The set of primitive terms of a universe of discourse consists of "string eigenvalues" that are mapped to the Gödel numbers of the corresponding eigenvalue operators of the universe. Each primitive (and defined) term maps to one eigenvalue operator. Then all statements in the universe of discourse can be expressed as symbolic sums of the corresponding eigenvalue operators.

The primitive terms are the key to the entire universe of discourse. We now consider primitive terms in human and symbolic languages.

Statements can be expressed in primitive terms in human languages such as English or in artificial symbolic languages. A problem that arises—particularly in human languages—is that there are different ways of expressing the same statement and different ways of specifying the same primitive terms. A particularly good example of this issue appears when one compares a primitive term in German, which can create words by combining other words, and English where the equivalent of a German primitive term may be a phrase.

In this section we point out that one can define a map between the primitive terms of two semantic universes of discourse that have the same calculus universe of discourse. Thus no problem can arise if we choose to use differing primitive terms in two semantic universes of discourse with the same calculus universe of discourse. A famous example of this situation is the calculus of Euclidean geometry *without* the fifth postulate. Depending on the definition of

[79] At a later point in the development they will be used to determine the truth or falsity of such statements.

"point", "straight line" and "congruence" one obtains Euclidean geometry or a non-Eucliean geometry.[80]

Thus the issue of primitive terms is important in the consideration of semantic universes of discourse. We will now consider general features of primitive terms.

A universe of discourse consists of

1. Primitive terms
2. Axioms expressed in terms of the primitive terms
3. Terms defined as combinations of primitive terms
4. A set of theorems derivable from the axioms

The primitive terms in a calculus universe of discourse are undefined. The primitive terms in a semantic universe of discourse are given a meaning external to the universe of discourse. Frequently the meaning is based on physical or mathematical entities.

If two semantic universes of discourse have the same calculus universe of discourse then the primitive terms in each semantic universe must map to the other on a one-to-one basis. (It is possible that a primitive term in one universe of discourse maps to a pair of (or several) primitive terms in the other universe of discourse. However the pair of (or several) terms always appear together in the other universe of discourse with the result that they are effectively one term.)

*6.10 The Relation Between Statements, Predicates, and Functions

Most scientists, Logicians and Mathematicians are familiar with mathematical functions that have numerical arguments and calculate a numerical value or set of numerical values. They are also familiar with Logic statements. Superficially statements and mathematical functions appear to be unrelated. However they are related at a deeper level. *We shall see that a statement is a form of function call. Paradoxes arise because the argument(s) of a statement function call is outside the domain of the statement function.*

6.10.1 Types of Functions

Those familiar with Computer Science know that a function, in general, has two types of output: its numeric value(s), and a status value(s) indicating whether the computation that the function performed was successfully done, or failed, or how it succeeded or failed. Typically the status value is a zero or one,[81] but it can be understood as true or false also. (True – computation successful; false

[80] Weyl (1950) p. 81.

[81] In many cases the status value could be one of a set of values indicating various stages of success and/or how the computation failed.

– computation failed for some reason) Thus we categorize functions as being in one of three broad categories:

Types of Functions

A Mathematical Function Producing A Value Only Example: sin function	A Function Producing a Value(s) and a Status Value(s) Example: a Programming Function	A Function Producing a Status Value Only Example: a logical statement

6.10.2 Predicates as Functions

Many years ago Leibniz noted that the predicate of a statement could be viewed as a function and the subject could be viewed as the argument of the function. Thus a statement can be described as a "function call" in a fashion similar to mathematical function calls such as sin(90°).

Based on this characterization of predicates and subjects we can assert that for a given predicate a decidable statement must have a subject that is a member of a set or domain of allowed subjects. An undecidable statement has a subject that is not a member of the predicate's domain of allowed subjects.

6.10.3 Undecidable Statements

Thus the interpretation of statements as function calls to functions with a domain of allowed subjects puts statements on a par with mathematical functions which also usually have restricted domains for their arguments. Undecidable statements then are on a par with function calls whose arguments have values that correspond to singularities, or other flaws, in the function.

The main difference between statements and mathematical function calls in this respect is that invalid mathematical function arguments are usually easily discerned from the form of the function whereas the invalidity of the subjects of undecidable statements can only be decided on a case by case basis – usually with a detailed analysis that reveals undecidability. Thus undecidability was regarded as a flaw in Logic when it was really a flaw in particular statements.

6.11 Predicate Domains and Paradox Avoidance

From the discussion of previous sections it is clear that each predicate has a domain of allowed subjects.[82] The subjects that are not in the domain of allowed

[82] In this respect our view is contrary to Leibniz's view of 1686, and other views, that the predicate is contained in the concept of the subject in each true statement. Our view does agree with Frege's use

subjects can be used to form statements. But these statements would have no truth value.[83] They would be neither true nor false and thus would properly be called undecidable or paradoxical statements.[84]

We will now discuss the resolution of most of the well-known paradoxes using our new paradigm.

Liar Paradox

The statement of the Liar Paradox has a subject such as Subject = "The statement that I am now making" and predicate "is a lie." If the subject is reflexive as it appears implicitly in the predicate, which might be better phrased as "is my lie", then a paradox is evident. The elimination of this paradox is accomplished by limiting the domain of the predicate to non-reflexive subjects. In the Operator Logic formalism the domain of the predicate operator is limited to non-reflexive subject operators. One implements this by assigning a projection P to the predicate and the projection (1 – P) to all reflexive subject operators. Thus paradoxical statements of this type have no truth value and the Operator Logic formalism excludes such paradoxes.

Grelling Paradox

Some adjectives have the property that they apply to themselves. Such adjectives can be called autological. Other adjectives do not apply to themselves. These adjectives can be called heterological. The Grelling paradox is:

"Heterological is heterological."

If we identify the primitive terms as the predicate "is heterological" and the subject "Heterological", then assigning eigenvalue operators to each term, and the projection operator P as the subject projection operator, and (1 – P) as the predicate projection operator gives the Grelling statement no truth value, thus eliminating the paradox.

Barber Pseudoparadox

The Town mayor issues an order: "The one town barber must shave those men in the village who do not shave themselves." In the simplest Operator Logic

---

of a functional notation wherein the predicate is the function and a subject is the variable. The set of subjects constitutes the domain of the function. It also does agree with Hilbert (1950), "To each predicate corresponds a certain 'class' of objects, consisting of all objects for which the predicate holds." p. 46.

[83] Truth value becomes provability in a calculus universe of discourse.

[84] Herbrand, Schmidt, Wang, Hintikka, Hailperin, Lightstone & Robinson, Gilmore, and Quine have considered approaches within the framework of earlier formalisms based on restricting the subject domain.

form the subject is "The one town barber" and the predicate is "must shave those men in the village who do not shave themselves." Assigning eigenvalue operators to each term, and the subject projection operator P and the predicate projection operator (1 – P) results in the statement not having a truth value. Thus the paradox is avoided.

Berry Paradox

The number of positive integers that can be named in English in less than a fixed number of syllables is finite. Thus there must be a least integer that cannot be so named. However, " the least integer that cannot be named in English in less than fifty syllables" is an English name of less than fifty syllables. Thus the least integer has a name contrary to the assumption and thus a paradox.

Resolution:

This superficial paradox is based on an ambiguity in the use of the word "named." If it means "numerically named" (in words representing a numerical value) there is no paradox. If it means "any type of name" then an apparent paradox appears until one realizes that it is then a tautology "The least integer that cannot be named in English in less than fifty syllables is the least integer that cannot be named in English in less than fifty syllables."

Russell's Paradox

From experience we know we can consider classes of things; classes of integer numbers, classes of cars, and so on. We can consider classes of classes such as the class of all classes of cars in the various big cities of America. There are two interesting varieties of classes: proper classes and improper classes. *Proper classes* are classes, which are not members of themselves. For example, the class of all cars in China is proper. *Improper classes* are classes, which are members of themselves. For example, the class of all classes is a member of itself and thus improper.

Let us define the Russell class R as the class of all proper classes. Paradox: if R is a proper class, it is a member of itself, and is thus by definition an improper class. If R is an improper class, it is not a member of itself and therefore, by definition, it is a proper class. Thus there is no resolution of this paradox according to conventional logic.

Resolution:

The primitive terms of this paradox are the subject, "the class of all proper classes" and the predicate, "is a proper class." After associating an eigenvalue operator and projection operator with each term we see that the statement "the class of all proper classes is a proper class" has no truth value if we specify a projection operator P for the subject and (1 – P) for the predicate. This is again a case where the paradox is avoided by the predicate having a non-reflexive domain.

Cantor Paradox

According to the theory of cardinal (infinite) numbers the set of all subsets of a set C has a higher cardinal number than C. If C is the set of all sets, then the preceding statement is a contradiction.

Resolution:

The Cantor paradox can be changed to the following by simple substitution for C:

> The set of all subsets of the set of all sets has a higher cardinal number than the set of all sets.

The subject can be simplified to give:

> The set of all sets has a higher cardinal number than the set of all sets.

which is a manifestly verbal paradox but not an Operator Logic paradox.

If we identify the subject term as "The set of all sets" and the predicate term as "has a higher cardinal number than the set of all sets", and define corresponding eigenvalue and projection operators, then the statement can be shown to have no truth value eliminating the paradox. We choose the projection operator for the subject as P and the projection operator for the predicate as (1 – P) with the result that the statement has no truth value within the framework of Operator Logic. Again the paradox is avoided by the predicate having a non-reflexive domain.

Burali-Forti Paradox

The Burali-Forti paradox is based on the theory of transfinite (infinite) numbers. The theory proves a) every well-ordered set has a unique ordinal number; b) any set of ordinals, that is placed in a natural order such that each element contains all its predecessors, has an ordinal number which is greater than any preceding element in the set; and c) the set A of all ordinals placed in natural order is well-ordered. Then by theorems a and c, A has an ordinal number n. Since n is in A we see $n < n$ by theorem b, thus establishing a contradiction.

Resolution

The paradoxical conclusion can be stated verbally as:

> The ordinal of the set of all ordinals is greater than the ordinal of the set of all ordinals.

In this statement we can identify the subject primitive term as "The ordinal of the set of all ordinals" and the predicate primitive term as "is greater than the ordinal of the set of all ordinals." Here again we see a reflexivity in the subject. If we restate the conclusion in terms of eigenvalue operators with the subject projection operator P and the predicate projection operator (1 – P) then the inner product $<\Omega_u|P(1 - P)|\Omega_u> = 0$ indicating that the statement has no truth value. Thus the statement is excluded from the set of statements with truth values in this universe of discourse calculus. The domain of the predicate does not include reflexive subjects such as the one given above.

Richard Paradox

The Richard paradox is concerned with the proposition: the set of all numerical functions is not enumerable. A commonly used argument to prove this proposition is the following. Suppose an enumeration existed symbolized by $f_n(m)$, which represents the nth function with argument m. Consider the function g defined by

$$g(n) = f_n(n) + 1$$

for any value of n. Let n_0 be the index number of g(n) in the enumeration:

$$g(n) = f_{n_0}(n) + 1$$

then

$$f_{n_0}(n) = g(n) = f_n(n) + 1 \qquad (6.27)$$

and

$$f_{n_0}(n_0) = g(n_0) = f_{n_0}(n_0) + 1 \qquad (6.28)$$

which is a contradiction. Thus the set of all numerical functions is not enumerable.

Contrarian argument: Consider the set of all definable functions. Definable is taken to mean definable in some specific language with a fixed dictionary and grammar. Since the number of words in the language is finite, then the number of expressions is enumerable. Thus the set of expressions that form the definitions of definable functions is enumerable. Thus the set of definable functions is enumerable. Since the set of numerical functions is a subset of the set of definable functions it also must be enumerable. Thus the set of all numerical functions is enumerable.

The result of the preceding two arguments is a contradiction (paradox).

Resolution

The projection operator inner product corresponding to eq. 6.27 has the form (see subsection 3.2.3 for a discussion of variables)

$$<\Omega_u| \ldots P_{f_{n_0}}(n) \ldots (1 - P_{f_n}(n)) \ldots |\Omega_u> \qquad (6.29)$$

which is non-zero (and thus the statement has a truth value) for $n \neq n_0$. However for $n = n_0$ eq. 6.29 is equal to zero[85] and thus eq. 6.28 has no truth value and therefore the paradox is avoided. The contrarian proof is not contradicted.

Gödel's Undecidability Theorem

Gödel's Undecidability Theorem, as he pointed out in his celebrated paper, is closely related to Richard's paradox (antinomy) and the Liar's paradox. Gödel provides an explicit undecidable statement, eq. 1, in his paper:

$$n \, \varepsilon \, K \equiv \overline{\text{Bew}}[R(n); n] \tag{6.30}$$

where

1. Bew Y means Y is a provable formula (statement)
2. The bar over Bew means "not"
3. K is a class of natural numbers
4. A class-sign is a formula[86] with one free natural number variable
5. The set of class-signs is arranged in a series using some rule with the nth class-sign denoted R(n).
6. Define [α, m] to be the formula derived upon replacing the free variable in the class-sign α by the sign of the natural number m. Thus the relation Y = [α, z] is definable in PM.
7. Since the above are all definable in PM, so is the class K that is formed from them.

Therefore there is a class-sign S such that [S; n] implies that n ε K. Thus S = R(q) for some specfic natural number q. If we assume [R(q); q] is provable, then q ε K. But, by eq. 6.30

$$\overline{\text{Bew}}[R(q); q] \tag{6.31}$$

would be true contrary to our assumption.

If we assume the contrary: the negation of

$$[R(q); q] \tag{6.32}$$

is provable, then

$$\overline{q \, \varepsilon \, K} \tag{6.33}$$

and

$$\text{Bew}[R(q); q] \tag{6.34}$$

---

[85] Note all projection operators commute as do all eigenvalue operators in the formulation of this chapter.

[86] Of the Principia Mathematica (denoted PM).

would be true. Thus [R(q); q] and its negation would be provable.

Gödel's Theorem (more precisely any statement conforming to eq. 6.30) is excluded from a mathematical deductive system if the subject is reflexive. Eq. 6.30 can be stated in a subject predicate form as

"n is contained in K which is equivalent to ..."

where "n" is the subject and "is contained ..." is the predicate. In this form we can associate the projection P with "n" and (1 − P) with the predicate part "[R(n); n]" yielding

$$<\Omega_u|P \ldots (1 - P)|\Omega_u> = 0 \quad (6.35)$$

which shows the statement has no truth value.

Gödel's more succinct undecidable statement[87] 17 Gen r is similar in character but disguised in a more condensed notation. The observant reader will note the close resemblance between Gödel's undecidablity result, Richard's paradox and the Liar paradox—as Gödel himself noted.[88]

Conclusion

The preceding examples show that Operator Logic provides a mechanism to systematically exclude statements with no truth value ("undecidable" or paradoxical statements) from a universe of discourse. The following chapters show further advantages of Operator Logic including a generalization to quantum and probabilistic logic, hierarchies of universes of discourse, and the formulation of a mathematical philosophy (metaphysics) relating Platonic Ideas to the physical universe, Reality, that is conceptually similar to the theories of various Platonic schools.

---

[87] Gödel (1992) p. 60.

[88] Gödel (1992) p. 40.

7. Sentential Calculus in Operator Logic

In this chapter we will extend our discussion of projection operators to the determination of the truth value of simple sentences, or sentences composed of clauses consisting of a subject and a predicate.

In their classic book, Hilbert and Ackermann[89] begin with a discussion of the sentential calculus[90] (calculus of sentences[91]). They point out that sentences can be combined in a number of ways to create compound sentences using connectives such as "and", "or", "if ... then", and "not". They specify five fundamental combinations of sentences using the following symbols:

Connective	Symbol
and	&
or	v
if ... then (modus ponens)	→
if and only if	~
not[92]	___

The symbols can be used to form basic combinations of sentences.

Denoting sentences with upper case letters: A, B, ... the five basic combinations are:

1. A & B is a compound sentence that is true if and only if both A and B are true.
2. A v B is true if and only if at least one of the sentences A or B is true.
3. A → B is false if and only if A is true and B is false.
4. A ~ B is true if and only if both A and B are true, or both A and B are false.
5. The sentence $\underline{A}$ is false if A is true, and is true if A is false.

7.1 Truth Values of Simple Sentences and Clauses

The projection operators, that we have used up to now, have determined the decidability of sentences or clauses. A *clause* consists of a subject(s) and predicate. A sentence or clause was undecidable if the product of the subject

89 Hilbert (1950).

90 We assume a semantic universe of discourse in this, and the following, subsections.

91 We use sentence and statement interchangeably.

92 For lexicographic reasons we use underscore ___ for "not" rather than the more commonly used overscore ‾‾ symbol.

projection operator and predicate projection operator were zero—usually in the simple form of $P(1 - P) = 0$.

Now we extend the formalism to the determination of the truth value of a sentence or clause using projection operators. The procedure begins by assigning a set of projections to each predicate—one truth projection for each subject in the domain of subjects for the predicate. For predicate p we denote the truth projection as P_{ptot} (eq. 6.19). For each subject s we associate a projection P_{stot} (eq. 6.23). Then the sentence

$$<\Omega_u|A_1 + A_2|\Omega_u> \tag{7.1}$$

where A_1 is the subject and A_2 is the predicate, has the truth value

$$\text{Truth Value} = <\Omega_u|P_{1tot}P_{2tot}|\Omega_u> = <\Omega_u|P'_1P'_2|\Omega_u> \tag{7.2}$$

where the form of P_{itot} for i = 1,2 is given by eq. 6.8. For example, if the sentence is "The sky is blue" then, when we assign truth projections for all possible subjects of the predicate "is blue", we could assign

$$\begin{aligned} P_1 &= P_{\text{"The sky"}} \\ P_{1tot} &= P_1P' \\ P_2 &= P_{\text{"is blue"}} \\ P_{2tot} &= P_2P' \end{aligned} \tag{7.3}$$

and let $P' = P_1$ using eq. 6.20. Consequently, the statement has a truth value.

We now extend Classical Operator Logic by assigning numeric values to truth values:

$$\begin{aligned} \text{True} &= +1 \\ \text{Undecidable} &= 0 \\ \text{False} &= -1 \end{aligned} \tag{7.4}$$

In order to obtain numeric truth values from inner products of projection operator expressions we must introduce a dependence on the subject in the truth value projection operator expression. To accomplish this we generalize eq. 6.20 to the operator expression

$$P'_{pk} = \prod_{i\,\varepsilon\,D}[(-1)^{n_{ik}}P'_i] \prod_{j \notin D} (1 - P'_j) \tag{7.5}$$

where the integers n_{ik} are 0 if subject i and predicate k form a true subject-predicate statement (or clause), and 1 if subject i and predicate k form a false subject-predicate statement (or clause). *The set of all predicates in the universe of discourse is labeled by the integer k.*

The set denoted $\{P'_s\}$ is the set of subject truth value projection operators for all subjects in the universe of discourse. The operator expression P'_{pk} satisfies

$$<\Omega_u|P'_s\ P'_{pk}|\Omega_u> = (-1)^{n_{sk}} \tag{7.6}$$

if subject s is contained in D and

$$<\Omega_u|P'_s\ P'_{pk}|\Omega_u> = 0 \tag{7.7}$$

if subject s is *not* contained in D.

For example, consider now the sentence "The sky is green." When we assign truth projections for all possible subjects of the predicate "is green" we choose to assign the subject $n_{\text{"The sky""is green"}} = 1$ so that

$$<\Omega_u|P'_{\text{"The sky"}}\ P'_{\text{"is green""The sky"}}|\Omega_u> = -1 \tag{7.8}$$

The truth value –1 indicates a false statement.

We now note that the operator P'_{pk} is not a projection operator. It satisfies

$$P'_{pk}\ P'_{pk} = P'_p \tag{7.9}$$

where P'_p is defined by eq. 6.20.

For a given predicate, the choice of subject truth values (i.e. the values of the integers n_{ik}) in eq. 7.5 is a semantic issue and thus is determined by semantic considerations.

If a given sentence has multiple subjects the determination of the truth value is slightly more complex. We will consider two special cases:

$$\text{Subject1 and subject2 predicate} \tag{7.10}$$

and

$$\text{Subject1 or subject2 predicate} \tag{7.11}$$

In the case of eq. 7.10 the truth value[93] is obtained by evaluating the expression:[94]

[93] In this example and the following examples we factor expectation values into products of expectation values of clauses since this procedure is both well defined and produces results consistent with our intuitive expectations. Interestingly, a somewhat similar procedure is followed in perturbation theory expansions in quantum field theory. In perturbation theory, terms with more than two operators in them are expanded in terms of vacuum expectation values of products of two quantum field operators.

$$\text{Truth Value} = \langle\Omega_u|P_{s1}P_{pa}|\Omega_u\rangle\langle\Omega_u|P_{s2}P_{pa}|\Omega_u\rangle[1 - 2\theta(-\langle\Omega_u|P_{s1}P_{pa}|\Omega_u\rangle - \langle\Omega_u|P_{s2}P_{pa}|\Omega_u\rangle)] \quad (7.12)$$

where the $\theta(x)$ is the step function:

$$\theta(x) = 1 \text{ if } x > 0 \text{ and } \theta(x) = 0 \text{ if } x \leq 0 \quad (7.13)$$

and where P_{si} for i = 1, 2 are the subject projection operators for Subject1 and Subject2 respectively. P_{pa} is the predicate status operator.
Eq. 7.10 is equivalent to

Subject1 predicate and subject2 predicate (7.14)

Eq. 7.12 for the truth value follows from its equivalent, eq. 7.14, taking account of the possible truth values and undecidability issue. Eq. 7.12 specifies that the subject-predicate expression must be true for both subjects for the statement to be true. If either factor in eq. 7.12 is zero then the truth value of the statement is zero meaning it is undecidable. If either factor, or both factors, in eq. 7.12 are false (and both decidable) then the truth value of the statement is –1 meaning false.

In the case of eq. 7.11 the truth value is obtained by evaluating the truth projection expression:

$$\text{Truth Value} = \text{Max}(\langle\Omega_u|P_{s1}P_{pa}|\Omega_u\rangle, \langle\Omega_u|P_{s2}P_{pa}|\Omega_u\rangle)\cdot |\langle\Omega_u|P_{s1}P_{pa}|\Omega_u\rangle\langle\Omega_u|P_{s2}P_{pa}|\Omega_u\rangle| \quad (7.15)$$

where | ... | indicates the numerical absolute value. The statement is true if the subject-predicate expression is true for either subject. The absolute value $|\langle\Omega_u|P_{s1}P_{pa}|\Omega_u\rangle\langle\Omega_u|P_{s2}P_{pa}|\Omega_u\rangle|$ guarantees the truth value of the statement is zero if either or both subject-predicate expressions is undecidable. The function Max(x, y) is the maximum of the quantities x and y.

The generalization to the case of multiple subjects is direct.[95]

---

[94] The statement is separated into an equivalent consisting of two parts each of which is evaluated using the universe state $|\Omega_u\rangle$. It is clearly the only way of obtaining truth values correctly in this formalism. Therefore we so define it.

[95] In the case of quantifiers such as "all" or "some" we make the quantifier part of the subject or predicate as the case may be. For example, in the sentence "All men are good" the subject is "all men" and the predicate is "are good". The truth value status of the sentence is determined by whether "all men" is a member of the domain of "are good". A similar procedure can be followed for "some". The appearance of "all" or "some" in a predicate also can be similarly treated. For example, "Peacemakers are all good men" has the predicate "are all good men" and the truth value status of the sentence is determined by whether "peacemakers" is in the domain of that predicate. Other quantifiers such as "there exists" and "for any" are also embodiable within Operator Logic.

7.2 Truth Value of Compound Sentences and Clauses

In the case of compound sentences and clauses, classical Operator Logic is based on factoring them into simple subject-predicate clauses as discussed in the previous section.

In this section we will consider some general cases of compound sentences to illustrate the procedure for obtaining a truth value.

Suppose we have a sentence that can be expressed as a series of "and" clauses such as

$$\text{Statement} = \text{clause1 and clause2 and} \ldots \tag{7.16}$$

Then the truth value of the statement is given by the expression:[96]

$$\begin{aligned}\langle\Omega_u|\text{statement}|\Omega_u\rangle = |\langle\Omega_u|\text{clause1}|\Omega_u\rangle\langle\Omega_u|\text{clause2}|\Omega_u\rangle\ldots|\{1 -\\ - 2\theta(|\langle\Omega_u|\text{clause1}|\Omega_u\rangle| - \langle\Omega_u|\text{clause1}|\Omega_u\rangle + |\langle\Omega_u|\text{clause2}|\Omega_u\rangle| -\\ - \langle\Omega_u|\text{clause2}|\Omega_u\rangle + \ldots)\}\end{aligned} \tag{7.17}$$

Note the following truth value cases implied by eq. 7.17:

Case 1: all clauses true (no undecidable clauses), Truth Value = +1
Case 2: One or more false clauses (with no undecidable clauses), Truth Value = –1
Case 3: One or more undecidable clauses, Truth Value = 0

Suppose we consider a sentence that can be expressed as a series of "or" clauses such as

$$\text{Statement} = \text{clause1 or clause2 or} \ldots \tag{7.18}$$

Then the truth value of the statement is given by the expression:[97]

$$\begin{aligned}\langle\Omega_u|\text{statement}|\Omega_u\rangle = [2\theta(\theta(\langle\Omega_u|\text{clause1}|\Omega_u\rangle) + \theta(\langle\Omega_u|\text{clause2}|\Omega_u\rangle) + \ldots) - 1]\cdot\\ \cdot|\langle\Omega_u|\text{clause}|\Omega_u\rangle\langle\Omega_u|\text{clause}|\Omega_u\rangle\langle\Omega_u|\text{clause}|\Omega_u\rangle\ldots|\end{aligned} \tag{7.19}$$

The following truth value cases are implied by eq. 7.19:

Case 1: Any clauses are true (no undecidable clauses), Truth Value = +1
Case 2: All clauses are false (with no undecidable clauses), Truth Value = –1

---

[96] Note that because the projection expressions do not change the universe state the expression $\langle\Omega_u|\text{clause1}|\Omega_u\rangle\langle\Omega_u|\text{clause2}|\Omega_u\rangle \equiv \langle\Omega_u|\text{clause1clause2}|\Omega_u\rangle$ and similarly elsewhere. The separation into "clause" factors is clearer.
[97] Note nested step functions appear in eq. 7.19.

Case 3: One or more undecidable clauses, Truth Value = 0

Lastly, let us consider the case of an "if ... then" statement:

$$\text{If A, then B} \tag{7.20}$$

This statement is equivalent to

$$\underline{A\underline{B}} \tag{7.21}$$

which in words is NOT(A and NOT B). In terms of Operator Logic the truth value of eqs. 7.20 and 7.21 is numerically

$$-[\langle\Omega_u|A|\Omega_u\rangle(-\langle\Omega_u|B|\Omega_u\rangle)] = \langle\Omega_u|A|\Omega_u\rangle\langle\Omega_u|B|\Omega_u\rangle \tag{7.22}$$

where $\langle\Omega_u|A|\Omega_u\rangle$ and $\langle\Omega_u|B|\Omega_u\rangle$ are evaluated in a manner similar to the previous cases (eqs. 7.17 and 7.19), or for other sentential forms of A and B.

7.3 Truth Value of the Normal Form

It is well known that all sentential calculus statements can be reduced to operations involving two logical connectives such as & and __ (not).[98] So our Operator Logic clearly suffices to handle all sentential calculus statements including undecidable statements.

Furthermore any set of sentences can be into a form called a *normal form*[99] by means of equivalence transformations.[100]

Since our Operator Logic also encompasses the predicate calculus and second level predicate calculus (See the following chapter.) we conclude we have a complete Operator Logic formalism[101] for Logic that decisively handles the issue of undecidability.

---

[98] Hilbert (1950) pp. 10-11.

[99] "consisting of a conjunction of disjunctions in which each component of the disjunction is either an elementary sentence or the negation of one" in the words of Hilbert (1950) p. 12.

[100] Hilbert (1950) p. 12.

[101] The rules of inference and proof procedures are analogous in Operator Logic and in conventional Logic, and so will not be considered here.

8. Why Operator Logic?

8.1 Operator Logic is the Correct Formalism for Logic

Operator Logic enables logical systems and statements to be constructed that exclude paradoxes and undecidable statements in a mathematically well defined way. In addition it has other important advantages described later. It provides a superior formalism for Logic without the known defects of conventional language/symbolic language formulations, which we now see as incomplete.

After we establish Operator Logic we can proceed to develop the intermediate mathematical-metaphysical transition to a theory of physical Reality

8.2 Advantages of Operator Logic

Operator Logic has some major advantages, which we will discuss in this chapter and in succeeding chapters:

1. Operator Logic provides a unified formulation of classical and quantum probabilistic logic.

2. Operator Logic is the only approach with a fundamental, non-intuitive basis in Nature (being somewhat similar conceptually to Quantum Theory) and thus is guaranteed to be consistent.[102] Otherwise, logic is futile in its foundations.

3. Raising and lowering operators can be defined that enable us to create subworlds (subuniverses) of Calculi and of Semantic Universes of Discourse.

4. Projection operators can also be used to generate subuniverses.

5. A semantic universe of discourse can be defined, which can be mapped to a physical universe realizing the Platonic concepts of Ideas

---

[102] G. Feinberg, Phys. Rev. **159**, 1089 (1967) quotes Dr. M. Tausner to the effect, "No [experimental] observations can be logically inconsistent." Thus no formalism truly embodying Nature can be inconsistent. This comment presumes nature to be consistent and to embody a well-defined set of laws. The regularity of nature has become more and more evident in recent centuries with the growth of scientific knowledge.

(pure forms), of an intermediate mathematical connection, and of the physical universe of Reality.

These possibilities will be discussed in the following sections and chapters.

8.3 Operator Logic Raising and Lowering Operators

8.3.1 Eigenvalue Operators that are not Functions of Variables

If we consider an eigenvalue operator A_E (without a dependence on variables) for a term E, and its corresponding eigenstates |0> and |E>,

$$A_E |E> = gn(E)|E> \quad (8.1)$$
$$A_E |0> = 0 \quad (8.2)$$

then we can[103] define raising and lowering operators that transform one eigenstate into the other. We define the raising operator $a_E^\dagger$ that satisfies

$$|E> = [gn(E)]^{-\frac{1}{2}} a_E^\dagger |0> \quad (8.3)$$

with <E|E> = 1, and, the lowering operator a_E that satisfies

$$|0> = [gn(E)]^{-\frac{1}{2}} a_E |E> \quad (8.4)$$

where † signifies hermitean conjugatation. In the present situation we do not wish repeated application of the raising operator to generate states with higher eigenvalues.[104] Therefore we define

$$a_E^\dagger |E> = 0 \quad (8.5)$$
$$a_E |0> = 0 \quad (8.6)$$

Consequently, raising and lowering operators must satisfy anticommutation relations[105] with the form:

$$\{a_E^\dagger, a_E\} = gn(E)$$
$$\{a_E, a_E\} = 0 \quad (8.7)$$
$$\{a_E^\dagger, a_E^\dagger\} = 0$$

---

[103] Although this is commonly done in the case of a harmonic oscillator in Quantum Mechanics, and also in Quantum Field Theory.

[104] Since such eigenstates would not correspond to a primitive or defined term in the universe of discourse.

[105] Anticommutation relations appear in the quantum field theory of fermions such as the electron.

for eqs. 8.3 – 8.6 to hold[106] where an anti-commutator is defined by

$$\{a, b\} = ab + ba$$

More generally if there are a number of semantic eigenvalue operators A_{E_k} where k = 1,2, …, q then the anticommutation relations are

$$\begin{aligned} \{a_{E_k}{}^{\dagger}, a_{E_{k'}}\} &= gn(E_k)\delta_{kk'} \\ \{a_{E_k}, a_{E_{k'}}\} &= 0 \\ \{a_{E_k}{}^{\dagger}, a_{E_{k'}}{}^{\dagger}\} &= 0 \end{aligned} \tag{8.8}$$

where $\delta_{kk'}$ is the Kronecker delta. The semantic eigenvalue operators can be represented in terms of these operators by

$$A_{E_k} = a_{E_k}{}^{\dagger} a_{E_k} \tag{8.9}$$

and the corresponding projection operators by

$$P_{E_k tot} \equiv P_{E_k} = P'_{E_k}\,[1 - [gn(E_k)]^{-1} a_{E_k} a_{E_k}{}^{\dagger}] \tag{8.10}$$

where P'_{E_k} is a truth value projection operator. (See section 7.1 for its definition and use.) The product of P'_{E_k} operators determines the truth value of expressions and precludes logical paradoxes.

The projection operator factor $[1 - [gn(E_k)]^{-1} a_{E_k} a_{E_k}{}^{\dagger}]$ in eq. 8.10 projects substates from the universe state whose kth eigenvalue is zero. Thus the projection operator P_{E_k} plays two roles: 1) it projects any state $|\Omega>$, whose kth eigenvalue is zero, to zero; and 2) it enables statements to avoid paradoxes by zeroing statements and clauses whose subjects or predicates are not terms in the universe of discourse.[107]

Subjects that are not in the subject domain of a predicate have a truth value projection operator $P'_{E_{k'}}$ such that when it is multiplied by the predicate's P'_{E_k} it gives zero identically. As a result such a statement's corresponding operator product (containing both subject and predicate operators) yields a truth value of zero indicating undecidable.

---

[106] This definition of raising and lowering operators leads to a simpler form for eq. 8.9.

[107] In the cases considered hitherto, this is not an issue since the universe state has a non-empty string factor for each term in the universe. Later when we develop subuniverses state projection operators become non-trivial. They are also relevant in Quantum Operator Logic as will be seen.

Calculus eigenvalue operators, and raising and lowering operators, have exactly the same definitions, forms and equations as the semantic analogues given above in this subsection. For definiteness we give their basic definitions:

$$\{b_{E_k}{}^{\dagger}, b_{E_{k'}}\} = gn(E_k)\delta_{kk'}$$
$$\{b_{E_k}, b_{E_{k'}}\} = 0 \qquad (8.11)$$
$$\{b_{E_k}{}^{\dagger}, b_{E_{k'}}{}^{\dagger}\} = 0$$

with

$$B_{E_k} = b_{E_k}{}^{\dagger} b_{E_k} \qquad (8.12)$$

and

$$Q_{E_k} = Q'_{E_k}\tfrac{1}{2}[1 - [gn(E_k)]^{-1} b_{E_k} b_{E_k}{}^{\dagger}] \qquad (8.13)$$

where Q'_{E_k} is the calculus equivalent of a semantic truth projection operator. The Q'_{E_k} projection operator can prevent logical paradoxes in a manner similar to the semantic case.

8.3.2 Eigenvalue Operators that are Functions of Variables

If we consider any eigenvalue operator that is dependent on free variables such as $A_E(x_1, x_2, \ldots)$ with corresponding eigenstates $|0, x_1, x_2, \ldots>$ and $|gn(E), x_1, x_2, \ldots>$ for each set of values of $x_1, x_2, \ldots$ then we can again define raising and lowering operators that transform between eigenstates. The raising operator becomes variable dependent:

$$|E, x_1, x_2, \ldots> = [gn(E_k)]^{-1/2} a_E{}^{\dagger}(x_1, x_2, \ldots)|0> \qquad (8.14)$$

as does the lowering operator:

$$a_E(x_1, x_2, \ldots)|E, x'_1, x'_2, \ldots > = [gn(E_k)]^{1/2}\delta(x_1, x'_1)\, \delta(x_2, x'_2)\ldots|0> \qquad (8.15)$$

where † signifies the hermitean conjugate, and $\delta(x_k, x'_k)$ is a Kronecker-type δ-function for discrete variables and a Dirac-type δ-function for continuous variables.

Again we do not wish repeated application of a raising operator to generate states with higher eigenvalues. Therefore

$$a_E{}^{\dagger}(x_1, x_2, \ldots)|E> = 0 \qquad (8.17)$$
$$a_E{}^{\dagger}(x_1, x_2, \ldots) \ldots a_E{}^{\dagger}(x_1, x_2, \ldots)|0> = 0 \qquad (8.16)$$
$$a_E(x_1, x_2, \ldots) \ldots a_E(x_1, x_2, \ldots)|E> = 0 \qquad (8.17)$$

Consequently the reader can verify that the raising and lowering operators must satisfy the anicommutation relations:

$$\begin{aligned}&\{a_E^\dagger(x_1, x_2, \ldots), a_E(x'_1, x'_2, \ldots)\} = gn(E)\delta(x_1, x'_1)\delta(x_2, x'_2)\ldots \\ &\{a_E(x_1, x_2, \ldots), a_E(x'_1, x'_2, \ldots)\} = 0 \\ &\{a_E^\dagger(x_1, x_2, \ldots), a_E^\dagger(x'_1, x'_2, \ldots)\} = 0\end{aligned} \tag{8.18}$$

where $\delta(x, y)$ is a Dirac delta function for continuous variables x and y, and a Kronecker delta function for discrete variables x and y.

More generally if there are a number of semantic eigenvalue operators A_{E_k} where $k = 1,2, \ldots , q$ then the anticommutation relations are

$$\begin{aligned}&\{a_{E_k}^\dagger(x_1, x_2, \ldots), a_{E_{k'}}(x'_1, x'_2, \ldots)\} = gn(E_k)\delta_{kk'}\delta(x_1, x'_1)\delta(x_2, x'_2)\ldots \\ &\{a_{E_k}(x_1, x_2, \ldots), a_{E_{k'}}(x'_1, x'_2, \ldots)\} = 0 \\ &\{a_{E_k}^\dagger(x_1, x_2, \ldots), a_{E_{k'}}^\dagger(x'_1, x'_2, \ldots)\} = 0\end{aligned} \tag{8.19}$$

where $\delta_{kk'}$ is a Kronecker delta. The semantic eigenvalue operators can be represented in terms of these operators by

$$A_{E_k}(x_1, x_2, \ldots) = a_{E_k}^\dagger(x_1, x_2, \ldots)a_{E_k}(x_1, x_2, \ldots) \tag{8.20}$$

and the projection operators are

$$P_{E_k}(x_1, x_2, \ldots) = P'_{E_k}(x_1, x_2, \ldots)\{1 - [gn(E_k)]^{-1}a_{E_k}(x_1, x_2, \ldots)a_{E_k}^\dagger(x_1, x_2, \ldots)\} \tag{8.21}$$

The calculus eigenvalue operators, and raising and lowering operators have a similar form

$$\begin{aligned}&\{b_{E_k}^\dagger(x_1, x_2, \ldots), b_{E_{k'}}(x'_1, x'_2, \ldots)\} = gn(E_k)\delta_{kk'}\delta(x_1, x'_1)\delta(x_2, x'_2)\ldots \\ &\{b_{E_k}(x_1, x_2, \ldots), b_{E_{k'}}(x'_1, x'_2, \ldots)\} = 0 \\ &\{b_{E_k}^\dagger(x_1, x_2, \ldots), b_{E_{k'}}^\dagger(x'_1, x'_2, \ldots)\} = 0\end{aligned} \tag{8.22}$$

with

$$B_{E_k}(x_1, x_2, \ldots) = b_{E_k}^\dagger(x_1, x_2, \ldots)b_{E_k}(x_1, x_2, \ldots) \tag{8.23}$$

and

$$Q_{E_k}(x_1, x_2, \ldots) = \{1 - [gn(E_k)]^{-1}b_{E_k}(x_1, x_2, \ldots)b_{E_k}^\dagger(x_1, x_2, \ldots)\}Q'_{E_k}(x_1, x_2, \ldots) \tag{8.24}$$

8.3.3 Effect of Raising and Lowering Operators on Universes of Discourse

Consider the effect of a lowering operator on a universe state:

$$a_{E_k}|E_1, E_2, E_3, \ldots , E_k, \ldots ,E_n> = [gn(E_k)]^{1/2}|E_1, E_2, E_3, \ldots, \overset{k^{th}}{0}, \ldots ,E_n> \tag{8.25}$$

The kth eigenvalue becomes zero, the Gödel number of the empty string.

We now examine a semantic universe of discourse (Parallel observations and comments can be made for a calculus universe of discourse.) with terms corresponding to Gödel numbers: $E_1, E_2, \ldots, E_n$.[108] The state $|\Omega_u>$ corresponding to this universe of discourse is given by eqs. 6.26 and 8.25.
A universe of discourse state $|\Omega_u>$ can be "restricted" using products of lowering operators. For example, from eq. 8.14 we see

$$a_{E_k}|\Omega_u> = [gn(E_k)]^{1/2}|\Omega'_u> \tag{8.26}$$

where $<\Omega'_u|\Omega'_u> = 1$, and where the kth eigenvalue of $|\Omega'_u>$ is 0:

$$|\Omega'_u> = |E_1, E_2, E_3, \ldots, \overset{k^{th}}{0}, \ldots ,E_n> \tag{8.27}$$

The axioms, theorems and proofs of the subuniverse of discourse state $|\Omega'_u>$ are those of the universe of discourse $|\Omega_u>$ which do not contain the kth term.

Thus a statement such as

$$<\Omega'_u| \sum_{n \in \{m'\}} A_{E_n}|\Omega'_u> \tag{8.28}$$

satisfies the corresponding operator product equation

$$F(P'_{E_1}, P'_{E_2}, \ldots) \prod_{n \in \{m'\}} P_{E_n}|\Omega'_u> = 0 \tag{8.29}$$

where $F(P'_{E_1}, P'_{E_2}, \ldots)$ is a function of P'_{E_1}, P'_{E_2} and so on, if k (from eq. 8.27) is among the set of integers {m'} since P_{E_k} brings the expression to zero. The expectation value[109] is also zero

$$<\Omega'_u|F(P'_{E_1}, P'_{E_2}, \ldots) \prod_{n \in \{m'\}} P_{E_n}|\Omega'_u> = 0 \tag{8.30}$$

---

[108] Zero is the Gödel number of the empty string.
[109] Inner product.

for each statement containing the kth term. Thus the set of statements with a possible non-zero truth value in the $|\Omega'_u\rangle$ universe of discourse consists of those statements not containing the kth term. $|\Omega'_u\rangle$ is properly regarded as a subuniverse of discourse.

Whether the subuniverse of discourse is of interest is another question. In the next chapter we consider a simple case where the creation of subuniverses is of interest.

To summarize: A statement in a subuniverse defined above has the form

$$\langle\Omega'_u| \sum_{n \in \{m''\}} A_{E_n}|\Omega'_u\rangle \tag{8.31}$$

for {m"}, the subset of the integers of {m} not containing k.
The operator product corresponding to this statement has the form[110]

$$F(P'_{E_1}, P'_{E_2}, \ldots)\prod_{n \in \{m''\}} P_{E_n} \tag{8.32}$$

Whether a statement in the subuniverse has a truth value is determined by the expectation value

$$\langle\Omega'_u| F(P'_{E_1}, P'_{E_2}, \ldots)\prod_{n \in \{m''\}} P_{E_n}|\Omega'_u\rangle \tag{8.33}$$

If it is zero then it has no truth value; if it is non-zero then it has a truth value. The decidable statements of this subuniverse do not contain statements of the parent universe containing the kth term.

We now have a method for generating subuniverses through the application of one or more lowering operators. We shall see examples of subuniverses in the next chapter.

8.3.4 Multiple Lowering Operator Case

In the case where we apply q lowering operators (whose indices form the set {p}) to a universe state[111], then eq. 8.25 generalizes directly to

$$\prod_{n \in \{p\}} a_{E_n}|\Omega_u\rangle = \prod_{n \in \{p\}} [gn(E_n)]^{1/2}|\Omega''_u\rangle \tag{8.34}$$

---

[110] The expression $F(P'_{E_1}, P'_{E_2}, \ldots)$ is assumed not to contain P'_{E_k}.

[111] $|\Omega_u\rangle$ is defined by eq. 6.26 and {m} is defined in the statement following it.

where $<\Omega''_u|\Omega''_u> = 1$, and members of the set of statements with truth values in this subuniverse have the form

$$<\Omega''_u| \sum_{n \in \{m-p\}} A_{E_n}|\Omega''_u> \tag{8.35}$$

where {m – p} is the subset of integers in {m} that does not include any integers in {p}.

The corresponding operator product expectation value has the form[112]

$$<\Omega''_u| F(P'_{E_1}, P'_{E_2}, \ldots) \prod_{n \in \{m-p\}} P_{E_n}|\Omega''_u> \tag{8.36}$$

The truth value of the statement is determined by the value of eq. 8.36: if the value is zero then the corresponding statement has no truth value - undecidable; if the value is non-zero then the corresponding statement has a truth value.

---

[112] The expression $F(P'_{E_1}, P'_{E_2}, \ldots)$ is assumed not to contain operators corresponding to terms in {p}.

9. Hierarchies of Universes of Discourse

In this chapter we will define hierarchies of universes and consider some simple examples. In addition we will show how to combine universes into "superuniverses." We will use these concepts when we establish a mathematical connection to the theory of Reality – The Standard Model of Elementary Particles.

9.1 Hierarchies of Subuniverses

In the previous chapter we examined the case of a two level hierarchy. One can construct multi-level hierarchies of universes using various sets of lowering operator products at each level. Fig. 9.1 shows a three level hierarchy diagramatically. The procedure is so simple that writing the explicit equations generating the hierarchy is unnecessary in the author's view. They are similar to those given for a two level hierarchy in section 8.3.4.

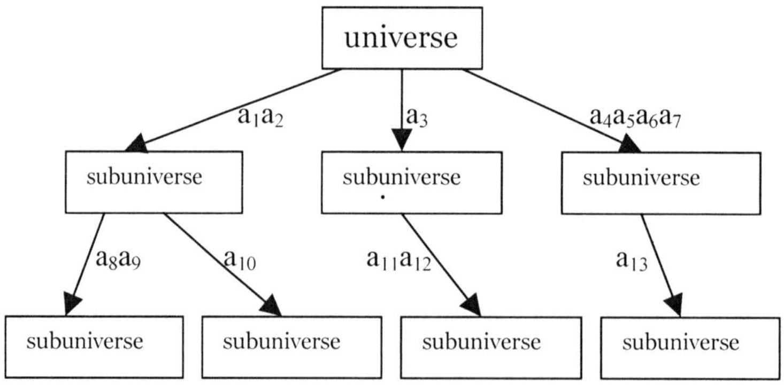

Figure 9.1. A hierarchy of universes with the lowering operators indicated for the transition to each subuniverse.

One primary purpose in defining universe hierarchies is to enable separate universes of discourse that have a significant overlap in their primitive terms, axioms and theorems to be united in a superuniverse through the reversal of the mechanism described for creating subuniverse hierarchies. We form a unified set of terms with one subset containing the primitive terms of one subuniverse and another subset containing the primitive terms of the other subuniverse. Then a combined set of axioms is constructed such that setting one set of terms to zero (empty strings) yields one of the initial subuniverses and setting a different set of terms to zero yields the other subuniverse.[113] Thus one achieves a unification of what had been two separate but similar universes of discourse.[114] The arithmetic universe considered next is an example of this process.

9.2 Subuniverses of a Simple Arithmetical Universe of Discourse

We will consider a simple universe of discourse, that is similar to Gödel's "formal system P"[115], to illustrate the utility of creating subuniverses of discourse and to illustrate combining universes to produce a super universe. The axioms in this universe are:

1. a + b = b + a Commutativity of Addition
2. a + (b + c) = (a + b) + c Associativity of Addition
3. a·b = b·a Commutativity of Multiplication
4. a·(b·c) = (a·b)·c Associativity of Multiplication
5. a·(b + c) = a·b + a·c Distributive Law

They are the fundamental arithmetic axioms of natural numbers (integers).

The arithmetic universe of discourse state $|\Omega_u>$ is:

$$|\Omega_u> = |\text{"a"}, \text{"b"}, \text{"c"}, \text{"+"}, \text{"·"}, \text{"="}, \text{"("}, \text{")"} > \qquad (9.1)$$

9.2.1 Additive Universe of Discourse

If we now apply the lowering operator for multiplication "·", namely $a_{\text{"·"}}$, to $|\Omega_u>$ then the universe of discourse is reduced to addition only:

1. a + b = b + a Commutativity of Addition
2. a + (b + c) = (a + b) + c Associativity of Addition

with the subuniverse state:

$$|\Omega_+> = |\text{"a"}, \text{"b"}, \text{"c"}, \text{"+"}, 0, \text{"="}, \text{"("}, \text{")"} > \qquad (9.2)$$

All statements containing a multiplication become statements without a truth value when their operator products are evaluated between $<\Omega_+|$ and $|\Omega_+>$. These

---

[113] Additional axioms might also be needed to unite the subuniverses. These axioms contain terms from both subuniverses and thus are only relevant for the unified subuniverses.

[114] J. C. Maxwell achieved this feat by successfully uniting electricity and magnetism in his theory of Electromagnetism in 1865.

[115] Gödel (1962) p. 41.

statements are undecidable statements since they are neither true nor false and excluded from the subuniverse of decidable statements.

9.2.2 Multiplicative Universe of Discourse

The multiplicative universe of discourse state $|\Omega_\bullet>$ is obtained by applying the lowering operator for addition "+", namely $a_{"+"}$, to $|\Omega_u>$. The subuniverse state $|\Omega_\bullet>$ is:

$$|\Omega_\bullet> = |"a", "b", "c", 0, "\cdot", "=", "(", ")" > \qquad (9.3)$$

The axioms in this subuniverse are the purely multiplicative axioms:

1. a·b = b·a	Commutativity of Multiplication
2. a·(b·c) = (a·b)·c	Associativity of Multiplication

Thus we obtain the multiplicative subuniverse. All statements containing an addition become undecidable (truth value = 0) when their projection operator products are evaluated between $<\Omega_\bullet|$ and $|\Omega_\bullet>$.

9.3 Subuniverses of Euclidean Geometry

Euclidean geometry was plagued for millenia by concerns over the fifth postulate. In the 19th Century Bolyai and Lobachevsky showed that omitting the fifth postulate opened the door to non-Euclidean geometries that were valid on certain curved surfaces. Euclidean geometry has five axioms plus five common notions including the notion of congruence. The primitive terms and axioms are:

Primitive terms:
point, line, straight line, circle, angle, congruent (in common notions)

Axioms:
1. Given two points there is a straight line that joins them.
2. Any straight line can be prolonged indefinitely.
3. A circle can be constructed when its center, and a point on it, are specified.
4. All right angles are equal.
5. If a straight line falling on two straight lines makes the interior angles on the same side sum to less than two right angles, the two straight lines if extended indefinitely, meet on that side on which the angles are less than two right angles.

They also include some common English words.

If we consider the semantic universe of discourse of Euclidean geometry, then the interpreted primitive terms, the axioms, and common notions comprise the knowledge base of the Euclidean universe of discourse.

If we define a semantic subuniverse by setting "congruent" to the empty string using a lowering operator, and reinterpreting the primitive term "straight line" so that axiom 5 is not necessarily true, then we obtain a family of geometries that includes Euclidean geometry and non-Euclidean geometries.
We conclude a subuniverse can be more general then its parent universe.

9.4 A Subuniverse of Electromagnetism

A more general form of the Maxwell equations (We will call it the pseudo-Maxwell equations.) in the absence of sources and currents can be defined by

$$\begin{aligned} &\nabla \times \mathbf{E} = -\beta\partial\mathbf{B}/\partial t \\ &\nabla \cdot \mathbf{E} = 0 \\ &\nabla \cdot \mathbf{B} = 0 \\ &\nabla \times \mathbf{B} = -\beta\partial\mathbf{E}/\partial t \end{aligned} \tag{9.4}$$

where β is a constant and $\times$ represents the vector cross product while $\cdot$ represents a vector inner product. If $\beta = 1$ then eqs. 9.4 are exactly the Maxwell equations of electrodynamics in the absence of sources and currents.

We now define the nine primitive terms of eqs. 9.4:

"$\nabla\times$" "$\nabla\cdot$" "=" "–" "β" "$\partial/\partial t$" "**E**" "**B**" "0"

which we associate with the eigenvalue operators $A_1, A_2, \ldots, A_9$. The universe state is

$$|\Omega_u\rangle = |\text{“}\nabla\times\text{”}, \text{“}\nabla\cdot\text{”}, \text{“=”}, \text{“–”}, \text{“}\beta\text{”}, \text{“}\partial/\partial t\text{”}, \text{“}\mathbf{E}\text{”}, \text{“}\mathbf{B}\text{”}, \text{“0”}\rangle \tag{9.5}$$

We apply the lowering operator to make the eigenvalue of A_5 (the term "β") zero producing the resulting universe state $|\Omega_{ul}\rangle$:

$$\begin{aligned} [gn(\text{“}\beta\text{”})]^{1/2}|\Omega_{ul}\rangle &= a_{\text{“}\beta\text{”}}|\Omega_u\rangle \\ &= |\text{“}\nabla\times\text{”}, \text{“}\nabla\cdot\text{”}, \text{“=”}, \text{“–”}, 0, \text{“}\partial/\partial t\text{”}, \text{“}\mathbf{E}\text{”}, \text{“}\mathbf{B}\text{”}, \text{“0”}\rangle \end{aligned} \tag{9.6}$$

Thus the first and fourth Maxwell equations become:

$$\begin{aligned} &\langle\Omega_{ul}|(A_1 + A_7 + A_3 + A_9)|\Omega_{ul}\rangle \equiv \nabla\times\mathbf{E} = 0 \\ &\langle\Omega_{ul}|(A_1 + A_8 + A_3 + A_9)|\Omega_{ul}\rangle \equiv \nabla\times\mathbf{B} = 0 \end{aligned} \tag{9.7}$$

The complete set of the subuniverse "Maxwell" equations is

$$
\begin{aligned}
\nabla \times \mathbf{E} &= 0 \\
\nabla \cdot \mathbf{E} &= 0 \\
\nabla \cdot \mathbf{B} &= 0 \\
\nabla \times \mathbf{B} &= 0
\end{aligned}
\tag{9.8}
$$

Eqs. 9.8 represent a "new" theory in which the electric and magnetic fields are decoupled and become separate "subtheories." An important result of this theory is the absence of electromagnetic waves.

The creation of subuniverses, and the reverse process of combining subuniverses to produce a superuniverse, provide a mechanism for relating theories, and a mechanism for unifying similar theories into a joint theory in some areas of physics or mathematics.

9.5 Products of Universes

All universes, as we have defined them, are linear vector spaces. Thus it is natural to consider direct products of universes. We will start by considering the case of the direct product of the additive and multiplicative subuniverses defined earlier in this chapter, namely $|\Omega_+\rangle|\Omega_\bullet\rangle$. Then the axioms

1. a + b = b + a   Commutativity of Addition
2. a + (b + c) = (a + b) + c   Associativity of Addition
3. a·b = b·a   Commutativity of Multiplication
4. a·(b·c) = (a·b)·c   Associativity of Multiplication

emerge directly for the product of subuniverses. For example, axiom 1 is generated from

$$\langle\Omega_\bullet|\langle\Omega_+|(A_{+\text{"a"}}I_\bullet + A_{+\text{"+"}}I_\bullet + A_{+\text{"b"}}I_\bullet + A_{+\text{"="}}I_\bullet + A_{+\text{"b"}}I_\bullet + A_{+\text{"+"}}I_\bullet + A_{+\text{"a"}}I_\bullet)|\Omega_+\rangle|\Omega_\bullet\rangle$$

where $I_\bullet$ is the identity operator in the multiplicative subuniverse. The other axioms are similarly generated. The plus subscript on the operators A_+ indicates they are operators in the additive subuniverse. Since we are taking the direct product of linear vector spaces the eigenvalue operators are paired from each space as shown.

The fifth axiom:

5. a·(b + c) = a·b + a·c   Distributive Law

which combines addition and multiplication can also be added to the set of axioms. The statement of the fifth axiom is slightly more complex since both subuniverses "interact" through this axiom:

$$<\Omega_\bullet|<\Omega_+|(A_{+\text{"a"}}I_\bullet + A_{\bullet\text{"."}}I_+ + A_{+\text{"("}}I_\bullet + A_{+\text{"b"}}I_\bullet + A_{+\text{"+"}}I_\bullet + A_{+\text{"c"}}I_\bullet + A_{+\text{")"}}I_\bullet + A_{+\text{"="}}I_\bullet + \\ + A_{+\text{"a"}}I_+ + A_{\bullet\text{"."}}I_+ + A_{+\text{"b"}}I_\bullet + A_{+\text{"+"}}I_\bullet + A_{+\text{"a"}}I_\bullet + A_{\bullet\text{"."}}I_+ + A_{+\text{"c"}}I_\bullet)|\Omega_+>|\Omega_\bullet>$$

The "•" subscript on the operators $A_\bullet$ indicates they are operators in the multiplicative subuniverse. The axiom can be restated in alternate ways as well since we identify the terms "a", "b", and "c" in $|\Omega_+>$ with the corresponding terms in $|\Omega_\bullet>$.

This example illustrates "direct" products of universes.[116] It also shows that an issue, that may arise, is duplicate terms in the universes. As shown, this issue is usually easily resolved.

The procedure of taking the "direct" product of two universes can result in the creation of a "superuniverse" that combines the features of both universes. Additional axioms can be added to the combined universe to provide inter-relationships between the universes. Axiom 5 above is an example of an axiom that interrelates (causes an interaction between) the component universes. Thus a combination of universes can lead to more than their initial combined content.

[116] Not quite a true direct product due to the overlap of terms in the subuniverses.

10. Matrix Form of Operator Logic

The eigenvalue operators A_{E_k} and B_{E_k} in semantic and calculus universes of discourse each have two eigenvalues: a Gödel number gn(E_k) and zero (which is the Gödel number of the empty string). For each k, the operator A_{E_k} or B_{E_k}, and each operator's two eigenstates |0> and |E_k>, can be expressed in terms of SU(2) operators and eigenstates in a 2×2 matrix formulation. This representation is also used to describe spin ½ particles such as electrons and quarks in physics theories. Later when we develop the Platonic realm of Ideas and their relation to Reality this correspondence will turn out to be of importance.

10.1 Spinor Representation of Operator Logic

The SU(2) spinor representation is based on the Pauli matrices:

$$\sigma_1 = \begin{bmatrix} 0 & 1 \\ 1 & 0 \end{bmatrix} \qquad \sigma_2 = \begin{bmatrix} 0 & -i \\ i & 0 \end{bmatrix} \qquad \sigma_3 = \begin{bmatrix} 1 & 0 \\ 0 & -1 \end{bmatrix} \tag{10.1}$$

Operator Logic can be expressed in a spinor representation. Consider an eigenvalue operator A_{E_k} for some value k.[117] We can represent this operator and its eigenvectors as a matrix and as column vectors (spinors):

$$A_{E_k} \rightarrow \begin{bmatrix} \mathrm{gn}(E_k) & 0 \\ 0 & 0 \end{bmatrix} \qquad |E_k> \rightarrow \begin{bmatrix} 1 \\ 0 \end{bmatrix} = s_{k\uparrow} \qquad |0> \rightarrow \begin{bmatrix} 0 \\ 1 \end{bmatrix} = s_{k\downarrow} \tag{10.2}$$

The raising and lowering operators of the spinor representation are

$$c_{E_k}^{\dagger} = \begin{bmatrix} 0 & 1 \\ 0 & 0 \end{bmatrix} \qquad c_{E_k} = \begin{bmatrix} 0 & 0 \\ 1 & 0 \end{bmatrix} \tag{10.3}$$

where † signifies hermitean conjugation, and where

[117] Calculus Logic Operators have a completely analogous formalism.

$$c_{E_k}^{\dagger} \equiv [gn(E_k)]^{-1/2} a_{E_k}^{\dagger} \quad (10.4)$$

$$c_{E_k} \equiv [gn(E_k)]^{-1/2} a_{E_k}$$

with the anti-commutation relations

$$\begin{aligned} c_{E_k}^{\dagger} c_{E_{k'}} + c_{E_{k'}} c_{E_k}^{\dagger} &\equiv \{c_{E_k}^{\dagger}, c_{E_{k'}}\} = \delta_{kk'} \\ \{c_{E_k}, c_{E_{k'}}\} &= 0 \\ \{c_{E_k}^{\dagger}, c_{E_{k'}}^{\dagger}\} &= 0 \end{aligned} \quad (10.5)$$

where $\delta_{kk'}$ is the Kronecker delta. Note the semantic eigenvalue operator A_{E_k} then becomes

$$A_{E_k} = gn(E_k) c_{E_k}^{\dagger} c_{E_k} \quad (10.6)$$

and the corresponding projection operator is

$$P_{E_k} = P'_{E_k}[1 - c_{E_k} c_{E_k}^{\dagger}] \quad (10.7)$$

where P'_{E_k} is a truth projection operator. (See section 7.1 for details on truth projection operators.)
The universe of discourse $|\Omega_u\rangle$ with q factors is generated by the direct product of the q individual eigenstates $s_{i\uparrow}$ of each primitive term in the universe:

$$|\Omega_u\rangle = \prod_i s_{i\uparrow} \quad (10.8)$$

Note the universe state is normalized to one, $\langle\Omega_u|\Omega_u\rangle = 1$. A statement has the form:

$$\text{statement} = \langle\Omega_u|A_1 + A_2 + A_3 + \ldots |\Omega_u\rangle \quad (10.9)$$

where the matrices A_i have the form specified in eq. 10.2. The corresponding truth value is given by

$$\text{Truth Value} = \langle\Omega_u| F(P'_1, P'_2, \ldots)P_1 P_2 P_3 \ldots |\Omega_u\rangle \quad (10.10)$$

A simple example of this formalism is a two term universe. In this universe of discourse

$$|\Omega_u\rangle = s_{1\uparrow} s_{2\uparrow}$$

The statement (with a symbolic + sign)

$$\text{statement} = <\Omega_u|A_1 + A_2|\Omega_u> \\ = gn(E_1) + gn(E_2)$$

can be transformed into a human language by mapping the Gödel numbers to their human language equivalent, dropping the + sign, and putting the statement into a grammatically correct form for that language. This might include add "a" and "the" (in English), and possibly reordering the words. (For example in German words often appear in a different order than English.)

10.2 A Symmetry of the Spinor Matrix Representation

Eq. 10.2 implies a discrete symmetry is present in the matrix form of a universe of discourse.[118] If we define the eigenvalue and projection operators

$$A''_{E_k} = \sigma_1^{-1} A_{E_k} \sigma_1 \quad (10.11)$$
$$P''_{E_k} = \sigma_1^{-1} P_{E_k} \sigma_1 \quad (10.12)$$
$$P'''_{E_k} = \sigma_1^{-1} P'_{E_k} \sigma_1 \quad (10.13)$$

for all k using $\sigma_1^{-1} = \sigma_1$[119], and if we define a new universe state[120]

$$|\Omega_u> = \sigma_1^{-1}|\Omega''_u> \quad (10.14)$$

then the transformed statements and their truth values in the new matrix representation are the same as that of the original matrix representation.

A statement and its truth value is transformed to

$$\text{statement} = <\Omega_u|A_1 + A_2 + A_3 \ldots |\Omega_u> = <\Omega''_u|A''_1 + A''_2 + \ldots |\Omega''_u> \quad (10.15)$$

and

$$\text{TV} = \text{Truth Value} = <\Omega_u|F(P'_{E_1}, P'_{E_2}, \ldots)\prod_{n \in \{m'\}} P_{E_n} |\Omega_u> \\ = <\Omega''_u|F(P''_{E_1}, P''_{E_2}, \ldots)\prod_{n \in \{m'\}} P''_{E_n} |\Omega''_u> \quad (10.16)$$

Thus we have invariance under the discrete transformation σ_1.

---

[118] Similar comments apply to the matrix form of a calculus universe of discourse.
[119] A σ_1 transformation is applied for each k.
[120] There is a σ_1 transformation applied to each factor in $|\Omega_u>$.

*10.3 The Direct Product of Spinor Representations of Universes – Dirac Matrices

In this section we will consider the direct product of the spinor representations of two universes that are, in general, different universes. In part, our goal is to construct a space whose eigenvectors are similar to 4-component Dirac spinors. We will use these eigenvectors later to develop the relation of the Platonic realm of Ideas to Reality.[121]

The universes that we will define have different terms, axioms and proof schema. We will define the direct product to join the two universes of discourse to create a combined universe. We denote the constructs of the first universe as

$$A_{E_{1k}} \quad P_{E_{1k}} \quad c_{E_{1k}} \quad c_{E_{1k}}^{\dagger} \quad s_{1k\uparrow} \quad s_{1k\downarrow} \quad |\Omega_{1u}> \qquad (10.17)$$

and the constructs of the second universe as

$$A_{E_{2k}} \quad P_{E_{2k}} \quad c_{E_{2k}} \quad c_{E_{2k}}^{\dagger} \quad s_{2k\uparrow} \quad s_{2k\downarrow} \quad |\Omega_{2u}> \qquad (10.18)$$

From these constructs we form four column vectors with four columns:

$$s_{\uparrow\uparrow k} = \begin{bmatrix} s_{1k\uparrow} \\ s_{2k\uparrow} \end{bmatrix} \quad s_{\uparrow\downarrow k} = \begin{bmatrix} s_{1k\uparrow} \\ s_{2k\downarrow} \end{bmatrix} \quad s_{\downarrow\uparrow k} = \begin{bmatrix} s_{1k\downarrow} \\ s_{2k\uparrow} \end{bmatrix} \quad s_{\downarrow\downarrow k} = \begin{bmatrix} s_{1k\downarrow} \\ s_{2k\downarrow} \end{bmatrix} \qquad (10.19)$$

For example the first 4-vector $s_{\uparrow\uparrow k}$ can be representd by the direct product form:

$$s_{\uparrow\uparrow k} = s_{1k\uparrow}\otimes(1\ 0) + (0\ 1)\otimes s_{2k\uparrow} \qquad (10.19a)$$

---

[121] There are four logic values in the direct product of spinor representations. Four valued logic is conceptually used in the description of the clockless system design of computer logic circuits. The 4-valued logic developed by Fant (2005) has the logic values TRUE, FALSE, NULL, and INTERMEDIATE. It is used to develop an extension of Boolean Logic that can accommodate time asynchronicities in asynchronous computer circuits. It enables circuits to avoid the use of system clocks to implement synchronization. There is also a nine-valued logic defined by IEEE Standard 1164 with data type std_ulogic (standard unresolved logic) having logic values 'U' – uninitialized; 'X' - strong drive, unknown logic value; '0' - strong drive, logic zero; '1' - strong drive, logic one; 'Z' - high impedance; 'W' - weak drive, unknown logic value; 'L' - weak drive, logic zero; 'H' - weak drive, logic one; '-' - don't care. This logic is used in VHDL hardware design descriptions. It is quite useful in CMOS logic design. A nine valued logic can be defined as a direct sum of the direct product of two 4-logics plus a singlet (uni-valued) logic.

The 4×4 matrices are composed of 2×2 pairs of matrix entries such as [122]

$$A_{E_k} = \begin{bmatrix} A_{E_{1k}} & 0 \\ 0 & A_{E_{2k}} \end{bmatrix} \qquad P_{E_k} = \begin{bmatrix} P_{E_{1k}} & 0 \\ 0 & P_{E_{2k}} \end{bmatrix} \qquad (10.20)$$

$$c_{E_k} = \begin{bmatrix} c_{E_{1k}} & 0 \\ 0 & c_{E_{2k}} \end{bmatrix} \qquad d_{E_k} = \begin{bmatrix} c_{E_{1k}} & 0 \\ 0 & c_{E_{2k}} \end{bmatrix} \qquad (10.21)$$

and also similarly for $c_{E_k}{}^\dagger$ and $d_{E_k}{}^\dagger$. These particular forms are obtained from expressions of the form

$$A_{E_k} = A_{E_{1k}} \otimes (I + \sigma_3)/2 + (I - \sigma_3)/2 \otimes A_{E_{2k}} \qquad (10.20a)$$

using the Pauli matrix σ_3. The direct product universe state is

$$| \Omega_u > = |\Omega_{1u}>|\Omega_{2u}> \qquad (10.22)$$

Since there are 4 independent matrices, the identity matrix and 3 Pauli matrices, in which to express matrices in each universe, and thus 16 linearly independent matrices in the direct product of the two universes (See section 16.5 for the list of these matrices.) we see any 4×4 matrix in the direct product can be expressed in terms of these 16 matrices. Eqns. 10.20 and 10.21 are some examples.

10.3.1 Independent Universes

We can express all the statements in the first universe by projecting out the parts associated with the second universe. The universe projection operator that plays this role is

$$U_1 = ½(I_4 + \gamma^0) \qquad (10.23)$$

where I_4 is the 4×4 identity matrix and γ^0 is one of the four Dirac γ matrices.[123] The four Dirac γ matrices and their product $\gamma^5 = \gamma_5 = i\gamma^0\gamma^1\gamma^2\gamma^3$ are:

---

[122] If the number of terms in universe one and universe two differ then one pads eq. 10.20 and 10.21 with 2×2 zero submatrices for the values of k with unpaired submatrices.

[123] We introduce Dirac γ matrices with a Platonic view of connecting Operator Logic to Elementary Particle Physics later in this volume.

$$\gamma^0 = \begin{bmatrix} I & 0 \\ 0 & -I \end{bmatrix} \quad \gamma^i = \begin{bmatrix} 0 & \sigma_i \\ -\sigma_i & 0 \end{bmatrix} \quad \gamma^5 = \begin{bmatrix} 0 & I \\ I & 0 \end{bmatrix}$$

(10.24)

for i = 1, 2, 3 where the submatrices are the 2×2 Pauli matrices, and I is the 2×2 identity matrix. We will use the γ^5 matrix in the next subsection.

A statement – strictly of the first universe – has the form

$$\text{Universe 1 statement} = \langle\Omega_u|U_1(A_1 + A_2 + A_3 \ldots)|\Omega_u\rangle \quad (10.25)$$

where the eigenvalue operators A_i have the form of eq. 10.20. Its corresponding truth value is given by

$$\text{Universe 1 Truth Value} = \langle\Omega_u|U_1F(P'_1, P'_2, \ldots)P_1\, P_2\, P_3 \ldots|\Omega_u\rangle \quad (10.26)$$

where the projection operators P_i and P'_i have the form of eq. 10.20.

Similarly if we define the universe projection operator

$$U_2 = \tfrac{1}{2}(I_4 - \gamma^0) \quad (10.27)$$

then we see that a statement – strictly of the second universe – has the form

$$\text{Universe 2 statement} = \langle\Omega_u|U_2(A_1 + A_2 + A_3 \ldots)|\Omega_u\rangle \quad (10.28)$$

where the eigenvalue operators A_i have the form of eq. 10.20. Its corresponding truth value is given by

$$\text{Universe 2 Truth Value} = \langle\Omega_u|U_2F(P'_1, P'_2, \ldots)P_1P_2P_3 \ldots|\Omega_u\rangle \quad (10.29)$$

where the operators P_i and P'_i have the form of eq. 10.20. Thus both universes are embodied in the direct product.

10.3.2 Mixing Between the Universes

In the preceeding subsection both universes in the direct product maintained their distinct identity with separate sets of statements. In this section we will consider statements that have terms from both universes. This possibility is of interest because it is analogous to the mixing of two component particle spinors in the free Dirac equation in Physics that is caused by a particle's mass term. Again we are looking towards the development of the Platonic Ideas – Reality relationship later. This possibility of mixing universes is also of interest in its own right.

In order to create mixed statements with terms that come from both universes we have to define eigenvalue and operator matrices that perform that task:[124]

$$\hat{A}_{E_k} = \gamma^5 A_{E_k} \qquad \text{and} \qquad \hat{S} = \gamma^5 P_{E_k} \tag{10.30}$$

using the matrices in eq. 10.20. The new matrices have the form

$$\hat{A}_{E_k} = \begin{bmatrix} 0 & A_{E_{2k}} \\ A_{E_{1k}} & 0 \end{bmatrix} \qquad \hat{S}_{E_k} = \begin{bmatrix} 0 & P_{E_{2k}} \\ P_{E_{1k}} & 0 \end{bmatrix} \tag{10.31}$$

Some examples of statements that mix both universes are

$$\text{Mixed Universe statement} = \langle\Omega_u|U_1(A_1 + \ldots + \hat{A}_i + \ldots + A_j + \ldots + \hat{A}_k + \ldots)|\Omega_u\rangle \tag{10.32}$$

and

$$\text{Mixed Universe statement} = \langle\Omega_u|U_2(A_1 + \ldots + \hat{A}_i + \ldots + A_j + \ldots + \hat{A}_k + \ldots)|\Omega_u\rangle \tag{10.33}$$

Their corresponding truth values having the form

$$\text{Mixed Universe Truth Value 1} = \\ = \langle\Omega_u|U_1F(P'_1, P'_2, \ldots, S'_1, S'_2, \ldots)P_1 \ldots \hat{S}_i \ldots P_j \ldots \hat{S}_k \ldots |\Omega_u\rangle \tag{10.34}$$

$$\text{Mixed Universe Truth Value 2} = \\ = \langle\Omega_u|U_2F(P'_1, P'_2, \ldots, S'_1, S'_2, \ldots)P_1 \ldots \hat{S}_i \ldots P_j \ldots \hat{S}_k \ldots |\Omega_u\rangle \tag{10.35}$$

respectively where the A_k in eq. 10.32 and 10.33 correspond to the operators P_k in eqs. 10.34 and 10.35; and the $\hat{A}_k$ in eq. 10.32 and 10.33 correspond to the operators $\hat{S}_k$ in eqs. 10.32 and 10.33.

A reader who is a logician may wonder why we have developed the concept of mixed universe statements. Again we note this concept is analogous to the Dirac equation for spin ½ elementary particles and will be used later as we attempt to construct a realization of Plato's theory of Ideas and Reality. The analogy is also evident in the two component form of the free spin ½ particle Dirac

---

[124] There is also an alternate approach based on the use of a mass term, which we will discuss later.

equation. The Dirac wave function ψ has four components that can be viewed as a pair of spinors:

$$\psi = \begin{bmatrix} \psi_+ \\ \psi_- \end{bmatrix} \tag{10.36}$$

that satisfy the two-component form of the Dirac equation:

$$i\partial\psi_+/\partial t = -i\boldsymbol{\sigma}\cdot\nabla\psi_+ - m\psi_- \tag{10.37}$$

$$i\partial\psi_-/\partial t = i\boldsymbol{\sigma}\cdot\nabla\psi_- - m\psi_+ \tag{10.38}$$

where m is the particle's mass, $\boldsymbol{\sigma}$ is a 3-vector whose components are the Pauli matrices, and ∇ is the grad differential operator. If we view these equations as statements, which is what they are, then the mass terms mix the "universe" of ψ_+ spinors with the "universe" of ψ_- spinors. Thus the analogy. We will discuss this in more detail at a later point.[125]

The fact that γ_5 appears in the Operator Logic mixing terms (eq. 10.30) is also significant since γ_5 also appears in the 4 component Dirac equation formalism for left-handed and right-handed neutrinos. (Neutrinos are spin ½ particles.)

10.4 Multi-valued Matrix Logics

It is possible to develop a matrix formalism for 3-valued Logic that might be appropriate for some types of logic universes. The SU(3) group is the natural choice for the matrix representation of 3-valued logic.

It is also possible to develop 4-valued, 5-valued, 9-valued and other multi-valued matrix formalisms for universes of discourse. Their development parallels the spin ½ case discussed in section 10.3. We defer that development to a later work since the development is relatively straightforward.

---

[125] Eqs. 10.38 and 10.39 also show we could create a spinor representation for spin ½ particles directly. Therefore forming the direct product (eqs. 10.19 – 10.23) was not necessary but it was helpful because it led directly to a Dirac matrix formulation that is conventional in spin ½ particle physics.

11. Probabilistic and Quantum Operator Logics

The Operator Logic that we have developed in the preceding chapters is not a probabilistic or quantum logic. The eigenvalue operators and the projection operators of all the terms in a universe of discourse commute with each other. The expectation values (inner products) defined within this formalism have a well-defined value: true, false or undecidable (+1, –1 or 0).

In this chapter we develop formalisms for Probabilistic Operator Logic and Quantum Operator Logic. These logics have expectation values that are probabilities and thus can be called probabilistic or quantum respectively.

The material in this chapter is not needed to develop Relativistic Quantum Metaphysics in the succeeding chapters. However it seemed important to the author to discuss probabilistic Operator Logics as a natural extension of the development of deterministic Operator Logic in the preceeding chapters.

11.1 Quantum Operator Logic is Not Fuzzy Logic

Fuzzy Logic in its simplest form is a type of classical probabilistic logic that is based on assigning probabilities to the members of the domain of a predicate that leads to a probability for each statement consisting of the predicate and one or more of the subjects of its domain.

We discuss Fuzzy Logic within the framework of our operator formalism. We call it Probabilistic Operator Logic.

Quantum Operator Logic which is the main subject of this chapter is probabilistic in nature but is not Fuzzy Logic. It describes a quantum probabilistic universe of discourse where terms may not commute and statements have a truth value that is a probability.

11.2 Propositions are Analogous to Experiments

There is a correspondence between the stages of an experiment and the sequence of parts of a statement. In part this analogy is based on Hilbert's[126] view of a predicate, "the use of the term 'predicate' is the one usual in philosophy, namely, that by which one subject can be more particularly characterized." Correspondingly the analysis of an experiment shows it generally to be a series of filtrations which particularize the input to an experiment through a sequence of

---

[126] Hilbert (1950) p. 44.

stages to reach a final stage of output.[127] A filtration in an experiment takes an input state and lets a specified part of it proceed to the next stage of the experiment – stage by stage. A simple example is the case of an input stream of particles of varying velocities. A filtration may select only those particles of a certain velocity to go through to the next stage of the experiment. Analogously the statement "The car is green" takes the general set of all cars and particularizes (filters) it to a green car. An example of a multiple particularization (multi-stage) is "The car is green and has a silver star on it." This statement first particularizes the subject with the predicate "is green" and then particularizes it with the predicate "has a silver star on it."

We thus come to the conclusion that there is a formal analogy between physical experiments and logic propositions. In a sense this observation is the beginning of the relation between the Platonic sphere of Ideas and physical Reality. *Ideas and Reality "map" to each other.*

11.3 Overview of the Relation Between Operator Logic, Probabilistic Operator Logic & Quantum Operator Logic

All three of these formalisms assume that any proposition has one of three forms:

1. A simple subject- predicate form.
2. A combination of clauses of the simple subject-predicate form that are joined by connectives such as "and" and "or".
3. A mathematical equation.

11.3.1 Operator Logic Formulation

In the previous chapters we developed the formalism of Operator Logic. In Operator Logic the eigenvalue operators commute with each other – subject state projection operators, predicate state pojection operators, predicate truth value operators, and subject truth value projection operators.[128] As a result we obtain deterministic propositions and deterministic well defined truth values for propositions.

11.3.2 Probabilistic Operator Logic Formulation

Probabilistic Operator Logic is a generalization of Operator Logic in which the eigenvalue operators commute with each other. Predicate state projection operators, and subject state projection operators also all commute with each other. As a result simultaneous eigenstates of all terms in a universe of discourse are allowed.

---

[127] See Mackey (1963), Messiah (1965), and Gottfried (1989).

[128] See the discussion of eqs. 6.19 – 6.23.

However, predicate *truth value* operators and subject *truth value* operators do not form a commuting set. For a given predicate each subject in the predicate's domain has a probability value associated with it that determines the truth value of a subject-predicate statement or clause. These probabilities are classical probabilities – not quantum probabilities.[129] The sum of the probability values of all the subjects in each predicate's domain is one.

11.3.3 Quantum Operator Logic Formulation

Quantum Operator Logic addresses a theoretical physics issue as well as more general issues. Logic, Operator Logic and Probabilistic Operator Logic all concern c-number terms – terms that are simply words in a language or mathematics – that commute with each other even in our Operator Logic formulations. The areas of Quantum Mechanics and Quantum Field Theory have statements in the form of equations that contain q-number terms – terms that are non-commuting operators. Because of the inherent quantum nature of the terms in these theories the truth or falsity of statements (equations) containing them cannot be truly addressed by Logic, Operator Logic or Probabilistic Operator Logic.

The normal approach of physicists is to postulate a lagrangian and derive equations through variational techniques ignoring the q-number nature of the terms.[130] While this approach works in physics[131] it is not satisfactory from the viewpoint of logic.

In this chapter we will define a Quantum Operator Logic for equations that enables the truth or falsity of q-number equations (statements) to be determined. Quantum Operator Logic lies at the basis of Relativistic Quantum Metaphysics.

11.4 Probabilistic Operator Logic Formalism

The Probabilistic Operator Logic formalism is an extension of the formalism developed for classical Operator Logic in earlier chapters to the case where the eigenvalue operators of the terms in a universe commute with each other[132] but *the set of truth value operators of the universe of discourse do not commute with each other in general.*

We will define the state of a probabilistic universe of discourse in the same manner as we did in the case of the classical universe of discourse using Dirac's

---

[129] The probability of each subject in a predicate's domain is determined by semantic considerations in the case of a semantic universe of discourse and arbitrarily in the case of a calculus universe of discourse.

[130] To some extent this issue is mitigated by path integral formulations of quantum theories that use c-number terms but lead to results identical to q-number quantum theories (if the q-number quantum theories are correctly calculated.)

[131] And avoids the issue of false equations generally.

[132] This enables the eigenvalue operators all terms in a universe of discourse to have a simultaneous eigenvector, namely, the universe state as defined previously.

bra-ket notation for states and operators. We will assume we have q primitive terms with which any statement in the universe can be expressed.

Thus the universe state for a probabilistic universe of discourse (either semantic or calculus) with q terms will be

$$|\Omega_u\rangle = \prod_{i=1}^{q} |E_i\rangle \tag{11.1}$$

$$\equiv |E_1, E_2, E_3, \ldots, E_q\rangle$$

with the set of q Gödel numbers E_i each corresponding to a term gi(E_i) as in eq. 6.26. Note $|\Omega_u\rangle$ is normalized to one: $\langle\Omega_u|\Omega_u\rangle = 1$.

In defining eq. 11.1 we are assuming that the set of q eigenvalue operators for the primitive terms, which we denote $\{A_i\}$, is compatible—the eigenvalue operators all commute with each other.

The elements of the set of state projection operators have the form[133]

$$P_{E_k} = [1 - c_{E_k} c_{E_k}^{\dagger}] \tag{11.2}$$

and also all commute with each other.

The operators $c_{E_k}^{\dagger}$ and c_{E_k} are raising and lowering operators defined by eq. 10.3 – 10.5. (We will not use the spinor formalism here to avoid mathematical complexity.)

The elements of the set of truth value operators are denoted P'_{E_k}. The set of truth value operators P'_{E_k} form an incompatible set—some, or all, of them do not commute with each other in general. For a predicate term A_{E_k} the truth value operator factor P'_{E_k} has the form[134]

$$P'_{E_k} = \sum_j a_{E_j E_k} |E_j\rangle\langle E_k| \tag{11.3}$$

where subject states are denoted E_j, and predicate states are denoted E_k; and where the sum is over all subjects in the universe of discourse. The number $a_{E_j E_k}$ is the probability that the statement "gi(E_j) gi(E_k)" is true. Each combination of subject, E_j, and predicate, E_k, in a statement or clause has the probability $a_{E_j E_k}$ of being true. If "gi(E_j) gi(E_k)" is undecidable or is nonsense or is outside the domain of subjects

---

[133] These operators, together with the eigenvalue operators, conform to the general character of linear vector space operators. Truth value operators are well-defined linear vector space operators.

[134] Eq. 11.3 is a sum of outer products of eigenvectors.

of the predicate, *then we define* $a_{E_jE_k}$ *such that* $0 > a_{E_jE_k} \geq -1$.[135] *The introduction of negative probabilities is necessary to distinguish false statements from undecidable statements as we will see below.*[136]

For a subject term A_{E_j} the truth value operator is

$$P'_{E_j} = |E_j\rangle\langle E_j| \quad (11.4)$$

The probability of the truth of a statement:

$$\text{statement} = \langle\Omega_u|A_1 + A_2 + A_3 + \ldots |\Omega_u\rangle \quad (11.5)$$

is determined by the evaluation of its probability

$$\text{Probability} = \langle\Omega_u|P_{1tot}\, P_{2tot}\, P_{3tot} \ldots |\Omega_u\rangle \quad (11.6)$$

where the $P_{itot} = P'_iP_i$ is the projection operator corresponding to the eigenvalue operator A_i.

A clause or subject-predicate statement has the form

$$\begin{aligned}\text{statement} &= \langle\Omega_u|A_{subject} + A_{predicate}|\Omega_u\rangle \\ &= gn(subject)gn(predicate)\end{aligned} \quad (11.7)$$

namely, a sequence of two Gödel numbers equivalent to two terms in the universe of discourse. The corresponding probability of the subject-predicate statement or clause in Probabilistic Operator Logic is

$$\begin{aligned}\text{Probability} &= \langle\Omega_u|P_{subject}P_{predicate}|\Omega_u\rangle \\ &= \langle\Omega_u|P'_{subject}P'_{predicate}|\Omega_u\rangle \\ &= \langle\Omega_u|\; |E_{subject}\rangle\langle E_{subject}|a_{subject,predicate}|E_{subject}\rangle\langle E_{predicate}|\; |\Omega_u\rangle \\ &= a_{subject,predicate}\end{aligned} \quad (11.8)$$

where $P'_{subject}$ and $P'_{predicate}$ are the eigenvalue operators' corresponding truth value operators.

Probabilities are assigned to all subjects in the domain of a predicate according to some semantic (or calculus) rule.

At this point we see we have a sentential calculus and a rule to reduce all compound statements to equivalent series of clauses as seen previously.

We note again the sentential connectives and their interpretation

---

[135] If the probability of undecidability is not known then one can define $a_{EjEk} = -1$.

[136] Note this differs from deterministic Operator Logic. See eq. 7.4.

Connective	Symbol
and	&
or	v
if ... then (modus ponens)	$\rightarrow$
if and only if	$\sim$
not	___

Denoting sentences with upper case letters: A, B, ... the five basic sentential combinations have the probabilities:

1. A & B is a compound sentence whose probability is the product of the probabilities of A and B.
2. A v B is a compound sentence whose probability is the sum of the probabilities of A and B.
3. A $\rightarrow$ B is a compound sentence whose probability = 1 – (probability of A)·(1 – (probability of B)). This is based on the non-probabilistic case: A $\rightarrow$ B is false if and only if A is true and B is false, which is equivalent to $A\underline{B}$ or in words is equivalent to NOT(A and NOT B).
4. A $\sim$ B is a compound sentence whose probability = 1 – |(probability of A) – (probability of B)|. This is based on the non-probabilistic case: A $\sim$ B is true if and only if both A and B are true, or both A and B are false.
5. The sentence $\underline{A}$ has the probability = 1 – (probability of A). This is based on the non-probabilistic case: the sentence $\underline{A}$ is false if A is true, and is true if A is false.

(11.9)

In specifying these probability rules we assume that if a statement has the probability $p \geq 0$ of being true then $p_f = 1 - p$ is the probability that the statement is false.

Rule 11.1: *Given a simple or compound statement with one or more clauses the probability of the statement follows the rules stated in eq. 11.9 which is an expression in the probabilities p_i of the clauses that we denote $F(p_1, p_2, \ldots)$ multiplied by the minimum value in the set of p_i divided by its absolute value:*

$$\text{Probability} = F(p_1, p_2, \ldots)\text{Min}(p_1, p_2, \ldots)/|\text{Min}(p_1, p_2, \ldots)| \qquad (11.10)$$

If the probability is positive or zero it is the probability of the statement being true.[137]

[137] If any clause is undecidable then the factor Min(p1, p2, ...)/|Min(p1, p2, ...)| is negative and the probability has a negative value indicating the statement is undecidable or has a probability of undecidability specified by eq. 11.10. See section 11.4.3 for the interpretation of negative probabilities.

11.4.1 Probabilistic Example "if ... then ..." Modus Ponens

We will now examine the case of a probabilistic modus ponens (if ... then ...) statement. Suppose we have a statement

$$A \rightarrow B \qquad \text{which is equivalent to} \qquad \underline{\overline{A}B} \tag{11.11}$$

and A can be expressed as a sequence of "anded" clauses:

$$A = \text{clauseA1 and clauseA2 and} \ldots \tag{11.12}$$

and B can also be expressed as a sequence of "anded" clauses:

$$B = \text{clauseB1 and clauseB2 and} \ldots \tag{11.13}$$

Then the probability of the statement eq. 11.10 according to eq. 11.9 is given by the expression:

$$\text{probability} = 1 - (\text{probability of A})\cdot(1 - (\text{probability of B})) \tag{11.14}$$

where the probability for A is

$$<\Omega_u|A|\Omega_u> = \{\, p_{A1}p_{A2} \ldots\}\text{Min}(p_{A1}, p_{A2}, \ldots)/|\text{Min}(p_{A1}, p_{A2}, \ldots)| \tag{11.15}$$

with $p_{Ai} = <\Omega_u|\text{clauseAi}|\Omega_u>$; and where the probability for B is

$$<\Omega_u|B|\Omega_u> =\{\, p_{B1}p_{B2} \ldots\}\text{Min}(p_{B1}, p_{B2}, \ldots)/|\text{Min}(p_{B1}, p_{B2}, \ldots)| \tag{11.16}$$

with $p_{Bi} = <\Omega_u|\text{clauseBi}|\Omega_u>$. The probability of A is

$$<\Omega_u|A|\Omega_u> \tag{11.17}$$

and the probability of B is

$$<\Omega_u|B|\Omega_u> \tag{11.18}$$

If, for the sake of example,

$$<\Omega_u|A|\Omega_u> = 3/4 \tag{11.19}$$

and

$$<\Omega_u|B|\Omega_u> = 1/4 \tag{11.20}$$

then the probability of $A \rightarrow B$

$$p(A \rightarrow B) = 1 - 3/4(1 - 1/4) = 7/16 \tag{11.21}$$

For the special case where the probability of A equals one and the probability of B is one then p(A → B) = 1. If the probability of A equals one and if the probability of B is zero then p(A → B) = 0. Both these special cases conform to our expectations based on classical Logic.[138] If the probability is negative then the statement is undecidable.

11.4.2 Probabilistic Operator Logic Example

Consider a mechanical process that causes a stream of particles to fill a circle of a certain radius on a screen. The probabilities of the various possible radii of the circle are calculated and found to yield the $a_{subject,predicate}$ values for the cases listed below.

Predicate = "is the radius"

Domain:

Subject1 = "Five cm" $a_{subject,predicate} = 1/4$
Subject2 = "Ten cm" $a_{subject,predicate} = 1/4$
Subject3 = "Fifteen cm" $a_{subject,predicate} = 1/4$
Subject4 = "Twenty cm" $a_{subject,predicate} = 1/4$

Based on this probabilistic universe of discourse we find the probabilities of each statement to be:

p("Five cm ", "is the radius") = ¼
p("Ten cm ", "is the radius") = ¼
p("Fifteen cm ", "is the radius") = ¼
p("Twenty cm ", "is the radius") = ¼

The probabilities sum to one as they should.

11.4.3 The Probability of Undecidability

Rule 11.1 defined probability and its interpretation in the case of non-negative probabilities. Negative probabilities indicate a statement is undecidable: completely undecidable, or to some degree undecidable. Since decidable statements have a probability of truth or falsity, consistency suggests, in principle, decidability should also be probabilistic in nature. In the case of deterministic Logic decidable statements are either true or false; statements are either decidable or undecidable. In a probabilistic logic decidability and truth should both be probabilistic.

---

[138] Note that this formalism differs from Operator Logic. Compare the above discussion to the discussion of section 7.1.

Therefore it is reasonable to define the absolute value of a negative probability as the probability that a statement is undecidable. One can envision situations where a proposition has a certain probability of undecidability. The common view is to regard a proposition as either decidable or undecidable. However, there could be situations, where terms are not fully defined or there are other sources of ambiguity, in which one can only estimate the probability that a proposition is undecidable.[139]

The thought process for the probability of undecidability of a proposition is analagous to the thought process for the probability of the truth of a proposition.

The issue of the probability of undecidability is also relevant for Quantum Operator Logic although we will not address it in this book except to say the previous statements in this subsection also apply in Quantum Operator Logic.

11.5 Quantum Operator Logic

Quantum Operator Logic deals with propositions containing terms that are q-numbers.[140] The label q-number comes from physics and designates a term that is inherently an operator.[141] Quantum Mechanics is based on q-numbers. For example, the position of a particle is represented in quantum mechanics as an operator – a q-number. The momentum of a particle is similarly an operator. These operators satisfy the Heisenberg Uncertainty Principle

$$[x, p] \equiv xp - px = i\hbar \qquad (11.22)$$

where $\hbar = h/(2\pi)$ with h being Planck's constant. Quantum Mechanics is best understood as an operator formalism. Operator expressions are evaluated between quantum states (inner products) to yield expectation values (probability amplitudes whose absolute values squared are probabilities.)

Quantum Electrodynamics, the most accurately measured theory in physics, is formulated in terms of operators for the electromagnetic fields. Unified Quantum Field theories such as ElectroWeak Theory[142] and The Standard Model of Elementary Particles[143] are also formulated in terms of operators for the various particle fields.

The operator equations of Quantum Mechanics and Quantum Field Theories are not propositions whose truth or falsity can be determined in Logic, Operator Logic or Probabilistic Operator Logic. This issue has been overlooked

---

[139] In science it is not uncommon to do experiments to measure something where the success of the measurement may be undecidable or only decidable with a certain probability.

[140] We use the label "Quantum" in this context because the propositions, which are usually equations, contain q-numbers – operators. The label does not limit the formalism to quantum physics.

[141] The label c-number, used earlier, also comes from physics and designates a term that is not an operator and has a strictly numerical value.

[142] The union of electromagnetism and the Weak Interactions.

[143] The union of electromagnetism, the Weak Interactions and the Strong Interactions.

because Quantum Mechanical Theories of specific phenomena and the various Quantum Field Theories have been normally defined using a lagrangian and a variational principle.[144] If the lagrangian is correct the quantum equations derived from it are true[145] and the issue of truth or falsity does not arise. So the logic analysis of operator equations has hitherto not been investigated.[146] In the author's opinion the logic analysis of operator equations, viewed as propositions, is of interest for physics, logic and metaphysics.

In the case of metaphysics we are confronted with a very successful theory of Reality, The Standard Model. Its basis in metaphysics is not susceptible to significant logic analysis because the "language" of the theory, its quantum operator equations, cannot be analyzed with the logic that has hitherto been developed. Metaphysics, as Leibniz observed, is based on logic. So we must develop a logic suitable for operator propositions. We call this logic Quantum Operator Logic. Having this logic in hand we can address the deeper issues of the foundations of The Standard Model in a meaningful way.

11.5.1 Quantum Operator Logic Formalism

The fundamental consideration of Quantum Operator Logic is the non-commutativity of some or all of the terms in a universe of discourse. The consequence of non-commuting terms is non-commuting eigenvalue operators as well as non-commuting truth value operators in general.

In the previously discussed formalisms of Operator Logic and Probabilistic Operator Logic we could define a single state for a universe of discourse (See eq. 6.1). *In the present case we must define a vector composed of a set of universe of discourse states: one state for each subset of commuting eigenvalue operators.* For example if term1 and term2 are operators that do not commute with each other while term3, term4, ... commute with each other and with term1 and term2[147] then there are two subuniverse states which together form the universe state $|\Omega_u\rangle$:

$$|\Omega_{u1}\rangle = |E_{term1}, E_{term3}, E_{term4}, \ldots\rangle \qquad (11.23)$$
$$|\Omega_{u2}\rangle = |E_{term2}, E_{term3}, E_{term4}, \ldots\rangle \qquad (11.24)$$

In general, the set of universe states consists of all states that are simultaneous eigenstates of the maximally commuting subsets of the set of all terms in the

---

[144] The variational principles of physics come from the Calculus of Variations which was originally formulated by Leibniz and others.

[145] In the case of Quantum Field Theories containing Yang-Mills fields a path integral formalism yields the correct results in calculations. For examples of this formalism see appendix 19-A.

[146] Of course experimental verification has been used to determine whether a theory is true or not. But that is not logic analysis.

[147] To keep the example simple we assume [term1, term2] equals a c-number and thus this commutator commutes with term3, term4, ... This is required by the commutator identity [A,[B,C]] + [C,[A,B]] + [B,[C,A]] = 0.

universe of discourse. Since the number of terms in a universe is finite by assumption the number of maximally commuting subsets of terms is finite. The form of the ith term is

$$|\Omega_{ui}\rangle = \prod_{j \in \{m_i\}} |E_j\rangle \qquad (11.25)$$

where $\{m_i\}$ is the ith maximally commuting set of terms and thus eigenvalue operators.

We will now consider the example of a universe of discourse composed of a coordinate operator x and its corresponding momentum operator p where x and p satisfy eq. 11.22 and thus do not commute. We define the corresponding eigenvalue operators as $A_{\text{"x"}}$ and $A_{\text{"p"}}$ with Gödel number eigenvalues $E_{\text{"x"}}$ and $E_{\text{"p"}}$ respectively. We also define the eigenvalue operator $A_{\text{"}i\hbar\text{"}}$ with eigenvalue $E_{\text{"}i\hbar\text{"}}$. $A_{\text{"}i\hbar\text{"}}$ commutes with both $A_{\text{"x"}}$ and $A_{\text{"p"}}$. It will be convenient to define the set of universes of discourse

$$|\Omega_{u\text{"x"}}\rangle = |E_{\text{"x"}}, E_{\text{"}i\hbar\text{"}}\rangle \quad \text{and} \quad |\Omega_{u\text{"p"}}\rangle = |E_{\text{"p"}}, E_{\text{"}i\hbar\text{"}}\rangle \qquad (11.26)$$

and form a two-vector (spinor) from them which we will call *term space*

$$|\Omega_u\rangle = \begin{bmatrix} |\Omega_{u\text{"x"}}\rangle \\ |\Omega_{u\text{"p"}}\rangle \end{bmatrix} \qquad (11.27)$$

where each universe state includes an "$i\hbar$" term as in eq. 11.26.

We now introduce a matrix notation that enables us to evaluate equations (statements) involving non-commuting terms. We define the two diagonal matrices I_t in term space as having the matrix elements:

$$I_t = [\delta_{ti}] \qquad (11.28)$$

where δ_{ik} is the Kronecker delta function with value one if i = k and zero otherwise. Thus I_t has only one non-zero entry, namely, a one at row t and column t – the diagonal matrix element for the tth term.

Next consider two examples that illustrate the basic nature of the Quantum Operator Logic formalism. First consider the additive commutation condition which in terms of the x and p operators is

$$x + p = p + x \qquad (11.29)$$

This operator equation can be expressed in terms of Quantum Operator Logic eigenvalue operators

$$A_{tot\text{"}x\text{"}} = A_{\text{"}x\text{"}}\, I_{\text{"}x\text{"}} \tag{11.30}$$

and

$$A_{tot\text{"}p\text{"}} = A_{\text{"}p\text{"}}\, I_{\text{"}p\text{"}} \tag{11.31}$$

where

$$A_{\text{"}x\text{"}}\, |\Omega_{u\text{"}x\text{"}}\rangle = gn(\text{"}x\text{"})|\Omega_{u\text{"}x\text{"}}\rangle \tag{11.32}$$

and

$$A_{\text{"}p\text{"}}\, |\Omega_{u\text{"}p\text{"}}\rangle = gn(\text{"}p\text{"})|\Omega_{u\text{"}p\text{"}}\rangle \tag{11.33}$$

The eigenvalue operator equation leading to eq. 11.29 is

$$A_{tot\text{"}x\text{"}} + A_{tot\text{"}p\text{"}} = A_{tot\text{"}p\text{"}} + A_{tot\text{"}x\text{"}} \tag{11.34}$$

Taking the expectation value (inner product) of eq. 11.34 using the universe of discourse two-vector (eq. 11.27), followed by the trace over the universe states, of

$$\langle\Omega_u|(A_{tot\text{"}x\text{"}} + A_{tot\text{"}p\text{"}} = A_{tot\text{"}p\text{"}} + A_{tot\text{"}x\text{"}})|\,\Omega_u\rangle$$

yields (after mapping Gödel numbers to their corresponding terms)

$$x + p = p + x \tag{11.35}$$

upon realizing that the matrices contain operators and the order of operators matters. We thus find eq. 11.29 follows.[148] Note that we do not express the plus sign or the equal sign as operators although we could have done so.[149]

We now consider the case of the commutation relation eq. 11.22. We wish to evaluate the expectation value (inner product) of the commutator $[A_{tot\text{"}x\text{"}}, A_{tot\text{"}p\text{"}}]$ between universe two-vectors (eq. 11.27). The only sensible way to proceed is to take the expectation value of each eigenvalue operator independently followed by the trace over the universe states of the result:

$$\begin{aligned}\langle\Omega_u|[A_{tot\text{"}x\text{"}}, A_{tot\text{"}p\text{"}}]|\Omega_u\rangle &= \text{Tr}\,\{\langle\Omega_u|A_{tot\text{"}x\text{"}}|\Omega_u\rangle\langle\Omega_u|A_{tot\text{"}p\text{"}}|\Omega_u\rangle - \\ &\quad - \langle\Omega_u|\, A_{tot\text{"}p\text{"}}|\Omega_u\rangle\langle\Omega_u|A_{tot\text{"}x\text{"}}|\Omega_u\rangle\} \\ &= E_{\text{"}x\text{"}}\, E_{\text{"}p\text{"}} - E_{\text{"}p\text{"}}\, E_{\text{"}x\text{"}}\end{aligned} \tag{11.36}$$

Upon mapping the Gödel numbers to the corresponding operators we obtain

[148] Addition is normally considered a commutative operation but it is not required to be commutative. Thus commutativity is often an axiom and sometimes a theorem or result.

[149] Since in earlier discussions of the Operator Logic formalism the plus sign was discarded. Note the + sign eigenvalue operator commutes with all the other operators in this universe.

$$<\Omega_u|[A_{tot\text{"}x\text{"}}, A_{tot\text{"}p\text{"}}]|\Omega_u> = xp - px \tag{11.37}$$

We define

$$A_{tot\text{"}i\hbar\text{"}} = IA_{\text{"}i\hbar\text{"}}/2 \tag{11.38}$$

where I is the 2 × 2 identity matrix and $A_{\text{"}i\hbar\text{"}}$ is the eigenvalue operator for the c-number quantity $i\hbar$. Then we obtain eq. 11.22 through a suitably evaluated Quantum Operator Logic statement:

$$<\Omega_u|[A_{tot\text{"}x\text{"}}, A_{tot\text{"}p\text{"}}]|\Omega_u> = <\Omega_u|A_{tot\text{"}i\hbar\text{"}}|\Omega_u> \tag{11.39}$$

The evaluation of the sequence of eigenvalue operators between universe states in the preceding examples requires:

1. Taking the expectation values of each eigenvalue operator in an operator product separately to obtain its Gödel number.
2. Using a matrix notation with 2-vectors and 2 × 2 matrices (or an equivalent).
3. Evaluating the inner product of a statement.
4. Mapping Gödel numbers to corresponding operator terms in a fashion similar to Operator Logic.
5. Using the mathematical operators for multiplication, addition, subtraction, and so on after substituting corresponding operators for Gödel numbers.

Thus we find the Quantum Operator Logic equation

$$[A_{tot\text{"}x\text{"}}, A_{tot\text{"}p\text{"}}] = A_{\text{"}i\hbar\text{"}} \tag{11.40}$$

when evaluated in the above formalism implies eq. 11.22.

The above examples evaluate the eigenvalue operator expression of a quantum statement. The question of the true/false/undecidable probability remains to be addressed.

11.5.2 Operator Term Space

Recap of the Preceding Discussion:

The non-commutativity of operators in Quantum Operator Logic statements leads to a set of universe states (eq. 11.25). Each universe state has a set of maximally commuting subsets, C, of the set of operator terms. Since the number of operator terms is finite by assumption, the number of sets in C is finite. We can construct a universe state for each member of C. The number of such universe states is finite. They can be used to form a vector space. Eq. 11.27 in the preceding subsection is an example of a two dimensional vector space of universe states.

It is easy to find larger, more complex examples, particularly in physics. For example the group SU(2) plays a role in many areas of quantum physics. Let us consider the general case of its three generators S^i. They satisfy the commutation relations:

$$[S^i, S^j] = i\varepsilon_{ijk} S^k \qquad (11.41)$$

where ε_{ijk} is the unit totally anti-symmetric tensor with $\varepsilon_{123} = 1$. Since there are three non-commuting operators, there are three subsets in C, and we can form a three-dimensional universe vector space:

$$|\,\Omega_{u3}> = \begin{bmatrix} |\Omega_{u\text{"}S^1\text{"}}> \\ |\Omega_{u\text{"}S^2\text{"}}> \\ |\Omega_{u\text{"}S^3\text{"}}> \end{bmatrix} \qquad (11.42)$$

We introduce three eigenvalue operators $A_{\text{"}S^i\text{"}}$ that correspond to the three SU(2) operators where $A_{\text{"}S^i\text{"}}$ has the Gödel number eigenvalues $E_i = gn(\text{"}S^i\text{"})$ or zero. We define projection operators (similarly to eq. 11.28) as the diagonal matrices I_t in term space with the matrix elements:

$$I_{3t} = [\delta_{ti}] \qquad (11.43)$$

for t = 1, 2, 3 where δ_{ik} is the Kronecker delta function.

Eigenvalue operators in term space have the more general form

$$A_{tot\text{"}S^i\text{"}} = A_{\text{"}S^i\text{"}}\, I_{3i} \qquad (11.44)$$

for i = 1, 2, 3.
Then eq. 11.41 is equivalent to the Quantum Operator Logic statement

$$[A_{tot\text{"}S^i\text{"}}, A_{tot\text{"}S^j\text{"}}] = i\varepsilon_{ijk} A_{tot\text{"}S^k\text{"}} \qquad (11.45)$$

If we evaluate

$$<\Omega_{u3}|[A_{tot\text{"}S^i\text{"}}, A_{tot\text{"}S^j\text{"}}] = i\varepsilon_{ijk} A_{tot\text{"}S^k\text{"}}\, |\Omega_{u3}> \qquad (11.46)$$

in the same manner as we did in the case of eqs. 11.36 – 11.40 then we find eq. 11.46 leads to eq. 11.41.

Similarly in the case of the SU(3) Lie algebra, which is also of great importance in elementary particle theory, we find that the eight generators of SU(3) lead us to define a seven dimensional space since two of the generators commute with each other. The commuting generators are usually designated F^3 and F^8. The *universe vector* is thus

$$|\Omega_{u7}> \;=\; \begin{bmatrix} |\Omega_{u\text{"F1"}}> \\ |\Omega_{u\text{"F2"}}> \\ |\Omega_{u\text{"F3", "F8"}}> \\ |\Omega_{u\text{"F4"}}> \\ |\Omega_{u\text{"F5"}}> \\ |\Omega_{u\text{"F6"}}> \\ |\Omega_{u\text{"F7"}}> \end{bmatrix} \tag{11.47}$$

where

$$|\Omega_{u\text{"F3", "F8"}}> = |E_{\text{"F3"}}, E_{\text{"F8"}}> \tag{11.48}$$

We define projection operators (similarly to eq. 11.28) as the diagonal matrices I_t in term space with the matrix elements:

$$I_{7t} = [\delta_{ti}] \tag{11.49}$$

for t = 1,2, ..., 7 where δ_{ik} is the Kronecker delta function.

Eigenvalue operators in term space have the general form

$$A_{tot\text{"Fi"}} = A_{\text{"Fi"}}\, I_{7i} \tag{11.50}$$

for i = 1, 2, ..., 7 and

$$A_{tot\text{"F8"}} = A_{\text{"F8"}} I_{73} \tag{11.51}$$

Thus

$$A_{tot\text{"Fi"}}|\Omega_{u7}> = E_{\text{"Fi"}}\, I_{7i}|\Omega_{u7}> \tag{11.52}$$

for i = 1, ...,7 excepting i = 3, and

$$A_{tot\text{"F3"}}|\Omega_{u7}> = E_{\text{"F3"}} I_{73}|\Omega_{u7}> = gn(\text{"F}^{3}\text{"})I_{73}|\Omega_{u7}> \tag{11.53}$$
$$A_{tot\text{"F8"}}|\Omega_{u7}> = E_{\text{"F8"}}\, I_{73}|\Omega_{u7}> = gn(\text{"F}^{8}\text{"})I_{73}|\Omega_{u7}> \tag{11.54}$$

The commutation relations for SU(3) are

$$[F^{i}, F^{j}] = ic_{ijk}\, F^{k} \tag{11.55}$$

Again following the procedure outlined above we see that the corresponding Quantum Operator Logic statement is

$$[A_{toti}, A_{totj}] = ic_{ijk}\, A_{totk} \tag{11.56}$$

11.5.3 The Term Space of Quantum Operator Logic

In the preceding subsections we have seen how to develop and evaluate Quantum Operator Logic statements. These statements are typically equations and constructs in quantum mechanics and quantum field theories in physics. Statements (propositions) in ordinary language usually do not contain operator (q-number) terms.[150] Such statements are handled within the framework of Operator Logic.

We will now consider a theory that consists of statements (axioms, theorems, and equations) that are constructed from a set of generally non-commuting operator terms. The examples in the preceding subsections illustrate the main features of the Quantum Operator Logic formulation of this type of theory.

One might ask why we should develop a reformulation of perfectly acceptable quantum theories? I believe the answer lies in the hope of finding a more fundamental theory of Reality – particularly in the case of The Standard Model of Elementary Particles. This theory is the closest experimentally verified theory of fundamental Reality.

We found the concept of *term space* was necessary to develop a Quantum Operator Logic formulation of a set of operator statements since one cannot formulate a simultaneous eigenstate of non-commuting operators. Thus we were led to define a universe of discourse state that is a vector[151] of eigenstates of maximally commuting subsets of terms $\{m_i\}$ of the form of eq. 11.25:

$$|\Omega_{ui}> = \prod_{j \in \{m_i\}} |E_{ij}> \qquad (11.25')$$

the universe state of the ith maximally commuting subset $\{m_i\}$ of set of operators where each E_{ij} is the non-zero Gödel number of a term operator. Since the number of terms in a universe is finite by assumption the number of maximally commuting subsets of terms is finite. If there are q maximally commuting subsets then the universe of discourse state vector has q components:

$$|\Omega_u> = \begin{bmatrix} |\Omega_{u1}> \\ \dots \\ |\Omega_{uq}> \end{bmatrix} \qquad (11.57)$$

---

[150] They could also contain a set of operator terms – all of which commute with each other.

[151] Eqs. 11.42 and 11.47 are examples of Quantum Operator Logic universe of discourse states.

as does its hermitean conjugate row vector $<\Omega_u|$. We call the q dimensional vector composed of these q universe states a vector in *term space*. To access each of the individual universe states it is necessary to define projection operators that are diagonal matrices I_i in term space with the matrix elements:

$$I_i = [\delta_{ij}] \tag{11.58}$$

for i, j = 1,2, ..., q where δ_{ij} is the Kronecker delta function.

The eigenvalue operators $A_{\text{"}T^{ij}\text{"}}$ satisfy

$$A_{\text{"}T^{ij}\text{"}}\,|E_{ij}> = E_{ij}|E_{ij}> = gn(\text{“}T^{ij}\text{”})|E_{ij}> \tag{11.59}$$

The form of a full eigenvalue operator in term space is

$$A_{tot\text{"}T^{ij}\text{"}} = A_{\text{"}T^{ij}\text{"}}\, I_i \tag{11.60}$$

where j designates an operator within the ith maximally commuting subset $\{m_i\}$. T^{ij} is the name of the jth operator term in the ith subset of commuting term operators with its eigenvalue specified by eq. 11.59. The eigenvalue equation in term space is

$$A_{tot\text{"}T^{ij}\text{"}}|\Omega_u> = E_{ij}\, I_i\, |\Omega_u> = gn(\text{“}T^{ij}\text{”})I_i\, |\Omega_u> \tag{11.61}$$

for i = 1, ..., q.

Non-operator terms (c-number terms such as numeric constants and mathematical operation signs) that appear in Quantum Operator Logic statements are basically treated like terms in Operator Logic. We can choose to add these terms to the largest subset of commuting operators (although this is not required). The steps of defining corresponding eigenvalue operators whose eigenvalues are the Gödel numbers of the c-number terms and then using them to express a statement in terms of eigenvalue operators is similar to the Operator Logic case – see chapter 6 – particularly section 6.1 and eq. 6.24.

To each c-number term C_a we associate an eigenvalue operator A_{C_a} and an eigenstate $|E_{C_a}>$ satisfying

$$A_{C_a}|E_{C_a}> = E_{C_a}|E_{C_a}> = gn(C_a)|E_{C_a}> \tag{11.62}$$

We can (arbitrarily) adjoin the c-number term eigenstates to the eigenvector of the largest maximally commuting subset of operators denoted $\{m_{i_1}\}$

$$|\Omega_{ui_1}> = \prod_{j\in\{m_{i_1}\}} |E_{i_1j}> \prod_{k\in\{m_C\}} |E_{C_k}> \tag{11.25"}$$

where $\{m_C\}$ is the set of all c-number terms in the universe of discourse. For the C_a^{th} term we define

$$A_{totC_a} = A_{C_a} I_{i_1} \qquad (11.63)$$

Thus we have a Quantum Operator Logic that encompasses a universe of discourse with both q-number and c-number terms.

As we saw in the previous section we can map between operator statements (equations) and eigenvalue operator statements in term space:

$$S(T^1, T^2, \ldots , T^n, C_1, C_2, \ldots , C_m) \leftrightarrow S'(A_{tot\text{"}T^1\text{"}}, A_{tot\text{"}T^2\text{"}}, \ldots , A_{tot\text{"}T^n\text{"}}, A_{totC_1}, A_{totC_2}, \ldots , A_{totC_m}) \qquad (11.64)$$

which are evaluated as traces of expectation values of the universe state eq. 11.57:

$$<\Omega_u| S'(A_{tot\text{"}T^1\text{"}}, A_{tot\text{"}T^2\text{"}}, \ldots , A_{tot\text{"}T^n\text{"}}, A_{totC_1}, A_{totC_2}, \ldots , A_{totC_m})|\Omega_u> \qquad (11.65)$$

to obtain the the operator statement $S(T^1, T^2, \ldots , T^n, C_1, C_2, \ldots , C_m)$.

We have thus succeeded in mapping operator statements (equations) of a theory into a term space of Quantum Operator Logic. At this point it seems appropriate to inquire as to the benefits of such a map. One primary goal seems evident: to uncover a deeper layer of Reality using the formalism and methodology of Quantum Operator Logic that we can the map back to the original operator theory. The key will be to develop a semantic theory for the origin and truth[152] of the statements in term space. We initiate a discussion of a semantic theory in term space in the next subsection.

The existence of a term space due to the operator nature of the original theory is provocative. We will now examine some features of term space. If we apply the projection operators I_i to the universe state we can generate q unit vectors in the various "directions" of term space

$$v_i = I_i |\Omega_u> \qquad (11.66)$$

where i = 1, 2, ... , q and where

$$v_i^\dagger v_j = \delta_{ij} \qquad (11.67)$$

due to the projection I_i appearing in eq. 11.66. If we sum the vectors v_i we obtain the universe of discourse state

$$|\Omega_u> = \sum_{i=1}^{q} v_i \qquad (11.68)$$

[152] Or probability amplitude of truth.

We can define lowering operators for the eigenstates within the product of eigenstates constituting each unit vector v_i. In a manner similar to subsections 8.3.1 and 8.3.2. For the jth eigenstate in the product of eigenstates that constitutes v_i we can define a lowering operator $a_{E_{ij}}$ by

$$|0\rangle = [gn(E_{ij})]^{-½} a_{E_{ij}} |E_{ij}\rangle \qquad (11.69)$$

and, the corresponding raising operator $a_{E_{ij}}^\dagger$ with

$$|E_{ij}\rangle = [gn(E_{ij})]^{-½} a_{E_{ij}}^\dagger |0\rangle \qquad (11.70)$$

where † signifies hermitean conjugatation. Thus

$$v_{i,-j} = [gn(E_{ij})]^{-½} a_{E_{ij}} v_i \qquad (11.71a)$$

is the vector obtained from v_i by setting the jth eigenvalue to zero.

The inverse operation of changing the eigenvalue of a term eigenvalue operator A_{ij} from zero to E_{ij} is

$$v_{i,+j} = [gn(E_{ij})]^{-½} a_{E_{ij}}^\dagger v_i \qquad (11.71b)$$

As in sections 8.3.1 and 8.3.2 we do not wish repeated application of raising operators to generate states with higher eigenvalues.[153] Therefore we define

$$a_{E_{ij}}^\dagger |E_{ij}\rangle = 0 \qquad (11.72)$$

$$a_{E_{ij}} |0\rangle = 0 \qquad (11.73)$$

for each individual eigenstate. Consequently the raising and lowering operators must satisfy the *anticommutation* relations with the form:

$$\begin{aligned} \{a_{E_{ij}}^\dagger, a_{E_{km}}\} &= \delta_{ik}\, \delta_{jm}\, gn(E_{ij}) \\ \{a_{E_{ij}}, a_{E_{km}}\} &= 0 \\ \{a_{E_{ij}}^\dagger, a_{E_{km}}^\dagger\} &= 0 \end{aligned} \qquad (11.74)$$

for eqs. 11.72 and 11.73.

Eigenvalue operators can be represented in terms of these operators by

---

153 Since such eigenstates would not correspond to a primitive or defined term in the universe of discourse.

$$A_{E_{ij}} = a_{E_{ij}}{}^{\dagger} a_{E_{ij}} \tag{11.75}$$

and the corresponding projection operators by

$$P_{E_{ij}} = P'_{E_{ij}}[1 - [gn(E_{ij})]^{-1} a_{E_{ij}} a_{E_{ij}}{}^{\dagger}] \tag{11.76}$$

where $P'_{E_{ij}}$ is a truth projection operator. (Truth projection operators are discussed in the next subsection.)

Lastly it is possible to define rotations in this q dimensional space having the group structure of U(q) – the unitary group in q dimensions. These rotations would lead to rotations amongst the eigenvalue operators and thus to their Gödel number eigenvalues. Then the mapping from Gödel numbers to operators in statements would be similarly rotated. However the statements would remain the same. U(q) rotations may eventually play some role in understanding the underpinnings of The Standard Model of Elementary Particles.

11.5.4 Quantum Operator Logic Projection Operators

In the previous subsections of 11.5 we have focussed on the mapping from operator terms in the operator equations of a theory to eigenvalue operators in term space. We have only briefly examined projection operators in term space (eq. 11.76) and have not addressed the issue of the truth value projection operators that figured prominently in our study of Operator Logic.

Determining truth value projection operators is essentially a semantic question. Bacon's famous opening line of his essay "On Truth" begins "What is truth ..." and we must confront the same issue. In Operator Logic and Probabilistic Operator Logic we resolved the issue for subject-predicate statements and statements reducible to combinations of subject-predicate clauses by defining a domain for each predicate in a universe of discourse.[154] The specification of a domain is a semantic issue for semantic universes of discourse and a matter of definition for a calculus universe of discourse.

Thus one approach to the determination of truth value projection operators in Quantum Operator Logic would be to separate each operator equation (statement) into a "subject" part and a "predicate" part in such a way as to minimize the number of predicates,[155] and then to determine the domain of subjects for each predicate. An example of this approach is to regard the SU(3) commutation relations (eq. 11.55) as a set of statements where the subject is "$[F^i$,

---

[154] With a probability for each subject-predicate pair in the case of Probabilistic Operator Logic. In the case of Quantum Operator Logic following one approach we could define a probability amplitude for each subject-predicate pair and calculate probabilities from the probability amplitudes. Statements that had negative probabilities would again be undecidable as discussed in subsection 11.4.3.

[155] For the sake of simplicity.

F^j]" and the predicate is "= ic_{ijk} F^k". The domain of this predicate is the subject "[F^i, F^j]".

This approach is viable but it assumes that one specifies the set of domains and thus requires a foreknowledge of all the domains. One obtains a well-defined universe of discourse within the framework of Quantum Operator Logic but one only gets that which one puts in. Perhaps in some cases that is all one really wants—a reformulation of a set of operator equations (statements) in the formalism of Quantum Operator Logic.

Another more interesting approach, in the view of this author, to the case of the metaphysics of ultimate Reality is to grow a Quantum Operator Logic formulation from a "secure" foundation using a set of principles that lead to a physical theory of ultimate Reality. Among the principles that seem to be most relevant are Ockham's Razor[156] and Leibniz's Principle of Perfection.[157] We will consider this approach in the next chapter. Its constructive approach to the creation of the theory of Reality is more intellectually satisfying then simply positing a theory either as a set of postulates or a lagrangian for the universe.

11.6 Quantum Operator Logic Subuniverses Generated by Lowering Operators

In the preceding sections we have considered Quantum Operator Logic universes of discourse. We now consider the possibility of subuniverses as we did similarly in the case of Operator Logic earlier.

The creation of subuniverses of discourse is described in the case of Operator Logic in chapter 9. The creation of a quantum subuniverse follows almost the same procedure of deleting primitive terms from a universe of discourse by using lowering operators as in those sections.

Suppose we wish to delete the operator term with eigenvalue operator A_{E_m}, eigenstate $|E_m>$ and lowering operator a_{E_m} from a quantum universe of discourse. The universe of discourse eigenstates are defined by eqs. 11.25, 11.25', 11.25", and 11.57. We must first identify the subset of the states $\{|\Omega_{ui}>\}$ containing the eigenstate $|E_m>$ as a factor. We will denote these states as $|\Omega_{umi}>$ and express the universe vector as

[156] William of Ockham – Law of Parsimony – "Pluralitas non est ponenda sine necessitate" or "Plurality should not be posited without necessity." First stated by Durand De Saint-Pourçain (1270-1334 AD). In simple terms the principle states the simplest solution to a problem is most likely to be the correct solution.

[157] This principle can be phrased for our purposes as, "The universe is based on the smallest set of properties or features that lead to the greatest variety of phenomena." This principle reflects the minimum/maximum criteria of the Calculus of Variations that play a central role in many physics theories. Leibniz was one of the founders of the Calculus of Variations.

$$|\Omega_u\rangle = \begin{bmatrix} |\Omega_{um1}\rangle \\ \cdots \\ |\Omega_{ump}\rangle \\ |\Omega_{up+1}\rangle \\ \cdots \\ |\Omega_{uq}\rangle \end{bmatrix} \qquad (11.77)$$

where $p \leq q$. Thus we have a q dimensional vector to which we can apply the lowering operator

$$a_{E_m}|\Omega_{um}\rangle = \begin{bmatrix} |\Omega_{um1}'\rangle \\ \cdots \\ |\Omega_{ump}'\rangle \\ |\Omega_{up+1}\rangle \\ \cdots \\ |\Omega_{uq}\rangle \end{bmatrix} \equiv |\Omega_u'\rangle \qquad (11.78)$$

The application of the lowering operator produces a state such that

$$A_{E_m}|\Omega_u'\rangle = 0 \qquad (11.79)$$

We can use $|\Omega_u'\rangle$ to evaluate the expectation value of an equation (statement)

$$\langle\Omega_u'|S|\Omega_u'\rangle \qquad (11.80)$$

It does not contain the term corresponding to A_{E_m} but does contain all the other terms that were originally in the equation. Thus we have generated a subuniverse of discourse with the term corresponding to A_{E_m} omitted from all statements. This procedure could be applied repeatedly to obtain "sub-sub-...-sub universes generating specialized subtheories from an initial general theory.

12. From Ideas to Reality: First Principles

In this chapter we will develop a framework for relating the realm of Ideas as conceived by the Platonists to ultimate physical Reality as understood today. The approach will be based on certain principles which are closer to physical thought than the principles that have been the subject of investigation and discussion by metaphysicians from early times to the present.

12.1 Six General Principles

In section 2.3 we discussed seven principles upon which Leibniz based his metaphysics. His 17th century principles furnished a framework for thought and discussion then but are not sufficient for our current understanding of physical Reality. Metaphysical thought since Leibniz also suffers from a lack of specificity about physical Reality.

In this section we will specify six principles which furnish a starting point for the development of a fundamental basis for physical Reality and the construction of a specific physical theory of ultimate Reality – The Standard Model. The principles seem reasonable, and sufficiently detailed, to provide useful guidance in the construction of a theory of physical Reality. They also are general in nature.

The six principles are:

1. Logic, particularly Operator Logic and Quantum Operator Logic as the case may be, applies to metaphysical and physical derivations and discussions. (This principle is clearly assumed by metaphysicians.)

2. In the case of several alternative choices Ockham's Razor[158] should be used to determine the correct choice.[159]

3. We assume the universe is based on the smallest set of properties or features that lead to the greatest variety of phenomena.[160] Alternately put, we choose

---

[158] See subsection 11.5.4.

[159] One very clear reason for this principle is that physics is very difficult and the simplest choice is usually the choice that enables further progress to occur. An example is Copernicus' theory that the planets circled the sun. This theory was nicely adapted to support Newton's work on the theory of gravitation. On the other hand, the Ptolemaic theory was more complex and not consistent with Newton's theory of gravitation. Copernicus' theory was far simpler than Ptolemaic theory.

[160] A version of Leibniz's Principle of Perfection.

the most minimal conditions necessary to lead to the known phenomena of nature.

4. Whatever physical entities exist in the universe they can only make their presence known by interacting (exerting forces) with other entities. Thus each physical entity interacts with at least one other type of entity.[161]

5. The behavior and properties of physical entities must be reproducible whenever the same circumstances occur. Behavior can be deterministic or can be specified in terms of probabilities. Consequently the behavior and properties must be governed directly or indirectly by mathematics.

6. We assume that space and time exist and that the properties of entities are local in the sense that the properties of an entity depend on the point in space and time where it is located.[162]

12.2 Logic and Metaphysics

The principles stated in the previous section in themselves do not constitute a physical theory. In order to progress further we need to use logic and appropriate constructs.

Metaphysics deals with many areas including physical Reality, the nature of God, attempts at proofs of God's existence, Theology, and so on. We focus entirely on physical Reality as stated earlier. We will begin with a foundation built on Logic – the only fundamental starting point of which we can be certain. Without Logic the construction and study of the implications of theories is impossible. From Logic, in particular Asynchronous Logic, we will develop constructs that lead us to a physical theory of matter and energy. In this author's view, this approach is sounder and more intellectually satisfying than simply positing a physical theory with a set of axioms or simply stating a lagrangian for "everything."

Physicists have spent many years developing a type of comprehensive "Theory of Everything" called SuperString Theory. This theory purports to be ultimate Reality. However at present there is no experimental evidence for it and no conclusive derivation of The Standard Model from it. It assumes that everything is ultimately composed of ultra-small mathematical strings. However it does not answer the question: Why is it *ultimate* Reality? One simply posits certain types of mathematical strings and then derives their consequences. But where and why do strings arise? We conclude that a SuperString Theory, even if found to be experimentally true, would reduce physics to a lower level of Reality but, in turn, raises questions as to its origin in a yet deeper Reality. Some metaphysicians have

---

[161] Consequently there is no knowable physical entity with no interactions with other types of entities.
[162] This principle is the basis of local symmetries such as Yang-Mills theories.

proposed that there is no ultimate Reality but only ever-increasing lower levels of Reality. We do not believe this to be true.[163]

We suggest that a better alternative, if it is viable,[164] is to found the ultimate theory of Reality on the bedrock of Logic. For if Logic is not a true science, then no science or mathematics is possible. And choosing Logic as the bedrock of ultimate Reality gives us an unimpeachable foundation below which no deeper foundation is necessary. Our experience of Reality in the past few hundred years very convincingly demonstrates that science and mathematics "work."[165]

12.3 Platonic Conception of the Relation of Ideas to Reality

Plato and subsequent Platonic philosophers postulated that there existed a realm of Ideas independent of the human mind but consisting of the concepts and thought processes (Logic) existent in the mind.[166] Each thing, both material and conceptual, in Reality had an abstract counterpart in the realm of Ideas that embodied its features. Thus there was an Idea of a plow, an Idea of Justice, and so on. The realm of Ideas was thought to be connected through mathematics (Number) to the realm of Reality. The nature of the connection was not known to the Platonic philosophers.

Plato and subsequent philosophers of course were not familiar with the modern view of Reality, which has only become apparent in the quantum and relativistic revolution of the twentieth century.

Quantum Theory has a formal method of viewing the world. It describes the process of quantum experimental measurement as a series of filtrations (or refinements that select certain properties or states) that occur at various stages of an experiment. Since all observations are experiments, it applies in principle to all the ways that we can obtain information about the universe in which we live. In the macroscopic world quantum effects are usually negligible. In the very small, quantum effects often dominate physical processes. Quantum Theory applies in principle in all cases.

So the gropings of Plato and other philosophers towards a total view of Reality had an element of truth in that there is a fundamental set of Ideas that govern Reality, but was wrong since Reality does not map directly to Ideas in the one-to-one fashion envisioned by the Platonists.

---

[163] As shown in section 1.4.
[164] And we believe it is.
[165] It is the author's belief that the enormous progress in the sciences – particularly in the past two hundred years – has mistakenly led us to believe that we are approaching a fairly complete knowledge of the physical and biological sciences. In the author's view nothing could be further from the truth. Mankind has barely begun to scratch the surface of the sciences and can look forward to a glorious further awakening over the next millenium as the scope and depth of its knowledge increases.
[166] Plato discusses the concept of Ideas in the dialogue Parmedides as well as other dialogues.

Instead, the set of Ideas is the set of laws of Physics: Quantum Theory, the Theory of Relativity and Elementary Particle Theory. The phenomena of Reality behave according to the laws of physics. The mathematics connecting Ideas to Reality will be shown to start from Logic and end in The Standard Model of Elementary Particles. Thus Platonic thought was correct in a certain sense; but the specifics were incorrect due to the primitive state of physical science.

12.4 Operator Logic and Quantum Operator Logic Exist Independent of Our Knowledge of Them

We have constructed Operator Logic and Quantum Operator Logic in the "language" of linear vector spaces, and, more specifically, based on the view that a proposition is analogous to a quantum experiment in that it proceeds to particularize subjects in the same manner that a quantum experiment filters quantum states to achieve a particular final state. Linear vector spaces appear in many areas of physical Reality. Since Reality cannot be internally inconsistent, incomplete or "incorrect", Operator Logics are well founded and exist independently of our knowledge of them.[167] Our understanding of Quantum Physics only emerged in the twentieth century yet the universe has followed the laws of quantum theory since its beginning. Thus we must attribute a conceptual reality to the concepts and laws of Quantum Theory. The extreme precision of the theory of Quantum Electrodynamics strongly indicates that Quantum Theory is correct.

Consequently the only reasonable philosophical stance is Platonist—the concepts and laws of Quantum Field Theory have a true existence outside of our knowledge of them.[168]

12.5 The Realm of Ideas

The preceding section shows that our formalisms for Operator Logic and Quantum Operator Logic exist independently of our awareness of them. One may ask what other Ideas exist in the Realm of Ideas. Confining ourselves to the physical, we see that the Realm of Ideas must contain Operator Logic, Quantum Operator Logic, Mathematics, and the fundamental theory of physical Reality.

---

[167] The deepest part of our current physical understanding of the universe is Quantum Field Theory. All branches of physics ultimately conform to it. It is extremely unlikely that the major known parts of Quantum Field Theory will change although part 3 of this book develops a major new formulation of perturbative calculations.

[168] This statement should not be confused with the quantum mechanical requirement of an observer to make quantum mechanical measurements. Quantum physical processes can proceed without an observer although experimental results, being probabilistic, can only be obtained by an observer.

12.6 Role of the Observer in the Realm of Reality

Section 12.4 addressed the issue of whether the Realm of Ideas exists independently of our knowledge of it and its contents. One might also ask the same question of the Realm of Reality: Do the events happening in Reality, and the manner in which they occur, require an Observer to detect them? Or do events in Reality happen irrespective of our presence? The answer to these questions lies in Quantum Theory.

In Quantum Theory there is an implicit Observer who "sets up" experiments, performs eigenvalue measurements (filters) at various stages in an experiment, and measures the end result(s) of an experiment.

12.6.1 The Observer in Quantum Theory

The role of the Observer in Quantum Theory is extremely significant. One indication of the importance of the Observer in quantum theory is the effect of an observation (without filtration) of the state of an intermediate stage of an experiment compared to the case of the non-observation of the state of the intermediate stage of an experiment.

Consider the experiment diagrammed in Fig. 12.1. The probability amplitude of the transition of the initial physical state from having eigenvalue x of X to having eigenvalue y of Y and thence to having eigenvalue z of Z is

$$<Z(z)|Y(y)><Y(y)|X(x)>$$

and thus the corresponding quantum probability is

$$|<Z(z)|Y(y)>|^2|<Y(y)|X(x)>|^2 \qquad (12.1)$$

If we do not filter at stage 2 of the experiment then we must sum over the Y eigenvalue states to obtain the probability of the transition from x to z:

$$p_Y(z, x) = \sum_y |<Z(z)|Y(y)>|^2|<Y(y)|X(x)>|^2 \qquad (12.2)$$

$$= \sum_y <Z(z)|Y(y)><Y(y)|Z(z)><Y(y)|X(x)><X(x)|Y(y)> \qquad (12.3)$$

Note the probability $p_Y(z, x)$ is dependent on the summation over intermediate (stage 2) Y eigenstates in the particular way specified by eq. 12.3.

If stage 2 were not present in the experiment and the experiment ran from stage 1 directly to stage 3 without intermediate observation then a different probability, which we will denote p(z, x), would result.

$$p(z, x) = |<Z(z)|X(x)>|^2 \tag{12.4}$$

$$= \sum_{y_1} <Z(z)|Y(y_1)><Y(y_1)|X(x)> \sum_{y_2} <X(x)|Y(y_2)>< Y(y_2)|Z(z)>$$

$$= \sum_{y_1, y_2} <Z(z)|Y(y_1)><Y(y_1)|X(x)><X(x)|Y(y_2)>< Y(y_2)|Z(z)> \tag{12.5}$$

which is manifestly different from $p_Y(z, x)$ as given in eq. 12.3.

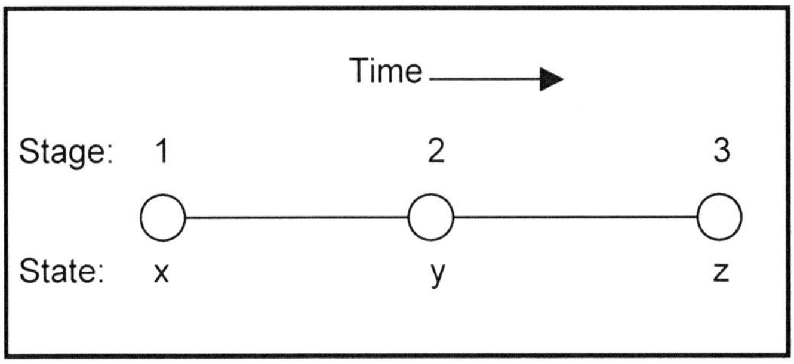

Figure 12.1. A three stage experiment beginning with a system in the state with eigenvalue x of observable X, having an intermediate state 2 with the eigenvalue y of observable Y, and ending in state 3 with eigenvalue z of observable Z. Time proceeds from left to right in the diagram.

Thus we conclude that the mere presence of an observation point (stage 2 in the present example) changes the probability for an experiment even if a filtration is not made.

We thus have concrete proof (substantiated by extensive experiments) that the passive observation of the state of an experiment in progress affects the probability of the outcome of the experiment.

Since an observation (passive, or active (i.e. a filtration)) necessarily requires an observer we see that observers influence the evolution of experiments, and thus have a more important role in quantum theory then they do in classical physics. Since the mathematics of Operator Logic, and Quantum Operator Logic, is analogous in many respects to Quantum Theory we conclude that a parallel observer role exists, in principle, for Operator Logic and Quantum Operator Logic.

12.6.2 The Observer in Operator Logic & Quantum Operator Logic

In Operator Logic and Quantum Operator Logic an implicit "observer" (a logician or theorist) determines the truth (semantic universe of discourse) or provability (calculus universe of discourse) of a statement.

An observer has three major roles in Operator Logic and Quantum Operator Logic:

1. Create statements.
2. Determine the truth (provability) of statements.
3. Create derivations of statements (theorems).

The effect of the observer on the truth (provability) of statements depends on whether the statement is a c-number statement (Operator Logic) or a q-number statement (Quantum Operator Logic). In the case of Operator Logic the observer has no effect on the truth value (true or false, provable or not, or probability of truth) of the statement. In the case of Quantum Operator Logic, if the observer determines the state of a proposition (statement) at an intermediate point(s) in the proposition (statement) then the probability of the statement is affected in the same manner as we saw in the previous subsection in the discussion of Quantum Measurement Theory.

12.7 Space and Time

In the previous section we began the process of constructing the transition from the Realm of Ideas to the Realm of Reality. An important idea in the Realm of Ideas is the concept of time. Time can be viewed as a subsidiary feature (although usually implicit) not necessarily contained in Logic, Operator Logic, and Quantum Operator Logic. However in multi-valued Logic such as 4-valued Logic time can be embodied within the Logic using the extra logic values beyond true and false. Space is also an important Idea in the Realm of Ideas. It also must somehow be embeddable in Logic since location is very important in determining the truth of many statements. We shall adjoin space (and time) to Logic in chapter 14 in the form of coordinate space.

12.7.1 The Necessity of Time, and an Arrow of Time, in Operator Logic and Quantum Operator Logic

A number of logicians have noted that a concept of time is implicit in conventional Logic. For example, proofs of theorems proceed step by step from initial postulates and theorems to a theorem's proof. Embedded in that process is a notion of discrete time steps, and a direction of the time steps. The directionality of the time steps[169] specifies an "arrow of time." The question of the Arrow of Time – why time proceeds forward and not backward – has been a subject of much discussion over the years. In the present situation Logic, Operator Logic, Quantum Operator Logic automatically embody an arrow of time.

Not only is this true for proofs but it is also true for statements. Although the order of the parts of a statement is language dependent the order is specific and consecutive within a given language and thus has a time order as well.

So we conclude that the various Logics that we have considered all embody discrete time steps and a definite concept of time ordering – an Arrow of Time.

Having ascertained that discrete time, and time ordering, is implicit in our Logics we now define physical time as the continuous limit of discrete time (with the understanding that physical time may be discrete and may possibly consist of very small time steps of time intervals of the order of the Planck time scale 5.39×10^{-44} seconds.) Discrete time intervals of that order of magnitude are not detectable experimentally at present. Thus the assumption of continuous time, with an arrow of time, is satisfactory.

12.7.2 Why add Space to Logic?

Space is necessarily a part of the Realm of Ideas because propositions often depend on a spatial location. (See subsection 8.3.2 where the variables of eigenvalue operators x_1, x_2, ... could specify spatial locations.) Thus we must add spatial (and time) dimensions to our specification of the Realm of Reality as well as the Realm of Ideas. Later we will connect the spin ½ matrix formulation of Operator Logic augmented with space-time dimensions to the Dirac equation for spin ½ particles. Thus we map Operator Logic spinors and the spinors in physical Reality. Fermion particles (spin ½ particles) in physical Reality emerge from a map from Operator Logic spinors.

Now we address the issue of the number of space-time dimensions. Clearly if spinors exist in Reality, as we know they do, then they must be "spinning" in spatial dimensions. The number of components of a spinor is related to the total number of time and space dimensions.[170] For the case of an even number of dimensions d a spinor has $2^{d/2}$ components. For the case of an odd number of dimensions d a spinor has $2^{(d-1)/2}$ components. Based on these formulas we find the results in the following Table 12.1.

The case of d = 1 is immediately ruled out because Operator Logic supports, at minimum, 2 component spinors or 4 component spinors. The case d = 2 is also ruled out because spinor particles in a one-dimensional space reduce to scalar particles, and Reality has true spinors. The case d = 3 is ruled out because in

---

[169] After all one does not proceed "backward" from a theorem through the proof steps to the initial postulats.

[170] Weinberg (1995) p. 216.

two spatial dimensions there is no difference between left-handedness and right-handedness. Thus the minimal number of spatial dimensions that yield true physical spinors and support "handedness" is three spatial dimensions. This case meets Leibniz's criteria: principle 3 of section 12.1. The simplest features associated with space are spin (represented by spinors) and handedness. They yield a rich spectrum of particle types and interaction types (maximal complexity).

Total Number of Space-Time Dimensions d	Number of Spinor Components
1	1
2	2
3	2
4	4

Table 12.1. The number of spinor components for various numbers of space-time dimensions.

Thus we have a rationale for the extension of Operator Logic to include one time and three spatial dimensions.

12.7.3 Truth is Generally Local – Space-Time Dependent

The extension of Operator Logic to include space-time is further buttressed by the dependence of the truth of statements on location and time in general. For example: "Today it rained in Concord, New Hampshire." is a space and time dependent statement.

Thus we find that statements are *local* in general – they depend on the time and spatial location.

The locality that we find in Logic naturally leads to locality in physical theories – particularly the locality of the Yang-Mills rotations in quantum field theories such as The Standard Model where the values of quantum numbers can vary from space-time point to space-time point but in such a way that their variation is compensated by local rotations of quantum fields. The locality of logical statements thus supports the connection of Logic to fundamental Physics.

12.8 Being and Existence of the Material World

Having established the Ideas of Quantum Theory, Time and Space and their counterparts in Reality we now turn to the Idea of Being.

The question of being or existence has been a subject of discussion in Philosophy and Metaphysics for millenia. In the absence of "experimental" information the discussions have centered on the definitions of being and the implications of these definitions for the "properties" of being.

For many scholars the state of Philosophy and Metaphysics was considered satisfactory in the 20th century with respect to issues such as Being. For example, Hans Bethe, a dominant figure in theoretical physics from the 1930's through the 1950's, and a Nobel Prize winner, stated that at the beginning of his graduate studies in the 1920's he considered the state of Philosophy and Metaphysics, and concluded that they were satisfactory. He then decided to become a physicist where he felt that he could make significant contributions (which he did by discovering the solar energy carbon cycle, and making notable contributions to nuclear physics and quantum field theory as well as guiding a generation of physicists including R. P. Feynman and M. Gell-Mann). In our view, Professor Bethe's comments on Metaphysics did not anticipate the impact of Quantum Theory and Relativity.

In this section we will consider a phenomenological (Realistic) view of Being and nothingness based on the experimental observations of the creation and annihilation of particles over the past eighty years. The experimental picture that we will use dates from the 1930's when the occurrence of particle creation and annihilation was first recognized. The mathematics of quantum field theory adequately describes particle creation and annihilation as seen experimentally. Since particle creation is the "creation" of Being and since particle annihilation is the "destruction" of Being one could say that the issues of Being and non-Being can be resolved by experimental observations and their theoretical analysis – thus making metaphysical speculations experimentally accountable. *Experiment can provide direct guidance on the nature of Being.*

Having said that, we recognize that the mathematics of creation and annihilation of particles somehow doesn't fully answer the question, "What is Being?" from a human perspective. Part of the problem is that we don't exactly "know" what a particle is. We know particles have particle-like properties, and wave-like properties as well. But statements about properties do not address the issue of what Being is in itself. Can we say particles are composed of a substance or substances? If so, what substance or substances? Or are particles merely form without substance?

12.8.1 The Substance of Particles

Philosophers have long argued (two and a half millenia) about the nature of the substances of Reality. These discussions have largely centered around the definition of substances followed by an analysis based on one definition or another. The question of their reality was also a recurring issue. In the earliest discussions the world was thought to consist of varying combinations of the four elements: earth, air, fire and water. So early philosophical discussions took the lead from "experimental" observation because these four elements are what we encounter in nature, and one can argue that various materials seen in nature are combinations of the four elements. More recently philosophers such as Bishop Berkeley have raised the question whether the world is real or perhaps some form of evolving "thought" in the mind of God, or a similar insubstantiality.

In this subsection we address the issue of Being based on experimental observations – particularly particle observations made from the 1930's onward. *We look to experiment/reality, as the first philosophers did, to uncover the nature of substances.*

The first major qualitative experimental reality that we find is that elementary particles can transform into each other in certain ways that are theoretically well described by the Standard Model of Elementary Particles.[171] Some typical transformations are diagrammed in Fig. 12.2.

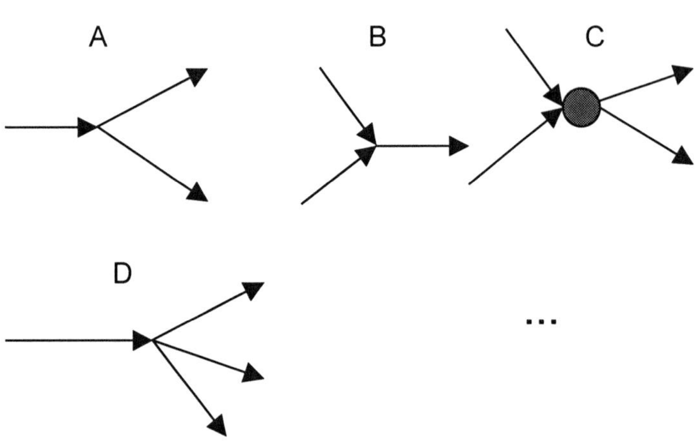

Figure 12.2. Diagrams showing some ways in which particles interact with each other and transform to other types of particles. In diagram A a particle "splits" into two particles. In B two particles combine to produce an "outgoing" particle. In C two particles interact in some complex way represented by the filled circle to produce two outgoing particles which may or may not be the same particles. In D a particle transforms (decays) into three particles.

Because of the nature of the Standard Model, particles of matter (quarks and leptons) and particles of "energy" (photons and massive vector bosons) can transform into each other in all sorts of ways if the incoming particle(s) have sufficient energy and momentum to generate the outgoing particles.

---

[171] There are some uncertainties with regard to a small number of special transformations, and the possibility of new particles and new types of transformations at higher energies is not excluded. But the overwhelming majority of known particle physics phenomena are very well described by The Standard Model.

Since, under the right energy and momentum conditions, it is possible for any incoming set of particles to transform into a set of different outgoing particles[172] it seems reasonable to assume that the substance of *all particles in nature is the same and differ only in their form.* This assumption follows from the principle of Ockham's Razor, which states that usually the simplest explanation of a phenomenon is usually the correct one.[173]

So instead of assuming each particle is composed of a different substance the well-substantiated Standard Model, which contains transformations (interactions) between all the known particles, suggests that particles are made of the same substance (or perhaps no substance) and that their differences are due to their internal form.[174] There is no experimental data that could completely clarify the nature and properties of this substance. There is a further problem in that the dominant form of matter in our universe is Dark Matter and the dominant form of energy is Dark Energy. Dark Matter and Dark Energy interact with "normal" matter and energy through the gravitational interaction, which is very weak.[175] So the substance of which Dark Matter and Dark Energy are made are open questions. But it would seem that they are different forms of the same substance as known particles as well.

It is also possible that all types of matter and energy are not made of any substance but instead consist of form imposed on nothingness. Then Dark Matter and Dark Energy would be "dark" only because they did not interact with normal matter and energy except possibly through very weak forces.

It is difficult to imagine form imposed on nothingness yet this possibility is not unreasonable. There is also the question of how nothingness relates to the vacuum. In quantum field theory the vacuum is a quantum thing and possesses properties. It possesses enormous energy. Vacuum fluctuations can occur in which particle – antiparticle pairs "pop out" of the vacuum for extremely short periods of time. The vacuum can "exert" Casimir forces that can be measured in the laboratory. So the vacuum is an extremely dynamic "substance" in quantum field theory.

If we identify the vacuum of quantum field theory with nothingness then nothingness is an extremely dynamic thing as well. Perhaps its major distinguishing feature is its non-conservation. Nothingness is unlimited in quantity. One thinks of the "sacred" law of 19th century physics, "Mass is conserved and cannot be created or destroyed," which was overturned by Einstein's theory of

---

172 With the proviso that charge and internal quantum numbers such as strangeness, color, and so on are properly conserved (or not conserved according to the dictates of The Standard Model).

173 Principle 2 of section 12.1. Richard P. Feynman expressed a similar view – that the simplest solution is often the correct one.

174 As stated earlier this concept is also implicit in SuperString Theories. All particles consist of extremely small mathematical strings in these theories and differ only in form. Thus the substance would be mathematical strings.

175 And possibly other ways.

Relativity and the transformation of matter into energy in fusion and fission. Nothingness being unlimited can easily be interpreted as the substance of particles and its unlimited nature enables transitions between the numbers and types of particles.

12.8.2 The Form of Particles

If, as the previous subsection suggested, particles are composed of one substance, and particles (and the vacuum) are differentiated from one another by their form, then the connection of the realm of Ideas with Reality becomes less difficult since form is based on Ideas – theoretical concepts. We have learned that the form of particles is in part based on space-time features such as their spin, their energy, their mass, and their momentum. In addition, their form is based on their electric charge and their internal quantum numbers. These quantities are determined by their group structure. Their group structure is $SU(2)\otimes U(1)\oplus U(1)$ (ElectroWeak symmetry) and color SU(3) (quark and gluon particle Strong interaction symmetry). There may be additional symmetries that will appear in the future as higher energy accelerators discover new phenomena.

But, from what we have learned, it is clear that space-time symmetry and internal group symmetries determine the form of particles.

12.8.3 Being as Form

The consequences of the considerations in the previous two subsections suggest that Being is form and that there is one unlimited substance in nature from which the particles of Reality and thus Reality is constructed.

If Being is form, then there is no difficulty in combining Being with Operator Logic to form Reality using the knowledge base of space-time features and internal symmetries that we have acquired experimentally.

Thus we have realized the Platonic scenario of a realm of Ideas connected mathematically to Reality.

12.8.4 Origin of Being

The components of Reality are tangible—perceivable by our senses (directly or indirectly), and the components of Reality interact with each other. But how did the components of Reality acquire existence?

Remarkably, there is strong evidence that the universe emerged from a point (or a "small" neighborhood of a point) at its beginning, The Big Bang. So the question of Being may degenerate to the question of "Being" at a single point without extension, without space or time, and perhaps without content. A number of theorists have suggested that the Big Bang is a quantum fluctuation—something emerging from nothing (the "vacuum") in such a way that the sum total of the emergent fluctuation has zero energy and thus is still nothing if considered in toto.[176]

In view of the major uncertainties, both physically and philosophically, in the understanding of existence and Being the question of the origin of Being is primarily work for the future. This section provides a preliminary discussion of Being based on our knowledge of particle physics – a subject which confronts existence and non-existence (Being and non-Being) on an everyday experimental and theoretical basis.

12.9 The Origin and Mechanism of Physical Laws

When we were developing the various Logics in earlier chapters we assumed a corpus of predicates and subjects, and also domains for each predicate. The specification of predicate domains was based on an implicit *knowledge base* that was used to determine the subjects that were in the domain of each given predicate and/or to determine the probability or probability amplitude that a subject was in the domain of a particular predicate. Thus a knowledge base is an essential ingredient of Operator Logic and Quantum Operator Logic.

In the case of physical Reality a physics knowledge base was implicit and operated between the various stages of an experiment. It affected the filtrations that took place as the experiment progressed. Physical quantities such as mass, energy, momentum and so on, and the laws that governed them, constitute a knowledge base that is implicit in the application of Quantum Theory experiment.

To make the transition from the realm of Ideas to Reality not only do we need the existence of physical things but we also need a source for physical laws and a mechanism for their execution. In human societies legislatures frame laws and the police enforce them. In Reality we must determine the origin of physical laws and the mechanism by which they are "enforced" on the components of Reality.

The typical methods, by which physical laws are determined, are 1) a fundamental lagrangian which through "canonical" methods leads to the equations embodying physical laws; or 2) a set of fundamental equations from which all physical laws are derived.

In either of these cases the question arises: "Where did the fundamental lagrangian or equations come from?" One could say that their origin is not knowable within a scientific framework. (Perhaps a deity specified them.) Or one could say they were the result of some as yet unknown process of self-organization (order emerging from chaos).

---

[176] Being then becomes an illusory artifact of our consciousness. Combinations of forms (particles) constitute matter and the forces between these particles en masse give matter its solidity or liquidity as the case may be. The forces between chunks of matter when they collide are the cumulative effect of the forces between the particles constituting each mass.

Once we have the laws then the question arises of how things in Reality are coerced into obeying them. Who are the police that enforce physical laws? The importance of this question is very apparent today in the case of quantum entanglement, and especially in the Einstein-Rosen-Podolsky thought experiment. In this experiment doing something to a particle that is quantum entangled with another particle at a great distance away—affects the distant particle. How? Clearly there is an enforcement issue for physical laws that is not as yet understood. Interestingly, the enforcement must be instantaneous and thus infinitely faster than the speed of light. One mechanism that could resolve this problem is the tachyon propagation of quantum effects.[177] Tachyons can travel at infinite speed and thus instantaneously "enforce" physical laws.

Thus we add the need for a source of physical laws, and a mechanism for their enforcement, to the needs for the origin of Reality.

---

[177] See Blaha (2008) for a discussion of tachyon quantum field theory and the possible role tachyons play in the Standard Model of elementary particles.

13. The Nature of the Physical Derivation/Construction of a Fundamental Theory of Physics

The further we pursue these inquiries, the fewer become the primitive truths to which we reduce everything; and this simplification is inself a goal worth pursuing. ... The aim of proof is, in fact, not merely to place the truth of a proposition beyond all doubt, but also to afford us an insight into the dependence of truths upon one another.
Gottlob Frege – *The Foundations of Arithmetic*

13.1 What is the Form of the Fundamental Theory of Physics?

Many individuals have speculated on the ultimate form of the fundamental theory of physics, often called the "Theory of Everything". Some say it is a form of string theory. Others have suggested a theory based on a discrete lattice-like space-time with, perhaps, extra dimensions. Yet others have suggested a self-organizing theory from a primordial chaos. And, lastly, a more conservative group has suggested theories based on quantum field theory.

In all cases the fundamental theory of physics must take the form of a set of postulates that use primitive terms that are either loosely defined or "intuitive" or both. The implications of the postulates are explored and experimentally testable consequences ultimately derived.

It is clear from the experimental success of the Standard Model of Elementary Particles (modulo a few discrepancies and some major open questions[178]) that the Standard Model must be "derivable" or "constructable", at least approximately, for currently accessible energies from THE fundamental theory.

13.2 A Theory for its Time

In the time of Charlemagne Western European scholars obtained access to the literature and philosophy of classical Greek times. A mini-Renaissance, the Carolingian Renaissance, followed that expired after a relatively short time

[178] Such as how Dark Matter and Dark Energy relate to The Standard Model.

because the intellectual climate of the times in Western Europe "was not ready for it." In contrast the Italian Renaissance flowered, and was culturally diffused throughout a receptive Western Europe, beginning the modern age of culture and science. For a cultural or intellectual advance to succeed, and not wither, the intellectual/cultural climate must be in a receptive state.

In the case of physics a similar situation holds. Consider the case of Newton's theory of motion, which survived unchanged for hundreds of years. If someone in that period had pointed out that Newton's transformation between reference frames in relative motion with respect to each other was a special case of the Lorentz transformation, the general reaction would have been vast disinterest since there was no experimental means to test the dynamics of relativistic velocities. Neither was there any theoretical motivation. And one could also argue that there were many possible transformation laws that approximated Newton's transformation at low velocities.[179] Thus the times were not ripe for the appearance of the Theory of Special Relativity.

More generally, if we trace the development of physics from Aristotelian times to the present then it is clear that, at all times, the major changes in physical theories were unanticipated and generally driven by experimental results. This was particularly true in the twentieth century. It is also true in each historical period of physics, that when a successful theory was constructed, it described experimental results as they were understood in that period. The transitions between physical theories evidenced a continuity that was reflected in the "old" theory being an approximation to the new theory in the region of the "old' theory's approximate validity.

The drift of the above considerations is clear. Having been driven to the Standard Model by over seventy years of experimental and theoretical work, it is incumbent on theoretical physics to make sense of its form and origin.

Countless negative comments about the peculiarity of its form have appeared since the general form of the Standard Model was established around 1975. Attempts to develop a more fundamental theory of which the Standard Model is an approximation, have been largely in the form of SuperString theories. Lacking experimental guidance, and limited by the complexity and variety of the mathematics of SuperString theories, progress along these lines has been somewhat limited and not experimentally substantiated.

Thus we still face the issue of explaining the Standard Model. It certainly accounts for the vast bulk of experimental data. And so we must ask for a rationale/derivation of the Standard Model within the context of the current time rather than attempt to leapfrog to a future physics without any significant experimental guidance.

We have two choices to explain the Standard Model. The first choice—although it seems not to be an appealing choice—is to take the Standard Model

[179] There was also a belief in that period that the speed of light was infinite.

lagrangian (with perhaps a few refinements and a unification with quantum gravity) as The Fundamental Theory and simply accept its peculiarities. The other choice is to develop a reasonable set of basic postulates from which the known particles and interactions of nature follow as well as the peculiar form of the Standard Model lagrangian.

Recently a series of books[180] by this author have appeared that have developed a set of postulates that lead to the known particles, interactions and the form of the Standard Model lagrangian. This book extends Blaha (2009) with some improvements and additions. The great virtue of this derivation, which is in fact a construction,[181] is that it leads directly to The Standard Model (possibly with a fourth generation) but without a plethora of new, unseen particles.

Thus in this volume we show

1. The origin of the spectrum of elementary fermions: differentiated into Leptons and Quarks with each quark occurring as a color triplet. Probably four generations of fermions.
2. The ElectroWeak symmetry is SU(2)⊗U(1)⊕U(1).
3. The Strong Interaction symmetry is SU(3).
4. A broken SU(2)⊗U(1)⊕U(1) symmetry of the fermion generations based on a new mass mechanism that explains particle masses in the event Higgs particles are not found experimentally.
5. Local gauge interactions for each of the known interactions.
6. A mass generation mechanism that does not require Higgs particles.
7. Fermion generation mixing is possible without Higgs particles.

Thus we have derived the known form of The Standard Model completely from space-time considerations based on GL(4)⊗GL(4)[182] – in particular to their complex Lorentz subgroup and superluminal Lorentz transformations, and a γ matrix symmetry associated with superluminal transformations in Dirac-like equations.

13.3 Derivation vs. Construction of a Physical Theory

Let us assume that we have a physics theory based on a lagrangian such as The Standard Model that describes physical phenomena such as the accumulated experimental data on elementary particles and their interactions.

---

180 Blaha (2006), (2007b), (2008), (2009), and (2010a).

181 A construction starts with a set of axioms and proceeds to develop a theory based on the addition of features that are consistent with the set of axioms.

182 GL(4)⊗GL(4) may result from broken GL(16). We ascribe it to an asynchronous time Logic GL(4) and an asynchronous space Logic – the other GL(4) factor.

Perhaps the lagrangian in itself describes the phenomena and a deeper layer of physical theory is not possible. (A simple example of this case is a classical particle in a conservative potential.)

But in other cases it is possible that the lagrangian is the result of a more fundamental physics. (An example of this situation is the Landau-Ginzberg theory of superconductivity, which is based on the more fundamental theory of Cooper pairing.)

It is almost universally believed that the Standard Model is a phenomenological theory that is a consequence of a more fundamental theory of elementary particles. The open question, of course, is the nature of the more fundamental theory. We have defined a set of postulates that leads to the Standard Model's form. But we realize from simple logic that the ability to derive a theory from a more fundamental theory does *not* prove the more fundamental theory is the one and only correct theory.

So in seeking to define a more fundamental theory that is *most likely* to be the true physical basis of an experimentally verified (phenomenological) theory, it seems reasonable to assume the six principles stated previously as a starting point.

Further, it will be seen that the development of the specific phenomenological theory of interest (The Standard Model of Elementary Particles) can be best viewed as a construction from first principles rather than as a strict derivation from axioms. A *construction* begins with a set of axioms, and then develops a theory based on the axioms, and on the addition of features that are consistent with, and share the spirit of, the set of axioms.

Euclid's geometry is an example of a construction although most students have viewed it as strictly derived from its five axioms. As logicians have pointed out the geometrical figures used to prove many geometrical theorems embody extensions of the five axioms, and are not implied by the axioms. Thus ***Euclid's gemoetry is a construction*** – not a strictly derived theory.

In the spirit of Euclid's approach we constructed The Standard Model's form from a set of initial postulates. As a result of the construction, the superficial peculiarities of form of The Standard Model, the various symmetries, are seen to follow from the initial postulates.

13.4 What Purpose does a Construction or Derivation Serve?

As the eminent logician Frege indicated in the introductory quotes there are several benefits in the reduction of a theory to a more fundamental theory (set of postulates). These benefits include:

i) The more fundamental theory is usually simpler and more comprehensive.

ii) The derivation of the phenomenological theory shows the origin of the various parts of the derived theory, and the interdependence of the parts of the derived theory and their derivation from a particular postulate or set of postulates.

iii) The more fundamental theory enables us to consider a deeper level of physics and perhaps find a path to a yet deeper level.

iv) The more fundamental theory enables us to consider alternative postulate sets and the universes to which they lead. (A good example of this possibility is the controversy over Euclid's Fifth Postulate and the non-Euclidean geometries that emerged from the controversy.)

13.5 The Rigor of a Derivation

In defining a set of postulates that lead to the Standard Model, either exactly or as an approximation, the question of the rigor of the derivation unavoidably appears. First there is the question of the rigor of differential and integral calculus which is still an issue despite the apparent rigorous development of calculus in the nineteenth century by Dedekind and others. To show the issue is still alive we simply mention the question, "If a point has no width or breadth, what does it mean to say the "next point" on a line?" This question inevitably leads to issues when one considers the definition of a derivative as a limit. Since "bare" particles are "point-like" the issue also surfaces in physics.

Secondly there is the issue of the path integral formulation of the Standard Model which (together with the Faddeev-Popov Mechanism) generates the correct perturbation theory in the ElectroWeak sector and is generally believed to generate correct results for Standard Model particle physics. The path integral formalism has not been put on a rigorous basis and it appears unlikely to be made rigorous for the foreseeable future.

These obstacles to a completely rigorous development mean that we must follow a procedure similar to the "traditional" physics approach of doing things "rigorously" and exercise reasonable moderation in rigor. In this approach we have the history of physics since Newton and Leibniz to support us. For approximately three hundred years physicists have used the differential and integral calculus successfully. These areas of mathematics were not put on a quasi-rigorous footing in the view of many mathematicians until the mid-nineteenth century and there remains more to be done to obtain a truly rigorous calculus.

Thus a quasi-rigorous physicist's approach to proofs and derivations appears to be fully justified.

13.6 Consistency and Completeness of a Set of Postulates

Since our goal is to define a set of postulates at a deeper level from which we can derive the Standard Model it is prudent to inquire about the consistency and completeness of the set of postulates so defined. It would be the height of hubris to believe that the set of postulates that we define completes the study of the fundamental nature of the universe(s). So we will take these postulates to be a step in the direction of deeper knowledge but realize that these postulates now become the subject of deeper theoretical and experimental investigation.

There probably will be more phenomena at higher energies that will extend the domain of particle physics beyond The Standard Model. Indeed the vast majority of particle physicists expects new phenomena as higher energy accelerators appear. And the mysteries of Dark Matter, Dark Energy, and other unusual cosmological features, also suggest there is much more to learn since we have found that the very small is intimately connected to the very large.

However taking the postulates seriously as an interim deeper theory of particle physics we have two "simple" issues to address: the consistency of the set of postulates and the completeness of the set of postulates.

The consistency issue can be resolved by two remarks:

i) The consistency of a set of postulates cannot be mathematically proved, in principle, within the framework of the theory it defines according to the celebrated Consistency Theorem of Gödel.

ii) From a physicist's point of view a theory is consistent if it predicts a unique result for all possible experiments within the domain of applicability of the theory. This criteria is, of course, impossible to meet since the number of possible experiments is infinite. (Presumably, the predictions are verified by experiment as well.)

The path integral formulation of the Standard Model appears to fulfill the second consistency criteria since it provides, in principle, a unique, well-defined approach to calculating any experimental result.

The completeness of the set of postulates is determined by the theoretic results it implies. If we assume the Standard Model is a complete description of particle physics (and, of course, we do not), then if the postulates lead to The Standard Model then the postulates can be considered complete from a physical point of view. If new phenomena are found that require the Standard Model be extended, then the set of postulates may well have to be extended as well: the extended set of postulates should imply the extended Standard Model. (The possibility exists that some of the postulates might have to be modified as well.)

13.7 The Difference Between a Mathematical-Deductive System and a Fundamental Scientific Theory

The eminent philosopher and logician R. B. Braithwaite[183] has said, "The irreducible difference between the propositions of logic and mathematics and those of a natural science are that the former are logically necessary and the latter contingent."[184] This observation is true, in the author's opinion, in the case of

---

[183] Braithwaite (1960) p. 353.

[184] Braithwaite is not using "necessary" and "contingent" in the sense of modality where one distinguishes between the necessary and the contingent. Rather it appears he is using these words to

interim scientific theories that are dependent upon experiment for clarification and further growth. However, a theory, which purports to be a complete "Theory of Everything", falls into the other category—a mathematical theory of a purely mathematical deductive type—that needs, in principle no further experimentation and thus is strictly mathematical-deductive in nature.[185] In this book we claim to have a theory that is deeper than the Standard Model and that implies the Standard Model. This theory, framed in terms of postulates, is a step in the direction of a mathematical-deductive theory and thus it is treated as such. However it is clearly not a complete "Theory of Everything" and does not pretend to be.

It is only a step—a very significant step—in the author's view that obviates certain other theoretical attempts and brings us closer to a truly fundamental theory—of great simplicity—and depth—that is based on the geometry of the universe.

13.8 Modality and Physical Theories

Superficially the issues raised in the study of modality: necessary vs. contingent would appear to be relevant to the physical sciences. However a consideration of the test for contingency shows it to be untenable for physical science: a property is *contingent* if it is not required in all possible worlds (universes); a property is *necessary* if it is required in all possible worlds (universes). For physics Modality requires physical knowledge of all possible worlds (universes). This knowledge is unattainable experimentally or theoretically.

We can only create a fundamental physical theory for our universe at best. Other universes are, by definition, beyond our ken. Thus we must conclude that all physical properties—all features of the fundamental physical theory—are contingent due to our inability to investigate all possible universes. We cannot create a set of postulates that form the basis of Reality and declare any of the postulates to be necessary.[186]

Thus in the derivation/construction of The Standard Model in the chapters that follow all postulates are contingent. Modality is a moot issue in the study of Reality for the foreseeable future.[187]

---

distinguish between mathematics which has a logical structure that follows from fundamental logic and mathematical principles, and natural science which is provisional and deduced from experiment.

[185] It should be noted that plane geometry was a deductive theory until Euclid's five postulates were recognized as implying all the theorems of plane geometry. At this point geometry transitioned to a mathematical-deductive theory. Thus one can say Euclid's geometry is a "Theory of Everything" for plane geometry. The ultimate goal of elementary particle theory is to accomplish the same feat for elementary particle physics.

[186] Unless they are trivially true.

[187] Note that a world (universe) is totally separate and distinct from any other world (universe). If one says that one can tunnel from one universe to another, then one must view these universes as branches of one all-encompassing universe. In this sense we cannot learn of other worlds (universes) through experiments. They are by definition totally separate from our universe.

If in some distant epoch Mankind can develop the mathematical and physical knowledge to declare a property is necessary then it can only result from a deep analysis that shows any world without the property is "inconsistent." Gödel's Inconsistency Theorem[188] indicates that a proof that a system is internally consistent is not possible in general. Consequently a proof that a system is internally *in*consistent is undecidable in general. (It can be proven to be consistent and can be proven to be inconsistent within a system in general.)

Lastly, a totally empty universe without quantum phenomena stands as an example of a property-less universe (world). *Therefore no physical property is necessary using the many worlds analysis of Modal Logic.*

---

[188] The theorem, in brief, states, "If a formal system A is consistent, then its consistency is not provable within A."

Appendix 13-A. Generalized Metaphysics

13-A.1 Goals of Generalized Metaphysics

In an earlier chapter we considered the metaphysics of our universe which we have called Relativistic Quantum Metaphysics. This metaphysics embodies the deepest known physical information about our universe and establishes a metaphysical view based on it (particularly in sections 12.6 – 12.9 as well as appendix 14-A).

We now turn to universes in general and inquire into their physical Realities and their metaphysics. The word universe[189] is a bit vague when it is used in the physics and metaphysics literature. One reads of worm holes between universes, quantum tunneling between universes, and so on. These concepts and gedanken (thought) experiment discussions raise the question – if one can connect "universes" are they really separate universes or are they parts of the same universe with difficult paths between them? We therefore will define a *universe* to be specifically a completely self-contained entity totally unconnected with any other entity in any way whatsoever. Thus beings within a universe cannot communicate/contact/obtain physical knowledge of or even be aware of other possible universes. Beings within a universe can speculate on possible other universes and their features. These other universes are "universes of the mind" – assemblages of thought and logic created by imagination.

The metaphysical studies and conceptualizations of aspects of universes of the mind we will call *Generalized Metaphysics*. In this chapter we will consider some fundamentals of Generalized Metaphysics.

13-A.2 Logic is Pan-Universal

The various logics that we have considered and developed in the earlier chapters of this book would apply in any universe since they stand independent of physical details and are based on the rules of rational thought. There is one type of exception to this principle. One can conceive of a universe – perhaps empty or perhaps with contents – in which there are no physical laws and all behavior is capricious and chaotic. In this type of universe there are two possibilities: the laws of probability, statistics and chaos theory may apply (the *chaotic* class of

[189] Metaphysicians use the word "world" where physicists use the word "universe" generally speaking.

universes)[190]; or no laws whatsover apply (the *lawless* class of universes). In the absence of physical laws logic is irrelevant for physical metaphysics.

Logic applies to all other types of universes.

13-A.3 All Nontrivial Postulates in all Universes are Contingent

Modality concerns the distinction between the necessary and the contingent. Properties and propositions that are necessary are present in all possible universes. Properties and propositions that are contingent are present in some universes and absent in other universes. (See subsection 3.2.4 for more detail.) Modalists phrase their discussions in terms of worlds; we will use the word universes instead. The assumption of multiple universes is required to meaningfully distinguish between the necessary and the contingent.

However in the case of properties and propositions of physical Reality it is not possible to examine the properties and propositions (laws) of other universes since by definition any other universe is completely separate from our universe and no communication or detection of the properties and features of other universes is possible. Thus the acid test of necessary/contingent of modalists is not possible for aspects of physical Reality – except conceptually.

Some counterexamples showing universes with no necessary properties are:

1. A point universe consisting of one point
2. An empty universe containing nothing but having extension
3. A non-quantum universe
1. A non-relativistic universe with a non-Lorentzian space-time transformation group

We conclude that all properties and postulates (propositions) are necessarily contingent. An examination of the postulates of The Standard Model listed earlier supports this view. Modality is physically irrelevant.

13-A.4 Towards a Generalized Metaphysics of Universes

It would seem that a Generalized Metaphysics would include Leibniz's seven principles listed in section 2.3. However the generality of these principles reduces their value to a minimum that almost amounts to triviality as far as contributing to an understanding of ultimate physical Reality.

One might also think of listing all possible properties (attributes) and bundles of properties that might be relevant for classes of universes. This exercise seems to be less than illuminating in grasping the range of ultimate Realities.

---

[190] Universes in which quantum theory holds are not part of this class of universes since quantum theory is a physical theory with physical laws albeit probabilistic.

Considerations of Modality which superficially seems relevant to a consideration of "many universes" also seem to be less than relevant in the study of possible ultimate Realities for reasons given in the preceding section.

Thus one is left with the approaches broached in chapter 28. One of the primary difficulties of studying an ultimate Reality[191] is the consistency of postulates required to define an ultimate Reality. Are there deep connections between postulates that make certain combinations of postulates inconsistent?

Another area of study is the range of ultimate Realities subject to certain specified constraints? For example a constraint could be that a set of postulates for an ultimate Reality lead to a set of stable chemical elements, or, more interestingly, a set of chemical elements that would support the evolution of some form of life. Clearly, these questions would require a Generalized Metaphysics that is mathematically deep and not qualitative in nature (as most past metaphysics studies have been.) Also clear, is our current inability to develop the consequences of a set of ultimate postulates. For example, given the postulates that lead to The Standard Model in this book, what mind or eye could see the chemical elements to which they lead, or to the combinations of chemical elements that led to life in this universe?

We conclude we are only at the beginning of our understanding of this universe's ultimate Reality, and not even at the beginning of understanding the ultimate Realities of other possible universes.

---

[191] Presumably different universes will have different ultimate Realities in general.

Appendix 13-B. Logic, Language and the Universe

A reading of this book brings us to the conclusion that the ultimate Reality of our universe is expressible in words in the general sense of both verbal language and mathematics. We also conclude that the stuff of the universe is insubstantial[192] – form imposed on nothingness. So in the end we can only view our universe's ultimate Reality as a language embodying logic whose words are particles since words, in a general sense, are forms. The known elementary particles constitute a limited vocabulary. Thus it is appropriate to end this part of our study with a brief word on language.

13-B.1 Logical Equivalence of Languages

All logic is expressed in human or symbolic languages. As a result there is an intimate connection between logic and language. Statements and deductive systems require a sufficiently robust language to express their content. A language usually must have the equivalent of predicates, subjects, connectives and quantifiers.

A question of some interest is the equivalence of languages. When are two languages equally capable of expressing the statements of a universe of discourse? Clearly they must have sets of equivalent terms although a term in one language might be a combination of terms in the other. However in some cases of human languages this type of equivalence is hard to achieve. A classic example is the Greek language in comparison to English. The Greek language has hundreds of words expressing various forms and nuances of love while English, in comparison, has few words for love. In other areas of the world some languages have a plethora of words for one aspect of nature, or another, which have no simple analogue in European languages.

---

[192] Evoking Shakespeare's prescient vision of Reality in *The Tempest*:

"The baseless fabric of this vision,
The cloud-capp'd towers, the gorgeous palaces,
The solemn temples, the great globe itself,
Yea, all which it inherit, shall dissolve,
And, like this insubstantial pageant faded,
Leave not a rack behind. We are such stuff
As dreams are made on ..."

One might think that the equivalence of languages is not of great importance. However, the growth of culture and science is directly tied to the growth in their terminology and the concepts that they embody. An example is the growth in the knowledge of quantum theory in the twentieth century, which introduced a host of new terms in physics. Thus the equivalence of languages reflects, to some degree, the equivalence of the range of universes of discourse (and their intellectual content) that the languages can support.

The requirements for two languages to be equivalent are:

1. Equivalent expressions in the two languages must have the same truth value.

2. Any expression in one language must have an expression with equivalent meaning in the other language.

3. The primitive terms of one language must be equivalent to the primitive terms in the other language, or to combinations of the primitive terms in the other language.

13-B.2 Operator Languages

Hitherto the languages that have been considered in studies of Logic have been human languages or symbolic languages that we will call c-number languages following the terminology of Quantum Theory. A c-number is a quantity, a number, character, or string of characters, that is not an operator. We have introduced the concept of a q-number language for logic in the study of Quantum Operator Logic. Primitive terms are represented by operators in a linear vector space. q-number languages are as valid as c-number languages. But they have not been studied previously because language studies have focussed on human languages. q-number languages only appear in quantum theories.

q-number languages have the added advantages of furnishing a unifying framework for deterministic logic and quantum probabilistic logic that is guaranteed to be well-formed and consistent since it is based on Quantum Measurement Theory, which, embodying Reality, cannot be inconsistent or incomplete.

13-B.3 Quantum Languages, Grammar, Turing Machines, Computers, and Computer Programs

This book develops Operator Logic and Quantum Operator Logic. Blaha (2005b) developed the concepts of Quantum Languages, Quantum Grammars, Quantum Turing Machines, Quantum Computers, and Quantum Computer Programs; and proved Gödel's Undecidability Theorem required the fundamental laws of Nature to be quantum.

Blaha (2005b) and this book complement each other by bringing the Quantum concept, which is undoubtedly the deepest knowledge that we have of Reality, to logic and language.

A reading of Blaha (2005b) displays a remarkably similarity in the concepts and mathematics of quantum languages with our present development of Quantum Operator Logic.

Together they give us a coherent weltanschauung of Thought and Reality. The major open question is the determination of the knowledge base of Reality. In this author's view this question will be resolved by an extension of our understanding of the formulation of the nature of space, time, and substance within the framework of Quantum Operator Logic and Quantum Language.

*14. The Transition from Ideas to Reality – The Principle of Asynchronicity

14.1 The Aspects of Reality

The four aspects of Reality, space, time, matter, and radiation, are intimately related. Without change in matter and/or radiation, time becomes unmeasurable and irrelevant. Without matter and/or radiation, distance and thus space becomes unmeasurable and irrelevant. So the presence of matter and radiation are necessary to make time and space meaningful and thus real.

In turn matter and radiation modify the properties of time and space as Einstein's Theory of General Relativity shows. Thus the components of Reality are highly interrelated.

14.2 Matter

Our understanding of matter has evolved tremendously in the past two millenia. For much of that time we viewed matter as concrete, substantial "stuff" that we could kick, as Dr. Johnson did to refute Bishop Berkeley's claim that matter was not "real". Dr. Johnson reputedly said, " I refute it thus!" as he kicked a rock. Starting in the nineteenth century it became clear that matter was composed of molecules, which in turn were composed of atoms, which in turn were composed of electrons circling a nucleus containing protons and neutrons (the solar system view of the atom.) In the nineteen seventies it became clear that protons, neutrons and other particles were composed of quark particles.

At present matter[193] appears to be consist of eighteen types of quarks and six particles[194] called leptons.[195] The major qualitative, distinguishing property between quarks and leptons is that quarks experience a type of force called the strong interaction while leptons do not.

This author has proposed a set of postulates from which various forms of the theory of these particles (the Standard Model of Elementary Particles) may be "derived."[196] We will examine the starting point of this derivation/construction in this chapter. Blaha (2008) contains the derivation, which we will provide in an

---

[193] There is also an unknown form of matter called Dark Matter that constitutes about 95% of the matter in the universe. The properties and nature of this form of matter are not known at present although there are a number of speculative theoretical proposals. One possibility, considered in this book, is that Dark Matter is composed of WIMPs.

[194] Electrons, muons, τ particles – each with a corresponding neutrino.

[195] Together with gluons generated by quarks.

[196] Blaha (2008).

expanded form later in this book. Chapters 15 – 23 of this book contain a more complete derivation.

14.3 Matter is Insubstantial

We now wish to get to the heart of matter.[197] What is a particle made of? It appears that a particle has form without substance (as we commonly understand substance). Its form is particulate in part and wave-like in part. We might think of it as "nothingness" upon which a semi-permanent (or permanent) form is imposed with the quality of "Being." Being consists of existence for some interval of time and implies observability. An entirely unobservable entity could have a persistent form, but, lacking observability, could not be considered real since there would be no manifestations of it in our Reality. Fortunately, for us Quantum Theory requires an observer, and observability, for all real entities and events. Thus all known particles interact and that makes them part of our Reality.[198]

Particles exist and possess Being. Particles can interact with each other to truly create new particles or to truly annihilate into radiation.[199] So the forms imposed on nothingness (particles) can transform to other forms but do so in such a way that certain features of these forms are preserved. These features satisfy what we call conservation laws such as the conservation of energy.

It is rather remarkable that particles can undergo true creation and annihilation because these features represent transformations between *Being* and *non-Being*. Earlier we developed Operator Logic and Quantum Operator Logic based on linear vector space operators similar to those of Quantum Theory. Noting that one thus has a framework applicable to both Logic and Reality, we raised the issue of the "addition" of the quality of Being (existence) to "create" Reality from Ideas. Seeing the process of true creation and annihilation in the laboratory it is clear that Being is an acquirable property. As such, the Big Bang, which seems to be the origin of the universe, might be a transition from nothingness to Being through an expansion from a point to space-time differentiated regions with a variety of forms of particles. Thus we perceive the possible nature of the process although the precise details remain to be determined.[200]

14.4 The Knowledge Base of Reality

As we have seen, a calculus universe of discourse consists of a set of axioms written in terms of undefined primitive terms and the theorems derived

---

[197] This section repeats parts of earlier chapters for completeness for those readers who might skip earlier chapters.

[198] This includes Dark Matter, which interacts with normal matter through gravitation and perhaps through other forces very weakly.

[199] Radiation is composed of particles as well: photons, ElectroWeak bosons, Strong Interaction gluons, and gravitons. Gravitons have not as yet been detected due to the weakness of the force of gravity.

[200] Blaha (2004) describes a theory of a quantum Big Bang.

therefrom. A semantic universe of discourse attributes meanings/interpretations to the primitive terms thereby giving meaning to the axioms and consequent theorems. A theory expressed in terms of Operator Logic usually is based on a set of axioms for either type of universe of discourse. A physical interpretation of the primitive terms is also required for a semantic universe of discourse.

We will call the set of primitive term definitions and the set of axioms the *knowledge base* of the semantic universe of discourse of Reality.

A knowledge base for physical Reality (at least the part of it with which we are familiar) was first expounded in a series of books by Blaha culminating in Blaha (2008). Blaha (2008) lists a set of axioms that lead directly, and *exactly*, to most of the established features of the form of the Standard Model of Elementary Particles. The major known features of The Standard Model of elementary particles were determined in a period of forty-five years (1930 – 1975 approximately). These features include parity violation, peculiar symmetries, and the somewhat complicated nature of the particle spectrum. These features are *exact* results of the axioms in Blaha (2008), and this book, unlike other theoretic attempts to explain the Standard Model. Other attempts view the Standard Model as a low energy approximation of a larger theory and have no inherent justification for parity violation but rather define their theories to incorporate it.

We suggest the form of The Standard Model *is* exact and directly based on the space-time properties of subgroups of GL(4)⊗GL(4).

*14.5 Principle of Asynchronicity - From 4-Valued Logic to Dirac-like Equations and Four Generations

It is difficult to discern fundamental general principles of Physics from our knowledge of physical Reality which is best expressed currently in The Standard Model. One new principle that this author feels is implicit in The Standard Model and in quantum Mechanics is a Principle of Asynchronicity. When processes take place in parallel whether it is Quantum Mechanical entanglement of processes at large distances from each other or in high order Feynman diagrams (or their old fashioned time ordered perturbation theory predecessor) the synchronicity of the processes is an issue. It is resolved by physical law. Asynchronicity issues, situations when parallel processes get "out of sync" resulting in the failure of an entire physical process to complete properly, do not arise.

In computation asynchronicity issues do arise. For example parallel computations or computer processes on a chip or set of chips have to be carefully managed for a parallel computer process to complete properly. In the case of computer chip design (VLSI chips and so on) techniques have been developed for the design of chips based on multi-valued logic. One conceptual approach (as remarked in section 10.3 on 4-valued matrix Logic) uses 4-valued logic to define clockless computer logic circuits. The 4-valued logic developed by Fant (2005) has the four logic values TRUE, FALSE, NULL, and INTERMEDIATE. It is an extension of Boolean Logic that can accommodate time asynchronicities in asynchronous computer circuits. It enables circuits to avoid the use of system clocks to implement synchronization.[201] Thus the coordination is explicit in the 4-valued logic and non-logical constructs are not needed. Concurrent transitions are coordinated solely by logical relationships with no need for any time constraints or relationships.

Now, realizing that The Standard Model, and physical theories ultimately derived from it such as Quantum Mechanics, contain asynchronicities that are controled by The Standard Model, we suggest that a Principle of Asynchronicity is embodied in The Standard Model that has major consequences in two areas: 1) it leads to Dirac-like equations for the fundamental fermions – leptons and quarks; 2) it implies four generations of fermions.

14.5.1 4-Valued Asynchronous Logic

The definitions of asynchronous circuits and asynchronous Logic are:

1. An *asynchronous circuit* is a circuit in which the component parts are autonomous and can act in parallel at various rates of time evolution. They are not controled by a clock mechanism but proceed or wait for signals indicating that they can proceed.

2. *Asynchronous logic* is the logic used in the design of asynchronous circuits. The logic embodies the asynchronicity and so the circuits built using the logic do not use a clock to control the execution speed of the various parts of an asynchronous circuit. Consequently logic elements do not necessarily have a distinct true or false state at any given point in time. 2-valued Boolean Logic is not sufficient and so asynchronous logic is multi-valued. The logic embodies states that allow for "stop and go" states within an executing asynchronous circuit.

In Fant's asynchronous 4-valued logic the four possible truth values of a state are:

True – status is true and all data is current
False – status is false and all data is current
Intermediate – status is indefinite with some data current
NULL – status is indefinite with no data present – results in a suspension of processing of the circuit part in a NULL state until current data becomes present

[201] Remarkably Bjorken (1965) pp 220-226 presents an analogy of Feynman diagrams with electrical circuits with momenta mapping to currents, coordinates to voltages, Feynman parameters to resistance, free particle eqn of motion to Ohm's Law plus the equivalent of Kirchhoff's Laws. Thus Feynman diagrams and compter circuits are analogous in some ways.

"Data" is the information flowing through all or part of a circuit. Using these truth values the evolution in time of the parts of an asynchronous circuit are effectively synchronized by the logic and without the use of a clock mechanism. A clock mechanism effectively is a subsidiary time constraint or set of time constraints. See Fant (2005) for further details.

An implicit aspect of asynchronous logic is the coordination of spatially separated parts of a circuit. Since spatial separations in a circuit can be mapped to time delays using the speed of data propagation between parts, spatial asynchronicites are normally subsumed under time asynchronicities. This is particularly true for computer chips which are kept small to minimize delays.

In the case of physical phenomena spatial separations canot be transformed unambiguously to time delays since propagation speeds range from infinite (for quantum entanglements) to varying finite values. Thus spatial asynchronicity is different in principle, and practice, from time asynchronicity.

14.5.2 Principle of Spatial and Time Asynchronicity

An obvious, and thus little thought of, feature of elementary particle phenomena is the coordination of the parts of a physical process in time and space. Complex Feynman diagrams embody the coordination of the parts of interacting particles over a period of time. Quantum entanglement phenomena embody the coordination of the parts of a physical phenomena separated by large distances. Examples of these types, which could be multiplied indefinitely, lead to the Principle of Spatial and Time Asynchronicity.

Principle: Nature requires time and space asynchronicity. This asynchronicity is coordinated by 4-valued physico-logical structures for matter (fermions).

Elaboration: Elementary particle physical phenomena must support extended coordinated physical phenomena in space and time. The fundamental laws of particle physics must be such as to permit coordinated physical phenomena with coordination between the parts of a physical phenomena at large distances and large time intervals. The coordination must be embodied within physical laws.

This principle will be applied below (and later in this book in more detail) to justify Dirac-like equations for particle dynamics, and to justify a four generation fermion spectrum. The origin of the fermion generations has been obscure since the 1970's. This new principle furnishes a general basis for their origin.

Coordination is an obvious feature of physical phenomena. This principle goes beyond that by asserting that extended coordinated physical phenomena must exist. So if particles exist, then their antiparticles must also exist to give asynchronous behavior in interaction regions. If only particles exist then all interactions proceed forward in time and the state of the interaction at any point in time is known. With the addition of antiparticles, asynchronicity is introduced and at various time slices of an interaction, the state can be ambiguous since antiparticles are negative energy particles moving backward in time.

*14.6 Time Asynchronicity – Dirac-like Equations with Particles and Antiparticles

Time asynchronicities are common in the many subcircuits of a computer chip. Time asynchronicities are also common in the many interaction subregions of a set of particles in interaction. Fant (2005) has a diagram on p. 7 of a circuit with a set of subcircuits with five time slices of the interacting subcircuits showing five states of the "'data' wavefront" at five points in time. This diagram is similar to the time sliced diagram of an interacting system of particles in "old fashioned" time-ordered perturbation theory. Blaha (2005b) p. 29 displays such a diagram (Fig. 5.1.4) in a description of a Standard Model Quantum Langauge Grammar – a language representation of particle physics. Blaha's diagram is remarkably similar to Fant's diagram in overall features as one might expect since they both address time asynchronicity..

The asynchronicity that appears in perturbation theory diagrams is intimately related to the appearance of antiparticles in diagrams. As noted earlier antiparticles are interpretable as negative energy particles traveling backwards in time. The time orderings which are implicit in the Feynman diagram approach, and explicit in old fashioned perturbation theory, show the time asynchronicity, and the effect of the dynamics to coordinate the asynchronicities so that meaningful results follow from perturbative calculations.

Thus we are led to propose:

1. Matter particles have four fundamental states
2. Their dynamics is governed by 4×4 dynamical equations
3. They are spin ½ particles in 4 dimensions[202]

We assume one time dimension. We find the complex Lorentz group in four dimensions is the initial physical space-time group subject to further restrictions stated later.

Section 14.9 below expands on the above points and provides a transition to Dirac-like equations for fundamental fermions. Particles are found to be described by two fundamental states. Antiparticles are found to result from the other two particle states.

Interestingly, our association of particle states with truth values complements recent efforts to use particle spins, up and down, as storage for true and false in advanced computer devices. Complex particle diagrams thus specify

---

[202] Weinberg (1995) p. 216. Discussed later in more detail.

symbolic Logic computations[203] as well as being physical computation specifications.

*14.7 Spatial Asynchronicity – Four Generations with Particle Oscillations

The Principle of Asynchronicity defined in subsection 14.5.2 also specifies the existence and coordination of spatial asynchronicities. An excellent example of spatial asynchronicities is the flow of solar neutrinos to the earth. After much effort experimenters have found that solar neutrinos oscillate between electron type neutrinos and muon type neutrinos during their travel from the sun. At present there are three known types of neutrinos as well as three known types of charged leptons, up-type quarks and down-type quarks. Taken together these four types of particles are grouped in three generations which appear to be duplicates of each other in most respects. Recently, preliminary evidence of a fourth generation of fermions has appeared.

The spatial oscillations of the stream of solar neutrinos between generations is a form of asynchronous behavior that is governed by (coordinated) by physical laws. On this basis we propose:

1. There are four levels – generations – of each fermion particle type corresponding to the four values in asynchronous 4-valued logic.

2. Transitions between generations can occur asynchronously in streams of particles. These transitions and the corresponding states at various points along the path of the stream are coordinated by the fundamental laws of physics embodied in The Standard Model.

Thus a major unexplained feature of the fermion species – generations – is explainable by the Principle of Asynchronicity. We return to this topic in more detail in chapter 20.

*14.8 Implications of Asynchronicities for Particle Interactions

Spatial translation generators do not explicitly contain interactions.[204] Time translation generators, hamiltonians do have interaction terms. Thus we suspect generations may not have interactions but rather "interact" via mixing terms.

---

[203] Blaha (2005).

[204] In the absence of interactions with derivative couplings. All known Standard Model interactions contain no derivative couplings.

*14.9 Dirac-like Equations of Matter from 4-valued Logic

In Blaha's derivation every truly fundamental particle of matter, whether quark or lepton, has spin ½. We have seen in chapter 10 that the basic algebra of Operator Logic eigenvalue operators, and that of the raising and lowering operators, is the same as the algebra of free spin ½ particles. Operator Logic is part of the Realm of Ideas. Our goal is to build a theory of Reality on the scaffolding of Operator Logic. The remainder of this chapter will be devoted to outlining the general idea of the construction of the theory of Reality. Blaha (2008) and the following chapters in this book, provide a detailed discussion. In building this theory we will have implemented the Platonic goal of mathematically connecting the Realm of Ideas and the Realm of Reality.

Before beginning we note as we did previously that Logic statements are generally local.[205] *Therefore the introduction of space-time coordinates and a space-time transformation group is a necessity.*

We will start with the direct product of two spinor universes of discourse as given in section 10.3. In particular the 4-vectors of four rows defined in section 10.3 were

$$s_{\uparrow\uparrow k} = \begin{bmatrix} s_{1k\uparrow} \\ \\ s_{2k\uparrow} \end{bmatrix} \quad s_{\uparrow\downarrow k} = \begin{bmatrix} s_{1k\uparrow} \\ \\ s_{2k\downarrow} \end{bmatrix} \quad s_{\downarrow\uparrow k} = \begin{bmatrix} s_{1k\downarrow} \\ \\ s_{2k\uparrow} \end{bmatrix} \quad s_{\downarrow\downarrow k} = \begin{bmatrix} s_{1k\downarrow} \\ \\ s_{2k\downarrow} \end{bmatrix} \tag{10.19}$$

We form linear combinations of these vectors to establish contact with the standard treatment of spin ½ free quantum fields in physics:

$$u(k, +\tfrac{1}{2}) = \begin{bmatrix} 1 \\ 0 \\ 0 \\ 0 \end{bmatrix} \quad u(k, -\tfrac{1}{2}) = \begin{bmatrix} 0 \\ 1 \\ 0 \\ 0 \end{bmatrix} \tag{14.1}$$

and

$$v(k, -\tfrac{1}{2}) = \begin{bmatrix} 0 \\ 0 \\ 1 \\ 0 \end{bmatrix} \quad v(k, +\tfrac{1}{2}) = \begin{bmatrix} 0 \\ 0 \\ 0 \\ 1 \end{bmatrix} \tag{14.2}$$

where k = 0. k will become non-zero and identified as the momentum of a spin ½ particle shortly. The second arguments of u and v are the values of particle spin:

[205] For example, "It will rain here today." is local in both time and space.

+½ represents an "up" spin state and −½ represents a "down" spin state.[206] Note the relations to the spinors in eq. 10.19 such as

$$u(k, +½) + v(k, -½) = s_{\uparrow\uparrow k} \qquad (14.3)$$

etc.

Having defined the spinor states for a particle in eqs. 14.1 and 14.2 we now introduce 4-dimensional space/time. The three real space dimensions and the one real time dimension of our experience are the residue of 4-dimensional complex space-time. We introduce the coordinates $x = (t, \mathbf{x})$ where $\mathbf{x}$ is a 3-vector and t is the time, and introduce the momentum $p = (p^0, \mathbf{p})$ where p^0 is the energy and $\mathbf{p}$ the 3-momentum of a particle. The manner of combination of Logic matrices and coordinate specifications is a matter of choice. We will combine them with a view to Physics; in a human language context we would undoubtedly combine them differently.

Then we form a Fourier representation of a spin ½ particle wave function:

$$\psi(x) = \sum_{\pm s} \int d^3p N(p)[b(p,s)u(p, s)e^{-ip\cdot x} + d^\dagger(p,s)v(p, s)e^{+ip\cdot x}] \qquad (14.4)$$

where N(p) is a normalization factor, b(p,s) is lowering operator that annihilates a particle of momentum p and spin s, $d^\dagger(p,s)$ is a creation operator, u(p, s) and v(p, s) are column vectors like those in eqs. 14.1 and 14.2 that have been "boosted" to momentum p, and $p\cdot x = p^0x^0 - \mathbf{p}\cdot\mathbf{x}$ is the inner product of 4-vectors.[207] The raising and lowering operators b(p,s) and $d^\dagger(p,s)$ (and their hermitean conjugates $b^\dagger(p,s)$ and d(p,s)) are mathematically similar to the raising and lowering operators of eq. 10.21. They satisfy the anticommutation relations

$$\{b(q,s), b^\dagger(p,s')\} = \delta_{ss'}\delta^3(\mathbf{q} - \mathbf{p}) \qquad (14.5)$$
$$\{d(q,s), d^\dagger(p,s')\} = \delta_{ss'}\delta^3(\mathbf{q} - \mathbf{p})$$

using Dirac δ-functions instead of Kronecker δ-functions due to the continuous nature of the momentum variables.

Thus we see spin ½ particle wave functions originating from the spinors, and raising and lowering operators of the spinor formulation of Operator Logic. We will see how this happens in some detail below and in more detail in subsequent chapters.

---

[206] Again we choose not to deviate from our line of discussion to discuss the details of particle spin and refer the interested reader to popular books on elementary particles. The point we are seeking to establish is that the direct product of the matrix representation of logical universes of discourse mathematically leads to the mathematics of spin ½ Dirac particles in a simple and direct manner. And that gives us the mathematical connection of Ideas to Reality of which the Platonists spoke.

[207] This equation is discussed in detail in the many good books on quantum field theory.

Another interesting point that emerges from this discussion is the nature of spin ½ particle states such as

$$|p, s\rangle = b^{\dagger}(p, s)|0\rangle \qquad (14.6)$$

This state is interpreted in Reality as a one particle state. It also has an analogous interpretation in Operator Logic as creating a one term universe of discourse – a construct which is in part linguistic and in part logic.

14.9.1 Boosts and Fermion Masses Related to Gödel Numbers?

In this section we will show how Dirac-like equations can be obtained by "boosting" the Dirac equation for a particle at rest to a particle of momentum p = $(p^0, \mathbf{p})$. *The novel feature of this derivation is that the particle mass turns out to be a scale factor with the dimension of [mass] times a Gödel number.*

Our starting point is A_{E_k} in eq. 10.20, which expanded to show its 4 × 4 rows and columns, is

$$A_E = \begin{bmatrix} gn(E_1) & 0 & 0 & 0 \\ 0 & 0 & 0 & 0 \\ 0 & 0 & gn(E_2) & 0 \\ 0 & 0 & 0 & 0 \end{bmatrix} \qquad (14.7)$$

where we surpress the subscript k. A_E can be put into a more physically relevant form using the unitary transformation[208]

$$U = \begin{bmatrix} 1 & 0 & 0 & 0 \\ 0 & 0 & 1 & 0 \\ 0 & 1 & 0 & 0 \\ 0 & 0 & 0 & 1 \end{bmatrix} = U^{\dagger} = U^{-1} \qquad (14.8)$$

$$A_E' = UA_EU^{-1} = \begin{bmatrix} gn(E_1) & 0 & 0 & 0 \\ 0 & gn(E_2) & 0 & 0 \\ 0 & 0 & 0 & 0 \\ 0 & 0 & 0 & 0 \end{bmatrix} \qquad (14.9)$$

---

[208] This transformation poses no physics or logic issues since 4×4 Dirac γ matrices are equivalent up to a unitary transformation. See p. 18 of Bjorken (1965) and also R. H. Good Jr., Rev. Mod. Phys. **27**, 187 (1955).

If we now let $E_2 = E_1$ and $A = A_E'$ then

$$A = \begin{bmatrix} gn(E_1) & 0 & 0 & 0 \\ 0 & gn(E_1) & 0 & 0 \\ 0 & 0 & 0 & 0 \\ 0 & 0 & 0 & 0 \end{bmatrix} \quad (14.10)$$

$$= gn(E_1)(\gamma^0 + I_4) \quad (14.11)$$

where I_4 is the 4×4 identity matrix and γ^0 is one of the four Dirac γ matrices defined in eq. 10.24. If we multiply eq. 14.11 by a mass scale m_0 and define the particle mass[209]

$$m = gn(E_1)m_0 \quad (14.12)$$

then we have

$$m_0A = m\gamma^0 + mI_4 \quad (14.13)$$

If we apply eq. 14.13 to a Dirac spinor U such as those in eqs. 14.1 or 14.2, introduce an exponentiated time factor, and set eq. 14.13 to zero we obtain the Dirac equation for a particle at rest:[210]

$$(m\gamma^0 + mI_4)Ue^{-imt} = 0 \quad (14.14)$$

where t is the time variable.[211] If we now perform a "Lorentz boost" – a Lorentz transformation to a reference frame moving with a velocity v with respect to the reference frame in which the particle is at rest and described by eq. 14.14 – we obtain[212]

---

[209] The reader may wonder whether the known quark and lepton masses – particularly their ratios have the Gödel number form of $2^a3^b5^c7^d$... where a, b, c, d, ... are integers. Unfortunately the masses of most particles are not well known, and we expect the Higgs Mechanism and other mechanisms that may modify particle masses to ruin the simple Gödel ratios. The known mass ratios of the charged leptons are not exact Gödel ratios. There were observations of factors of 3 in the mass spectrum of "constituent" quarks some years ago with the mass ratios of s::c::b quarks being 3. m_s = 0.5 Gev, m_c = 1.5 Gev, and m_b = 4.5 Gev. (One can also approximately relate top and bottom quark masses to the masses of other quarks using Gödel ratios.) (See "On A Possible Similarity Between The Heavy Lepton And Heavy Constituent Quark Mass Spectra", S. Blaha, Phys.Lett. **B84**, 116 (1979)) Since then the masses of these quarks have been adjusted to experimentally more "acceptable" values and the almost exact ratio of 3 has disappeared.

[210] This approach leads one to speculate that the fundamental fermions form an alphabet in the sense that each particle's mass is proportional to a Gödel number and thus constitutes a language symbol. Blaha (2005c) and earlier books discuss particle physics as a language.

[211] From the viewpoint of Operator Logic eq. 14.14 can be viewed as a proposition or statement in the Operator Logic formalism.

[212] Blaha (2008) p. 29.

$$(\not{p} - m)e^{-ip\cdot x}U(p) = 0 \qquad (14.15)$$

where the exponential factor, mt, is also "boosted" to $p\cdot x = p^0 x^0 - \mathbf{p}\cdot\mathbf{x}$, and where $\not{p} = \gamma^0 p^0 - \boldsymbol{\gamma}\cdot\mathbf{p}$. The 4-momentum $p = (p^0, \mathbf{p})$ satisfies the mass condition

$$(p^0)^2 - |\mathbf{p}|^2 = m^2 \qquad (14.16)$$

where m is given by eq. 14.12. A Fourier transformation of eq. 14.15 yields the free, coordinate space Dirac equation:[213]

$$(i\gamma^\mu\partial/\partial x^\mu - m)\psi(x) = 0 \qquad (14.17)$$

with the general solution given by eq. 14.4. The Dirac equation,[214] generalized to include internal symmetries associated with the ElectroWeak Interaction and the Strong Interaction, incorporating the three known generations of quarks and leptons, and with interaction terms introduced; has the form of the matter sector of the Standard Model of Elementary Particles.

Thus we have found a mathematical path from Operator Logic implemented by a spin ½ matrix formalism to the basic equations of matter. The Platonists, of course, had no knowledge of Operator Logic, or The Standard Model, or the mathematical path that we have outlined between them. But they were able to develop a conceptual outline of the realm of Ideas and a conceptual mathematical bridge to Reality.

In subsequent chapters we will develop an expanded version of Blaha (2008) and (2009) that leads to the form of The Standard Model.

The steps from Operator Logic to the matter sector of the Standard Model may be briefly summarized:

1) The γ matrices being 4×4 matrices imply a 4-dimensional space-time (Weinberg (1995) p. 216).

2) Assume the complex Lorentz subgroup L(C) of GL(4) is the space-time group of particles we find four types of spin ½ Dirac-like equations follow through boosts from the rest state to a state with one real time variable. These correspond to the four known types of spin ½ particles: up-quark, down-quark, electron-like and neutrino-like.

---

[213] Three other Dirac-like equations can be obtained using complex Lorentz boosts. The set of four equations corresponds naturally to the four known types of spin ½ particles and leads naturally to the form of the Standard Model.

[214] Fant (2005) also develops 2-valued and 3-valued logics to deal with asynchronous circuits. These logics are less appealing as they require the use of additional constructs. However there is an analogue in physics. The Dirac equation has an equivalent two dimensional formulation using 2×2 Pauli matrices.

3) The four free Dirac-like equations are naturally grouped in SU(2)⊗U(1)⊕U(1) Weak triplets to obtain L(C)[215] covariant dynamic equations with WIMPs. Parity violation naturally appears as does color SU(3) in the quark sector.

4) Upon the introduction of weak interaction coupling terms and particle mass splitting, L(C) is broken to Lorentz covariance.[216] With the addition of another GL(4) group the four (three known) generations of spin ½ particles and mass mixing follow yielding the known particle matter sector of the Standard Model.

There is a mathematical bridge between the Realm of Ideas and the Realm of Reality.

14.10 Particles as Monads

The combination of a physical interpretation and an Operator Logic interpretation is somewhat similar to Leibniz's concept of a *monad.* However it differs in that no spiritual aspect is introduced,[217] and also in that, states, such as that defined in eq. 14.6, do not have "perception" in the usual sense of the word. But they do have something analogous to perception through their interactions since they are affected by interactions, and thus may be said to 'perceive" other monads (particles) interacting with them.

---

[215] L_c is the complex Lorentz group.

[216] Lorentz covariance of the real Lorentz group sort.

[217] A religious reader might say that since everything in Reality emanates from God an implicit spiritual aspect is present. Possible spiritual extensions of this development are beyond the scope of this book.

Part 2. The Construction/Derivation of The Standard Model

15. Time, Space, and the Complex Lorentz Group

15.1 Are Time and Space Contingent or Necessary?

This section, 15.1, points out the connection between Operator Logic and the foundations of the derivation of the Standard Model. It is optional in the sense that it is not necessary for the Physics derivation of The Standard Model if the reader is willing to accept space-time and the existence of fundamental spin ½ fermions as the basis of The Standard Model. These topics are discussed in chapters 1 – 14 wherein a new formulation of Logic is developed called Operator Logic that has an SU(2) representation as well as multi-valued Logic representations that naturally lead to Dirac-like equations for fermions and four generations of fermions.

In section 12.7 we gave arguments for the existence of time and a direction for time based on the consecutive nature and directionality of proofs in logic. Since logical statements may be true in one locale and not true in another the existence of space, and locality, is also required. Upon further examination of space, and the types of spinors that it supports, we found that one time dimension and three space dimensions were required[218] at minimum to have Dirac spinors.

We then associated space and time with a generic term in Operator Logic in section 14.6 and found that the simplest generic 4×4 Operator Logic statement (eq. 14.14) could be transformed into the Dirac equation (eq. 14.17) by performing a Lorentz (boost) transformation on the statement.

Thus we found that we can construct a fermion basis for a theory of Reality by taking a simple, generic Operator Logic statement, introducing space and time, and proceding to apply a Lorentz transformation.

In section 3.2.4 we discussed modal considerations: Is a property necessary or contingent?[219] In the case of time it appears that time is a contingently necessary[220] property of any universe (world) supporting logical analysis, or logical proofs, which are sequential in time, would not be possible. To paraphrase Descartes: "I prove, therefore time exists." In the case of space one could argue that space is a contingent property if the common conception of the Big Bang is correct. The universe at the time (and before the time) of the Big Bang is thought

---

[218] Table 12.1 and the discussion following it.

[219] A contingent property is one which may or may not be possessed by an entity.

[220] Contingently necessary is intermediate between contingent and necessary. A property is contingently necessary if it is necessary when another property is present. See Blaha (2011a), where this term is first introduced, for more detail.

to have consisted of one point without extension. Thus we have an example of a spaceless universe demonstrating the contingency of space.

Space may be a contingent property of universes. But our universe at present does have extension (space). So in the construction of our theory of Reality we introduce a space of three dimensions. The issue that now arises is the structure of space and its relation to time.

15.2 Transformations Between Coordinate Systems

The measurement of time and space is simple in practice but raises weighty questions when their underlying basis is examined. We shall begin by measuring spatial distances with a ruler, and by measuring time with a clock. Earlier we determined that four dimensions: one space dimension and three space dimensions were required for a non-trivial universe (section 12.7). We now define rectangular coordinate systems with x, y, and z axes as pictured in Fig. 15.1 below. We then assert the postulate:

Postulate 15.1. Any observer can define a set of time and space coordinates called a coordinate system in which the observer is at rest. One can define a transformation that relates the coordinate systems of two observers traveling at a any constant velocity with respect to each other.

One can always relate the coordinates of two coordinate systems by having an observer in each coordinate system specify the coordinates of objects located at each spatial point, and then creating a map between the coordinates of corresponding spatial locations. Consequently the coordinates of one coordinate system are related to the coordinates of the other coordinate system.

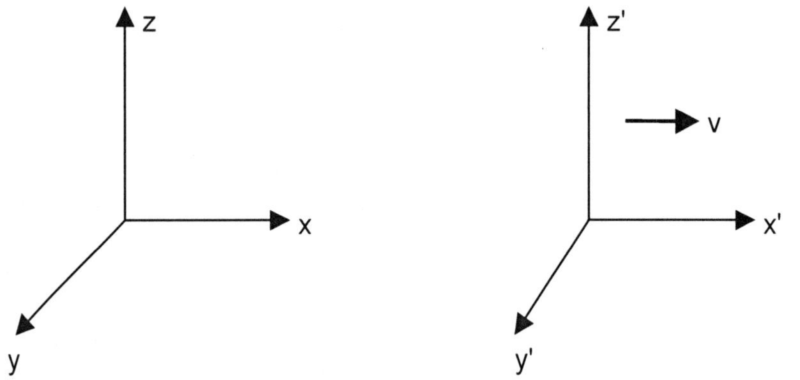

Figure 15.1. Depiction of two coordinate systems. The “primed” coordinate system is moving with velocity **v** in the positive x direction with respect to the “unprimed” coordinate system. We choose parallel axes for convenience.

If space is flat the relation between the respective coordinates is linear. (One could reverse the logic of that statement by defining a flat space to be one in which the coordinates of a point in any coordinate system are linearly related to the coordinates of any other coordinate system moving at a constant velocity with respect to it.) Thus we can express the relation between the coordinates in the "unprimed" system to the coordinates in the "primed" system as the transformation between coordinate systems

$$\begin{aligned} \mathbf{a}' &= A\mathbf{a} + \mathbf{B}t + \mathbf{C} \\ t' &= Dt + \mathbf{E}\cdot\mathbf{a} \end{aligned} \tag{15.1}$$

where A is a 3×3 matrix, **B**, **C** and **E** are 3-vectors, and D is a number (scalar value).

Having restricted the set of transformations between coordinate systems to the form of eq. 15.1 we now assert postulates that restrict the form of the transformation to Lorentz transformations and transformations similar to Lorentz transformations.

Postulate 15.2. The speed of light, c, is the same in all coordinate systems.

Postulate 15.3. The invariant interval or distance $d\tau$ is defined by

$$d\tau^2 = g_{\mu\nu}dx^\mu dx^\nu \tag{15.2}$$

It is invariant under a change of coordinate systems. The 16 quantities $g_{\mu\nu}$ are known as the metric tensor.[221] *The four quantities dx^μ are infinitesimal displacements in space and time.*

Postulates 15.2 and 15.3 are assumptions that we can only categorize as contingent because we do not have an a priori reason to motivate these assumptions.[222] Since we can conceive of other universes without these postulates they cannot be viewed as necessary.

If we expand eq. 15.2 in rectangular coordinates it is equivalent to

$$d\tau^2 = g_{00}dx^0dx^0 + g_{11}dx^1dx^1 + g_{22}dx^2dx^2 + g_{33}dx^3dx^3 \tag{15.3}$$

which equals

$$d\tau^2 = c^2dt^2 - dx^2 - dy^2 - dz^2 \tag{15.4}$$

---

[221] The repeated indices indicate a summation. In this case from 0 to 3 as shown in eq. 15.3

[222] There is an a postiori reason pointed out by Einstein and Poincaré that the equations of electromagnetism imply them. But since our goal is to construct electromagetism these postulates can only be asserted.

using the familiar form of the time and rectangular space coordinates.

15.3 The Lorentz Group

The metric tensor $g_{\mu\nu}$ for rectangular coordinates has the matrix form G = diag(1, −1, −1, −1):

$$G = \begin{bmatrix} 1 & 0 & 0 & 0 \\ 0 & -1 & 0 & 0 \\ 0 & 0 & -1 & 0 \\ 0 & 0 & 0 & -1 \end{bmatrix} \tag{15.5}$$

The invariant interval under a transformation between rectangular coordinate systems (with "primed" and "unprimed" coordinates) has the form of eq. 15.4 for the unprimed coordinates and the same form for the primed coordinates:

$$d\tau^2 = c^2dt'^2 - dx'^2 - dy'^2 - dz'^2 \tag{15.6}$$

In matrix form we can define an "unprimed" coordinate column vector with

$$a = \begin{bmatrix} t \\ x \\ y \\ z \end{bmatrix} \tag{15.7a}$$

and its corresponding "primed" coordinate with

$$a' = \begin{bmatrix} t' \\ x' \\ y' \\ z' \end{bmatrix} \tag{15.7b}$$

If a and a' are the coordinates of the same point in the respective coordinate systems then, by postulates 15.2 and 15.3, they are related by a boost Lorentz transformation $\Lambda(\mathbf{v})$ with the form

$$a' = \Lambda(\mathbf{v})a \tag{15.8}$$

and possibly a spatial rotation matrix factor, where **v** is the relative velocity of the coordinate systems. The form of the transformation eq. 15.8, which is called a Lorentz *boost*, is constrained by postulates 15.2 and 15.3 to be[223]

$$\Lambda(\mathbf{v}) = \begin{bmatrix} \gamma & -\gamma v_x & -\gamma v_y & -\gamma v_z \\ -\gamma v_x & 1 + (\gamma - 1)v_x^2/v^2 & (\gamma - 1)v_x v_y/v^2 & (\gamma - 1)v_x v_z/v^2 \\ -\gamma v_y & (\gamma - 1)v_x v_y/v^2 & 1 + (\gamma - 1)v_y^2/v^2 & (\gamma - 1)v_y v_z/v^2 \\ -\gamma v_z & (\gamma - 1)v_x v_z/v^2 & (\gamma - 1)v_y v_z/v^2 & 1 + (\gamma - 1)v_z^2/v^2 \end{bmatrix} \quad (15.9)$$

where $\gamma = (1 - v^2)^{-1/2}$, $\mathbf{v} = (v_x, v_y, v_z)$, $v = |\mathbf{v}|$ and we set c = 1 for convenience.[224] The set of all matrices of the form of $\Lambda(\mathbf{v})$, or $\Lambda(\mathbf{v})\mathcal{R}(\boldsymbol{\theta})$ or $\mathcal{R}(\boldsymbol{\theta})\Lambda(\mathbf{v})$ where $\mathcal{R}(\boldsymbol{\theta})$ is a spatial rotation with angle vector $\boldsymbol{\theta}$, for v < c form a matrix representation of the Lorentz group. Elements, $\Lambda(\mathbf{v}, \boldsymbol{\theta})$, of the Lorentz group satisfy the defining relation of the Lorentz group:

$$\Lambda(\mathbf{v}, \boldsymbol{\theta})^T G \Lambda(\mathbf{v}, \boldsymbol{\theta}) = G \quad (15.10)$$

where the superscript ^{T} specifies the transpose of the matrix.

The Lorentz group, with which we are familiar, relates the coordinates of an event in two coordinate systems that differ by a a spatial rotation, and a relative velocity whose magnitude is less than the speed of light. The inhomogenous Lorentz group includes coordinate displacements.[225]

The group elements of the homogeneous Lorentz group can be expressed in terms of the generators **K** of boosts to coordinate systems moving at a constant velocity **v** and the generators **J** of purely spatial rotations by

$$\Lambda(\mathbf{v}, \boldsymbol{\theta}) = \exp[i\omega\hat{\mathbf{u}}\cdot\mathbf{K} + i\boldsymbol{\theta}\cdot\mathbf{J}] \quad (15.11)$$

where the vector $\boldsymbol{\theta}$ is a 3-vector specifying the rotation angles, and where $\mathbf{v} = \hat{\mathbf{u}} \tanh\omega$, $\hat{\mathbf{u}}\bullet\hat{\mathbf{u}} = 1$.

The boost transformation $\Lambda(\mathbf{v}) = \Lambda(\mathbf{v}, \mathbf{0})$ has the form

$$\Lambda(\mathbf{v}) = \exp[i\omega\hat{\mathbf{u}}\cdot\mathbf{K}] \quad (15.12)$$

---

[223] We shall consider only the proper, orthochronous Lorentz group at this point. We assume that the primed and unprimed coordinate systems have parallel axes. So there is no rotation of axes embodied in eq. 15.9.

[224] One can set c = 1 by an appropriate choice of time and spatial distance scales. The demonstration that $\Lambda(\mathbf{v})$ has the form given by eq. 15.9 can be found in many textbooks.

[225] See Weinberg (1995) for a discussion of the inhomogeneous Lorentz group.

Its matrix form is eq. 15.9. The matrix form can be expressed in terms of the unit normalized velocity vector $\mathbf{u} = (u_x, u_y, u_z)$ and ω as

$$\Lambda(\omega, \mathbf{u}) = \Lambda(\mathbf{v}) \tag{15.13}$$

$$= \begin{bmatrix} \cosh(\omega) & -\sinh(\omega)u_x & -\sinh(\omega)u_y & -\sinh(\omega)u_z \\ -\sinh(\omega)u_x & 1 + (\cosh(\omega) - 1)u_x^2 & (\cosh(\omega) - 1)u_xu_y & (\cosh(\omega) - 1)u_xu_z \\ -\sinh(\omega)u_y & (\cosh(\omega) - 1)u_xu_y & 1 + (\cosh(\omega) - 1)u_y^2 & (\cosh(\omega) - 1)u_yu_z \\ -\sinh(\omega)u_z & (\cosh(\omega) - 1)u_xu_z & (\cosh(\omega) - 1)u_yu_z & 1 + (\cosh(\omega) - 1)u_z^2 \end{bmatrix}$$

where $\Lambda(\omega, \mathbf{u}) = \Lambda(\omega, \mathbf{u}, \boldsymbol{\theta} = \mathbf{0})$ in the previous notation.
This definition of the general form of proper, orthochronous, Lorentz boost matrices $\Lambda(\omega, \mathbf{u})$ will be used in subsequent sections to define faster-than-light boost transformations.

The vector form of a Lorentz boost transformation is

$$\begin{aligned} \mathbf{x}' &= \mathbf{x} + (\gamma - 1)\mathbf{x}\cdot\mathbf{v}\ \mathbf{v}/v^2 - \gamma\mathbf{v}t \\ t' &= \gamma(t - \mathbf{v}\cdot\mathbf{x}/c^2) \end{aligned} \tag{15.14}$$

where $\gamma = (1 - \beta^2)^{-1/2}$ with $\beta = v/c = v$ (since we set c = 1).

15.4 The Nature of Λ(ω, u) for Complex ω

We now turn to the case of complex ω where we will see superluminal (faster-than-light) Lorentz transformations can be defined using complex values of ω (section 15.6). Since, for any complex value z

$$\cosh^2(z) - \sinh^2(z) = 1 \tag{15.15}$$

it follows that for any complex value of ω, $\Lambda(\omega, \mathbf{u})$ is a member of the Lorentz group, or of the complex Lorentz group[226] for complex ω:

$$\Lambda(\omega, \mathbf{u})^T G \Lambda(\omega, \mathbf{u}) = G \tag{15.16}$$

For certain values of the imaginary part of ω the matrix $\Lambda(\omega, \mathbf{u})$ has a particularly simple form, similar to that of $\Lambda(\omega, \mathbf{u})$ for real ω, but which generates boosts to relative velocities greater than the speed of light. Among these values are:

---

[226] The complex Lorentz group is defined as the group of all complex transformations that satisfy eq. 15.16.

$$\omega = \omega_{\pm} = \omega \pm i\pi/2 \tag{15.17}$$

Later we will see that these alternate choices $\omega_{\pm}$ correspond to specific choices of parity.

15.5 Complex Lorentz Group

In the preceding section we saw that the parameter ω can be complex and the boost transformation will still satisfy the Lorentz condition eq. 15.10. More generally we can consider complex homogeneous Lorentz transformations $\Lambda(\mathbf{v}, \boldsymbol{\theta})$ which can be represented by eq. 15.11 with complex parameters ω, $\hat{\mathbf{u}}$, and $\boldsymbol{\theta}$ where $\hat{\mathbf{u}}$ and $\boldsymbol{\theta}$ are complex 3-vectors. $\boldsymbol{\theta}$ specifies a rotation angle.

In general $\Lambda(\mathbf{v}, \boldsymbol{\theta})$ is then a transformation between coordinate systems that have complex coordinates. One coordinate system is moving at a constant complex velocity with respect to the other. And the coordinate systems do not have parallel spatial axes in general.

Within the complex Lorentz group, denoted L(C),[227] *there are subsets of boosts with important physical roles in the derivation of the form of The Standard Model. In particular we will see that certain classes of boosts generate faster-than-light transformations. These transformations can be further divided into subclasses of "left-handed" and "right-handed" transformations based on the quantum field theories to which they lead. Further within each subclass there are subclasses of transformations that naturally lead to Dirac-like free field equations that can be described as lepton-like and quark-like.*

Thus these boosts are a key ingredient to understanding the form of The Standard Model.

15.6 Faster-than-Light Transformations

In this section we will substitute $\omega_{\pm}$ for ω in $\Lambda(\omega, \mathbf{u})$ and then show that we obtain two sets of possible transformations from sublight reference frames to faster-than-light reference frames. One set of transformations, where $\omega_L = \omega + i\pi/2$, will be called *left-handed superluminal boosts*. They eventually lead to the "left-handed" Standard Model. We denote members of this set, $\Lambda_L(\omega, \mathbf{u})$, with the subscript "L" for left-handed.

The other set of boosts where $\omega_R = \omega - i\pi/2$ will be called *right-handed superluminal boosts*. They eventually lead to a right-handed, unphysical,[228] version of The Standard Model. We denote members of this set of boosts, $\Lambda_R(\omega, \mathbf{u})$, with the subscript "R" for right-handed.

---

[227] Streater (2000) points out that the complex Lorentz group is essential to the proof of the CPT theorem.

[228] Currently the case. If a right-handed counterpart to the current Standard Model surfaces at higher energies then the features emerging from right-handed superluminal boosts then become physically important.

Before considering faster-than-light boosts we note the relation between a real valued ω in a *conventional* Lorentz boost $\Lambda(\omega, \mathbf{u})$, and the magnitude of the relative velocity v for $v < 1$, is

$$\mathbf{v} = \hat{\mathbf{u}} \tanh\omega \qquad \text{with} \qquad \hat{\mathbf{u}}\bullet\hat{\mathbf{u}} = 1$$

$$\begin{aligned} \cosh(\omega) &= \gamma = (1 - v^2)^{-1/2} \\ \sinh(\omega) &= v\gamma = \beta\gamma \end{aligned} \tag{15.18}$$

where $\beta = v = |\mathbf{v}|$.

15.7 Left-Handed Superluminal Transformations

Left-handed (proper orthochronous) superluminal boost transformations $\Lambda_L(\mathbf{v})$ have the same form as eq. 15.9 for ordinary (proper orthochronous) Lorentz boost transformations. However the magnitude of the relative velocity **v** is greater than the speed of light. Thus $\gamma = (1 - v^2)^{-1/2}$ is pure imaginary and $\Lambda_L(\mathbf{v})$ is complex.

$$\Lambda_L(\mathbf{v}) = \begin{bmatrix} \gamma & -\gamma v_x & -\gamma v_y & -\gamma v_z \\ -\gamma v_x & 1 + (\gamma - 1)v_x^2/v^2 & (\gamma - 1)v_xv_y/v^2 & (\gamma - 1)v_xv_z/v^2 \\ -\gamma v_y & (\gamma - 1)v_xv_y/v^2 & 1 + (\gamma - 1)v_y^2/v^2 & (\gamma - 1)v_yv_z/v^2 \\ -\gamma v_z & (\gamma - 1)v_xv_z/v^2 & (\gamma - 1)v_yv_z/v^2 & 1 + (\gamma - 1)v_z^2/v^2 \end{bmatrix} \tag{15.19}$$

This transformation raises several issues – the most prominent of which is the interpretation of the imaginary coordinates generated by the transformation. Imaginary coordinates would appear at first glance to be unphysical. However we view the measurement of these quantities operationally: an observer measures distances with "rulers", and time with clocks, which both give real numeric values. Thus an observer *in any coordinate system* will always measure real numbers for time and space distances. However an observer *in another coordinate system* that is related to the first coordinate system by a superluminal transformation will view the coordinates in the first system as complex as eq. 15.19 indicates.

The reconciliation of these points of view requires the introduction of a new transformation in addition to a superluminal Lorentz transformation for the case of faster than light transformations. This transformation maps the complex coordinates generated by a Lorentz transformation to the real coordinates seen by the observer[229] in the "faster than light" reference frame. We describe this transformation in section 15.12. We show how this transformation, an

[229] The linearity of the superluminal transformation makes this secondary transformation physically possible.

SU(2)⊗U(1)⊕U(1) transformation, leads to the SU(2)⊗U(1)⊕U(1) ElectroWeak symmetry in chapter 18.[230]

15.7.1 Cosh-Sinh Representation of Left-Handed Superluminal Boosts

We will now develop the representation of left-handed superluminal boost transformations in terms of cosh(ω) and sinh(ω) for later use in our discussion of tachyons. We find that we must use a complex $\omega_L \equiv \omega + i\pi/2$ to properly describe left-handed superluminal boosts. The relation between ω_L and v is different from eq. 15.18 for the case of left-handed superluminal boosts:

$$\cosh(\omega_L) = i\sinh(\omega) = -\gamma = i\gamma_s \qquad (15.20)$$
$$\sinh(\omega_L) = i\cosh(\omega) = -\beta\gamma = i\beta\gamma_s$$

where $\beta = v > 1$, $\boldsymbol{\omega \geq 0}$, and

$$\gamma_s = (\beta^2 - 1)^{-1/2} \qquad (15.21)$$

Note that for the real part of ω_L namely Re $\omega_L \equiv \omega$

$$\sinh(\omega) = \gamma_s \qquad (15.22)$$
$$\cosh(\omega) = \beta\gamma_s$$

Upon substituting $\boldsymbol{\omega_L}$ for ω in eq. 15.13 we obtain another form for a left-handed superluminal transformation (equivalent to that of eq. 15.19):

$$\Lambda_L(\omega, \mathbf{u}) = \Lambda(\omega + i\pi/2, \mathbf{u})$$

$$= \begin{bmatrix} \cosh(\omega_L) & -\sinh(\omega_L)u_x & -\sinh(\omega_L)u_y & -\sinh(\omega_L)u_z \\ -\sinh(\omega_L)u_x & 1+(\cosh(\omega_L)-1)u_x^2 & (\cosh(\omega_L)-1)u_xu_y & (\cosh(\omega_L)-1)u_xu \\ -\sinh(\omega_L)u_y & (\cosh(\omega_L)-1)u_xu_y & 1+(\cosh(\omega_L)-1)u_y^2 & (\cosh(\omega_L)-1)u_yu_z \\ -\sinh(\omega_L)u_z & (\cosh(\omega_L)-1)u_xu_z & (\cosh(\omega_L)-1)u_yu_z & 1+(\cosh(\omega_L)-1)u_z^2 \end{bmatrix}$$

$$= \begin{bmatrix} i\gamma_s & -i\beta\gamma_su_x & -i\beta\gamma_su_y & -i\beta\gamma_su_z \\ -i\beta\gamma_su_x & 1+(i\gamma_s-1)u_x^2 & (i\gamma_s-1)u_xu_y & (i\gamma_s-1)u_xu_z \\ -i\beta\gamma_su_y & (i\gamma_s-1)u_xu_y & 1+(i\gamma_s-1)u_y^2 & (i\gamma_s-1)u_yu_z \\ -i\beta\gamma_su_z & (i\gamma_s-1)u_xu_z & (i\gamma_s-1)u_yu_z & 1+(i\gamma_s-1)u_z^2 \end{bmatrix} = \Lambda_L(\mathbf{v}) \qquad (15.23)$$

[230] The extra U(1) symmetry leads to the introduction of WIMPs (Weakly Interacting Massive Particles) which causes the SU(2)⊗U(1) doublets of particles to become SU(2)⊗U(1)⊕U(1) triplets with a WIMP as the third member of each triplet. See appendix 18-A for details. Preliminary experimental evidence for WIMPs has recently been found. While the introduction of WIMPs to form triplets is natural, a restriction of the theory to particle doublets is allowed by the form of the theory if WIMPs are not found experimentally.

A simple case that illustrates a left-handed superluminal boost is to assume the relative velocity is in the x direction. Then eq. 15.23 becomes

$$\Lambda_L(\omega, \mathbf{u} = (1,0,0)) = \begin{bmatrix} i\gamma_s & -i\beta\gamma_s & 0 & 0 \\ -i\beta\gamma_s & i\gamma_s & 0 & 0 \\ 0 & 0 & 1 & 0 \\ 0 & 0 & 0 & 1 \end{bmatrix} \quad (15.24)$$

implementing the coordinate transformation:

$$X' = \Lambda_L(\omega, \mathbf{u} = (1,0,0))X$$

or

$$\begin{aligned} t' &= i\gamma_s(t - \beta x) \\ x' &= i\gamma_s(x - \beta t) \\ y' &= y \\ z' &= z \end{aligned} \quad (15.25)$$

The addition rule for the x-component of velocity can be computed for infinitesimal displacements in space and time:

$$v_x' = \Delta x' / \Delta t' = (\Delta x\ \gamma_s - \Delta t\ \beta\gamma_s)/(\Delta t\ \gamma_s - \Delta x\ \beta\gamma_s)$$

$$= (v_x - \beta)/(1 - \beta v_x) \quad (15.26)$$

(where $\beta = w$ is the relative speed) in the limit $\Delta t \to 0$ where the x component of a particle's velocity in the unprimed frame is $v_x = \Delta x/\Delta t$. $\Delta t'$ is determined by

$$\Delta t' = i\Delta t\ \gamma_s(1 - \beta v_x) \quad (15.27)$$

Note the velocity of light is the same in the primed and unprimed reference frames. (If $v_x = 1$ then $v_x' = 1$.) *Thus left-handed superluminal transformations preserve the constancy of the speed of light in all reference frames.* (Postulate 15.2)

Further note that increasing the value of ω in $\Lambda_L(\omega, \mathbf{u})$ corresponds to decreasing the magnitude of the relative velocity v since

$$v = \text{cotanh}(\omega) \quad (15.28)$$

by eq. 15.22. Thus when $\omega = 0$ then $v = \infty$ and when $\omega = \infty$ then $v = 1$. This is the reverse of the sublight case: By eq. 15.18 $v = \tanh(\omega)$. Thus when $\omega = 0$ then $v = 0$ and when $\omega = \infty$ then $v = \infty$.

15.7.2 General Velocity Transformation Law – Left-Handed Superluminal Boosts

The general velocity transformation law for a particle moving with velocity **v** in the unprimed reference frame and velocity **v'** in the primed reference frame is

$$\mathbf{v}' = [\mathbf{v} + (\gamma - 1)\mathbf{w}\cdot\mathbf{v}\ \mathbf{w}/w^2 - \gamma\mathbf{w}]/[\gamma(1 - \mathbf{w}\cdot\mathbf{v})] \qquad (15.29)$$

where **w** is the relative velocity of the primed reference frame with respect to the unprimed reference frame, and $\gamma = (1 - w^2)^{-1/2}$. Eq. 15.29 is obtained by calculating the derivative d**x**'/dt' using eqs. 15.14. The relative velocity **w** can be greater or less than the speed of light. Eq. 15.29 implies

$$v'^2 = 1 + (v^2 - 1)(1 - w^2)/(1 - \mathbf{w}\cdot\mathbf{v})^2 \qquad (15.30)$$

The relation of the velocities (eq. 15.30) will be used to determine the multiplication rules for subluminal and superluminal Lorentz transformations (next subsection).

15.7.3 Left-Handed Transformations Multiplication Rules

In this subsection we will determine the multiplication rules of left-handed subluminal and superluminal Lorentz boosts. To do this we will consider three reference frames: an "unprimed" frame, a "primed" frame moving with velocity **w** with respect to the unprimed frame, and a "double-primed" frame moving with velocity **v** with respect to the unprimed frame and velocity **v'** with respect to the primed frame. See Fig.15.2.

The velocity **v'** is related to **v** by eqs. 15.29 and 15.30. Think of the double-primed coordinate system as attached to a particle. In addition note that the transformation law from the unprimed to the double-primed reference frame can be viewed as the product of consecutive transformations (boosts) from the unprimed to the primed reference frames and then from the primed to the double-primed reference frames.

Thus the transformations have the general form:

$$\Lambda_?(\mathbf{v}) = \Lambda_?(\mathbf{v}')\Lambda_?(\mathbf{w}) \qquad (15.31)$$

where the "?" subscripts indicate subluminal or superluminal transformations (boosts) depending on the magnitude of the relative velocity in the transformation's parentheses.

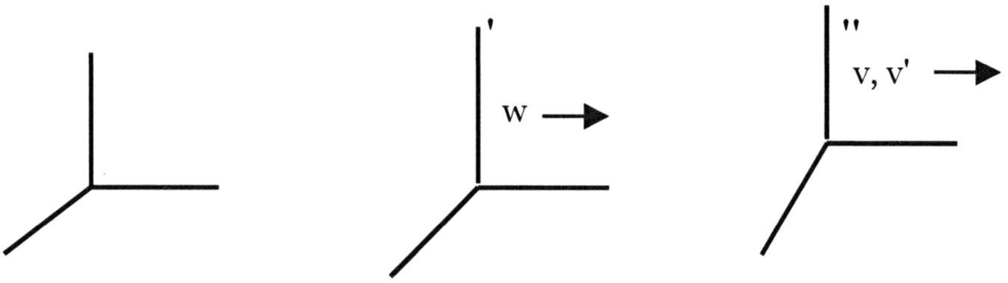

Figure 15.2. Three reference frames used to establish transformation multiplication rules.

We now consider the various cases using eq. 15.30:

1) If w > 1 and v' > 1

   then eq. 15.30 implies v < 1 and thus the left $\Lambda_?(\mathbf{v})$ is a subluminal transformation

$$\Lambda(\mathbf{v}) = \Lambda_L(\mathbf{v'})\Lambda_L(\mathbf{w}) \qquad (15.32)$$

2) If w > 1, v' < 1

   then eq. 15.30 implies v > 1 and thus the left $\Lambda_?(\mathbf{v})$ is a superluminal transformation

$$\Lambda_L(\mathbf{v}) = \Lambda(\mathbf{v'})\Lambda_L(\mathbf{w}) \qquad (15.33)$$

3) If w < 1, v' > 1

   then eq. 15.30 implies v > 1 and thus the left $\Lambda_?(\mathbf{v})$ is a superluminal transformation

$$\Lambda_L(\mathbf{v}) = \Lambda_L(\mathbf{v'})\Lambda(\mathbf{w}) \qquad (15.34)$$

4) If w < 1, v' < 1

   then eq. 15.30 implies v < 1 and thus the left $\Lambda_?(\mathbf{v})$ is a Lorentz transformation

$$\Lambda(\mathbf{v}) = \Lambda(\mathbf{v'})\Lambda(\mathbf{w}) \qquad (15.35)$$

where, in each above case, the transformation on the left side of the equation may be a boost or a combination of a boost and a spatial rotation. Thus we have obtained the multiplication rules for left-handed subluminal and superluminal Lorentz transformations.

15.7.4 Inverse of Left-Handed Transformations

The inverse of a Lorentz boost is

$$\Lambda^{-1}(\omega, \hat{\mathbf{u}}) = \exp[-i\omega\hat{\mathbf{u}}\cdot\mathbf{K}] \tag{15.36}$$

where $\boldsymbol{\omega} \geq \mathbf{0}$. Thus the inverse is generated by letting $\omega \rightarrow -\omega$. Note that since v = tanhω, the effect of $\omega \rightarrow -\omega$ is to let $v \rightarrow -v$. In the case of superluminal left-handed boosts, since

$$\Lambda_L(\omega, \mathbf{u}) = \Lambda(\omega + i\pi/2, \mathbf{u}) = \exp[i(\omega + i\pi/2)\hat{\mathbf{u}}\cdot\mathbf{K}] \tag{15.37}$$

we find the inverse is

$$\Lambda_L^{-1}(\omega, \mathbf{u}) = \Lambda(-(\omega + i\pi/2), \mathbf{u}) = \exp[-i(\omega + i\pi/2)\hat{\mathbf{u}}\cdot\mathbf{K}] \tag{15.38}$$

where $\boldsymbol{\omega} \geq \mathbf{0}$. Since $\Lambda_L^{-1}(\omega, \mathbf{u})$ is not the hermitean conjugate of $\Lambda_L(\omega, \mathbf{u})$, superluminal boosts are not unitary. However unitarity is not required since complex Lorentz group elements satisfy the defining relation of the Lorentz group (eq. 15.10).

15.8 Right-Handed Superluminal Transformations

When we transform between reference frames using a *right-handed*[231] superluminal boost the relation between ω and v is different. The variable ω becomes $\omega_R = \omega - i\pi/2$ and

$$\cosh(\omega_R) = -i\sinh(\omega) = \gamma = -i\gamma_s \tag{15.39}$$
$$\sinh(\omega_R) = -i\cosh(\omega) = \beta\gamma = -i\beta\gamma_s \tag{15.40}$$

where $\beta = v > 1$ and $\boldsymbol{\omega} \geq \mathbf{0}$. Note that $\omega = \text{Re}\,\omega_R$

$$\sinh(\omega) = \gamma_s \tag{15.41}$$
$$\cosh(\omega) = \beta\gamma_s \tag{15.42}$$

with

$$\gamma_s = (\beta^2 - 1)^{-1/2} \tag{15.43}$$

Upon substituting ω_R for ω in eq. 15.13 we obtain the form of the right-handed superluminal boost:[232]

[231] We call these transformations right-handed because they lead eventually to an alternate right-handed Standard Model This alternate right-handed Standard Model does not appear to correspond to current experimental reality.

$$\Lambda_R(\omega, \mathbf{u}) = \Lambda(\omega - i\pi/2, \mathbf{u}) \tag{15.44}$$

$$= \begin{bmatrix} -i\gamma_s & i\beta\gamma_s u_x & i\beta\gamma_s u_y & i\beta\gamma_s u_z \\ i\beta\gamma_s u_x & 1 + (-i\gamma_s - 1)u_x^2 & (-i\gamma_s - 1)u_x u_y & (-i\gamma_s - 1)u_x u_z \\ i\beta\gamma_s u_y & (-i\gamma_s - 1)u_x u_y & 1 + (-i\gamma_s - 1)u_y^2 & (-i\gamma_s - 1)u_y u_z \\ i\beta\gamma_s u_z & (-i\gamma_s - 1)u_x u_z & (-i\gamma_s - 1)u_y u_z & 1 + (-i\gamma_s - 1)u_z^2 \end{bmatrix}$$

A simple case that illustrates right-handed superluminal transformations is to assume a relative velocity in the x direction. Then eq. 15.44 becomes

$$\Lambda_R(\omega, \mathbf{u} = (1,0,0)) = \begin{bmatrix} -i\gamma_s & i\beta\gamma_s & 0 & 0 \\ i\beta\gamma_s & -i\gamma_s & 0 & 0 \\ 0 & 0 & 1 & 0 \\ 0 & 0 & 0 & 1 \end{bmatrix} \tag{15.45}$$

implementing the coordinate transformation:

$$X' = \Lambda_R(\omega, \mathbf{u})X$$

or

$$\begin{aligned} t' &= -i\gamma_s(t - \beta x) \\ x' &= -i\gamma_s(x - \beta t) \\ y' &= y \\ z' &= z \end{aligned} \tag{15.46}$$

Comparing eq. 15.45 with eq. 15.24 for a left-handed superluminal boost we see that

$$PT\Lambda_L(\omega, \mathbf{u} = (1,0,0)) = \begin{bmatrix} -i\gamma_s & i\beta\gamma_s & 0 & 0 \\ i\beta\gamma_s & -i\gamma_s & 0 & 0 \\ 0 & 0 & -1 & 0 \\ 0 & 0 & 0 & -1 \end{bmatrix}$$

where P is the parity operator and T is the time reversal operator. If we now apply a spatial rotation $\mathfrak{R}$ of π radians around the x axis then we obtain

$$\mathfrak{R}PT\Lambda_L(\omega, \mathbf{u} = (1,0,0))\mathfrak{R}^{-1} = \begin{bmatrix} -i\gamma_s & i\beta\gamma_s & 0 & 0 \\ i\beta\gamma_s & -i\gamma_s & 0 & 0 \\ 0 & 0 & 1 & 0 \\ 0 & 0 & 0 & 1 \end{bmatrix} \tag{15.47}$$

---

[232] We note the singularities at $\beta = \pm 1$ or $\omega = \pm\infty$. As a result we have a branch cut in the complex ω-plane consisting of the entire real ω axis. Therefore three left-handed boosts are not equivalent to a right-handed boost but rather appear on a different Riemann sheet.

$$= \Lambda_R(\omega, \mathbf{u} = (1,0,0))$$

Since P and T commute with spatial rotations we find

$$\Lambda_R(\omega, \mathbf{u} = (1,0,0)) = PT\mathfrak{R}\Lambda_L(\omega, \mathbf{u} = (1,0,0))\mathfrak{R}^{-1} \tag{15.48}$$

or, more generally, performing additional spatial rotations:

$$\Lambda_R(\omega, \mathbf{u}) = PT\mathfrak{R}_u\mathfrak{R}\mathfrak{R}_w\Lambda_L(\omega, \mathbf{w})\mathfrak{R}_w^{-1}\mathfrak{R}^{-1}\mathfrak{R}_u^{-1} \tag{15.49}$$

or,

$$\Lambda_R(\omega, \mathbf{u}) = PT\mathfrak{R}_{tot}\Lambda_L(\omega, \mathbf{w})\mathfrak{R}_{tot}^{-1} \tag{15.50}$$

where **u** and **w** are unit vectors, and $\mathfrak{R}_{tot} = \mathfrak{R}_u\mathfrak{R}\mathfrak{R}_w$. Alternately,

$$\Lambda_L(\omega, \mathbf{w}) = PT\mathfrak{R}_{tot}^{-1}\Lambda_R(\omega, \mathbf{u})\mathfrak{R}_{tot} \tag{15.51}$$

or

$$\Lambda_L(\omega, \mathbf{w}) = PT\Lambda_R(\omega, \mathbf{u}') \tag{15.52}$$

for some unit vector **u'**.

Thus we have shown that PT can be used to relate left-handed and right-handed boosts in a one-to-one fashion. *The appearance of the* parity *operator P takes on great significance when we derive features of the Standard Model. The appearance of left-handed form of The Standard Model stems directly from the implicit parity dependence of the left-handed sector of the superluminal part of the complex Lorentz group.*

For a right-handed boost the addition rule for the x-component of velocity can be computed for infinitesimal displacements in space and time:

$$\begin{aligned} v_x' = \Delta x' / \Delta t' &= (\Delta x\ \gamma_s - \Delta t\ \beta\gamma_s)/(\Delta t\ \gamma_s - \Delta x\ \beta\gamma_s) \\ &= (v_x - \beta)/(1 - \beta v_x) \end{aligned} \tag{15.53}$$

in the limit $\Delta t \to 0$ where the x component of a particle's velocity in the unprimed frame is $v_x = \Delta x/\Delta t$. Note if $v_x = 1$ then $v_x' = 1$. *Thus right-handed superluminal transformations also preserve the constancy of the speed of light in all reference frames.*

15.9 Inhomogeneous Left-Handed Lorentz Group Transformations

The *Left-Handed transformations of the complex Lorentz group* consist of the elements of the real Lorentz group plus left-handed superluminal boost transformations, and combinations of boosts and spatial rotations. Thus the homogeneous left-handed superluminal transformations have the general form:

$$\Lambda_L(\mathbf{v}, \boldsymbol{\theta}) = \exp[i\omega_L \hat{\mathbf{u}}\cdot\mathbf{K} + i\boldsymbol{\theta}\cdot\mathbf{J}] \tag{15.54}$$

where $\omega_L' = \omega + i\pi/2$, $\boldsymbol{\theta}$ is the angular vector, and **J** is the angular momentum operator vector. Inhomogeneous left-handed superluminal transformations, which include displacements, can be expressed as

$$\Lambda_L(\mathbf{v}, \boldsymbol{\theta}, \mathbf{d}) = \exp[i\omega_L \hat{\mathbf{u}}\cdot\mathbf{K} + i\boldsymbol{\theta}\cdot\mathbf{J} - i\mathbf{d}\cdot\mathbf{P}] \tag{15.55}$$

where **P** is the momentum operator vector and **d** is a displacement vector.

We note

$$\det \Lambda_L(\omega, \mathbf{u}) = \pm 1 \tag{15.56}$$

The ordinary Lorentz group is divided into four disjoint subgroups that are often denoted:

$$\begin{aligned}
L_+^\uparrow &: \det \Lambda(\omega, \mathbf{u}) = +1; \ \text{sgn}\, \Lambda(\omega, \mathbf{u})^0{}_0 = +1 \\
L_-^\uparrow &: \det \Lambda(\omega, \mathbf{u}) = -1; \ \text{sgn}\, \Lambda(\omega, \mathbf{u})^0{}_0 = +1 \\
L_+^\downarrow &: \det \Lambda(\omega, \mathbf{u}) = +1; \ \text{sgn}\, \Lambda(\omega, \mathbf{u})^0{}_0 = -1 \\
L_-^\downarrow &: \det \Lambda(\omega, \mathbf{u}) = -1; \ \text{sgn}\, \Lambda(\omega, \mathbf{u})^0{}_0 = -1
\end{aligned} \tag{15.57}$$

where $\text{sgn}\, \Lambda(\omega, \mathbf{u})^0{}_0$ is the sign of the 00 component of the $\Lambda(\omega, \mathbf{u})$ matrix. The various subgroups are related by the discrete transformations of parity P and time reversal T:

$$\begin{aligned}
L_+^\uparrow &\xrightarrow{P} L_-^\uparrow \\
L_+^\uparrow &\xrightarrow{PT} L_+^\downarrow \\
L_+^\uparrow &\xrightarrow{T} L_-^\downarrow
\end{aligned}$$

The left-handed superluminal transformations are disjoint in a somewhat different way. By eq. 15.56 the determinants are ±1. However the 0-0 matrix element of eq. 15.16 gives

$$\Lambda_L{}^0{}_0{}^2 - \Sigma_i (\Lambda_L{}^i{}_0)^2 = 1 \tag{15.58}$$

The representation of superluminal boosts shows that each factor in eq. 15.58 is imaginary. Thus eq. 15.58 implies

$$\Sigma_i |\Lambda_L{}^i{}_0|^2 \geq 1 \tag{15.59}$$

$$|\Lambda_L{}^0{}_0| \geq 0 \qquad \text{(not } \geq 1) \tag{15.60}$$

where || indicates absolute value since the quantities in eq. 15.58 are squares – not in absolute value. Thus the magnitude of $\Lambda_L{}^0{}_0$ does not have a gap. Therefore left-handed superluminal transformations can be divided into two categories:

$$\begin{aligned} {}_L L_+ &: \quad \det \Lambda_L(\omega, \mathbf{u}) = +1 \\ {}_L L_- &: \quad \det \Lambda_L(\omega, \mathbf{u}) = -1 \end{aligned} \tag{15.61}$$

as one expects for complex Lorentz group transformations.[233]

Earlier we saw that under a PT transformation a left-handed superluminal transformation becomes a right handed superluminal transformation. Again, as in the left-handed case, the various disjoint pieces are related by the discrete transformations of parity P and time reversal T:

$$\begin{aligned} {}_L L_+ &\xrightarrow{P} {}_L L_- \\ {}_L L_+ &\xrightarrow{T} {}_L L_- \end{aligned} \tag{15.62}$$

15.10 Inhomogeneous Right-Handed Extended Lorentz Group

The inhomogeneous right-handed part of the complex Lorentz group[234] consists of the real Lorentz group plus right-handed superluminal transformations plus rotations and displacements that have the form:

$$\Lambda_R(\mathbf{v}, \boldsymbol{\theta}, \mathbf{d}) = \exp[i\omega_R \hat{\mathbf{u}}\cdot\mathbf{K} + i\boldsymbol{\theta}\cdot\mathbf{J} - i\mathbf{d}\cdot\mathbf{P}] \tag{15.63}$$

in general where $\omega_R = \omega - i\pi/2$.

15.11 General Forms of Superluminal Boosts

The group elements of the homogeneous complex Lorentz group L(C) can be expressed in terms of the group generators as

---

[233] Streater (2000) p. 13.

[234] Since γ_s has branch points at $v = \pm 1$ (which corresponds to $\omega = \pm\infty$ for both the left-handed and right-handed groups) there is a cut along the real ω axis between $-\infty$ and $+\infty$ in the ω complex plane. Therefore, we note, the product of three left-handed Lorentz transformations does not yield a right-handed transformation (as might be supposed from eqs. 2.43 and 2.51) but rather a left-handed transformation on the second sheet. A transformation with $\omega + 3i\pi/2$ is not equivalent to a transformation with $\omega - i\pi/2$.

$$\Lambda_C = \exp[i(\omega_r\hat{\mathbf{u}}_r + i\omega_i\hat{\mathbf{u}}_i)\cdot\mathbf{K} + i\boldsymbol{\theta}_c\cdot\mathbf{J}] \tag{15.64}$$

where the vector $\boldsymbol{\theta}_c$ is a complex 3-vector, $\omega_r \geq 0$ and $\omega_i \geq 0$ are real numbers, and $\hat{\mathbf{u}}_r$ and $\hat{\mathbf{u}}_i$ are real normalized 3-vectors such that $\hat{\mathbf{u}}_r\cdot\hat{\mathbf{u}}_r = 1 = \hat{\mathbf{u}}_i\cdot\hat{\mathbf{u}}_i$. The generators of the homogeneous complex Lorentz group are **K**, and **J** just as for the homogeneous real Lorentz group.

We now focus on boosts because they will be crucial in the determination of the equations of motion of various types of spin ½ particles. A boost has the form

$$\Lambda_C(\mathbf{v}_c) = \exp[i\omega\hat{\mathbf{w}}\cdot\mathbf{K}] \tag{15.65}$$

where

$$\omega = (\omega_r^2 - \omega_i^2 + 2i\omega_r\omega_i\,\hat{\mathbf{u}}_r\cdot\hat{\mathbf{u}}_i)^{1/2} \tag{15.66}$$

and

$$\hat{\mathbf{w}} = (\omega_r\hat{\mathbf{u}}_r + i\omega_i\hat{\mathbf{u}}_i)/\omega \tag{15.67}$$

Since $\hat{\mathbf{u}}_r\cdot\hat{\mathbf{u}}_r = 1 = \hat{\mathbf{u}}_i\cdot\hat{\mathbf{u}}_i$ we see

$$\hat{\mathbf{w}}\cdot\hat{\mathbf{w}} = 1 \tag{15.68}$$

The complex relative velocity is

$$\mathbf{v}_c = \hat{\mathbf{w}}\tanh(\omega) \tag{15.69}$$

Having placed boost transformations in the form of eq. 15.12 we can take advantage of the form of real proper orthchronous Lorentz boost transformations, eq. 15.13, and analytically continue to complex ω and complex unit vectors $\hat{\mathbf{w}}$ provided eq. 15.69 is satisfied. The resulting complex generalization will be the matrix form of proper boosts:

$$\Lambda_C(\mathbf{v}_c) = \exp[i\omega\hat{\mathbf{w}}\cdot\mathbf{K}] \equiv \Lambda_C(\omega, \hat{\mathbf{w}})$$

$$= \begin{bmatrix} \cosh(\omega) & -\sinh(\omega)\hat{w}_x & -\sinh(\omega)\hat{w}_y & -\sinh(\omega)\hat{w}_z \\ -\sinh(\omega)\hat{w}_x & 1+(\cosh(\omega)-1)\hat{w}_x^2 & (\cosh(\omega)-1)\hat{w}_x\hat{w}_y & (\cosh(\omega)-1)\hat{w}_x\hat{w}_z \\ -\sinh(\omega)\hat{w}_y & (\cosh(\omega)-1)\hat{w}_x\hat{w}_y & 1+(\cosh(\omega)-1)\hat{w}_y^2 & (\cosh(\omega)-1)\hat{w}_y\hat{w}_z \\ -\sinh(\omega)\hat{w}_z & (\cosh(\omega)-1)\hat{w}_x\hat{w}_z & (\cosh(\omega)-1)\hat{w}_y\hat{w}_z & 1+(\cosh(\omega)-1)\hat{w}_z^2 \end{bmatrix} \tag{15.70}$$

Since analytic continuations are unique, the above form for $\Lambda_C(\mathbf{v}_c)$ is well-defined and unique. It spans the complete set of proper complex Lorentz boosts.

We now will study six classes of boosts that have the property that they boost from a coordinate system with real time and space coordinates to a coordinate system with either a purely real or purely imaginary time, and real, imaginary or complex spatial coordinates. These boosts produce left-handed lepton-like and "quark-like" free Dirac-like equations. They also produce right-handed lepton-like and "quark-like" free Dirac-like equations. We will discuss these Dirac-like equations in detail later. First we describe the four categories of boosts that have the property that they transform the reference frame of a particle at rest to a reference frame where the energy is either purely real or purely imaginary – the distinguishing feature of these four sets of transformations.

15.11.1 "Lepton-like" Left-Handed Boosts

If we let

$$\hat{\mathbf{u}}_i = \hat{\mathbf{u}}_r \equiv \hat{\mathbf{u}} \tag{15.71}$$

so that the vector $\hat{\mathbf{u}}_i$ is parallel to $\hat{\mathbf{u}}_r$, and let

$$\omega_i = \pi/2 \tag{15.72}$$

then $\Lambda_C(\mathbf{v}_c)$ becomes a lepton-like left-handed boost:[235]

$$\Lambda_C = \exp[i(\omega_r + i\,\pi/2)\hat{\mathbf{u}}_r\cdot\mathbf{K}] \tag{15.73}$$

15.11.2 "Lepton-like" Right-Handed Boosts

If we let

$$\hat{\mathbf{u}}_i = -\hat{\mathbf{u}}_r \equiv -\hat{\mathbf{u}} \tag{15.74}$$

so that the vector $\hat{\mathbf{u}}_i$ is anti-parallel to $\hat{\mathbf{u}}_r$, and

$$\omega_i = -\pi/2 \tag{15.75}$$

then $\Lambda_C(\mathbf{v}_c)$ becomes a right-handed boost:

$$\Lambda_C = \exp[i(\omega_r - i\,\pi/2)\hat{\mathbf{u}}_r\cdot\mathbf{K}] \tag{15.76}$$

15.11.3 "Quark-like" Left-Handed Boosts

If the real and imaginary relative vectors parts of $\hat{\mathbf{w}}$, namely $\hat{\mathbf{u}}_r$ and $\hat{\mathbf{u}}_i$, are perpendicular, $\hat{\mathbf{u}}_r\cdot\hat{\mathbf{u}}_i = 0$, then by eq. 15.66

$$\omega = (\omega_r^2 - \omega_i^2)^{1/2} \tag{15.77}$$

Thus ω is either pure real ($\omega_r \geq \omega_i$) or pure imaginary ($\omega_r < \omega_i$). We choose ω real, and then reset

---

[235] We say "lepton-like" because we obtain a lepton-like Dirac-like equation using these boosts later. Similarly for "quark-like.'

$$\omega = (\omega_r^2 - \omega_i^2)^{1/2} \rightarrow \omega' = (\omega_r^2 - \omega_i^2)^{1/2} + i\pi/2 = \omega + i\pi/2 \qquad (15.78)$$

by adding $i\pi/2$ to the ω factor in eq. 15.65 since ω is a free parameter. Then the resulting Lorentz transformation then becomes a "quark-like" left-handed boost:[236]

$$\Lambda_C = \exp[i((\omega_r^2 - \omega_i^2)^{1/2} + i\pi/2)(\omega_r\hat{\mathbf{u}}_r + i\omega_i\hat{\mathbf{u}}_i)\cdot\mathbf{K}/\omega] \qquad (15.79)$$

15.11.4 "Quark-like" Right-Handed Boosts

If the real and imaginary relative vectors parts of $\hat{\mathbf{w}}$, namely $\hat{\mathbf{u}}_r$ and $\hat{\mathbf{u}}_i$, are perpendicular, $\hat{\mathbf{u}}_r\cdot\hat{\mathbf{u}}_i = 0$, then by eq. 15.66

$$\omega = (\omega_r^2 - \omega_i^2)^{1/2} \qquad (15.80)$$

Thus ω again starts out either pure real ($\omega_r \geq \omega_i$) or pure imaginary ($\omega_r < \omega_i$). In this case we also choose ω real, and then reset

$$\omega = (\omega_r^2 - \omega_i^2)^{1/2} \rightarrow \omega' = (\omega_r^2 - \omega_i^2)^{1/2} - i\pi/2 \qquad (15.81)$$

by subtracting $i\pi/2$ from ω in eq. 15.65 since ω is a free parameter. The resulting Lorentz boost

$$\Lambda_C = \exp[i((\omega_r^2 - \omega_i^2)^{1/2} - i\pi/2)(\omega_r\hat{\mathbf{u}}_r + i\omega_i\hat{\mathbf{u}}_i)\cdot\mathbf{K}/\omega] \qquad (15.82)$$

becomes a quark-like right-handed boost.[237]

15.11.5 "Quark-like" Boosts

If the real and imaginary relative vectors parts of $\hat{\mathbf{w}}$, namely $\hat{\mathbf{u}}_r$ and $\hat{\mathbf{u}}_i$, are perpendicular, $\hat{\mathbf{u}}_r\cdot\hat{\mathbf{u}}_i = 0$, then by eq. 15.66

$$\omega = (\omega_r^2 - \omega_i^2)^{1/2} \qquad (15.83)$$

Thus ω again starts out either pure real ($\omega_r \geq \omega_i$) or pure imaginary ($\omega_r < \omega_i$). In this case choose ω_r real and use ω as defined by eq. 15.83.

Then the resulting Lorentz boost

$$\Lambda_C = \exp[i(\omega_r^2 - \omega_i^2)^{1/2}(\omega_r\hat{\mathbf{u}}_r + i\omega_i\hat{\mathbf{u}}_i)\cdot\mathbf{K}/\omega] \qquad (15.84)$$

becomes a quark-like boost without handedness.[238]

---

[236] We say "quark-like" because we obtain a quark-like left-handed Dirac-like equation with complex spatial momentum terms using these boosts later.

[237] We say "quark-like" because we obtain a quark-like right-handed Dirac-like equation with complex spatial momentum terms using these boosts later.

15.11.6 Conventional "Dirac" Boosts

If we let

$$\hat{\mathbf{u}}_i = \hat{\mathbf{u}}_r \equiv \hat{\mathbf{u}} \tag{15.85}$$

so that the vector $\hat{\mathbf{u}}_i$ is parallel to $\hat{\mathbf{u}}_r$, and let

$$\omega_i = 0 \tag{15.86}$$

then $\Lambda_C(\mathbf{v}_c)$ becomes a Dirac boost:[239]

$$\Lambda = \exp[i\omega_r \hat{\mathbf{u}}_r \cdot \mathbf{K}] \tag{15.87}$$

This boost can be used to generate the free Dirac equation.

15.12 Changing Complex Coordinates After a Superluminal Boost to Real Coordinates – A New Hidden Transformation

In the discussions up to this point in this chapter we have not considered imaginary (and more generally complex) coordinates resulting from a superluminal Lorentz transformation. In this section we show that they require us to introduce another transformation that maps complex coordinates to real coordinates. This transformation will be of significance because it leads to a hitherto unstated SU(2)⊗U(1)⊕U(1) symmetry that emerges when we consider superluminal transformations but is not required for ordinary sublight Lorentz transformations. The SU(2)⊗U(1) part of this new symmetry can be identified as the source of the SU(2)⊗U(1) symmetry of the ElectroWeak sector of The Standard Model. (See appendix 18-A.)

We introduce this new transformation by reconsidering the previous simple example wherein one coordinate system is traveling at a speed v in the x direction with respect to the "laboratory" system. (Fig. 15.1) The (left-handed[240]) Lorentz transformation is given by eq. 15.24 and the coordinates in the two reference frames are related by eq. 15.25. We now define a transformation that maps the real coordinates of the unprimed reference frame to real coordinates in the primed reference frame.

$$\Pi_L(\mathbf{u}) = \begin{bmatrix} -i & 0 & 0 & 0 \\ 0 & -i & 0 & 0 \\ 0 & 0 & 1 & 0 \\ 0 & 0 & 0 & 1 \end{bmatrix} \tag{15.88}$$

---

[238] We again say "quark-like" because we obtain a quark-like Dirac equation with complex spatial momentum terms using these boosts later.

[239] We say "Dirac" because we obtain a Dirac equation using this boost later.

[240] The right-handed Lorentz transformation case is analogous.

where **u** is the unit vector corresponding to the direction of **v** (the positive x direction in this example). Inserting $\Pi_L(\mathbf{u})$ in eq. 15.25 we obtain an overall transformation from real coordinates to real coordinates:

$$X'' = \Pi_L(\mathbf{u})\Lambda_L(\omega, \mathbf{u} = (1,0,0))X$$

or

$$\begin{aligned} t'' &= \gamma_s(t - \beta x) \\ x'' &= \gamma_s(x - \beta t) \\ y'' &= y \\ z'' &= z \end{aligned} \tag{15.89}$$

An observer in the primed reference frame would consider his/her time to be real when measured on a clock, and distances along the x axis to be real when measured with a ruler. Thus eq. 15.89 makes good sense physically because in any reference frame, observers measure real distances and real times. For this reason we will call transformations of the type of eq. 15.89 – from real coordinates to real coordinates – *physical* superluminal transformations.

This simple example generalizes to arbitrary relative velocities **v**. First we note that the Lorentz transformation for a velocity **v** that is a rotation of the velocity in the x-direction ($\mathbf{v} = |\mathbf{v}|R\mathbf{u}$ where R is the relevant rotation matrix) has the form

$$\Lambda_L(\omega, \mathbf{v}) = \mathcal{R}(\mathbf{v}/v, \mathbf{u})\Lambda_L(\omega, \mathbf{u} = (1,0,0))\mathcal{R}^{-1}(\mathbf{v}/v, \mathbf{u}) \tag{15.90}$$

where $\mathcal{R}(\mathbf{v}/v, \mathbf{u})$ is a rotation from the velocity direction **u** to direction **v**/v.

The original transformation (eq. 15.89) can be written as

$$\Pi_L(\mathbf{u})\Lambda_L(\omega, \mathbf{u} = (1,0,0)) = \Pi_L(\mathbf{u})\mathcal{R}^{-1}(\mathbf{v}/v, \mathbf{u})\Lambda_L(\omega, \mathbf{v})\mathcal{R}(\mathbf{v}/v, \mathbf{u}) \tag{15.91}$$

Consequently the combined transformation for velocity **v** is

$$\begin{aligned} \mathcal{R}(\mathbf{v}/v, \mathbf{u})\Pi_L(\mathbf{u})\Lambda_L(\omega, \mathbf{u} = (1,0,0))\mathcal{R}^{-1}(\mathbf{v}/v, \mathbf{u}) \\ = \mathcal{R}(\mathbf{v}/v, \mathbf{u})\Pi_L(\mathbf{u})\mathcal{R}^{-1}(\mathbf{v}/v, \mathbf{u})\Lambda_L(\omega, \mathbf{v}) \\ = \Pi_L(\mathbf{v}/v)\Lambda_L(\omega, \mathbf{v}) \end{aligned} \tag{15.92}$$

Thus for a Lorentz transformation $\Lambda_L(\omega, \mathbf{v})$ for velocity **v** we see that we can define a subsidiary transformation $\Pi_L(\mathbf{v}/v)$ of the form

$$\Pi_L(\mathbf{v}/v) = \mathcal{R}(\mathbf{v}/v, \mathbf{u})\Pi_L(\mathbf{u})\mathcal{R}^{-1}(\mathbf{v}/v, \mathbf{u}) \tag{15.93}$$

The general form of $\mathcal{R}(\mathbf{v}/v, \mathbf{u})$, is

$$\mathfrak{R}(\mathbf{v}/v, \mathbf{u}) = \begin{bmatrix} 1 & 0 & 0 & 0 \\ 0 & & & \\ 0 & & \mathfrak{R}_3(\mathbf{v}/v, \mathbf{u}) & \\ 0 & & & \end{bmatrix} \quad (15.94)$$

where $\mathfrak{R}_3(\mathbf{v}/v, \mathbf{u})$ is a 3×3 rotation matrix that can be expressed in terms of the generators of the 3-dimensional rotation group as

$$\mathfrak{R}_3(\mathbf{v}/v, \mathbf{u}) = \exp(i\boldsymbol{\theta}\cdot\mathbf{J}) \quad (15.95)$$

The rotation angles $\boldsymbol{\theta}$ are real numbers since we are rotating the real vector **u** to the real number **v**/v. Given the form of eq. 15.95 then we see that the form of $\Pi_L(\mathbf{v}//v)$ is

$$\Pi_L(\mathbf{v}/v) = \begin{bmatrix} -i & 0 & 0 & 0 \\ 0 & & & \\ 0 & & \mathfrak{R}_3(\mathbf{v}/v, \mathbf{u})\Pi_{L3}(\mathbf{u})\mathfrak{R}_3^{-1}(\mathbf{v}/v, \mathbf{u}) & \\ 0 & & & \end{bmatrix} \quad (15.96)$$

where

$$\Pi_{L3}(\mathbf{u}) = \begin{bmatrix} -i & 0 & 0 \\ 0 & 1 & 0 \\ 0 & 0 & 1 \end{bmatrix} \quad (15.97)$$

If we consider the case of an infinitesimal rotation $\boldsymbol{\theta}$ to first order in $\boldsymbol{\theta}$

$$\mathfrak{R}_3(\mathbf{v}/v, \mathbf{u}) \simeq I + i\boldsymbol{\theta}\cdot\mathbf{J} \quad (15.98)$$

then

$$\Pi_{L3}(\mathbf{v}/v) = \mathfrak{R}_3(\mathbf{v}/v, \mathbf{u})\Pi_{L3}(\mathbf{u})\mathfrak{R}_3^{-1}(\mathbf{v}/v, \mathbf{u}) \simeq \Pi_{L3}(\mathbf{u}) + i\boldsymbol{\theta}\cdot\mathbf{J}\Pi_{L3}(\mathbf{u}) - i\Pi_{L3}(\mathbf{u})\boldsymbol{\theta}\cdot\mathbf{J}$$
$$\simeq \Pi_{L3}(\mathbf{u})[I + i\Pi_{L3}^{-1}(\mathbf{u})[\boldsymbol{\theta}\cdot\mathbf{J}, \Pi_{L3}(\mathbf{u})] \quad (15.99)$$

where $\Pi_{L3}^{-1}(\mathbf{u})$ is the inverse of $\Pi_{L3}(\mathbf{u})$ and [...] represents the commutator. Thus for arbitrary rotations eq. 15.99 implies

$$\Pi_{L3}(\mathbf{v}/v) = \mathfrak{R}_3(\mathbf{v}/v, \mathbf{u})\Pi_{L3}(\mathbf{u})\mathfrak{R}_3^{-1}(\mathbf{v}/v, \mathbf{u}) = \Pi_{L3}(\mathbf{u})\exp\{i\Pi_{L3}^{-1}(\mathbf{u})[\boldsymbol{\theta}\cdot\mathbf{J}, \Pi_{L3}(\mathbf{u})]\} \quad (15.100)$$

We can find the general form of $\Pi_{L3}(\mathbf{v}/v)$ by considering the case of eq. 15.89 in more detail. The exponentiated matrix expression in 15.100 can written

$$\Pi_{L3}^{-1}(\mathbf{u})[\boldsymbol{\theta}\cdot\mathbf{J}, \Pi_{L3}(\mathbf{u})] = \Pi_{L3}^{-1}(\mathbf{u})\boldsymbol{\theta}\cdot\mathbf{J}\Pi_{L3}(\mathbf{u}) - \boldsymbol{\theta}\cdot\mathbf{J} = \boldsymbol{\theta}\cdot\mathbf{Q} \quad (15.101)$$

where

$$\mathbf{Q} = \Pi_{L3}^{-1}(\mathbf{u})\mathbf{J}\Pi_{L3}(\mathbf{u}) - \mathbf{J} = \mathbf{Q}' - \mathbf{J} \quad (15.102)$$

The matrices Q_i can be evaluated using eq. 15.97 and the matrix representations for the rotation generators J_i: which are equivalent in form to the SU(2) generators T_i:

$$J_1 = \begin{bmatrix} 0 & 0 & 0 \\ 0 & 0 & -i \\ 0 & i & 0 \end{bmatrix} = T_1 \qquad (15.103)$$

$$J_2 = \begin{bmatrix} 0 & 0 & i \\ 0 & 0 & 0 \\ -i & 0 & 0 \end{bmatrix} = T_2 \qquad (15.104)$$

$$J_3 = \begin{bmatrix} 0 & -i & 0 \\ i & 0 & 0 \\ 0 & 0 & 0 \end{bmatrix} = T_3 \qquad (15.105)$$

The rotation generators satisfy the commutation relations

$$[J_i, J_j] = i\epsilon_{ijk}J_k \qquad (15.106)$$

as do the SU(2) generators:

$$[T_i, T_j] = i\epsilon_{ijk}T_k \qquad (15.107)$$

We can calculate Q' from eqs. 15.97 and 15.102 – 15.104 and obtain

$$Q'_1 = \begin{bmatrix} 0 & 0 & 0 \\ 0 & 0 & -i \\ 0 & i & 0 \end{bmatrix} \qquad (15.108)$$

$$Q'_2 = \begin{bmatrix} 0 & 0 & -1 \\ 0 & 0 & 0 \\ -1 & 0 & 0 \end{bmatrix} \qquad (15.109)$$

$$Q'_3 = \begin{bmatrix} 0 & 1 & 0 \\ 1 & 0 & 0 \\ 0 & 0 & 0 \end{bmatrix} \qquad (15.110)$$

We note that each Q'_i is hermitean and the Q'_i satisfy the commutation relations:

$$[Q'_i, Q'_j] = i\epsilon_{ijk}Q'_k \qquad (15.111)$$

Consequently the set of Q'_i are also equivalent to SU(2) generators. As a result the exponential factor in eq. 15.100

$$\Pi_{L3}(\mathbf{v}/v) = \Pi_{L3}(\mathbf{u})\exp\{i\boldsymbol{\theta}\cdot(\mathbf{Q}' - \mathbf{J})\} \quad (15.112)$$

is equivalent to a combination of SU(2) rotations not only in this case but in general for superluminal transformations. The factor $\Pi_{L3}(\mathbf{u})$ is not an SU(2) matrix since its determinant is not +1 but

$$\Pi'_{L3}(\mathbf{u}) = -i\Pi_{L3}(\mathbf{u}) \quad (15.112a)$$

is an SU(2) matrix since

$$\Pi'^{-1}_{L3}(\mathbf{u}) = \Pi'^{\dagger}_{L3}(\mathbf{u}) \quad (15.112b)$$

$$\det \Pi'_{L3}(\mathbf{u}) = 1 \quad (15.112c)$$

and

$$\Pi'_{L3}(\mathbf{v}/v) = \Pi'_{L3}(\mathbf{u})\exp\{i\boldsymbol{\theta}\cdot(\mathbf{Q}' - \mathbf{J})\} \quad (15.112d)$$

is similarly an SU(2) rotation.

Thus the general form of superluminal transformation from a real set of coordinates to a real set of coordinates is[241]

$$\Pi_L(\mathbf{v}/v)\Lambda_L(\omega, \mathbf{v}) \quad (15.113)$$

where

$$\Pi_L(\mathbf{v}/v) = \begin{bmatrix} -i & 0 & 0 & 0 \\ 0 & & & \\ 0 & & \Pi_{L3}(\mathbf{u})\exp\{i\,\boldsymbol{\theta}\cdot(\mathbf{Q}' - \mathbf{J})\} & \\ 0 & & & \end{bmatrix} \quad (5.114)$$

by eqs. 15.92, 15.96 and 15.112. The Lorentz condition for real to real physical transformations generalizes to

$$\Lambda(\mathbf{v})^T\Pi_L(\mathbf{v}/v)^\dagger G\, \Pi_L(\mathbf{v})\Lambda(\mathbf{v}/v) = G \quad (15.115)$$

Since superluminal transformations $\Lambda_L(\omega, \mathbf{v})$ transform real coordinates to complex coordinates in general, we can generalize the form of a superluminal transformation to

$$e^{i\varphi}\Pi_L(\mathbf{v}'/v')\Lambda_L(\omega, \mathbf{v}) \quad (15.116)$$

where φ is a constant phase and $\mathbf{v}'$ is an arbitrary velocity. This generalization will satisfy the generalized Lorentz condition

$$\Lambda(\mathbf{v})^T\Pi_L(\mathbf{v}'/v')^\dagger e^{-i\varphi}G\, e^{i\varphi}\Pi_L(\mathbf{v}'/v')\Lambda(\mathbf{v}) = G \quad (15.117)$$

---

[241] The choice of the unit vector **u** and the angle vector **θ** must be such that applying eq. 15.113 to a real set of coordinates yields a real set of coordinates.

but the transformation will, in general, yield a complex set of coordinates when applied to a set of real coordinates.

These considerations imply:

1. Any observer in a coordinate system will treat a complex 4-dimensional coordinate system as if it were a real 4-dimensional coordinate system with complex-valued straight lines along each dimension (assuming rectangular coordinates).

2. The transformation $e^{i\varphi}\Pi'_{L3}(\mathbf{v}/v)$ is a SU(2)⊗U(1) transformation that takes complex 3-dimensional spatial coordinates to complex 3-dimensional spatial coordinates. In particular straight lines map to straight lines.

3. Physical observations in the observer's coordinate system are invariant under SU(2)⊗U(1) rotations of the spatial coordinates and the multiplication of the time component by an arbitrary phase.

4. The matrix

$$\Pi'_L(\mathbf{v}/v, \chi, \varphi) = \begin{bmatrix} e^{i\chi} & 0 \quad 0 \quad 0 \\ 0 & \\ 0 & e^{i\varphi}\Pi'_{L3}(\mathbf{u})\exp\{i\,\boldsymbol{\theta}\cdot(\mathbf{Q}' - \mathbf{J})\} \\ 0 & \end{bmatrix} \quad (15.118)$$

(where χ and φ are real numbers and **u** is a unit vector along any convenient coordinate axis) is an SU(2)⊗U(1)⊕U(1) transformation that transforms complex 4-dimensional coordinates to complex 4-dimensional coordinates. Note, $\Pi_L(\mathbf{v}/v) = \Pi'_L(\mathbf{v}/v, 3\pi/2, \pi/2)$ is a special case of $\Pi'_L(\mathbf{v}/v, \chi, \varphi)$. Due to the manifest form of 15-118 we see

$$\Pi'_L{}^{\mu}{}_{\alpha}{}^{*}\Pi'_L{}^{\mu}{}_{\beta} = [\Pi'_L{}^{\dagger}\Pi'_L]_{\alpha\beta} = I_{\alpha\beta} \quad (15.119)$$

(with an implied sum over μ) or, in matrix form,

$$\Pi'_L{}^{\dagger}\,\Pi'_L = I \quad (15.120)$$

and also[242]

$$\Pi'_L{}^{\dagger}G\Pi'_L = G \quad (15.121)$$

5. Complex coordinate values of the type generated by superluminal transformations are transformable to real coordinates by a transformation of the form of eq. 15.118. The complex coordinates are thus physically equivalent to

[242] Eq. 15.121 is close to the defining condition for a Lorentz group element but the presence of complex conjugation rather than a transpose means Π'_L is outside the real and complex Lorentz groups.

corresponding real coordinate values in the sense that an observer in that frame would automatically use the real coordinates so obtained since rulers and clocks always measure real spatial coordinates and times. *Therefore physical theory is invariant under global SU(2)⊗U(1)⊕U(1) coordinate transformations since complex coordinates, so generated, can be rotated back to real coordinates.*

6. The complex coordinates of any point obtained through a superluminal transformation can be transformed to a real set of coordinates by a transformation of the form of eq. 15.118 – a SU(2)⊗U(1)⊕U(1) transformation.

This SU(2)⊗U(1)⊕U(1) invariance will be shown to lead to the SU(2)⊗U(1)⊕U(1) symmetry of the ElectroWeak interactions (which can be restricted, if desired, to the conventional SU(2)⊗U(1) symmetry). See appendix 18-A.

Appendix 15-A. Superluminal Boosts of Momentum

15-A.1 Superluminal Transformations from Real to Complex Values

In chapter 15 we considered a simple superluminal boost from a "laboratory" reference frame to a moving reference frame (from the unprimed frame to the primed frame in eq. 15.46) as well as more complex superluminal transformations.

Eq. 15.46 shows that the time and x coordinate in the primed reference frame are imaginary. We showed how to remedy this situation in section 15.12 by introducing a subsidiary transformation $\Pi_L(\mathbf{u})$, applied after the superluminal Lorentz transformation, that yielded real values for the time and x coordinate in the primed reference frame (eqs. 15.88 and 15.89).

The question we now address is the physical meaning of transformations such as $\Pi_L(\mathbf{u})$ to obtain real values for the time and space variables in the primed reference frame. The answer is simple enough: an observer in the primed reference frame would use clocks and rulers to measure real values for the time and space cariables. So the transformation $\Pi_L(\mathbf{u})$ (and its more general forms) transforms the complex values yielded by the superluminal transformation to the physical values an observer would measure in the primed reference frame.

15-A.2 Superluminal Transformation of a Particle's Momentum

A particle's 4-momentum also undergoes superluminal transformations and becomes complex in general under an arbitrary superluminal transformation. The transformation law is the same as that of coordinates since both transform as 4-vectors. We use the same example as that represented by the transformation in eq. 15.46 (Fig. 15.1) for two frames moving at a relative velocity **v** in the positive x direction.

We assume a particle is moving with momentum p^μ in the unprimed frame. Then in the primed frame

$$p' = \Lambda_R(\omega, \mathbf{u})p$$

or

$$p'^0 = -i\gamma_s(p^0 - \beta p^x)$$

$$p'^x = -i\gamma_s(p^x - \beta p^0) \qquad (15\text{-A.1})$$
$$p'^y = p^y$$
$$p'^z = p^z$$

Using the subsidiary transformation $\Pi_L(\mathbf{u})$ we obtain a real 4-momentum in the primed frame (as in eq. 15.89)

$$p'' = \Pi_L(\mathbf{u})\Lambda_L(\omega, \mathbf{u} = (1,0,0))p$$

or

$$p''^0 = \gamma_s(p^0 - \beta p^x)$$
$$p''^x = \gamma_s(p^x - \beta p^0) \qquad (15\text{-A.1})$$
$$p''^y = p^y$$
$$p''^z = p^z$$

This 4-momentum p''^μ has purely real values and corresponds to the particle momentum that would be measured by an observer in the primed frame.

Thus we consider the *physical* superluminal transformation to be

$$\Pi_L(\mathbf{v}/v)\Lambda_L(\omega, \mathbf{v}) \qquad (15\text{-A.2})$$

For later use we will now explicitly list the form of the 4-momentum of a particle of mass m at rest in the unprimed (laboratory) reference frame, and the form of the 4-momentum of a particle in the primed reference frame:

Unprimed Reference Frame

$$p^0 = m$$
$$\mathbf{p} = \mathbf{0} \qquad (15\text{-A.3})$$

Primed Reference Frame

$$p'^0 = -im\gamma_s$$
$$p'^x = -imv\gamma_s \qquad (15\text{-A.4})$$
$$p'^y = 0$$
$$p'^z = 0$$

where $\gamma_s = (v^2 - 1)^{-\frac{1}{2}}$ and after applying $\Pi_L(\mathbf{u})$ we obtain the physical 4-momentum

$$p''^0 = m\gamma_s$$
$$p''^x = mv\gamma_s \qquad (15\text{-A.5})$$
$$p''^y = 0$$
$$p''^z = 0$$

which is the 4-momentum measured by an observer in the primed reference frame.

One immediate reason for stating the forms of the 4-momenta above is to specify the mass shell conditions.

Unprimed Reference Frame

$$p^{0\,2} - \mathbf{p}^2 = m^2 \qquad (15\text{-A.6})$$

Primed Reference Frame

$$p'^{0\,2} - \mathbf{p}'^2 = m^2 \qquad (15\text{-A.7})$$

which follows from Lorentz invariance. The *physical* 4-momentum yields the mass shell condition (resulting from applying $\Pi_L(\mathbf{u})$)

$$p''^{0\,2} - \mathbf{p}''^2 = -m^2 \qquad (15\text{-A.8})$$

indicating the particle is a tachyon from the viewpoint of an observer in the primed reference frame.

In the general case of a *physical* superluminal transformation of the form of eq. 15-A.2 from a rest frame we find the resulting 4-momentum p''^{μ} of a particle is real and satisfies eq. 15-A.8. A tachyon looks like a "normal" sublight particle in its own reference frame or a frame related to it by an ordinary subluminal Lorentz transformation.

Appendix 15-B. Phenomena Beyond the Light Barrier

15-B.1 Superluminal (Faster-than-Light) Transformations

In this Appendix we will briefly survey some of the very different features of faster-than light physical phenomena. We will frame our discussion in terms of the two simple reference frames depicted in Fig. 15-B.1. The primed frame is moving at a speed $v > c$ (the speed of light) in the positive x direction with respect to the unprimed reference frame.

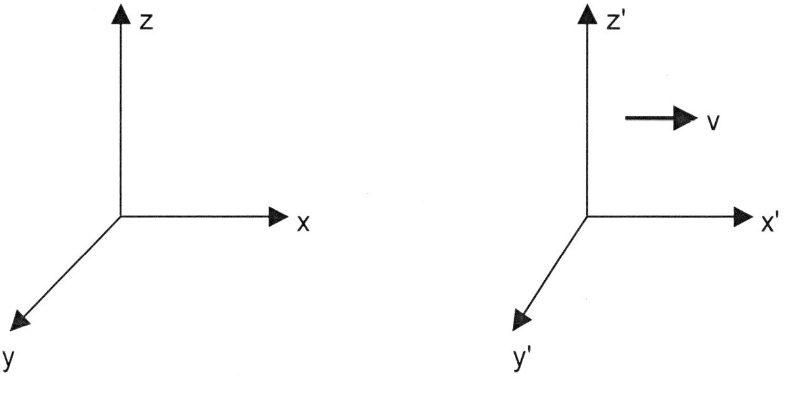

Figure 15-B.1. Two coordinate systems having a relative speed v in the x direction.

As previously in eq. 15.25 we define a physical superluminal (faster-than-light) transformation between coordinates in these reference frames with

$$\begin{aligned} t' &= \gamma_s(t - \beta x/c) \\ x' &= \gamma_s(x - \beta ct) \\ y' &= y \\ z' &= z \end{aligned} \tag{15.89}$$

where

$$\gamma_s = (\beta^2 - 1)^{-1/2} \tag{15.21}$$

and $\beta = v/c > 1$.

The energy and momentum of a tachyon (faster-than-light) particle of mass m traveling at a speed $v > c$ is

$$E = \gamma_s mc^2 \quad (15\text{-B.1})$$

and

$$\mathbf{p} = m\gamma_s \mathbf{v} \quad (15\text{-B.2})$$

Note that the tachyon defining condition is satisfied:

$$E^2 - c^2\mathbf{p}^2 = -m^2c^4 \quad (15\text{-B.3})$$

Also note that in the limit $\beta = v/c \rightarrow \infty$ that

$$E = 0 \quad (15\text{-B.4})$$

and

$$p = mc \quad (15\text{-B.5})$$

where $p = |\mathbf{p}|$. Tachyons are always in motion. The minimal momentum of a tachyon is given by eq. 15-B.5. It corresponds to zero energy. It is the tachyon equivalent of Einstein's famous $E = mc^2$.

15-B.2 Length Dilations and Time Contractions

In ordinary Lorentz transformations a moving ruler will appear to be shorter in the direction of its motion when measured in another reference frame. This phenomenon is called *Lorentz contraction.*

Superluminal Length Dilation/Contraction

In the case of a superluminal transformation we find precisely the opposite effect, *superluminal length dilation*, is a possibility. Consider the case of the transformation of eq. 15.25 above (coresponding to Fig. 15-B.1), which relates the primed reference frame traveling at speed v in the positive x direction to the unprimed reference frame. A ruler perpendicular to the x-axis will have the same length in both reference frames if its endpoints are simultaneously measured – perhaps by photographing it. The y and z equations in eqs. 15.25 specify this fact.

If the ruler is at rest in the primed reference frame and parallel to the x' axis, then a simultaneous measurement of its endpoints at the same time t_0 by an observer in the unprimed reference frame (perhaps by photographing it) will reveal both *length contraction and dilation* depending on the value of β. If the length is $L' = x'_2 - x'_1$ in the prime frame and $L = x_2 - x_1$ in the unprimed frame, then the equations:

$$x'_1 = \gamma_s(x_1 - \beta ct_0) \quad (15\text{-B.7})$$

$$x'_2 = \gamma_s(x_2 - \beta c t_0) \qquad (15\text{-B.8})$$

imply

$$L' = \gamma_s\, L = (\beta^2 - 1)^{-1/2}\, L \qquad (15\text{-B.9})$$

Thus we have three cases:

Case 1: $\beta \in <1, \sqrt{2}>$: $L < L'$ Contraction (15-B.10)

Case 2: $\beta = \sqrt{2}$: $L = L'$ Equality (15-B.11)

Case 3: $\beta \in <\sqrt{2}, \infty>$: $L > L'$ Dilation (15-B.12)

Superluminal Time Contraction/Dilation

In the case of a superluminal transformation we find *superluminal time contraction* is a possibility. Consider again the case of the transformation of eq. 15.89 above coresponding to Fig. 15-B.1 relating the primed reference frame traveling at speed v in the positive x direction to the unprimed reference frame. Consider the time interval between two events occurring at the same point x'_0 in the primed reference frame. From the viewpoint of an observer in the unprimed frame the events take place at different points x_1 and x_2. If the time interval is $T' = t'_2 - t'_1$ in the primed frame and $T = t_2 - t_1$ in the unprimed frame, then the inverse of eqs. 15.89 give:

$$t_1 = \gamma_s(t'_1 + \beta x'_0/c) \qquad (15\text{-B.13})$$
$$t_2 = \gamma_s(t'_2 + \beta x'_0/c) \qquad (15\text{-B.14})$$

and imply

$$T = \gamma_s T' = (\beta^2 - 1)^{-1/2} T' \qquad (15\text{-B.15})$$

Again we have three cases:

Case 1: $\beta \in <1, \sqrt{2}>$: $T > T'$ Dilation (15-B.16)

Case 2: $\beta = \sqrt{2}$: $T = T'$ Equality (15-B.17)

Case 3: $\beta \in <\sqrt{2}, \infty>$: $T < T'$ Contraction (15-B.18)

The time interval in the unprimed frame can be less than, equal to, or greater than the time interval in the frame where the events take place at the same spatial point.

Thus superluminal transformations are more complex than Lorentz transformations with respect to space and time, dilation and contraction.

15-B.3 Tachyon Fission to More Massive Particles – Reverse Fission

Another way in which faster-than-light phenomena differ from sublight phenomena is particle fission. Normally when a particle or nucleus decays or fissions the masses of the particles produced by the decay are smaller than the mass of the original particle or nucleus. And energy is released. We are familiar with fission as the source of nuclear energy.

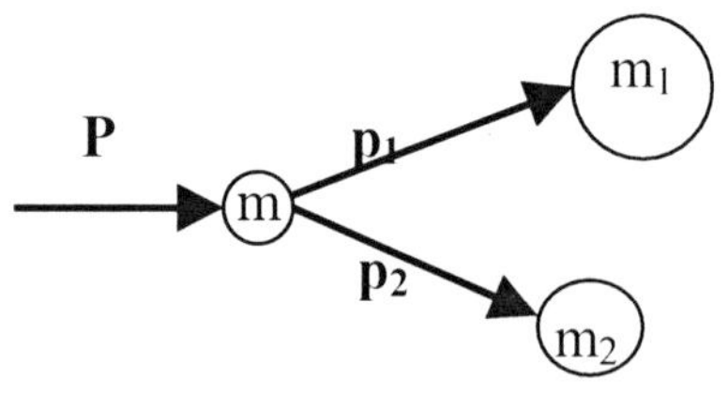

Figure 15-B.2. Two particle decay of a tachyon.

In the case of faster-than-light particles, tachyons, a much different possibility appears: a tachyon can decay into heavier tachyons: *a particle's spatial 3-momentum can be transformed into mass*. We will consider the specific case of a tachyon decaying into two particles to illustrate this possibility. (See Fig. 15-B.2.)

We will assume the initial tachyon has zero energy[243] and thus the tachyons emerging from the decay also have zero energy. The analysis is based on conservation of total energy and momentum.

Momentum conservation implies

$$\mathbf{P} = \mathbf{p}_1 + \mathbf{p}_2 \qquad (15\text{-B}.19)$$

Since all energies are zero

$$(cP)^2 = (c\mathbf{P})^2 = m^2$$
$$(cp_1)^2 = (c\mathbf{p}_1)^2 = m_1^2 \qquad (15\text{-B}.20)$$
$$(cp_2)^2 = (c\mathbf{p}_2)^2 = m_2^2$$

where $P = |\mathbf{P}|$, $p_1 = |\mathbf{p}_1|$, and $p_2 = |\mathbf{p}_2|$. If we now square eq. 15-B.19 and use eqs. 15-B.20 we obtain

$$m^2 = m_1^2 + m_2^2 + 2m_1m_2 \cos\theta \qquad (15\text{-B}.21)$$

[243] If a particle has zero energy its velocity is infinite so the case considered is somewhat artificial. However the results would still be approximately true for very large velocities. The simplicity of the kinematics leads us to consider this case.

where θ is the angle between the emerging particles' momenta $\mathbf{p}_1$ and $\mathbf{p}_2$.

Eq. 15-B.21 has a number of interesting cases:

Case θ = 0:

$$m = m_1 + m_2 \qquad (15\text{-B.22})$$

The masses of the outgoing tachyons sum to the mass of the original tachyon.

Case θ = π/2:

$$m^2 = m_1^2 + m_2^2 \qquad (15\text{-B.23})$$

The masses of each outgoing tachyon is less than the mass of the original tachyon.

Case θ = π:

$$m^2 = (m_1 - m_2)^2 \qquad (15\text{-B.24})$$

In this case either $m_1 > m$ or $m_2 > m$. Thus one of the outgoing tachyons has a greater mass than the original tachyon. Mass is effectively created from the spatial momentum of the particle. This process is the inverse of normal particle decay or fission where the sum of the outgoing masses is always less than the original particle's mass and the difference is mass converted into energy in the form of additional photons via $E = mc^2$.

This last case, where one of the outgoing particles is more massive than the original particle, is not just for θ = π. Since

$$\cos\theta = (m^2 - m_1^2 - m_2^2)/(2m_1m_2) \qquad (15\text{-B.25})$$

we see that *the sum of the outgoing tachyon masses is always greater than the original tachyon mass (except when θ = 0)* since

$$\cos\theta = 1 + [m^2 - (m_1 + m_2)^2]/(2m_1m_2) \le 1 \qquad (15\text{-B.26})$$

and thus

$$[m^2 - (m_1 + m_2)^2]/(2m_1m_2) \le 0 \qquad (15\text{-B.27})$$

Note $m = m_1 + m_2$ only if θ = 0.

Since we can transform the above discussion to the case of tachyons with a non-zero energy using an ordinary Lorentz transformation the above discussion in this subsection is general.

We therefore conclude that when a tachyon decays into two tachyons the sum of the masses of the produced tachyons is greater than the mass of the original tachyon except if the angle between the momenta of the produced tachyons is zero. In that case the sum of the masses of the produced tachyon equals the mass of the original tachyon.

Thus tachyons can engage in ***reverse fission*** *in which* ***momentum is converted into mass so the outgoing particles have a total mass greater than the incoming particle.*** In the case of "normal" fission part of the mass of a particle is converted to energy and the sum of the masses of the decay product particles is less than the mass of the original particle.

15-B.4 Light Chasing Faster-than-Light Particles?

Einstein told a story that he imagined positioning himself in a (Galilean) reference frame moving at the speed of light and seeing electromagnetic waves "frozen" in time so that they were no longer vibrating. This vision inspired him to reconsider the transformation laws between coordinate systems and to derive the theory of Special Relativity. In Special Relativity the speed of light is the same in all reference frames.

In this subsection we will consider a light pulse from the points of view of two reference frames whose relative speed v is greater than the speed of light. We will use the example considered earlier and add a pulse of light traveling in the positive x direction. (See Fig. 15-B.3.)

Eq. 15.29 contains the general law for the addition of velocities in a situation such as depicted in Fig. 15-B.3. If we adapt it to the present example and let u be the speed of the pulse in the unprimed frame (temporarily forgetting it is a light pulse) we find eq. 15.30 implies

$$u' = (u - \beta c)/(1 - \beta u/c) \qquad (15.30a)$$

where $\beta = v/c > 1$, and u' is the speed of the pulse in the prime frame. Then if we set u = c we see that u' = c as well. *Thus our superluminal transformations preserve the constancy of the speed of light just like Lorentz transformations.*

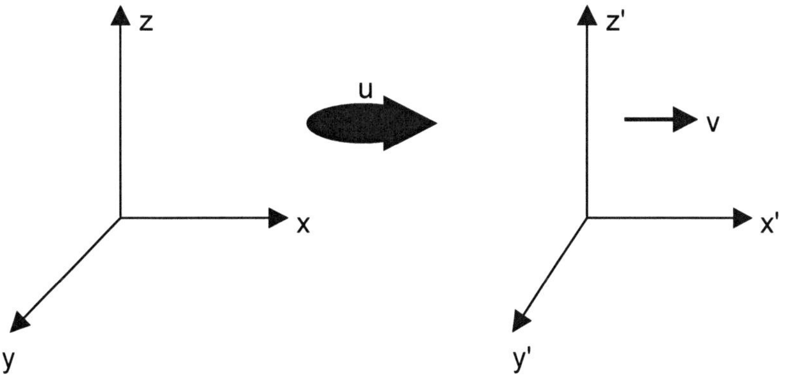

Figure 15-B.3. Two coordinate systems having a relative speed v in the x direction. A pulse of light is displayed as a thick arrow.

As a result the pulse of light will intersect the z' axis eventually. However if superluminal transformations did not preserve the speed of light in all frames the pulse might never reach the z' axis. For example under a Galilean transformation the speed of the pulse would be u' = u – v = c – v and the pulse would actually be falling further and further behind the z' axis.

15-B.5 Electromagnetic Field of a Charged Tachyon – A Pancake Effect?

The electric field of a charge q at rest in a reference frame is:

$$\mathbf{E} = (q/(4\pi\varepsilon_0)\check{\mathbf{r}}/r^2 \qquad (15\text{-B.28})$$

in spherical coordinates where $\check{\mathbf{r}}$ is a unit vector in the radial direction.

Sublight Charged Particle

The electric and magnetic fields of a charge q moving in the positive x direction with speed v < c are

$$\mathbf{E} = (q/(4\pi\varepsilon_0)\check{\mathbf{r}}(1 - \beta^2)/[r^2(1 - \beta^2\sin^2\theta)^{\frac{3}{2}}] \qquad (15\text{-B.29})$$

$$\mathbf{B} = (q/(4\pi\varepsilon_0)\check{\mathbf{r}}\beta(1 - \beta^2)\sin\theta/[r^2(1 - \beta^2\sin^2\theta)^{\frac{3}{2}}] \qquad (15\text{-B.30})$$

where $\check{\mathbf{r}}$ is the radial unit vector, β = v/c, and θ is measured with respect to the polar axis which is taken to be the x axis. As $\beta \to 1$ the electric and magnetic fields develop a "pancake" form with large field strengths in the directions perpendicular to the direction of motion similar to the transverse fields of electromagnetic quanta. This feature is the basis of the Weizsäcker-Williams method of virtual quanta.

Charged Tachyon

The electric and magnetic fields of a tachyon of charge q moving in the positive x direction with speed v > c are

$$\mathbf{E} = (q/(4\pi\varepsilon_0)\check{\mathbf{r}}(\beta^2 - 1)/[r^2(\beta^2\sin^2\theta - 1)^{\frac{3}{2}}] \qquad (15\text{-B}.31)$$

$$\mathbf{B} = (q/(4\pi\varepsilon_0)\check{\mathbf{r}}\beta(\beta^2 - 1)\sin\theta/[r^2(\beta^2\sin^2\theta - 1)^{\frac{3}{2}}] \qquad (15\text{-B}.32)$$

where $\beta = v/c > 1$, and θ is again measured with respect to the polar axis which is taken to be the x axis. In the case of tachyons there are three cases of interest.

Case $\beta^2\sin^2\theta - 1 < 0$:
The electric and magnetic fields are pure imaginary and are excluded from the forward and backward cones surrounding the x axis defined by $|\sin\theta| < \beta^{-1}$.

Case $\beta^2\sin^2\theta - 1 = 0$:
The electric and magnetic fields are "infinite." Thus the field strengths are infinite on a cone at the angle θ with respect to the x-axis. By comparison, a magnetic monopole only has a one-dimensional, singularity line extending from the monopole to infinity.

Case $\beta^2\sin^2\theta - 1 > 0$:
The electric and magnetic fields decrease in strength as $\sin^2\theta$ increases. Thus the region of maximum field strength are the forward and backward cones where $|\sin\theta|$ is greater than, but near, β^{-1} in value. The pancake picture of the sublight charged particle does not hold for charged tachyons.

Are Tachyonic Cones in the Au-Au Scattering Quark-Gluon Plasma?

Cones have been observed in high energy Au-Au scattering in which a quark-gluon plasma is created. These cones have been attributed to a variety of causes such as hydrodynamically generated Mach cones, and Cherenkov radiation. The possibility exists that tachyonic excitations may transiently exist in the quark-gluon plasma and may, in part, explain the observed cones and dips. The above described cones in the case of a moving charged tachyon are remarkably similar in character. See the CERES collaboration paper arXiv:nucl-ex/0701023, and references therein, for experimental findings.

15-B.6 Superluminal (Tachyon) Physics is Different

The simple classical examples presented in this appendix demonstrate that superluminal physics has many interesting new features that are worthy of interest. Since tachyons exist in Black Holes, and in other contexts, their study is a worthwhile endeavor.

16. GL(4)⊗GL(4) Basis of Space-Time and Symmetries

Our examination of Asynchronous Logic, and its implications for a fundamental physics theory with asynchronous time and spatial phenomena, leads us to consider two four dimensional "space-times" which have Dirac-like equations and are best initially viewed as complex space-times. Thus the group GL(4)⊗GL(4) appears to be the best initial starting point for the group structure of "space-time." The coordinates of one GL(4) become the coordinates of the space-time that we experience although the coordinates are restricted to real values except within hadrons. And the transformation group is restricted to the complex Lorentz group denoted L_C if superluminal transformations are allowed. The other GL(4), denoted GL'(4), describes another space-time that can be viewed as either inaccessible or, perhaps, as contracted to a point yet having physical effects in our space-time – notably being the source of 4 generations and also of mass mixing matrices between generations.

While it would be pleasing to embed GL(4)⊗GL(4) within GL(16) and use GL(16) as the fundamental space-time-symmetries group there does not as yet appear to an experimental ot theoretical reason to do so at current or foreseeable energies.

In this chapter we develop a provisional space-time formalism with 16 complex dimensions broken to GL(4)⊗GL'(4) that leads to a derivation of The Standard Model. This approach seems reasonable and enables us to develop a complete framework for the derivation.

However we might have started directly from GL(4)⊗GL'(4). The reader who wishes to skip to the beginning postulates for the derivation based on the broken(?) symmetry GL(16)→GL(4)⊗GL(4) can skip to section 16.3.

16.1 Complex, Sixteen Dimensional General Relativity (Optional)

The current climate of physics envisions a unified theory of the Standard Model and Quantum Gravity. Therefore it makes sense to define a General Relativistic framework for The Standard Model derivation. However since space-time is almost flat the derivation/construction of the Standard Model should be possible in flat space-time. A General Relativistic framework is not required. It will become apparent that assuming a flat space-time does not impede/prevent the derivation of Standard Model features such as Parity Violation, the (broken) symmetries of the ElectroWeak sector, and color SU(3).

In this section we will indicate aspects of a 16-dimensional, General Relativistic framework[244] for the sake of completeness before going to a flat space-time framework initially based on 4 complex dimensions in section 16.3.

Postulate 16.1. The universe has sixteen complex dimensions and is described by a curved, 16-dimensional complex analytic extension of 4-dimensional General Relativity. Complex, 16-dimensional General Relativity is invariant under complex coordinate transformations of the form:

$$X^{\mu} = f^{\mu}(X'^{0}, X'^{1}, \ldots, X'^{15}) \qquad (16.1)$$

where X'^{μ} denotes an alternate complex coordinate system that is related to X^{μ} by continuous, holomorphic functions f^{μ}.

Postulate 16.2. We define an invariant interval[245] with a metric $g_{\mu\nu}$ where

$$d\tau^2 = g_{\mu\nu}dX^{\mu}dX^{\nu} \qquad (16.2)$$

One important consequence of eq. 16.2 is that $g_{\mu\nu}$ is symmetric, $g_{\mu\nu} = g_{\nu\mu}$, in complex 16-dimensional space-time.[246] Under complex coordinate transformations (eq. 16.1) the complex interval $d\tau^2$ is invariant. The transformed vectors dX^{μ} are given by

$$dX^{\mu} = dX'^{\nu}\, \partial f^{\mu}(X'^{0}, X'^{1}, X'^{2}, X'^{3})/\partial X'^{\nu} \qquad (16.3)$$

or, more simply,

$$dX^{\mu} = dX'^{\nu}\, \partial X^{\mu}/\partial X'^{\nu} \qquad (16.4)$$

by the rules of partial differentiation. (In this book we define holomorphic in a conventional manner to mean analytic in *some* domain. Thus we will say $g_{\mu\nu}$, as we

---

[244] We should use a vierbein description but in view of the length of this book we will use the usual Einsteinian approach.

[245] Note that we use complex variable coordinates (and not their complex conjugates) throughout the development. This enables $g_{\mu\nu}$ (and other tensors such as the Riemann-Christoffel curvature tensor) to be a holomorphic (analytic) function of the coordinates – a feature that survives a change of coordinate system. Thus we obtain the benefits of analyticity. See Blaha (2004) for details of four complex dimensional General Relativity, particularly chapter 3.

[246] Another important consequence of eq. 16.2 is the matrix form of the Lorentz condition, $\Lambda_{16}{}^{T}G_{16}\Lambda_{16} = G_{16}$, uses the transpose just like the 4-dimensional Lorentz condition. In contrast the often used complex space-time interval $d\tau^2 = g_{\mu\nu}dX^{\mu}dX^{\nu *}$ does not imply $g_{\mu\nu}$ is symmetric and must use a different "Lorentz condition" $\Lambda_{16}{}^{\dagger}G_{16}\Lambda_{16} = G_{16}$ where $\Lambda_{16}{}^{\dagger}$ is the hermitian conjugate of Λ_{16}.

define it, is an holomorphic function of X^μ. Nevertheless $g_{\mu\nu}$ can have singularities, branch points, etc.) Since the components of each X^μ vector are complex variables, we can define the partial derivative of any function, or tensor, with respect to a coordinate X^ν to be:

$$\partial f(X)/\partial X^\nu = \lim_{w^\nu \to 0} [f(X^\nu + w^\nu) - f(X^\nu)]/w^\nu \tag{16.5}$$

with no implied summation over ν and with the other coordinates X^μ held fixed when $\mu \neq \nu$. The only non-zero component of the vector w^μ is the νth component w^ν which is an arbitrary complex quantity that approaches zero in the limit. From a geometric viewpoint, the point $X^\nu + w^\nu$ can approach X^ν along any arbitrary curve in the complex X^ν plane in eq. 16.5. The limit in eq. 16.5 must exist and must be independent of the path along which $X^\nu + w^\nu$ approaches X^ν. The quantity $f(X^\mu)$ is *differentiable* at the point X^μ if all first order partial derivatives with respect to all components of X^μ exist at the point X^μ.

Since the definition of a complex partial derivative is formally identical to that of a partial derivative with respect to a real variable we can make the following observation.

Observation: *All the formal rules of complex partial differentiation are the same as those of real partial differentiation. Thus the differentiation of sums, products, quotients and so on are formally the same as in the real variable case. In particular, a complex partial derivative will coincide with the corresponding real partial derivative on the "real axis" (i.e. for all X^ν real).*

Thus we are saved the labor of recalculating the equations of Riemannian geometry for the complex case although we now have to contend with sixteen dimensions.

Under complex coordinate transformations the metric tensor transforms as a covariant second rank tensor:

$$g_{\mu\nu} = g'_{\alpha\beta}\, \partial X'^\alpha/\partial X^\mu\, \partial X'^\beta/\partial X^\nu \tag{16.6}$$

$d\tau^2$ is invariant under complex coordinate transformations:

$$d\tau^2 = g_{\mu\nu} dX^\mu dX^\nu = d\tau'^2 = g'_{\mu\nu} dX'^\mu dX'^\nu \tag{16.7}$$

The inverse of $g_{\mu\nu}$, denoted $g^{\mu\nu}$, satisfies

$$g_{\alpha\nu}\, g^{\nu\beta} = \delta_\alpha^{\ \beta} \tag{16.8}$$

where δ_α^β is the Kronecker delta function ($\delta_\alpha^\beta = 1$ if $\alpha = \beta$ and is zero otherwise), and transforms as a contravariant tensor under a complex coordinate transformation:

$$g'^{\mu\nu} = g^{\alpha\beta}\,\partial X'^{\mu}/\partial X^\alpha\ \partial X'^{\nu}/\partial X^\beta \qquad (16.9)$$

The Principle of Equivalence in sixteen complex dimensions is somewhat problematic. It requires a definition of a local,[247] flat space-time metric that is the analog of the 4-dimensional real space-time metric of Einstein's General Relativity Theory. The motion of a particle, moving under the influence of gravitational forces only, must be described as straight line motion in a free falling coordinate system x^μ where

$$d^2x^\mu/d\tau^2 = 0 \qquad (16.10)$$

with

$$d\tau^2 = \eta_{16\mu\nu}dx^\mu dx^\nu \qquad (16.11)$$

It is reasonable to assume the metric tensor $\eta_{16\mu\nu}$ in rectangular coordinates consists of diagonal entries with the values ± 1. In light of the reduction of the 16-dimensional Lorentz group L_{16} eventually to the exterior product of two Lorentz groups in four dimensions $L_4 \otimes L_4$ in section 3.3, and the known flat space-time metric $\eta_{4\mu\nu}$ of our 4-dimensional space-time, we *define* the 16-dimensional flat space-time metric to be the outer product of the 4-dimensional metric $\eta_{4\mu\nu}$ with itself. (Note: $\eta_{400} = 1$; $\eta_{4ij} = -\delta_{ij}$ for i, j = 1, 2, 3; and $\eta_{4\mu\nu} = 0$ if $\mu \neq \nu$.) Thus,

Postulate 16.3. We assume a 16-dimensional Equivalence Principle: At any point one can transform to the coordinates of a locally flat space-time where

$$g'^{\mu\nu} = \eta_{16}^{\alpha\beta}\,\partial X'^{\mu}/\partial X^\alpha\ \partial X'^{\nu}/\partial X^\beta \qquad (16.9a)$$

with

$$\eta_{16\mu\nu} = diag(1, -1, -1, -1, -1, 1, 1, 1, -1, 1, 1, 1, -1, 1, 1, 1) \equiv \eta_{16}^{\mu\nu} \qquad (16.12)$$

The development of complex, 16-dimensional General Relativity proceeds from this point along the lines of the development of complex 4-dimensional General Relativity in Blaha (2004). We will not pursue it here because of the similarity and also because our goal is the derivation of the flat space-time Standard Model (for which this development is irrelevant.)

---

[247] The locality of General Relativity meshes with the locality of statements in Logic as described in section.12.7.3.

16.2 Flat Complex 16-Dimensional Space-time; GL(16), U(16), and L(16) (Optional)

The general linear group GL(16) in sixteen complex dimensions consists of all non-singular linear coordinate transformations with complex coefficients. The $\gamma\lambda$-matrix elements of the 256 generators $M_{\alpha\beta}$ of this group have the form:

$$(M_{\alpha\beta})_{\gamma\lambda} = \delta_{\alpha\gamma}\delta_{\beta\lambda} \tag{16.13}$$

where $\delta_{\alpha\beta} = 1$ if $\alpha = \beta$ and is zero otherwise, and where $\alpha, \beta, \gamma, \lambda = 0, \ldots, 15$. The generators satisfy the commutation relation

$$[M_{\alpha\beta}, M_{\kappa\lambda}] = \delta_{\beta\kappa}M_{\alpha\lambda} - \delta_{\alpha\lambda}M_{\kappa\beta} \tag{16.14}$$

GL(16) group transfomations can be expressed in the form:

$$W(a) = \exp(ia^{\alpha\beta}M_{\alpha\beta}) = \exp(i\eta_{16}{}^{\alpha\gamma}\eta_{16}{}^{\beta\delta}a_{\alpha\beta}M_{\gamma\delta}) \tag{16.15}$$

where the constants $a^{\alpha\beta}$ are complex numbers.

16.2.1 Relation of GL(16) to U(16)

The unitary group in 16 dimensions U(16) is based on the unitary condition $U^{\dagger}U = I$. Its 256 generators can be defined in terms of the GL(16) generators by the symmetric matrices:

$$U^{s}{}_{\alpha\beta} = M_{\alpha\beta} + M_{\alpha\beta} \tag{16.16a}$$

and by the anti-symmetric matrices:

$$U^{a}{}_{\alpha\beta} = i(M_{\alpha\beta} - M_{\alpha\beta}) \tag{16.16b}$$

for $\alpha \neq \beta$, and, for diagonal elements,

$$U_{\alpha\alpha} = M_{\alpha\alpha} \tag{16.17}$$

Linear combinations of the matrices, eqs. 16.16 – 16.17, with real coefficients generate all elements of the Lie algebra of U(16) while linear combinations of the matrices, eqs. 16.16 – 16.17, with *complex* coefficients generate all elements of the Lie algebra of GL(16). Thus the groups U(16) and GL(16) are intimately related.

The general form of a unitary transformation is

$$U(a) = \exp(ia^{k}U_{k}) \tag{16.18}$$

where the constants a^k are real and k is a numbering of the generators defined by eqs. 16.16 – 16.17. If the a^k are allowed to be complex, then the set of U(a) are the elements of the GL(16) group.

16.2.2 The Complex, 16-Dimensional Lorentz Group

The coordinates of inertial reference frames in sixteen flat complex dimensions are related by transformations of the complex, 16-dimensional inhomogeneous Lorentz group, denoted L(16). To determine the form of the L(16) generators we consider infinitesimal "rotations" and translations. We denote translations by d^μ and homogeneous "rotations" by $\Lambda^\alpha{}_\beta$. The identity rotation is $\Lambda_{16}{}^\alpha{}_\beta = \delta^\alpha{}_\beta$. An infinitesmal rotation can be expresssed in the form

$$\Lambda_{16}{}^\alpha{}_\beta = \delta^\alpha{}_\beta + \omega^\alpha{}_\beta \tag{16.19}$$

The defining feature of Lorentz transformations is

$$\Lambda_{16}{}^T G_{16}\, \Lambda_{16} = G_{16} \tag{16.20}$$

where Λ_{16} is a "Lorentz" transformation expressed in matrix form, $\Lambda_{16}{}^T$ is the transpose of Λ_{16}, and G_{16} is the metric $\eta_{16\mu\nu}$ expressed as a (diagonal) matrix (eq. 16.12). In the case of infinitesimal rotations eq. 16.20 becomes

$$(\delta^\alpha{}_\mu + \omega^\alpha{}_\mu)\eta_{16\alpha\beta}\,(\delta^\beta{}_\nu + \omega^\beta{}_\nu) = \eta_{16\mu\nu} \tag{16.21}$$

Using $\eta_{16\alpha\beta}$ to lower indices and $\eta_{16}{}^{\alpha\beta}$ to raise indices we have

$$\omega_{\beta\mu} = \eta_{16\alpha\beta}\omega^\alpha{}_\mu \tag{16.22}$$

so that eq. 16.20 implies the antisymmetry of $\omega_{\mu\beta}$

$$\omega_{\mu\beta} + \omega_{\beta\mu} = 0 \tag{16.23}$$

to leading order in ω.

The infinitesimal, unitary, linear Hilbert space transformation corresponding to eq. 16.19, to which we add a translation by d_μ, is

$$U(\omega, d) = I + ½\, i\, \omega_{\alpha\beta}J^{\alpha\beta} - id_\mu P^\mu \tag{16.24}$$

where the unitarity of U requires the 120 operators $J^{\alpha\beta}$ and 16 operators P^μ to be hermitian, and eq. 16.23 requires $J^{\alpha\beta}$ to be antisymmetric in α and β. After some further conventional considerations the following complex, 16-dimensional

Poincaré group generator algebra commutation relations result with the metric $\eta_{16\mu\nu}$ and its inverse $\eta_{16}^{\ \mu\nu}$.

$$[J^{\alpha\beta}, J^{\kappa\lambda}] = i(\eta_{16}^{\ \alpha\kappa} J^{\beta\lambda} + \eta_{16}^{\ \lambda\alpha} J^{\kappa\beta} - \eta_{16}^{\ \beta\kappa} J^{\alpha\lambda} - \eta_{16}^{\ \lambda\beta} J^{\kappa\alpha})$$
$$[P^{\alpha}, J^{\kappa\lambda}] = i(\eta_{16}^{\ \alpha\lambda} P^{\kappa} - \eta_{16}^{\ \alpha\kappa} P^{\lambda}) \tag{16.25}$$
$$[P^{\alpha}, P^{\kappa}] = 0$$

The momentum operators are

$$\mathbf{P} = (P^0, P^1, \ldots, P^{15}) \tag{16.26}$$

The boost 15-vector is

$$\mathbf{K} = (J^{10}, J^{20}, \ldots, J^{150}) \tag{16.27}$$

The angular momentum 15-vector is

$$\mathbf{J} = (J^{1415}, J^{151}, J^{12}, J^{23}, J^{34}, \ldots, J^{1314}) \tag{16.28}$$

The form of an inhomogeneous complex 16-dimensional Lorentz, Hilbert space transformation is

$$U(\mathbf{v}, \boldsymbol{\theta}, \mathbf{d}) = \exp[i\mathbf{a}\cdot\mathbf{K} + i\boldsymbol{\theta}\cdot\mathbf{J} - i\mathbf{d}\cdot\mathbf{P}] \tag{16.29}$$

where **a**, **θ** and **d** are complex.

We leave the development of the commutation relations of the Lorentz transformation generators as an exercise for the reader.

The remainder of the development of Special and General Relativity in 16 complex dimensions is a generalization of the development of complex, 4-dimensional General and Special Relativity of Blaha (2004) to which the reader is referred.

16.3 GL(4)⊗GL(4) Subgroup

The postulates, and discussions, of sections 16.1 and 16.2 are not necessary to our derivation of the Standard Model. However they seem to furnish a natural setting for the derivation.

The following postulate appears to be the current proper starting point for the postulates required for the derivation.

Postulate 16.4. We assume that GL(16) is broken to GL(4)⊗GL'(4) through some mechanism and this group is the basis of the form of the Standard Model.[248]

Variations[249] on this postulate are also possible that would lead to the Standard Model.

The mechanism for the breakdown to GL(4)⊗GL'(4), if such is the case, remains to be determined. We prefer to attribute the GL(4)⊗GL'(4) structure to The Standard Model's basis in 4-valued time and spatial Asynchronous Logic as described in chapter 14. Knowledge of its nature is not required for our derivation. So we leave its determination to the future.

Taking the simplest approach to the breakdown of GL(16) to GL(4)⊗GL'(4) we assume:

Postulate 16.5. After breakdown, a complex 16-dimensional vector can be represented as the outer product of two complex 4-dimensional vectors

$$x^{\mu} \rightarrow y^{i}y'^{j} \tag{16.30}$$

where μ ranges from 0 to 15, and i, j range from 0 through 3. The 4-dimensional metrics are assumed to be the outer product of the Lorentz metric

$$\eta_{16\mu\nu} \rightarrow diag(1, -1, -1, -1) \otimes diag(1, -1, -1, -1) \tag{16.31}$$

$$= (\eta_4) \otimes (\eta'_4)$$

using a matrix notation.

The generators and group elements of each GL(4) group can be expressed in the form of eqs. 16.13 – 16.16 using 4×4 matrices; and can be related to U(4) with expressions similar to eqs. 16.16 – 16.18 with 4×4 matrix rows and columns.

16.4 The Complex, 4-Dimensional Lorentz Group, L(C)

In this section we will consider a subgroup of GL(4) – the complex 4-dimensional Lorentz group that is conventionally labeled L(C). We will define a 4×4 matrix representation for GL(4)'s sixteen generators:

---

[248] The prime character will be used to distinguish the GL(4) factors. The GL(4) factor's coordinate space representation becomes the Minkowski space-time with which we are familiar. GL'(4) denotes another GL(4) group. GL'(4)'s coordinate space representation collapses to a point completely leaving only the fermion generations and mixing matrices as its residue.

[249] For example, alternate postulate: GL(16) is broken to GL(4)⊗U(4). This postulate seems less satisfactory to the author since the groups in the product are different, and the author feels that there ought to be an initial symmetry between the group leading to our space-time and the group leading to the generations.

$$(M_{\alpha\beta})_{\gamma\lambda} = \delta_{\alpha\gamma}\delta_{\beta\lambda} \tag{16.32}$$

where $\delta_{\alpha\beta} = 1$ if $\alpha = \beta$ and is zero otherwise, and where $\alpha, \beta, \gamma, \lambda = 0, \ldots, 3$. The generators satisfy the commutation relation (as before)

$$[M_{\alpha\beta}, M_{\kappa\lambda}] = \delta_{\beta\kappa}M_{\alpha\lambda} - \delta_{\alpha\lambda}M_{\kappa\beta} \tag{16.33}$$

We define the standard boost generator vector **K**'s components with

$$K^i = -i(M^{0i} + M^{i0}) \tag{16.34}$$

for i = 1, 2, 3 where $\eta_4^{\mu\nu}$ is used to raise indices in the conventional way. The rotation generator vector **J**'s components are

$$J^i = -i\epsilon^{ijk}M_{jk} \tag{16.35}$$

where i, j, k = 1,2,3.

These generators satisfy the familiar Lorentz generator commutation relations

$$\begin{aligned}
[K_i, K_j] &= -i\varepsilon_{ijk}J_k \\
[J_i, J_j] &= i\varepsilon_{ijk}J_k \\
[J_i, K_j] &= i\varepsilon_{ijk}K_k
\end{aligned} \tag{16.36}$$

The L(C) group transformations in 4-dimensional complex coordinate space are

$$\Lambda_C = \exp[i\mathbf{a}\cdot\mathbf{K} + i\boldsymbol{\theta}_c\cdot\mathbf{J}] \tag{16.37}$$

where **a** and $\boldsymbol{\theta}_c$ are complex vectors. Upon adding complex translations **d** we obtain complex Poincaré group transformations:

$$P_C = \exp[i\mathbf{a}\cdot\mathbf{K} + i\boldsymbol{\theta}_c\cdot\mathbf{J} + i\mathbf{d}\cdot\mathbf{P}] \tag{16.38}$$

The fact that L(C) group transformations have a complex parameter **a** will be crucial in the next chapter when we define the four species of fermions: charged leptons, neutral leptons, up-type quarks and down-type quarks.

16.5 New Notation for GL(4) Generators Bringing Out the Subgroups: L(C), U(3) and U(1)

It is possible to define the 16 generators of GL(4) using a notation based on Pauli matrices. This approach brings out the subgroups contained within GL(4) in a direct way. The Pauli matrices σ_i are

$$\sigma_1 = \begin{bmatrix} 0 & 1 \\ 1 & 0 \end{bmatrix} \qquad \sigma_2 = \begin{bmatrix} 0 & -i \\ -i & 0 \end{bmatrix} \qquad \sigma_3 = \begin{bmatrix} 1 & 0 \\ 0 & -1 \end{bmatrix} \tag{16.39}$$

The outer product of two Pauli matrices can be used to produce 4×4 matrices which we will express using a functional notation:

$$\sigma_i\sigma''_k = \sigma_i(\sigma''_k) \tag{16.40}$$

where the double prime on the Pauli matrix differentiates between the two Pauli matrices in the outer product. (Note the Pauli matrices are <u>not</u> matrix multiplied in eq. 16.40. Rather they represent the matrix elements $[\sigma_i]_{ab}[\sigma'_k]_{cd}$ where a, b, c, and d equal 1 or 2.) The functional form in eq. 16.40 implements the concept that σ''_k appears in each of the four matrix elements of σ_i multiplied by the value of that σ_i matrix element. For example

$$\sigma_2\sigma''_3 = \sigma_2(\sigma''_3)$$

$$= \begin{bmatrix} 0 & 0 & -I & 0 \\ 0 & 0 & 0 & i \\ i & 0 & 0 & 0 \\ 0 & -i & 0 & 0 \end{bmatrix} \tag{16.41}$$

We also define $\sigma_0 = I$, the two-dimensional identity matrix, and use the notation $\sigma_0(\sigma''_i) = \sigma_0\sigma''_i$. For example,

$$\sigma_0(\sigma''_3) = \sigma_0\sigma''_3$$

$$= \begin{bmatrix} 1 & 0 & 0 & 0 \\ 0 & -1 & 0 & 0 \\ 0 & 0 & 1 & 0 \\ 0 & 0 & 0 & -1 \end{bmatrix} \tag{16.42}$$

Similarly we define $\sigma_0(\sigma''_0) = \sigma_0\sigma''_0 = 4 \times 4$ identity matrix, and $\sigma_i(\sigma''_0) = \sigma_i\sigma''_0$.

The $\sigma_a(\sigma''_b)$ representation for a, b = 0, 1, 2, 3 can be expressed in terms of Dirac matrices as

$$\sigma_0(\sigma''_k) = \tfrac{1}{2}\varepsilon_{ijk}\sigma^{ij} = \tfrac{1}{4}\, i\varepsilon_{ijk}[\gamma^i,\gamma^j]$$
$$\sigma_1(\sigma''_k) = \gamma^0\gamma^k$$

$$\sigma_2(\sigma''_k) = -i\gamma^k$$
$$\sigma_3(\sigma''_k) = ½\varepsilon_{ijk}\gamma^0\sigma^{ij} \quad (16.43)$$
$$\sigma_1(\sigma''_0) = \gamma^5$$
$$\sigma_2(\sigma''_0) = -i\gamma^5\gamma^0$$
$$\sigma_3(\sigma''_0) = \gamma^0$$

for k = 1, 2, 3. Furthermore the $\sigma_i(\sigma''_k)$ matrices are hermitean and their square is the 4×4 identity matrix:

$$[\sigma_i(\sigma''_k)]^\dagger = \sigma_i(\sigma''_k) \quad (16.44)$$
$$[\sigma_i(\sigma''_k)]^2 = \sigma_0(\sigma''_0)$$

for i, k = 0,1,2,3.

Let us define ten symmetric matrices

$$\Omega_{+cd} = ½[\sigma_c(\sigma''_d) + \sigma_d(\sigma''_c)] \quad (16.45)$$

(with $c \leq d$) in terms of $\sigma_i(\sigma''_k)$ (eq. 16.40). It also will be important to define the six antisymmetric matrices (with $c \leq d$)

$$\Omega_{-cd} = ½[\sigma_c(\sigma''_d) - \sigma_d(\sigma''_c)] \quad (16.46)$$

We now show that linear combinations of the matrices defined by eqs. 16.45 and 16.46 form representations of the generators of the Lorentz group, the U(3) group and a U(1) group. There are sixteen matrices with six matrices serving as Lorentz group generators, one matrix generator of U(1), and nine matrices serving as U(3) generators.

16.5.1 L(C) Lorentz Group Generators

Six matrices defined by eqs. 16.45 – 16.46 can be used to define boost and rotation generator matrices of the Lorentz group. We define the Lorentz boost generators (in our notation):

$$K_i = i\Omega_{-0i} \quad (16.47)$$

and the three rotation generators

$$J_i = \Omega_{+0i} \quad (16.48)$$

(Note the hermitean conjugate relations $K_i^\dagger = -K_i$, and $J_i^\dagger = J_i$ are the same as those of the corresponding Lorentz generators in our notation.)

Their commutation relations are

$$[K_i, K_j] = -i\varepsilon_{ijk}J_k$$

$$[J_i, J_j] = i\varepsilon_{ijk}J_k \tag{16.49}$$
$$[J_i, K_j] = i\varepsilon_{ijk}K_k$$

16.5.2 U(1) Generator

The matrix $\sigma_0(\sigma'_0) = \Omega_{+00}$ serves as the generator of a U(1) symmetry.

16.5.3 U(3) Generators

The nine matrices Ω_{-ik} and Ω_{+ik} (for i, k = 1, 2, 3), together with linear combinations of Ω_{+0k} and Ω_{-0k}, form a 4×4 representation of the U(3) generators with the 4th row and 4th column consisting of 0's.

16.5.4 Gell-Mann SU(3) Generators

$$\begin{aligned}
F_1 &= \tfrac{1}{2}(\Omega_{+13} - \Omega_{-13} + \Omega_{+01} + \Omega_{-01}) \\
F_2 &= \tfrac{1}{2}(\Omega_{+23} - \Omega_{-23} + \Omega_{+02} + \Omega_{-02}) \\
F_3 &= \tfrac{1}{2}(\Omega_{+33} + \Omega_{+03} + \Omega_{-03}) \\
F_4 &= \tfrac{1}{2}(\Omega_{+10} + \Omega_{-10} + \Omega_{+13} + \Omega_{-13}) \\
F_5 &= \tfrac{1}{2}(\Omega_{+20} + \Omega_{-20} + \Omega_{+23} + \Omega_{-23}) \\
F_6 &= \tfrac{1}{2}(\Omega_{+11} + \Omega_{+22}) \\
F_7 &= \tfrac{1}{2}(\Omega_{+21} + \Omega_{-21} - \Omega_{+12} - \Omega_{-12}) \\
F_8 &= (2\sqrt{3})^{-1}(-\Omega_{+00} + 2\Omega_{+33} + \Omega_{+03} - 5\Omega_{-03})
\end{aligned} \tag{16.50}$$

They satisfy the standard SU(3) commutation relations:

$$[F_i, F_j] = i\, f_{ijk}F_k \tag{16.51}$$

where f_{ijk} are the totally antisymmetric SU(3) structure constants for i, j, k = 1, 2, ...,8. Note: the Lorentz generators do not commute with the F_i.

$$[F_i, K_j] \neq 0 \tag{16.52}$$
$$[F_i, J_j] \neq 0$$

16.5.5 Additional U(1) Generator needed for U(3)

$$F_9 = \tfrac{1}{4}(3\Omega_{+00} - \Omega_{+33} + 2\Omega_{+03}) \tag{16.53}$$

This generator consists of a 3×3 identity matrix with the element in the fourth row and column having the value 0.

We note also

$$[F_i, F_9] = 0 \tag{16.54}$$
$$[F_9, K_j] \neq 0$$
$$[F_9, J_j] \neq 0$$

for j = 1, 2, 3 and i = 1, 2, …, 8.

16.6 GL(4) Transformations have Complex Parameters

The group elements of the L(C), U(3) and U(1) algebras defined in section 16.5 are transformations with complex parameters since these are subgroups of GL(4) whose transformations have the general form of eq. 16.15 where $a^{\alpha\beta}$ are complex numbers.

Thus the L(C) subgroup of GL(4) has the form of eq. 16.37 with complex constant vectors **a** and $\boldsymbol{\theta}_c$. We now postulate

Postulate 16.6. The GL(4) group has a familiar coordinate space representation except that it is a complex space-time with a complex Lorentz group denoted L(C). The elements of L(C), Λ_C, satisfy $\Lambda_C^T G_+ \Lambda_C = G_+$. L(C) is in turn broken to the real Lorentz group L of space-time that we physically observe.

The U(3) algebra of generators defined in section 16.5 is used with nine complex parameters a_i to define elements of GL(3)

$$V_C(a) = \exp[ia_k F_k] \tag{16.55}$$

(Note we don't distinguish between a_k and a^k in this discussion.) Alternately these same generators can be used to define elements of the unitary group U(3) if the parameters a_i are real:

$$U_3(a) = \exp[ia_k F_k] \tag{16.56}$$

Similarly the U(1) generator Ω_{+00} defines the U(1) group element

$$U_1(b) = \exp[ib\Omega_{+00}] \tag{16.57}$$

if the constant b is real.

Since

$$[U_3(a), U_1(b)] = 0 \tag{16.58}$$

for all real a and b, these elements can be used to form a U(3)⊗U(1) group consisting of the elements

$$U(a,b) = U_3(a)U_1(b) \tag{16.59}$$

with the constants a_i and b real.

16.7 GL(4) and GL'(4) Combine to Yield Four Fermion Species and Four Fermion Generations

Since the elements of the L(C) algebra and the U(3)⊗U(1) algebra do not commute, and since one cannot combine internal symmetries with space-time symmetries without violating "no go' theorems, we will use L(C) boost transformations of GL(4) to generate the four fermion species: charged leptons, neutral leptons, up-type quarks and down-type quarks. We will use the the other GL(4) (denoted GL'(4)) to obtain the fermion generations. Thus we postulate the source of the fermion generations:

Postulate 16.7. The GL'(4) group has a space-time (that may be reduced to a point.) However it implies four fermion generations, (OR possibly it may have only the three known generations due to the imposition of an additional constraint(s).)

The pattern of the "breakdowns" of these GL(4) groups is indicated in Fig. 16.1.

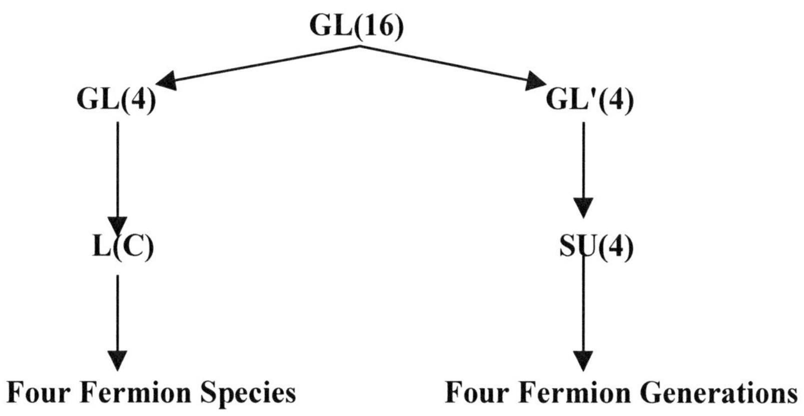

Figure 16.1. The hierarchy of GL(4)⊗GL'(4) subgroups that lead to a four generation fermion spectrum and four fermion species.

By having two GL(4) groups which separately generate the four species and the fermion generations we avoid issues associated with an approach based solely on the subgroups of one GL(4): we saw the L(C) and SU(4) (or SU(3)) subgroups of GL(4) do not commute with each other.

17. Fermion Particle Species

The derivation of the four fermion species is provided in much more detail in part 4 of this book.

17.1 Four Fermion Species

One of the important features of the Standard Model is the presence of four fermion species: charged lepton, neutral lepton, up-type quark, and down-type quark species. The distinguishing features of these fermion species is their appearance in ElectroWeak multiplets, the parity violating form of ElectroWeak interactions, and the presence of Strong interactions for quarks – the main qualitative difference between quarks and leptons.

In this chapter we will show how these features follow from L(C) and an SU(3) group which we identify with color SU(3).[250]

We identify the free Dirac equation as the free dynamical equation of charged leptons such as electrons; a free tachyon equation as the free dynamical equation of neutral leptons (neutrinos); a free Dirac equation with complex 3-momenta as the free dynamical equation of up-type quarks; and a tachyon Dirac equation with complex 3-momenta as the free dynamical equation of down-type quarks.[251] These equations will be derived from L(C) boosts applied to the Dirac equation of a particle at rest. *The key feature of these four types of boosts is that the boost from a rest state yields an energy that is either purely real or purely imaginary.*

17.2 Deriving the Free Dynamical Equations of Motion for Fermions

The starting point for our derivation of the four species' dynamical equations is the homogeneous L(C) subgroup of GL(4). Its group elements can be expressed as

$$\Lambda_C = \exp[i\mathbf{a}\cdot\mathbf{K} + i\boldsymbol{\theta}_c\cdot\mathbf{J}] \qquad (17.1)$$

$$= \exp[i(\omega_r\hat{\mathbf{u}}_r + i\omega_i\hat{\mathbf{u}}_i)\cdot\mathbf{K} + i\boldsymbol{\theta}_c\cdot\mathbf{J}] \qquad (17.2)$$

---

[250] The detailed discussion and proofs of the statements and equations in this chapter and the following chapters may be found in Blaha (2007b) and part 4 of this book.

[251] An alternate fomulation wherein quarks have real energies and 3-momenta will also be considered later.

where the vectors $\mathbf{a}$ and $\boldsymbol{\theta}_c$ are complex 3-vectors. Eq. 17.2 expresses $\mathbf{a}$ in terms of complex constants: real unit vectors with real constants $\omega_r \geq 0$ and $\omega_i \geq 0$; and with $\hat{\mathbf{u}}_r$ and $\hat{\mathbf{u}}_i$ being real normalized 3-vectors such that $\hat{\mathbf{u}}_r \cdot \hat{\mathbf{u}}_r = 1 = \hat{\mathbf{u}}_i \cdot \hat{\mathbf{u}}_i$. The generators of the homogeneous group are the same $\mathbf{K}$ and $\mathbf{J}$ as the real Lorentz group generators. Their matrix form is given by eqs. 16.34 and 16.35.

The general form of a pure L(C) boost can be expressed in terms of a complex relative velocity $\mathbf{v}_c$ between reference frames as

$$\Lambda_C(\mathbf{v}_c) \equiv \Lambda_C(\omega, \mathbf{v}_c) = \exp[i\omega\hat{\mathbf{w}}\cdot\mathbf{K}] \tag{17.3}$$

where

$$\omega = (\omega_r^2 - \omega_i^2 + 2i\omega_r\omega_i\, \hat{\mathbf{u}}_r \cdot \hat{\mathbf{u}}_i)^{1/2} \tag{17.4}$$

and

$$\hat{\mathbf{w}} = (\omega_r \hat{\mathbf{u}}_r + i\omega_i \hat{\mathbf{u}}_i)/\omega \tag{17.5}$$

Since $\hat{\mathbf{u}}_r \cdot \hat{\mathbf{u}}_r = 1 = \hat{\mathbf{u}}_i \cdot \hat{\mathbf{u}}_i$

$$\hat{\mathbf{w}}\cdot\hat{\mathbf{w}} = 1 \tag{17.6}$$

and the complex relative velocity is

$$\mathbf{v}_c = \hat{\mathbf{w}}\, \tanh(\omega) \tag{17.7}$$

The matrix form of proper ($\det \Lambda_C = 1$) L(C) coordinate boosts is

$$\Lambda_C(\mathbf{v}_c) = \begin{bmatrix} \cosh(\omega) & -\sinh(\omega)\hat{w}_x & -\sinh(\omega)\hat{w}_y & -\sinh(\omega)\hat{w}_z \\ -\sinh(\omega)\hat{w}_x & 1+(\cosh(\omega)-1)\hat{w}_x^2 & (\cosh(\omega)-1)\hat{w}_x\hat{w}_y & (\cosh(\omega)-1)\hat{w}_x\hat{w}_z \\ -\sinh(\omega)\hat{w}_y & (\cosh(\omega)-1)\hat{w}_x\hat{w}_y & 1+(\cosh(\omega)-1)\hat{w}_y^2 & (\cosh(\omega)-1)\hat{w}_y\hat{w}_z \\ -\sinh(\omega)\hat{w}_z & (\cosh(\omega)-1)\hat{w}_x\hat{w}_z & (\cosh(\omega)-1)\hat{w}_y\hat{w}_z & 1+(\cosh(\omega)-1)\hat{w}_z^2 \end{bmatrix} \tag{17.8}$$

The free dynamical equations of the four fermion species will be generated by L(C) boosts of the free Dirac equation of a particle at rest with the *requirement that the time (or energy) variable is real in the resulting field equations*. The procedure can most easily be performed in momentum space and the coordinate space version of the generated equation determined from the momentum space version. We therefore postulate:

Postulate 17.1. The dynamical equations of the four species of free fermions can be generated by L(C) boosts from the Dirac equation for a fermion at rest subject to the requirement that there be only one real time variable in each resulting dynamical differential equation.

Postulate 17.2. Each free fermion field has the simplest possible lagrangian from which the canonically conjugate momentum, the energy-momentum tensor, the equal-time (or equal light-front) anti-commutation relations, and other physically important quantities can be derived.

The consequence of Postulate 17.1 is that there are exactly four types of boosts from rest. Each type of boost corresponds to a fermion species. Details are provided below and in part 4 of the book.

17.2.1 Dirac Equation – Free Charged Lepton Dynamical Equation

In this section we describe a well-known method of obtaining the Dirac equation by conventional Lorentz boosts[252] of the spinor wave function of a particle at rest. This section amplifies the discussion of section 14.6 (and eq. 14.14 in particular) where we showed that the Operator Logic spinor formalism leads to the Dirac equation if we introduce coordinates and perform Lorentz boosts. *Thus the most fundamental aspects of Physics can be viewed as Operator Logic statements combined with space-time coordinates and Lorentz transformations. We will see that the four fermion species naturally emerge from this basis and that The Standard Model follows upon the further introduction of local[253] vector boson interactions.*

A generic positive energy plane wave solution of the Dirac equation for a particle at rest with rest energy m is

$$\psi(x) = e^{-imt}w(0) \tag{17.9}$$

with w(0) a four component spinor column vector. It satisfies the momentum space Dirac equation for a particle at rest:[254]

$$(m\gamma^0 - m)e^{-imt}w(0) = 0 \tag{17.10}$$

The 4 x 4 spinor matrix form of a Lorentz transformation with relative velocity **v** of the Dirac matrices is

$$S^{-1}(\Lambda(\mathbf{v}))\gamma^\nu S(\Lambda(\mathbf{v})) = \Lambda^\nu{}_\mu(\mathbf{v})\gamma^\mu \tag{17.11}$$

where $S(\Lambda(\mathbf{v}))$ is

$$S(\Lambda(\mathbf{v})) = \exp(-i\omega\sigma_{0i}v_i/(2|\mathbf{v}|)) = \exp(-\omega\gamma^0\boldsymbol{\gamma}\cdot\mathbf{v}/(2|\mathbf{v}|))$$

---

[252] A Lorentz boost is a special case of an L_C boost with $\omega_i = 0$ and thus a real $\omega = \omega_r$ and a real $\hat{\mathbf{w}}$.

[253] The locality of vector boson sectors meshes well with the locality of statements in Logic as described in section.12.7.3.

[254] Compare to eq. 14.14.

$$= \cosh(\omega/2)I + \sinh(\omega/2)\gamma^0\boldsymbol{\gamma}\cdot\mathbf{p}/|\mathbf{p}| \qquad (17.12)$$

with real ω = arctanh($|\mathbf{v}|$) and real $\mathbf{v}$ where $\cosh(\omega/2) = [(E+m)/(2m)]^{1/2}$ and $\sinh(\omega/2) = |\mathbf{p}|[2m(E+m)]^{-1/2}$. Also

$$S^{-1}(\Lambda(\mathbf{v})) = \gamma^0 S^\dagger(\Lambda(\mathbf{v}))\gamma^0 = \exp(\omega\gamma^0\boldsymbol{\gamma}\cdot\mathbf{v}/(2|\mathbf{v}|))$$

$$= \cosh(\omega/2)I - \sinh(\omega/2)\gamma^0\boldsymbol{\gamma}\cdot\mathbf{p}/|\mathbf{p}| \qquad (17.13)$$

If we now apply $S(\Lambda(\mathbf{v}))$ to the momentum space Dirac equation of a particle at rest (eq. 17.10) we find

$$0 = S(\Lambda(\mathbf{v}))(m\gamma^0 - m)e^{-imt}w(0)$$
$$= [mS(\Lambda(\mathbf{v}))\gamma^0 S^{-1}(\Lambda(\mathbf{v})) - m]S(\Lambda(\mathbf{v}))w(0)$$

A straightforward evaluation shows

$$mS(\Lambda(v))\gamma^0 S^{-1}(\Lambda(v)) = g_{\mu\nu}p^\mu\gamma^\nu = \not{p} \qquad (17.14)$$

where $p^0 = (p^2 + m^2)^{1/2}$, $\mathbf{p} = \gamma m\mathbf{v}$, and $p = |\mathbf{p}|$. In addition we define

$$S(\Lambda(v))w(0) = w(p) \qquad (17.15)$$

a positive energy Dirac spinor. Therefore the Dirac equation in momentum space has the familiar form:

$$(\not{p} - m)e^{-ip\cdot x}w(p) = 0 \qquad (17.16)$$

where the exponential factor, mt, is also boosted to $p\cdot x$. Eq. 17.16 implies the free, coordinate space Dirac equation:

$$(i\gamma^\mu\partial/\partial x^\mu - m)\psi(x) = 0 \qquad (17.17)$$

We identify this equation as the dynamical equation of a free charged lepton.

17.2.2 L(C) Complex Boost for a Dirac Spinor

The form of the L(C) spinor boost transformation corresponding to the coordinate transformation eq. 17.8 is:

$$S_C(\omega, \mathbf{v}_c) \equiv S_C = \exp(-i\omega\sigma_{0k}\hat{w}_k/2) = \exp(-\omega\gamma^0\boldsymbol{\gamma}\cdot\hat{\mathbf{w}}/2)$$
$$= \cosh(\omega/2)I + \sinh(\omega/2)\gamma^0\boldsymbol{\gamma}\cdot\hat{\mathbf{w}} \qquad (17.18)$$

with $\mathbf{v}_c$ and $\hat{\mathbf{w}}$ defined by eqs. 17.7 and 17.5 respectively. The inverse transformation is

$$S_C^{-1}(\omega, \mathbf{v}_c) = \gamma^2\gamma^0 K^{-1} S_C^{\dagger} K \gamma^0\gamma^2 = \gamma^2\gamma^0 S_C^{\ T}\gamma^0\gamma^2 = \exp(\omega\gamma^0\boldsymbol{\gamma}\cdot\hat{\mathbf{w}}/2)$$

$$= \cosh(\omega/2)I - \sinh(\omega/2)\gamma^0\boldsymbol{\gamma}\cdot\hat{\mathbf{w}} \qquad (17.19)$$

where the superscript T denotes the transpose and K is the complex conjugation operator (that also appears in the time-reversal operator). Note that S_C is not unitary just as $S(\Lambda(\mathbf{v}))$ of eq. 17.12 is not unitary.

We now apply an L(C) spinor boost (eq. 17.18) to the Dirac equation for a particle at rest (eqs. 14.14 - an Operator Logic statement, and 17.10) in this more general case of complex ω and $\hat{\mathbf{w}}$.

$$\begin{aligned} 0 &= S_C(\omega, \mathbf{v}_c))(m\gamma^0 - m)e^{-imt}w(0) \\ &= [mS_C\gamma^0 S_C^{-1} - m]e^{-imt}S_C w(0) \end{aligned} \qquad (17.20)$$

where $S_C = S_C(\omega, \mathbf{v}_c)$. After some algebra we find

$$mS_C\gamma^0 S_C^{-1} = m[\cosh(\omega)\gamma^0 - \sinh(\omega)\boldsymbol{\gamma}\cdot\hat{\mathbf{w}}] \qquad (17.21)$$

We will use these complex boosts to generate the other fermion species' Dirac-like equations.

17.2.3 Tachyon Dirac Equation – Free Neutral Lepton Dynamical Equation

The development of the L(C) spinor boost transformation (eqs. 17.18 – 17.21) leads to two possible forms of the tachyon Dirac equation. One form will lead to a free lagrangian theory with physical left-handed neutrinos. The other form leads to a free lagrangian theory with physical right-handed neutrinos.

17.2.3.1 Form Leading to a Left-Handed Neutrino Theory

If the real and imaginary relative vectors parts of $\hat{\mathbf{w}}$, namely $\hat{\mathbf{u}}_r$ and $\hat{\mathbf{u}}_i$, are parallel, then $\hat{\mathbf{u}}_r\cdot\hat{\mathbf{u}}_i = 1$ and

$$\omega = \omega_r + i\omega_i \qquad (17.22)$$

Eqs. 17.18 and 17.19 then imply

$$mS_C\gamma^0 S_C^{-1} = m[\cosh(\omega_r)\cos(\omega_i) + i\sinh(\omega_r)\sin(\omega_i)]\gamma^0 -$$

$$- m[\sinh(\omega_r)\cos(\omega_i) + i\cosh(\omega_r)\sin(\omega_i)]\boldsymbol{\gamma}\cdot\hat{\mathbf{u}}_r \qquad (17.23)$$

or

$$mS_C\gamma^0 S_C^{-1} = \cos(\omega_i)\gamma\cdot p_r + i\sin(\omega_i)\gamma\cdot p_i \qquad (17.24)$$

where

$$p_r^{\,0} = m\cosh(\omega_r) \qquad p_i^{\,0} = m\sinh(\omega_r) \qquad (17.25)$$

and

$$\mathbf{p}_r = m\hat{\mathbf{u}}_r\sinh(\omega_r) \qquad \mathbf{p}_i = m\hat{\mathbf{u}}_r\cosh(\omega_r) \qquad (17.26)$$

If $\omega_i = 0$, then we recover the momentum space Dirac equation eq. 17.16 (or Operator Logic statement eq. 14.14). If $\omega_i = \pi/2$, then we obtain the left-handed momentum space tachyon equation:

$$mS_C\gamma^0 S_C^{-1} = i\gamma\cdot p_i \qquad (17.27)$$

and the tachyon energy and momentum expressions

$$\mathbf{p} = m\mathbf{v}\gamma_s \qquad E = m\gamma_s \qquad (17.28)$$

where $\sinh(\omega) = \gamma_s = (\beta^2 - 1)^{-1/2}$ with $\beta = v/c > 1$. Also

$$S_C w(0) = w_C(p) \qquad (17.29)$$

is a tachyon spinor. (See Blaha (2007b).)

The momentum space tachyonic Dirac equation is

$$(i\not{p} - m)e^{ip\cdot x}w_T(p) = 0 \qquad (17.30)$$

where $p\cdot x = Et - \mathbf{p\cdot x}$ after performing a corresponding L(C) boost in the exponential factor. Thus the positive energy wave is transformed into a negative energy wave by the superluminal boost transformation.

If we apply $i\not{p}$ to we find the tachyon mass condition is satisfied

$$-E^2 + \mathbf{p}^2 = m^2 \qquad (17.31)$$

Transforming back to coordinate space we obtain the "left-handed" *tachyonic Dirac equation*:

$$(\gamma^\mu\partial/\partial x^\mu - m)\psi_T(x) = 0 \qquad (17.32)$$

17.2.3.2: Form Leading to a Right-handed Neutrino Theory

If the real and imaginary relative vectors parts of $\hat{\mathbf{w}}$, $\hat{\mathbf{u}}_r$ and $\hat{\mathbf{u}}_i$, are anti-parallel $\hat{\mathbf{u}}_r = -\hat{\mathbf{u}}_i$, then $\hat{\mathbf{u}}_r\cdot\hat{\mathbf{u}}_i = -1$ and

$$\omega = \omega_r - i\omega_i \tag{17.33}$$

Then

$$mS_C\gamma^0 S_C^{-1} = m[\cosh(\omega_r)\cos(\omega_i) - i\sinh(\omega_r)\sin(\omega_i)]\gamma^0 -$$

$$- m[\sinh(\omega_r)\cos(\omega_i) - i\cosh(\omega_r)\sin(\omega_i)]\boldsymbol{\gamma}\cdot\hat{\mathbf{u}}_r \tag{17.34}$$

or

$$mS_C\gamma^0 S_C^{-1} = \cos(\omega_i)\gamma\cdot p_r - i\sin(\omega_i)\gamma\cdot p_i \tag{17.35}$$

where

$$p_r^0 = m\cosh(\omega_r) \qquad p_i^0 = m\sinh(\omega_r) \tag{17.36}$$

and

$$\mathbf{p}_r = m\hat{\mathbf{u}}_r\sinh(\omega_r) \qquad \mathbf{p}_i = m\hat{\mathbf{u}}_r\cosh(\omega_r) \tag{17.37}$$

If $\omega_i = \pi/2$, then we obtain the right-handed momentum space tachyon equation.[255]

$$(-\gamma^\mu\partial/\partial x^\mu - m)\psi_T(x) = 0 \tag{17.38}$$

The sign difference between eqs. 17.32 and 17.38 is significant taking account of the required sign of the mass term in the lagrangian due to positivity requirements of the associated hamiltonian. It leads to the difference between a theory with physical interacting, left-handed neutrinos and a theory with physical interacting, right-handed neutrinos as we shall see.

17.2.4 Quark Dirac Equation – Free Up-type Quark Dynamical Equation

There are two other cases where we can obtain fermion dynamical equations with a *real* time variable and real energy. In one case we set $\hat{\mathbf{u}}_r\cdot\hat{\mathbf{u}}_i = 0$ and have a real ω.

If the real and imaginary relative vectors parts of $\hat{\mathbf{w}}$, namely $\hat{\mathbf{u}}_r$ and $\hat{\mathbf{u}}_i$, are perpendicular, $\hat{\mathbf{u}}_r\cdot\hat{\mathbf{u}}_i = 0$, then by eq. 17.4

$$\omega = (\omega_r^2 - \omega_i^2)^{1/2} \tag{17.39}$$

Thus ω is either pure real ($\omega_r \geq \omega_i$) or pure imaginary ($\omega_r < \omega_i$).

The momentum space equation generated by the corresponding L(C) spinor boost (eqs. 17.12 and 17.13) is

$$\{m\cosh(\omega)\gamma^0 - m\sinh(\omega)\boldsymbol{\gamma}\cdot(\omega_r\hat{\mathbf{u}}_r + i\omega_i\hat{\mathbf{u}}_i)/\omega - m\}e^{-ip\cdot x}w_c(p) = 0 \tag{17.40}$$

---

[255] We note that $\gamma_s = (\beta^2 - 1)^{-1/2}$, if expressed in terms of ω, has a branch cut extending from <-∞, +∞> in the complex ω plane. Thus values of ω with positive imaginary parts are physically different from values of ω with negative imaginary parts.

Defining the momentum 4-vector

$$p = (p^0, \mathbf{p}) \qquad (17.41)$$

where

$$p^0 = m\cosh(\omega) \qquad \mathbf{p} = \mathbf{p_r} + i\mathbf{p_i} \qquad (17.42)$$

with

$$\mathbf{p_r} = m\omega_r\hat{\mathbf{u}}_r \sinh(\omega)/\omega \qquad \mathbf{p_i} = m\omega_i\hat{\mathbf{u}}_i \sinh(\omega)/\omega \qquad (17.43)$$

$$\mathbf{p_r}\cdot\mathbf{p_i} = 0 \qquad (17.44)$$

then we obtain a positive energy Dirac-like equation

$$[p\cdot\gamma - m]e^{-ip\cdot x}w_c(p) = 0$$

or

$$[p^0\gamma^0 - (\mathbf{p_r} + i\mathbf{p_i})\cdot\gamma - m]e^{-ip\cdot x}w_c(p) = 0 \qquad (17.45)$$

with a complex 3-momentum $\mathbf{p}$ and the 4-momentum mass shell condition:

$$p^2 = p^{0\,2} - \mathbf{p_r}\cdot\mathbf{p_r} + \mathbf{p_i}\cdot\mathbf{p_i} = m^2 \qquad (17.46)$$

Note

$$|\mathbf{v_c}| = |\mathbf{p}|/p^0 = [(\mathbf{p_r} + i\mathbf{p_i})\cdot(\mathbf{p_r} + i\mathbf{p_i})]^{1/2}/p^0 = \tanh(\omega) \qquad (17.47)$$

and so the Lorentz factor is

$$\gamma = \cosh(\omega) \qquad (17.48)$$

Eq. 17.45 is the momentum space equivalent of the wave equation

$$[i\gamma^0\partial/\partial t + i\gamma\cdot(\nabla_r + i\nabla_i) - m]\psi_{Cu}(t, \mathbf{x_r}, \mathbf{x_i}) = 0 \qquad (17.49)$$

where $\mathbf{x} = \mathbf{x_r} - i\mathbf{x_i}$, and where the grad operators ∇_r and ∇_i are with respect to $\mathbf{x_r}$ and $\mathbf{x_i}$ respectively. Since $\hat{\mathbf{u}}_r\cdot\hat{\mathbf{u}}_i = 0$, which in turn implies eq. 17.44, we see that there is a subsidiary condition on the wave function

$$\nabla_r\cdot\nabla_i\,\psi_{Cu}(t, \mathbf{x_r}, \mathbf{x_i}) = 0 \qquad (17.50)$$

We will call the particles satisfying eqs. 17.49 and 17.50 *complexons*.

We note that eq. 17.49 is covariant under the real Lorentz group and eq. 17.50 can be easily put into covariant form as the difference of two 4-vectors squared (which is a real Lorentz group invariant):

$$[\gamma^0\partial/\partial t + i\gamma\cdot(\nabla_r + i\nabla_i)]^2 - [\gamma^0\partial/\partial t + i\gamma\cdot(\nabla_r - i\nabla_i)]^2 = 4\nabla_r\cdot\nabla_i.$$

We identify eq. 17.49 as the dynamical equation of a free, up-type quark.[256] Later we will show that the fourier solution of the equation has a SU(3) global symmetry.

17.2.5 Left-handed Quark Dirac Equation – Free Down-type Quark Dynamical Equation

In this case we set $\hat{\mathbf{u}}_r \cdot \hat{\mathbf{u}}_i = 0$. Then by eq. 17.4

$$\omega = (\omega_r^2 - \omega_i^2)^{1/2}$$

Thus ω again starts out either pure real (if $\omega_r \geq \omega_i$) or pure imaginary (if $\omega_r < \omega_i$). In this case we also choose ω real, and then change ω to

$$\omega = (\omega_r^2 - \omega_i^2)^{1/2} \rightarrow \omega' = (\omega_r^2 - \omega_i^2)^{1/2} + i\pi/2 = \omega + i\pi/2$$

by adding $i\pi/2$ to ω in eq. 15.65 since ω is a free parameter and proceed as we did in the prior tachyon case.[257]. The resulting Lorentz boost

$$\Lambda_C = \exp[i((\omega_r^2 - \omega_i^2)^{1/2} + i\pi/2)(\omega_r\hat{\mathbf{u}}_r + i\omega_i\hat{\mathbf{u}}_i)\cdot\mathbf{K}/\omega] \tag{17.51}$$

becomes a left-handed quark-like boost. The resulting tachyon dynamical equation is

$$[\gamma^0\partial/\partial t + \boldsymbol{\gamma}\cdot(\nabla_r + i\nabla_i) - m]\psi_{Cd}(x) = 0 \tag{17.52}$$

with the constraint equation

$$\nabla_r\cdot\nabla_i\, \psi_{Cd}(t, \mathbf{x}_r, \mathbf{x}_i) = 0 \tag{17.53}$$

We will call the particles satisfying eqs. 17.52 and 17.53 *tachyonic complexons*.

We identify eq. 17.52 as the dynamical equation of a free, down-type, left-handed quark. Later we will show that the fourier solution of this equation also has an SU(3) global symmetry.

Thus we have seen how to generate the dynamical equations of the four fermion species from eq. 14.14 – a generic Operator Logic statement – which is

---

[256] As we point out later in the chapter on experimental questions we have chosen to consider the possibility that the free dynamical equations for up-type and down-type quarks are the same as those for charged leptons and neutral leptons respectively, and the possibility that quarks are complexons. The global SU(3) symmetry implicit in complexons strongly suggested that quarks be identified with complexons.

[257] Here again the choice of ω in eq. 17.51 leads to a left-handed quark ElectroWeak sector while the choice $\omega' = \omega - i\pi/2$ leads to a right-handed quark ElectroWeak sector.

equivalent to the dynamical equation of a fermion at rest. This transition is the key to the transition from Plato's Realm of Ideas to the Realm of Reality.

17.3 Free Lagrangians of the Four Fermion Species

In defining the lagrangians for the four species that lead to their dynamical equations in the usual canonical way, we require the conventional quantum field theory feature that the hamiltonian derived from the lagrangian is hermitean. We will develop a separate lagrangian for each species and combine them later in the derivation of the form of the Standard Model. The interested readers can derive these lagrangians for themselves or see their derivation in Blaha (2007b) or part 4.

Since the Standard Model is formulated as a quantum field theory and calculations are often based on its path integral formulation it appears reasonable to make the following essentially mathematical postulates[258] (assumptions):

Postulate 17.3. The Standard Model is a quantum field theory with a lagrangian and a path integral formulation. The mathematics of these constructs is assumed to be correct.

Postulate 17.4. The free field approximation to the Standard Model is a canonical quantum field theory but not necessarily quantized on light-like surfaces.

17.3.1 Free Charged Lepton Lagrangian

The Dirac equation lagrangian is

$$\mathcal{L} = \bar{\psi}(i\gamma^\mu\partial/\partial x^\mu - m)\psi(x) \tag{17.54}$$

where

$$\bar{\psi} = \psi^\dagger\gamma^0$$

and $\psi^\dagger$ is the hermitean conjugate of ψ.

17.3.2 Free Neutral Lepton Lagrangian

$$\mathcal{L}_T = \psi_T^{\ S}(\gamma^\mu\partial/\partial x^\mu - m)\psi_T(x) \tag{17.55}$$

where

$$\psi_T^{\ S} = \psi_T^{\ \dagger}\, i\gamma^0\gamma^5 \tag{17.56}$$

The peculiar form of the tachyon lagrangian is necessitated by the hermiticity of the hamiltonian calculated from it.

---

[258] Postulates will be numbered and appear in bold type as seen above.

17.3.3 Free Up-type Quark Lagrangian

The lagrangian density of a free up-type complexon quark is

$$\mathcal{L}_{Cu} = \bar{\psi}_{Cu}(i\gamma^{\mu}D_{\mu} - m)\psi_{Cu}(x) \tag{17.57}$$

where $\bar{\psi}_{Cu} = \psi_{Cu}{}^{\dagger}\gamma^0$ and

$$\psi_{Cu}{}^{\dagger} = [\psi_{Cu}(\mathbf{x_r}, \mathbf{x_i})]^{\dagger}\,|_{\mathbf{x_i} = -\mathbf{x_i}} \tag{17.58}$$

$$D_0 = \partial/\partial x^0$$
$$D_k = \partial/\partial x_r{}^k + i\,\partial/\partial x_i{}^k \tag{17.59}$$

for k = 1, 2, 3. The action

$$I = \int d^7x \mathcal{L}_{Cu} \tag{17.60}$$

is invariant under real Lorentz transformations involving $(x^0, \mathbf{x_r})$ with $\mathbf{x_i}$ held constant. It is easy to show that this action is also real.

17.3.4 Free Down-type Quark Lagrangian

The simplest, physically acceptable, free, spin ½ tachyon lagrangian density[259] for ψ_{Cd} is:

$$\mathcal{L}_{Cd} = \psi_{Cd}{}^{C}(x)(\gamma^{0}\partial/\partial t + \boldsymbol{\gamma}\cdot(\nabla_{\mathbf{r}} + i\nabla_{\mathbf{i}}) - m)\psi_{Cd}(x) \tag{17.61}$$

where

$$\psi_{Cd}{}^{C}(x) = [\psi_{Cd}(x)]^{\dagger}|_{\mathbf{x_i} = -\mathbf{x_i}}\, i\gamma^0\gamma^5 \tag{17.62}$$

In words, eq. 17.62 states: take the hermitean conjugate of $\psi_{Cd}(x)$; change $\mathbf{x_i}$ to $-\mathbf{x_i}$; and then post-multiply by the indicated factors.

The free tachyon complexon, action

$$I = \int d^7x \mathcal{L}_{Cd} \tag{17.63}$$

is invariant under real Lorentz transformations involving $(x^0, \mathbf{x_r})$ with $\mathbf{x_i}$ held constant. And the action can be shown to be real.

---

[259] Leading to a left-handed quark ElectroWeak sector.

17.4 Canonical Commutation Relations of the Four Fermion Species

17.4.1 Free Charged Lepton Commutation Relations

The canonically conjugate momentum determined by the lagrangian density eq. 17.54 is

$$\pi_a = \partial\mathcal{L}/\partial\dot{\psi}_a \equiv \partial\mathcal{L}/\partial(\partial\psi_a/\partial t) = i\psi^\dagger_a \tag{17.64}$$

The resulting non-zero, canonical, equal time, anti-commutation relations are

$$\{\pi_a(x), \psi_b(x')\} = i\,\delta_{ab}\,\delta^3(x - x')$$

or

$$\{\psi^\dagger_a(x), \psi_b(x')\} = \delta_{ab}\,\delta^3(x - x') \tag{17.65}$$

17.4.2 Free Neutral Lepton Commutation Relations

We identify neutral leptons with tachyons. Having defined a suitable tachyon lagrangian we can now proceed to its canonical quantization. The conjugate momentum calculated from the lagrangian density eq. 17.55 is

$$\pi_{Ta} = \partial\mathcal{L}_T/\partial\dot{\psi}_{Ta} \equiv \partial\mathcal{L}_T/\partial(\partial\psi_{Ta}/\partial t) = -i(\psi_T^{\dagger}\gamma^5)_a \tag{17.66}$$

The resulting non-zero, canonical anti-commutation relations are

$$\{\pi_{Ta}(x), \psi_{Tb}(x')\} = i\,\delta_{ab}\,\delta^3(x - x')$$

or

$$\{\psi_T{}^\dagger{}_a(x), \psi_{Tb}(x')\} = -\,[\gamma^5]_{ab}\,\delta^3(x - x') \tag{17.67}$$

The presence of the γ^5 in eq. 17.67 indicates that parity differentiates between the left-handed and right-handed field anti-commutation relations. We define left-handed and right-handed fields using a transformed set of Dirac matrices:

$$\gamma^0 = \begin{bmatrix} 0 & -I \\ -I & 0 \end{bmatrix} \quad \gamma^i = \begin{bmatrix} 0 & \sigma_i \\ -\sigma_i & 0 \end{bmatrix} \quad \gamma^5 = \begin{bmatrix} I & 0 \\ 0 & -I \end{bmatrix} \tag{17.68}$$

which are obtained from the usual Dirac matrices by applying the unitary transformation $U = 2^{-1/2}(I + \gamma^5\gamma^0)$. Note I is the 4×4 identity matrix. The γ^5 chirality operator's eigenvalues define handedness: +1 corresponds to right-handed; and –1 corresponds to left-handed:

$$\gamma^5\psi_{TL} = -\psi_{TL} \qquad \gamma^5\psi_{TR} = \psi_{TR} \tag{17.69a}$$

The projection operators:

$$\begin{aligned} C^{\pm} &= \tfrac{1}{2}(I \pm \gamma^5) \\ C^{+} + C^{-} &= I \\ C^{\pm\,2} &= C^{\pm} \\ C^{+}C^{-} &= 0 \end{aligned} \tag{17.69b}$$

can be used to define left and right handed tachyon fields

$$\begin{aligned} \psi_{TL} &= C^{-}\psi_{T} \\ \psi_{TR} &= C^{+}\psi_{T} \end{aligned} \tag{17.70}$$

Eq. 17.67 implies the anti-commutation relations

$$\{\psi_{TLa}{}^{\dagger}(x), \psi_{TLb}(x')\} = C^{-}{}_{ab}\delta^3(x - x') \tag{17.71}$$

$$\{\psi_{TRa}{}^{\dagger}(x), \psi_{TRb}(x')\} = -\,C^{+}{}_{ab}\delta^3(x - x') \tag{17.72}$$

$$\{\psi_{TLa}{}^{\dagger}(x), \psi_{TRb}(x')\} = \{\psi_{TRa}{}^{\dagger}(x), \psi_{TLb}(x')\} = 0 \tag{17.73}$$

The lagrangian density of eq. 17.55 decomposes into left-handed and right-handed parts (modulo the mass term):

$$\mathcal{L}_T = \psi_{TL}{}^{\dagger}\gamma^0 i\gamma^{\mu}\partial_{\mu}\psi_{TL} - \psi_{TR}{}^{\dagger}\gamma^0 i\gamma^{\mu}\partial_{\mu}\psi_{TR} - im[\psi_{TR}{}^{\dagger}\gamma^0\psi_{TL} - \psi_{TL}{}^{\dagger}\gamma^0\psi_{TR}] \tag{17.74}$$

Noting the sign difference between the left-handed and right-handed kinetic terms in eq. 17.74, we see that the right-handed kinetic term (having the wrong sign) leads to an unphysical quantum field theory. This difference is the source of parity violation in the Standard Model, and the source of its left-handedness, as we will show in a later section.

17.4.3 Free Up-type Quark Commutation Relations

There are two choices for the up-type free quark (and corrrespondingly two choices in the interacting quark case). The first choice is to assume a free up-type quark theory similar to the theory of the free charged lepton considered earlier in this section based on the lagrangian eq. 17.54. Pursuing this approach would lead to the conventional Standard Model quark up-type sector.[260]

However it is clear there is a profound difference between quarks and leptons in nature. Quarks are always bound; leptons are not necessarily bound. Quarks experience the strong interaction; leptons do not. Consequently it is of interest to consider a second case embodied in the third type of fermion as specified in eqs. 17.57 – 17.60. In this case we shall see that the field solutions contain an embedded global SU(3) symmetry. This symmetry leads us to postulate that quarks are in the $\underline{3}$ representation of an SU(3) symmetry that we identify with color.

In this second case the conjugate momentum is

$$\pi_{\text{Cua}} = \partial\mathcal{L}/\partial\dot{\psi}_{\text{Cua}} \equiv \partial\mathcal{L}/\partial(\partial\psi_{\text{Cua}}/\partial x^0) = i\psi_{\text{Cu}\,a}^{\dagger} \tag{17.75}$$

where a is a spinor index. The non-zero anti-commutation relation, *at first glance*, appears to be

$$\{\psi_{\text{Cu}\,a}^{\dagger}(x), \psi_{\text{Cub}}(y)\} = \delta_{ab}\,\delta^3(x_r - y_r)\delta^3(x_i - y_i) \tag{17.76}$$

However the constraint eq. 17.50 *requires* the anti-commutator to be

$$\{\psi_{\text{Cu}\,a}^{\dagger}(x), \psi_{\text{Cub}}(y)\} = -\,\delta_{ab}\delta'(\nabla_r\cdot\nabla_i/m^2)[\delta^3(x_r - y_r)\delta^3(x_i - y_i)] \tag{17.77}$$

where all ∇_r and ∇_i are ∇ derivatives with respect to x, and where $\delta'(\nabla_r\cdot\nabla_i)$ is the derivative of a delta function with the argument being differential operators such as those in eq. 17.50. The minus sign is due to the presence of a *derivative* of a delta-function and is not an issue.

17.4.4 Free Down-type Quark Commutation Relations

The conjugate momentum can be calculated from the lagrangian density eq. 17.61:

$$\pi_{\text{Cda}} = \partial\mathcal{L}_{\text{Cd}}/\partial\dot{\psi}_{\text{Cda}} \equiv \partial\mathcal{L}_{\text{Cd}}/\partial(\partial\psi_{\text{Cda}}/\partial t) = -i([\psi_{\text{Cd}}(x)]^{\dagger}|_{\mathbf{x}_i = -\mathbf{x}_i}\gamma^5)_a \tag{17.78}$$

The resulting non-zero, canonical anti-commutation relations are presumably

$$\{\pi_{\text{Cda}}(x), \psi_{\text{Cdb}}(y)\} = i\,\delta_{ab}\,\delta^3(x_r - y_r)\delta^3(x_i - y_i) \tag{17.79}$$

---

[260] We consider this possibility later and call the resulting theory the Standard Standard Model.

based on locality in both the real and imaginary coordinates. However, or taking account of the constraint eq. 17.53 it must be modified to

$$\{\psi_{Cd}{}^{\dagger}{}_{a}(x)|_{\mathbf{x_i} = -\mathbf{x_i}'}\, \psi_{Cdb}(y)\} = [\gamma^5]_{ab}\delta'(\nabla_r \cdot \nabla_i/m^2)[\delta^6(x-y)] \tag{17.80}$$

As in the case of the neutral lepton commutation relations the presence of a γ^5 in eq. 17.80 differentiates between left and right-handed field anti-commutators:

$$\{\psi_{CdLa}{}^{\dagger}(x)|_{\mathbf{x_i} = -\mathbf{x_i}'}\, \psi_{CdLb}(y)\} = -C^{-}{}_{ab}\delta'(\nabla_r \cdot \nabla_i/m^2)[\delta^6(x-y)] \tag{17.81}$$

$$\{\psi_{CdRa}{}^{\dagger}(x)|_{\mathbf{x_i} = -\mathbf{x_i}'}\, \psi_{CdRb}(y)\} = C^{+}{}_{ab}\delta'(\nabla_r \cdot \nabla_i/m^2)[\delta^6(x-y)] \tag{17.82}$$

$$\{\psi_{CdLa}{}^{\dagger}(x)|_{\mathbf{x_i} = -\mathbf{x_i}'}\, \psi_{CdRb}(y)\} = \{\psi_{CdRa}{}^{\dagger}(x)|_{\mathbf{x_i} = -\mathbf{x_i}'}\, \psi_{CdLb}(x')\} = 0 \tag{17.83}$$

The left-handed fields again have a physically acceptable sign and lead to the left-handedness of the Standard Model quark sector.

17.5 Fourier Expansions of Free Fermion Fields

In this section we will simply list the fourier expansions, and some salient features, of the four species of fermion fields. The interested reader can read about the details of the fourier expansions in Blaha (2007b). We will express the fourier expansions in terms of light front variables[261] for the Dirac and tachyon species although they are required only for a correct canonical *tachyon* quantization.[262] (If equal-time canonical quantization is attempted, then non-localities,[263] and other problems, appear.) The light front variables are:

$$\begin{gathered} x^{\pm} = (x^0 \pm x^3)/\surd 2 \\ \partial/\partial x^{\pm} \equiv \partial^{\mp} \equiv (\partial/\partial x^0 \pm \partial/\partial x^3)/\surd 2 \end{gathered} \tag{17.84}$$

---

[261] L. Susskind, Phys. Rev. **165**, 1535 (1968); K. Bardakci and M. B. Halpern Phys. Rev. **176**, 1686 (1968), S. Weinberg, Phys. Rev. **150**, 1313 (1966); J. Kogut and D. Soper, Phys. Rev. **D1**, 2901 (1970); J. D. Bjorken, J. Kogut, and D. Soper, Phys. Rev. **D3**, 1382 (1971); R. A. Neville and F. Rohrlich, Nuov. Cim. **A1**, 625 (1971); F. Rohrlich, Acta Phys Austr. Suppl. **8**, 277 (1971); S-J Chang, R. Root, and T-M Yan, Phys. Rev. **D7**, 1133 (1973); S-J Chang, and T-M Yan, Phys. Rev. **D7**, 1147 (1973); T-M Yan, Phys. Rev. **D7**, 1761 (1973); T-M Yan, Phys. Rev. **D7**, 1780 (1973); C. Thorn, Phys. Rev. **D19**, 639 (1979); and references therein.

[262] This was first demonstrated in Blaha (2006) who showed tachyon fields must be separated into left-handed and right-handed parts, and then second quantized using light-front coordinates, to obtain local, equal light-front anti-commutators.

[263] See G. Feinberg, Phys. Rev. **159**, 1089 (1967) for example.

with the "transverse" coordinate variables, x^1 and x^2, unchanged.

17.5.1 Free Charged Lepton Field Fourier Expansion

The fourier expansion of the free charged lepton field is the conventional Dirac field expansion. In light-front coordinates the free, "+" wave function Fourier expansion[264] of a Dirac field is:

$$\psi^+(x) = \sum_{\pm s} \int d^2p\, dp^+ N^+(p)\theta(p^+)[b^+(p, s)u^+(p, s)e^{-ip\cdot x} + d^{+\dagger}(p, s)v^+(p, s)e^{+ip\cdot x}] \tag{17.85}$$

where

$$N^+(p) = [m/((2\pi)^3p^+)]^{1/2} \tag{17.86}$$

with $u^+(p, s)$ and $v^+(p, s)$ being projections of the conventional spinors u and v; and with the non-zero creation and annihilation operator anti-commutators:

$$\begin{aligned}\{b^+(q,s), b^{+\dagger}(p,s')\} &= \delta_{ss'}\delta^2(\mathbf{q} - \mathbf{p})\delta(q^+ - p^+)\\ \{d^+(q,s), d^{+\dagger}(p,s')\} &= \delta_{ss'}\delta^2(\mathbf{q} - \mathbf{p})\delta(q^+ - p^+)\end{aligned} \tag{17.87}$$

17.5.2 Free Neutral Lepton Field Fourier Expansion

The free neutral lepton field fourier expansion is somewhat more involved since it is a tachyon field. The free, "+" light-front, *left-handed* tachyon wave function Fourier expansion is:

$$\psi_{TL}^+(x) = \sum_{\pm s} \int d^2p\, dp^+ N_{TL}^+(p)\theta(p^+)[b_{TL}^+(p, s)u_{TL}^+(p, s)e^{-ip\cdot x} + d_{TL}^{+\dagger}(p, s)v_{TL}^+(p, s)e^{+ip\cdot x}] \tag{17.88}$$

with

$$N_{TL}^+(p) = [2m|\mathbf{p}|/((2\pi)^3(p^+(p^+ - p^-) + p_\perp^2))]^{1/2} \tag{17.89}$$

and where the non-zero anti-commutators of the Fourier coefficient operators are

$$\begin{aligned}\{b_{TL}^+(q,s), b_{TL}^{+\dagger}(p,s')\} &= \delta_{ss'}\delta^2(\mathbf{q} - \mathbf{p})\delta(q^+ - p^+)\\ \{d_{TL}^+(q,s), d_{TL}^{+\dagger}(p,s')\} &= \delta_{ss'}\delta^2(\mathbf{q} - \mathbf{p})\delta(q^+ - p^+)\end{aligned} \tag{17.90}$$

The spinors are

$$\begin{aligned}u_{TL}^+(p, s) &= C^- R^+ S_C(\omega, \mathbf{v_c})w^1(0)\\ u_{TL}^+(p, -s) &= C^- R^+ S_C(\omega, \mathbf{v_c})w^2(0)\\ v_{TL}^+(p, s) &= C^- R^+ S_C(\omega, \mathbf{v_c})w^3(0)\end{aligned}$$

---

[264] See S-J Chang and T- M. Yan Phys. Rev. D7, (1973) for a detailed presentation on light-front (infinite momentum frame) quantization of Dirac fields as well as Blaha (2007b).

$$v_{TL}^{+}(p, -s) = C^{-} R^{+} S_C(\omega, \mathbf{v_c}) w^4(0)$$

where the superscript "T" indicates the transpose, $R^{\pm} = ½(I \pm \gamma^0\gamma^3)$ are ± field light-front projection operators[265], $S_C(\omega, \mathbf{v_c})$ is an L(C) spinor boost, and the column 4-vector components $[w^i(0)]^k = \delta^{ik}$.

The case of *right-handed* free neutral lepton fields (tachyons) is similar to the left-handed case with only two differences: a minus sign in the creation and annihilation operator anti-commutation relations, and the use of right-handed projection operators. The right-handed tachyon wave function light-front Fourier expansion has the form:

$$\psi_{TR}^{+}(x) = \sum_{\pm s} \int d^2p dp^+ N_{TR}^{+}(p)\theta(p^+)[b_{TR}^{+}(p, s)u_{TR}^{+}(p, s)e^{-ip\cdot x} + d_{TR}^{+\dagger}(p, s)v_{TR}^{+}(p, s)e^{+ip\cdot x}]$$

with the non-zero anti-commutators of the Fourier coefficient operators:

$$\{b_{TR}^{+}(q,s), b_{TR}^{+\dagger}(p,s')\} = -\delta_{ss'}\delta^2(\mathbf{q} - \mathbf{p})\delta(q^+ - p^+)$$
$$\{d_{TR}^{+}(q,s), d_{TR}^{+\dagger}(p,s')\} = -\delta_{ss'}\delta^2(\mathbf{q} - \mathbf{p})\delta(q^+ - p^+)$$

The right-handed spinors are

$$u_{TR}^{+}(p, s) = C^{+}R^{+} S_C(\omega, \mathbf{v_c})w^1(0)$$
$$u_{TR}^{+}(p, -s) = C^{+}R^{+} S_C(\omega, \mathbf{v_c})w^2(0)$$
$$v_{TR}^{+}(p, s) = C^{+}R^{+} S_C(\omega, \mathbf{v_c})w^3(0)$$
$$v_{TR}^{+}(p, -s) = C^{+}R^{+} S_C(\omega, \mathbf{v_c})w^4(0)$$

17.5.3 Free Up-Type Complexon Quark Field Fourier Expansion

We believe quarks are complexons although this is an experimental question that remains to be resolved. Otherwise the quark field fourier expansions would be the same as the lepton fourier expansions described in the previous two sub-sections.

We will express the fourier expansion of the up-type quark field in conventional coordinates for the sake of illustration and because they are more familiar to most physicists. The form of the up-type complexon quark field, ignoring generations and color indices temporarily, and taking account of the subsidiary condition is[266, 267]

---

265 Blaha (2007b) p. 48.

266 Note that when $|\mathbf{p_i}| \geq |\mathbf{p_r}|$ (for imaginary $\omega = (\omega_r^2 - \omega_i^2)^{½}$) the 3-momentum becomes imaginary $\mathbf{p}\cdot\mathbf{p} < 0$. However, since we will be identifying confined quarks with this type of particle – much modified by a confining color quark interaction – the issue of an imaginary 3-momentum in the hypothetical free quark case becomes moot. We note the energy gap between positive and negative energy states disappears so E = 0 is possible. Thus real Lorentz transformations can mix positive and

$$\psi_{Cu}(x_r, x_i) = \sum_{\pm s} \int d^3p_r d^3p_i\, N_C(p)\delta(\mathbf{p}_r\cdot\mathbf{p}_i/m^2)[b_{Cu}(p,s)u_{Cu}(p, s)e^{-i(p\cdot x + p^*\cdot x^*)/2} + $$
$$+ d_{Cu}^{\dagger}(p,s)v_{Cu}(p, s)e^{+i(p\cdot x + p^*\cdot x^*)/2}] \qquad (17.91)$$

where $\mathbf{p} = \mathbf{p}_r + i\mathbf{p}_i$, $\mathbf{x} = \mathbf{x}_r - i\mathbf{x}_i$, $p\cdot x = p^0x^0 - \mathbf{p}\cdot\mathbf{x}$, and where we use

$$(p\cdot x + p^*\cdot x^*)/2 = p^0x^0 - \mathbf{p}_r\cdot\mathbf{x}_r - \mathbf{p}_i\cdot\mathbf{x}_i \qquad (17.92)$$

in the exponentials in order to avoid divergences that would appear in the calculation of the equal-time commutator, the Feynman propagator and other quantities of interest after second quantization. Note that

$$(\nabla_r + i\nabla_i)e^{-i(p\cdot x + p^*\cdot x^*)/2} = i(\mathbf{p}_r + i\mathbf{p}_i)e^{-i(p\cdot x + p^*\cdot x^*)/2} \qquad (17.93)$$

and

$$(\nabla_r + i\nabla_i)e^{-ip^*\cdot x^*} = 0 \qquad (17.94)$$

for all p.

Further we note

$$N_C(p) = [2m/((2\pi)^6p^0)]^{1/2} \qquad (17.95)$$

and

$$u_{Cu}(p, s) = S_C(\omega, \mathbf{v}_c)w^1(0)$$
$$u_{Cu}(p, -s) = S_C(\omega, \mathbf{v}_c)w^2(0)$$
$$v_{Cu}(p, s) = S_C(\omega, \mathbf{v}_c)w^3(0)$$
$$v_{Cu}(p, -s) = S_C(\omega, \mathbf{v}_c)w^4(0) \qquad (17.96)$$

The momentum 4-vector $p = (p^0, \mathbf{p})$ is related to the other quantities by

$$p^0 = m\cosh(\omega) \qquad \mathbf{p} = \mathbf{p}_r + i\mathbf{p}_i$$
$$\mathbf{p}_r = m\omega_r\hat{\mathbf{u}}_r \sinh(\omega)/\omega \qquad \mathbf{p}_i = m\omega_i\hat{\mathbf{u}}_i \sinh(\omega)/\omega$$
$$\omega = (\omega_r^2 - \omega_i^2)^{1/2}$$
$$\mathbf{p}_r\cdot\mathbf{p}_i = 0$$
$$\hat{\mathbf{w}} = (\omega_r\hat{\mathbf{u}}_r + i\omega_i\hat{\mathbf{u}}_i)/\omega$$
$$\hat{\mathbf{w}}\cdot\hat{\mathbf{w}} = 1$$
$$\mathbf{v}_c = \hat{\mathbf{w}}\tanh(\omega) \qquad (17.97)$$

The non-zero anti-commutators of the Fourier coefficient operators are

$$\{b_{Cu}(p,s), b_{Cu}^{\dagger}(p'^*,s')\} = \delta_{ss'}\delta^3(\mathbf{p}_r - \mathbf{p'}_{r'})\delta^3(\mathbf{p}_i + \mathbf{p'}_{i'})$$

---

negative energy states. The solution is to do all calculations in the light-front frame as we do for tachyons. Then the mixing issue is resolved. In the present case we second quantize on the "time-front" for illustrative purposes.

[267] We scale $\mathbf{p}_r\cdot\mathbf{p}_i$ with m^2 in the delta function for convenience. In the case of a zero mass particle some other scale could be used.

$$\{d_{Cu}(p,s),\ d_{Cu}^{\dagger}(p'^*,s')\} = \delta_{ss'}\,\delta^3(\mathbf{p_r} - \mathbf{p'}_{r'})\delta^3(\mathbf{p_i} + \mathbf{p'}_{i'}) \qquad (17.98)$$

17.5.4 Free Down-Type Complexon Quark Field Fourier Expansion

It is necessary that we perform the fourier expansion of the down-type quark field in light-front coordinates in order to obtain a local canonical quantization.

17.5.4.1 Left-Handed Down-Type Complexon Quark Field Fourier Expansion

The independent, left-handed down-type complexon quark field Fourier expansion (free, "+" light-front, left-handed tachyonic complexon), ignoring fermion generations and color indices temporarily, and taking account of the subsidiary condition, is

$$\begin{aligned}\psi_{CdL}^{+}(x_r, x_i) = \sum_{\pm s} \int d^2p_r dp^+ d^3p_i\, N_{CdL}^{+}(p)\theta(p^+)\cdot\\ \cdot\delta((p_i^{3}(p^+ - p^-)/\surd 2 + \mathbf{p_{r\perp}}\cdot\mathbf{p_{i\perp}})/m^2)\cdot\\ \cdot[b_{CdL}^{+}(p, s)u_{CdL}^{+}(p, s)e^{-i(p\cdot x\, +\, p^*\cdot x^*)/2} +\\ + d_{CdL}^{+\dagger}(p, s)v_{CdL}^{+}(p, s)e^{+i(p\cdot x\, +\, p^*\cdot x^*)/2}]\end{aligned} \qquad (17.99)$$

where

$$N_{CdL}^{+}(p) = (2\pi)^{-3}(2m/p^+)^{1/2}$$

and

$$\begin{aligned}u_{CdL}^{+}(p, s) &= C^- R^+ S_C(\omega',\mathbf{v_c})w^1(0)\\ u_{CdL}^{+}(p, -s) &= C^- R^+ S_C(\omega',\mathbf{v_c})w^2(0)\\ v_{CdL}^{+}(p, s) &= C^- R^+ S_C(\omega',\mathbf{v_c})w^3(0)\\ v_{CdL}^{+}(p, -s) &= C^- R^+ S_C(\omega',\mathbf{v_c})w^4(0)\end{aligned} \qquad (17.100)$$

where

$$\omega' = \omega + i\pi/2 \qquad (17.101)$$

The momentum 4-vector $p = (p^0, \mathbf{p})$ is related to the other quantities by eq. 17.97 with ω replaced by $\omega' = \omega + i\pi/2$ in all the relations of eq. 17.97.

The non-zero anti-commutators of the Fourier coefficient operators are:

$$\begin{aligned}\{b_{CdL}(p,s),\ b_{CdL}^{\dagger}(p'^*,s')\} &= 2^{-1/2}\,\delta_{ss'}\delta(p^+ - p'^+)\delta^2(\mathbf{p_r} - \mathbf{p'}_{r'})\delta^3(\mathbf{p_i} + \mathbf{p'}_{i'})\\ \{d_{CdL}(p,s),\ d_{CdL}^{\dagger}(p'^*,s')\} &= 2^{-1/2}\delta_{ss'}\,\delta(p^+ - p'^+)\delta^2(\mathbf{p_r} - \mathbf{p'}_{r'})\delta^3(\mathbf{p_i} + \mathbf{p'}_{i'})\end{aligned} \qquad (17.102)$$

17.5.4.2 Right-Handed Down-Type Complexon Quark Field Fourier Expansion

The independent, right-handed down-type complexon quark field Fourier expansion (free, "+" light-front, right-handed tachyonic complexon), ignoring fermion generations and color indices temporarily, and taking account of the subsidiary condition, is

$$\psi_{CdR}{}^{+}(x_r, x_i) = \sum_{\pm s} \int d^2p_r dp^+ d^3p_i\, N_{CdR}{}^{+}(p)\theta(p^+)\cdot$$
$$\cdot\delta((p_i{}^3(p^+ - p^-)/\sqrt{2} + \mathbf{p}_{r\perp}\cdot\mathbf{p}_{i\perp})/m^2)\cdot$$
$$\cdot[b_{CdR}{}^{+}(p, s)u_{CdR}{}^{+}(p, s)e^{-i(p\cdot x + p^*\cdot x^*)/2} +$$
$$+ d_{CdR}{}^{+\dagger}(p, s)v_{CdR}{}^{+}(p, s)e^{+i(p\cdot x + p^*\cdot x^*)/2}] \qquad (17.103)$$

where

$$N_{CdR}{}^{+}(p) = (2\pi)^{-3}(2m/p^+)^{1/2}$$

and

$$u_{CdR}{}^{+}(p, s) = C^+ R^+ S_C(\omega',\mathbf{v_c})w^1(0)$$
$$u_{CdR}{}^{+}(p, -s) = C^+ R^+ S_C(\omega',\mathbf{v_c})w^2(0)$$
$$v_{CdR}{}^{+}(p, s) = C^+ R^+ S_C(\omega',\mathbf{v_c})w^3(0)$$
$$v_{CdR}{}^{+}(p, -s) = C^+ R^+ S_C(\omega',\mathbf{v_c})w^4(0) \qquad (17.104)$$

where

$$\omega' = \omega + i\pi/2 \qquad (17.101)$$

The momentum 4-vector $p = (p^0, \mathbf{p})$ is related to the other quantities by eq. 17.97 with ω replaced by $\omega' = \omega + i\pi/2$ in all the relations of eq. 17.97.

The non-zero anti-commutators of the Fourier coefficient operators are:

$$\{b_{CdR}(p,s), b_{CdR}{}^{\dagger}(p'^*,s')\} = -2^{-1/2}\delta_{ss'}\delta(p^+ - p'^+)\delta^2(\mathbf{p_r} - \mathbf{p'_{r'}})\delta^3(\mathbf{p_i} + \mathbf{p'_{i'}})$$
$$\{d_{CdR}(p,s), d_{CdR}{}^{\dagger}(p'^*,s')\} = -2^{-1/2}\delta_{ss'}\delta(p^+ - p'^+)\delta^2(\mathbf{p_r} - \mathbf{p'_{r'}})\delta^3(\mathbf{p_i} + \mathbf{p'_{i'}})$$
(17.105)

17.6 A Global SU(3) Symmetry of Complexon Quarks

We will now consider a global SU(3) covariance implicit in eqs. 17.49, 17.50, 17.52, and 17.53. The defining property of the group SU(3) is that it preserves the invariance of inner products of complex 3-vectors of the form:

$$u^*\cdot v = u^{1*}v^1 + u^{2*}v^2 + u^{3*}v^3 \qquad (17.106)$$

If we examine the dynamical equation eq. 17.49 (the case of eq. 17.52 differs only in some details) we see that the differential operator is covariant under a global SU(3) transformation U of the complex spatial 3-coordinates:

$$[i\gamma^0\partial/\partial t + i\mathbf{D}_c{}^*\cdot\gamma - m] = [i\gamma^0\partial/\partial t + i\mathbf{D}_c{}'^*\cdot\gamma' - m] \tag{17.107}$$

where

$$\mathbf{D}_c{}^* = \nabla_c = \nabla_r + i\nabla_i$$

and

$$\gamma^a = U^{ab}\gamma'^b \tag{17.108a}$$
$$D_c{}^{*a} = D_c{}'^{*b}U^{ab*} \tag{17.108b}$$

where $U^\dagger = U^{-1}$. We now wish to exhibit the covariance of eq. 17.49. Since we can view the three spatial γ-matrices as SU(3) 3-vectors, we can express eq. 17.108 as the result of an SU(3) rotation V of the γ-matrices (on the spinor indices)

$$V\gamma^a V^{-1} = U^{ab}\gamma'^b \tag{17.109}$$

where V is a 4×4 reducible representation[268] of SU(3), namely, $\underline{3} + \underline{1}$. Since V commutes with γ^0 in the Pauli matrix representation of the γ matrices (eq. 17.68) we see that V can have the form

$$V = \begin{bmatrix} A\exp(i\alpha_i\sigma_i) & 0 \\ 0 & B\exp(i\beta_i\sigma_i) \end{bmatrix}$$

where A, B, α_i and β_i are constants, and the zeroes represent 2 by 2 zero matrices. The inverse of V is $V^\dagger$. Eq. 17.109 is

$$V\gamma^a V^{-1} = \begin{bmatrix} 0 & AB^*\exp(i\alpha_i\sigma_i)\sigma_a\exp(-i\beta_i\sigma_i) \\ -A^*B\exp(i\beta_i\sigma_i)\sigma_a\exp(-i\alpha_i\sigma_i) & 0(-i\beta_i\sigma_i) \end{bmatrix}$$

A $\underline{3} + \underline{1}$ reducible representation of V is described in section 17.6.1.

In a manner similar to the covariance proof of the Dirac equation[269] we see that eq. 17.49 is covariant under SU(3) transformations:

---

[268] Eqs. 16.50, and 16.53 furnish a reducible 4×4 matrix representation described later in section 17.6.1. The generators of this reducible representation are ${}_4F_i = F_i + \Omega_{+00} - F_9$ with F_i being the Gell-Mann SU(3) generators for i = 1, 2, …, 8.

[269] For example see Bjorken (1964) pp. 18 – 20.

$$V[i\gamma^0\partial/\partial t + i\gamma\cdot\mathbf{D}_c^* - m]V^{-1}V\psi_C(t, \mathbf{x}_r, \mathbf{x}_i) = 0$$

or

$$[i\gamma^{0\prime}\partial/\partial t' + i\mathbf{D}_c'^*\cdot\boldsymbol{\gamma}' - m]V\psi_C(t, \mathbf{x}_r, \mathbf{x}_i) = 0 \qquad (17.110)$$

(Note $\gamma^{0\prime} = V\gamma^0V^{-1}$ and $t' = t$.) The SU(3) transformed wave function $\psi_C'(t, \mathbf{x}')$ is

$$\psi_C'(t', \mathbf{x}') = V\psi_C(t, \mathbf{x}) = V\psi_C(t', U\mathbf{x}') \qquad (17.111)$$

Thus the complexon Dirac equation is covariant under global SU(3).

The subsidiary condition, eq. 17.50, is also covariant under an SU(3) rotation:

$$\nabla_r'^*\cdot\nabla_i'\psi_C'(t, \mathbf{x}') = \nabla_r\cdot\nabla_i\, V\psi_C(t, \mathbf{x}) = V\nabla_r^*\cdot\nabla_i\, \psi_C(t, \mathbf{x}) = 0 \qquad (17.112)$$

We now examine the transformation of the wave function eq. 17.111 under the SU(3) transformation U. If we define

$$q^{*\mu} = (q^0, \mathbf{q}^*) = (p^0, \mathbf{p}_r + i\mathbf{p}_i) = (p^0, \mathbf{p}) = p^\mu \qquad (17.113)$$

then $\psi_C(t, \mathbf{x})$ will be seen to be covariant form under a SU(3) transformation:

$$\psi_C(t, x) = \sum_{\pm s} \int d^3q_r d^3q_i\, N_C(p^0)\delta(\mathbf{q}_r^*\cdot\mathbf{q}_i/m^2)[b_C(q^*,s)u_C(q^*,s)e^{-i(q^*\cdot x + q\cdot x^*)/2} + $$
$$+ d_C^\dagger(q^*,s)v_C(q^*,s)e^{+i(q^*\cdot x + q\cdot x^*)/2}] \qquad (17.114)$$

Note both terms in each exponential are separately invariant under global SU(3). ($\mathbf{q}_r^* = \mathbf{q}_r$ since $\mathbf{q}_r$ is real.)

Eq. 17.111 implies that the spinors appearing in eq. 17.114 are covariant under SU(3) transformations

$$u_C'(q'^*,s') = Vu_C(q^*,s) \qquad (17.115)$$
$$v_C'(q'^*,s') = Vv_C(q^*,s) \qquad (17.116)$$

The fourier coefficients, if second quantized in a complex spatial coordinate generalization of the usual way, also have covariant anti-commutation relations under an SU(3) transformation:

$$\{b_C(q,s), b_C^\dagger(q'^*,s')\} = \delta_{ss'}\delta^3(q_r - q'_{r'})\delta^3(q_i - q'_{i'}) \qquad (17.117)$$

Under an SU(3) transformation, $z = Uq$ and $z' = Uq'$, the right side of eq. 17.117 transforms to

$$\delta^3(q_r - q'_{r'})\delta^3(q_i - q'_{i'}) \rightarrow \delta^3(z_r - z'_{r'})\delta^3(z_i - z'_{i'})/|\partial(q)/\partial(z)| = \delta^3(z_r - z'_{r'})\delta^3(z_i - z'_{i'}) \qquad (17.118)$$

where

$$|\partial(q)/\partial(z)| = |\partial(q_r^{\,1},q_r^{\,2},q_r^{\,3},q_i^{\,1}, q_i^{\,2}, q_i^{\,3})/\partial(z_r^{\,1},z_r^{\,2},z_r^{\,3},z_i^{\,1}, z_i^{\,2}, z_i^{\,3})| = 1 \qquad (17.119)$$

is the Jacobian of the transformation U. The fourier coefficients transform trivially under SU(3):

$$b_C(q^*,s) \rightarrow b_C(z^*,s) \qquad (17.120)$$

Since the integrand transforms as

$$\int d^3q_r d^3q_i \rightarrow \int d^3z_r d^3z_i\, |\partial(q)/\partial(z)| = \int d^3z_r d^3z_i \qquad (17.121)$$

we see that the wave function $\psi_C(t, \mathbf{x})$ transforms covariantly.

17.6.1 Global SU(3) Generators

The generators of the global SU(3) symmetry under discussion have a 4×4 matrix reducible representation ($\underline{3}$ + $\underline{1}$) based on eqs. 16.50, and 16.53. The generators of this reducible representation are

$$_4F_i = F_i + \Omega_{+00} - F_9 \qquad (17.122)$$

with F_i being the Gell-Mann SU(3) generators for i = 1, 2, …, 8. The form of these matrices is

$$_4F_i = \left[\begin{array}{ccc|c} & & & 0 \\ & F_i & & 0 \\ & & & 0 \\ \hline 0 & 0 & 0 & 1 \end{array}\right] \qquad (17.123)$$

Projection operators can be defined to project out the $\underline{3}$ and the $\underline{1}$ representations of complexon spinor fields:

$$P_1 = \Omega_{+00} - F_9$$

and

$$P_3 = F_9 \qquad (17.124)$$

where Ω_{+00} is the identity matrix. Thus the $\underline{3}$ complexon field is

$$\psi_{C3}(t, \mathbf{x}_r, \mathbf{x}_i) = P_3\psi_C(t, \mathbf{x}_r, \mathbf{x}_i) \tag{17.125}$$

while the $\underline{1}$ complexon field is

$$\psi_{C1}(t, \mathbf{x}_r, \mathbf{x}_i) = P_1\psi_C(t, \mathbf{x}_r, \mathbf{x}_i) \tag{17.126}$$

Since P_1 and P_3 do not commute with Lorentz transformations, a Lorentz transformation mixes ψ_{C1} and ψ_{C3}.[270] Since P_1 and P_3 do not commute with $\gamma 5$, left-handed and right-handed complexons would also be mixed by these projection operators.

However the appearance of an SU(3) covariance in complexon dynamics is suggestive of the color local SU(3) of the Strong interactions.

In chapter 19 we show that the global SU(3) symmetry described in this subsection can be interpreted as related to the local color SU(3) symmetry of quarks. This point gives us a strong theoretical inclination to regard color SU(3) as having a geometrical origin that is ultimately due to GL(4).[271]

---

[270] At this point it is worth noting that the construction of complexon fields, based on a boost from a particle rest state, guarantees that a reference frame exists in which any complexon particle has a single real time variable. Similarly a reference frame exists for a set of complexon particles (that is within a Lorentz of the center of momentum frame) with a single real time variable. The time variables of the individual complexon particles in the set are complex in general but are functions of the center of momentum real time variable. So there is only one real time variable for each complexon in the set although the time variable of an individual particle may be a complex function of the real center of momentum time variable.

[271] Some may feel that associating color SU(3) with this essentially GL(4) space-time SU(3) leads to issues similar to those that led to the demise of SU(6) in the 1960's. (See Coleman, S., Phys. Rev. **138,** 1262 (1965) for a discussion of these issues.) However, we evade these issues as will be seen later.

18. ElectroWeak Sector of the Standard Model

In this chapter we establish the conventional SU(2)⊗U(1) symmetry of the Standard Model with doublets of leptons and quarks. In Appendix 18-A we find a natural symmetry SU(2)⊗U(1)⊕U(1) associated with the Lorentz group when extended to include superluminal transformations – the complex Lorentz group. The extended symmetry supports triplets of leptons and quarks. The extra member of each triplet is a neutral fermion with Weak interactions only. It is natural to identify these extra fermions as leptonic and color singlet quark WIMPs (Weakly Interacting Massive Particles). WIMPs have been suggested as the constituents of Dark Matter.

As we progress in our derivation of the Standard Model we will begin to see that many Standard Model features follow from the boost properties of the L(C) group and the requirement that the free dynamical equation of each of the fermion species results from a boost from a free fermion at rest to a state with either purely real or purely imaginary energy.

Thus we will show that extended space-time geometrical considerations are directly the source of much of the form of the Standard Model. We believe that all Standard Model features (those known and those to be discovered in future experiments) will be found to originate in the geometry of extended space-time and another "internal" GL'(4) space-time.

This chapter begins the connection of the form of the Standard Model with dynamics based on the L(C) group. In this chapter we will construct a lepton doublet consisting of a Dirac field and a tachyon field, develop the dynamical equations of the doublet as well as its transformation under the L(C) and the real Lorentz groups, and then show how parity violation and the other features of the dynamical part of the ElectroWeak sector follow. Then we develop the quark ElectroWeak sector of the Standard Model for both complexon quarks and "normal" quarks.

An important part of the derivation of the form of the ElectroWeak sector is the covariance of the equations of motion under a SU(2)⊗U(1) symmetry that follows from the equivalence of 4-dimensional complex coordinates to 4-dimensional real coordinates under a SU(2)⊗U(1)⊕U(1) transformation. (See section 15.12.) *SU(2)⊗U(1)⊕U(1) covariance is demonstrated in appendix 18-A together with an extension of ElectroWeak theory to SU(2)⊗U(1)⊕U(1). The extended theory has triplets which consist of the SU(2)⊗U(1) doublets discussed in this chapter, and a U(1) singlet that we identify with WIMPs (Weakly Interacting Massive Particles). It is possible to restrict the SU(2)⊗U(1)⊕U(1) ElectroWeak theory to SU(2)⊗U(1) and recover the standard ElectroWeak Theory. This chapter describes the standard SU(2)⊗U(1) ElectroWeak Theory while appendix 18-A describes the full SU(2)⊗U(1)⊕U(1) ElectroWeak Theory.*

18.1 L(C) Transformations of Dirac and Tachyon Equations

An L(C) boost of a Dirac field (of the type discussed in chapter 17) can transform it into a spin ½ tachyon field, and vice versa:

$$S_C(\omega, \mathbf{v_c})\psi(x) \rightarrow \psi_T'(x') \qquad (18.1a)$$
$$S_C(\omega, \mathbf{v_c})\psi_T(x) \rightarrow \psi'(x')$$

if $\omega = \omega_r + i\pi/2$ and $\mathbf{v_c}$ is real. The dynamical equations' operators then transform as

$$S_C(\omega, \mathbf{v_c})(\gamma^\mu\partial/\partial x^\mu - m)S_C^{-1}(\omega, \mathbf{v_c}) = (i\gamma^\mu\partial/\partial x'^\mu - m) \qquad (18.1b)$$
$$S_C(\omega, \mathbf{v_c})\gamma^5(i\gamma^\mu\partial/\partial x^\mu - m)\gamma^5 S_C^{-1}(\omega, \mathbf{v_c}) = (\gamma^\mu\partial/\partial x'^\mu - m)$$

where

$$x'^\mu = \Lambda_C{}^\mu{}_\nu(\omega, \mathbf{v_c})x^\nu \qquad (18.1c)$$
$$\partial/\partial x'^\mu = \Lambda_C{}^\nu{}_\mu(\omega, \mathbf{v_c})\partial/\partial x^\nu$$

Eq. 18.1c can be written in matrix form as

$$X' = \Lambda_C(\mathbf{v_c})X$$

Eqs. 18.1a – 18.1c imply

$$S_C(\omega, \mathbf{v_c})(\gamma^\mu\partial/\partial x^\mu - m)\psi_T(x) = (i\gamma^\mu\partial/\partial x'^\mu - m)S_C(\omega, \mathbf{v_c})\psi_T(x)$$
$$= (i\gamma^\mu\partial/\partial x'^\mu - m)\psi'(x') \qquad (18.1d)$$

and

$$S_C(\omega, \mathbf{v_c})\gamma^5(i\gamma^\mu\partial/\partial x^\mu - m)\psi(x) = (\gamma^\mu\partial/\partial x'^\mu - m)S_C(\omega, \mathbf{v_c})\gamma^5\psi(x)$$
$$= (\gamma^\mu\partial/\partial x'^\mu - m)\psi_T'(x') \qquad (18.1e)$$

where

$$\psi'(x') = S_C(\omega, \mathbf{v_c})\psi_T(x) \qquad (18.1f)$$

and

$$\psi_T'(x') = S_C(\omega, \mathbf{v_c})\gamma^5\psi(x) \qquad (18.1g)$$

We conclude that both the Dirac equation and the tachyon equation are not covariant under an L(C) boost of the type discussed in chapter 17.

18.2 Extension of Dirac & Tachyon Equations to a Doublet Format

We will now consider the issue of generalizing the Dirac equation to include a tachyon equation in such a way that the generalized equation is covariant under both real Lorentz transformations and L(C) boost transformations.

The direct method to obtain a generalized equation that is covariant under both types of transformations is to define an 8 × 8 matrix generalization with a doublet consisting of a Dirac field and a tachyon field. Let

$$đ(x) = \begin{bmatrix} (\gamma^\mu\partial/\partial x^\mu - m) & 0 \\ 0 & (i\gamma^\mu\partial/\partial x^\mu - m) \end{bmatrix} \tag{18.2}$$

be an 8 × 8 matrix differential operator with the 4 × 4 matrix elements shown, and let

$$\Psi(x) = \begin{bmatrix} \psi_T(x) \\ \psi(x) \end{bmatrix} \tag{18.3}$$

be an 8 row field composed of a Dirac field and a tachyon field. Then the extended L(C) covariant equation is

$$đ(x)\Psi(x) = 0 \tag{18.4}$$

We now define the 8 × 8 L(C) transformation

$$S_{L8}(\Lambda_C(\omega, \mathbf{v_c})) = \begin{bmatrix} 0 & S_C(\omega, \mathbf{v_c})\gamma^5 \\ S_C(\omega, \mathbf{v_c}) & 0 \end{bmatrix} \tag{18.5}$$

with inverse transformation

$$S_{L8}^{-1}(\Lambda_C(\omega, \mathbf{v_c})) = \begin{bmatrix} 0 & S_C^{-1}(\omega, \mathbf{v_c}) \\ \gamma^5 S_C^{-1}(\omega, \mathbf{v_c}) & 0 \end{bmatrix} \tag{18.6}$$

Applying S_{L8} to eq. 18.4 yields

$$0 = S_{L8}(\Lambda_C(\omega, \mathbf{v_c}))đ(x)\Psi(x) = đ(x')\Psi'(x') \tag{18.7}$$

where

$$\Psi'(x') = \begin{bmatrix} S_C\gamma^5\psi(x) \\ \\ S_C\psi_T(x) \end{bmatrix} = \begin{bmatrix} \psi_T'(x') \\ \\ \psi'(x') \end{bmatrix} \qquad (18.8)$$

This exercise leads us to the postulate:

Postulate 18.1a. A Dirac field and a tachyon field form a doublet whose dynamical equation is covariant under the 8 ×8 L(C) transformation eq. 18.5.

Note: L(C) covariance requires the tachyon and the Dirac particle must have the same absolute value for the bare mass.

It is easy to show that the extended Dirac equation eq. 18.4 is also covariant under conventional Lorentz transformations in the 8 × 8 representation:

$$S_8(\Lambda(v)) = \begin{bmatrix} S(\Lambda(v)) & 0 \\ \\ 0 & S(\Lambda(v)) \end{bmatrix} \qquad (18.9)$$

with inverse

$$S_8^{-1}(\Lambda(v)) = \begin{bmatrix} S^{-1}(\Lambda(v)) & 0 \\ \\ 0 & S^{-1}(\Lambda(v)) \end{bmatrix} \qquad (18.10)$$

and also the alternate Lorentz transformation:

$$S_{8A}(\Lambda(v)) = \begin{bmatrix} 0 & S(\Lambda(v)) \\ \\ S(\Lambda(v)) & 0 \end{bmatrix} \qquad (18.11)$$

with inverse

$$S_{8A}^{-1}(\Lambda(v)) = \begin{bmatrix} 0 & S^{-1}(\Lambda(v)) \\ \\ S^{-1}(\Lambda(v)) & 0 \end{bmatrix} \qquad (18.12)$$

Under a conventional Lorentz transformation we find

$$0 = S_8(\Lambda(v))d\!\!\!/(x)\Psi(x) = d\!\!\!/(x')\Psi'(x') \qquad (18.13)$$
$$0 = S_{8A}(\Lambda(v))d\!\!\!/(x)\Psi(x) = d\!\!\!/(x')\Psi'(x')$$

The lagrangian density that corresponds to our 8-dimensional construction is

$$\mathcal{L}_8 = \overline{\Psi}(x)d\!\!\!/(x)\Psi(x) \qquad (18.14)$$

where

$$\overline{\Psi}(x) = \Psi^\dagger\Gamma^0 \qquad (18.15)$$

and

$$\Gamma^0 = \begin{bmatrix} i\gamma^0\gamma^5 & 0 \\ 0 & \gamma^0 \end{bmatrix} \qquad (18.16)$$

The action

$$I = \int d^4x\mathcal{L}_8 \qquad (18.17)$$

is invariant under the 8 × 8 Lorentz transformations S_8 and S_{8A}.

The Hamiltonian density for the 8 × 8 formulation is

$$\mathcal{H}_8(x) = \begin{bmatrix} i\psi_T^\dagger\gamma^5(\boldsymbol{\alpha}\cdot\nabla + \beta m)\psi_T & 0 \\ 0 & \psi^\dagger(-i\boldsymbol{\alpha}\cdot\nabla + \beta m)\psi \end{bmatrix} \qquad (18.18)$$

18.3 Non-Invariance of the Extended Equation's Action under an L(C) Transformation

The action eq. 18.17 is *not* invariant under 8 × 8 L(C) transformations. The fundamental cause of this non-invariance is the three dimensional nature of space. In the case of Dirac particles one can define a Lorentz invariant action because time is one-dimensional. Thus one can use $\psi^\dagger\gamma^0 = \bar{\psi}$ to form the Dirac field lagrangian and action. A key factor in the Lorentz invariance is the relation between the inverse and hermitean conjugate of the spinor boost operator

$$\gamma^0 S^{-1}\gamma^0 = S^\dagger \qquad (18.19)$$

In the case of the tachyon lagrangian and action, L(C) boost invariance is not possible because the tachyonic equivalent to eq. 18.19 is[272]

[272] This relation is derivable from eq. 17.19.

$$S_L^{-1}(\Lambda_C(\omega, \mathbf{v}_c))\boldsymbol{\gamma}\cdot \mathbf{v}_c/|\mathbf{v}_c| = S_L^{-1}(\Lambda_C(\omega, \mathbf{v}_c))\boldsymbol{\gamma}\cdot\mathbf{p}/|\mathbf{p}|$$
$$= i\gamma^0\, S_L^{\dagger}(\Lambda_C(\omega, \mathbf{v}_c)) \qquad (18.20)$$

where $\mathbf{p} = m\gamma_s\mathbf{v}_c$. The appearance of $\boldsymbol{\gamma}\cdot\mathbf{p}/|\mathbf{p}|$ in eq. 18.20 precludes the invariance of the free tachyon action under L(C).

We will now show the effect of an L(C) boost transformation (eqs. 18.5 and 18.6) on the lagrangian density eq. 18.14. The two non-zero parts of the lagrangian density $\mathcal{L}_8$ (eq. 18.14) are

$$\mathcal{L}_1 = \psi_T^{\dagger} i\gamma^0\gamma^5(\gamma^\mu\partial/\partial x^\mu - m)\psi_T(x) \qquad (18.21)$$
$$\mathcal{L}_2 = \psi^{\dagger}\gamma^0(i\gamma^\mu\partial/\partial x^\mu - m)\psi(x) \qquad (18.22)$$

The effect of the transformation eqs. 18.5 and 18.6 on these terms is

$$\mathcal{L}_1' = \psi_T^{\dagger} i\gamma^0\gamma^5 S_L^{-1}S_L(\gamma^\mu\partial/\partial x^\mu - m)\, S_L^{-1}S_L\psi_T(x)$$
$$= \psi_T^{\dagger} i\gamma^0\gamma^5 S_L^{-1}(i\gamma^\mu\partial/\partial x'^\mu - m)S_L\psi_T(x)$$
$$= -\psi_T^{\dagger} S_L^{\dagger}\gamma^5(\boldsymbol{\gamma}\cdot\mathbf{p}/|\mathbf{p}|)(i\gamma^\mu\partial/\partial x'^\mu - m)S_L\psi_T(x)$$
$$= \psi'^{\dagger}(x')(\boldsymbol{\gamma}\cdot\mathbf{p}/|\mathbf{p}|)\gamma^5(i\gamma^\mu\partial/\partial x'^\mu - m)\psi'(x') \qquad (18.23)$$

using eqs. 18.20, 18.1f and 18.1g; and

$$\mathcal{L}_2' = \psi^{\dagger}\gamma^0\gamma^5 S_L^{-1}S_L\gamma^5(i\gamma^\mu\partial/\partial x^\mu - m)\gamma^5 S_L^{-1}S_L\gamma^5\psi(x)$$
$$= \psi^{\dagger}\gamma^0\gamma^5 S_L^{-1}(\gamma^\mu\partial/\partial x'^\mu - m)S_L\gamma^5\psi(x)$$
$$= i\psi^{\dagger}\gamma^5 S_L^{\dagger}(\boldsymbol{\gamma}\cdot\mathbf{p}/|\mathbf{p}|)(\gamma^\mu\partial/\partial x'^\mu - m)S_L\gamma^5\psi(x)$$
$$= i\psi_T'^{\dagger}(x')(\boldsymbol{\gamma}\cdot\mathbf{p}/|\mathbf{p}|)(\gamma^\mu\partial/\partial x'^\mu - m)\psi_T'(x') \qquad (18.24)$$

where $\psi'(x')$ is a solution of the Dirac equation obtained by a superluminal L(C) boost (by $\mathbf{v}_c = \mathbf{p}/(\gamma_s m)$) of the tachyon field and where $\psi_T'(x')$ is a solution of the tachyon equation obtained by the same boost of the Dirac field. Eqs. 18.23 and 18.24 clearly show that $\mathcal{L}_8$ is *not* invariant under L(C) transformations in general.

Consequently the action of eq. 18.17 is only invariant under inhomogeneous Lorentz transformations. *This state of affairs is actually an advantage when we derive features of The Standard Model because it will be seen to prevent any interplay between unbroken ElectroWeak SU(2)⊗U(1) rotations and L(C) transformations.*

Postulate 18.2a. A lepton doublet in any generation consists of a charged Dirac field and a neutral tachyon field (a neutrino). By convention the neutrino field is *the topmost component in a lepton doublet. The equations of motion of the lepton doublet are eqs. 18.4 with $\Psi(x) = \Psi_{\ell e}(x)$ (eq. 18.25 below).*

Postulate 18.3a. A quark doublet in any generation consists of an up-type quark Dirac field and a down-type quark tachyon field. By convention the up-type quark field is the topmost component in a quark doublet. (The quark fields may be "normal" or complexon – an experimental issue.)

18.3.1 The Lepton Doublet

Thus the electron generation doublet is:

$$\Psi_{\ell e}(x) = \begin{bmatrix} \psi_{T\nu} \\ \psi_e \end{bmatrix} = \begin{bmatrix} \nu_e \\ e \end{bmatrix} \tag{18.25}$$

With this interpretation we can introduce SU(2)⊗U(1) gauge interactions and develop leptonic ElectroWeak theory naturally. SU(2)⊗U(1) transformations, and covariance, are a consequence of a set of transformations and an associated symmetry SU(2)⊗U(1)⊕U(1) originating in a consideration of superluminal transformations. (See section 15.12 for a description of the coordinate transformations associated with SU(2)⊗U(1)⊕U(1) and appendix 18-A for a discussion of the consequent SU(2)⊗U(1)⊕U(1) symmetry.) The SU(2)⊗U(1) symmetry of Weinberg-Salam ElectroWeak Theory is a subgroup of SU(2)⊗U(1)⊕U(1) – the full SU(2)⊗U(1)⊕U(1) theory includes WIMPs and uses triplets consisting of SU(2)⊗U(1) doublets and a WIMP.

We have identified two of the four types of spin ½ fermions as leptons. The remaining two types of spin ½ fermions – complexons – ψ_C and ψ_{CT} seem to naturally correspond with quarks since their equations of motion and wave functions have a natural global SU(3) symmetry.[273]

We therefore associate a color SU(3) symmetry with these two types of spin ½ complexons. The Electroweak doublet of first generation quarks is then

$$\Psi_{q1}{}^{a}(x) = \begin{bmatrix} \psi_C{}^{a} \\ \psi_{CT}{}^{a} \end{bmatrix} = \begin{bmatrix} u^a \\ d^a \end{bmatrix} \tag{18.26}$$

---

[273] The global SU(3) symmetry of complexons makes their identification with quarks reasonable. However, the complexon theory that we develop and use for quark dynamics in the Standard Model is <u>not</u> required. So we will consider both cases: non-complexon quarks and complexon quarks.

where u is the up quark and d is the down quark[274] and a is a color index with a = 1, 2, or 3.

The rationale for constructing quark doublets is the same as the leptonic case: We wish to define a generalization of the "Dirac-like" equations of motion that is covariant under general L(C) boosts.

18.4 L(C) Covariance of Complexon (Quark) Doublet Dynamical Equations

An L(C) spinor boost of a complexon can change it into a tachyonic complexon and vice versa.[275] We will denote an L(C) spinor boost that generates an ordinary complexon (from a free quark at rest) as S_{CLu} (eqs. 17.39-17.50) and an L(C) spinor boost that generates a tachyon complexon (from a free quark at rest) as S_{CLd} (eqs. 17.51-17.53).[276] We will define the up-type complexon quark field as $\psi_{Cu}(x)$ and the down-type (tachyon) quark field as $\psi_{Cd}(x)$. S_{CLd} is a superluminal boost that transforms a Dirac-like quark into a tachyon quark and vice versa:

$$S_{CLd}\psi_{Cu}(x) \rightarrow \psi_{Cd}'(x') \qquad (18.27)$$
$$S_{CLd}\psi_{Cd}(x) \rightarrow \psi_{Cu}'(x')$$

Similarly the differential operator used in the equations of motion can also be transformed.

$$S_{CLd}(\gamma^{\mu}D_{\mu} - m)S_{CLd}^{-1} = (i\gamma^{\mu}D'_{\mu} - m)$$
$$S_{CLd}\gamma^5(i\gamma^{\mu}D_{\mu} - m)\gamma^5 S_{CLd}^{-1} = (\gamma^{\mu}D'_{\mu} - m) \qquad (18.28)$$

where D_{μ} is given by eq. 17.59 and where

$$x'^{\mu} = \Lambda_{CLd}{}^{\mu}{}_{\nu}(\omega, \mathbf{v_c})x^{\nu} \qquad (18.29)$$

---

[274] The lepton situation is clear in the sense that charged leptons cannot be tachyons since their masses are known. Thus tachyonic neutrinos are the only currently allowed possibility. The quark situation is somewhat unclear. We have provisionally chosen the "down" type of quark (d, s, and b) as tachyonic. The association of bound states of these quarks such as the K^0 and B^0 systems which are known to have CP violation, and the CP violation engendered by tachyons, encourages this interpretation.

In addition, $W^{\pm}$ charge asymmetry in $p\bar{p}$ collisions indicate the d sea in a proton is greater than the u sea (K. Abe et al, PRL **74**, 850 (1995)) as does the asymmetry of Drell-Yan production in deep inelastic scattering on p and n targets (A. Baldit et al, Phys. Lett. **B332**, 244 (1994)). These results are to be expected since there is no mass gap for a d tachyon sea while there is a mass gap for a u Dirac particle sea. Complexon quarks may explicate the discrepancies between theory and experiment in the spin structure functions of the parton model for nucleons.

[275] The considerations of this section (and chapter) would still apply if quarks were not complexons but rather Dirac fields and tachyon fields.

[276] The reader might think that the generation of both "normal" fermions and tachyonic fermions from a Dirac particle at rest makes them equivalent. This would be true if L_C covariance were not broken to Lorentz covariance as it is. The L_C breakdown makes "normal" fermions different from tachyonic fermions.

$$D'_\mu = \Lambda_{CLd}{}^{\nu}{}_{\mu}(\omega, \mathbf{v_c})D_\nu$$

In matrix form

$$X' = \Lambda_{CLd}(\omega, \mathbf{v_c})X \tag{18.30}$$

Eqs. 18.27 – 18.29 imply

$$\begin{aligned} S_{CLd}(\gamma^\mu D_\mu - m)\psi_{Cd}(x) &= (i\gamma^\mu D'_\mu - m)S_{CLd}\psi_{Cd}(x) \\ &= (i\gamma^\mu D'_\mu - m)\psi_{Cu}'(x') \end{aligned} \tag{18.31}$$

and

$$\begin{aligned} S_{CLd}\gamma^5(i\gamma^\mu D_\mu - m)\psi_{Cu}(x) &= (\gamma^\mu D'_\mu - m)S_{CLd}\gamma^5\psi_{Cu}(x) \\ &= (\gamma^\mu D'_\mu - m)\psi_{Cd}'(x') \end{aligned} \tag{18.32}$$

where

$$\psi_{Cu}'(x') = S_{CLd}\psi_{Cd}(x) \tag{18.33}$$

and

$$\psi_{Cd}'(x') = S_{CLd}\gamma^5\psi_{Cu}(x) \tag{18.34}$$

Thus neither free complexon dynamical equation is L(C) covariant.

18.4.1 Doublet Dynamical Equation for Complexon Quarks

We will now generalize the complexon quark dynamical equations as we did for lepton dynamical equations so that the generalized equation is covariant under both Lorentz transformations and L(C) boosts. The approach is the same except that the spatial derivatives have an imaginary part and the quark tachyon field is now the low component of the doublet.

So we define an 8 × 8 matrix generalization:

$$đ_C(x) = \begin{bmatrix} (i\gamma^\mu D_\mu - m) & 0 \\ 0 & (\gamma^\mu D_\mu - m) \end{bmatrix} \tag{18.35}$$

with D_μ given by eq. 17.59 and

$$\Psi_C(x) = \begin{bmatrix} \psi_{Cu}(x) \\ \psi_{Cd}(x) \end{bmatrix} \tag{18.36}$$

where $\Psi_C(x)$ is an 8 component field composed of a up-type quark complexon field and a down-type (tachyonic) complexon field.

Then the generalized complexon equation is

$$đ_C(x)\Psi_C(x) = 0 \tag{18.37}$$

We now define the 8 × 8 L(C) spinor boost transformation

$$S_{CL8} = \begin{bmatrix} 0 & S_{CLd} \\ S_{CLd}\gamma^5 & 0 \end{bmatrix} \tag{18.38}$$

with inverse transformation

$$S_{CL8}^{-1} = \begin{bmatrix} 0 & \gamma^5 S_{CLd}^{-1} \\ S_{CLd}^{-1} & 0 \end{bmatrix} \tag{18.39}$$

Applying S_{CL8} to eq. 18.37 yields

$$0 = S_{CL8}\not{d}_C(x)\Psi_C(x) = \not{d}_C(x')\Psi_C'(x') \tag{18.40}$$

where

$$\Psi_C'(x') = \begin{bmatrix} S_{CL8}\psi_{Cd}(x) \\ S_{CL8}\gamma^5\psi_{Cu}(x) \end{bmatrix} = \begin{bmatrix} \psi_{Cu}'(x') \\ \psi_{Cd}'(x') \end{bmatrix} \tag{18.41}$$

Thus the generalized complexon equation is covariant under L(C) boosts. Covariance requires the complexon, and tachyonic complexon, must have the same absolute value for the mass.

It is easy to show that the generalized complexon equation is also covariant under conventional real Lorentz transformations represented as 4 × 4 diagonal blocks in an 8 × 8 matrix representation. (The demonstration is analogous to eqs. 18.9 – 18.13.)

The lagrangian density that corresponds to our 8-dimensional construction is

$$\mathcal{L}_{C8} = \overline{\Psi}_C(x)\not{d}_C(x)\Psi_C(x) \tag{18.42}$$

where

$$\overline{\Psi}_C(x) = \Psi_C^{\dagger}|_{\mathbf{x_i} = -\mathbf{x_i}}\,\Gamma_C^{\;0} \tag{18.43}$$

and

$$\Gamma_C^{\ 0} = \begin{bmatrix} \gamma^0 & 0 \\ 0 & i\gamma^0\gamma^5 \end{bmatrix} \tag{18.44}$$

The action

$$I = \int d^4x \mathcal{L}_{C8} \tag{18.45}$$

is invariant under Lorentz transformations S_8 and S_{8A} (eqs. 18.9 – 18.12).

The Hamiltonian density for the 8-dimensional theory is

$$\mathcal{H}_{C8}(x) = \begin{bmatrix} \psi_{Cu}^{\dagger}(-i\boldsymbol{\alpha}\cdot\nabla_C + \beta m)\psi_{Cu} & 0 \\ 0 & i\psi_{Cd}^{\dagger}\gamma^5(\boldsymbol{\alpha}\cdot\nabla_C + \beta m)\psi_{Cd} \end{bmatrix} \tag{18.46}$$

where

$$\nabla_C = \mathbf{D} \tag{18.47}$$

is the spatial vector part of D^{μ}.

18.4.2 Non-Invariance of the Generalized Free Quark Complexon Action under an L(C) Boost

The action 18.45 is not invariant under L(C) boosts. The reason is similar to that of section 18.3 for the "leptonic" type of particle: there is no simple relation between the hermitean conjugate of an L(C) spinor boost and its inverse that is similar to eq. 18.19 for the Dirac boost case.

Consequently the action of eq. 18.45 is only invariant under inhomogeneous real Lorentz transformations. *This state of affairs is again an advantage when we derive features of the Standard Model because it prevents any interplay between unbroken ElectroWeak SU(2) rotations and* L(C) *transformations in the quark complexon sector.*

18.5 Postulates of the Derivation of the Leptonic ElectroWeak Sector

In this section we will derive the leptonic sector of The Standard Model based on the L(C) covariance of the doublet equations of motion. *The naturalness of the derivation, and its close connection to the L(C) group, strongly suggest the Standard Model, which was grown by theorists from experiment, has an undeniable truth that will likely survive the passage of time. The basis of the derivation in a more fundamental theoretical framework raises the hope that we have found a newer, deeper level of understanding of elementary particle dynamics.*

We will add certain new postulates that provide a basis for the derivation of the form of the leptonic sector of the Standard Model:

Postulate 18.4a. The form of the equations of motion of the unbroken leptonic sector of the Standard Model are determined by covariance under the L(C) group, and local gauge symmetries.[277] *Electron number is not gauged.*

Postulate 18.5a. In each generation leptonic particles consist of spin ½ Dirac particles (electron-like) with charge -1 and tachyons (neutrinos) of charge zero as well as their anti-particles.[278]

Postulate 18.6. A neutrino is a tachyon with a non-zero bare mass.

Postulate 18.7a. The interacting extensions of doublet equations of motion have interactions consisting of local SU(2)⊗U(1) Yang-Mills fields[279]*. Gauge fields are massless "before" spontaneous symmetry breaking and are thus conventional local Yang-Mills gauge fields without a tachyon equivalent.*

Postulate 18.8. L(C) group covariance, and gauge symmetries of the dynamical equations of motion, are spontaneously broken through the appearance of additional fermion and vector boson mass terms generated by a mechanism such as the Higgs mechanism or the Dimensional Mass Mechanism (sections 21.3 and 21.4 below).

18.6 Derivation of the Leptonic ElectroWeak Sector of the Standard Model

The steps of the derivation are:

A. L(C) group covariance of the free dynamical equations of motion requires that leptons be described by a generalization of the Dirac equation to an 8×8 matrix form in section 18.2 and eq. 18.25 based on a doublet consisting of a charged Dirac particle and a tachyon neutrino.

---

[277] The locality of gauge symmetries is consonant with the locality of statements in Logic as described in section.12.7.3.

[278] If neutrinos are Majorana particles then the derivation must be modified.

[279] SU(2)⊗U(1) symmetry is a subgroup of SU(2)⊗U(1)⊕U(1) which appendix 18-A suggests is the true ElectroWeak symmetry.

B. The bare masses of these doublet particles have the same numeric value (before symmetry breaking).

C. The leptonic sector free field lagrangian (without gauge fields introduced yet) is explicitly

$$\mathcal{L}_{\text{freelep}} = \Psi^\dagger(x) \begin{bmatrix} \gamma^0\gamma^5 i(\gamma^\mu\partial/\partial x^\mu - m) & 0 \\ 0 & \gamma^0(i\gamma^\mu\partial/\partial x^\mu - m) \end{bmatrix} \Psi(x) \qquad (18.48)$$

Focussing on the derivative term we see that it can be put in the form

$$\Psi^\dagger(x)\gamma^0 \begin{bmatrix} \gamma^5 & 0 \\ 0 & I_4 \end{bmatrix} i\gamma^\mu\partial/\partial x^\mu\, \Psi(x) = \Psi^\dagger(x)\gamma^0[C^+I - C^-\sigma_3]i\gamma^\mu\partial/\partial x^\mu\Psi(x) \qquad (18.49)$$

where I_4 is a 4×4 identity matrix, where I (an identity matrix) and σ_3 are 2 × 2 matrices, and where C^+ and C^- are defined in eq. 17.69. The Pauli matrices σ_i are defined by eq. 16.39.

Expression 18.49 can be re-expressed in terms of left-handed and right-handed fields as

$$\Psi_L^\dagger(x)\gamma^0 i\gamma^\mu\partial/\partial x^\mu\Psi_L(x) - \Psi_R^\dagger(x)\gamma^0 i\gamma^\mu\partial/\partial x^\mu\sigma_3\Psi_R(x) \qquad (18.50)$$

D. At this point we are in a position to introduce couplings to gauge fields. In view of the doublet nature of the fields $\Psi_L(x)$ and $\Psi_R(x)$ it would appear, at first glance, that the symmetry group of the gauge fields would be $SU(2)_L \otimes SU(2)_R$. However the right-handed tachyon field in expression 18.50 has the wrong sign in the lagrangian, as has been noted in the previous discussion of the free tachyon lagrangian and anti-commutator. Consequently the right-handed tachyon field *cannot* have trilinear or higher order couplings. If it did have such interactions then it would rapidly degrade to lower and lower energy by the emission of particles since right-handed leptonic tachyons can exist in principle as free particles (modulo possible Higgs interaction terms). (In this regard the situation is similar to that of time-like photons, except that the set of tachyon physical states cannot be defined in a manner analogous to Gupta-Bleuler electrodynamics where the timelike and longitudinal photons "cancel" each other so that only transverse photons have physical effects.) *Thus there can be no right-handed trilinear or higher leptonic tachyon interactions.*[280] The doublet nature of the left-handed sector implies at least local SU(2) symmetries implemented with a covariant derivative.

The restricted nature of the right-handed leptonic sector indicates that *only* the Dirac particle in the "right-handed doublet" can have an interaction. Also the appearance of σ_3 in the right-handed term in expression 18.50 breaks SU(2) invariance if the left-handed covariant derivative (eq. 18.51 below) were substituted for $\partial/\partial x^\mu$ in the right-handed term. Thus a U(1) local gauge field interaction, restricted to the Dirac field member of the right-handed doublet and coupling to both members of the left-handed doublet, is the only allowed possibility. Without the U(1) interaction, the "right-handed doublet" would have no trilinear or higher order elementary particle interactions and would be physically irrelevant (except gravitationally) in the unbroken gauge theory before spontaneous breakdown.

Putting these symmetries together we obtain a left-handed covariant derivative implementing local SU(2)⊗U(1) invariance:

$$D_{L\mu} = \partial/\partial x^\mu + \tfrac{1}{2} i g_2 \boldsymbol{\sigma}\cdot\mathbf{W}_\mu + \tfrac{1}{2} i g' B_\mu \tag{18.51}$$

and a right-handed covariant derivative[281]

$$\begin{aligned} D_{R\mu} &= \partial/\partial x^\mu \sigma_3 + \tfrac{1}{2} i g' B_\mu |Q| \sigma_3 \\ &= \partial/\partial x^\mu \sigma_3 + \tfrac{1}{2} i g' B_\mu |Q| \end{aligned} \tag{18.52}$$

where Q is the charge operator using the relation $|Q|\sigma_3 = |Q|$ for leptons. We use the absolute value of Q in order to achieve consistency in form with the right-handed quark sector described in the next chapter. As a result expression 18.50 becomes

$$\Psi_L^\dagger(x)\gamma^0 i\gamma^\mu D_{L\mu}\Psi_L(x) - \Psi_R^\dagger(x)\gamma^0 i\gamma^\mu D_{R\mu}\Psi_R(x) \tag{18.53}$$

Thus the non-Strong, leptonic sector of the lagrangian[282] (modulo mass/Higgs terms or the Dimensional Mass Mechanism (sections 21.3 and 21.4 below)) is

$$\begin{aligned} \mathcal{L}_{lepl} &= \Psi_L^\dagger\gamma^0 i\gamma^\mu D_{L\mu}\Psi_L - \Psi_R^\dagger\gamma^0 i\gamma^\mu D_{R\mu}\Psi_R \\ &= \Psi_L^\dagger\gamma^0 i\gamma^\mu D_{L\mu}\Psi_L + \bar{\psi}_{eR}\gamma^0 i\gamma^\mu D_{R\mu}\psi_{eR} - \bar{\psi}_{\upsilon R}\gamma^0 i\gamma^\mu \partial/\partial x^\mu \psi_{\upsilon R} \end{aligned} \tag{18.54}$$

---

[280] Right-handed neutrinos must interact with gravitons due to their mass-energy and the universality of the gravitational interaction. The extreme weakness of the gravitational interaction mitigates this effect.

[281] The coupling constants are defined by $e = -g'\cos\theta_W = g_2\sin\theta_W$.

[282] Note again that the gauge fields do not have a tachyon equivalent since they are initially massless prior to spontaneous symmetry breaking.

$$= \Psi_L^{\dagger}\gamma^0 i\gamma^\mu D_{L\mu}\Psi_L + \bar{\psi}_{eR}\gamma^0 i\gamma^\mu(\partial/\partial x^\mu + \tfrac{1}{2}ig'B_\mu)\psi_{eR} -$$
$$- \bar{\psi}_{\upsilon R}\gamma^0 i\gamma^\mu\partial/\partial x^\mu\psi_{\upsilon R}$$

where we identify the tachyon as a neutrino and the Dirac particle as its charged lepton partner. Our leptonic sector lagrangian is now the usual leptonic sector ElectroWeak lagrangian with a tachyon neutrino.

E. Local gauge invariance prior to symmetry breaking: The gauge field sector has the usual local Yang-Mills lagrangian terms and the B field has lagrangian terms similar to that of the QED lagrangian.

F. Spontaneous symmetry breaking of gauge symmetry, and of L(C) group covariance, via the Higgs mechanism (or an alternative mechanism such as the Dimensional Mass Mechanism (sections 21.3 and 21.4)) can be implemented in such a way as to give the electron its known mass as well as the massive vector bosons. Since spontaneous symmetry breaking breaks L(C) covariance to Lorentz covariance it is a moot point whether the Higgs sector (if there is one) exhibits a similar covariance.

This concludes the derivation of leptonic sector of the Standard Model. We have shown that the form of the leptonic sector of the Standard Model is fundamental in nature and based on L(C) group covariance of the equations of motion and the postulate that leptons can be L(C) boosted from a rest state subject to the restriction of one real time. *Thus the leptonic ElectroWeak sector is directly based on geometry.*

18.7 Postulates of the Quark ElectroWeak Sector

The derivation of the form of the quark ElectroWeak sector of the Standard Model is very similar to the preceding derivation of the form of the leptonic ElectroWeak sector – but with some important points of difference. The primary difference is that we can identify quarks with complexons[283] or with a "normal" doublet. We consider both cases in this derivation. (As we will see it is plausible to assume complexon quarks in view of the appearance of global SU(3) symmetry in complexon field dynamics. Chapter 19 shows that complexons have a local SU(3) group (color) symmetry. Since complexons have very different spin characteristics compared to Dirac fields, complexon quarks may also resolve difficulties in parton spin studies.)

The quark ElectroWeak sector postulate is:

---

[283] This choice seemed to be the appropriate choice in view of the global SU(3) invariance of complexons. However we could have opted for a Dirac particle and ordinary tachyon combined in a doublet and developed a quark ElectroWeak sector in exactly the same form as the conventional Standard Model ElectroWeak sector.

Postulate 18.9a. The equations of motion of free quark doublets in the unbroken form of the ElectroWeak sector of the Standard Model are determined by covariance under the complex Lorentz group L(C) and SU(2)⊗U(1), and local gauge symmetries. Baryon number is not gauged.

18.8 Derivation of the Quark ElectroWeak Sector of the Standard Model

The derivation:

A. L(C) covariance of the dynamical equations of motion requires that spin ½ particles be described by equations[284] generalized to an 8 × 8 matrix form (eqs 18.35 – 18.37) based on a doublet consisting of a Dirac-type particle and a tachyon. (They may or may not be complexons.) In in the case of quarks the Dirac-type particle is the top component in the doublet and the tachyon is the bottom component.

B. Thus the 8 × 8 quark matrix formalism is:

$$đ_q(x)\Psi_C(x) = 0 \tag{18.37q}$$

where[285]

$$đ_q(x) = \begin{bmatrix} (i\gamma^\mu D_{q\mu} - m_0) & 0 \\ 0 & (\gamma^\mu D_{q\mu} - m_0) \end{bmatrix} \tag{18.35q}$$

and

$$\Psi_C(x) = \begin{bmatrix} \psi_{Cu}(x) \\ \psi_{Cd}(x) \end{bmatrix} \tag{18.36q}$$

---

[284] The global SU(3) symmetry of complexons makes their identification with quarks reasonable. However, the complexon theory that we develop and use for quark dynamics in the Standard Model is not required. Our Standard Model could use Dirac fermion dynamics for the up-type quarks and tachyon dynamics for down-type quarks. We choose to use complexon dynamics for quarks because they have an SU(3)-like structure suggestive of color SU(3). More importantly, their spin dynamics is different and this difference may resolve the differences between theory and experiment for the deep inelastic parton spin-dependent structure functions. Nevertheless, quarks could be similar to leptons in this regard and form a doublet of a Dirac fermion and an ordinary tachyon. Whether quarks are complexons or not is an experimental question!

[285] If we wish to obtain the Standard Model without complexon quarks (which is the conventional Standard Model) then $D_{q\mu} = \partial/\partial x^\mu$. If we wish to obtain the Standard Model with complexon quarks (which is a new form of the Standard Model) then $D_{q\mu}$ is given by eq. 17.59.

The upper 4-component field is a u-type field and the lower 4-component field is a d-type tachyonic field. The generalized equation eq. 18.37q is covariant under 8 × 8 L(C) transformations similar to eqs. 18.38 – 18.41. The generalized quark fermion equation eq. 18.35q is also covariant under conventional Lorentz transformations in the 8 × 8 representation.

The free quark sector lagrangian density that corresponds to the 8-dimensional fermion equation eq. 18.37q is

$$\mathcal{L}_{\text{freeQuark}} = \overline{\Psi}_C(x)\not{d}_q(x)\Psi_C(x) \tag{18.55}$$

where

$$\overline{\Psi}_C(x) = \Psi_C{}^{\dagger}\Gamma_C{}^0 \tag{18.56}$$

with

$$\Psi_C{}^{\dagger} = (\Psi_C{}^{\dagger})|_{\mathbf{x_i} = -\mathbf{x_i}} \tag{18.57a}$$

for complexon quarks, and with

$$\Psi_C{}^{\dagger} = \text{the hermitean conjugate of } \Psi_C \tag{18.57b}$$

in the case of non-complexon quarks. The † in the parentheses on the right side of eq. 18.57a indicates hermitean conjugation. Also

$$\Gamma_C{}^0 = \begin{bmatrix} \gamma^0 & 0 \\ 0 & i\gamma^0\gamma^5 \end{bmatrix} \tag{18.58}$$

C. The free quark field lagrangian is explicitly

$$\mathcal{L}_{\text{freeQuark}} = \Psi_C{}^{\dagger} \begin{bmatrix} \gamma^0(i\gamma^\mu D_{q\mu} - m_0) & 0 \\ 0 & \gamma^0\gamma^5 i(\gamma^\mu D_{q\mu} - m_0) \end{bmatrix} \Psi_C \tag{18.59}$$

Focussing on the derivative term we see that it can be put in the form

$$\Psi_C{}^{\dagger}\gamma^0 \begin{bmatrix} I_4 & 0 \\ 0 & \gamma^5 \end{bmatrix} i\gamma^\mu D_{q\mu}\Psi_C = \Psi_C{}^{\dagger}\gamma^0[C^+I + C^-\sigma_3]i\gamma^\mu D_{q\mu}\Psi_C \tag{18.60}$$

where I_4 is a 4×4 identity matrix, where I, and the Pauli matrix σ_3, are 2 × 2 matrices, and C^+ and C^- are defined in eq. 17.69. Expression 18.60 can be expressed in terms of left-handed and right-handed fields as

$$\Psi_{CL}^{\dagger}\gamma^0 i\gamma^\mu D_{q\mu}\Psi_{CL} + \Psi_{CR}^{\dagger}\gamma^0 i\gamma^\mu D_{q\mu}\sigma_3\Psi_{CR} \quad (18.61)$$

D. At this point we are in a position to introduce couplings to gauge fields. In view of the doublet nature of the fields $\Psi_{CL}(x)$ and $\Psi_{CR}(x)$ it would again appear, at first glance, that the symmetry group of the gauge fields would be $SU(2)_L \otimes SU(2)_R$. However the right-handed tachyonic field has the wrong sign in the lagrangian, as has been noted in the earlier discussion of the free tachyon lagrangian and anti-commutator. Consequently, **if quarks were *not* confined,** the right-handed quark tachyon field could not have trilinear or higher order couplings. **If free tachyon quarks existed, and had interactions, then they would rapidly degrade to lower and lower energy by the emission of particles.**

However, because of quark confinement in bound states with discrete energy levels, a bound tachyon quark can only emit particles if a lower energy bound state exists. As a result right-handed tachyon quarks can have interactions, such as the electromagnetic interaction, because quark confinement "tames" their propensity to emit particles due to the "wrong sign" in the lagrangian.

Again there is an analogy to Gupta-Bleuler QED quantization. In Gupta-Bleuler quantization physical states are required to have equal numbers of time-like and longitudinal photons thus canceling their physical effects. Similarly, right-handed tachyon quarks are required to be bound to other quarks by quark confinement to avoid continuous emission of particles.[286] **Since interactions are allowed for right-handed tachyon quarks the Higgs mechanism (or the Dimensional Mass Mechanism (sections 21.3 and 21.4)) can be used to change their mass.**

The doublet nature of the left-handed sector implies at least local SU(2) symmetries. The appearance of σ_3 in the right-handed term in expression 18.61 explicitly breaks SU(2) invariance if the left-handed covariant derivative (eq. 18.62 below) were substituted for $D_{q\mu}$ in the right-handed term. Thus the right-handed fields can only have a U(1) local gauge field interaction, and are

---

[286] Therefore quark confinement is required in order to have a properly formulated quark sector. Another interaction – the strong interaction – is required for quark confinement. Presently there is only one accepted mechanism for quark confinement – through a non-abelian gauge coupling. (Higher derivative theories with quark confinement are in disfavor.) An additional non-abelian symmetry must be introduced for quarks. As discussed in chapter 19, SU(3) appears to be a natural choice.

SU(2) singlets. We thus obtain a left-handed covariant derivative implementing local SU(2)⊗U(1) covariance:

$$D_{qL\mu} = D_{q\mu} + \tfrac{1}{2}ig_2\boldsymbol{\sigma}\cdot\mathbf{W}_\mu + ig_1B_\mu/6 \tag{18.62}$$

where $D_{q\mu}$ is given by eq. 17.59 and a right-handed covariant derivative[287]

$$D_{qR\mu} = D_{q\mu}\sigma_3 + ig_1B_\mu|Q| \tag{18.63}$$

where |Q| is the absolute value of the charge operator (with u eigenvalue 2/3 and d eigenvalue 1/3). The absolute value is used in order to compensate for the minus sign in front of the right-handed tachyon (d quark) term. As a result expression 18.61 becomes

$$\Psi_{CL}{}^\dagger(x)\gamma^0 i\gamma^\mu D_{qL\mu}\Psi_{CL}(x) + \Psi_{CR}{}^\dagger(x)\gamma^0 i\gamma^\mu D_{qR\mu}\Psi_{CR}(x) \tag{18.64}$$

Thus the quark sector of the lagrangian[288] (modulo possible mass/Higgs/Dimensional Mass Mechanism (sections 21.3 and 21.4) and Strong interaction terms) is

$$\mathcal{L}_{quark1} = \Psi_{CL}{}^\dagger\gamma^0 i\gamma^\mu D_{qL\mu}\Psi_{CL} + \Psi_{CR}{}^\dagger\gamma^0 i\gamma^\mu D_{qR\mu}\Psi_{CR} \tag{18.65}$$

$$= \Psi_{CL}{}^\dagger\gamma^0 i\gamma^\mu D_{qL\mu}\Psi_{CL} + \bar\psi_{CuR} i\gamma^\mu D_{qR\mu}\psi_{CuR} - \bar\psi_{CdR} i\gamma^\mu D_{qR\mu}\psi_{CdR}$$

$$= \Psi_{CL}{}^\dagger\gamma^0 i\gamma^\mu D_{qL\mu}\Psi_{CL} + \bar\psi_{CuR} i\gamma^\mu(D_{q\mu} + \tfrac{2}{3}ig_1B_\mu)\psi_{CuR} -$$

$$- \bar\psi_{CdR} i\gamma^\mu(D_{q\mu} + \tfrac{1}{3}ig_1B_\mu)\psi_{CdR}$$

where we *provisionally* identify the tachyon as a d-type quark and the Dirac-type particle as a u-type quark. Our quark sector lagrangian is now the usual Standard Model quark sector lagrangian (modulo possible mass/Higgs/Dimensional Mass Mechanism (sections 21.3 and 21.4), modulo complex 3-momenta, and modulo the strong interaction, terms) except that d-type quarks are tachyons.

E. The SU(2) gauge field sector has the usual Yang-Mills lagrangian terms, and the B field is a U(1) abelian gauge field.

---

[287] The quark SU(2) coupling constant is, by gauge invariance, required to have the same value as the leptonic SU(2) coupling constant. The U(1) coupling constants are not required to be the same in both sectors and, in fact, are different. The coupling constants here are defined by $e = g_1\cos\theta_W = g_2\sin\theta_W$.

[288] Note that the gauge fields do not have a tachyon equivalent since they are initially massless prior to spontaneous symmetry breaking.

F. Spontaneous symmetry breaking of gauge symmetry, and of L(C) covariance, via the Higgs mechanism or the Dimensional Mass Mechanism (sections 21.3 and 21.4) can be implemented in such a way as to give the quarks, and massive vector bosons, their "known" masses. Since spontaneous symmetry breaking breaks L(C) covariance of the dynamical equations of motion to Lorentz covariance it is a moot point whether the Higgs sector (if there is one) is manifestly L(C) covariant or not.

Thus the initial L(C) covariance of the doublet quark equations of motion results in the "unusual" features of the ElectroWeak quark sector of the Standard Model.

18.9 Geometrical Basis of Parity Violation and ElectroWeak Theory

Ockham's Razor[289] (principle 2 of section 12.1) embodies the precept that, of all the proposed solutions to a problem, the simplest is preferred. The essence of the derivation of the form of the ElectroWeak sector of the Standard Model is 1) in the extension of the Lorentz group (which has real parameters) to the complex Lorentz group L(C); and 2) in the assumption that the four fermion species follow from L(C) boosts plus SU(2)⊗U(1) transformations of a spin ½ particle at rest to a particle with a real energy (and thus a real time coordinate).[290] The rest consists of canonical quantum field theory considerations and Yang-Mills gauge fields. No other approach to the derivation/understanding of ElectroWeak theory has such a simple basis. And thus this approach is to be preferred not only because of Ockham's Razor but also because its naturalness connotes its truth.

---

[289] William of Ockham – Law of Parsimony – "Pluralitas non est ponenda sine necessitate" or "Plurality should not be posited without necessity." First stated by Durand De Saint-Pourçain (1270-1334 AD).

[290] As appendix 18-A points out an SU(2)⊗U(1)⊕U(1) symmetry also implies SU(2)⊗U(1) ElectroWeak Theory as a part of a larger SU(2)⊗U(1)⊕U(1) ElectroWeak Theory.

*Appendix 18-A. SU(2)⊗U(1))⊕U(1) ElectroWeak Theory: Leptons, Quarks and WIMPs

In section 15.12 we introduced an SU(2)⊗U(1)⊕U(1) transformation that initially was used to change complex coordinates generated by a superluminal transformation to a real set of coordinates. We then noted that this transformation could be used to change a 4-vector to another complex 4-vector, and that this transformation could not affect the physical situation since an observer – also similarly transformed – would view the new 4-vector of coordinates as real. Consequently we found an SU(2)⊗U(1)⊕U(1) symmetry that we will now show leads to a new global SU(2)⊗U(1)⊕U(1) symmetry of the fermion dynamical equations, and thence by gauging to an SU(2)⊗U(1)⊕U(1) ElectroWeak sector of the Standard Model.

In this appendix we extend the SU(2)⊗U(1) ElectroWeak discussion of chapter 18 to SU(2)⊗U(1)⊕U(1) ElectroWeak Theory. One important consequence is the natural occurrence of WIMPs (Weakly Interacting Massive Particles) in SU(2)⊗U(1)⊕U(1) ElectroWeak Theory. Dark Matter has been conjectured to be composed of WIMPs. The naturalness of this extended ElectroWeak Theory, and its firm grounding in a fundamental symmetry, provide theoretical support for the Dark Matter proposal.

18-A.1 SU(2) ⊗U(1) Coordinate Transformations and Symmetry

In section 15.12 we introduced an SU(2)⊗U(1)⊕U(1) coordinate transformation in conclusion 4 of that section:

$$\Pi'_L(\boldsymbol{\theta}, \chi, \varphi) = \begin{bmatrix} e^{i\chi} & 0 & 0 & 0 \\ 0 & & & \\ 0 & & e^{i\varphi}\Pi'_{L3}(\mathbf{u})\exp\{i\boldsymbol{\theta}\cdot(\mathbf{Q}' - \mathbf{J})\} & \\ 0 & & & \end{bmatrix} \tag{15.118}$$

where χ and φ are real numbers, $\boldsymbol{\theta}$ is a vector specifying a spatial rotation, and $\mathbf{u}$ is any convenient coordinate axis. Π'_L is an SU(2)⊗U(1)⊕U(1) transformation that transforms complex 4-dimensional coordinates to complex 4-dimensional coordinates in general.

If we apply $\Pi'_L(\boldsymbol{\theta}, \chi, \varphi)$ to the superluminal coordinates of a point X^ν we can produce another expression for the point that has either real or complex coordinates.

$$x'^\mu = [\Pi'_L(\boldsymbol{\theta}, \chi, \varphi)]^\mu{}_\nu x^\nu \qquad (18\text{-A.1})$$

In conclusion 5 of section 15.12 we pointed out that

5. Complex coordinate values of the type generated by superluminal transformations are transformable to real coordinates by a transformation of the form of eq. 15.118. The complex coordinates are thus physically equivalent to corresponding real coordinate values in the sense that an observer in that frame would automatically use the real coordinates, so obtained, since rulers and clocks always measure real spatial coordinates and times.

Conclusion 5 thus identifies a new space-time symmetry which manifestly has the group structure $SU(2)\otimes U(1)\oplus U(1)$ and leads to the $SU(2)\otimes U(1)\oplus U(1)$ equivalence principle stated in conclusion 6 of section 15.12.

An $SU(2)\otimes U(1)\oplus U(1)$ Equivalence Principle
6. The complex coordinates of any point obtained through a superluminal transformation can be transformed to a real set of coordinates by a transformation of the form of eq. 15.118 – an $SU(2)\otimes U(1)\oplus U(1)$ transformation.

18-A.2 Impact of $SU(2)\otimes U(1)\oplus U(1)$ Symmetry on Dirac-like Equations

If we abbreviate $[\Pi'_L(\boldsymbol{\theta}, \chi, \varphi)]^\mu{}_\nu$ as

$$\Pi^\mu{}_\nu = [\Pi'_L(\boldsymbol{\theta}, \chi, \varphi)]^\mu{}_\nu \qquad (18\text{-A.2})$$

then eq. 18-A.1 becomes

$$x'^\mu = \Pi^\mu{}_\nu x^\nu \qquad (18\text{-A.3})$$

and the 4-derivative $D^\mu = \partial/\partial x^\mu$ (or $D^\mu = (\partial/\partial x^0, \mathbf{D}_c{}^* = \nabla_r + i\nabla_i)$ for complexons) transforms similarly as

$$D^\mu = \Pi^\mu{}_\nu D'^\nu \qquad (18\text{-A.4})$$

We can transform the Dirac γ matrices with Π.[291] Π 's inverse is:

[291] The 4×4 matrix form of $\Pi^\mu{}_\nu$.

$$\Pi^{-1} = \Pi^{\dagger} \tag{18-A.5}$$

since Π is unitary. The γ matrices transformation can be written

$$\gamma^{\mu} = \Pi_{\nu}{}^{\mu*}\gamma'^{\nu} \tag{18-A.6}$$

Combining eqs. 18-A.4 and 18-A.6 we find

$$[iD\cdot\gamma - m] = [iD'\cdot\gamma' - m] \tag{18-A.7}$$

due to the unitarity of Π.

We now note that the four Dirac γ-matrices can be treated as an SU(2)⊗U(1)⊕U(1) 4-vector. An SU(2)⊗U(1)⊕U(1) rotation V of the γ-matrices (on the spinor indices) can be defined that satisfies

$$V\gamma^{\mu}V^{-1} = \Pi_{\nu}{}^{\mu*}\gamma'^{\nu} \tag{18-A.8}$$

where V is a 4×4 representation of SU(2)⊗U(1)⊕U(1). Thus the free fermion equations of motion are covariant under global SU(2)⊗U(1)⊕U(1) transformations:

$$V[iD\cdot\gamma - m]V^{-1}V\psi(x) = 0 \tag{18-A.9}$$

or

$$[iD'\cdot\gamma' - m]V\psi(x) = 0 \tag{18-A.10}$$

and the free tachyon equations of motion (eq. 17.49) are also covariant under global SU(2)⊗U(1)⊕U(1) transformations:

$$V[D\cdot\gamma - m]V^{-1}V\psi(x) = 0 \tag{18-A.11}$$

or

$$[D'\cdot\gamma' - m]V\psi(x) = 0 \tag{18-A.12}$$

with the SU(2)⊗U(1)⊕U(1) transformed wave function $\psi_{SUU}'(t, \mathbf{x}')$ being

$$\psi_{SUU}'(x') = V\psi(x) = V\psi(\Pi^{-1}x') \tag{18-A.13}$$

Thus the four types of free Dirac-like equations are covariant under global SU(2)⊗U(1)⊕U(1).

The subsidiary condition for complexons, eq. 17.50, is also covariant under an SU(2)⊗U(1)⊕U(1) rotation:

$$\nabla_{\mathbf{r}}'^{*}\cdot\nabla_{i}'\psi'_{SUU}(x') = \nabla_{\mathbf{r}}\cdot\nabla_{i}V\psi_{C}(x) = V\nabla_{\mathbf{r}}^{*}\cdot\nabla_{i}\psi_{C}(x) = 0 \tag{18-A.14}$$

where we take advantage of the block diagonal form of Π and the unitarity of the SU(2) ⊗U(1) spatial block $e^{i\varphi}\Pi'_{L3}(\mathbf{u})\exp\{i\,\boldsymbol{\theta}\cdot(\mathbf{Q}' - \mathbf{J})\}$ within Π.

18-A.3 SU(2)⊗U(1)⊕U(1) Triplet Field

We now introduce a three component field consisting of an SU(2) doublet and a U(1) singlet of the form

$$\Psi_{\ell e3}(x) = \begin{bmatrix} \psi_{eS} \\ \psi_{T\nu} \\ \psi_e \end{bmatrix} = \begin{bmatrix} e_{eS} \\ \nu_e \\ e \end{bmatrix} \tag{18-A.15}$$

for a U(1) neutral scalar lepton[292] ψ_{eS} (as yet undiscovered but possibly a particle of Dark Matter) and an SU(2)⊗U(1) leptonic doublet where the electron doublet and e_S are displayed for concreteness. Similar triplets can be defined for the other lepton generations.

Having generalized ElectroWeak doublets to SU(2)⊗U(1)⊕U(1) triplets we now must address the question of the superluminal transformation covariance of the equations of motion for a triplet. The new element in the equations of motion, and the superluminal transformation matrix, is the field ψ_{eS}. If we assume ψ_{eS} is a Dark Matter field satisfying a Dirac equation then the following generalizations of the doublet case discussed in chapter 18 are the only reasonable choice. We define the differential operator matrix assuming the U(1) singlet ψ_{eS} is initially massless:[293]

$$đ_3(x) = \begin{bmatrix} i\gamma^\mu\partial/\partial x^\mu & 0 & 0 \\ 0 & (\gamma^\mu\partial/\partial x^\mu - m) & 0 \\ 0 & 0 & (i\gamma^\mu\partial/\partial x^\mu - m) \end{bmatrix} \tag{18-A.16}$$

and the equations of motion

$$đ_3(x)\Psi_{\ell e3}(x) = 0 \tag{18-A.17}$$

---

[292] We denote this particle e_S based on the possibility that it is a "dark" electron.

[293] All three fields will have their masses changed due to spontaneous symmetry breaking. In particular the U(1) singlet could acquire a large mass making it a candidate as a WIMP (See section 18-A.3.1).

The generalization of the superluminal transformation of the dynamical equations is

$$S_{L8,3}(\Lambda_C(\omega, \mathbf{v}_c)) = \begin{bmatrix} S_C(\omega, \mathbf{v}_c) & 0 & 0 \\ 0 & 0 & S_C(\omega, \mathbf{v}_c)\gamma^5 \\ 0 & S_C(\omega, \mathbf{v}_c) & 0 \end{bmatrix} \tag{18-A.18}$$

with inverse transformation

$$S_{L8,3}{}^{-1}(\Lambda_C(\omega, \mathbf{v}_c)) = \begin{bmatrix} S_C{}^{-1}(\omega, \mathbf{v}_c) & 0 & 0 \\ 0 & 0 & S_C{}^{-1}(\omega, \mathbf{v}_c) \\ 0 & \gamma^5 S_C{}^{-1}(\omega, \mathbf{v}_c) & 0 \end{bmatrix}$$

(18-A.19)

Applying $S_{L8,3}$ to eq. 18-A.17 yields

$$\begin{aligned} 0 &= S_{L8,3}(\Lambda_C(\omega, \mathbf{v}_c))\text{đ}_3(x)\Psi(x) \\ &= S_{L8,3}\text{đ}_3(x)S_{L8,3}{}^{-1}S_{L8,3}\Psi(x) = \text{đ}_3(x')\Psi'(x') \end{aligned} \tag{18-A.20}$$

upon rearrangement of the rows of $S_{L8,3}\text{đ}_3(x)S_{L8,3}{}^{-1}$, where we surpress the $\Lambda_C(\omega, \mathbf{v}_c)$ in the second line for brevity, and where

$$\Psi'_{\ell e3}(x') = \begin{bmatrix} \psi'_{eS}(x') \\ \psi'_{Tv}(x') \\ \psi'_e(x') \end{bmatrix} = \begin{bmatrix} S_C\psi_{eS}(x) \\ S_C\gamma^5\psi_e(x) \\ S_C\psi_{Tv}(x') \end{bmatrix}$$

(18-A.21)

We note again that we can use the equivalence of a massless fermion and tachyon equation of motion for the U(1) singlet ψ_{eS} (the subscript S indicates the WIMP Singlet) to obtain superluminal covariance of the equations of motion.

The formulation of the superluminal transformation covariance of the quark equations of motion is analogous to the preceding lepton formulation. We can directly generalize from section 18.4.1 and define

$$\Psi_{C3}(x) = \begin{bmatrix} \psi_{CuS}(x) \\ \psi_{Cu}(x) \\ \psi_{Cd}(x) \end{bmatrix} \qquad (18\text{-A}.22)$$

for the u-d generation, and similarly for the other quark generations, and then proceed to establish superluminal covariance.

The covariance of the lepton and quark triplet equations of motion under Lorentz transformations is easily established along the lines of eqs. 18.9 – 18.13.

Postulate 18.1b. A free Dirac field, tachyon field, and singlet field form a triplet whose dynamical equation is covariant under L(C) transformations.

Postulate 18.2b. A lepton triplet in any generation consists of a charged Dirac field, a neutral tachyon field (the neutrino), and a neutral singlet lepton (WIMP). The equations of motion of the lepton triplet are eqs. 18-A.17.

Postulate 18.3b. A quark triplet in any generation consists of a singlet quark, an up-type quark Dirac field, a down-type quark tachyon field. (The quark fields may be "normal" or complexon – an experimental issue – but complexon is theoretically favored.)

Postulate 18.4b. The form of the equations of motion of the unbroken leptonic sector of the Standard Model are determined by covariance under L(C) group boosts, and local SU(2)⊗U(1)⊕U(1) gauge symmetries. Electron number is not gauged.

Postulate 18.5b. In each generation leptonic particles consist of a spin ½ Dirac particle (electron-like) with charge –1, a tachyon (neutrinos) of charge zero, and a singlet (WIMP) of charge zero as well as their anti-particles.[294] *(Redundant for clarity.)*

Postulate 18.7b. The interacting extensions of free triplet equations of motion have interactions consisting of local Yang-Mills fields. Gauge fields are massless "before" spontaneous symmetry breaking and are thus conventional local Yang-Mills gauge fields without a tachyon equivalent.

---

[294] If neutrinos are Majorana particles then the derivation must be modified.

Postulate 18.9b. The equations of motion of free quark triplets in the unbroken part of the ElectroWeak sector of the Standard Model are determined by covariance under the complex Lorentz group L(C), and local SU(2)⊗U(1)⊕U(1) gauge symmetries. Baryon number is not gauged.

18-A.4 The U(1) Singlet Members of Triplets – WIMPs – Dark Matter?

In the preceding section we introduced a third member in each lepton and quark generation as a result of the additional U(1) symmetry of SU(2)⊗U(1)⊕U(1). Thus in a four (three) generation theory there are four (three) U(1) lepton singlets and four (three) U(1) quark singlets[295] plus their antiparticles.

18-A.4.1 SU(2)⊗U(1)⊕U(1) WIMPs

The form of the lepton and quark triplets suggests the set of singlets are WIMPs – Weakly Interacting Massive Particles. WIMPS have been a candidate for Dark Matter since 1985.[296] However WIMPs were introduced in an *ad hoc* manner to satisfy the few known characteristics of Dark Matter – not on the basis of fundamental principles. This development introduces WIMPs for good reason: the SU(2)⊗U(1)⊕U(1) Equivalence Principle stated earlier.

18-A.4.2 Alternate Possibility: Singlets With a Non-Electric Charge

The lepton singlets cannot possess electric charge or they would have been detected. They could have an electric charge of a different type, a "dark" electric charge, and thus exert "dark" electromagnetic forces on each other via a coupling to "dark" photons. The quark singlets could also possess dark electric charge. Consequently "dark" hydrogen consisting of a singlet quark and singlet lepton bound together by dark electromagnetism would be possible. This possibility is not favored since singlet Physics in this case would lack SU(2)⊗U(1) ElectroWeak symmetry unless a "dark" Weak Interaction sector were also present.

It seems unlikely that Nature would duplicate ElectroWeak Theory in this manner. WIMPs, in contrast, fit well within an extended ElectroWeak framework.

---

[295] Quark singlets have complexon fields as do the known quarks. However they must be color SU(3) singlets, for, if they had color, they would bind to the known quarks and exotic quark bound states with fractional electric charge would exist. No such bound states have been found although it is possible the singlet quarks are so massive that they are not accessible at current accelerator energies. We view this possibility as unlikely.

[296] C. A. de S. Pires, F. S. Queiroz, P. S. Rodrigues da Silva, and J. K. Mizukoshi, arXiv/Hep-ph/1010.4097 (2010); J. Faulkner and R. L. Gilliland, *Astrophys. J.* **299**, 994 (1985); A. de Rùjula, S. L. Glashow, and L. Hall, *Nature* **320**, 38 (1986); W. H. Press and D. N. Spergel, *Astrophys. J.* **296**, 679 (1985); D. N. Spergel and W. H. Press, *Astrophys. J.* **294**, 663 (1985).

18-A.5 Derivation of ElectroWeak Sector Symmetry

The SU(2)⊗U(1)⊕U(1) equivalence principle and symmetry, which is ultimately based on superluminal transformations is the source of the SU(2)⊗U(1)⊕U(1) ElectroWeak sector.[297] We begin by deriving the SU(2)⊗U(1)⊕U(1) covariance of the free dynamical equations of the triplet starting from eqs. 18-A.1 – 18-A.17.

Eq. 15.118 for SU(2)⊗U(1)⊕U(1) coordinate transformations corresponds to a 3×3 SU(2)⊗U(1)⊕U(1) transformation matrix for a triplet consisting of an SU(2)⊗U(1) doublet and a U(1) singlet. The matrix has the block diagonal form

$$S_\Pi(\boldsymbol{\theta}, \chi, \varphi) = [V] \begin{bmatrix} e^{i\chi} & 0 \quad 0 \\ \begin{matrix} 0 \\ 0 \end{matrix} & e^{i\varphi} S_2(\boldsymbol{\theta}) \end{bmatrix} \qquad (18\text{-A}.23)$$

where $S_2(\boldsymbol{\theta})$ is a 2×2 SU(2) doublet transformation matrix, and where $[V] = \mathrm{diag}(V, V, V)$ and $[V^{-1}] = \mathrm{diag}(V^{-1}, V^{-1}, V^{-1})$ are diagonal matrices with each V and V^{-1} (defined by eq. 18-A.8) operating on spinor indices. The inverse of $S_\Pi(\boldsymbol{\theta}, \chi, \varphi)$ is

$$S_\Pi^{-1}(\boldsymbol{\theta}, \chi, \varphi) = \begin{bmatrix} e^{-i\chi} & 0 \quad 0 \\ \begin{matrix} 0 \\ 0 \end{matrix} & e^{-i\varphi} S_2^{-1}(\boldsymbol{\theta}) \end{bmatrix} [V^{-1}] \qquad (18\text{-A}.24)$$

The lepton and the quark triplets dynamical equations transform under the SU(2)⊗U(1)⊕U(1) matrix $S_\Pi(\boldsymbol{\theta}, \chi, \varphi)$ as

$$\begin{aligned} 0 &= S_\Pi(\boldsymbol{\theta}, \chi, \varphi)\not{d}_3(x)\Psi(x) \\ &= S_\Pi(\boldsymbol{\theta}, \chi, \varphi)\not{d}_3(x)\, S_\Pi^{-1}(\boldsymbol{\theta}, \chi, \varphi)S_\Pi(\boldsymbol{\theta}, \chi, \varphi)\Psi(x) = \not{d}_3(x')\Psi'(x') \end{aligned} \qquad (18\text{-A}.25)$$

where $\Psi(x)$ is $\Psi_{\ell e3}(x)$ for leptons or $\Psi_C(x)$ for quarks, and $x'^\mu = \Pi^\mu{}_\nu x^\nu$. The application of $S_\Pi(\boldsymbol{\theta}, \chi, \varphi)$ to a lepton triplet yields

[297] The theory could be reformulated to SU(2)⊗U(1) by excluding the singlet particle from the equations. In view of the importance of resolving the Dark Matter issue we will keep the singlet in the discussion of this appendix.

$$S_\Pi \Psi_{\ell e3}(x) = [V] \begin{bmatrix} e^{i\chi}\psi_{eS}(x') & - \\ & \\ e^{i\varphi}S_2(\boldsymbol{\theta}) & \begin{pmatrix} \psi_{T\nu}(x') \\ \psi_e(x') \end{pmatrix} \end{bmatrix} \qquad (18\text{-A}.26)$$

where $x'^\mu = \Pi^\mu{}_\nu x^\nu$.

Note that the SU(2)⊗U(1)⊕U(1) rotation factors into an SU(2)⊗U(1)⊕U(1) rotation of the three fields in each triplet and an SU(2)⊗U(1)⊕U(1) rotation of the four spinor components of each individual field by V. This factorization enables us to consider a global SU(2)⊗U(1)⊕U(1) rotation $S_{\Pi g}$ of the fields while holding the coordinates fixed:

$$S_{\Pi g} \Psi_{\ell e3}(x) = \begin{bmatrix} e^{i\chi}\psi_{eS}(x) & - \\ & \\ e^{i\varphi}S_2(\boldsymbol{\theta}) & \begin{pmatrix} \psi_{T\nu}(x) \\ \psi_e(x) \end{pmatrix} \end{bmatrix} \qquad (18\text{-A}.27)$$

The equations of motion are covariant under this global transformation:

$$0 = S_{\Pi g}(\boldsymbol{\theta}, \chi, \varphi)\,đ_3(x)\Psi(x) = đ_3(x)\Psi'(x) \qquad (18\text{-A}.28)$$

Thus we have a global SU(2)⊗U(1)⊕U(1) symmetry of the free dynamical equations for each generation of free lepton and quark triplets.

We now note the form of eq. 18-A.27 is the same as that of a *local* Yang-Mills rotation:

$$\psi_C{}'^a(x) = \Theta^{ab}(x)\psi_C{}^b(x) \qquad (18\text{-A}.29)$$

Therefore we can introduce five local SU(2)⊗U(1)⊕U(1) Yang-Mills fields: $W_i{}^\mu(x)$ for i = 1, 2, 3 and $B^\mu(x)$ having SU(2)⊗U(1) symmetry, and $B'^\mu(x)$ for the singlet (WIMP) U(1) sector. We define a covariant derivative for leptons and quarks that generalizes

$$đ_3(x)\Psi(x) = 0 \qquad (18\text{-A}.30)$$

to the case of local SU(2)⊗U(1)⊕U(1) if we do <u>not</u> perform the spinor rotation V.[298]

---

[298] This approach enables us to avoid the dilemmas associated with mixing coordinate and internal symmetries as described by Coleman, S., Phys. Rev. **138** B1262 (1965) and others in the case of

$$\mathcal{D}_\nu = D_\nu - (igW_{i\nu}(x)\tau^i + ig'B_\nu(x))(1 - Q') - ig''B'_\nu(x)Q' \qquad (18\text{-A}.31)$$

where D_ν is as in eq. 18-A.4, and

$$Q' = \text{diag}(1, 0, 0) \qquad (18\text{-A}.32)$$

Thus we have the covariant derivative of the Weinberg-Salam ElectroWeak Theory and a U(1) WIMP adddition $ig''B'_\nu(x)$.[299] The dynamical equations of the gauge fields are conventional as required by Yang-Mills local gauge invariance.

This extension of ElectroWeak symmetry to SU(2)⊗U(1)⊕U(1) results in only minor changes to the discussions of chapter 18.

SU(6) in the 1960's. Note that the spinor rotation V is expressed in terms of numerical matrices while, in the second quantized formulation, the $S_{\Pi g}(\boldsymbol{\theta}, \chi, \varphi)$ rotation is expressed in terms of second quantized fields as well as numeric matrices. Thus the factorization is reflected in the form of the transformation.

[299] It is possible that WIMPs interact with Z bosons. We have not placed the corresponding interaction term in the lagrangian. If experiments indicate the existence of this interaction then the above covariant derivative would have to be modified by adding the interaction term.

19. Color SU(3)

19.1 Two Possible Approaches to Color SU(3)

There are two approaches to obtaining the Strong interaction and Color SU(3) symmetry:

1. Assume up-type and down-type quarks are in $\underline{3}$ representations of Color SU(3). This assumption sheds no light on a deeper origin of the Strong interaction and Color SU(3). It simply assumes the color SU(3) of the Strong interaction sector of the Standard Model. Thus our understanding is not deepened. A postulate corresponding to this assumption is:

Postulate 19.1a. Quarks are in the $\underline{3}$ representation of Color SU(3). The SU(3) symmetry is gauged with local Yang-Mills SU(3) fields called gluons that constitute the Strong interaction of the quark sector. Quarks are minimally coupled to the gluons in a gauge covariant fashion.

2. In the preceding chapter, and appendix, the ElectroWeak sector of the Standard Model (modulo generations and their mixing) was shown to essentially follow from the L(C) covariance of the doublet equations of motion for leptons and quarks, minimal coupling to gauge fields and stability requirements for tachyons, and from a previously unrecognized SU(2)⊗U(1)⊕U(1) symmetry. Thus we have a significant geometrical basis for the form of the ElectroWeak sector. We would like to establish a similar geometrical basis for the Strong interaction and Color SU(3). In this regard we note that the generators of GL(4) were shown (section 16.5) to include the generators of the Lorentz group and the generators of U(3) as well as a U(1) generator. Since GL(4) has complex parameters, and has a complex 4-dimensional representation which we take to be an extension of our real coordinate space, GL(4) has L(C) as a subgroup. We used L(C) and a coordinate symmetry to extract the form of the ElectroWeak sector in the preceding chapter. *We now suggest that the subset of SU(3) generators of GL(4) is the source of Color SU(3) when we limit the parameter set of the subgroup defined with these generators to real values. If we now extend the parameters to be real functions of the space-time coordinates (i.e. local SU(3) transformations), then we obtain a geometrical basis for color SU(3). A key factor in this interpretation is the global covariance of complexon equations of motion under global SU(3).* We thus establish a basis for ElectroWeak symmetry and for color SU(3) symmetry in

geometry. (See Fig. 19.1.) This line of development leads to the following alternative postulate.

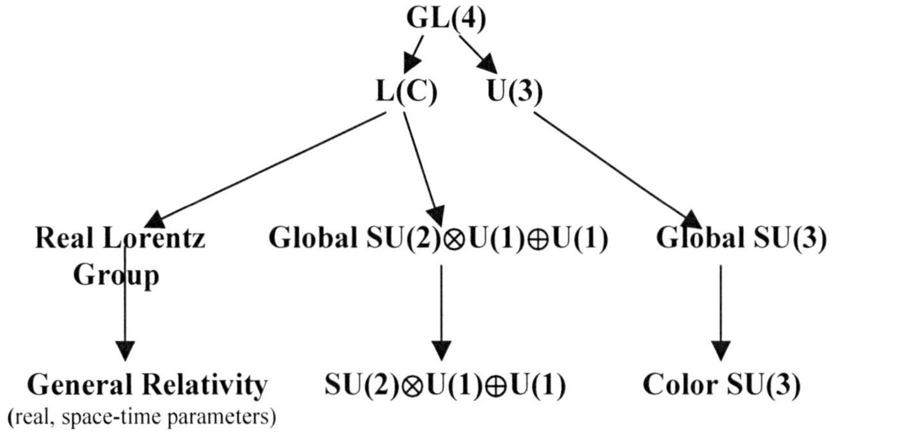

Figure 19.1. Generation of ElectroWeak SU(2)⊗U(1)⊕U(1) and color SU(3) from L(C). This diagram is symbolic. For example, General Relativity actually may actually originate in 16-dimensional space-time which then breaks down to the real 4-dimensional space-time we experience.

Postulate 19.1b. Quarks are complexons.

We now proceed to derive local Color SU(3).

19.2 Complexon Basis of Color SU(3) and the Strong Interactions

In section 17.6 we showed that the equations of motion of free Dirac-like, complexon, up-type quarks are covariant under a SU(3) subgroup of GL(4). The free, tachyon, complexon, down-type quark equations of motion are also easily seen to be covariant under this SU(3) subgroup. In this section we will show this covariance is the "source" of local Color SU(3) symmetry of quarks, and then we will introduce the Strong interaction via minimal coupling to SU(3) Yang-Mills gluons in gauge covariant derivatives.

In section 17.6 we considered a global SU(3) rotation of the derivative vector:

$$\mathbf{D}_c{}^* = \nabla_c = \nabla_r + i\nabla_i$$

namely,

$$D_c{}^{*a} = D_c{}'^{*b}U^{ab*}$$

where U^{ab} is a global SU(3) rotation. We saw

$$[i\gamma^0\partial/\partial t + i\mathbf{D}_c{}^*\cdot\gamma - m] = [i\gamma^0\partial/\partial t + i\mathbf{D}_c{}'^*\cdot\gamma' - m] \qquad (17.107)$$

where

$$\gamma^a = U^{ab}\gamma'^b \qquad (17.108)$$

due to the unitarity of U. ($U^\dagger = U^{-1}$) We then noted the three spatial γ-matrices can be treated as SU(3) 3-vectors with eq. 17.108 representing the result of an SU(3) rotation V of the γ-matrices (on the spinor indices)

$$V\gamma^a V^{-1} = U^{ab}\gamma'^b \qquad (17.109)$$

where V is a matrix in a 4×4 reducible representation[300] of SU(3), namely, $\underline{3} + \underline{1}$. Thus we see that the free complexon equations of motion (eq. 17.49) are covariant under global SU(3) transformations:

$$V[i\gamma^0\partial/\partial t + i\gamma\cdot\mathbf{D}_c{}^* - m]V^{-1}V\psi_C(t, \mathbf{x}_r, \mathbf{x}_i) = 0$$

or

$$[i\gamma^{0\prime}\partial/\partial t' + i\mathbf{D}_c{}'^*\cdot\gamma' - m]V\psi_C(t, \mathbf{x}_r, \mathbf{x}_i) = 0 \qquad (17.110)$$

with the SU(3) transformed wave function $\psi_C'(t, \mathbf{x}')$ being

$$\psi_C'(t', \mathbf{x}') = V\psi_C(t, \mathbf{x}) = V\psi_C(t', U\mathbf{x}') \qquad (17.111)$$

(Note $\gamma^{0\prime} = V\gamma^0V^{-1}$ and t' = t.) Note the relation of the SU(3) vectors $\mathbf{x} = U\mathbf{x}'$ can be expressed in component form by:

$$x^a = U^{ab}x'^b \qquad (19.1)$$

The subsidiary condition, eq. 17.50, is also covariant under an SU(3) rotation:

$$\nabla_r'^*\cdot\nabla_i'\psi_C'(t', \mathbf{x}') = \nabla_r\cdot\nabla_i V\psi_C(t, \mathbf{x}) = V\nabla_r{}^*\cdot\nabla_i\,\psi_C(t, \mathbf{x}) = 0 \qquad (17.112)$$

We now introduce a complexon field with a global SU(3) index a which takes values from 1 to 3 making the field a member of the $\underline{3}$ representation of global SU(3):

---

[300] Eqs. 16.50, and 16.53 furnish a reducible 4×4 matrix representation. The generators of this reducible representation are ${}_4F_i = F_i + \Omega_{+00} - F_9$ with F_i being the Gell-Mann SU(3) generators for i = 1, 2, ..., 8. See eqs. 17.122 and 17.123.

$$\psi_C{}^a(t, \mathbf{x}) \qquad (19.2)$$

Due to the SU(3) index the transformation property of $\psi_C{}^a(t, \mathbf{x})$ changes from eq. 17.111 to

$$\psi_C{}''^a(t, \mathbf{x}') = U^{ab}V\psi_C{}^b(t, \mathbf{x}) = U^{ab}V\psi_C{}^b(t, U\mathbf{x}') \qquad (19.3)$$

where U^{ab} is an SU(3) rotation of 3 representation "vectors" such as $\psi_C{}^b$ and **x.** V is the corresponding rotation of the spinor indices of $\psi_C{}^b(t, \mathbf{x})$.

Note that the SU(3) rotation of the field factorizes into an SU(3) rotation of the three fields $\psi_C{}^b$ by U^{ab} and an SU(3) rotation of the four spinor components of each individual field $\psi_C{}^b$ by V.

This factorization enables us to consider a global SU(3) rotation of the $\psi_C{}^b$ fields while holding the coordinates fixed:

$$\psi_C{}'^a(t, \mathbf{x}) = U^{ab}\psi_C{}^b(t, \mathbf{x}) \qquad (19.4)$$

The equations of motion are covariant under this global transformation

$$0 = U^{ab}[i\gamma^0\partial/\partial t + i\gamma\cdot\mathbf{D}_c{}^* - m]\psi_C{}^b(t, \mathbf{x}_r, \mathbf{x}_i)$$

$$= [i\gamma^0\partial/\partial t + i\gamma\cdot\mathbf{D}_c{}^* - m]\psi'_C{}^a(t, \mathbf{x}_r, \mathbf{x}_i) \qquad (19.5)$$

We now note the form of eq. 19.4 is the same as that of a *local* Yang-Mills rotation:

$$\psi_C{}'^a(t, \mathbf{x}) = \Theta^{ab}(t, \mathbf{x})\psi_C{}^b(t, \mathbf{x}) \qquad (19.6)$$

where $\mathbf{x} = \mathbf{x}_r + i\mathbf{x}_i$. Therefore if we introduce a local SU(3) Yang-Mills field $A_{C\nu}(t, \mathbf{x}_r, \mathbf{x}_i)$ and define a covariant derivative we can generalize eq. 19.5 to the case of local, color SU(3) if we do not perform the spinor rotation V.[301]

$$\mathcal{D}_\nu = D_\nu - igA_{C\nu} \qquad (19.7)$$

where

---

[301] This approach enables us to avoid the dilemmas associated with mixing coordinate and internal symmetries as described by Coleman, S., Phys. Rev. **138** B1262 (1965) and others in the case of SU(6) in the 1960's. Note that the spinor rotation V is expressed in terms of numerical matrices while, in the second quantized formulation, the U^{ab} rotation is expressed in terms of second quantized fields as well as numeric matrices. Thus the factorization is reflected in the form of the transformation.

$$A_{C\nu} = A_C{}^a{}_\nu t^a \quad (19.8)$$

and where D_ν is given by eq. 17.59. The SU(3) 3×3 matrix generators satisfy

$$[t^a, t^b] = if^{abc}t^c \quad (19.9)$$

The locality of the SU(3) gauge group (and of the ElectroWeak interactions) can be viewed as following from the locality of Logic statements in general. (See section 12.7.3.)

We can represent $\Theta_{ab}(x)$ in the form:

$$\Theta_{ab}(x) = [\exp(-i\varphi_c(x)t^c)]_{ab} \quad (19.10)$$

where $\varphi_c(x)$ is a local parameter dependent on $x = (x^0, \mathbf{x} = \mathbf{x_r} + i\mathbf{x_i})$, and t^c is an SU(3) generator.

Applying a gauge transformation to the gauge covariant derivative of a complexon fermion field $\mathcal{D}_\nu\psi_C(x)$:

$$\Theta\mathcal{D}_\nu\psi_C(x) = \Theta D_\nu\psi_C(x) - ig\Theta A_{C\nu}\Theta^{-1}\Theta\psi_C(x) \quad (19.11)$$

$$= D_\nu\psi_C'(x) - igA_C{}'_\nu\psi_C'(x) = (\mathcal{D}_\nu\psi_C(x))'$$

where

$$\psi_C'(x) = \Theta(x)\psi_C(x) \quad (19.12)$$

we find

$$A_C{}'_\nu = (-i/g)(D_\nu\Theta(x))\Theta^{-1}(x) + \Theta(x)A_{C\nu}(x)\Theta^{-1}(x) \quad (19.13)$$

The reader will note that the form of eqs. 19.6 – 19.12 is identical to those associated with a conventional non-abelian gauge interaction with the replacement:

$$\partial/\partial x^\nu \rightarrow D_\nu \quad (19.14)$$

with D_ν given by eq. 17.59. Note that $\varphi_c(x)$, the local parameter in eq. 19.10 is dependent in general, on time, and the real and imaginary parts of the complex spatial 3-vector.

Introducing the SU(3) gauge covariant derivative transforms eq. 19.5 to

$$0 = [i\gamma^\nu\mathcal{D}_\nu - m]\psi_C{}^a(t, \mathbf{x_r}, \mathbf{x_i}) \quad (19.15)$$

The preceding argumentation supports the following postulates (assuming Postulate 19.1b):

Postulate 19.2. Quarks are in a $\underline{3}$ *representation of a global SU(3) subgroup of GL(4). Their transformation law under SU(3) transformations is eq. 19.6 wherein the space-time coordinates are not transformed.*

Postulate 19.3. The covariance of the quark equations of motion under global SU(3) transformations becomes covariance under local color SU(3) transformations when the equations of motion are generalized to gauge covariant derivatives. We assume the equations of motion are so generalized. The interaction terms introduced constitute the Strong interaction.

We note the case of tachyon complexon quarks differs only in small details from the above discussion of Dirac-type complexon quarks.

19.3 The One Generation "Standard" Standard Model

If we assume postulate Postulate 19.1a so that quarks are not complexons then we obtain the conventional form of the Standard Model modulo the values of coupling constants, the introduction of masses (Higgs or Dimensional Mass Mechanisms (sections 21.3 and 21.4)), the introduction of generations and the mixing of the generations.

The lagrangian density that we have derived for the conventional Standard Model[302] with ElectroWeak triplets (and with one generation at this point in the derivation) is

$$\mathcal{L}_{SSM} = \Psi_{3L}{}^{\dagger}\gamma^0 i\gamma^\mu D_{L\mu}\Psi_{3L} - \Psi_{3R}{}^{\dagger}\gamma^0 i\gamma^\mu D_{R\mu}\Psi_{3R} +$$

$$+ \Psi_{q3L}{}^{\dagger}\gamma^0 i\gamma^\mu D_{qL\mu}\Psi_{q3L} + \Psi_{q3R}{}^{\dagger}\gamma^0 i\gamma^\mu D_{qR\mu}\Psi_{q3R} -$$

$$- \mathcal{L}_{BareMasses} + \mathcal{L}_{Gauge} + \mathcal{L}_{Higgs} \qquad (19.16)$$

where

$$D_{L\mu} = \partial/\partial x^\mu - (\tfrac{1}{2} i g_2 \boldsymbol{\sigma}\cdot\mathbf{W}_\mu - i g_1 B_\mu/2)(1 - Q') - i g'' Q' B'_\nu(x)/2 \qquad (19.17a)$$

$$D_{R\mu} = (\sigma_3 \partial/\partial x^\mu + i g_1 B_\mu |Q|)(1 - Q') + (\partial/\partial x^\mu + i g'' B'_\nu(x))Q' \qquad (19.17b)$$

$$D_{qL\mu} = \mathfrak{D}_\mu + (\tfrac{1}{2} i g_2 \boldsymbol{\sigma}\cdot\mathbf{W}_\mu + i g_1 B_\mu/6)(1 - Q') + i g'' B'_\nu(x) Q'/6 \qquad (19.18)$$

$$D_{qR\mu} = (\sigma_3 \mathfrak{D}_\mu + i g_1 B_\mu |Q|)(1 - Q') + (\mathfrak{D}_\mu + i g'' B'_\nu(x))Q' \qquad (19.19)$$

---

302 We call this theory the one generation Standard Standard Model. The ElectroWeak sector has fermion triplets and broken SU(2)⊗U(1)⊕U(1) symmetry. This theory with fermion triplets can easily be restricted to fermion doublets to obtain the conventional SU(2)⊗SU(1) ElectroWeak sector.

where $B'_v(x)$ is the U(1) WIMP ElectroWeak field, where Q' is given by eq. 18-A.32, where $\mathfrak{D}_v$ is given by eq. 19.7 with D_v given by

$$D_0 = \partial/\partial x^0$$
$$D_k = \partial/\partial x_r^{\ k} + i\,\partial/\partial x_i^{\ k} \tag{19.20}$$

for k = 1, 2, 3, where $\Psi_{q3}^{\ \dagger}$ is the hermitean conjugate of the quark triplet with $\Psi_{q3L}^{\ \dagger}$ and $\Psi_{q3R}^{\ \dagger}$ being its left-handed and right-handed parts respectively, where A_v are local SU(3) Yang-Mills fields, and where

$$\mathcal{L}_{Gauge} = \mathcal{L}_{GaugeEW} + \mathcal{L}_{GaugeColor} \tag{19.21}$$

with

$$\mathcal{L}_{GaugeEW} = -\tfrac{1}{4}\,F_W^{\ a\mu\nu}F_{W\ \mu\nu}^{\ a} - \tfrac{1}{4}\,F_B^{\ \mu\nu}F_{B\mu\nu} - \tfrac{1}{4}\,F_{B'}^{\ \mu\nu}F_{B'\mu\nu} + \mathcal{L}_{EW}^{\ ghost} \tag{19.22}$$

and

$$\mathcal{L}_{GaugeColor} = -\tfrac{1}{4}\,F_{Color}^{\ a\mu\nu}F_{Color\ \mu\nu}^{\ a} + \mathcal{L}_{Color}^{\ ghost} \tag{19.23}$$

$\mathcal{L}_{BareMasses}$ contains the fermion bare mass terms. The ElectroWeak gauge bosons $W_\mu^{\ a}$, B'_μ and B_μ field tensors are:

$$F_{W\ \mu\nu}^{\ a} = \partial W^a_{\ \mu}/\partial x^\nu - \partial W^a_{\ \nu}/\partial x^\mu + g_2 f^{abc} W^b_{\ \mu} W^c_{\ \nu} \tag{19.24}$$

$$F_{B\mu\nu} = \partial B_\mu/\partial x^\nu - \partial B_\nu/\partial x^\mu \tag{19.25a}$$

$$F_{B'\mu\nu} = \partial B'_\mu/\partial x^\nu - \partial B'_\nu/\partial x^\mu \tag{19.25b}$$

$\mathcal{L}_{EW}^{\ ghost}$ are the Faddeev-Popov ghost terms for the ElectroWeak $W_\mu^{\ a}$ gauge bosons. The color gauge field tensors are

$$F_{Color\ \mu\nu}^{\ a} = \partial A^a_{\ \mu}/\partial x^\nu - \partial A^a_{\ \nu}/\partial x^\mu + g_2 f_{su(3)}^{\ abc} A^b_{\ \mu} A^c_{\ \nu} \tag{19.26}$$

$\mathcal{L}_{Color}^{\ ghost}$ is the color SU(3) ghost terms defined using the Faddeev-Popov mechanism. The Higgs sector(?) $\mathcal{L}_{Higgs}$ or the Dimensional Mass Mechanism (sections 21.3 and 21.4) generates fermion masses and ElectroWeak vector boson masses.

The effective action for the path integral formulation is

$$I_{SSM} = \int dx^0 d^3x\ \mathcal{L}_{SSM} \tag{19.27}$$

19.4 The One Generation Complexon Standard Model

If we assume Postulates 19.1b, 19.2, and 19.3 then we obtain the one generation Complexon Standard Model[303] which differs from the "Standard Standard Model" by having WIMPs, and complexon quarks and gluons. Which of these Standard Models is the correct one chosen by Nature? That is a question that only experiment can answer. However the naturalness of color SU(3) within the complexon formulation suggests that the Complexon Standard Model is the correct Standard Model.

In this section we show the results of the derivation of previous sections for the Complexon Standard Model: its lagrangian density and action.

$$\mathcal{L}_{CSM} = \Psi_{3L}{}^{\dagger}\gamma^0 i\gamma^{\mu} D_{L\mu}\Psi_{3L} - \Psi_{3R}{}^{\dagger}\gamma^0 i\gamma^{\mu} D_{R\mu}\Psi_{3R} +$$
$$+ \Psi_{C3L}{}^{\dagger}\gamma^0 i\gamma^{\mu}\mathcal{D}_{qL\mu}\Psi_{C3L} + \Psi_{C3R}{}^{\dagger}\gamma^0 i\gamma^{\mu}\mathcal{D}_{qR\mu}\Psi_{C3R} -$$
$$- \mathcal{L}_{BareMasses} + \mathcal{L}_{Gauge} + \mathcal{L}_{Mass} \tag{19.28}$$

using eq. 18.65 where $D_{L\mu}$ and $D_{R\mu}$ are given by eq. 19.17, and where the left-handed complexon quark covariant derivative[304] is

$$D_{qL\mu} = \mathcal{D}_{\mu} + (\tfrac{1}{2} i g_2 \boldsymbol{\sigma}\cdot\mathbf{W}_{\mu} + i g_1 B_{\mu}/6)(1 - Q') + i g'' B'_{\nu}(x) Q'/6 \tag{19.29}$$

and the right-handed complexon quark covariant derivative is

$$D_{qR\mu} = (\sigma_3\, \mathcal{D}_{\mu} + i g_1 B_{\mu}|Q|)(1 - Q') + (\mathcal{D}_{\mu} + i g'' B'_{\nu}(x))Q' \tag{19.30}$$

where $\mathcal{D}_{\mu}$ is given by eq. 19.7 with D_{ν} given by eq. 17.59, where $\Psi_{C3}{}^{\dagger}$ is given by eq. 18-A.22, and $\Psi_{C3L}{}^{\dagger}$ and $\Psi_{C3R}{}^{\dagger}$ are its left-handed and right-handed parts respectively. $\mathcal{L}_{BareMasses}$ contains the fermion bare mass terms. Also,

$$\mathcal{L}_{Gauge} = \mathcal{L}_{GaugeEW} + \mathcal{L}_{GaugeC} \tag{19.31}$$

with

$$\mathcal{L}_{GaugeEW} = -\tfrac{1}{4} F_W{}^{a\mu\nu} F_W{}^{a}{}_{\mu\nu} - \tfrac{1}{4} F_B{}^{\mu\nu} F_{B\mu\nu} - \tfrac{1}{4} F_{B'}{}^{\mu\nu} F_{B'\mu\nu} + \mathcal{L}_{EW}{}^{ghost} \tag{19.32}$$

and

---

[303] This theory with fermion triplets can easily be restricted to fermion doublets to obtain the conventional SU(2)⊗U(1) ElectroWeak sector but with complexon quarks and gluons.

[304] The quark SU(2) coupling constant is, by gauge invariance, required to have the same value as the leptonic SU(2) coupling constant. The U(1) coupling constants are not required to be the same in both sectors and, in fact, are different. The coupling constants here are defined by $e = g_1\cos\theta_W = g_2\sin\theta_W$.

$$\mathcal{L}_{\text{GaugeC}} = \mathcal{L}_{\text{CCG}} + \text{Faddeev-Popov terms} \tag{19.33}$$

See eqs. 19-A.17, 19-A.18, 19-A.26 and 19-A.28 appearing later for the Faddeev-Popov terms. The ElectroWeak gauge bosons $W_\mu{}^a$, B_μ and B'_ν field tensors are:

$$F_{W}{}^{a}{}_{\mu\nu} = \partial W^a{}_\mu/\partial x^\nu - \partial W^a{}_\nu/\partial x^\mu + g_2 f^{abc} W^b{}_\mu W^c{}_\nu \tag{19.24}$$

$$F_{B\mu\nu} = \partial B_\mu/\partial x^\nu - \partial B_\nu/\partial x^\mu \tag{19.25a}$$

$$F_{B'\mu\nu} = \partial B'_\mu/\partial x^\nu - \partial B'_\nu/\partial x^\mu \tag{19.25b}$$

$\mathcal{L}_{EW}{}^{ghost}$ contains the Faddeev-Popov ghost terms for the ElectroWeak $W_\mu{}^a$ gauge bosons. The complexon color gluon lagrangian term $\mathcal{L}_{CCG}$ is defined by

$$\mathcal{L}_{\text{CCG}} = -\tfrac{1}{4} F_{CC}{}^{a\mu\nu}(x) F_{CC}{}^{a}{}_{\mu\nu}(x) \tag{19.34}$$

where

$$F_{CC}{}^{a}{}_{\mu\nu} = D_\nu A_C{}^a{}_\mu - D_\mu A_C{}^a{}_\nu + g f_{su(3)}{}^{abc} A_C{}^b{}_\mu A_C{}^c{}_\nu \tag{19.35}$$

where $A_C{}^a{}_\nu$ is the color gluon complexon gauge field, and D_μ is defined by eq. 17.59, g is the color coupling constant, and the $f_{su(3)}{}^{abc}$ are the SU(3) structure constants.

In addition $\mathcal{L}_C{}^{ghost}$ contains the color SU(3) Faddeev-Popov ghost terms defined in appendix 19-A for the complexon Lorentz gauge. $\mathcal{L}_{CC}{}^{ghost}$ contains the complexon color SU(3) constraint (eq. 19-A.8) ghost terms also defined through the Faddeev-Popov mechanism. The mass sector terms $\mathcal{L}_{Mass}$ are Higgs terms or the Dimensional Mass Mechanism (sections 21.3 and 21.4) terms. They generates masses and generation mixing.

The coordinates are

$$x = (x^0, \mathbf{x}_r + i\mathbf{x}_i) \tag{19.36}$$

and the action is therefore

$$I_{CSM} = \int dx^0 d^3x_r d^3x_i \, \mathcal{L}_{CSM} \tag{19.37}$$

The lagrangian is supplemented with the following condition on all complexon fields $\Phi_{\dots}$:

$$\nabla_r \cdot \nabla_i \Phi\ldots = 0 \tag{19.38}$$

Non-complexon fields Ω... in the left-handed formulation under consideration satisfy the subsidiary condition:

$$[\nabla_r \cdot \nabla_i - (\nabla_r^2 \nabla_i^2)^{1/2}]\Omega\ldots = 0 \tag{19.39}$$

which guarantees a complexon's real momentum is parallel to its imaginary momentum.

19.4.1 Renormalization and Divergence Issues

The theory derived from $\mathcal{L}_{CSM}$ when calculated perturbatively in conventional quantum field theory has divergences that are not renormalizable due to its 7-dimensional nature. This issue does not appear to be curable by known renormalization methods.[305] However if the theory is reformulated in the two-tier formalism (next section and part 5) developed by Blaha[306] then a finite theory results. This is also true for the three or four generation Complexon Standard Model derived in the next chapter.

19.5 Two-Tier Formulation of the One Generation Complexon Standard Model

The general formulation of Two-Tier quantum field theory, and its application to the Standard Model, were presented in Blaha (2003) and Blaha (2005a).[307] In this section we will develop a Two-Tier formulation of the one generation Complexon Standard Model described in the previous section. This formulation goes beyond our earlier books by using complex spatial coordinates which necessitate a slightly more complicated Two-Tier formulation. The original Two-Tier formulation is described in detail in part 5. It applies unchanged to the part of the Complexon Standard Model that does not contain complexons.

The basic ansatz of the Two-Tier formalism is to replace every appearance of a coordinate x with a variable that is in part a quantum field:

$$x^\mu \to X^\mu = (y^0, \mathbf{y} + \mathbf{Y}(y^0, \mathbf{y})) \tag{19.40}$$

where $\mathbf{Y}(y^0, \mathbf{y})$ is the spatial part of a free massless vector field with features that are identical to the free QED field.

Then one finds that the momentum space free field Feynman propagators G(k) of all particles acquires a Gaussian factor exp(h(k)):

$$G(k) \to G(k) \exp(h(k)) \tag{19.41}$$

---

[305] The known methods pioneered by t'Hooft and Veltmann do renormalize the "Standard Standard Model" *IF Higgs particles are found experimentally.*

[306] Blaha (2005a) and Blaha (2003).

[307] See Part 5 for a detailed discussion. It contains Blaha (2005a).

so that all perturbation theory diagrams are finite. The result is a finite perturbative result for all calculations to any order in perturbation theory. Blaha (2005a) shows that Two-Tier theories are finite, Poincare covariant, and unitary. (See Blaha (2005a), part 5, for a complete discussion.)

19.5.1 Simple Two-Tier Formalism

In this subsection we will describe the basic Two-Tier formalism. Taking the lagrangian described in Blaha (2005A):[308]

$$\mathscr{L}(y) = \mathscr{L}_F(X_\mu(y))J + \mathscr{L}_C(X^\mu(y), \partial X^\mu(y)/\partial y^\nu, y) \qquad (19.42)$$

where

$$X_\mu(y) = y_\mu + i\, Y_\mu(y)/M_c^2 \qquad (19.43)$$

with M_C being a large mass scale, $Y_\mu(y)$ a vector quantum field, and where J is the absolute value of the Jacobian of the transformation from X to y coordinates:

$$J = |\partial(X)/\partial(y)| \qquad (19.44)$$

The lagrangian term $\mathscr{L}_C$ is

$$\mathscr{L}_C = +\tfrac{1}{4}\, M_c^4 F^{\mu\nu} F_{\mu\nu} \qquad (19.45)$$

with

$$F_{\mu\nu} = \partial X_\mu/\partial y^\nu - \partial X_\nu/\partial y^\mu \qquad (19.46)$$
$$\equiv i\,(\partial Y_\mu/\partial y^\nu - \partial Y_\nu/\partial y^\mu)/M_c^2$$

The sign in eq. 19.45 is not negative – contrary to the conventional electromagnetic Lagrangian. The reason for this difference is that the quantum field part of X^μ is imaginary. Thus $\mathscr{L}_C$ ends up having the correct sign after taking account of the factor of i in the field strength $F_{\mu\nu}$.

Defining

$$F_{Y\mu\nu} = (\partial Y_\mu/\partial y^\nu - \partial Y_\nu/\partial y^\mu) \qquad (19.47)$$

we see the Lagrangian assumes the form of the conventional electromagnetic Lagrangian:

$$\mathscr{L}_C = -\tfrac{1}{4}\, F_Y^{\mu\nu} F_{Y\mu\nu} \qquad (19.48)$$

The action of this theory has the form

[308] Eq. 7.1.

$$I = \int d^4y\, \mathscr{L}(y) \tag{19.49}$$

The further development of this theory is described in Part 3 (Blaha (2005a).)

19.5.2 Two-Tier L(C)-based Standard Model Theory

In the present case of the Complexon Standard Model we will need two variables X_r^μ and X_i^μ since we have complex spatial 3-coordinates. We define them similarly to the previous case:

$$X_{r\mu}(y_r) = y_{r\mu} + i\, Y_{r\mu}(y_r)/M_c^2 \tag{19.50}$$

$$X_{i\mu}(y_i) = y_{i\mu} + i\, Y_{i\mu}(y_i)/M_c^2 \tag{19.51}$$

where we choose the same mass scale for both the "real" and "imaginary" variables. The Two-Tier, single generation, version of the Complexon Standard Model given in eq. 19.37 then has an action of the form

$$I_{CSMtt} = \int dy^0 d^3y_r d^3y_i\, (\mathscr{L}_{CSM}(X_r^\mu(y_r), \mathbf{X}_i^k(y_i))J_2)|_{y_i^0 = 0,\, Y_r^0 = Y_i^0 = 0} +$$

$$+ \int dy_r^0 d^3y_r\, \mathscr{L}_C(X_r^\mu(y_r), \partial X_r^\mu(y_r)/\partial y_r^\nu, y_r) +$$

$$+ \int dy_i^0 d^3y_i\, \mathscr{L}_C(X_i^\mu(y_i), \partial X_i^\mu(y_i)/\partial y_i^\nu, y_i) \tag{19.52}$$

where the replacements

$$x^\mu \equiv x_r^\mu \rightarrow X_r^\mu(y_r) \tag{19.53}$$

$$x_i^k \rightarrow X_i^k(y_i) \tag{19.54}$$

for $\mu = 0, 1, 2, 3$ and $k = 1, 2, 3$ are made in $\mathscr{L}_{CSM}$ in eq. 19.37 followed by defining $y_r^0 = y^0$ and making an L(C) transformation to a frame where $y_i^0 = 0$. J_2 is the absolute value of the Jacobian of the transformation from (X_r, X_i) to (y_r, y_i) coordinates:

$$J_2 = |\partial(X_r, X_i)/\partial(y_r, y_i)| \tag{19.55}$$

We also choose gauges where $Y_r^0 = Y_i^0 = 0$. These types of transformations and gauge choices are discussed in detail in Blaha (2005a). The lagrangian terms $\mathscr{L}_C(X_r^\mu(y_r), \partial X_r^\mu(y_r)/\partial y_r^\nu, y_r)$ and $\mathscr{L}_C(X_i^\mu(y_i), \partial X_i^\mu(y_i)/\partial y_i^\nu, y_i)$ have the same form:

$$\mathscr{L}_C = +\tfrac{1}{4} M_c^4 F^{\mu\nu} F_{\mu\nu} \tag{19.56}$$

with

$$F_{\mu\nu} = \partial X_\mu/\partial y^\nu - \partial X_\nu/\partial y^\mu \tag{19.57}$$

$$\equiv i\,(\partial Y_\mu/\partial y^\nu - \partial Y_\nu/\partial y^\mu)/M_c^2 \tag{19.58}$$

or defining

$$F_{Y\mu\nu} = (\partial Y_\mu/\partial y^\nu - \partial Y_\nu/\partial y^\mu) \tag{19.59}$$

we see each lagrangian assumes the form of the conventional electromagnetic Lagrangian:

$$\mathscr{L}_C = -\tfrac{1}{4} F_Y^{\mu\nu} F_{Y\mu\nu} \tag{19.60}$$

The lagrangian is supplemented with the following condition on all complexon fields $\Phi_{...}$:

$$(\partial/\partial X_r^k(y_r))\,(\partial/\partial X_i^k(y_i))\Phi... = 0 \tag{19.61}$$

summed over k = 1, 2, 3. Non-complexon fields Ω... in our left-handed formulation satisfy the subsidiary condition:

$$\{(\partial/\partial X_r^k(y_r))(\partial/\partial X_i^k(y_i)) - [(\partial/\partial X_r^k(y_r))^2(\partial/\partial X_i^m(y_i))^2]^{1/2}\}\Omega... = 0 \tag{19.62}$$

summed over k = 1, 2, 3 and over m = 1, 2, 3 separately in each of the two terms.

19.5.3 Two-Tier Feynman Propagators for Real Space-time

The momentum space, free field, Feynman propagators G...(k) of all particles and ghosts in Two-Tier QFT acquire a Gaussian factor exp(h(k)):

$$G...(k) \rightarrow G...(k)\exp(h(k)) \tag{19.63}$$

so that all perturbation theory diagrams are finite. The consequence is finite results in all calculations to any order in perturbation theory. Blaha (2005a) shows that Two-Tier theories are finite, Poincaré covariant, and unitary.

An example of the Two-Tier effect on propagators is the Two-Tier photon propagator. The Two-Tier photon propagator[309] in Two-Tier QED is:

---

[309] Blaha (2005a).

$$iD_F^{TT}(y_1 - y_2)_{\mu\nu} = -i\int \frac{d^4p\, e^{-ip\cdot z}\, g_{\mu\nu}\, R(\mathbf{p}, z)}{(2\pi)^4\,(p^2 + i\varepsilon)} \qquad (19.64)$$

(since the imaginary parts can be taken to be zero: $y_{1i}^{\mu} - y_{2i}^{\mu} = 0$) where

$$z^{\mu} = y_{1r}^{\mu} - y_{2r}^{\mu} \qquad (19.65)$$

$$R(\mathbf{p}, \mathbf{z}) = \exp[-p^i p^j \Delta_{Tij}(z)/M_c^4] \qquad (19.66)$$
$$= \exp\{ -\mathbf{p}^2[A(v) + B(v)\cos^2\theta] / [4\pi^2 M_c^4|\mathbf{z}|^2]\} \qquad (19.67)$$

with i, j = 1, 2, 3, and with $\Delta_{Tij}(z)$ the commutator of the positive frequency part $Y^+_k(y)$ and the negative frequency part $Y^-_k(y)$ of $Y_k(y)$:

$$\Delta_{Tij}(z) = [Y^+_j(y_{1r}), Y^-_k(y_{2r})] = \int d^3k\, e^{ik\cdot(y_{1r} - y_{2r})}\,(\delta_{jk} - k_j k_k/\mathbf{k}^2)/[(2\pi)^3 2\omega_k] \qquad (19.68)$$

and

$$v = |z^0|/|\mathbf{z}|$$
$$A(v) = (1 - v^2)^{-1} + .5v\,\ln[(v - 1)/(v + 1)] \qquad (19.69)$$
$$B(v) = v^2(1 - v^2)^{-1} - 1.5v\,\ln[(v - 1)/(v + 1)] \qquad (19.70)$$
$$\mathbf{p}\cdot\mathbf{z} = |\mathbf{p}|\,|\mathbf{z}|\cos\theta \qquad (19.71)$$

with $|\mathbf{p}|$ denoting the length of a spatial vector $\mathbf{p}$, $|\mathbf{z}|$ denoting the length of a spatial vector $\mathbf{z}$, and with $|z^0|$ being the absolute value of z^0.

The gaussian factors R(**p**, z) which appear in all Two-Tier propagators damp the large momentum behavior of all perturbation theory integrals producing a completely finite perturbation theory and *yet give the usual results of perturbation theory at energies that are small compared to the mass scale M_c.*

19.5.4 Complexon Feynman Propagators

In the case of complexons the Two-Tier Feynman propagator differs from the non-complexon case by having an integration over imaginary spatial 3-momenta, a derivative of a delta function embodying the orthogonality of the real and imaginary 3-momenta, and two factors of R(**p**, z): one factor being $R(\mathbf{p_r}, z_r)$ and the other factor being $R(\mathbf{p_i}, z_i)$ (where the time components $z_r^0 = z^0$ and $z_i^0 = 0$ since there is only one real time coordinate[310]). Thus we obtain large momentum convergence for both real and imaginary 3-momentum integrations.

---

[310] We can arrange for $z_i^0 = 0$ by making a L_C transformation to an inertial frame where z is real.

For a scalar complexon particle in conventional quantum field theory we find the Feynman propagator:

$$i\Delta_{CTF}(x - y) = \theta(x^+ - y^+)\langle 0|\phi_{CT}(x)\,\phi_{CT}(y)|0\rangle + \theta(y^+ - x^+)\langle 0|\phi_{CT}(y)\phi_{CT}(x)|0\rangle$$

$$= i\int \frac{d^4p_r d^3p_i (2\pi)^{-7}\delta'(\mathbf{p}_r\cdot\mathbf{p}_i/m^2)e^{-ip^+(x^- - y^-) - ip^-(x^+ - y^+) + ip_\perp\cdot(x_\perp - y_\perp) - ip_i\cdot(x_i - y_i)}}{(p^2 + m^2 + i\varepsilon)}$$

(19.72)

In the case of Two-Tier quantum field a scalar complexon has the the Feynman propagator

$$i\Delta_{CTFtt}(x - y) = i\int d^4p_r d^3p_i (2\pi)^{-7}\delta'(\mathbf{p}_r\cdot\mathbf{p}_i/m^2)\, R(\mathbf{p}_r, z_r)R(\mathbf{p}_i, z_i)\cdot$$
$$\cdot\, e^{-ip^+(x^- - y^-) - ip^-(x^+ - y^+) + ip_\perp\cdot(x_\perp - y_\perp) - ip_i\cdot(x_i - y_i)}/(p^2 - m^2 + i\varepsilon) \qquad (19.73)$$

where the time components $z_r^0 = z^0$ and $z_i^0 = 0$ since there is only one time coordinate and where $p^2 = p^{0\,2} - p_r^2 + p_i^2$. Note both the real and imaginary integrations have gaussian damping at high momenta.

Propagators for other types of particles are similarly modified in the Two-Tier formalism (See Blaha 2005a).

The above discussion of Two-Tier theory appears to be quite complicated but it is simple enough in practice. One important feature of Two-Tier theory is that it produces the same results as conventional perturbation theory at low energies ($E << M_c$.) The mass scale M_c is assumed to be very large – perhaps of the order of the Planck mass.

19.6 Reason for Parity Violating Form of the Standard Model

Parity violation[311] that was discovered 50 years ago is now "explained" by the necessary appearance of tacyonic leptons and quarks in all forms of the derived Standard Model. Tachyonic particles require a left-handed (or right-handed) second quantization procedure. See part 4 for details.

---

[311] T. D. Lee and C. N. Yang, Phys. Rev. **104**, 254 (1956).

Appendix 19-A. Complexon Gauge Field Dynamics

19-A.1 Color SU(3)

Complexon quarks have a complex spatial 3-momentum. There are Dirac-type complexons and tachyonic complexons. If we think of quark gluon interactions, and in particular, perturbative diagrams with quark-gluon loops such as the simplest quark self-energy diagram, then imaginary momentum should "flow" around a loop as well as real momentum.

Gluons should be massless spin 1 complexon SU(3) gauge fields if quarks are complexons for the sake of consistency. More importantly, this is required because the color gluon fields appearing in the gauge covariant derivatives in the fermion field equations are necessarily functions of time and complex spatial 3-coordinates. The color gluon lagrangian density is

$$\mathcal{L}_{CCG} = -\tfrac{1}{4} F_{CC}^{a\mu\nu}(x) F_{CC\,\mu\nu}^{a}(x) \qquad (19\text{-A.1})$$

where

$$F_{C\,\mu\nu}^{a} = D_\nu A_{C\,\mu}^{a} - D_\mu A_{C\,\nu}^{a} + g f^{abc} A_{C\,\mu}^{b} A_{C\,\nu}^{c} \qquad (19\text{-A.2})$$

where D_μ is defined by eq. 17.59, g is the color coupling constant, and the f^{abc} are the SU(3) structure constants.

The theory of the strong color interaction must be developed within a path integral framework that takes account of Faddeev-Popov gauge fixing. Before doing so we must consider the effect of complex spatial 3-vectors on the non-abelian gauge formalism.

19-A.2 Pure Complexon Gauge Groups

The commutator of of gauge covariant complexon derivatives is

$$F_{C\mu\nu} = F_{C\,\mu\nu}^{a} t^a = (i/g)[\mathfrak{D}_\mu, \mathfrak{D}_\nu] \qquad (19\text{-A.3})$$

where $\mathfrak{D}_\mu$ is given by eq. 19.7. $F_{CC\mu\nu}$ is covariant under gauge transformations:

$$F_{CC\,\mu\nu}' = \Theta F_{CC\mu\nu} \Theta^{-1} \qquad (19\text{-A.4})$$

The strong interaction part of the action has terms of the form

$$I = \int d^7x[\mathcal{L}_{CCG} + \overline{\Psi}_q i\gamma^\mu \mathcal{D}_\mu \Psi_q + \ldots] \qquad (19\text{-A.5})$$

where the "extra" three integrations are over the imaginary spatial coordinates, and Ψ_q represents a Dirac quark complexon field. The gauge field part of the action must be supplemented with a constraint similar to those seen earlier for complexons, which ensures the reality of the lagrangian term $\mathcal{L}_{CCG}$ (and thus the corresponding Hamiltonian terms, and ultimately yields the unitarity of the theory.) The constraint simply specifies the imaginary part of $\mathcal{L}_{CCG}$ is zero:

$$[\partial A_C^{a\mu}/\partial x_{ik}]\mathrm{Re}\, F_C{}^a{}_{\mu k} = [\partial A_C^{a\mu}/\partial x_{ik}](\partial A_C{}^a{}_\mu/\partial x_r{}^k - \partial A_C{}^a{}_k/\partial x_r{}^\mu + g f^{abc} A_C{}^b{}_\mu A_C{}^c{}_k)$$

$$= 0 \qquad (19\text{-A.6})$$

where r indicates the real part of the coordinate $\mathbf{x} = \mathbf{x}_r + i\mathbf{x}_i$, and $\partial/\partial x_i^k$ is the derivative with respect to the kth component of the imaginary coordinate 3-vector $\mathbf{x}_i$. This restriction is not a choice of gauge but rather a restriction on the dependence of the $A_{C\nu}$ field on the real and imaginary spatial 3-vectors.

If we assume that we can integrate by parts in the imaginary coordinates then we can re-express the constraint as

$$A_C^{a\mu}\ \mathrm{Re}\ \partial F_C{}^a{}_{\mu k}/\partial x_i^k = 0 \qquad (19\text{-A.7})$$

which is explicitly

$$A_C^{a\mu}[\partial^2 A_C{}^a{}_\mu/\partial x_r{}^k \partial x_i{}^k - \partial^2 A_C{}^a{}_k/\partial x_r{}^\mu \partial x_i{}^k + g f^{abc}\partial(A_C{}^b{}_\mu A_C{}^c{}_k)/\partial x_i{}^k] = 0 \qquad (19\text{-A.8})$$

where $x_r^0 = x^0$. This restriction can be implemented within the framework of the path integral formalism by using the Faddeev-Popov Method with the introduction of ghosts in a manner similar to the use of the Faddeev-Popov Method to implement gauge conditions. The form of the restriction stated in eqs. 19-A.7 and 19-A.8 leads to second order derivative Faddeev-Popov ghost terms in the path integral while eq. 19-A.6 would lead to a fourth order derivative ghost terms in the path integral with attendant unitarity (and possibly other) issues.

19-A.3 Pure Gauge Complexon Path Integral Formulation and Faddeev-Popov Method

The path integral formalism for complexon, non-abelian, pure, Yang-Mills fields differs significantly from the conventional gauge field path integral formalism. The path integral for a complexon gauge field can be written symbolically as:

$$Z(J^{\mu})=N\int DA_C\Delta_{FP}(A_C)\delta(F(A_C))\Delta_C(A_C)\delta(F_C(A_C))\exp\{i\int d^7y[\mathscr{L}+J^{\mu}(y)A_{C\mu}(y)]\}$$
(19-A.8a)

where $\delta(F(A_C))$ specifies the gauge, $\Delta_{FP}(A_C)$ is its Faddeev-Popov determinant; and $\delta(F_C(A_C))$ specifies the complexon condition (eq. 19-A.8) with $\Delta_C(A_C)$ the Faddeev-Popov determinant for the complexon condition. In both cases the Faddeev-Popov determinant can be calculated in the standard way.[312]

First we consider the gauge fixing delta function. Note that it can be written as a delta function in the gauge times a determinant:

$$\delta(F(A_C{}^{\omega})) = \delta(\omega - \omega_0)|\det \delta F(A_{C\mu}{}^{\omega}(x))/\delta\omega(x)|^{-1}|_{F(A_C)=0} \tag{19-A.9}$$

where ω_0 is a reference gauge, where

$$\begin{aligned} A_C{}^{a\,\omega}_{\mu}(x) &= A_C{}^{a}_{\mu}(x) - g^{-1}D_{\mu}\omega^a + f^{abc}\,\omega^b(x)A_C{}^{c}_{\mu}(x) \\ &= A_C{}^{a}_{\mu}(x) + \delta A_C{}^{a\,\omega}_{\mu}(x) \end{aligned} \tag{19-A.10}$$

and

$$\begin{aligned} \text{Re } F_C{}^{a}_{\mu k}{}^{\omega} &= \text{Re } F_C{}^{a}_{\mu k} + f^{abc}\,\omega^b(x)\text{Re } F_C{}^{c}_{\mu k} \\ &= \text{Re } F_C{}^{a}_{\mu k} + \delta(\text{Re } F_C{}^{a}_{\mu k}{}^{\omega}) \end{aligned} \tag{19-A.11}$$

under an infinitesimal gauge transformation, and where

$$\Delta_{FP}(A_C) = |\det \delta F(A_{C\mu}{}^{\omega}(x))/\delta\omega(x)||_{F(A_C)=0,\ \omega=0} \tag{19-A.12}$$

We will choose the complexon Lorentz gauge to evaluate the Faddeev-Popov determinant:

$$F^a(A_C) = D_{\mu}A_C{}^{a\mu}(x) = 0 \tag{19-A.13}$$

We find

$$\begin{aligned} F^a(A_{C\mu}{}^{\omega}(x)) &= D^{\mu}(A_C{}^{a}_{\mu}(x) - g^{-1}D_{\mu}\omega^a(x) + f^{abc}\,\omega^b(x)A_C{}^{c}_{\mu}(x)) \\ &= -g^{-1}D^{\mu}D_{\mu}\omega^a(x) + f^{abc}A_C{}^{c}_{\mu}(x)D^{\mu}\omega^b(x) \end{aligned} \tag{19-A.14}$$

Thus

$$\delta F^a(A_{C\mu}{}^{\omega}(x))/\delta\omega^b(x) = -g^{-1}\,\delta^{ab}D^{\mu}D_{\mu} + f^{abc}A_C{}^{c\mu}(x)D_{\mu} \tag{19-A.15}$$

---

[312] See for example Huang (1992).

and

$$\Delta_{FP}(A_C) = |\det(g^{-1}\delta^{ab}D^\mu D_\mu - f^{abc}A_C{}^{c\mu}(x)D_\mu)| \tag{19-A.16}$$

where | ... | represent absolute value.

We can rewrite the Faddeev-Popov determinant as a path integral over anti-commuting c-number fields with a ghost Lagrangian:

$$\Delta_{FP}(A_C) = \int D\chi^* D\chi \exp[\, i\int d^7x\, \mathscr{L}_C^{\,ghost}(x)] \tag{19-A.17}$$

where

$$\mathscr{L}_C^{\,ghost}(x) = \chi^{a*}(x)[\delta^{ab}D^\mu D_\mu - gf^{abc}A_C{}^{c\mu}(x)D_\mu]\chi^b(x) \tag{19-A.18}$$

19-A.3.1 Faddeev-Popov Application to the Complexon Condition

The complexon condition can also be implemented within the path integral formalism using the Faddeev-Popov Mechanism. Using the identity

$$1 = \int DA_C\Delta_C(A_C)\delta(F_C(A_C)) \tag{19-A.19}$$

we see that an infinitesimal gauge transformation yields eqs. 19-A.10 and 19-A.11. This enables us to relate $\Delta_C(A)$ to the determinant

$$\delta(F_C(A_C{}^\omega)) = |\det \delta F_C(A_{C\mu}{}^\omega(x))/\delta\omega(x)|^{-1}|_{F_C(A_C)=0,\ \omega=0} \tag{19-A.20}$$

and

$$\Delta_C(A_C) = |\det \delta F_C(A_{C\mu}{}^\omega(x))/\delta\omega(x)||_{F_C(A_C)=0,\ \omega=0} \tag{19-A.21}$$

From eq. 19-A.8 we see

$$F_C(A_{C\mu}(x)) = A_C{}^{a\mu}[\partial^2 A_C{}^a{}_\mu/\partial x_r{}^k\partial x_i{}^k - \partial^2 A_C{}^a{}_k/\partial x_r{}^\mu\partial x_i{}^k + \\ + gf^{abc}\partial(A_C{}^b{}_\mu A_C{}^c{}_k)/\partial x_i{}^k] \tag{19-A.22}$$

with

$$F_C{}^a(A_{C\mu}) = 0 \tag{19-A.23}$$

Inserting eq. 19-A.10 and 19-A.11 we find

$$[\delta F_C(A_{C\mu}{}^\omega(x))/\delta\omega^a(x)]|_{F_C(A_C)=0,\ \omega=0} = \delta[\delta A_C{}^{b\mu\omega}\,\mathrm{Re}\,\partial F_C{}^b{}_{\mu k}/\partial x_{ik} + \\ + A_C{}^{b\mu}\,\partial\delta(\mathrm{Re}\,F_C{}^b{}_{\mu k}{}^\omega)/\partial x_{ik}\,]/\delta\omega^a(x)|_{\omega=0} \\ = -g^{-1}(\mathrm{Re}\,\partial F_C{}^a{}_{\mu k}/\partial x_{ik})D^\mu - f^{abc}A_C{}^{b\mu}(\mathrm{Re}\,F_C{}^c{}_{\mu k})\partial/\partial x_{ik} \tag{19-A.24}$$

Thus

$$\Delta_C(A_C) = |\det (g^{-1}(\mathrm{Re}\ \partial F_{C\ \mu k}^{\ a}/\partial x_{ik})D^\mu + f^{abc}A_C^{\ b\mu}(\mathrm{Re}\ F_{C\ \mu k}^{\ c})\partial/\partial x_{ik}| \quad (19\text{-A}.25)$$

where | … | represents absolute value.

We can rewrite this Faddeev-Popov determinant as a path integral over anti-commuting c-number fields with a ghost Lagrangian:

$$\Delta_C(A_C) = \lim_{r \to \infty} \int D\chi_C^{\ *}D\chi_C \exp[ir^{-2}\int d^7x\ \mathscr{L}_{CC}^{\ \ ghost}(x)] \quad (19\text{-A}.26)$$

where r is a constant that is taken to the limit ∞, and where

$$\mathscr{L}_{CC}^{\ ghost}(x) = \chi_C^{\ *}(x)\{D^\mu D_\mu + r^2t^a[(\mathrm{Re}\ \partial F_{C\ \mu k}^{\ a}/\partial x_{ik})D^\mu + \\ + gf^{abc}A_C^{\ b\mu}(\mathrm{Re}\ F_{C\ \mu k}^{\ c})\partial/\partial x_{ik}]\}\chi_C(x) \quad (19\text{-A}.27)$$

where t^a is a 3 × 3 matrix of the 3 representation of color SU(3) and $\chi_C(x)$ is a three row field in the 3 representation. *The introduction of $D^\mu D_\mu$ is based on consistency with the complexon formalism. It is needed to establish a perturbative expansion of the path integral. Its effect vanishes in the limit $r \to \infty$ reducing ghost loops of this type to point interactions.* The reader will note that second order and third order derivative terms appear in the interaction in $\mathscr{L}_C^{\ ghost}(x^\mu)$ and raise the issue of non-renormalizable divergences. If one uses the Two-Tier approach to quantum field theory developed by Blaha (2005a) then all potential divergences disappear. *The Two-Tier formulation of the pure, complexon, Yang-Mills theory that we are discussing is finite in perturbation theory.*

The complete pure complexon, Yang-Mills path integral is

$$Z(J^\mu) = N\int DA_C D\chi^* D\chi D\chi_C^{\ *} D\chi_C \Delta_{FP}(A_C)\delta(F(A_C))\Delta_C(A_C)\delta(F_C(A_C))\cdot \\ \cdot\exp\{i\int d^7y\ [\mathscr{L}_{CCG} + J^\mu A_{C\mu}]\} \quad (19\text{-A}.28)$$

Eq. 19-A.28 appears previously in eq. 19.33 in the definition of the complexon color gauge lagrangian terms.

19-A.4 Complexon Quark-Gluon Perturbation Theory

In this section we will give an impression of complexon strong interaction theory by considering a simple diagram. Fig. 19-A.1 is a self-energy diagram for a quark in which a gluon is emitted and absorbed. Complexon particles have complex spatial momentum. From the form of the propagators and wave functions previously considered it is clear that the real part of the 3-momentum and the imaginary part of the 3-momentum will be separately conserved at each vertex. The general form of the self-energy integral corresponding to Fig. 19-A.1 is

$$I_{ab}(p) = \int d^7q\, P_{ab}(q, p)\exp(\text{Gaussian})/Q(q, p) \qquad (19\text{-A.29})$$

where a and b are SU(3) color indices, $P_{ab}(q, p)$ is a polynomial in q and p together with Dirac gamma matrix factors and color SU(3) matrix factors, and Q(q, p) is the product of a quark and a gluon propagator denominator factor. A Gaussian exponential factor appears if we use a Two-Tier formulation[313] of the complexon quark and gluon theory. This exponential factor guarantees the convergence of the integral yielding a finite result.

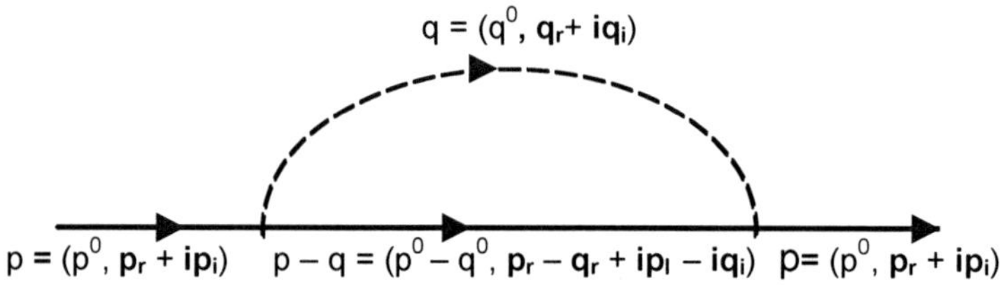

Figure 19-A.1. Quark self-energy diagram in which a quark emits and subsequently absorbs a gluon. The quark momentum is $p = (p^0, \mathbf{p}_r + i\mathbf{p}_i)$ and the gluon momentum is $q = (q^0, \mathbf{q}_r + i\mathbf{q}_i)$ with conservation of energy, and separate conservation of real spatial momentum and imaginary spatial momentum at each vertex.

Thus we see complexon perturbation theory is well defined, finite, and sensible if one uses Two-Tier quantum field theory.

The integral corresponding to Fig. 19-A.1 has the form:

$$I_{1ab}(p) = \int dq^0 d^3q_r d^3q_i\, P_{ab}(q, p)(\text{Gaussian factors})/Q(q, p) \qquad (19\text{-A.30})$$

and includes integrations over both real and imaginary 3-momenta.

19-A.5 Complexon Quark ElectroWeak Interactions

In the Complexon Standard Model quarks and gluons are complexons with complex spatial 3-momenta. On the other hand, the ElectroWeak bosons: $W^{\pm}_{\mu}$, Z_{μ}

313 See Blaha (2005a).

and A_μ are not complexons and so can be viewed as having totally real 3-momenta (or complex 3-momenta with parallel real and imaginary parts). In this section we address the issue of quark ElectroWeak perturbation theory. *The general perturbation theory rule for this situation again is that real and imaginary momenta are separately conserved at each vertex.*

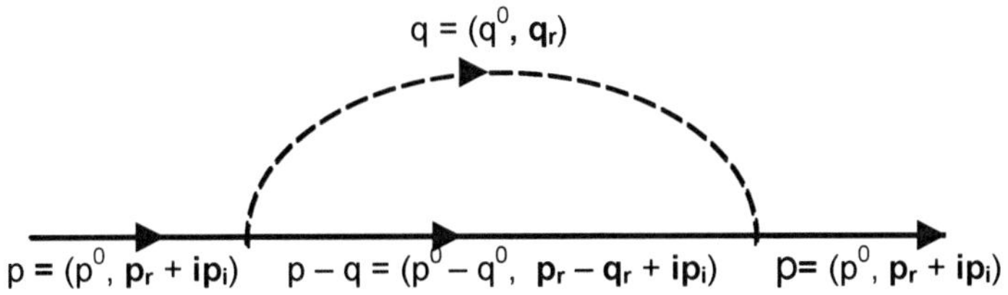

Figure 19-A.2. Quark self-energy diagram in which a quark emits and subsequently absorbs a photon. The quark momentum is $p = (p^0, \mathbf{p}_r + i\mathbf{p}_i)$ and the photon momentum is $q = (q^0, \mathbf{q}_r)$ with conservation of energy, and separate conservation of real spatial momentum and imaginary spatial momentum at each vertex.

Fig. 19-A.2 shows a quark self-energy diagram in which a quark emits a photon and subsequently reabsorbs it. Since ElectroWeak bosons can be viewed as having real energies and 3-momenta, and since real and imaginary momenta are separately conserved at each vertex by the above stated rule, the photon momentum is real while the quark spatial 3-momentum is complex.

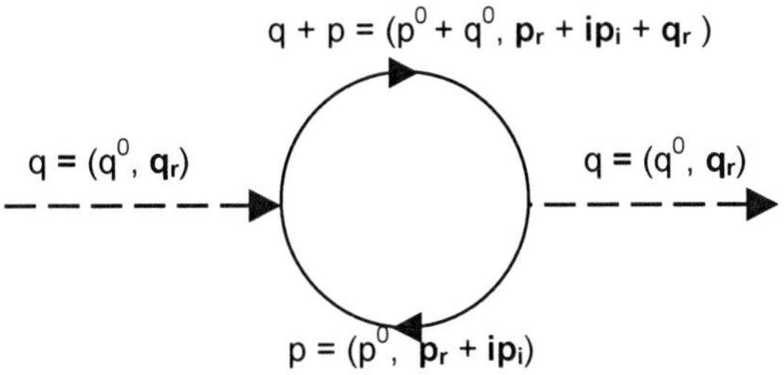

Figure 19-A.3. Photon self-energy diagram with a quark loop. The photon momentum is $q = (q^0, \mathbf{q}_r)$ and quark loop momentum is $p = (p^0, \mathbf{p}_r + i\mathbf{p}_i)$.

The integral corresponding to Fig. 19-A.2 has the form:

$$I_{2ab}(p) = \int dq^0 d^3q_r d^3q_i \delta^3(\mathbf{q}_i) P_{ab}(q, p)(\text{Gaussians})/Q(q, p)$$
$$= \int dq^0 d^3q_r P_{ab}(q, p)(\text{Gaussians})/Q(q, p) \qquad (19\text{-A}.31)$$

In cases where there are complexon quark loops or gluon loops the imaginary quark or gluon 3-momenta affects the result. For example, Fig. 19-A.3 shows a complexon quark loop contribution to the photon self-energy in which the imaginary quark 3-momenta integration contributes.

Thus complexon imaginary 3-momenta can appear in many calculations in the Complexon Standard Model.

*20. Origin of the Generations

In the preceding chapters we have seen how most of the features of the Standard Model originate in the L(C) group: Parity Violation, the four species of fermions, the ElectroWeak sector, WIMPs, and color SU(3).

The three known generations of each species appear to have a different origin. Unlike the ElectroWeak vector bosons, whose masses and mixing angle (the Weinberg angle) are fairly well known, the masses of the fermions are only well known for the charged leptons, and the masses and mixing angles are only partly known for the quark species. Neutrino masses are known to be small and some mixing angles are approximately known.[314]

This experimental situation opens the door to many theoretical approaches to the determination of the values of physical particle masses and mixing angles between generations.

In this chapter, based on the 4-valued spatial asynchronous logic described in sections 14.7–14.9 we will assume there are 4 generations initially describable by the 4 truth value matrix formalism of section 10.3. The 4×4 matrices of 4-valued matrix logic are mathematically equivalent to Dirac matrices. If we assume a map to Dirac spinors, then the space-time of the spinors must be a 4-dimensional space-time which we can assume initially has the group structure of GL(4), denoted GL'(4). In analogy with our space-time physics we assume that Dirac-like equations describe the dynamics of particles in this space-time. Thus GL'(4) space-time and its features are to some extent a mirror of our space-time and its features. We suggest (in chapter 21) that similar interactions may exist in the GL'(4) space-time. We also suggest the GL'(4) space-time may be restricted to a point. As far as we know (for the current highest energies) the GL'(4) space-time is inaccessible.

The above discussion can be summarized by the following postulates:

Postulate 20.1. Spatial asynchronicity of physical phenomena requires a 4-valued logic that can defined in a 4×4 matrix formulation.

Postulate 20.2. The 4×4 matrix formulation can be mapped to Dirac spinors in a different four dimensional space-time (not the space-time of our experience) and, upon introducing coordinates, serve as the starting point for the definition of Dirac-like equations.

---

[314] E. Ma, arXiv:hep-ph/0612013 (2006) and references therein; P. F. Harrison, D. H. Perkins, and W. G. Scott, Phys. Lett. **B530**, 167 (2002).

Postulate 20.3 For each fermion species the four solutions of each Dirac-like equation in GL'(4) *are the four generations of the species.*

Comment: The mixing of the four generations forces the ElectroWeak and color SU(3) interactions to be the same for the four generations.

In the next chapter we will describe the dynamics of the GL'(4) (space-time) generations in some detail. In addition we describe a non-Higgsian mechanism, Dimensional Mass Generation, that gives masses to particles and can produce mixing angles between generations.[315]

---

[315] The four generation theory presented could alternately accommodate Higgs mass generation should Higgs particles be found.

21. The Dimensional Mass Mechanism

21.1 The Origin of Mass

The origin of mass is both a physics question and a metaphysical question. On the metaphysical question we defer consideration. On the physical question, the Higgs Mechanism is a solution that is widely accepted and meets the needs for mass terms in the Standard Model. However, the Higgs particles that are required by this mechanism have not been found experimentally.

In this chapter we will consider the possibility of a different mechanism for generating the known fermion and vector boson particle masses. This mechanism has several advantages over the Higgs Mechanism:

1. Economy – no extra particles beyond the known Standard Model particles are required.
2. It fits very nicely within the GL'(4) explanation of the origin of the fermion generations as introduced in the previous chapter.
3. It is well motivated by the method through which the curvature constant, usually denoted k, with the dimensions of $[\text{mass}]^2$ arises in the derivation of the Robertson-Walker metric of General Relativity.

Thus this new approach to the origin of mass has both an experimental and theoretical advantage over the Higgs Mechanism at the present time.

21.2 The Origin of k (with Dimension $[\text{mass}]^2$) in the Robertson-Walker Metric

The Robertson-Walker metric:

$$d\tau^2 = dt^2 - R^2(t)[dr^2/(1 - kr^2) + r^2(d\theta^2 + \sin^2\theta\, d\varphi^2)] \qquad (21.1)$$

contains the constant k which is a measure of the cuvature of the universe. This constant has the dimensions of $[\text{mass}]^2$ and arises in the solution of the Einstein equations of General Relativity. As Weinberg (1972) and Blaha (2004) show[316] the constant k arises as a dimensionful constant in the separation of the Einstein equations. In particular Blaha (2004) obtained the solution for a generalization of the generalized Robertson-Walker metric equation

[316] Weinberg (1972) p. 343 and Blaha (2004) p. 174.

$$\dot{a}^2 - 8\pi G\rho a^2/3 - (\check{r}b^2)^{-1}\,\partial(\check{r}\,b'/b)/\partial\check{r} = 0 \qquad (21.2)$$

where a (the scale factor) and b are certain functions of time and a radial coordinate ř, by noting that the first two terms in eq. 21.2 are solely functions of t while the third term is solely a function of a radial coordinate ř. Thus we can separate the differential equation in the usual way to obtain:

$$\dot{a}^2 - 8\pi G\rho a^2/3 = -k \qquad (21.3)$$

and

$$k + (\check{r}b^2)^{-1}\,\partial(\check{r}\,b'/b)/\partial\check{r} = 0 \qquad (21.4)$$

where k is the curvature parameter k in eq. 21.1 after some manipulations.

Thus we find that, in perhaps the most fundamental of theories, that a dimensionful parameter arises as a separation constant. This is the key to an alternate mechanism to generating particle masses.

*21.3 Origin of the Fermion Masses and Mass Mixing Matrices

Having seen that a parameter k with the dimensions of mass squared naturally appears in certain separable solutions of the Einstein equations (in particular Robertson-Walker solutions), we now consider the possibility that the masses of fermions appear as separation constants in GL(4)⊗GL'(4) Dirac-like equations. We will call this method of giving particles masses the *Dimensional Mass Mechanism*.[317]

In chapter 20 we saw how a Standard Model with four generations generations of fermions could be derived by assuming that each species of fermion were 4-vectors – four generations of each species including WIMPs..

A free Dirac-like dynamical equation is linear in coordinate derivatives and has the general form

$$(\gamma^\mu \mathcal{D}_\mu(x) - m)\psi(x) = 0 \qquad (21.5)$$

In the case of GL(4)⊗GL'(4), if we denote the GL(4) 4-vector coordinates as x^ν and the GL'(4) 4-vector coordinates as y^μ then each species[318] Dirac-like equation becomes a 4 generation first order Dirac-like equation of the form[319]

---

[317] We use the phrase "dimensional mass" since the origin of masses in this approach follows from the separation of the part of the dynamical equations dependent on the coordinates (dimensions) of GL'(4).

[318] The species of fermions are charged leptons, uncharged leptons, leptonic WIMPs, up-type quarks, down-type quarks, and quark-like WIMPs.

$$[D_{GL(4)}\otimes I_{GL'(4)} + I_{GL(4)}\otimes D_{GL'(4)} - m_0 I_{GL(4)}\otimes I_{GL'(4)}]\psi(x, y) = 0 \qquad (21.6a)$$

where I represents the identity matrix, or explicitly with indices displayed:

$$(\gamma^{\nu}{}_{ab}\mathcal{D}_{\nu}(x)\delta_{cd} + \gamma'^{\mu}{}_{cd}\mathcal{D}'_{\mu}(y)\delta_{ab} - m_0\delta_{ab}\delta_{cd})\psi_{bd}(x, y) = 0 \qquad (21.6b)$$

where the γ'^{μ} matrices are Dirac matrices with the spinor indices c and d, of GL'(4). The γ^{μ} matrices are Dirac matrices with spinor indices a and b of GL(4). In addition, $\mathcal{D}'_{\mu}(y)$ is the GL'(4) differential operator, m_0 is the bare mass, and d will become the generational index whereas b is a spinor index of our space-time.

Eq. 21.6 can be derived from a lagrangian density

$$\mathcal{L} = \psi_{ge}^{\dagger}(x, y)\gamma^{0}{}_{gc}\gamma^{0}{}_{ea}(\gamma^{\nu}{}_{ab}\mathcal{D}_{\nu}(x)\delta_{cd} + \gamma'^{\mu}{}_{cd}\mathcal{D}'_{\mu}(y)\delta_{ab} - m_0\delta_{ab}\delta_{cd})\psi_{bd}(x, y)$$

or

$$\mathcal{L} = \bar{\psi}(x, y)[D_{GL(4)}\otimes I_{GL'(4)} + I_{GL(4)}\otimes D_{GL'(4)} - m_0 I_{GL(4)}\otimes I_{GL'(4)}]\psi(x, y) \qquad (21.6c)$$

Eqs. 21.6 has similar $\mathcal{D}_{\nu}(x)$ and $\mathcal{D}'_{\mu}(y)$ differential operators. These operators are of the forms corresponding to eqs. 17.17, 17.32, 17.49, and 17.52. Thus, for example, for neutrinos both $\mathcal{D}_{\nu}(x)$ and $\mathcal{D}'_{\mu}(y)$ have the form $i\gamma^{\mu}\partial/\partial z^{\mu}$ where z = x and z = y respectively. Consequently there are four equations of the form of eq. 21.6 – one equation for each fermion species. The 4 solutions $\psi_{bd}(x, y)$ for d = 1, 2, 3, 4 correspond to the four generations of a species. Taking account of WIMPs there are 4 generations of six species and thus there will be 6 mass matrices when we separate the equations.

Eq. 21.6 is a manifestly separable equation. We can express the wave function of a species with 4 generations as[320]

$$\psi_{bd}(x, y) = \psi_{bd}(x)\psi'_{d}(y) \qquad (21.7)$$

An index for each of the 6 species is not displayed for simplicity and also because eq. 21.6 does not mix species. If we introduce a matrix of separation constants $\mathcal{M}_{cd}$ then we find that eq. 21.6 is solved using the solutions of the separated equations:[321]

---

[319] The form is largely determined by the requirement that the Dirac equation be a first order equation.

[320] The reader who is concerned about treating the wave function classically rather than as a second quantized field should remember that the manipulations that we are performing can be redone in a path integral formalism in which the manipulation of fields as classical fields is perfectly acceptable.

[321] The appearance of m_0 in eq. 21.8 rather than eq. 21.9 is arbitrary. It could have been otherwise with no change in results by simply shifting the $\mathcal{M}_{cd}$ diagonal values by m_0.

$$[\gamma'^{\mu}{}_{cd}\mathcal{D}'_{\mu}(y) - \mathcal{M}_{cd} - m_0\delta_{cd}]\psi'_d(y) = 0 \tag{21.8}$$

$$(\gamma^{\nu}{}_{ab}\mathcal{D}_{\nu}(x)\delta_{cd} - \mathcal{M}_{cd}\delta_{ab})\psi_{bd}(x) = 0 \tag{21.9a}$$

or, in matrix notation,

$$(\gamma^{\nu}\mathcal{D}_{\nu}(x)\mathrm{I} - \mathcal{M})\psi(x) = 0 \tag{21.9b}$$

where $\psi(x)$ is a vector composed of the four generations of a fermion species:

$$\psi(x) = \begin{bmatrix} \psi_{d=1} \\ \ldots \\ \psi_{d=4} \end{bmatrix} \tag{21.10}$$

with each ψ_d a four component coordinate space Dirac spinor.[322]

Eq. 21.9 exhibits the generation mixing of the separation constants $\mathcal{M}_{cd}$ which, in general, constitute a complex, non-diagonal mass matrix *for each species* of fermions. Thus there are six fermion mass matrices $\mathcal{M}$ that lead to the mixing matrices of the charged ElectroWeak currents.[323] This includes two mass matrices for the quark-like and lepton-like WIMPs.

If the coordinate space for GL'(4) is compressed to a point, then eq. 21.9 becomes a relation between the field and its derivatives in that limit.

$\mathcal{M}$ must be a non-singular generation mixing matrix. This mixing matrix may reflects the manner of the breakdown of the original symmetry from GL(16) to GL(4)⊗GL'(4). In the absence of detailed information about the cause and nature of this breakdown mechanism, if there is a breakdown, we will not speculate on its form and merely assume it appears in the broken theory.

The separation constants $\mathcal{M}$ must have the dimensions of [Mass] according to eq. 21.6. Thus we obtain particle masses as separation constants[324] just as the constant k with the dimension of $[\mathrm{Mass}]^2$ appears in the solution for the Robertson-Walker metric. This simple mechanism—the Dimensional Mass

---

[322] We note that $\psi'_d(y)$ is integrated out of any expectation value/scattering amplitude if there are no interactions in y-space. Thus its only role is to support generations and the construction of the mass matrices.

[323] The neutral ElectroWeak currents do not have generational mixing.

[324] The philosophically inclined reader may balk at thrusting the origin of the known physical fermion masses onto an unseen point-like universe, which is, in the present state of physical knowledge, unknown. However this assumption is in reality a challenge to examine that universe in more detail—something which is beyond the scope of the present work.

Mechanism—for fermion mass generation does not require the introduction of hitherto unseen scalar fields.

Once we have obtained equations of the form of eq. 21.9 for each of the fermion species then we can derive the leptonic and quark ElectroWeak charge changing currents. Before discussing ElectroWeak currents we discuss the transformation of each mass matrix $\mathcal{M}$ to diagonal form. We rely on a mathematical theorem to the effect that a complex matrix can be diagonalized (with non-negative real eigenvalues) using a transformation of the form

$$M_s = \mathcal{A}_{sL}\mathcal{M}_s\mathcal{A}_{sR}^{-1} \qquad (21.11)$$

where $\mathcal{A}_{sL}$ and $\mathcal{A}_{sR}$ are special unitary matrices and where M_s is diagonal. The subscript s represents the fermion species. Consequently if we apply the twelve $\mathcal{A}$ matrices to the left-handed and right-handed generation 4-vectors for each of the fermion species: s = e (charged leptons), ν (neutral leptons), leptonic WIMPs, u (up-type quarks), d (down-type quarks), and quark-type WIMPs:

$$\begin{bmatrix} s_{1L} \\ s_{2L} \\ s_{3L} \\ s_{4L} \end{bmatrix} \rightarrow \mathcal{A}_{sL} \begin{bmatrix} s_{1L} \\ s_{2L} \\ s_{3L} \\ s_{4L} \end{bmatrix} = s_L \qquad (21.12)$$

$$\begin{bmatrix} s_{1R} \\ s_{2R} \\ s_{3R} \\ s_{4R} \end{bmatrix} \rightarrow \mathcal{A}_{sR} \begin{bmatrix} s_{1R} \\ s_{2R} \\ s_{3R} \\ s_{4R} \end{bmatrix} = s_R \qquad (21.13)$$

then we obtain the physical species 4-vector states s_L and s_R. Also the six diagonalized mass matrices M_s appear in the ElectroWeak lagrangian for the physical particle 4-vectors e, ν, w_l, u, d and w_q:[325]

$$\mathcal{L}_{mass} = \bar{e}M_e e + \bar{\nu}M_\nu \nu + \bar{u}M_u u + \bar{d}M_d d + \bar{w}_l M_{wl} w_l + \bar{w}_q M_{wq} w_q \qquad (21.14)$$

With the transformation of generation 4-vectors to physical particle mass eigenstates we find the leptonic charge changing current has the form:

---

[325] w_l and w_q are the leptonic and quark-like WIMP species respectively.

$$J_{leptonicCC}{}^{\alpha} \sim \bar{e}_L\gamma^{\alpha}C_{\ell}\nu_L + \bar{e}_L\gamma^{\alpha}C_{ew}w_{lL} \tag{21.15a}$$

and the neutral leptonic current has the form:

$$J_{leptonicNeut}{}^{\alpha} \sim \bar{\nu}_L\gamma^{\alpha}C_{\nu w}w_{lL} \tag{21.15b}$$

where e, ν and w are physical mass eigenstate generation 4-vectors and the special unitary leptonic mixing matrices are

$$C_{\ell} = \mathcal{A}_{eL}\mathcal{A}_{\nu L}{}^{-1} \tag{21.16a}$$
$$C_{ew} = \mathcal{A}_{eL}\mathcal{A}_{wL}{}^{-1} \tag{21.16b}$$
$$C_{\nu w} = \mathcal{A}_{\nu L}\mathcal{A}_{wL}{}^{-1} \tag{21.16c}$$

The quark charge changing current has the form:

$$J_{quark}{}^{\alpha} \sim \bar{d}_L\gamma^{\alpha}C_q u_L \tag{21.17}$$

where u and d are physical mass eigenstate generation 4-vectors and the quark mixing matrix is

$$C_q = \mathcal{A}_{dL}\mathcal{A}_{uL}{}^{-1} \tag{21.18}$$

All the mixing matrices C_i are special unitary:

$$C_i\, C_i^{\dagger} = I \tag{21.19}$$
$$C_i^{-1} = C_i^{\dagger}$$
$$|\det C_i|^2 = 1$$

for i = ℓ, q.

Thus we find that the generalization of our four Dirac-like fermion equations to GL(4)⊗GL'(4) in eq. 21.6 naturally leads to fermion generations, differing fermion masses, and fermion generation mixing in general. Consequently there are now two possible origins of particle masses and mixing angles:

Postulate 21.1a. The masses, and mixing angles, of fermions result from the Higgs mechanism.

Postulate 21.1b. The masses, and mixing angles, of fermions appear as separation constants in GL(4)⊗GL'(4) fermion dynamical equations.

*21.4 A Group Structure of Generational Mixing Matrices

In preceding chapters we developed the implications of the complex Lorentz group for the form of the one generation fermion spectrum. We found that SU(2)⊗U(1)⊕U(1) group transformations were physically required to relate the real coordinates of one reference frame to the real physical coordinates of a reference frame related to it via a superluminal transformation. Then we found that this coordinate transformation group led to a symmetry of the fundamental fermion fields yielding ElectroWeak triplets of leptons and quarks (at the level of one generation.) In this derivation in Appendix 18-A we found that the Dirac matrices had the transformation law

$$V\gamma^{\mu}V^{-1} = \Pi_{\nu}{}^{\mu *}\gamma'^{\nu} \qquad (18\text{-A}.8)$$

where V is a 4×4 SU(2)⊗U(1)⊕U(1) matrix. Thus the free fermion equations of motion are covariant under the global SU(2)⊗U(1)⊕U(1) transformations:

$$V[iD\cdot\gamma - m]V^{-1}V\psi(x) = 0 \qquad (18\text{-A}.9)$$

or

$$[iD'\cdot\gamma' - m]V\psi(x) = 0 \qquad (18\text{-A}.10)$$

In section 21.3 we considered the general form of free Dirac-like equations including both the Dirac equation, the tachyon equation and the two types of complexon equations:

$$[\gamma'^{\mu}{}_{cd}\mathcal{D}'_{\mu}(y) - \mathcal{M}_{cd} - m_0\delta_{cd}]\psi'_d(y) = 0 \qquad (21.8)$$

in the space of GL'(4) with a corresponding Dirac-like equation (eq. 21.9) in our GL(4) space-time.

Let us assume that the structure of the GL'(4) space mirrors the structure of our space-time with three spatial coordinates and one time coordinate (although its space-time extent may have been shrunk to a point.) Thus the complex Lorentz group transformations as well as the SU(2)⊗U(1)⊕U(1) transformation group are the relevant physical transformation groups. Let us now apply the transformation V to eq. 21.8 introducing double prime characters " for the transformed quantities:

$$V_{ec}[\gamma'^{\mu}{}_{cd}\mathcal{D}'_{\mu}(y) - \mathcal{M}_{cd} - m_0\delta_{cd}]V_{da}{}^{-1}V_{an}\psi'_n(y) = 0$$
$$[\gamma''^{\nu}{}_{ea}\Pi_{\nu}{}^{\mu *}\mathcal{D}'_{\mu}(y) - V_{ec}\mathcal{M}_{cd}V_{da}{}^{-1} - m_0\delta_{ea}]\psi''_a(y) = 0$$

or

$$[\gamma''^{\nu}{}_{ea}\mathcal{D}''_{\nu}(y) - V_{ec}\mathcal{M}_{cd}V_{da}{}^{-1} - m_0\delta_{ea}]\psi''_a(y) = 0 \qquad (21.20)$$

where $V_{an}\psi'_n(y) = \psi''_a(y)$ and $\mathcal{D}''_{\nu}(y) = \Pi_{\nu}{}^{\mu *}\mathcal{D}'_{\mu}(y)$.

If we require that eq. 21.20 be covariant under SU(2)⊗U(1)⊕U(1) then the mass matrix is required to be invariant under SU(2)⊗U(1)⊕U(1) transformations:

$$\mathcal{M} = V\mathcal{M}V^{-1} \qquad (21.21)$$

as is $\mathcal{M} + m_0 I$ for any SU(2)⊗U(1)⊕U(1) transformation V. One possible form for V is

$$V = \begin{bmatrix} a & b & 0 & 0 \\ c & d & 0 & 0 \\ 0 & 0 & e & 0 \\ 0 & 0 & 0 & f \end{bmatrix} \qquad (21.22)$$

where we assume V has a block structure with the upper 3×3 block is an SU(2) block corrsponding to a reducible representation of SU(2) consisting of a doublet and a singlet and with the lower 1×1 block a U(1) block.

Then eq. 21.21 implies that the mass matrix M has the diagonal form

$$M = \text{diag}(m_1, m_1, m_2, m_3) \qquad (21.23)$$

This mass structure for the generations of each species approximately corresponds to the observed fermion mass spectrum. The first two generations of each known family has roughly equal masses: u and c, d and s, e and μ. The third generation: t, b and τ masses are much greater than those of the corresponding first two generations. The fourth generation of these species is may just be becoming known experimentally. Their masses must be much greater than the third generation masses. Thus the overall pattern of masses of the three known species masses is consistent with the general pattern we have obtained from SU(2)⊗U(1)⊕U(1) symmetry in the space-time of GL'(4).

However it is clear that there is more to the physics of the generations than the above derivation. Firstly, the masses of the first and second generation of each species differ. This fact argues for a symmetry breaking interaction. Secondly the issue of generation mixing is not determined by the above considerations. Again the resolution may lie in a new interaction.

But it is encouraging that the overall pattern of the mass spectrum arises so clearly based on the assumed SU(2)⊗U(1)⊕U(1) symmetry of GL'(4) space.

We conclude with postulates embodying the above discussion:

Postulate 21.2a. The GL'(4) Dirac equation is covariant under SU(2)⊗U(1).
OR
Postulate 21.2b. The GL'(4) Dirac equation is covariant under SU(2)⊗U(1)⊕U(1).

*21.5 Origin of the Massive Vector Boson Masses

The Higgs Mechanism was needed to give masses to ElectroWeak vector bosons in the t'Hooft-Veltmann renormalization program for ElectroWeak Theory. Since experiment has been unable to discover Higgs scalar bosons up to this point in time, and may well not find them at future accelerators, it is useful to develop other mechanisms for mass generation for massive vector bosons.

21.5.1 Renormalizable Massive Gauge Boson Theories

We could choose to use massive vector bosons ab initio and obtain a *fully renormalizable* (indeed finite) ElectroWeak Theory by using the Two-Tier quantization procedure (Blaha (2003) and (2005a) – Part 3). A massive vector boson lagrangian would then have the form:

$$\mathcal{L}_{\text{Y-M}} = -\tfrac{1}{4} F^{a\mu\nu}(x)F^{a}_{\mu\nu}(x) + \tfrac{1}{2} m^2 A^{a\mu} A^{a}_{\mu} \qquad (21.24)$$

where

$$F^{a}_{\mu\nu}(x) = D_\nu(x)A^{a}_{\mu}(x) - D_\mu(x)A^{a}_{\nu}(x) + gf^{abc}A^{b}_{\mu}(x)A^{c}_{\nu}(x) \qquad (21.25)$$

The form of the differential operator D_ν depends on whether the Yang-Mills field is "normal" or complexon. For a normal Yang-Mills Field $D_\nu = \partial_\nu$ while for a complexon Yang-Mills field D_ν is specified by eq. 17.59. The constants f^{abc} are the algebra's structure constants.

21.5.2 Dimensional Mass Generation in Gauge Theories

Since the structure of the ElectroWeak sector that we derived in earlier chapters comes from the interplay of GL(4) and GL'(4) it is reasonable to consider generating ElectroWeak vector boson masses through dimensional mass generation in a manner similar to the fermion case considered above.

The dynamical field equation of a conventional 4-dimensional massless Yang-Mills field is

$$G^{a\nu}(x) = (D_\mu(x)\delta^{ac} + gf^{abc}A^{b}_{\mu}(x))F^{c\mu\nu}(x) = 0 \qquad (21.26)$$

where $F^{a\mu\nu}(x)$ is defined by eq. 21.25. Following a parallel development to the fermion case that started from eq. 21.6 we consider a GL(4)⊗GL'(4) generalization of eq. 21.26:[326]

$$G^{a\nu}(x)_{GL(4)}A^{b\mu}(y)_{GL'(4)} + A^{a\nu}(x)_{GL(4)}G^{b\mu}(y)_{GL'(4)} = 0 \qquad (21.27a)$$

This equation results from the lagrangian density terms

$$V_{a\nu b\mu}(x, y)[G^{a\nu}(x)_{GL(4)}A^{b\mu}(y)_{GL'(4)} + A^{a\nu}(x)_{GL(4)}G^{b\mu}(y)_{GL'(4)}] \qquad (21.27b)$$

---

[326] The form is determined by requiring the equation be second order, and that both the GL(4) and the GL'(4) factors are vectors.

where $V_{av b\mu}(x, y)$ is a subsidiary field. Variation with respect to $V_{av b\mu}(x, y)$ yields eq. 21.27a. Variation with respect to $A^{av}(x)_{GL(4)}$ and $A^{b\mu}(y)_{GL'(4)}$ yield the other dynamical equations for this subsector. The linearity in $V_{av b\mu}(x, y)$ in these equations imply $V_{av b\mu}(x, y)$ is separable and the product of two gauge fields:

$$V_{av b\mu}(x, y) = V_{1av}(x)V_{2b\mu}(y) \qquad (21.27c)$$

Dividing eq. 21.27a by $A^{b\mu}(y)_{GL'(4)}A^{av}(x)_{GL(4)}$ yields a separable equation with the results:

$$G^{av}(x)_{GL(4)}/A^{av}(x)_{GL(4)} = -m^2 \qquad (21.28)$$

and

$$G^{av}(y)_{GL'(4)}/A^{av}(y)_{GL'(4)} = -m^2 \qquad (21.29)$$

where the right side of each equation is a constant mass squared (required on dimensional grounds). Eqs. 21.27 are not covariant uder local gauge transformations. This conclusion is consistent with the acquisition of mass by the fields (eqs. 21.28 and 21.29) breaking gauge symmetry.

Consequently the GL(4) Yang-Mills field (as well as the GL'(4) field) becomes a massive Yang-Mills field and satisfies

$$(D_\mu(x)\delta^{ac} + gf^{abc}A^b{}_\mu(x))F^{c\mu\nu}(x) + m^2A^{a\nu}(x) = 0 \qquad (21.30)$$

Thus we have an alternate mechanism to generate masses for Yang-Mills fields. The theories embodying these fields are renormalizable (as noted earlier) using Two-Tier quantization. As a result there are two possible alternate postulates for massive Yang-Mills fields.

*21.6 The Universe of the GL'(4) Space-Time

We have suggested that the GL'(4) space-time may be the origin of fermion and ElectroWeak vector boson masses. This origin implies that there are interactions in the GL'(4) space-time. The "size" of this space-time, hidden as it is, from direct observation is not presently know. It could be any size from a point to infinite. It could be flat or curved. It may have gravitation. It may not.

In view of these uncertainties we will *assume* that the GL'(4) space-time features are the same as that of the familiar space-time of our experience. Thus we will assume that there is a complex Lorentz group structure. We will assume that the the initially free equation of motion for each fermion species has corresponding boosts in the part corresponding to our space-time and the part corresponding to the GL'(4) space-time:

$$(\gamma^\nu{}_{ab}\mathcal{D}_{s\nu}(x)\delta_{cd} + \gamma'^\mu{}_{cd}\mathcal{D}'_{s\,\mu}(y)\delta_{ab} - m_0\delta_{ab}\delta_{cd})\psi_{bd}(x, y) = 0 \qquad (21.31)$$

for species index s. Thus each of the four "free" Dirac-like equations (chapter 17) holds for the GL'(4) space-time. Eqs. 21.8 and 21.9 thus acquire a species index:

$$[\gamma'^{\mu}{}_{cd}\mathcal{D}_s{}'_{\mu}(y) - \mathcal{M}_{cd} - m_0\delta_{cd}]\psi'_d(y) = 0 \tag{21.32}$$

$$(\gamma^{\nu}{}_{ab}\mathcal{D}_{s\nu}(x)\delta_{cd} - \mathcal{M}_{cd}\delta_{ab})\psi_{bd}(x) = 0 \tag{21.33}$$

Now the GL'(4) space-time Dirac-like equations (eq. 21.32) have a major difference from the GL(4) equations (eq. 21.33). The mass term in eq. 21.32 is manifestly not a multiple of the identity matrix. Therefore its covariance properties in its "spinor" space are different from the covariance properties in the "spinor" space of eq. 21.33. However since the mass term appears in the dynamic equations (eq. 21.33) for our space-time it must be constant on physical grounds. Therefore if we perform a boost in GL'(4):

$$S'_{sec}[\gamma'^{\mu}{}_{cd}\mathcal{D}_s{}'_{\mu}(y) - \mathcal{M}_{cd} - m_0\delta_{cd}]\, S'_{sda}{}^{-1}S'_{saf}\psi'_f(y) = 0 \tag{21.34}$$

then the mass term must be invariant under the transformation S'_s

$$S'_{sec}\mathcal{M}_{cd}S'_{sda}{}^{-1} = \mathcal{M}_{ea} \tag{21.35}$$

or

$$S'_s\mathcal{M}S'_s{}^{-1} = \mathcal{M} \tag{21.36}$$

The set of complex Lorentz transformations satisfying eq. 21.36 are the only physically allowed complex Lorentz transformations in the GL'(4) space-time. For each species the set of physically allowed transformations forms a group. We note that eq. 21.36 is satisfied by the identity transformation. In addition it is easy to show the product of two allowed transformations is an allowed transformation. Lastly the inverse of a transformation is an allowed transformation. Thus, for each species, the set of allowed transformations satisfying eq. 21.36 forms a group.

Next one might ask if there are there any transformations in the sets for each species? If we consider the <u>example</u>:

$$\mathcal{M} = a\gamma'^5 \tag{21.37}$$

where a is constant then from the form of the three types of boosts in sections 15.11 and 17.2 it is clear that eq. 21.36 will be satisfied for by any Lorentz or superluminal transformation. Thus eq. 21.37 would appear to be the "best" choice for $\mathcal{M}$ since one would like invariance under the full complex Lorentz group. However this feature is not physically required for the GL'(4) space-time. If $\mathcal{M}$ is expanded in terms the 16 4×4 linearly independent Γ matrices[327] it will be seen that all terms except the γ'^5 term violate invariance.

Eq. 21.36 is satisfied by the expression for $\mathcal{M}$ in eq. 21.37 for all Lorentz and superluminal transformations. However it does not conform to reality for any of the particle species. Another possibility is to specify a value for $\mathcal{M}$ that may be closer to a realistic expression. In doing this we assert a Mach principle for the GL'(4) space-time that there are preferred reference frames: in this case the preferred frames are a subset of the set of all inertial reference frames. To illustrate this possiblity we consider the example:

$$\mathcal{M} = a\boldsymbol{\gamma}'.\mathbf{p} \qquad (21.38)$$

Then the set of spinor boost transformations that satisfy eq. 21.36 are those which have a relative velocity vector perpendicular to the vector **p** as one can see from the form of spinor boosts in sections 15.11 and 17.2. Thus a subset of Lorentz and superluminal boosts satisfy eq. 21.36.

These speculations suggest that it may not be correct to treat the GL'(4) space-time as if it had the structure of our space-time with Lorentz transformations linking reference frames. Thus the separation constant matrix may well be fixed by breakdown conditions that are as yet unknown.

---

[327] Bjorken (1964) p. 25 and R. h. Good, Jr., Rev. Mod. Phys. **27**, 187 (1955).

22. The "Standard" Standard Model

22.1 Postulates

If we assume Postulate 19.1a stating that quarks are not complexons then we derive a conventional form of the Standard Model which we call the "Standard" Standard Model to distinguish it from the Complexon Standard Model that is described in the following chapter. The postulates of the Standard Standard Model are:

Postulate 15.1. Any observer can define a set of time and space coordinates called a coordinate system in which the observer is at rest. One can define a transformation that relates the coordinate systems of two observers traveling at a constant velocity with respect to each other.

Postulate 15.2. The speed of light, c, is the same in all coordinate systems.

Postulate 15.3. The invariant interval or distance $d\tau$ is defined by

$$d\tau^2 = g_{\mu\nu}dx^\mu dx^\nu \tag{15.2}$$

Postulate 16.1. The universe has sixteen complex dimensions and is described by a curved, 16-dimensional complex extension of 4-dimensional General Relativity. Complex, 16-dimensional General Relativity is invariant under complex coordinate transformations of the form:

$$X^\mu = f^\mu(X'^0, X'^1, \ldots, X'^{15}) \tag{16.1}$$

where X'^μ denotes an alternate complex coordinate system that is related to X^μ by continuous, holomorphic functions f^μ.

Postulate 16.2. We define an invariant interval with a metric $g_{\mu\nu}$ where

$$d\tau^2 = g_{\mu\nu}dX^\mu dX^\nu \tag{16.2}$$

Postulate 16.3. We assume a 16-dimensional Equivalence Principle: At any point one can transform to the coordinates of a locally flat space-time where

$$g'^{\mu\nu} = \eta_{16}{}^{\alpha\beta}\, \partial X'^{\mu}/\partial X^{\alpha}\ \partial X'^{\nu}/\partial X^{\beta} \qquad (16.9a)$$

with

$$\eta_{16}{}^{\mu\nu} = \text{diag}(1, -1, -1, -1, -1, 1, 1, 1, -1, 1, 1, 1, -1, 1, 1, 1) \qquad (16.12)$$
$$\equiv \eta_{16}{}^{\mu\nu}$$

Postulate 16.4. We assume that GL(16) is broken to GL(4)⊗GL'(4) through some mechanism and this group is the basis of the form of the Standard Model.

Postulate 16.5. After breakdown, a complex 16-dimensional vector can be represented as the outer product of two complex 4-dimensional vectors

$$x^{\mu} \to y^{i} y'^{j} \qquad (16.30)$$

where μ ranges from 0 to 15, and i, j range from 0 through 3. The 4-dimensional metrics are assumed to be the outer product of the Lorentz metric

$$\eta_{16\mu\nu} \to \text{diag}(1, -1, -1, -1) \otimes \text{diag}(1, -1, -1, -1) \qquad (16.31)$$
$$= (\eta_4) \otimes (\eta'_4)$$

using a matrix notation.

Postulate 16.6. The GL(4) group has a familiar coordinate space representation except that it is a complex space-time with a complex Lorentz group denoted L(C) whose elements Λ_C satisfy $\Lambda_C{}^T G_+ \Lambda_C = G_+$. L(C) is in turn broken to the real Lorentz group L of space-time that we physically observe.

Postulate 16.7. The GL'(4) group has a space-time (that may be reduced to a point.) However it implies four fermion generations, (OR possibly it may have only the three known generations due to the imposition of an additional constraint(s).)

Postulate 17.1. The dynamical equations of the four species of free fermions can be generated by L(C) boosts from the Dirac equation for a fermion at rest subject to the requirement that there be only one real time variable in each resulting dynamical differential equation.

Postulate 17.2. Each free fermion field has the simplest possible lagrangian from which the canonically conjugate momentum, the energy-momentum tensor, the equal-time (or equal light-front) anti-commutation relations, and other physically important quantities can be derived.

Postulate 17.3. The Standard Model is a quantum field theory with a lagrangian and a path integral formulation. The mathematics of these constructs is assumed to be correct.

Postulate 17.4. The free field approximation to the Standard Model is a canonical quantum field theory but not necessarily quantized on light-like surfaces.

SU(2)⊗U(1)
{

Postulate 18.1a. A Dirac field and a tachyon field form a doublet whose dynamical equation is covariant under the 8 × 8 L(C) transformation eq. 18.5.

Postulate 18.2a. A lepton doublet in any generation consists of a charged Dirac field and a neutral tachyon field (a neutrino). By convention the neutrino field is the topmost component in a lepton doublet. The equations of motion of the lepton doublet are eqs. 18.4 with $\Psi(x) = \Psi_{\ell e}(x)$ (eq. 18.25 below).

Postulate 18.3a. A quark doublet in any generation consists of an up-type quark Dirac field and a down-type quark tachyon field. By convention the up-type quark field is the topmost component in a quark doublet. (The quark fields may be "normal" or complexon – an experimental issue.)

Postulate 18.4a. The form of the equations of motion of the unbroken leptonic sector of the Standard Model are determined by covariance under L(C) group boosts, and local SU(2)⊗U(1) gauge symmetries. Electron number is not gauged.

Postulate 18.5a. In each generation leptonic particles consist of a spin ½ Dirac particle (electron-like) with charge –1, and a tachyon (neutrinos) of charge zero. as well as their anti-particles.[328] *(Redundant for clarity.)*

Postulate 18.7a. The interacting extensions of free doublet equations of motion have interactions consisting of local Yang-Mills fields. Gauge fields are massless "before" spontaneous symmetry breaking and are thus conventional local Yang-Mills gauge fields without a tachyon equivalent.

Postulate 18.9a. The equations of motion of free quark doublets in the unbroken part of the ElectroWeak sector of the Standard Model are determined by covariance under the complex Lorentz group L(C), and local SU(2)⊗U(1) gauge symmetries. Baryon number is not gauged.

Postulate 21.2a. The GL'(4) Dirac equation is covariant under SU(2)⊗U(1).

---

[328] If neutrinos are Majorana particles then the derivation must be modified.

}

OR <u>*SU(2)⊗U(1)⊕U(1)*</u>

{

Postulate 18.1b. A free Dirac field, tachyon field, and singlet field form a triplet whose dynamical equation is covariant under L(C) transformations.

Postulate 18.2b. A lepton triplet in any generation consists of a charged Dirac field, a neutral tachyon field (the neutrino), and a neutral singlet lepton (WIMP). The equations of motion of the lepton triplet are eqs. 18-A.17.

Postulate 18.3b. A quark triplet in any generation consists of a singlet quark, an up-type quark Dirac field, a down-type quark tachyon field. (The quark fields may be "normal" or complexon – an experimental issue – but complexon is theoretically favored.)

Postulate 18.4b. The form of the equations of motion of the unbroken leptonic sector of the Standard Model are determined by covariance under L(C) group boosts, and local SU(2)⊗U(1)⊕U(1) gauge symmetries. Electron number is not gauged.

Postulate 18.5b. In each generation leptonic particles consist of a spin ½ Dirac particle (electron-like) with charge –1, a tachyon (neutrino) of charge zero, and a singlet (WIMP) of charge zero as well as their anti-particles.[329] *(Redundant for clarity.)*

Postulate 18.7b. The interacting extensions of free triplet equations of motion have interactions consisting of local Yang-Mills fields. Gauge fields are massless "before" spontaneous symmetry breaking and are thus conventional local Yang-Mills gauge fields without a tachyon equivalent.

Postulate 18.9b. The equations of motion of free quark triplets in the unbroken part of the ElectroWeak sector of the Standard Model are determined by covariance under the complex Lorentz group L(C), and local SU(2)⊗U(1)⊕U(1) gauge symmetries. Baryon number is not gauged.

Postulate 21.2b. The GL'(4) Dirac equation is covariant under SU(2)⊗U(1)⊕U(1).

}

Postulate 18.6. A neutrino is a tachyon with a non-zero bare mass that may be changed by symmetry breaking effects.

---

[329] If neutrinos are Majorana particles then the derivation must be modified.

Postulate 18.8. L(C) group covariance, and ElectroWeak gauge symmetries of the dynamical equations of motion, are spontaneously broken through the appearance of additional fermion and vector boson mass terms generated by a mechanism such as the Higgs mechanism or the Dimensional Mass Mechanism (sections 21.3 and 21.4).

Postulate 19.1a. Quarks are in the $\underline{3}$ representation of Color SU(3). The SU(3) symmetry is gauged with local Yang-Mills SU(3) fields called gluons that constitute the Strong interaction of the quark sector. Quarks are minimally coupled to the gluons in a gauge covariant fashion.

Postulate 20.1. Spatial asynchronicity of physical phenomena requires a 4-valued logic that can defined in a 4×4 matrix formulation.

Postulate 20.2. The 4×4 matrix formulation can be mapped to Dirac spinors in a different four dimensional space-time (not the space-time of our experience) and, upon introducing coordinates, can serve as the starting point for the definition of Dirac-like equations.

Postulate 20.3 For each fermion species the four solutions of each Dirac-like equation in GL'(4) *are the four generations of the species.*

Postulate 21.1a. The masses, and mixing angles, of vector bosons result from the Higgs mechanism.
OR
Postulate 21.1b. The masses, and mixing angles, of fermions appear as separation constants in GL(4)⊗GL'(4) fermion dynamical equations.

The number of postulates, and their complexity, far exceeds Euclid's formulation of geometry. However, the subject is also far more complex.

The author's view is that a smaller, more compact set of postulates will eventually emerge from the set upon which this derivation is based. While the ability to derive the Standard Model is in itself a worthwhile pursuit because it informs us of the cohesion of the theory, it is also important in considering possible generalizations of the theory. Theorists have proposed a plethora of generalizations (and variants) of the Standard Model.

It is clear that the basic Standard Model has a strong geometric foundation in GL(16) or GL(4)⊗GL'(4). Can one develop a similarly strong geometric basis for any proposed extensions? If not, then the probability of the validity of a proposed extension is not as high as it would be if were geometrically based in a way similar to the proposed geometric basis of the Standard Model.

Perhaps the most important reason to establish the Standard Model on the basis of fundamental postulates is that we thereby obtain a deeper level of understanding of the physics of the Standard Model. Working at this deeper level is in itself an advance in our understanding of the fundamental basis of physical reality – a deeper layer in the onion of fundamental physics.

22.2 The Standard Standard Model Lagrangian

The lagrangian density that we have derived for the conventional Standard Model with four generations, and quark and lepton triplets is[330]

$$\begin{aligned}\mathcal{L}_{SSM} = \Psi_{3L}^{a\dagger}\gamma^0 i\gamma^\mu D_{L\mu}\Psi_{3L}^{a} - \Psi_{3R}^{a\dagger}\gamma^0 i\gamma^\mu D_{R\mu}\Psi_{3R}^{a} + \\ + \Psi_{q3L}^{a\dagger}\gamma^0 i\gamma^\mu D_{qL\mu}\Psi_{q3L}^{a} + \Psi_{q3R}^{a\dagger}\gamma^0 i\gamma^\mu D_{qR\mu}\Psi_{q3R}^{a} - \\ - \mathcal{L}_{BareMasses} + \mathcal{L}_{Gauge} + \mathcal{L}_{Mass}\end{aligned} \tag{22.1}$$

where a is the generation index, and where

$$D_{L\mu} = D_\mu - (\tfrac{1}{2}ig_2\boldsymbol{\sigma}\cdot\mathbf{W}_\mu - ig_1B_\mu/2)(1 - Q') - ig''Q'B'_\nu(x)/2 \tag{22.2}$$

$$D_{R\mu} = (\sigma_3\, D_\mu + ig_1B_\mu|Q|)(1 - Q') + (D_\mu + ig''B'_\nu(x))Q' \tag{22.3}$$

$$D_{qL\mu} = D_{q\mu} + (\tfrac{1}{2}ig_2\boldsymbol{\sigma}\cdot\mathbf{W}_\mu + ig_1B_\mu/6)(1 - Q') + ig''B'_\nu(x)Q'/6 \tag{22.4}$$

$$D_{qR\mu} = (\sigma_3\, D_{q\mu} + ig_1B_\mu|Q|)(1 - Q') + (D_{q\mu} + ig''B'_\nu(x))Q' \tag{22.5}$$

where Q' is given by eq. 18-A.32, where D_ν is $\partial/\partial x^\nu$, where $D_{q\nu}$ is the color covariant derivative

$$D_{q\nu} = \partial/\partial x^\nu - igA_\nu \tag{22.6}$$

where $\Psi_{q3}^{\dagger}$ is the hermitean conjugate of the quark triplet Ψ_{q3} with $\Psi_{q3L}^{\dagger}$ and $\Psi_{q3R}^{\dagger}$ being its left-handed and right-handed parts respectively, and where

$$\mathcal{L}_{Gauge} = \mathcal{L}_{GaugeEW} + \mathcal{L}_{GaugeColor} \tag{22.7}$$

with

$$\mathcal{L}_{GaugeEW} = -\tfrac{1}{4}\, F_W^{a\mu\nu}F_{W\,\mu\nu}^{a} - \tfrac{1}{4}\, F_B^{\mu\nu}F_{B\mu\nu} - \tfrac{1}{4}\, F_{B'}^{\mu\nu}F_{B'\mu\nu} + \mathcal{L}_{EW}^{ghost} \tag{22.8}$$

and

$$\mathcal{L}_{GaugeColor} = -\tfrac{1}{4}\, F_{Color}^{a\mu\nu}F_{Color\,\mu\nu}^{a} + \mathcal{L}_{Color}^{ghost} \tag{22.9}$$

---

[330] This theory with fermion triplets can easily be restricted to fermion doublets to obtain the conventional SU(2)⊗SU(1) ElectroWeak sector. The four generations can similarly be restricted to three generations.

$\mathcal{L}_{\text{BareMasses}}$ contains the fermion bare mass terms. The ElectroWeak gauge bosons $W_\mu{}^a$, B_μ and B'_μ field tensors are:

$$F_{W}{}^{a}{}_{\mu\nu} = \partial W^a{}_\mu/\partial x^\nu - \partial W^a{}_\nu/\partial x^\mu + g_2 f^{abc} W^b{}_\mu W^c{}_\nu \qquad (22.10)$$

$$F_{B\mu\nu} = \partial B_\mu/\partial x^\nu - \partial B_\nu/\partial x^\mu \qquad (22.11)$$

$$F_{B'\mu\nu} = \partial B'_\mu/\partial x^\nu - \partial B'_\nu/\partial x^\mu \qquad (19.25b)$$

$\mathcal{L}_{\text{EW}}{}^{\text{ghost}}$ are the Faddeev-Popov ghost terms for the ElectroWeak $W_\mu{}^a$ gauge bosons. The color gauge field tensors are

$$F_{\text{Color}}{}^{a}{}_{\mu\nu} = \partial A^a{}_\mu/\partial x^\nu - \partial A^a{}_\nu/\partial x^\mu + g_2 f_{su(3)}{}^{abc} A^b{}_\mu A^c{}_\nu \qquad (22.12)$$

$\mathcal{L}_{\text{Color}}{}^{\text{ghost}}$ is the color SU(3) ghost terms defined via the Faddeev-Popov mechanism. The mass sector $\mathcal{L}_{\text{Mass}}$ (using the Higgs Mechanism or the Dimensional Mass Mechanism (sections 21.3 and 21.4)) gives masses to the ElectroWeak vector bosons and the fermions as well as specifying generational mixing.

The effective action for the path integral formulation is

$$I_{SSM} = \int dx^0 d^3x \, \mathcal{L}_{SSM} \qquad (22.13)$$

Thus we have derived the *form* of the conventional Standard Model enhanced with WIMPs and a fourth generation (modulo the mass terms which may be generated by the Higgs Mechanism or Dimensional Mass Generation). Chapter 22 describes the missing pieces in this derivation: coupling constant values, generation mixing, and the mass spectrum that are not derived.

It is a somewhat amazing fact that we can precisely derive the form of the Standard Model with the constants appearing in the theory undetermined, and thus in a sense free, from strictly geometrical considerations within the framework of quantum field theory.

22.3 Renormalization

Since neutrinos and d-type quarks are tachyons in this theory the t'Hooft-Veltmann renormalization program should be reconsidered for this case if one wishes to use conventional renormalization.

However as pointed out in Blaha (2005a) and earlier books (as well as section 19.5 and Part 3) one can use the Two-Tier formalism to obtain a finite, renormalized theory in the present case. Consequently renormalization is not an issue.

23. The Complexon Standard Model

23.1 Postulates

If we assume postulate Postulate 19.1b is correct so that quarks and gluons *are* complexons then we obtain the Complexon Standard Model The postulates of the Complexon Standard Model are:

Postulate 15.1. Any observer can define a set of time and space coordinates called a coordinate system in which the observer is at rest. One can define a transformation that relates the coordinate systems of two observers traveling at a constant velocity with respect to each other.

Postulate 15.2. The speed of light, c, is the same in all coordinate systems.

Postulate 15.3. The invariant interval or distance $d\tau$ is defined by

$$d\tau^2 = g_{\mu\nu}dx^\mu dx^\nu \qquad (15.2)$$

Postulate 16.1. The universe has sixteen complex dimensions and is described by a curved, 16-dimensional complex extension of 4-dimensional General Relativity. Complex, 16-dimensional General Relativity is invariant under complex coordinate transformations of the form:

$$X^\mu = f^\mu(X'^0, X'^1, \ldots, X'^{15}) \qquad (16.1)$$

where X'^μ denotes an alternate complex coordinate system that is related to X^μ by continuous, holomorphic functions f^μ.

Postulate 16.2. We define an invariant interval with a metric $g_{\mu\nu}$ where

$$d\tau^2 = g_{\mu\nu}dX^\mu dX^\nu \qquad (16.2)$$

Postulate 16.3. We assume a 16-dimensional Equivalence Principle: At any point one can transform to the coordinates of a locally flat space-time where

$$g'^{\mu\nu} = \eta_{16}{}^{\alpha\beta}\, \partial X'^\mu/\partial X^\alpha \; \partial X'^\nu/\partial X^\beta \qquad (16.9a)$$

with

$$\eta_{16\mu\nu} = \text{diag}(1, -1, -1, -1, -1, 1, 1, 1, -1, 1, 1, 1, -1, 1, 1, 1) \qquad (16.12)$$
$$\equiv \eta_{16}{}^{\mu\nu}$$

Postulate 16.4. We assume that GL(16) is broken to GL(4)⊗GL'(4) through some mechanism and this group is the basis of the form of the Standard Model.

Postulate 16.5. After breakdown, a complex 16-dimensional vector can be represented as the outer product of two complex 4-dimensional vectors

$$x^\mu \rightarrow y^i y'^j \qquad (16.30)$$

where μ ranges from 0 to 15, and i, j range from 0 through 3. The 4-dimensional metrics are assumed to be the outer product of the Lorentz metric

$$\eta_{16\mu\nu} \rightarrow \text{diag}(1, -1, -1, -1) \otimes \text{diag}(1, -1, -1, -1) \qquad (16.31)$$

$$= (\eta_4) \otimes (\eta'_4)$$

using a matrix notation.

Postulate 16.6. The GL(4) group has a familiar coordinate space representation except that it is a complex space-time with a complex Lorentz group denoted L(C) whose elements Λ_C satisfy $\Lambda_C{}^T G_4 \Lambda_C = G_4$. L(C) is in turn broken to the real Lorentz group L of space-time that we physically observe.

Postulate 16.7. The GL'(4) group has a space-time (that may be reduced to a point.) However it implies four fermion generations, (OR possibly it may have only the three known generations due to the imposition of an additional constraint(s).)

Postulate 17.1. The dynamical equations of the four species of free fermions can be generated by L(C) boosts from the Dirac equation for a fermion at rest subject to the requirement that there be only one real time variable in each resulting dynamical differential equation.

Postulate 17.2. Each free fermion field has the simplest possible lagrangian from which the canonically conjugate momentum, the energy-momentum tensor, the equal-time (or equal light-front) anti-commutation relations, and other physically important quantities can be derived.

Postulate 17.3. The Standard Model is a quantum field theory with a lagrangian and a path integral formulation. The mathematics of these constructs is assumed to be correct.

Postulate 17.4. The free field approximation to the Standard Model is a canonical quantum field theory but not necessarily quantized on light-like surfaces.

SU(2)⊗U(1)

{

Postulate 18.1a. A Dirac field and a tachyon field form a doublet whose dynamical equation is covariant under the 8 × 8 L(C) transformation eq. 18.5.

Postulate 18.2a. A lepton doublet in any generation consists of a charged Dirac field and a neutral tachyon field (a neutrino). By convention the neutrino field is the topmost component in a lepton doublet. The equations of motion of the lepton doublet are eqs. 18.4 with $\Psi(x) = \Psi_{\ell e}(x)$ (eq. 18.25 below).

Postulate 18.3a. A quark doublet in any generation consists of an up-type quark Dirac field and a down-type quark tachyon field. By convention the up-type quark field is the topmost component in a quark doublet. (The quark fields may be "normal" or complexon – an experimental issue.)

Postulate 18.4a. The form of the equations of motion of the unbroken leptonic sector of the Standard Model are determined by covariance under L(C) group boosts, and local SU(2)⊗U(1) gauge symmetries. Electron number is not gauged.

Postulate 18.5a. In each generation leptonic particles consist of a spin ½ Dirac particle (electron-like) with charge –1, and a tachyon (neutrinos) of charge zero. as well as their anti-particles.[331] *(Redundant for clarity.)*

Postulate 18.7a. The interacting extensions of free doublet equations of motion have interactions consisting of local Yang-Mills fields. Gauge fields are massless "before" spontaneous symmetry breaking and are thus conventional local Yang-Mills gauge fields without a tachyon equivalent.

Postulate 18.9a. The equations of motion of free quark doublets in the unbroken part of the ElectroWeak sector of the Standard Model are determined by covariance under the complex Lorentz group L(C), and local SU(2)⊗U(1) gauge symmetries. Baryon number is not gauged.

Postulate 21.2a. The GL'(4) Dirac equation is covariant under SU(2)⊗U(1).

---

[331] If neutrinos are Majorana particles then the derivation must be modified.

}

OR <u>*SU(2)⊗U(1)⊕U(1)*</u>
{
Postulate 18.1b. A free Dirac field, tachyon field, and singlet field form a triplet whose dynamical equation is covariant under L(C) transformations.

Postulate 18.2b. A lepton triplet in any generation consists of a charged Dirac field, a neutral tachyon field (the neutrino), and a neutral singlet lepton (WIMP). The equations of motion of the lepton triplet are eqs. 18-A.17.

Postulate 18.3b. A quark triplet in any generation consists of a singlet quark, an up-type quark Dirac field, a down-type quark tachyon field. (The quark fields may be "normal" or complexon – an experimental issue – but complexon is theoretically favored.)

Postulate 18.4b. The form of the equations of motion of the unbroken leptonic sector of the Standard Model are determined by covariance under L(C) group boosts, and local SU(2)⊗U(1)⊕U(1) gauge symmetries. Electron number is not gauged.

Postulate 18.5b. In each generation leptonic particles consist of a spin ½ Dirac particle (electron-like) with charge –1, a tachyon (neutrinos) of charge zero, and a singlet (WIMP) of charge zero as well as their anti-particles.[332] *(Redundant for clarity.)*

Postulate 18.7b. The interacting extensions of free triplet equations of motion have interactions consisting of local Yang-Mills fields. Gauge fields are massless "before" spontaneous symmetry breaking and are thus conventional local Yang-Mills gauge fields without a tachyon equivalent.

Postulate 18.9b. The equations of motion of free quark triplets in the unbroken part of the ElectroWeak sector of the Standard Model are determined by covariance under the complex Lorentz group L(C), and local SU(2)⊗U(1)⊕U(1) gauge symmetries. Baryon number is not gauged.

Postulate 21.2b. The GL'(4) Dirac equation is covariant under SU(2)⊗U(1)⊕U(1).
}

Postulate 18.6. A neutrino is a tachyon with a non-zero bare mass that may be changed by symmetry breaking effects.

---

[332] If neutrinos are Majorana particles then the derivation must be modified.

Postulate 18.8. L(C) group covariance, and ElectroWeak gauge symmetries of the dynamical equations of motion, are spontaneously broken through the appearance of additional fermion and vector boson mass terms generated by a mechanism such as the Higgs mechanism or the Dimensional Mass Mechanism (sections 21.3 and 21.4).

Postulate 19.1a. Quarks are in the $\underline{3}$ *representation of Color SU(3). The SU(3) symmetry is gauged with local Yang-Mills SU(3) fields called gluons that constitute the Strong interaction of the quark sector. Quarks are minimally coupled to the gluons in a gauge covariant fashion.*

Postulate 19.1b. Quarks are complexons. (Implies quarks are in the $\underline{3}$ *representation of Color SU(3). Partly redundant with 19.1a)*

Postulate 19.2. Quarks are in a $\underline{3}$ *representation of a global SU(3) subgroup of GL(4). Their transformation law under SU(3) transformations is eq. 19.6 wherein the space-time coordinates are not transformed. (Redundant with 19.1a)*

Postulate 19.3. The covariance of the quark equations of motion under global SU(3) transformations becomes covariance under local color SU(3) transformations when the equations of motion are generalized to gauge covariant derivatives. We assume the equations of motion are so generalized. The interaction terms so introduced constitute the Strong interaction.

Postulate 20.1. Spatial asynchronicity of physical phenomena requires a 4-valued logic that can defined in a 4×4 matrix formulation.

Postulate 20.2. The 4×4 matrix formulation can be mapped to Dirac spinors in a different four dimensional space-time (not the space-time of our experience) and, upon introducing coordinates, can serve as the starting point for the definition of Dirac-like equations.

Postulate 20.3 For each fermion species the four solutions of each Dirac-like equation in GL'(4) *are the four generations of the species.*

Postulate 21.1a. The masses, and mixing angles, of vector bosons result from the Higgs mechanism.
OR
Postulate 21.1b. The masses, and mixing angles, of fermions appear as separation constants in GL(4)⊗GL'(4) fermion dynamical equations.

If the LHC accelerator (at CERN) does not find Higgs particles then Dimensional Mass Generation becomes more probable in this author's view. One cannot chase rainbows forever.

The preceding list of postulates is purposefully redundant in the interest of clarity.

23.2 The Complexon Standard Model Lagrangian

If we assume Postulates 19.1b, 19.2, and 19.3 then we obtain the Complexon Standard Model which differs from the "Standard Standard Model" by having complexon quarks and gluons. Which of these Standard Models is the correct one chosen by Nature? That is a question that only experiment can answer – perhaps deep inelastic electron-nucleon spin experiments.

In this section we show the results of the derivation of the four generation Complexon Standard Model with ElectroWeak triplets: its lagrangian density and action.[333]

$$\begin{aligned}\mathcal{L}_{CSM} = {}& \Psi_{3L}{}^{a\dagger}\gamma^0 i\gamma^\mu D_{L\mu}\Psi_{3L}{}^{a} - \Psi_{3R}{}^{a\dagger}\gamma^0 i\gamma^\mu D_{R\mu}\Psi_{3R}{}^{a} + \\ & + \Psi_{C3L}{}^{a\dagger}\gamma^0 i\gamma^\mu \mathcal{D}_{qL\mu}\Psi_{C3L}{}^{a} + \Psi_{C3R}{}^{a\dagger}\gamma^0 i\gamma^\mu \mathcal{D}_{qR\mu}\Psi_{C3R}{}^{a} - \\ & - \mathcal{L}_{BareMasses} + \mathcal{L}_{Gauge} + \mathcal{L}_{Mass}\end{aligned} \tag{23.1}$$

where a is the generation index, using eq. 18.65 where $D_{L\mu}$ and $D_{R\mu}$ are given by eq. 19.17, and where the left-handed complexon quark covariant derivative is

$$\mathcal{D}_{qL\mu} = \mathcal{D}_\mu + (\tfrac{1}{2}ig_2\boldsymbol{\sigma}\cdot\mathbf{W}_\mu + ig_1B_\mu/6)(1 - Q') + ig''B'_\nu(x)Q'/6 \tag{23.2}$$

where $\mathcal{D}_\mu$ is given by eq. 19.7, where Q' is given by eq. 18-A.32, and the right-handed complexon quark covariant derivative is

$$\mathcal{D}_{qR\mu} = (\sigma_3\mathcal{D}_\mu + ig_1B_\mu|Q|)(1 - Q') + (\mathcal{D}_\mu + ig''B'_\nu(x))Q' \tag{23.3}$$

where $D_\mu = \partial/\partial x^\mu$ in

$$D_{L\mu} = D_\mu - (\tfrac{1}{2}ig_2\boldsymbol{\sigma}\cdot\mathbf{W}_\mu - ig_1B_\mu/2)(1 - Q') - ig''Q'B'_\nu(x)/2 \tag{23.4a}$$

$$D_{R\mu} = (\sigma_3 D_\mu + ig_1B_\mu|Q|)(1 - Q') + (D_\mu + ig''B'_\nu(x))Q' \tag{23.4b}$$

[333] This theory with fermion triplets can easily be restricted to fermion doublets to obtain the conventional SU(2)⊗SU(1) ElectroWeak sector. The four generations can similarly be restricted to three generations.

and where $\Psi_{C3}^{\dagger}$ is given by eq. 18-A.22, and $\Psi_{C3L}^{\dagger}$ and $\Psi_{C3R}^{\dagger}$ are its left-handed and right-handed parts respectively. $\mathcal{L}_{BareMasses}$ contains the fermion bare mass terms. Also,

$$\mathcal{L}_{Gauge} = \mathcal{L}_{GaugeEW} + \mathcal{L}_{GaugeC} \qquad (23.5)$$

with

$$\mathcal{L}_{GaugeEW} = -\tfrac{1}{4}\, F_W^{a\mu\nu} F_{W\,\mu\nu}^{\,a} - \tfrac{1}{4}\, F_B^{\mu\nu} F_{B\mu\nu} - \tfrac{1}{4}\, F_{B'}^{\mu\nu} F_{B'\mu\nu} + \mathcal{L}_{EW}^{ghost} \qquad (23.6)$$

and

$$\mathcal{L}_{GaugeC} = \mathcal{L}_{CCG} + \mathcal{L}_{C}^{ghost} + \mathcal{L}_{CC}^{ghost} \qquad (23.7)$$

The ElectroWeak gauge bosons W_μ^a, B_μ and B'_μ field tensors are:

$$F_{W\,\mu\nu}^{\,a} = \partial W^a_\mu/\partial x^\nu - \partial W^a_\nu/\partial x^\mu + g_2 f^{abc} W^b_\mu W^c_\nu \qquad (23.8)$$

$$F_{B\mu\nu} = \partial B_\mu/\partial x^\nu - \partial B_\nu/\partial x^\mu \qquad (23.9)$$

$$F_{B'\mu\nu} = \partial B'_\mu/\partial x^\nu - \partial B'_\nu/\partial x^\mu$$

$\mathcal{L}_{EW}^{ghost}$ is the Faddeev-Popov ghost terms for the ElectroWeak W_μ^a gauge bosons. The complexon color gluon lagrangian $\mathcal{L}_{CCG}$ is defined by

$$\mathcal{L}_{CCG} = -\tfrac{1}{4}\, F_{CC}^{a\mu\nu}(x) F_{CC\,\mu\nu}^{\,a}(x) \qquad (23.10)$$

where

$$F_{CC\,\mu\nu}^{\,a} = D_\nu A_{C\,\mu}^{\,a} - D_\mu A_{C\,\nu}^{\,a} + g f_{su(3)}^{\;abc} A_{C\,\mu}^{\,b} A_{C\,\nu}^{\,c} \qquad (23.11)$$

where $A_{C\,\nu}^{\,a}$ is the color gluon gauge field, and D_μ is defined by eq. 17.59, g is the color coupling constant, and the $f_{su(3)}^{\;abc}$ are the SU(3) structure constants.

In addition $\mathcal{L}_{C}^{ghost}$ is the color SU(3) Faddeev-Popov ghost terms defined in appendix 19-A for the complexon Lorentz gauge and $\mathcal{L}_{CC}^{ghost}$ is the complexon color SU(3) constraint ghost terms defined through the Faddeev-Popov mechanism. The mass sector $\mathcal{L}_{Mass}$ is based on the Higgs Mechanism or the Dimensional Mass Mechanism (sections 21.3 and 21.4) which creates the fermion and ElectroWeak vector boson masses, and generation mixing.

The coordinates are

$$x = (x^0, \mathbf{x}_r + i\mathbf{x}_i) \qquad (23.12)$$

and the action is therefore

$$I_{CSM} = \int dx^0 d^3x_r d^3x_i \, \mathcal{L}_{CSM} \tag{23.13}$$

The lagrangian is supplemented with the following condition on all complexon fields $\Phi_{...}$:

$$\nabla_r \cdot \nabla_i \Phi ... = 0 \tag{23.14}$$

Non-complexon fields Ω... in the left-handed formulation under consideration satisfy the subsidiary condition:

$$[\nabla_r \cdot \nabla_i - (\nabla_r^2 \nabla_i^2)^{1/2}]\Omega ... = 0 \tag{23.15}$$

which guarantees a complexon's real momentum is parallel to its imaginary momentum.

23.3 Renormalization

Since quarks are complexons, since neutrinos and d-type quarks are tachyons, and since we have effectively a 7-dimensional lagrangian theory, the t'Hooft-Veltmann renormalization program is not applicable to the Complexon Standard Model.

However as pointed out in Blaha (2005a) and earlier books (as well as section 19.5 and part 3 in some detail) one can use a Two-Tier formalism to obtain a finite, renormalized Complexon Standard Model thus eliminating renormalization as an issue.

*23.4 A Minimal Set of Postulates

The preceding sets of postulates contain many (often redundant) references to the particles, interactions and symmetries. Also the full impact of the proposed complexon property of quarks and gluons is not evident in the stated postulates. The use of both ElectroWeak cases: SU(2)⊗U(1)⊕U(1) ElectroWeak theory and SU(2)⊗U(1) ElectroWeak theory, makes the list of postulates lengthy.

Therefore we felt it would be interesting to state a minimal set of postulates based on GL(4)⊗GL(4) space-times, SU(2)⊗U(1)⊕U(1) ElectroWeak symmetry and complexon quarks. The statement of these postulates relies on the definition of terms made earlier. In defining these postulates we *assume* that future experiments will resolve current experimental issues in a certain way. The next few chapters describe open issues. Chapter 26 lists the eight variations allowed by current experimental uncertainties. The initial beginning of our derivation/construction in time and space asynchronicity is not embodied in the postulates. We view the origin of the GL(4)⊗GL(4) space-times to lie in asynchronicity – an obvious, but usually unremarked, property of physical phenomena.

Postulate 1. Any observer can define a set of time and space coordinates called a coordinate system in which the observer is at rest. One can define a transformation that relates the coordinate systems of two observers traveling at a constant velocity with respect to each other.

Postulate 2. The speed of light, c, is the same in all coordinate systems.

Postulate 3. The invariant interval or distance $d\tau$ *is defined by*

$$d\tau^2 = g_{\mu\nu}dx^\mu dx^\nu$$

Postulate 4. We assume that there are two initially complex space-times which are representations of GL(4)⊗GL'(4).

Postulate 5. The GL(4) group has a familiar coordinate space representation except that it is a complex space-time with a complex Lorentz group denoted L(C) whose elements Λ_C *satisfy* $\Lambda_C{}^T G_+ \Lambda_C = G_+$*. L(C) is in turn broken to the real Lorentz group L of space-time that we physically observe.*

Postulate 6. A physical superluminal transformation consists of an L(C) transformation followed by an SU(2)⊗U(1)⊕U(1) transformation. The combined transformation ttransforms real coordinates to real coordinates.

Postulate 7. The free part of the dynamical equations of the four species of initially free fermions can be generated by L(C) boosts from the Dirac equation for a fermion at rest subject to the requirement that there be only one real time variable in each resulting dynamical differential equation.

Postulate 8. Each free fermion field has the simplest possible lagrangian from which the canonically conjugate momentum, the energy-momentum tensor, the equal-time (or equal light-front) anti-commutation relations, and other physically important quantities can be derived.

Postulate 9. The Standard Model is a canonical quantum field theory with a lagrangian and a path integral formulation. The mathematics of these constructs is assumed to be correct.

Postulate 10. A fermion triplet consists of an L(C) doublet consisting of a Dirac field and a tachyon field, and an L(C) singlet with all fields of the same particle generation where L(C) field transformations are defined by eqs. 18-A.17-18. Each triplet is also a representation of a broken local (upon introduction of gauge covariant derivatives) SU(2)⊗U(1)⊕U(1) ElectroWeak symmetry (broken partly because the local SU(2)⊗U(1) rotation of Dirac fields and tachyon fields is not physically acceptable.)

Postulate 11. The three species of leptons are charged leptons (satisfying a Dirac equation), neutrinos (satisfying a tachyon dynamical equation) and leptonic neutral WIMPs satisfying a Dirac equation.

Postulate 12. A lepton triplet in any generation consists of a charged Dirac field, a neutral tachyon field (the neutrino), and a neutral singlet lepton (WIMP). The equations of motion of the lepton triplet are eqs. 18-A.17.

Postulate 13. The three species of quarks are up-type quarks (satisfying a Dirac equation), down-type quarks (satisfying a tachyon dynamical equation) and color singlet quark WIMPs satisfying a Dirac equation. All quarks of all species have complex 3-momenta and real energies.

Postulate 14. A quark triplet in any generation consists of a singlet WIMP quark, an up-type quark Dirac field, a down-type quark tachyon field.

Postulate 15. The interacting extensions of free triplet equations of motion have interactions consisting of local Yang-Mills fields. ElectroWeak gauge fields are massless "before" spontaneous symmetry breaking and are thus conventional local Yang-Mills gauge fields without a tachyon equivalent.

Postulate 16. L(C) group covariance, and ElectroWeak gauge symmetries of the dynamical equations of motion, are spontaneously broken through the appearance of additional fermion and vector boson mass terms generated by the Dimensional Mass Mechanism.

Postulate 17. Quarks are in the $\underline{3}$ representation of Color SU(3). The SU(3) symmetry is gauged with local Yang-Mills SU(3) fields called gluons that constitute the Strong interaction of the quark sector. Quarks are minimally coupled to the gluons in a gauge covariant fashion.

Postulate 18. The GL'(4) group has a 4-dimensional complex space-time that supports four fermion generations and generates the fermion mass matrices.

Postulate 19. The GL'(4) Dirac-like equations are covariant under SU(2)⊗U(1)⊕U(1).

Postulate 20. The masses, and mixing angles, of fermions appear as separation constants in GL(4)⊗GL'(4) fermion dynamical equations.

24. Resolved and Unresolved Issues

24.1 Resolved Issues in the Standard Standard Model and the Complexon Standard Model

We have seen that the forms of the various types of the Standard Model appear to be ultimately based on the "broken" geometry of 16-dimensional space-time. Thus we can derive parity violation, the Strong interaction – color SU(3), the form of the Electroweak sector with SU(2)⊗U(1)⊕U(1) symmetry and WIMPs (or SU(2)⊗U(1) without WIMPs), the existence of six fermion species (charged leptons, neutral leptons, leptonic WIMPs, up-type quarks, down-type quarks, and quark WIMPS), and the existence of four fermion generations – *all from geometrical considerations*.

Remarkably the forms of all these features are determined by 16-dimensional space-time broken to the direct product of a four dimensional space-time by another four dimensional space-time. We base the choice of two 4-dimensional space-times on time and space asynchronicity. However the values of constants: masses, mixing matrices and coupling constants are not determined by these considerations. One thus has two choices: 1) the constants are not amenable to calculation and can only be determined experimentally, or 2) there is a deeper level of theory that remains to be determined in which the values of the constants can be determined theoretically. The author favors the second possibility knowing that this deeper level is as yet unattainable.

24.2 Unresolved Issues in the Standard Standard Model and the Complexon Standard Model

The unresolved issues that are associated with the derivation in this book can be divided into conceptual issues and experimental issues. These issues overlap to some extent.

The conceptual issues center around

- The breakdown mechanism(s) presumably from GL(16) to the Lorentz group in several conceptual stages, and ElectroWeak symmetry breaking.

- The values of coupling constants, mixing parameters, and particle masses.

The significant experimental issues that fix the structure of the Standard Model are:

1. The nature of quarks: Are they complexons?
2. Are neutrinos and down-type quarks tachyons?
3. Is there a fourth generation of fermions, and does it mix with the three known generations.
4. Do WIMPs exist and can ElectroWeak symmetry be shown to be SU(2)⊗U(1)⊕U(1)? Does Dark Matter consist of WIMPs?
5. Are there other subtle effects such as CP violation that give further insight into realistic possible extensions of the Standard Model.

Experimental issues are further addressed in chapter 25.

24.3 The Breakdown Mechanism

We have been constantly referring to the breakdown(s) of symmetries ultimately from GL(16). While a mechanism such as the Higgs Mechanism may account for these breakdowns, at least in part, we will not speculate as to the breakdown mechanism in the absence of decisive experimental evidence. The fact that the Higgs Mechanism offers a framework for breakdowns with the Standard Model is encouraging. However, the "proof of the pudding" is the experimental discovery of Higgs particles, which at the moment, is lacking.

The investigation of other possible methods of symmetry breaking such as the Dimensional Mass Mechanism (sections 21.3 and 21.4) should be pursued. It would be somewhat surprising if the Higgs Mechanism were the unique, one and only, non-trivial, symmetry breaking theoretical mechanism.

24.4 Determination of the Values of the Coupling Constants and Mass Matrices

While the derivation has determined the form of the possible Standard Model lagrangians, the values of the Electroweak coupling constants and Weinberg angle, the strong interaction coupling constant, (and the gravitational coupling constant upon unification with quantum gravity), the masses of ElectroWeak vector bosons, and the mass matrices (and mixing angles) of the four fermion species remain to be determined.

There are two general methods to pin down the values of constants within the framework of Quantum Field Theory: 1) imposition of a symmetry group structure that relate the values of constants; and 2) the use of consistency conditions that set the values of some or all of the theory's constants. There are many model theories illustrating approach 1.

The only major 4-dimensional quantum field theory illustrating approach 2 is the Johnson-Baker-Willey model[334] of massless Quantum Electrodynamics which used the requirement of cancellation of all divergences as a consistency condition on QED. Johnson, Baker, and Willey showed that the consistency condition would be met if a certain function (called the eigenvalue condition) had a zero[335] at the physical value of the fine structure constant α, which is approximately 1/137.

This author[336] developed an approximate solution of the Schwinger-Dyson equations to calculate the eigenvalue function for α.

The Johnson-Baker-Willey model has been made obsolete by the success of ElectroWeak theory and thus does not itself appear to be of interest any longer. Nevertheless it illustrates the possibility of using consistency conditions to determine the value of constants appearing in a quantum field theory.

---

[334] M. Baker and K. Johnson, Phys. Rev. **D8**, 1110 (1973) and references therein.

[335] Adler later showed the zero, if it exists, must be an essential zero.

[336] S. Blaha, Phys. Rev. **D9**, 2246 (1974). Although this solution summed an infinite number of diagrams' contributions – more than any other calculation in QED known to this author, and although the solution agreed with the exact calculated values to 4th order in α, an essential singularity was not found (although a complex analytic structure was found). The first zero was at $\alpha = 1$. For this reason the Johnson-Baker-Willey program for creating a finite QED seems to have failed.

25. Experimental Tests

25.1 Experimental Tests to Nail Down the Form of the Standard Model

Two general possible forms of the Standard Model have been derived based on two possible sets of postulates. The postulates embody several experimental tests. In this chapter we state the tests that would prove or disprove various postulates.

25.2 Tachyonic Neutrinos and Down-type Quarks

In both general types of Standard Models, the Standard Standard Model and the Complexon Standard Model, fermions are assumed to appear in triplets. In both Standard Models each triplet is assumed to have a tachyon member. A major experimental test of this derivation is to determine whether neutrinos are tachyons and down-type quarks are tachyons. In the case of neutrinos the known smallness of the neutrino masses makes an experimental test difficult. In the case of down-type quarks, quark confinement makes the determination of whether they are tachyons indirect.

One possible area of experimental tests are parton models of nucleons in electron-nucleon scattering. Another experimental test area would be accelerator experiments that create quark-gluon plasmas such as Au-Au scattering experiments.

25.3 Quarks and Gluons: Complexons or "Normal"

In the study of the implications of our postulates we have found two natural general forms of the Standard Model: the Standard Standard Model and the Complexon Standard Model. A major difference between these theories is in the 3-momenta of quarks and gluons. Are their 3-momenta real? If so, the Standard Standard Model is the correct model. Or are their 3-momenta complex? If so, then the Complexon Standard Model is the correct model of elementary particles.

How can we detect the difference between complexon and "normal" quarks and gluons? Color confinement makes this determination difficult. And since the Strong Interaction does not have good, accurate, perturbative calculation predictions like QED the determination of the nature of quarks and gluons is that much more difficult. At the moment it appears that fits to emerging quark-gluon plasma data from Au-Au (and similar) experiments may offer a means of distinguishing complexon from non-complexon particles.

In addition the difficulties in the analysis of the spin-dependent structure functions of electron-nucleon scattering may be explainable by complexon quark constituents since complexons have a richer spin structure than normal spin ½ particles.

25.4 A 4th Generation of Fermions?

Section 20.3 raised the possibility of a fourth generation of fermions possibly with its own one generation Standard Model. If a fourth generation of fermions exist the overall pattern of masses of the three known generations suggest that the masses of the charged fourth generation fermions would be extremely large.

There are three possible scenarios for a fourth fermion generation:

a. The fourth generation mixes with the three known generations. In this case they would constitute an extension of the set of particles of the Standard Model and could possibly be found at high energies.
b. The fourth generation does not mix with the three known generations. In this case the fourth generation could constitute a separate one generation Standard Model, which would then would be a candidate for Dark Matter. Experimentally fourth generation particles could not be created in accelerators from known particle interactions except possibly to an undetectable extent from gravitons.
c. The fourth generation does not mix with the three known generations but has the same Weak interactions as the known three generations. In this case they would have to be very massive to have escaped detection as part of unusual baryons created in accelerator experiments.

Possibility a is currently the most favored based on preliminary experimental data. It appears Dark Matter does decay into normal matter.

25.5 Higgs Mechanism or Dimensional Mass Mechanism

If Higgs particles are not found, then the “God-particle” is dead and an alternate mechanism for mass generation will be required. The alternate mechanism can be as simple as inserting mass terms in the Standard Model lagrangian and thus deferring the solution of the mystery of mass to the future. Or, in the case of fermion masses, the Dimensional Mass Mechanism described in section 21.3, which neatly combines the concepts of fermion generations, fermion masses, and mixing matrices between the generations. It appears to satisfactorily resolve the issue for fermions. In the case of ElectroWeak vector bosons the direct insertion of mass terms in the Standard Model lagrangian is completely acceptable if one uses Two-Tier quantization.[337] A fully renormalizable theory results. The Dimensional Mass Mechanism is also possble for non-abelian vector bosons. See section 21.5.2. It offers consistency with the fermion Dimensional Mass Mechanism as well as avoiding the use of Higgs particles which have not been found.

The primary experimental question in this case is then do Higgs particles exist? Hopefully the question will be resolved by the new LHC (Linear Hadron Collider) that is currently searching. (July, 2011).

If Higgs particles are not found then the Dimensional Mass Mechanism presents itself as an alternative. If there are four generations of fermions then, in this author's opinion, there would be fairly strong presumptive evidence for GL'(4) and the Dimensional Mass Mechanism. A more decisive test is not apparent at the present time.

25.6 Do WIMPs Exist And Are They Dark Matter?

The SU(2)⊗U(1)⊕U(1) ElectroWeak symmetry that we have newly derived predicts four leptonic WIMPs and four quark WIMPs (assuming four fermion generations). Dark Matter may be composed of WIMPs. Some recent astrophysical studies have suggested that certain astronomical sources that produce an excess of positrons may do so due to the weak decays of WIMPs.

A determination of the existence of WIMPs would be an important confirmation of the nature of our derivation of our various forms of Standard Models.

---

[337] The Two-Tier formalism has the added advantage of producing a fully renormalizable theory of quantum gravity. Thus a unified theory of the Standard Model and Quantum Gravity can be realized in a straightforward manner.

26. Summary of the Eight Possible Variations of the Derived Standard Model

Due to experimental uncertainties with respect to the features of the Standard Model there are eight possible variations of the derived "basic" Standard Model. In addition there are certain partially known detailed features of the "full" Standard Model such as CP violation that remain to be understood. Lastly, there are undoubtedly additional, currently unknown, features of the Standard Model that will become apparent as experiments are done at much higher energies in the future.

The eight variants of the "basic" Standard Model that we have explored that are derivable from reasonable variations of our set of postulates are summarized in the table of Fig. 26.1.

Standard Model	Number of Generations	Masses: Higgs (H) Dimensional (D)	Complexon Quarks/Gluons
Standard Model 1	4	D	Yes
Standard Model 2	4	H	Yes
Standard Model 3	4	D	No
Standard Model 4	4	H	No
Standard Model 5	3	D	Yes
Standard Model 6	3	H	Yes
Standard Model 7	3	D	No
Standard Model 8	3	H	No

Figure 26.1. A Table of the possible variants of the "basic" Standard Model. Experiment is required to determine which of these models is the correct model..

Standard Model 1 is the Complexon Standard Model and Standard Model 8 is the Standard Standard Model.

Tachyons and WIMPs appear naturally in all these models.

Obviously there are an unlimited number of possible extensions of these eight forms of Standard Model. *The close relation of the above Standard Models with space-time geometry suggests that Standard Model extensions should also have a close relation to space-time geometry.*

27. The Layer Below the Standard Model Lagrangian

At this point it is clear that we have uncovered a deeper level of particle physics that provides a rationale for the general form of the Standard Model (plus some variants) with all its unusual features and "peculiarities." We have phrased the description of this deeper level in terms of sets of postulates. The correct set of postulates remains to be specified due to experimental uncertainties detailed in the preceding chapters.

But we have good reason to believe that the key symmetry and space-time structures that lies at the base of the Standard Model are based on GL(16) or GL(4)⊗GL'(4).

The 4-dimensional, complex representation of the group that we denote as GL(4) is the source of the six species of fermions (two of which are tachyonic), parity violation, ElectroWeak triplets, the symmetry of the ElectroWeak sector of the Standard Model, and the Strong interaction symmetry SU(3) of quarks and gluons. This complex, 4-dimensional representation is further broken to the 4-dimensional real space-time that we experience. Again the breakdown mechanism remains to be determined.

The 4-dimensional, complex representation of the group that we denote as GL'(4) is the source of the four generations of fermions, and, if Higgs particles are not found, the source of the fermion masses and mixing matrices of each of the four fermion species via the Dimensional Mass Mechanism. This space may collapse to a point through an unknown mechanism.

Our purpose is to develop a ever-deeper basis for the Standard Model in the form of an increasingly more fundamental set of postulates:

Lagrangian → postulate set → refined postulate set → ?...

The success of our approach in deriving the *exact form* (and variants) of the Standard Model suggests that we are on the right track to a deeper level of physics. The major questions that currently confront this approach are: the source of the breakdown mechanisms, and the reason for values of the various constants in the theory: coupling constants, masses, and mixing matrices. It is somewhat amazing that we can derive the *form* of the Standard Model and yet have the various constants and masses completely "free." This situation opens the door to speculation on universes with a Standard Model of the same form but differing fundamental constants.

Another open question is whether more physics will be found at higher energies. The fact that our derivation follows so directly from geometry suggests that what we have found will remain true. What remains to be found will extend these results. *The Standard Model now has a firm foundation in theory as well as experiment.*

28. Beyond the Standard Model

28.1 Approaches to a Deeper Level of Reality

While we have shown that Operator Logic, particularly Asynchronous Logic, furnishes a basis for the development of the Standard Model of Particles, the additional assumptions – what we call the Knowledge Base – remain to be determined. It may be that, like Euclidean geometry, we can go no further then a set of postulates/axioms (assumptions). The number of explicit axioms needed to obtain the Standard Model in Blaha (2008) was twenty-three.[338] There were also a number of implicit assumptions in the derivation in Blaha (2008) just as there are a number of implicit assumptions in Euclid's'geometry which are only evident when figures are drawn in the process of proving theorems.

Irrespective of the question of the number, and content of the axioms required to prove the Standard Model of Elementary Particles and thus establish the basis of Reality as we know it, the question that immediately arises is the source/reason for these particular axioms.

The *possible* sources for the axioms leading to the Standard Model are:

1. There is no reason. They are just the basis of Reality.
2. They follow from an unknown deeper unifying principle(s).
3. They follow from a deeper known theory such as Superstring Theory.
4. They follow from an unknown mechanism that establishes order in the form of the axioms from chaos. A mechanism or explanation for the persistence of such order over billions of years must also be found.
5. They follow because they are the only totally consistent set of physical axioms.
6. They are the result of chance and one of many possible sets of axioms.
7. They follow because they are required for life, as we know it, to exist. (The Anthropic Principle) This case is clearly a subcase of cases 1 and 6.

While case 1 is possible it is intellectually unsatisfying and so we will not pursue it. We will discuss case 2 in the following subsections. We will discuss case 3 in section 28.4.

Case 4 seems unlikely to the author because of the secondary requirement that order persists indefinitely.

---

[338] The number of postulates in the Complexon Standard Model in this book is 29.

Case 5 brings us to Gödel's Consistency Theorem, which roughly states that a set of axioms cannot be proved to be consistent within the mathematical-deductive system that they define. Thus case 5 is unprovable.

Case 6 introduces an infinity of possible choices for the set of axioms for physics. The broadness of the set of choices makes this an unattractive possibility.

Case 7 is certainly true for humanity but recent studies have shown there is an extremely wide range of environments that can support life, and possibly, intelligent life. Thus, unless one posits religious reasons, it is not particularly clear that there is a compelling case for the Anthropic Principle. Case 7 then degenerates to case 6 – mere chance. And mere chance is not susceptible to intelligent discussion unless one can show that there is a "probability distribution" for sets of axioms and the set of axioms that governs our universe is amongst the most likely sets of axioms. Then case 7 becomes a subcase of case 2.

Therefore we find that case 2 is the most attractive case to study although case 3 (discussed elsewhere) is also of interest.

28.2 Extremum Rule for the Lagrangian of Physical Reality?

In Blaha (2005b) we developed a classification scheme for the "space" of all possible lagrangians and divided the space into subsets based on a definition of a Gödel number for lagrangians.[339] This classification scheme is a beginning towards developing a theory of lagrangian selection that determines the lagrangian of our universe or at least a set of lagrangians that lead to physically reasonable, non-trivial universes. Blaha (2005b) developed a map from lagrangians to Gödel numbers. Subsequent to the discussion below we will begin to consider the possibility of more physical principles to select a suitable "world lagrangian."

28.2.1 A Gödel Number Map for Lagrangians

Blaha (2005b):

> Definition: Suppose the Lagrangian L is an expression (string) of symbols, β_1, β_2, β_3, β_4, ... β_n (including field operator symbols, mathematical symbols, integral signs, the number of space dimensions, the number of time dimensions, space-time indices, and any other quantities that appear directly or indirectly in the Lagrangian); and v_1, v_2, v_3, v_4, ... v_n is the sequence of tokens corresponding to these symbols, then the Gödel number of a Lagrangian L is the integer
>
> $$\text{gn}(L) = \text{Min} \prod_{m=1}^{n} \text{Pr}(m)^{v_m}$$
>
> where $\text{Pr}(m)^{v_m}$ is the mth prime number raised to the power v_m with the 1st prime number taken to be 2; and where Min indicates the minimum is taken over all possible permutations of the order of symbols in the

[339] Blaha (2005b) pp. 151-154.

lagrangian that maintain its well-formed nature and all possible permutations of the assignment of tokens to the symbols in the lagrangian. The token numbers of the n symbols consist of a set of odd numbers ranging from 3 + p to 3 + p + 2(n – 1) where p is an even number greater than or equal to zero. (We set p = 0 for the sake of convenience.) *The set of token numbers consists of all odd integers ≥ 3.* A *well-formed lagrangian* has its symbols ordered in such a way that their order conforms to the syntax rules of the operators in the lagrangian. If L is an empty string (containing no symbols) then gn(L) = 1.

Since any number has a unique decomposition as a product of powers of prime numbers the following corollaries hold.

21. Corollary: If L_1 and L_2 are expressions such that $gn(L_1) = gn(L_2)$, then $L_1 = L_2$ (the Lagrangians are the same Lagrangian.)

22. Corollary: If the Gödel numbers of two Lagrangians are the same then they generate the same physical theory.

If we take an arbitrary lagrangian and generate all possible lagrangians from it through substitutions, definitions of new variables (and field operators if the lagrangian is for a field theory), and other Rules of Inference that are used to transform lagrangians to a new form, then the set of lagrangians so constructed constitutes a set of physically equivalent theories. (Certain additional changes are made in the case of theories with path integral formulations.) As stated, we denote these sets of physically equivalent theories as LPset_i for i = 1, 2, ... These sets are subsets of the set of all lagrangians Lset. The use of physically equivalent lagrangians is exemplified by a Higgs particle lagrangian, which is transformed to a different form to show the effects of its constant ground state value.

Those subsets of Lset, LPset_i, that contain classical (deterministic) physical theories such as theories in classical mechanics require no special treatment. Goldstein (1965) describes various forms of transformations: Legendre transformations and other transformations between equivalent physical theories.

In the case of subsets of Lset, LPset_i, that contain quantum theories, changes of variables (fields) require an additional modification in the path integral formulation of the theories. The path integral integrand must have a Jacobian factor that reflects the change of variables in the lagrangian in order for the original lagrangian theory and the transformed lagrangian theory to have the same physical implications. For example if the path integral for a certain lagrangian in a certain subset of lagrangian theories is

$$Z(J) = N \int D\phi \exp\{i\int d^4y\, [\mathscr{L}(\phi) + J(y)\phi(y)]\}$$

where N is a normalization factor. Then a change of field variable from ϕ to ψ

$$\phi = \phi(\psi)$$

in the lagrangian necessitates the introduction of a Jacobian factor $\mathcal{J}(\psi)$ in the path integral integrand (except for linear changes of variable)

$$Z(J) = N \int D\psi\, \mathcal{J}(\psi)\exp\{\, i\int d^4y\, [\mathcal{L}(\psi) + J(y)\phi(\psi(y))]\}$$

See Huang (1965) or Weinberg (1998) for examples in the case of Faddeev-Popov gauge fixing.

Thus we have shown that it is possible to define subsets of theories for deterministic and quantum lagrangians where each lagrangian in the subset embodies the same physical theory.

We can define a Gödel number for a subset $LPset_k$ as:

23. Definition: The Gödel number of a subset $LPset_k$ is the minimum of the Gödel numbers of the members of the subset.

Due to the unique decomposition of a whole number into its prime factors we have the lemma:

24. Lemma: If the Gödel numbers of two subsets of Lpset are the same then they are the same subset.

In the next part we will define a criteria for selecting the "simplest" lagrangian in a given subset. This definition will be of some importance since potential Theories of Everything are judged by their simplicity as well as their physical implications.

The criteria of simplicity and elegance, which are much talked about in discussions of the Theory of Everything, are subjective, ambiguous, and anthropomorphic. The assignment of Gödel numbers to lagrangians enables us to create a mathematical definition of simplicity and, to some extent, of elegance since these criteria are usually subjectively related. A lagrangian's Gödel number is a measure of the number of symbols and the number of times each symbol appears in the lagrangian. Therefore lagrangians in fewer dimensions, with fewer species of particles or fields, with lower spin entities, and with fewer terms will have a smaller Gödel number. Based on this observation we define the simplicity of a lagrangian as:

25. Definition: For any two lagrangians L_1 and L_2 if $gn(L_1) < gn(L_2)$ then L_1 is simpler than L_2.

which implies the following lemma:

26. Lemma: The Gödel number of the subset $LPset_{TE}$ that contains the Theory of Everything is the Gödel number of the simplest lagrangian in the subset.

28.2.2 The Natural Selection of Physical Lagrangians

The preceding quote gives us a method of finding the Gödel number of the set of equivalent lagrangians that define the "Theory of Everything". However an extremum method for determining this set of lagrangians is lacking. Clearly it is not defined by the least Gödel number. And, in fact, *there does not appear to be an obvious connection between Gödel numbers and lagrangians that would lead to an extremum method.*

Leibniz, who invented calculus including extending extremal (minimax) methods,[340] suggested perhaps the most concrete concept for an extremum rule to determine the fundamental laws of nature: *the laws of nature are of maximal simplicity yet capable of producing a universe of maximal complexity.*[341]

The union of these two features was one of the motivations that led Leibniz to describe this universe as "the best of all possible worlds." The mathematical framework needed to express this elegant, but currently unquantifiable, extremum principle, or any extremum principle that leads to the laws of space, time and matter, remains to be developed. One can hope that someday we will reach that level of understanding if finding the "right" lagrangian with this approach is the correct approach. It seems difficult to see how this approach can determine the features of space-time and of internal symmetries and interactions.

If we do, then we will perhaps be able to create "universes of the mind" in which we can explore the implications of other designs for a universe. And perhaps we may be able in that latter age to experimentally create bubbles of those other possibilities for a universe in the laboratory.

Whether there may be other sister universes that realize these other designs is also a question of importance. Today physicists speculate about such universes but if the history of physics has taught us anything, it has taught us that very few speculations about future physical theory turn out to be correct.

28.2.3 Using the Information↔Energy transformation to Select a Lagrangian

It has been known for a long time that information can be transferred into energy. Indeed the relation between energy and information theoretically is known to be

---

[340] Nova Methodus pro Maximis et Minimis ("New Method for the Greatest and the Least") 1684.

[341] In Leibniz's words, "at the same time the simplest in hypotheses [i.e., its laws] and the richest in phenomena." Quoted in Rescher (1967) p.19.

$$\text{one information bit} \equiv 3 \times 10^{-21} \text{ joules at room temperature}$$

Recently this equivalence has been confirmed experimentally for the first time by E. Muneyuki and M. Sano.[342]

Consider now a lagrangian for a theory of everything or a set of postulates that constitutes a theory of everything. These constructs contain information. This information is equivalent to energy. If we can construct a measure of the information in the theory of everything then we may be able to apply well-known minimax techniques to determine the "correct" theory.

Since an empty string constitutes a minimal physical theory it is clear that the minimax calculation must be subject to constraints that preclude a trivial physical theory. Leibniz's caveat *"capable of producing a universe of maximal complexity"* is a constraint of this type. But a mathematical definition of a universe of maximal complexity is not obvious.

Therefore we have to overcome the lack of a measure of the information content of a theory[343] and the lack of a set of physical constraints that the minimax solution must satisfy.

28.3 Axiomatic Approach to the Knowledge Base of Reality

In a series of books Blaha developed an axiomatic approach that led *exactly* to the established form of the lagrangian of the Standard Model of Particles. In this approach an axiomatic method was used and the established parts of the Standard Model derived.

The success of this approach leads us to consider physics to ultimately be the result of a set of fundamental axioms that lead to the Standard Model lagrangian. It is possible to define Gödel numbers for a set of axioms in a manner similar to that of the preceding subsection for lagrangians. Blaha (2005b) defined Gödel numbers for sets of axioms:[344]

> 27. Definition: If A is a non-Lagrangian, axiomatic theory, and a_1, a_2, a_3, ... , a_n are the *sequences* of symbols of the n (different) axioms of A, then the Gödel number of the theory can be defined to be
>
> $$gn(A) = \text{Min} \prod_{m=1}^{n} Pr(m)^{gn(a_m)}$$
>
> where $Pr(m)^{gn(a_m)}$ is the mth prime number raised to the power $gn(a_m)$, where $gn(a_m)$ is the Gödel number of the axiom a_m and where Min indicates the minimum is taken over all possible permutations of the

---

[342] E. Muneyuki and M. Sano, *Nature Physics*, DOI: 10.1038/NPHYS1821.

[343] It is possible that a Gödel number classification scheme might play a role in defining the information content of a theory.

[344] Blaha (2005b) pp. 154-156.

order of the axioms, all possible orderings of the symbols in the axioms that result in well-formed axioms, and all possible permutations of the assignment of tokens to the symbols in the production rules.

Thus if there are n axioms rules that use m symbols, then there at least n!m! Gödel numbers that correspond to the production rules (assuming the token numbers consist of the set of odd numbers ranging from 3 + k to 3 + k + 2(m – 1) where k is an even number greater than or equal to zero. (We set k = 0 for the sake of convenience.) A unique Gödel number for A is obtained by taking the prescribed minimum.

28. Theorem: If the Gödel numbers of two axiomatic theories are the same then they generate the same physical theory.

15.6.2 Subsets of NLset Containing Physically Equivalent Theories

Given a set of primitives and a set of axioms that define a physics theory, we can create a mathematical-deductive system that describes the physical theory in a manner similar to that used to define a mathematical-deductive system for a mathematics theory such as the theory of whole numbers.

Suppose we now take the primitive terms p_1, p_2, ... of the theory A and define a new set of primitive terms p_i' with

$$p_k = p_k(p')$$

and substitute the new primitives in the axioms creating a new theory A'. The new theory is physically equivalent to the original theory A but its mathematical-deductive system will look different.

Now consider the set of physical theories generated by all possible redefinitions of the primitive terms as above. This set of physical theories will be a subset of the set of all non-Lagrangian theories NLset. All theories in this subset are physically equivalent.

A moment's consideration leads to the realization that NLset is composed of subsets of equivalent axiomatic theories which we will denote NLset_i for i = 1, 2, ... The elements of each subset have different Gödel numbers in general. Therefore we will define the Gödel number of a subset by:

29. Definition: The Gödel number of a subset NLset_k is the minimum of the Gödel numbers of the members of the subset.

Due to the unique decomposition of a whole number into its prime factors we have the lemma:

30. Lemma: If the Gödel numbers of two NLset subsets are the same then they are the same subset.

In the next section we will define a criteria for selecting the "simplest" axiomatic theory in a given subset.

15.6.3 Simplicity Criteria for Subsets of NLset Containing Physically Equivalent Theories

A Gödel number is a measure of the number of symbols and the number of times each symbol appears in the set of axioms. Therefore sets of axioms for physics theories in fewer dimensions, with fewer species of particles or fields, with lower spin entities, and with fewer terms will have a smaller Gödel number. Based on this observation we define the simplicity of an axiomatic theory by:

31. Definition: For any two axiomatic theories A_1 and A_2 if $gn(A_1) < gn(A_2)$ then A_1 is simpler than A_2.

which implies the following lemma:

32. Lemma: The Gödel number of the subset $NLset_{TE}$ that contains the Theory of Everything, if it is axiomatic in nature, is the Gödel number of the simplest theory in the subset.

15.7. Escaping the Trap of Anthropomorphism

The form of Theories of Everything, and of physical theories in general, normally has anthropomorphic trappings: the choice of symbols, the form of the expressions, the characterization of the symmetries, and so on.

If we use the Gödel-number-based approach to defining lagrangian and non-lagrangian theories using the definitions of simplicity that we have developed we arrive at a non-anthropomorphic characterization of physical theories.

Thus we can hope that we can arrive at a universal characterization of the Theory of Everything that would pass this test: if we meet an intelligent alien scientist we could compare our theories of everything based on Gödel numbers! If the theories are the same their Gödel numbers should be equal. (We assume that they define tokens in the same way.)

As the passage shows one can define Gödel numbers for sets of axioms as well as for lagrangians. Unfortunately the same issue arises: there is no principle(s) that can be used to select the set of axioms actually used by Nature. Blaha's axioms are quite simple, relatively speaking, and lead directly to the Standard Model of Particles. It seems the best approach to progress in this area is to refine, and to reduce the number of, this set of axioms as well as to develop physical motivations for the axioms. If progress is made in this area then perhaps the Gödel number classification scheme, or some other scheme, may lead to a deeper understanding of fundamental physical theory.

28.4 What About String Theory?

String Theory has been the favored fundamental theory of elementary particles for many years. The major roadblocks to its acceptance is the complete absence of experimental evidence supporting it, and the theoretical difficulty of constructing the "correct" string theory.

Our derivation of the major features of the form of The Standard Model from symmetries based on space-time suggest that The Standard Model, in itself, is closely grounded on Reality and thus *a String Theory basis is not needed to understand particle physics.* Postulating infinitesimal strings in order to derive The Standard Model's form is clearly unnecessary and consequently excluded by Ockham's Razor!

Our space-time considerations have yielded: the fermion spectrum (four generations of four species), parity violation, SU(2)⊗U(1)⊕U(1) symmetry resulting in ElectroWeak triplets that include WIMPs, color SU(3), and the Dimensional Mass Mechanism. Locality (a space-time feature) has yielded the local gauge interactions. These results are a complete list of the experimentally certain features of elementary particles at the present time. And our derivation is based on initially simple "point-like" particles – not more complicated two-dimensional strings.

28.5 Constructing "The Theory of Everything" from Logic

In the preceding sections we have considered possible sources for the axioms that underlie The Theory of Everything for our universe. These approaches are not satisfactory for one reason or another.

An approach that does appear to be satisfactory starts from the fundamental principles of Logic. Part 2 describes the basis of Logic in a new formalism, Operator Logic, that leads to an understanding of the origin of Dirac-like spin ½ dynamical equations. From this starting point and GL(16), suitably broken, we derived the forms of the Standard Model from the geometry of space-time.

Unlike previously described approaches there is no need for unknown minimax principles, unseen strings or other artifacts that would provide a basis for elementary particle physics. Logic is the most critical factor in any attempt at a physical or mathematical theory. The foundation of Physics in Logic is the surest foundation for success!

Part 3. SuperLight Travel To The Stars Based On The Standard Model

The only possible approach to faster than light (superluminal) starships is based on The Standard Model with complex quark and gluon 3-momenta presented in this book. Thus the importance of experimental efforts to verify complex quark 3-momenta. All other approaches require resources greater than the earth's capacity, or crush starship passengers, or take generations of travel. Trips requiring generations of travel restrict starship travel to a small number of expeditions rather than large volume traffic back and forth to the stars envisioned by this author.

These two chapters describing superluminal starship dynamics and engines are a major benefit of the complexon version of The Standard Model. They are a slightly modified and extended part of the author's book *All The Universe* – Blaha (2011b).

3. Superluminal Starship Dynamics

3.1 Exceeding the Speed of Light

In standard sublight relativistic dynamics the speed of a massive object cannot exceed the speed of light if the force applied to the object is real. In this section we will consider the case of a *complex-valued* force applied to an object (complex thrust) that causes the object to attain a complex velocity whose real part can exceed the speed of light. Complex valued forces (tachyonic forces) have not been experimentally found in Nature as yet. However the motion of a particle inside a Black Hole is tachyonic. Its motion is determined by the gravitational curvature of space with the Black Hole. The Equivalence Principle of General Relativity tells us that the particle's gravitationally based motion can be viewed as a force acting on the particle in an inertial reference frame. (We experience the Equivalence Principle in this fashion every day – the force of gravity is equivalent to the gravitational curvature of space-time.) Now, the gravitational equivalent force on a tachyon particle in a Black Hole must be a complex valued force or the tachyon's motion would not be tachyonic. Thus Black Holes, and consequently Nature, do have complex valued forces.

In Blaha (2010a) we show that neutrinos and d type quarks are tachyons. Up-type quarks also have complex 3-momenta and can be accelerated to a speed faster than light. If we can harness quarks to create rocket thrust then we will have a mechanism for complex-valued thrust that can power starships faster than the speed of light.

Since a complex-valued rocket thrust will generate a complex-valued velocity and movement in space, the physical interpretation of complex velocities and distances must be addressed. In the previous chapter we showed that a superluminal transformation maps points with real coordinate values in one coordinate system to points with complex coordinate values in the target coordinate system. We then showed that the complex-valued coordinate points in the target coordinate system could be "rotated" to real valued coordinates using $\Pi_L(\mathbf{v}/v)$ (eq. 2.34). Thus the combined superluminal transformation and $\Pi_L(\mathbf{v}/v)$ transformation $\Pi_L(\mathbf{v}/v)\Lambda(\mathbf{v}/v)$ maps real coordinates to real coordinates. Complex coordinates are then merely an artifact of superluminal transformations.

However when we consider the path of a rocket with complex thrust that starts from a spatial point with real coordinates and, as it accelerates, traverses complex-valued spatial points a new issue arises: What is the physical meaning of these complex-valued spatial coordinates. Unlike the previous case of superluminal transformations one cannot simply use a global transformation to change the complex-valued points to points with real coordinate values. This is particularly clear if one considers a three point configuration: the earth, the rocket and the destination star. The earth and star have real valued coordinates in the earth coordinate system. The rocket in transit has complex coordinates at each point of its journey. In general, there is no global transformation that will make the coordinates of all three points real-valued. Therefore we conclude that complex coordinates are physically meaningful in this type of situation where one "point" is moving with a complex velocity. On this basis we will assume that space is three-dimensional with complex coordinate values in general.

Why haven't the complex values of coordinates been noticed before? Because objects with complex velocities have not been created and/or seen. To give an object a complex velocity we need either a highly curved space-time region (such as a Black Hole) with an event horizon that encloses the object so that we can't see it; or the movement of an object by a complex-valued force or thrust that would make the object traverse complex spatial points.

The second possibility can only be achieved with tachyon thrust or force generated by accelerating quarks. Neutrinos only interact via the Weak interaction and have strictly real momenta. They are not capable of generating a complex thrust since they only experience the weak force. Quarks are confined within protons and neutrons. To create macroscopic regions containing quarks we would need collisions at enormous energies. We are just entering the experimental stage where this possibility can be realized. RHIC at Brookhaven National Laboratory and LHC at CERN have started creating quark-gluon fluids by colliding heavy ions such as gold and lead ions. Evidence for tachyonic quarks within the collision regions will hopefully be forthcoming soon.

Then a superluminal, tachyon drive starship with complex thrust becomes possible.

3.2 Superluminal Starship Dynamics

In this section we will consider a constant, propulsive force in a starship's rest frame that drives the starship from a sublight velocity to a superluminal velocity. The key factor in achieving a superluminal speed is evading the singularity in γ at $v/c = 1$. We accomplish this goal by having a complex force – a force with a real and imaginary part – that generates a complex acceleration, and thus a complex velocity, that "goes around" the singularity in γ in the complex velocity plane. We assume that an "instantaneous" Lorentz transformation relates the earth reference frame and the starship reference frame.

We assume a constant, complex force exists in the rest frame of the starship due to the starship's thrust in the direction of the positive x' (and x) axis. The starship (primed coordinates) and earth (unprimed coordinates) coordinates have parallel axes as in Fig. 2.1. The spatial force in the positive x direction is

$$\mathbf{F'} = g\hat{\mathbf{x}} \qquad (3.1)$$

where g is assumed to now be a complex constant.

The fourth component of the force (since force is a Lorentz 4-vector) is zero in the rocket's rest frame:

$$F'^0 = 0 \qquad (3.2)$$

Applying the inverse of the Lorentz transformation eq. 2.1 we find the force in the earth rest frame is

$$\begin{aligned} F^0 &= \gamma(F'^0 + \beta F'^x/c) = \gamma\beta F'^x/c = \gamma v g/c^2 \\ F^x &= \gamma(F'^x + \beta c F'^0) = \gamma F'^x = \gamma g \\ F^y &= F^z = 0 \end{aligned} \qquad (3.3)$$

where $\beta = v/c$, c is the speed of light, and $\gamma = (1 - \beta^2)^{-1/2}$ as before. We again use the superscripts x, y, and z to identify the components of the spatial force. The spatial momentum of an object of mass m is

$$\mathbf{p} = \gamma m \mathbf{v} \qquad (3.4)$$

and the dynamical equation of motion is

$$d\mathbf{p}/dt = \mathbf{F} \qquad (3.5)$$

in the "earth" coordinate system resulting in

$$dp^x/dt = \gamma g \qquad (3.6)$$

with[345]

$$dp^y/dt = dp^z/dt = 0 \qquad (3.7)$$

The differential equation resulting from eq. 3.5 is[346]

$$d(\gamma v)/dt = \gamma g/m \qquad (3.8)$$

Assuming initially that g is real we must use $\gamma = (1 - \beta^2)^{-1/2}$ for $v < c$ and $\gamma = (\beta^2 - 1)^{-1/2}$ for $v > c$ based on the need for real coordinates for faster than light travel as expressed in eqs. 2.1 and 2.2. The solutions for real v are[347]

---

[345] There is thrust in the y and z direction as well. To avoid getting distracted by the details of an exact calculation we approximate the force in those directions as zero.

[346] Note that we neglect the decrease in mass of the starship as it uses its fuel. We assume that the effect of fuel consumption on the starship's dynamics will be of the order of 10%.

<u>v < c, Re v_0 < c</u>

$$v = c\{1 - 2/(1 + ((c + v_0)/(c - v_0))\exp[2g(t - t_0)/(mc)])\} \quad (3.9a)$$

<u>v > c, Re $\acute{\upsilon}_0 \geqslant$ c</u>

$$v = c\{1 - 2/(1 + ((c + \acute{\upsilon}_0)/(c - \acute{\upsilon}_0))\exp[2\breve{g}(t - t_0)/(mc)])\} \quad (3.9b)$$

where the velocity is v_0 at time t_0, $\acute{\upsilon}_0$ is the velocity[348] at t = t_0 and $\breve{g}$ is the acceleration for Re v $\geqslant$ c.[349]

Analytically continuing eqs. 3.9 to complex v with a complex force constant g we obtain the starship equation of motion. We require continuity when the real part of v = c by requiring that when Re v(t) of eq. 3.9a equal c, that t_0 = t and v(t_0) of eq. 3.9a equal $\acute{\upsilon}_0$. These conditions fix t_0 and $\acute{\upsilon}_0$.

<u>Note:</u>

Eqs. 3.9 can easily be integrated to give the distance traveled in the x direction.

<u>v < c, Re v_0 < c</u>

$$x = x_0 + (mc^2/g)\ln((1 - v_0/c + (1 + v_0/c)\exp[2g(t - t_0)/(mc)])/2) - c(t - t_0) \quad (3.10a)$$

<u>v $\geqslant$ c, Re $\acute{\upsilon}_0 \geqslant$ c</u>

$$x = x_0 - (mc^2/g)\ln((1 - \acute{\upsilon}_0/c + (1 + \acute{\upsilon}_0/c)\exp[2\breve{g}(t - t_0)/(mc)])/2) - c(t - t_0) \quad (3.10b)$$

or, correspondingly,

<u>v < c, Re v_0 < c</u>

$$x = x_0 + (mc^2/g)\ln[(1 - v_0/c)/(1 - v/c)] - c(t - t_0) \quad (3.11a)$$

<u>v $\geqslant$ c, Re $\acute{\upsilon}_0 \geqslant$ c</u>

$$x = x_0 + (mc^2/\breve{g})\ln[(1 - \acute{\upsilon}_0/c)/(1 - v/c)] - c(t - t_0) \quad (3.11b)$$

The complexity of g and thus of the velocity causes x to be complex. The starship is thus generally at a point x in complex space.

As a result superluminal travel to a distant star (or galaxy eventually) requires three phases in general. In the first phase the starship accelerates with a value for the thrust g that enables it to reach a high complex velocity whose real part was much greater than the speed of light. In the second phase the starship

---

[347] The velocity is entirely in the x-direction in this calculation. It can, and does, have complex values in this example. See footnote 45 in the discussion of our starship engine to see how the complexity of the value arises.

[348] It is greater than c by assumption in the calculation of eq. 3.9b.

[349] Although eqs. 3.9a and 3.9b have the same form, they have to be separately analytically continued to the complex velocity plane and the solutions matched on the vertical line Re v = c. (The acceleration can be changed at any point in the journey.)

coasts to a point “not far” from the destination. At this point the starship is located in complex space. In the third phase the starship engines are turned on and the thrust set at a value that will bring the starship to its destination (located at a real valued coordinate position.) See Figs. 3.1 – 3.3.

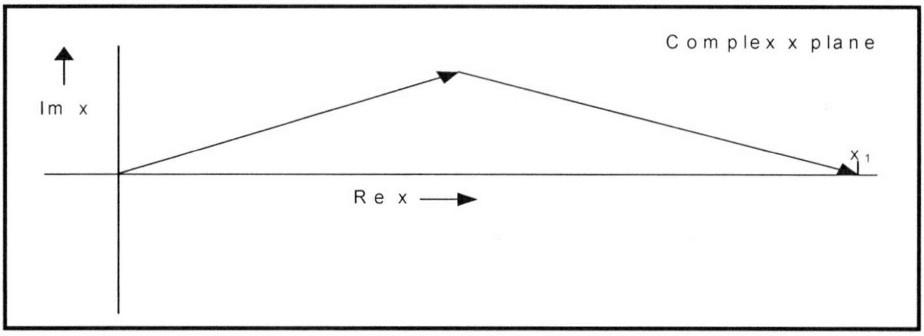

Figure 3.1. Rough depiction of the travel of a starship in the complex x dimension. The starship starts out in real space at x = 0. While accelerating, the starship is at a complex distance from its origin. After reaching cruising speed it turns off its superluminal engines until near its destination x_1. When nearing its destination it turns the superluminal engines back on, which brings it to its destination at the real distance x_1 at zero imaginary velocity and small real velocity.

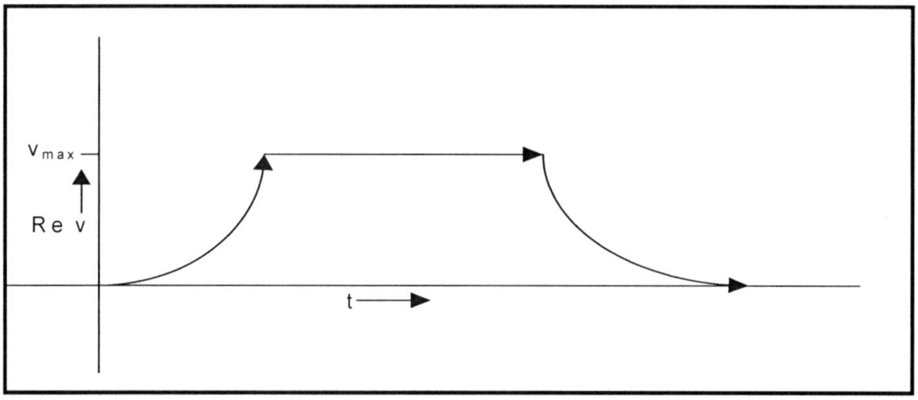

Figure 3.2. Depiction of the real part of a starship speed. There is an acceleration part to a desired maximum speed. Then the starship cruises at that speed until it reaches the vicinity of the destination. Then the starship drive decelerates it to a speed near zero so the starship can enter orbit around a star or planet.

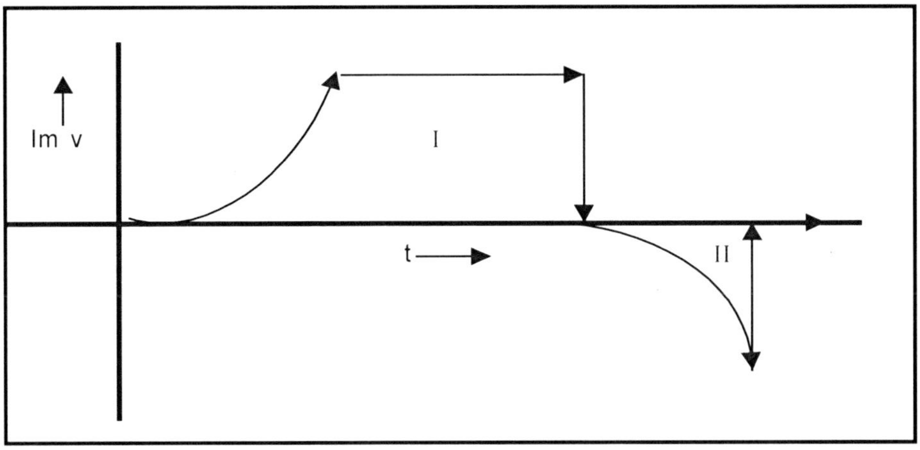

Figure 3.3. Rough depiction of a possible imaginary part of a starship speed. There is an acceleration part to high speed. Then the starship cruises at that speed until it reaches the vicinity of the destination. Then the starship drive decelerates causing the imaginary speed to decrease to zero so the starship has zero imaginary speed at the end of the deceleration. The net imaginary distance traveled is also forced to zero so that the starship ends in real space. The areas of regions I and II are of equal magnitude to bring the imaginary part of the starship's location to zero.

*3.3 Novel New Method of Achieving High Superluminal Starship Velocities

To achieve the type of motion depicted in Figs. 3.1 – 3.3 the constant force value $\breve{g}$ required after Re v $\geqslant$ c must satisfy a special set of conditions. These conditions emerge from a consideration of the denominator of eq. 3.9b:

$$1 + ((c + \acute{\upsilon}_0)/(c - \acute{\upsilon}_0))\exp[2\breve{g}(t - t_0)/(mc)] \qquad (3.12)$$

where $\acute{\upsilon}_0 \geqslant c$. If this denominator approaches zero then the speed v becomes infinite if g has an appropriate complex value. *This new, and the only known, method of achieving high superluminal speeds was proposed in Blaha (2011b) and earlier editions.* Let

$$\breve{g} = g_1 + ig_2 \qquad (3.13)$$

If we wish the velocity to get very large (approach infinity) after some acceleration time interval $\Delta t = t_1 - t_0$ we set

$$1 + ((c + \acute{\upsilon}_0)/(c - \acute{\upsilon}_0))\exp[2\breve{g}\Delta t/(mc)] = 0 \qquad (3.14)$$

with the result

$$g_2 = (mc/(2\Delta t))\{n\pi + \text{Im}\ \ln[(c - \acute{\upsilon}_0)/(c + \acute{\upsilon}_0)]\} \qquad (3.15)$$

and

$$g_1 = (mc/(2\Delta t))\ \text{Re}\ \ln[(c - \acute{\upsilon}_0)/(c + \acute{\upsilon}_0)] \qquad (3.16)$$

where n is an odd, positive integer, since $\acute{\upsilon}_0$ is complex in general. Eqs. 3.15 and 3.16 enable the real part of the velocity to become infinite in the time interval Δt. We assume n = 1 in the following discussions. Substituting in eq. 3.9b we obtain

$$v = c\{1 - 2/[1 + ((c + \acute{\upsilon}_0)/(c - \acute{\upsilon}_0))^{1 - (t - t_0)/\Delta t}\, e^{in\pi(t - t_0)/\Delta t}]\} \qquad (3.17)$$

3.4 Examples of a Starship Accelerating from Sublight speed to the Speed of Light

We will now consider a specific simple example of the acceleration phase of a starship. *This example has a conservative, slow acceleration. Much faster accelerations are possible with much shorter travel times of the order of months as shown in earlier editions of Blaha(2011b).* First we note that the acceleration of a rocket of mass m with a propellant exhaust speed v_e in the rocket's rest frame is given by

$$dv'/dt' = (v_e/m)\ dm/dt' \qquad (3.18)$$

and thus the constant g of eq. 3.1 is[350]

$$g = m dv'/dt' = v_e\ dm/dt' \qquad (3.19a)$$

Since we intend to generate the thrust with a quark-gluon plasma producing an extremely high-energy exhaust we will *choose* the value of the starship acceleration to be equal to twice the acceleration due to gravity at the earth's surface times (1 + i):[351]

$$g/m = 2(1 + i)g_E = (1 + i)1960\ \text{cm/sec}^2 \qquad (3.19b)$$

---

[350] In the illustrative example we will consider the acceleration g changes to $\breve{g}$ when Re v = c. The acceleration in a real world case would vary in such a way as to maximize speed while minimizing energy consumption.

[351] Astronauts can withstand up to $4g_E$ acceleration for several hours. On a starship with passengers in suspended animation even higher accelerations would appear to be acceptable.

where m is the mass of the starship. If we further choose the exhaust velocity v_e

$$v_e = -2c - 2ci \qquad (3.20)$$

which is a reasonable choice for the exit speed thrust of the quark fireball then

$$dm/dt' = -3.27\times10^{-8}m \qquad (3.21a)$$

in our example. If we envision a starship of 10,000 metric tons[352] then

$$dm/dt' = -327 \text{ gm/sec} \qquad (3.21b)$$

We can reach light speed starting from $v = 0$ in $t = 1.3\times10^7$ sec = 150 days measured in earth time using eq. 3.9:

$$v = 1.02c - 0.368ci \qquad (3.22)$$

The amount of fuel expended in this example is nominally 4,250 metric tons. Actually it would be a lesser amount if we took account of the declining mass of the starship as fuel is expended. This value can be reduced significantly by doubling or tripling[353] the exhaust velocity v_e. Tripling the exhaust velocity would reduce the fuel expended to 1420 metric tons – about 15% of the starship mass. Also there could be a significant transformation of energy into matter in the quark-gluon fireball that would also reduce the amount of stored fuel that was expelled.

3.5 From Light Speed to Enormous Speeds

In the example given in section 3.4 we considered an illustrative example of a starship accelerating to light speed. In this section we consider the second part of the acceleration: from light speed to enormous speeds taking advantage of the mechanism described in section 3.3. We will use an approximation to eq. 3.9b as its denominator approaches zero. Letting $t = t_1 + \tau$ where τ is small, and letting $\Delta t = t_1 - t_0$ then eq. 3.9b becomes

$$\begin{aligned} v &= c\{1 - 2/(1 + ((c + \acute{\upsilon}_0)/(c - \acute{\upsilon}_0))\exp[2\breve{g}(\Delta t + \tau)/(mc)])\} \\ &= c\{1 - 2/(1 - \exp[2\breve{g}\tau/(mc)])\} \\ &\simeq c\{1 - 2/(1 - (1 + 2\breve{g}\tau/(mc))\} \\ &\simeq c\{1 + (mc/\breve{g})(1/\tau)\} \\ &\simeq (\breve{g}^*mc^2/|\breve{g}|^2)(1/\tau) \end{aligned} \qquad (3.23a)$$

Continuing the preceding illustrative example with $\acute{\upsilon}_0 = 1.02c - 0.368ci$, and m = 10,000 metric tons, and choosing $\Delta t = 30$ days we find

352 Equals 10^{10} gm. About one-fifth the mass of the ship Queen Elizabeth.
353 An expelled mass decrease by a factor of ½ or 1/3.

$$\breve{g} = -9.95\times10^{13} + i2.86\times10^{14} \text{ gm-cm/sec}^2 \tag{3.23b}$$

Given the signs of g_1 and g_2 we see that

- For small negative τ both the real and imaginary parts of v approach $+\infty$ as $\tau \rightarrow 0$ from below.
- **For small positive τ both the real and imaginary parts of v approach $-\infty$ as $\tau \rightarrow 0$ from above.**

as displayed in Figs. 3.5 and 3.6. A starship can decide to switch off engines and "coast" at high speed towards the destination at some time close to the singularity point.

At t = 30 days – 1 sec ($\tau = -1$ sec) earth time we find using eq. 3.23a that the starship velocity in the earth's reference frame is

$$v = 324{,}690c + i934{,}885c \tag{3.23b}$$

At speeds of this order any of the 100 or so known stars within 21 light years are easily accssible. Any star in the galaxy can be easily reached. *Thus Milky Way travel times become comparable to 16th century oceanic travel times via ships to various parts of the world!*

The amount of fuel consumed in achieving these extremely high speeds is an important question. For very large speeds, integrating dm/dt (using $t' \sim vt$ and eq. 3.11b) gives

$$\text{Fuel Consumption} \sim (mc/\breve{g})\, dm/dt' \ln(v/c)$$

where dm/dt' is assumed constant. The logarithmic dependence on velocity is an encouraging factor. For the above $\breve{g}$ and velocity v, the fuel consumption is roughly the same order of magnitude as the mass of the 10,000 metric ton starship. Choosing a higher exhaust velocity would reduce the fuel consumption significantly.

3.6 The Acceleration Experienced on the Starship

The rapid acceleration, particularly in the neighborhood of $\tau = 0$ raises the question of the inertial forces that would be experienced by passengers on the starship.

The calculation of the maximum acceleration begins with the inverse of the relativistic transformation from earth coordinates to starship coordinates (eq. 3.3):

$$F'^0 = \gamma(F^0 - \beta F^x/c) \tag{3.24}$$

$$F'^{x} = \gamma(F^{x} - \beta c F^{0})$$
$$F'^{y} = F'^{z} = 0$$

which implies the apparent acceleration of the starship calculated in the starship's reference frame is

$$a' = F'^{x}/m = \breve{g}/m \tag{3.25}$$

whereas in the earth's reference frame, the acceleration a is given by the derivative of eq. 3.9b.

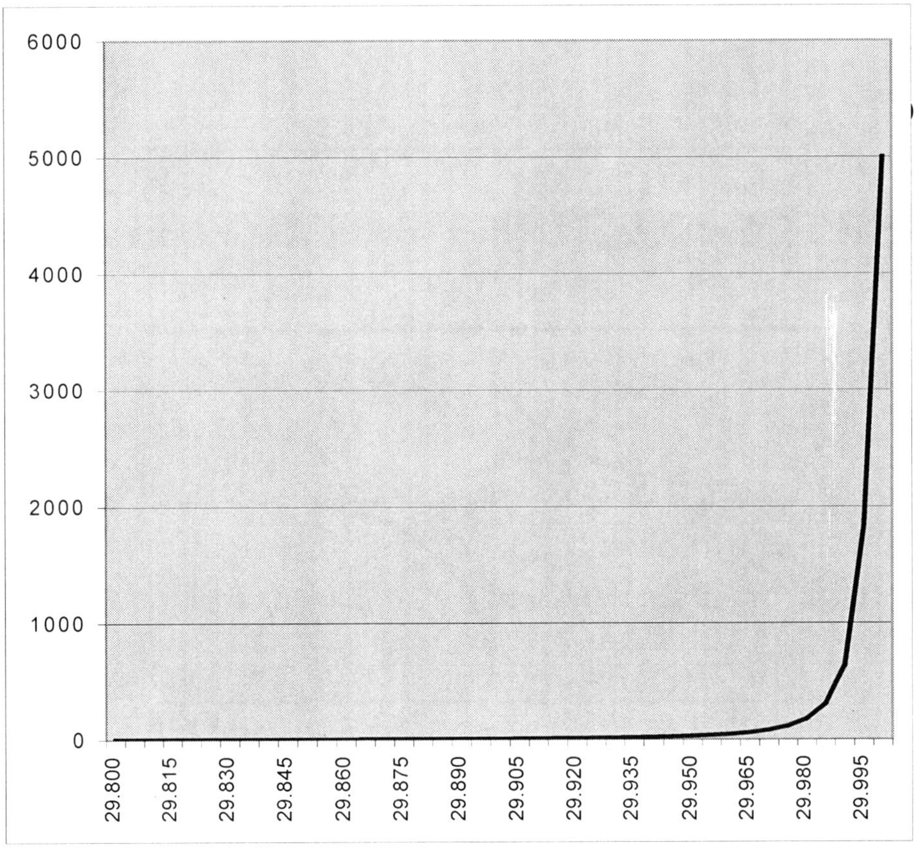

Figure 3.4. A plot of the <u>real</u> part of the velocity of a starship on its 29th and 30th earth day of travel up to 5,000c. The dynamics of this case are described in the text where the real speed reaches 8547c and beyond. Time is measured in earth days. <u>Note: as the speed of the starship increases rapidly near the singularity point, time on the starship also passes more quickly so that the starship occupants do</u>

not experience very high acceleration. Starship time t' ≈ βt when β ≫ 1 where t is earth time.

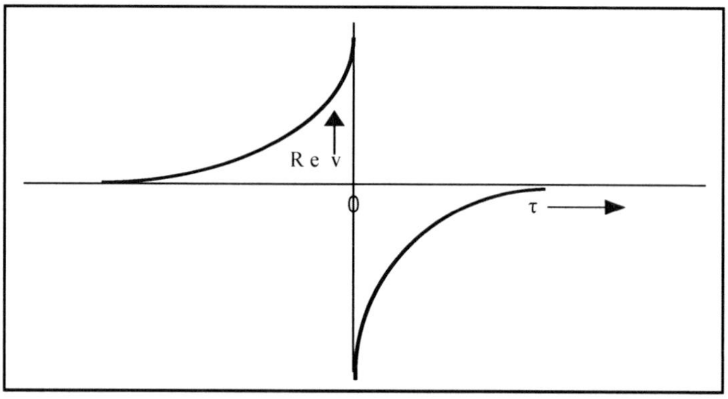

Figure 3.5. Qualitative plot of Re v from eq. 3.23 near the singularity at τ =0.

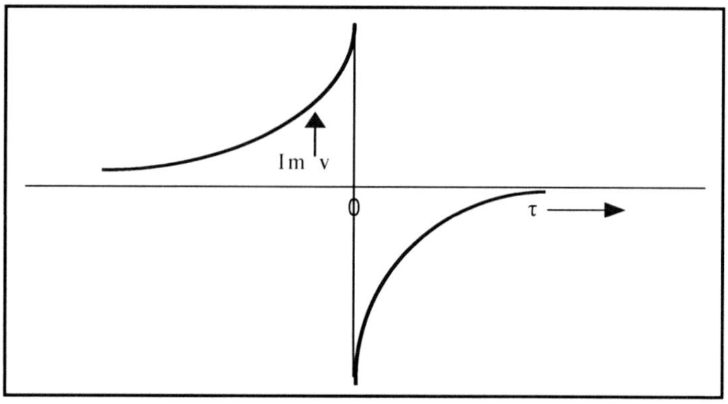

Figure 3.6. Qualitative plot of Im v from eq. 3.23 near the singularity at τ =0.

At $t = t_1 + \tau$ we see

$$a \simeq -mc^2/(\breve{g}\tau^2)[1 + 2\breve{g}\tau/mc] \approx -mc^2/(\breve{g}\tau^2) \qquad (3.26a)$$

in the earth's reference frame.

The acceleration experienced by the starship occupants, the relevant acceleration for the occupants, is

$$a' = \breve{g}/m = -9.945\times10^3 + i28.635\times10^3 \text{ gm-cm/sec}^2$$
$$= (-10.15 + i29.22)g_E \quad (3.27)$$

by eq. 3.25 and 3.23b. This acceleration (experienced by the starship occupants) would be too high if they were not in suspended animation. In suspended animation they might be able to survive in somewhat high accelerations of the magnitude of eq. 3.27.

If not, then the one month acceleration time of this example must be made somewhat longer. A longer acceleration time of, for example, 5 months[354] *would still be acceptable for human travel in months or years to stars in our celestial neighborhood.* It is also possible that a form of robotic ship could use high accelerations, such as the above, to explore the far reaches of the galaxy.

Having reached an enormous *real* speed such as a speed between 5000c and 30,000c we can turn off the superluminal engines. If we didn't, then the real speed would rapidly decline after passing "through" the singularity generated when the thrust is complex.[355]

The starship then moves at this constant speed in the absence of forces (and neglecting gravity and other minor perturbative forces). At a real speed of 5000c any place in the galaxy is a short travel time away. And nearby galaxies are reachable as well. Figure 3.7 shows the time required to reach various interesting destinations at 30,000c.

Destination	Distance (ly)	Approximate Travel Time (years)
To the other end of the Milky Way Galaxy	**100,000**	**3**
To the Center of the Milky Way	**30,000**	**1**
Large Magellenic Galaxy	**150,000**	**5**
Small Magellenic Galaxy	**200,000**	**7**
Andromeda Galaxy	**2,000,000**	**70**

Figure 3.7. "Coasting" part of travel time to various destinations at a real velocity of 30,000c.

Since much, much higher "coasting" velocities are also possible, almost the entire visible universe becomes accessible to Mankind. Mankind then has an incredible future if it has the will to seize it.

Starship travel will be in complex space. That could be an advantage. In real space, with which we are familiar, there are asteroids, planets, stars, nebulae, black holes, and dust. If we travel in real space in a straight line we will have a

---

[354] Increasing Δt by a factor of 5 would reduce a' by a factor of five.

[355] Due to the size of the starship, and other factors, the starship's velocity will not ever be precisely that of the singularity. Thus it is not an issue.

significant probability of colliding with some of these objects. We certainly would collide with interstellar and intergalactic dust. The dust alone would severely damage a starship traveling at high speed. Over long distances it would also reduce the speed of the starship significantly.

In complex space it is not known if there are any objects similar to those found in our real space. If it is empty then the starship will avoid all the pitfalls of travel in real space. Thus the starship speed will not be impeded and it will not be damaged or destroyed by collisions with matter. We can hope that this is the case.[356]

3.11 Quarks with Orthogonal Real and Imaginary 3-momenta

In chapter 7 we will consider a quark drive engine based on an exploding guided fireball that is confined by magnetic fields to "explode" in the "rear" direction to generate starship thrust.

In Blaha (2010a) quarks have complex 3-momenta. The real part of the 3-momenta of each quark was orthogonal to the imaginary part of its 3-momenta. This differs from the considerations presented earlier in this chapter. However if the average thrust of the engine is a rearward pointing, inverted v with a 90° angle between the parts of the "v" (Fig. 3.8 shows two examples) then a component of the average real and of the imaginary 3-momenta will point directly to the "rear" and yield a complex thrust. In chapter 5 we will discuss manipulating the real and imaginary parts of quark 3-momenta using magnetic and electric fields to change their magnitudes, directions and relative proportions. Thus the considerations of this chapter are relevant. Complex quark thrust can point to the "rear" and generate superluminal motion.

For example, it is possible that the quark exhaust might be manipulated to cancel thrust components perpendicular to the desired thrust direction by having the transverse thrust cancel in pairs on average. Fig. 3.8 illustrates this cancellation possibility.

Thus the total thrust, added in pairs as displayed in Fig. 3.8, yields a complex thrust to the "rear" for the starship as discussed earlier in this chapter.

---

[356] This hope may not be realized. As pointed out in Blaha (2004) p. 78-9 matter might exist in complex space. The "Great Attractor", a seemingly empty region of space has recently been shown to be drawing an enormous number of galaxies towards it. The Great Attractor may be an extraordinary large mass located in complex space – not the real space of our experience. As pointed out in Blaha (2008), and his earlier books, it appears that the universe began as a complex space that, due to symmetry breaking, broke into disjoint real and complex parts with no interaction between them except gravity, and except in localized regions such as inside hadrons in which quarks and gluons enjoy a small region of complex space.

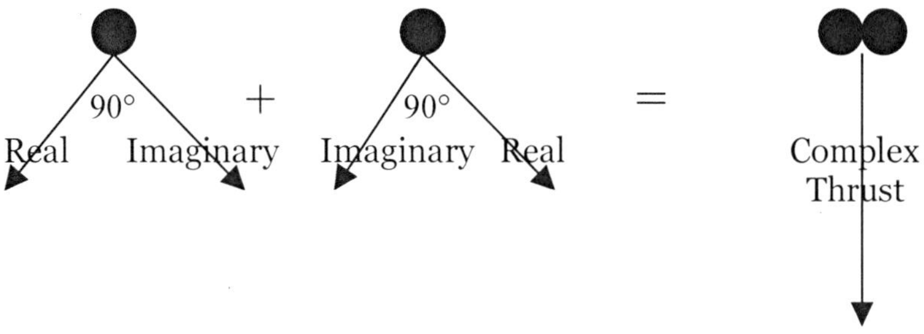

Figure 3.8. The sum of two quarks momenta to give a complex total 3-momenta (thrust) directly to the rear.

3.12 Solution for Starship with Decreasing Mass

The solution of the starship dynamical equations, eqs. 3.9-11, were based on the assumption that the starship mass was constant during the flight. In this section we assume the starship mass decreases due to its thrust. Specifically, defining $f = -g/v_e$ we find the mass m satisfies

$$m = m_0 - ft \tag{3.28}$$

by eq. 3.19a where m_0 is the initial starship mass. Substituting in eq. 3.8 we find the speed v is

$$v = c\{1 - 2/(1 + ((c + v_0)/(c - v_0))((m_0 - ft)/m_0)^{2v_e/c})\} \tag{3.29}$$

The singularity condition corresponding to eq. 3.12 is clearly

$$1 + ((c + v_0)/(c - v_0))((m_0 - ft_0)/m_0)^{2v_e/c} = 0 \tag{3.30}$$

for some time t_0. Thus the acceleration g must satisfy

$$g = (v_e m_0/t_0)\{\exp[((c - v_0)/(c + v_0))^{-c/(2v_e)}] - 1\} \tag{3.31}$$

If the argument of the exponential is small, then expanding in a power series followed by expanding the argument as a power series (after placing it in exponential form) yields the approximate singularity condition:

$$g = (m_0 c/(2t_0))\{\ln[(c - v_0)/(c + v_0)] + in\pi\} \tag{3.32}$$

where the $in\pi$ term is added (for n odd) since the argument of the logarithm may be negative. This condition is the same as eqs. 3.15 and 3.16 for constant mass.

7.2 Colliding Spherule Starship Thrust

Our starship is based on the R&D efforts expended to develop fusion power reactors based on Inertial Confinement Fusion and the extensive experience gained in building ultra-high energy colliding beam accelerators.[357]

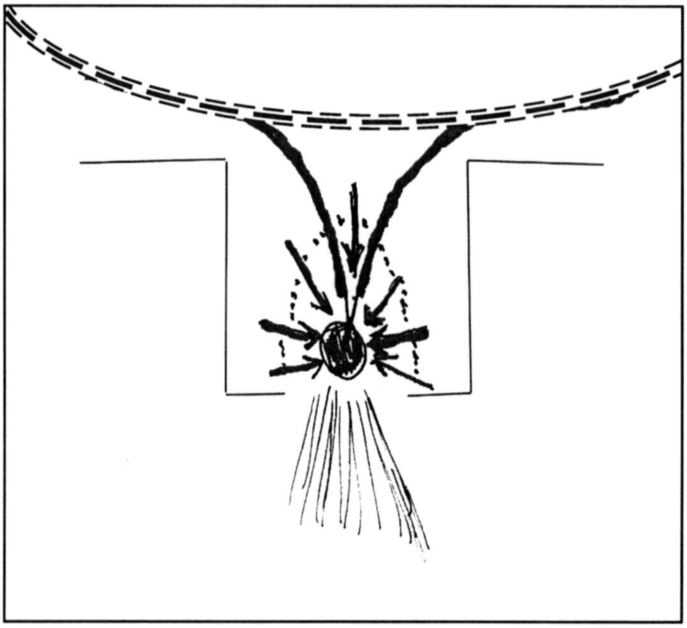

Figure 7.2. Diagram of the collision region of the intersecting spherules. The spherules (thick lines) that are extracted from the rings are thick until they are near the point of intersection where each spherule is compressed by high power lasers or particle beams (not shown) to nuclear density. They then collide generating a fireball that is constrained from expanding by multiple beams except towards the rear from which the fireball of quark-gluon plasma streams to generate the starship thrust. The straight lines enclosing the "combustion chamber" indicate the enclosing sides of the starship hull. The arrows represent some of the laser or particle beams used to enclose the fireball allowing expansion only towards the rear. The "dots" between the arrows indicate that there is likely to be an array

[357] Since the mathematical analysis of the starship propulsion mechanism requires complex computer codes that have yet to be developed we shall only be able to describe starship propulsion in qualitative terms. A mathematical presentation would require a significant programming effort, and extensive experimental input, that is beyond the current state of the art.

of many beams confining the expanding fireball to expansion to the rear of the starship as shown.

The steps of the engine cycle are:

1. Highly charged spherules are accelerated in "standard" colliding rings with new, more powerful magnets and rf accelerator modules. Two streams of spherules are diverted by magnets into the collision module as illustrated by Fig. 7.2. They are bent from their circular trajectory *to almost the thrust direction but retain their momentum component towards collision.*

2. The about-to-collide spherules are compressed by sets of laser beams similar to those used for inertial confinement fusion to nuclear density in the shape of thin ellipsoids. The long axis of each spheroid is oriented in the "thrust" direction to give a maximal collision surface for the colliding spherules. Each of the colliding spherules are compacted to ultra-high density as if being prepared for fusion (at perhaps twice nuclear density) using an array of powerful lasers or particle beams focussed on each spherule. After compaction the 2 spherules collide in a few fm/c to create a macroscopic fireball. The lasers/particle beams may vaporize part or all of a spherule but the ultra high density of the resulting stream makes that issue irrelevant.

3. The colliding fireballs in the multi-femtometer-size starship "combustion" chamber are confined to the chamber by laser or particle beams on the sides and in front of such strength as to only allow the fireballs to exit to the rear providing a quark-gluon thrust. The strength of the confinement beams must be orders of magnitude stronger than beams used in Inertial Confinement Fusion (ICF) since the density and explosive force is so much more than in ICF devices. The complex velocities of the quarks within the quark-gluon thrust enable complex starship mass and the speed of light to be exceeded.[358]

4. The thrust should have a high complex velocity of at least the order of $c(1 + i)$. The thrust will accelerate the starship to a speed well in excess of the speed of light. Since the exhaust speed is so high, small amounts of matter in the thrust will cause the starship speed to increase to high velocities.

5. The energy of the colliding spherule rings should be as high as possible to produce maximal thrust velocity.

---

358 The fireball expanding to the rear due to the confining effect of beams assumes an ellipsiodal shape as it exits from the starship. Therefore the real and imaginary parts of the velocities of the expanding fireball conform to Fig. 3.8. The real and imaginary parts of the quark velocities are directed to the "rear." Consequently, the rearward component of the quark velocities will be complex-valued.

The success of the starship engine requires carefully synchronized events at the fm/c level. The effort expended in developing this starship engine will also result in advances in ICF reactors for fusion power and in elementary particle accelerators for deeper studies of elementary particle physics and cosmology. Thus the R&D payoff will be very substantial.

7.3 Fireball Complex Spatial Momentum "Bubble" in Real Space-Time

The fireball created by the high energy collision of heavy ions or heavy atom spherules is substantially different from the collision region in nucleon-nucleon collisions. Fig. 7.3 shows a visualization of the fireball. Inside the fireball quarks and gluons form a perfect fluid and have complex spatial momenta. Outside the fireball the normal real-valued spatial dimensions prevail.

A fireball is created by the collision of heavy ions,[359] and rapidly (in a fm/c or so) becomes a perfect fluid described by thermodynamics. The fireball explosively expands as described in section 6.1. After 6 – 7 fm/c or so, its energy density becomes low enough to enter the "freeze out" stage where the quarks and gluons in the residue of the fireball combine (including quark-antiquark pairs created from the vacuum) to produce "normal particles" such as pions, protons , neutrons, and so on.

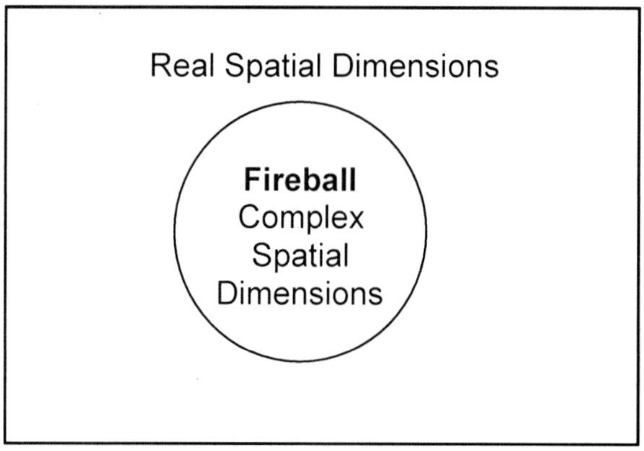

Figure 7.3. A depiction of a quark-gluon fireball in which the quarks and gluons have complex spatial momenta. The fireball is a complex "bubble" within the real space of our experience.

In the case of colliding spherules a much larger fireball will be produced if the colliding spherules are first compacted to nuclear density. Each compacted spherule can then be viewed as a "super-nucleus."

[359] Other remnants of the colliding ions are also produced.

The surface of a fireball has a certain surface tension – otherwise it would not be approximately spherical in shape. It does have some ellipticity at the 15% - 20% level. However the speed of fireball expansion shows the surface tension is much less than the pressure of expansion and thus may be ignored to leading order.

The existence of surface tension, and the freeze out of ion-ion collision fireballs after 6 – 7 fm/c, does raise the issue of whether the starship thrust would experience a drag due to these effects.

7.4 Freeze Out Stage – Impact on Starship Thrust

The laser or particle beams defining the "combustion chamber" sides confine the expanding fireball to expand only to the rear of the starship and thus provide thrust. The expanding fireball's perfect fluid elongates as shown in Fig. 7.4 and the fluid explosively expands out the rear of the chamber. In a short time period of perhaps a few fm/c freeze out occurs and the fireball "dissipates" with the quarks and gluons combining (together with quark-antiquark pairs excited from the vacuum) to transform into "normal" elementary particles.

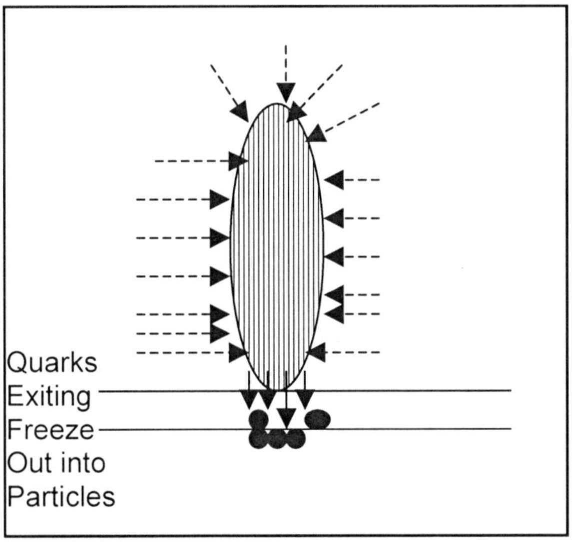

Figure 7.4. Diagram of the expanding, beam guided fireball exiting the "combustion" chamber. Dotted line arrows represent the confining laser or particle beams. The emerging quark-gluon fluid (indicated by the first horizontal line) dissipates (the freeze out) and forms "normal" elementary particles (indicated by the second horizontal line).

One may ask if the freeze out, which represents the breakup of the fireball surface, causes a "drag" to occur that reduces the thrust.[360] It appears that the quark thrust, which has a complex velocity, is the true thrust of the engine. The later recombination of the quarks and gluons outside the engine produces normal particles with real velocities. The imaginary part of the quark velocities is subsumed within the produced normal particles. This process of thrust transforming to another form occurs in chemical rockets as well. The molecules in the chemical rocket thrust, which often contains some free radicals, often recombine in the tail of the rocket thrust in a manner analogous to freeze out. The chemical recombination in the thrust tail does not affect rocket performance.

[360] Similar drag effects occur in fluid dynamics near surfaces.

REFERENCES

Bjorken, J. D., Drell, S. D., 1964, *Relativistic Quantum Mechanics* (McGraw-Hill, New York, 1965).

Bjorken, J. D., Drell, S. D., 1965, *Relativistic Quantum Fields* (McGraw-Hill, New York, 1965).

Blaha, S., 2004, *Quantum Big Bang Cosmology: Complex Space-time General Relativity, Quantum Coordinates™ Dodecahedral Universe, Inflation, and New Spin 0, ½, 1 & 2 Tachyons & Imagyons* (Pingree-Hill Publishing, Auburn, NH, 2004).

______ *2005a, Quantum Theory of the Third Kind: A New Type of Divergence-free Quantum Field Theory Supporting a Unified Standard Model of Elementary Particles and Quantum Gravity based on a New Method in the Calculus of Variations* (Pingree-Hill Publishing, Auburn, NH, 2005).

______, 2005b, *The Metatheory of Physics Theories, and the Theory of Everything as a Quantum Computer Language* (Pingree-Hill Publishing, Auburn, NH, 2005).

Blaha, S., 2005c, *The Equivalence of Elementary Particle Theories and Computer Languages: Quantum Computers, Turing Machines, Standard Model, Superstring Theory, and a Proof that Gödel's Theorem Implies Nature Must Be Quantum* (Pingree-Hill Publishing, Auburn, NH, 2005).

______, 2006, *A Derivation of ElectroWeak Theory based on an Extension of Special Relativity; Black Hole Tachyons; & Tachyons of Any Spin*. (Pingree-Hill Publishing, Auburn, NH, 2006).

______, 2007b, *The Origin of the Standard Model: The Genesis of Four Quark and Lepton Species, Parity Violation, the ElectroWeak Sector, Color SU(3), Three Visible Generations of Fermions, and One Generation of Dark Matter with Dark Energy* (Pingree-Hill Publishing, Auburn, NH, 2007).

______, 2008, *A Complete Derivation of the Form of the Standard Model With a New Method to Generate Particle Masses Second Edition* (Pingree-Hill Publishing, Auburn, NH, 2008)

______, 2009, *The Algebra of Thought & Reality: The Mathematical Basis for Plato's Theory of Ideas, and Reality Extended to Include A Priori Observers and Space-Time Second Edition* (Pingree-Hill Publishing, Auburn, NH, 2009).

______, 2010a, *Operator Metaphysics: A New Metaphysics Based on a New Operator Logic and a New Quantum Operator Logic that Lead to a Mathematical Basis for Plato's Theory of Ideas and Reality* (Pingree-Hill Publishing, Auburn, NH, 2010).

______, 2010b, *The Standard Model's Form Derived from Operator Logic, Superluminal Transformations and GL(16)* (Pingree-Hill Publishing, Auburn, NH, 2010).

______, 2011a, *21st Century Natural Philosophy Of Ultimate Physical Reality* (McMann-Fisher Publishing, Auburn, NH, 2011).

______, 2011b, *All the Universe! Faster Than Light Tachyon Quark Starships & Particle Accelerators with the LHC as a Prototype Starship Drive Scientific Edition* (Pingree-Hill Publishing, Auburn, NH, 2011).

Braithwaite, R. B., 1960, *Scientific Explanation* (Harper Torchbook, New York, 1960).

Carnap, R., 1956, *Meaning and Necessity* (Univ. Chicago Press, Chicago, 1956).

Carnap, R., (Ed. M. Gardner), 1995, A*n Introduction to the Philosophy of Science* (Dover Publications, New York, 1995).

Curry, H. B., 1976, *Foundations of Mathematical Logic* (Dover Publications, New York, 1976).

Davis, M., 1982, *Computability and Unsolvability* (Dover Publications, New York, 1982).

Devados, S., Ghosh, A., and Keutzer, K., 1994, *Logic Synthesis* (McGraw-Hill, New York, 1994).

Dirac, P. A. M., 1931, *Quantum Mechanics* Third Edition (Oxford University Press, Oxford, 1947).

Fant, Karl M., 2005, *Logically Determined Design: Clockless System Design With NULL Convention Logic* (John Wiley and Sons, Hoboken, NJ, 2005).

Frege, G., (Ed. M. Beaney), 1997, *The Frege Reader* (Blackwell Publishing, Malden, MA, 1997).

Garson, J. W., 2006, *Modal Logic for Philosophers* (Cambridge University Press, Cambridge, 2006).

Gödel, K., 1992, Tr. Meltzer, B., Introduction by R. B. Braithwaite, *On Formally Undecidable Propositions of Principia Mathematica and Related Systems* (Dover Publications, New York, 1992).

Gottfried, K., 1989, *Quantum Mechanics I Fundamentals* (Addison-Wesley, Reading, MA, 1989).

Hilbert, D. and Ackermann, W. (Tr. L. M. Hammond et al), 1950, *Principles of Mathematical Logic* (Chelsea Publishing Co., New York, 1950).

Kleene, S. C., 1967, *Mathematical Logic* (Dover Publications, New York, 1967).

Konyndyk, K., 1986, *Introductory Modal Logic* (University of Notre Dame Press, Notre Dame, Indiana, 1986).

Lavine, S., 1994, *Understanding the Infinite* (Harvard University Press, Cambridge, MA, 1994).

Loux, M. J., 2006, *Metaphysics: A Contemporary Introduction* (Routledge, New York, 2006).

Lowe, E. J., 2002, *A Survey of Metaphysics* (Oxford University Press, Oxford, 2002).

Mackey, G. W., 1963, Mathematical Foundations of Quantum Mechanics (W. A. Benjamin, New York, 1963).

Messiah, A., 1965, *Quantum Mechanics* Volume I (John Wiley & Sons, New York, 1965).

Potter, M., 2004, *Set Theory and its Philosophy* (Oxford University Press, Oxford, 2004).

Quine, W. van O., 1962, *Mathematical Logic* (Harper Torchbooks, New York, 1962).

Rescher, N., (1967), *The Philosophy of Leibniz* (Prentice-Hall, Englewood Cliffs, NJ, 1967).

Révész, G. E., 1983, *Introduction to Formal Languages* (Dover Publications, New York, 1983).

Smullyan, R. M., 1995, *First-Order Logic* (Dover Publications, New York, 1995).

Tarski, A., 1995, *Introduction to Logic and to the Methodology of Deductive Sciences* (Dover Publications, New York, 1995).

Tiles, M., 1989, *The Philosophy of Set Theory* (Dover Publications, New York, 1989).

Van Inwagen, P., 2009, *Metaphysics* (Westview Press, Boulder, CO, 2009).

Weinberg, S., 1995, *The Quantum Theory of Fields Volume I* (Cambridge University Press, New York, 1995).

Weyl, H., 1950, *Space, Time, Matter* (Dover, New York, 1950).

Weyl, H., (Tr. S. Pollard et al), 1987, *The Continuum* (Dover Publications, New York, 1987).

INDEX – Parts 1, 2 and 3

Part 4. “Normal” and Tachyon Quantum Field Theory: The “Origin” of the Fermion Species – Chapters 3 - 5 of

The Origin of the Standard Model

Chapters 3 – 5 of

The Origin of the Standard Model:

The Genesis of Four Quark and Lepton Species, Parity Violation, the ElectroWeak Sector, Color SU(3), Three Visible Generations of Fermions, and One Generation of Dark Matter with Dark Energy

Published in 2007

ISBN: 0-9746958-8-2

3. Free Spin ½ Particles – Leptons & Quarks

3.0 Chapter Overview

In this chapter we begin by developing dynamical equations for spin ½ particles based on the Extended Lorentz groups, and L_C, described in chapter 2. These spin ½ particles are conventional Dirac particles (Majorana particles are also allowed), spin ½ tachyons, and "color" versions of both types totalling four species. We will identify leptons and quarks with these fields.

3.1 Introduction

Tachyons are particles that move faster than the speed of light. As we saw in chapter 0 tachyons exist inside Black Holes, and within current theories – particularly SuperString theories.

Attempts to create canonical tachyon quantum field theories began in the 1960's. These attempts were made within the framework of the Lorentz group and, consequently, were limited to spin 0 theories since there are no finite dimensional representations of the Lorentz group for negative m^2 except for the one-dimensional representation. None of these attempts, or attempts since then, succeeded in creating a canonically quantized spin 0 tachyon quantum field theory.[1]

In this chapter we will formulate a free spin ½ tachyon Quantum Field theory. We choose to develop a spin ½ tachyon theory first because spin ½ particles (quarks and leptons) play an extraordinary role in the Standard Model. In chapter 4 we will consider bosonic tachyons.

We will develop our spin ½ tachyon theory from the "ground up" by applying a Left-Handed Extended Lorentz boost to the Dirac equation, and the Dirac spinor wave function, for a particle at rest. This procedure will give a tachyon spinor wave function, and the momentum space tachyon equation equivalent of the Dirac equation. Then we will obtain the coordinate space tachyon Dirac equation, define a lagrangian, and proceed to create a canonical quantum field theory for spin ½ tachyons.

---

[1] Except Blaha (2006), the first edition of this book.

3.2 A Method for Deriving the Dirac Equation

In this section we will review a method of obtaining the Dirac equation by Lorentz boosts of the spinor wave function of a particle at rest. In the case of a Lorentz transformation the 4 x 4 matrix form of a Lorentz transformation of the Dirac matrices is

$$S^{-1}(\Lambda(v))\gamma^{\nu}S(\Lambda(v)) = \Lambda^{\nu}{}_{\mu}(v)\gamma^{\mu} \qquad (3.1)$$

where $S(\Lambda(v))$ is

$$S(\Lambda(v)) = \exp(-i\omega\sigma_{0i}v_i/(2|\mathbf{v}|)) = \exp(-\omega\gamma^0\boldsymbol{\gamma}\cdot\mathbf{v}/(2|\mathbf{v}|))$$

$$= \cosh(\omega/2)I + \sinh(\omega/2)\gamma^0\boldsymbol{\gamma}\cdot\mathbf{p}/|\mathbf{p}| \qquad (3.2)$$

with $\omega = \text{arctanh}(|\mathbf{v}|)$, $\cosh(\omega/2) = [(E+m)/(2m)]^{1/2}$ and $\sinh(\omega/2) = |\mathbf{p}|[2m(E+m)]^{-1/2}$. Also

$$S^{-1}(\Lambda(v)) = \gamma^0 S^{\dagger}(\Lambda(v))\gamma^0 = \exp(\omega\gamma^0\boldsymbol{\gamma}\cdot\mathbf{v}/(2|\mathbf{v}|))$$

$$= \cosh(\omega/2)I - \sinh(\omega/2)\gamma^0\boldsymbol{\gamma}\cdot\mathbf{p}/|\mathbf{p}| \qquad (3.3)$$

A generic positive energy plane wave solution of the Dirac equation for a particle at rest with rest energy m is

$$\psi(x) = e^{-imt}w(0) \qquad (3.4)$$

with w(0) a four component spinor column vector. It satisfies the momentum space Dirac equation for a particle at rest:

$$(m\gamma^0 - m)e^{-imt}w(0) = 0 \qquad (3.5)$$

If we now apply $S(\Lambda(v))$ we find

$$0 = S(\Lambda(v))(m\gamma^0 - m)e^{-imt}w(0) = [mS(\Lambda(v))\gamma^0 S^{-1}(\Lambda(v)) - m]S(\Lambda(v))w(0)$$

A straightforward evaluation shows

$$mS(\Lambda(v))\gamma^0 S^{-1}(\Lambda(v)) = g_{\mu\nu}p^{\mu}\gamma^{\nu} = \not{p} \qquad (3.6)$$

where $p^0 = (p^2 + m^2)^{1/2}$, $\mathbf{p} = \gamma m\mathbf{v}$, and $p = |\mathbf{p}|$. In addition

$$S(\Lambda(v))w(0) = w(p) \qquad (3.7)$$

is a positive energy Dirac spinor. Therefore the Dirac equation in momentum space has the form:

$$(\not{p} - m)e^{-ip\cdot x}w(p) = 0 \tag{3.8}$$

where the exponential factor, mt, is also boosted to $p\cdot x$. Eq. 3.8 implies the free, coordinate space Dirac equation:

$$(i\gamma^\mu\partial/\partial x^\mu - m)\psi(x) = 0 \tag{3.9}$$

3.3 Derivation of the Tachyonic Dirac Equation

The Left-handed Extended Lorentz boost has the form:

$$\Lambda_L(\omega, \mathbf{u}) = \Lambda(\omega + i\pi/2, \mathbf{u}) = \exp[i\omega_L\hat{\mathbf{u}}\cdot\mathbf{K}] \tag{3.10}$$

where $\omega_L = \omega + i\pi/2$ and

$$\cosh(\omega_L) = i\sinh(\omega) = -\gamma = i\gamma_s \qquad (2.11)$$
$$\sinh(\omega_L) = i\cosh(\omega) = -\beta\gamma = i\beta\gamma_s$$

with, $\beta = v > 1$, $\gamma_s = (\beta^2 - 1)^{-1/2}$, and $\boldsymbol{\omega} \geq \mathbf{0}$. Thus

$$\sinh(\omega) = \gamma_s \qquad (2.12)$$
$$\cosh(\omega) = \beta\gamma_s$$

The corresponding spinor transformation is:

$$S_L(\Lambda_L(\omega, \mathbf{u})) = \exp(-i\omega_L\sigma_{0i}v_i/(2|\mathbf{v}|)) = \exp(-\omega_L\gamma^0\boldsymbol{\gamma}\cdot\mathbf{v}/(2|\mathbf{v}|))$$

$$= \cosh(\omega_L/2)I + \sinh(\omega_L/2)\gamma^0\boldsymbol{\gamma}\cdot\mathbf{p}/|\mathbf{p}| \tag{3.11}$$

The inverse transformation is

$$S_L^{-1}(\Lambda_L(\omega, \mathbf{u})) = \gamma^2\gamma^0 K^{-1}S_L^{\dagger}K\gamma^0\gamma^2 = \gamma^2\gamma^0 S_L^{T}\gamma^0\gamma^2 = \exp(\omega_L\gamma^0\boldsymbol{\gamma}\cdot\mathbf{v}/(2|\mathbf{v}|))$$

$$= \cosh(\omega_L/2)I - \sinh(\omega_L/2)\gamma^0\boldsymbol{\gamma}\cdot\mathbf{p}/|\mathbf{p}| \tag{3.12}$$

where the superscript T denotes the transpose and K is the complex conjugation operator (that also appears in the time-reversal operator). Note that S_L is not unitary just as the equivalent spinor Lorentz transformation $S(\Lambda(v))$ is not unitary.

We can now apply a left-handed superluminal transformation to the generic positive energy plane wave solution of the Dirac equation for a particle of mass m at rest. The result is

$$0 = S_L(\Lambda_L(\omega, \mathbf{u}))(m\gamma^0 - m)e^{-imt}w(0)$$

$$= [mS_L\gamma^0 S_L^{-1} - m]e^{-imt}S_L w(0)$$

where $S_L = S_L(\Lambda_L(\omega, \mathbf{u}))$. After some algebra

$$mS_L\gamma^0 S_L^{-1} = m[\cosh(\omega_L)\gamma^0 - \sinh(\omega_L)\boldsymbol{\gamma}\cdot\mathbf{p}/|\mathbf{p}|]$$

$$= i\gamma^0 E - i\boldsymbol{\gamma}\cdot\mathbf{p} = i\not{p} \qquad (3.13)$$

using eqs. 2.11 and the tachyon energy and momentum expressions

$$\mathbf{p} = m\mathbf{v}\gamma_s \qquad\qquad E = m\gamma_s \qquad (3.14)$$

Also

$$S_L w(0) = w_T(p') \qquad (3.15)$$

is a tachyon spinor. See Appendix 3-A (at the end of this chapter) for a discussion of tachyon spinors.

The momentum space tachyonic Dirac equation is

$$(i\not{p} - m)e^{ip\cdot x}w_T(p) = 0 \qquad (3.16)$$

where $p\cdot x = Et - \mathbf{p}\cdot\mathbf{x}$ after performing a corresponding left-handed superluminal coordinate transformation in the exponential factor based on eq. 2.10d. Thus the positive energy wave is transformed into a negative energy wave by the superluminal transformation.

If we apply $i\not{p}$ to we find the tachyon mass condition is satisfied

$$-E^2 + \mathbf{p}^2 = m^2 \qquad (3.17)$$

Transforming back to coordinate space we obtain the *tachyonic Dirac equation*:

$$(\gamma^\mu\partial/\partial x^\mu - m)\psi_T(x) = 0 \qquad (3.18)$$

The "missing" factor of i in the first term of eq. 3.18 requires the lagrangian to be different from the conventional Dirac lagrangian in order for the lagrangian to be real. The simplest, physically acceptable, free spin ½ tachyon lagrangian density is:

$$\mathcal{L}_T = \psi_T^S(\gamma^\mu\partial/\partial x^\mu - m)\psi_T(x) \qquad (3.19)$$

where

$$\psi_T^S = \psi_T^\dagger i\gamma^0\gamma^5 \tag{3.20}$$

The corresponding action is

$$I = \int d^4x \mathcal{L}_T \tag{3.21}$$

Appendix 3-B proves I is real. The Hamiltonian density is

$$\mathcal{H} = \pi_T\dot{\psi}_T - \mathcal{L} = i\psi_T^\dagger\gamma^5(\boldsymbol{\alpha}\cdot\nabla + \beta m)\psi_T = -i\psi_T^\dagger\gamma^5\dot{\psi}_T \tag{3.22}$$

using the tachyon Dirac equation to obtain the last equality. The reader will note that the tachyon hamiltonian is hermitean by explicit calculation up to an irrelevant total spatial divergence.

Probability Conservation Law

The tachyon Dirac equation implies a probability conservation law:

$$\partial\rho_5/\partial t = \nabla\cdot\mathbf{j}_5 \tag{3.23}$$

where

$$\rho_5 = \psi_T^\dagger\gamma^5\psi_T \qquad \mathbf{j}_5 = \psi_T^\dagger\gamma^5\boldsymbol{\alpha}\psi_T \tag{3.24}$$

We are thus led to define the conserved axial charge Q_5

$$Q_5 = \int d^3x\, \psi_T^\dagger\gamma^5\psi_T \tag{3.25}$$

Energy-Momentum Tensor

The tachyon energy-momentum tensor is

$$\mathfrak{T}_{T\mu\nu} = -g_{\mu\nu}\mathcal{L}_T + \partial\mathcal{L}_T/\partial(\partial\psi_T/\partial x_\mu)\,\partial\psi_T/\partial x^\nu \tag{3.26}$$

$$= i\psi_T^\dagger\gamma^0\gamma^5\gamma_\mu\partial\psi_T/\partial x^\nu \tag{3.27}$$

and thus the conserved energy and momentum are

$$P^0 = H = \int d^3x\, \mathfrak{T}_T^{00} = i\int d^3x\psi_T^\dagger\gamma^5(\boldsymbol{\alpha}\cdot\nabla + \beta m)\psi_T \tag{3.28}$$

and

$$P^i = \int d^3x\, \mathfrak{T}_T^{0i} = -i\int d^3x\, \psi_T^\dagger\gamma^5\partial\psi_T/\partial x_i \tag{3.29}$$

Both the energy and momentum differ significantly from the corresponding quantities for conventional Dirac fields.

3.4 Tachyon Canonical Quantization

Having defined a suitable tachyon lagrangian we can now proceed to its canonical quantization. The conjugate momentum can be calculated from the lagrangian density eq. 3.19:

$$\pi_{Ta} = \partial \mathcal{L}_T/\partial\dot{\psi}_{Ta} \equiv \partial\mathcal{L}_T/\partial(\partial\psi_{Ta}/\partial t) = -i(\psi_T^{\dagger}\gamma^5)_a \tag{3.30}$$

The resulting non-zero, canonical anti-commutation relations are

$$\{\pi_{Ta}(x), \psi_{Tb}(x')\} = i\,\delta_{ab}\,\delta^3(x - x')$$

or

$$\{\psi_{T\,a}^{\dagger}(x), \psi_{Tb}(x')\} = -[\gamma^5]_{ab}\,\delta^3(x - x') \tag{3.31}$$

At this point we might attempt to complete the canonical quantization procedure in the conventional manner by fourier expanding the field and specifying anti-commutation relations for the fourier component amplitudes. However the incompleteness of the set of plane waves, which are limited by the restriction $|\mathbf{p}| \geq m$, causes the anti-commutator of the fields *not* to yield a $\delta^3(x - x')$. Thus the conventional approach fails to yield the required anti-commutation relations.[2]

Other approaches: 1) decompose the tachyon field into left-handed and right-handed parts and then second quantize each part; and 2) second quantize in light-front coordinates ($x^{\pm} = (x^0 \pm x^3)/\sqrt{2}$). These approaches also both fail.[3]

The only approach that does succeed[4] is to decompose the tachyon field into left-handed and right-handed parts and then second quantize in light-front coordinates. We follow that procedure in the following subsections.

Separation into Left-Handed and Right-Handed Fields

We will use a transformed set of Dirac matrices to develop our left-handed and right-handed tachyon formulations:

---

[2] See G. Feinberg, Phys. Rev. **159**, 1089 (1967) for example.

[3] See the first edition Blaha (2006) where these possibilities were considered and found to fail.

[4] Blaha (2006) discusses this case in detail.

$$\gamma^0 = \begin{bmatrix} 0 & -I \\ -I & 0 \end{bmatrix} \qquad \gamma^i = \begin{bmatrix} 0 & \sigma_i \\ -\sigma_i & 0 \end{bmatrix} \qquad \gamma^5 = \begin{bmatrix} I & 0 \\ 0 & -I \end{bmatrix} \tag{3.32}$$

which are obtained from the usual Dirac matrices by applying the unitary transformation $U = 2^{-1/2}(I + \gamma^5\gamma^0)$. I is the 4 × 4 identity matrix. The γ^5 chirality operator's eigenvalues define handedness: +1 corresponds to right-handed; and −1 corresponds to left-handed:

$$\gamma^5\psi_L = -\psi_L \qquad \gamma^5\psi_R = \psi_R \tag{3.33}$$

Consequently, we can define left-handed and right-handed tachyon fields with the projection operators:

$$\begin{gathered} C^{\pm} = \tfrac{1}{2}(I \pm \gamma^5) \\ C^+ + C^- = I \\ C^{\pm\,2} = C^{\pm} \\ C^+C^- = 0 \end{gathered} \tag{3.34}$$

with the result

$$\begin{aligned} \psi_{TL} &= C^-\psi_T \\ \psi_{TR} &= C^+\psi_T \end{aligned} \tag{3.35}$$

We can calculate the commutation relations of the left-handed and right-handed tachyon fields from eq. 3.31 by pre-multiplying and post-multiplying by $\frac{1}{2}(1 - \gamma^5)$ and $\frac{1}{2}(1 + \gamma^5)$. The results are:

$$\{\psi_{TLa}^\dagger(x), \psi_{TLb}(x')\} = \tfrac{1}{2}(1 - \gamma^5)_{ab}\,\delta^3(x - x') \tag{3.36}$$

$$\{\psi_{TRa}^\dagger(x), \psi_{TRb}(x')\} = -\tfrac{1}{2}(1 + \gamma^5)_{ab}\,\delta^3(x - x') \tag{3.37}$$

$$\{\psi_{TLa}^\dagger(x), \psi_{TRb}(x')\} = \{\psi_{TRa}^\dagger(x), \psi_{TLb}(x')\} = 0 \tag{3.38}$$

The lagrangian density of eq. 3.19 decomposes into left-handed and right-handed parts:

$$\mathcal{L}_T = \psi_{TL}^\dagger\gamma^0 i\gamma^\mu\partial_\mu\psi_{TL} - \psi_{TR}^\dagger\gamma^0 i\gamma^\mu\partial_\mu\psi_{TR} - im[\psi_{TR}^\dagger\gamma^0\psi_{TL} - \psi_{TL}^\dagger\gamma^0\psi_{TR}] \tag{3.39}$$

Further Separation into + and – Light-Front Fields

There have been many studies of light-front (infinite momentum frame) physics in the past forty years.[5] Light-front coordinates cannot be obtained by a Lorentz transformation, or by a superluminal transformation, from a standard set of coordinate system variables even in a limiting sense. Instead they are a defined set of variables that have been used to develop quantum field theories that have been shown to be equivalent to quantum field theories based on conventional coordinates. In particular, light-front quantum field theories have been shown to yield fully Lorentz covariant S matrix elements that are the same as S matrix elements calculated in the conventional way.

Light-front variables can be defined by:

$$x^{\pm} = (x^0 \pm x^3)/\sqrt{2}$$
$$\partial/\partial x^{\pm} \equiv \partial^{\mp} \equiv (\partial/\partial x^0 \pm \partial/\partial x^3)/\sqrt{2} \qquad (3.40)$$

with the "transverse" coordinate variables, x^1 and x^2, unchanged.

The inner product of two 4-vectors has the form

$$x\cdot y = x^+y^- + y^+x^- - x^1y^1 - x^2y^2 \qquad (3.41)$$

and the light-front definition of Dirac matrices is:

$$\gamma^{\pm} = (\gamma^0 \pm \gamma^3)/\sqrt{2} \qquad (3.42)$$

with transverse matrices γ^1 and γ^2 defined as usual. Note the useful identity:

$$\gamma^{\pm\,2} = 0$$

We define "+" and "–" tachyon fields with the projection operators:

$$R^{\pm} = \tfrac{1}{2}(I \pm \gamma^0\gamma^3) \qquad (3.43)$$

Left-handed, ± light-front fields: $\psi_{TL}{}^{\pm} = R^{\pm}C^-\psi_T$ (3.44)

Right-handed, ± light-front fields: $\psi_{TR}{}^{\pm} = R^{\pm}C^+\psi_T$

---

[5] L. Susskind, Phys. Rev. **165**, 1535 (1968); K. Bardakci and M. B. Halpern Phys. Rev. **176**, 1686 (1968), S. Weinberg, Phys. Rev. **150**, 1313 (1966); J. Kogut and D. Soper, Phys. Rev. **D1**, 2901 (1970); J. D. Bjorken, J. Kogut, and D. Soper, Phys. Rev. **D3**, 1382 (1971); R. A. Neville and F. Rohrlich, Nuov. Cim. **A1**, 625 (1971); F. Rohrlich, Acta Phys Austr. Suppl. **8**, 277 (1971); S-J Chang, R. Root, and T-M Yan, Phys. Rev. **D7**, 1133 (1973); S-J Chang, and T-M Yan, Phys. Rev. **D7**, 1147 (1973); T-M Yan, Phys. Rev. **D7**, 1761 (1973); T-M Yan, Phys. Rev. **D7**, 1780 (1973); C. Thorn, Phys. Rev. **D19**, 639 (1979); and references therein.

Now if we transform to light-front variables and fields as above we obtain the light-front free tachyon lagrangian:

$$\mathcal{L}_T = 2^{1/2}\psi_{TL}{}^{+\dagger}i\partial^-\psi_{TL}{}^{+} + 2^{1/2}\psi_{TL}{}^{-\dagger}i\partial^+\psi_{TL}{}^{-} - \psi_{TL}{}^{+\dagger}\gamma^0 i\gamma^j\partial^j\psi_{TL}{}^{-} - \psi_{TL}{}^{-\dagger}\gamma^0 i\gamma^j\partial^j\psi_{TL}{}^{+} -$$

$$-2^{1/2}\psi_{TR}{}^{+\dagger}i\partial^-\psi_{TR}{}^{+} - 2^{1/2}\psi_{TR}{}^{-\dagger}i\partial^+\psi_{TR}{}^{-} + \psi_{TR}{}^{+\dagger}\gamma^0 i\gamma^j\partial^j\psi_{TR}{}^{-} + \psi_{TR}{}^{-\dagger}\gamma^0 i\gamma^j\partial^j\psi_{TR}{}^{+} -$$

$$- im[\psi_{TR}{}^{+\dagger}\gamma^0\psi_{TL}{}^{-} - \psi_{TL}{}^{+\dagger}\gamma^0\psi_{TR}{}^{-} + \psi_{TR}{}^{-\dagger}\gamma^0\psi_{TL}{}^{+} - \psi_{TL}{}^{-\dagger}\gamma^0\psi_{TR}{}^{+}] \qquad (3.45)$$

with implied sums over j = 1,2. In contrast to the light-front tachyon lagrangian we note the corresponding light-front Dirac fermion lagrangian is

$$\mathcal{L}_D = 2^{1/2}\psi_L{}^{+\dagger}i\partial^-\psi_L{}^{+} + 2^{1/2}\psi_L{}^{-\dagger}i\partial^+\psi_L{}^{-} - \psi_L{}^{+\dagger}\gamma^0 i\gamma^j\partial^j\psi_L{}^{-} - \psi_L{}^{-\dagger}\gamma^0 i\gamma^j\partial^j\psi_L{}^{+} -$$

$$+ 2^{1/2}\psi_R{}^{+\dagger}i\partial^-\psi_R{}^{+} + 2^{1/2}\psi_R{}^{-\dagger}i\partial^+\psi_R{}^{-} - \psi_R{}^{+\dagger}\gamma^0 i\gamma^j\partial^j\psi_R{}^{-} - \psi_R{}^{-\dagger}\gamma^0 i\gamma^j\partial^j\psi_R{}^{+} -$$

$$- im[\psi_R{}^{+\dagger}\gamma^0\psi_L{}^{-} + \psi_L{}^{+\dagger}\gamma^0\psi_R{}^{-} + \psi_R{}^{-\dagger}\gamma^0\psi_L{}^{+} + \psi_L{}^{-\dagger}\gamma^0\psi_R{}^{+}] \qquad (3.46)$$

The difference in signs between eqs. 3.45 and 3.46 will turn out to be a crucial factor in the derivation of features of the Standard Model later.

Returning to the tachyon lagrangian eq. 3.45 we obtain equations of motion through the standard variational techniques:

$$\begin{aligned} 2^{1/2}i\partial^-\psi_{TL}{}^{+} - \gamma^0 i\gamma^j\partial^j\psi_{TL}{}^{-} + im\gamma^0\psi_{TR}{}^{-} &= 0 \\ 2^{1/2}i\partial^-\psi_{TR}{}^{+} - \gamma^0 i\gamma^j\partial^j\psi_{TR}{}^{-} + im\gamma^0\psi_{TL}{}^{-} &= 0 \\ 2^{1/2}i\partial^+\psi_{TL}{}^{-} - \gamma^0 i\gamma^j\partial^j\psi_{TL}{}^{+} + im\gamma^0\psi_{TR}{}^{+} &= 0 \\ 2^{1/2}i\partial^+\psi_{TR}{}^{-} - \gamma^0 i\gamma^j\partial^j\psi_{TR}{}^{+} + im\gamma^0\psi_{TL}{}^{+} &= 0 \end{aligned} \qquad (3.47)$$

Eqs. 3.47 show that $\psi_{TL}{}^{-}$ and $\psi_{TR}{}^{-}$ are dependent fields that are functions of $\psi_{TL}{}^{+}$ and $\psi_{TR}{}^{+}$ on the light-front where x^+ equals a constant. They can be expressed in an integral form as well. (The independent fields $\psi_{TL}{}^{+}$ and $\psi_{TR}{}^{+}$ play a fundamental role in tachyon theory and are used to define "in" and "out" tachyon states in perturbation theory.)

The conjugate momenta implied by eq. 3.45 are

$$\pi_{TL}{}^{+} = \partial\mathcal{L}/\partial(\partial^-\psi_{TL}{}^{+}) = 2^{1/2}i\psi_{TL}{}^{+\dagger} \qquad (3.48)$$

$$\pi_{TL}{}^{-} = \partial\mathcal{L}/\partial(\partial^-\psi_{TL}{}^{-}) = 0$$

$$\pi_{TR}{}^{+} = \partial\mathcal{L}/\partial(\partial^-\psi_{TR}{}^{+}) = -2^{1/2}i\psi_{TR}{}^{+\dagger} \qquad (3.49)$$

$$\pi_{TR}^{\;-} = \partial\mathcal{L}/\partial(\partial^- \psi_{TR}^{\;-}) = 0$$

Quantization on surfaces of constant x^+ (light-front surfaces) has been shown to support satisfactory formulations of Quantum Electrodynamics and other quantum field theories. Thus x^+ plays the role of the "time" variable in light-front quantized theories. So we will define canonical equal x^+ anti-commutation relations for spin ½ tachyons.

The resulting canonical equal-light-front ($x^+ = y^+$) anti-commutation relations of the independent fields are:

$$\{\psi_{TL\;\;a}^{\;+\dagger}(x), \psi_{TL\;\;b}^{\;+}(y)\} = 2^{-1}[C^-R^+]_{ab}\,\delta(x^- - y^-)\delta^2(x - y) \qquad (3.50)$$

$$\{\psi_{TR\;\;a}^{\;+\dagger}(x), \psi_{TR\;\;b}^{\;+}(y)\} = -2^{-1}[C^+R^+]_{ab}\,\delta(x^- - y^-)\delta^2(x - y) \qquad (3.51)$$

$$\{\psi_{TL\;\;a}^{\;+\dagger}(x), \psi_{TR\;\;b}^{\;+}(y)\} = \{\psi_{TR\;\;a}^{\;+\dagger}(x), \psi_{TL\;\;b}^{\;+}(y)\} = 0 \qquad (3.52)$$

$$\{\psi_{TL\;\;a}^{\;+}(x), \psi_{TR\;\;b}^{\;+}(y)\} = \{\psi_{TR\;\;a}^{\;+\dagger}(x), \psi_{TL\;\;b}^{\;+\dagger}(y)\} = 0 \qquad (3.53)$$

where the factors of 2^{-1} are the result of the $2^{½}$ factor in eqs. 3.48 and 3.49, and the factor of $2^{-½}$ in the definition of x^- in eq. 3.40.

If we compare eqs. 3.50 and 3.51 with the corresponding anti-commutation relations of conventional Dirac quantum fields:

$$\{\psi_{L\;\;a}^{\;+\dagger}(x), \psi_{L\;\;b}^{\;+}(y)\} = 2^{-1}[C^-R^+]_{ab}\,\delta(x^- - y^-)\delta^2(x - y) \qquad (3.54)$$

$$\{\psi_{R\;\;a}^{\;+\dagger}(x), \psi_{R\;\;b}^{\;+}(y)\} = 2^{-1}[C^+R^+]_{ab}\,\delta(x^- - y^-)\delta^2(x - y) \qquad (3.55)$$

we see that the right-handed tachyon anti-commutation relation (eq. 3.51) has a minus sign relative to the corresponding right-handed conventional anti-commutation relation (eq. 3.55). The right-handed tachyon anti-commutation relation (eq. 3.51) with its minus sign will require compensating minus signs in its creation and annihilation Fourier component operators' anti-commutation relations.

The sign differences between the lagrangian terms in eqs. 3.47 and 3.48 ultimately lead to parity violating features in the Standard Model lagrangian and thus resolve the long-standing question: Why parity violation? Answer: Nature chooses the Left-handed Extended Lorentz group. Thus the source of parity violation, and much of the form of the Standard Model, is in superluminal physics.

Left-Handed Tachyons

The free, "+" light-front, left-handed tachyon wave function Fourier expansion is:

$$\psi_{TL}^{+}(x) = \sum_{\pm s} \int d^2p dp^+ N_{TL}^{+}(p)\theta(p^+)[b_{TL}^{+}(p, s)u_{TL}^{+}(p, s)e^{-ip\cdot x} + $$
$$+ d_{TL}^{+\dagger}(p, s)v_{TL}^{+}(p, s)e^{+ip\cdot x}] \qquad (3.56)$$

and its hermitean conjugate is

$$\psi_{TL}^{+\dagger}(x) = \sum_{\pm s} \int d^2p dp^+ N_{TL}^{+}(p)\theta(p^+) [b_{TL}^{+\dagger}(p, s)u_{TL}^{+\dagger}(p,s)e^{+ip\cdot x} +$$
$$+ d_{TL}^{+}(p, s)v_{TL}^{+\dagger}(p, s)e^{-ip\cdot x}] \quad (3.57)$$

where † indicates hermitean conjugate, where

$$N_{TL}^{+}(p) = [2m|\mathbf{p}|/((2\pi)^3(p^+(p^+ - p^-) + p_\perp^{\,2}))]^{1/2} \qquad (3.57a)$$

where the anti-commutation relations of the Fourier coefficient operators are

$$\begin{aligned}
\{b_{TL}^{+}(q,s), b_{TL}^{+\dagger}(p,s')\} &= \delta_{ss'}\delta^2(\mathbf{q} - \mathbf{p})\delta(q^+ - p^+)\\
\{d_{TL}^{+}(q,s), d_{TL}^{+\dagger}(p,s')\} &= \delta_{ss'}\delta^2(\mathbf{q} - \mathbf{p})\delta(q^+ - p^+)\\
\{b_{TL}^{+}(q,s), b_{TL}^{+}(p,s')\} &= \{d_{TL}^{+}(q,s), d_{TL}^{+}(p,s')\} = 0\\
\{b_{TL}^{+\dagger}(q,s), b_{TL}^{+\dagger}(p,s')\} &= \{d_{TL}^{+\dagger}(q,s), d_{TL}^{+\dagger}(p,s')\} = 0\\
\{b_{TL}^{+}(q,s), d_{TL}^{+\dagger}(p,s')\} &= \{d_{TL}^{+}(q,s), b_{TL}^{+\dagger}(p,s')\} = 0\\
\{b_{TL}^{+\dagger}(q,s), d_{TL}^{+\dagger}(p,s')\} &= \{d_{TL}^{+}(q,s), b_{TL}^{+}(p,s')\} = 0
\end{aligned} \qquad (3.58)$$

and where the spinors are

$$\begin{aligned}
u_{TL}^{+}(p, s) &= C^- R^+ S_L(\Lambda_L(\mathbf{p}))w^1(0)\\
u_{TL}^{+}(p, -s) &= C^- R^+ S_L(\Lambda_L(\mathbf{p}))w^2(0)\\
v_{TL}^{+}(p, s) &= C^- R^+ S_L(\Lambda_L(\mathbf{p}))w^3(0)\\
v_{TL}^{+}(p, -s) &= C^- R^+ S_L(\Lambda_L(\mathbf{p}))w^4(0)\\
u_{TL}^{+\dagger}(p, s) &= w^{1T}(0)S_L^{\dagger}(\Lambda_L(\mathbf{p}))R^+C^-\\
u_{TL}^{+\dagger}(p, -s) &= w^{2T}(0)S_L^{\dagger}(\Lambda_L(\mathbf{p}))R^+C^-\\
v_{TL}^{+\dagger}(p, s) &= w^{3T}(0)S_L^{\dagger}(\Lambda_L(\mathbf{p}))R^+C^-\\
v_{TL}^{+\dagger}(p, -s) &= w^{4T}(0)S_L^{\dagger}(\Lambda_L(\mathbf{p}))R^+C^-
\end{aligned} \qquad (3.59)$$

where the superscript "T" indicates the transpose. (These spinors are described in Appendix 3-A.)

The canonical left-handed, light-front anti-commutation relation (eq. 3.50) follows from eqs. 3.56 – 3.59:

$$\{\psi_{TL}{}^{+}{}_{a}(x), \psi_{TL}{}^{+\dagger}{}_{b}(y)\} = \sum_{\pm s,s'} \int d^2p dp^+ \int d^2p' dp'^+ \, N_{TL}{}^{+}(p) N_{TL}{}^{+}(p') \theta(p^+)\theta(p'^+) \cdot$$

$$\cdot [\{b_{TL}{}^{+\dagger}(p',s'), b_{TL}{}^{+}(p,s)\} u_{TL}{}^{+}{}_{a}(p,s) u_{TL}{}^{+\dagger}{}_{b}(p',s') e^{+ip'\cdot y - ip\cdot x} +$$

$$+ \{d_{TL}{}^{+}(p',s'), d_{TL}{}^{+\dagger}(p,s)\} v_{TL}{}^{+}{}_{a}(p,s) v_{TL}{}^{+\dagger}{}_{b}(p',s') e^{-ip'\cdot y + ip\cdot x}]$$

$$= \sum_{\pm s} \int d^2p dp^+ N_{TL}{}^{+2}(p)\theta(p^+)[u_{TL}{}^{+}{}_{a}(p,s) u_{TL}{}^{+\dagger}{}_{b}(p,s) e^{+ip\cdot(y-x)} +$$

$$+ v_{TL}{}^{+}{}_{a}(p,s) v_{TL}{}^{+\dagger}{}_{b}(p,s) e^{-ip\cdot(y-x)}]$$

$$= -i\int d^2p dp^+ \theta(p^+) N_{TL}{}^{+2}(p)(2m|\mathbf{p}|)^{-1}\{[C^- R^+(i\not{p} - m)\gamma\cdot\mathbf{p}R^+C^-]_{ab} e^{+ip\cdot(y-x)} +$$
$$+ [C^- R^+(i\not{p} + m)\gamma\cdot\mathbf{p}R^+C^-]_{ab} e^{-ip\cdot(y-x)}\}$$

$$= -i\int d^2p_\perp \int_0^\infty dp^+ N_{TL}{}^{+2}(p)\{[C^- R^+(ip^+(p^+ - p^-) + ip_\perp{}^2 - mp_\perp\cdot\gamma_\perp)C^-]_{ab} \cdot$$

$$\cdot e^{+ip^+(y^- - x^-) - ip_\perp\cdot(y_\perp - x_\perp)} -$$

$$- [C^- R^+(-ip^+(p^+ - p^-) - ip_\perp{}^2 - mp_\perp\cdot\gamma_\perp)C^-]_{ab} e^{-ip^+(y^- - x^-) + ip_\perp\cdot(y_\perp - x_\perp)}\}/(2m|\mathbf{p}|)$$

$$= \int d^2p_\perp \int_{-\infty}^{\infty} dp^+ N_{TL}{}^{+2}(p)[C^- R^+(p^+(p^+ - p^-) + p_\perp{}^2)]_{ab} \cdot$$

$$\cdot e^{+ip^+(y^- - x^-) - ip_\perp\cdot(y_\perp - x_\perp)}/(2m|\mathbf{p}|)$$

upon letting $p^+ \to -p^+$ and $\mathbf{p}_\perp \to -\mathbf{p}_\perp$ in the second term after using $N_{TL}{}^{+2}(p)(p^+(p^+ - p^-) + p_\perp{}^2) = 1$. The result

$$= \tfrac{1}{2}\int d^2p_\perp \int_{-\infty}^{\infty} dp^+ (2\pi)^{-3}[C^- R^+]_{ab} e^{+ip^+(y^- - x^-) - ip_\perp\cdot(y_\perp - x_\perp)}$$

$$= 2^{-1}[C^- R^+]_{ab}\delta(y^- - x^-)\delta^2(\mathbf{y} - \mathbf{x}) \qquad (3.60)$$

Therefore we have left-handed, light-front quantized tachyons with canonical commutation relations and localized tachyons. As a result we have a canonical tachyon Quantum Field Theory.

Right-Handed Tachyons

The case of right-handed tachyons is similar to the left-handed case with only two differences: a minus sign in the creation and annihilation operator anti-commutation relations, and the use of right-handed projection operators. The right-handed tachyon wave function light-front Fourier expansion is:

$$\psi_{TR}{}^{+}(x) = \sum_{\pm s} \int d^2p dp^+ N_{TR}{}^{+}(p)\theta(p^+)[b_{TR}{}^{+}(p, s)u_{TR}{}^{+}(p, s)e^{-ip\cdot x} + d_{TR}{}^{+\dagger}(p, s)v_{TR}{}^{+}(p, s)e^{+ip\cdot x}] \qquad (3.61)$$

and its hermitean conjugate is

$$\psi_{TR}{}^{+\dagger}(x) = \sum_{\pm s} \int d^2p dp^+ N_{TR}{}^{+}(p)\theta(p^+)\,[b_{TR}{}^{+\dagger}(p, s)u_{TR}{}^{+\dagger}(p, s)e^{+ip\cdot x} + d_{TR}{}^{+}(p, s)v_{TR}{}^{+\dagger}(p, s)e^{-ip\cdot x}] \qquad (3.62)$$

where $N_{TR}{}^{+}(p) = N_{TL}{}^{+}(p)$, where the anti-commutation relations of the Fourier coefficient operators are

$$\{b_{TR}{}^{+}(q,s), b_{TR}{}^{+\dagger}(p,s')\} = -\delta_{ss'}\delta^2(\mathbf{q} - \mathbf{p})\delta(q^+ - p^+) \qquad (3.63)$$
$$\{d_{TR}{}^{+}(q,s), d_{TR}{}^{+\dagger}(p,s')\} = -\delta_{ss'}\delta^2(\mathbf{q} - \mathbf{p})\delta(q^+ - p^+)$$
$$\{b_{TR}{}^{+}(q,s), b_{TR}{}^{+}(p,s')\} = \{d_{TR}{}^{+}(q,s), d_{TR}{}^{+}(p,s')\} = 0$$
$$\{b_{TR}{}^{+\dagger}(q,s), b_{TR}{}^{+\dagger}(p,s')\} = \{d_{TR}{}^{+\dagger}(q,s), d_{TR}{}^{+\dagger}(p,s')\} = 0$$
$$\{b_{TR}{}^{+}(q,s), d_{TR}{}^{+\dagger}(p,s')\} = \{d_{TR}{}^{+}(q,s), b_{TR}{}^{+\dagger}(p,s')\} = 0$$
$$\{b_{TR}{}^{+\dagger}(q,s), d_{TR}{}^{+\dagger}(p,s')\} = \{d_{TR}{}^{+}(q,s), b_{TR}{}^{+}(p,s')\} = 0$$

and where the spinors are

$$u_{TR}{}^{+}(p, s) = C^+R^+u_T(p,s) \qquad (3.64)$$
$$v_{TR}{}^{+}(p, s) = C^+R^+v_T(p,s) \qquad (3.65)$$

by Appendix 3-A (eq. 3-A.7).

The right-handed anti-commutation relation (eq. 3.51) with the minus sign follows in particular because of the minus signs in eqs. 3.63.

Interpretation of Tachyon Creation and Annihilation Operators

To properly discuss the physical interpretation of tachyon creation and annihilation operators we must first determine the hamiltonian and momentum operators in terms of creation and annihilation operators.

The energy-momentum tensor density is the symmetrized version of

$$\mathcal{T}^{\mu\nu} = \sum_i \partial\mathcal{L}/\partial(\partial\chi_i/\partial x_\mu)\, \partial\chi_i/\partial x_\nu - g^{\mu\nu}\mathcal{L} \qquad (3.66)$$

where the sum over i is over the fields. The light-front hamiltonian is

$$H \equiv P^- = T^{+-} = \int dx^- d^2x\, \mathcal{T}^{+-} \qquad (3.67)$$

and the "momenta" are

$$P^+ = T^{++} = \int dx^- d^2x\, \mathcal{T}^{++} \qquad (3.68)$$

$$P^i = T^{+i} = \int dx^- d^2x\, \mathcal{T}^{+i} \qquad (3.69)$$

for i = 1,2.

The light-front, left-handed and right-handed tachyon lagrangian $\mathcal{L}_T$ is eq. 3.45 and its equations of motion are eqs. 3.47. They imply

$$H = i2^{-1/2}\int dx^- d^2x\, [\psi_{TL}^{+\dagger}\partial^-\psi_{TL}^{+} - \partial^-\psi_{TL}^{+\dagger}\psi_{TL}^{+} + \psi_{TL}^{-\dagger}\partial^+\psi_{TL}^{-} - \partial^+\psi_{TL}^{-\dagger}\psi_{TL}^{-} -$$

$$- \psi_{TR}^{+\dagger}\partial^-\psi_{TR}^{+} + \partial^-\psi_{TR}^{+\dagger}\psi_{TR}^{+} - \psi_{TR}^{-\dagger}\partial^+\psi_{TR}^{-} + \partial^+\psi_{TR}^{-\dagger}\psi_{TR}^{-} + \text{mass terms}] \qquad (3.70)$$

After substituting for the various fields we find the *independent fields* (which create the in and out particle states) have the hamiltonian terms:

$$H = \sum_{\pm s}\int d^2p\,dp^+\, p^-[b_{TL}^{+\dagger}(p,s)b_{TL}^{+}(p,s) - d_{TL}^{+}(p,s)d_{TL}^{+\dagger}(p,s) -$$

$$- b_{TR}^{+\dagger}(p,s)b_{TR}^{+}(p,s) + d_{TR}^{+}(p,s)d_{TR}^{+\dagger}(p,s)] \qquad (3.71)$$

$$= \sum_{\pm s}\int d^2p\,dp^+\, p^-[b_{TL}^{+\dagger}(p,s)b_{TL}^{+}(p,s) + d_{TL}^{+\dagger}(p,s)d_{TL}^{+}(p,s) -$$

$$- b_{TR}^{+\dagger}(p,s)b_{TR}^{+}(p,s) - d_{TR}^{+\dagger}(p,s)d_{TR}^{+}(p,s)] \qquad (3.72)$$

up to the usual infinite constants due to left-handed operator rearrangement and right-handed operator rearrangement that are discarded. Eq. 3.72 is the basis for our particle interpretation of tachyon creation and annihilation operators based on Dirac's hole theory. Dirac hole theory as applied in light-front coordinates assumes all negative p^- ("energy") states are filled.

Left-Handed Tachyon Creation and Annihilation Operators

1. We identify $b_{TL}^{+\dagger}(p,s)$ and $d_{TL}^{+}(p,s)$ as creation operators for left-handed tachyons. $b_{TL}^{+\dagger}(p,s)$ creates a positive p^- ("energy") state and $d_{TL}^{+}(p,s)$ creates a negative p^- ("energy") state.

2. $b_{TL}^{+}(p,s)$ and $d_{TL}^{+\dagger}(p,s)$ are the corresponding annihilation operators for left-handed tachyons. $b_{TL}^{+}(p,s)$ annihilates a positive p^- ("energy") state and $d_{TL}^{+\dagger}(p,s)$ annihilates a negative p^- ("energy") state.

3. We assume Dirac hole theory holds for the left-handed tachyon vacuum with all negative energy states filled. There is no tachyon energy gap as there is for Dirac fermions. There is also the problem that the left-handed tachyon vacuum is not invariant under ordinary Lorentz transformations or Superluminal transformations. *However if we confine ourselves to light-front coordinates for computations no ambiguity can result and the Lorentz covariant quantities that we calculate, such as the S matrix, are well-defined.*

4. Using tachyon hole theory we identify $b_{TL}^{+}(p,s)$ and $d_{TL}^{+\dagger}(p,s)$ as annihilation operators for left-handed tachyons. $b_{TL}^{+}(p,s)$ annihilates a positive p^- ("energy") state and $d_{TL}^{+\dagger}(p,s)$ annihilates a negative p^- ("energy") state – thus creating a hole in the tachyon sea that we view as the creation of a positive p^- ("energy"), left-handed antitachyon. $d_{TL}^{+}(p,s)$ annihilates a positive p^- ("energy"), left-handed antitachyon.

Right-Handed Tachyons

The anti-commutation relations of right-handed tachyon creation and annhilation operators (eqs. 3.63) and the right-handed hamiltonian terms have the "wrong" sign compared to corresponding Dirac operators and left-handed tachyon operators. This situation is completely analogous to the situation of time-like photons in the covariant formulation of quantum Electrodynamics.[6] In the case of time-like photons it was possible to introduce an indefinite metric (Gupta-Bleuler formulation), and then to use the subsidiary condition $\partial A^{\nu}/\partial x^{\nu} = 0$ to reduce the dynamics of QED to the transverse components. Thus the time-like photons were intermediate artifacts needed to have a manifestly covariant formulation while QED observables depended solely on the transverse components of the electromagnetic field.

In the present case of free tachyons, and in leptonic ElectroWeak Theory there is no evident "subsidiary condition" to eliminate the right-handed tachyon

---

6 Bogoliubov (1959) pp. 130-136.

fields. But since the only manner in which the right-handed leptonic tachyon fields[7] interact is through mass terms, which can be easily 'integrated out", right-handed leptonic tachyon fields are removed from the observable part of the leptonic ElectroWeak Theory by their "lack of interaction" with left-handed fields.

In the case of quark ElectroWeak Theory right-handed tachyon quark fields have charge (–1/3) and thus experience an electromagnetic interaction as well as a Z interaction. However, since quarks are totally confined, right-handed tachyon quarks will not be able to continuously emit photons or Z's due to energy conservation and their confinement to bound states of fixed positive energy. Starting in section 2.6, when we consider complex Lorentz group boosts, we will suggest that quarks may not consist of Dirac particles or tachyons of the type considered up to this point in this chapter. Rather they may be variants on Dirac particles and tachyons satisfying different dynamical equations. However, the preceding comments on quarks would still apply.

Thus right-handed tachyons are analogous to time-like photons – necessary theoretically but prevented from causing a negative energy disaster by the forms of their interactions. We discuss this subject in more detail in the following chapters.

3.5 Tachyon Feynman Propagator

In this section we develop the light-front propagator for tachyons. We begin with a subsection describing the light-front propagators of Dirac fields.

Dirac Field Light-Front Propagators

The light-front Feynman propagator for the ψ^+ field of a Dirac fermion is

$$iS^+_F(x,y)\gamma^0 = \theta(x^+ - y^+)<0|\psi^+(x)\psi^{+\dagger}(y)|0> - \theta(y^+ - x^+)<0|\psi^{+\dagger}(y)\psi^+(x)|0> \quad (3.73)$$

and does not contain a non-covariant piece due to the projection operators:

$$\begin{aligned} iS^+_F(x,y) = \int d^2p dp^+\theta(p^+)[1/(2(2\pi)^3p^+)]\{&\theta(x^+ - y^+)[R^+(\not p + m)R^-]\, e^{-ip\cdot(x-y)} + \\ &+ \theta(y^+ - x^+)[R^+(-\not p + m)R^-]e^{+ip\cdot(x-y)}\} \\ = R^+ iS_F(x,y)R^- \end{aligned} \quad (3.74)$$

where $S_F(x,y)$ is the usual Feynman propagator.

The light-front Feynman propagator for a *left-handed* <u>Dirac</u> field ψ^+ is

$$iS^+_{LF}(x,y) = \int d^2p dp^+\theta(p^+)[1/(2(2\pi)^3p^+)]\{\theta(x^+ - y^+)[C^-R^+(\not p + m)R^-C^-]e^{-ip\cdot(x-y)} +$$

---

[7] The tachyon fields are provisionally assumed to be neutrino fields in the leptonic sector, and d, s and b quarks in the quark sector.

$$+ \theta(y^+ - x^+)[C^-R^+(-\not p + m)R^-C^-]e^{+ip\cdot(x-y)}\}$$

$$= C^-R^+ iS_F(x,y)R^-C^- \qquad (3.75)$$

Tachyon Field Feynman Propagator

Turning now to tachyons, the light-front Feynman propagator for the left-handed ψ_{TL}^{+} *tachyon* field is (using the Fourier expansion of the left-handed tachyon field eqs. 3.56 and 3.57):

$$iS^+_{TLF}(x,y) = \theta(x^+ - y^+)<0|\psi_{TL}{}^+(x)\psi_{TL}{}^{+\dagger}(y)\gamma^0|0> -$$
$$- \theta(y^+ - x^+)<0|\psi_{TL}{}^{+\dagger}(y)\gamma^0\psi_{TL}{}^+(x)|0>$$
$$= -i\int d^2p dp^+\theta(p^+)N_{TL}^{+2}(2m|\mathbf{p}|)^{-1}C^-R^+\{\theta(x^+ - y^+)[(i\not p - m)\gamma\cdot\mathbf{p}]e^{-ip\cdot(x-y)} +$$
$$+ \theta(y^+ - x^+)[(i\not p + m)\gamma\cdot\mathbf{p}]e^{+ip\cdot(x-y)}\}R^+C^-\gamma^0$$

If we define the on-shell momentum variable $p_0^- = (p_0{}^1p_0{}^1 + p_0{}^2p_0{}^2 - m^2)/(2p_0{}^+)$, $p_0{}^+ = p^+$, $p_0{}^j = p^j$ (for j = 1, 2), $p_{\perp 0}{}^2 = p_0{}^jp_0{}^j$ and $\not p_0 = p_0\cdot\gamma$ then the above equation can be rewritten as

$$= -iC^-R^+\int d^4p[32\pi^4(p_0{}^+(p_0{}^+ - p_0{}^-) + p_{0\perp}{}^2)]^{-1}e^{-ip\cdot(x-y)}\cdot$$

$$\cdot\{\theta(p^+)(i\not p_0 - m)\gamma\cdot\mathbf{p}_0]/[p^- - p_0{}^- + i\varepsilon] +$$

$$+ \theta(-p^+)(i\not p_0 + m)\gamma\cdot\mathbf{p}_0]/[p^- + p_0{}^- - i\varepsilon]\}R^+C^-\gamma^0$$

$$= -\tfrac{1}{2}\, i\int d^4p(2\pi)^{-4}[C^-R^+(i\not p - m)\gamma\cdot\mathbf{p}R^+C^-\gamma^0]e^{-ip\cdot(x-y)}.$$
$$\cdot[(p^2 + m^2 + i\varepsilon)(p^+(p^+ - p^-) + p_\perp{}^2))]^{-1}$$

and using $C^-R^+(i\not p - m)\gamma\cdot\mathbf{p}R^+C^- = i\, C^-R^+(p^+(p^+ - p^-) + p_\perp{}^2)$ we find

$$= \tfrac{1}{2}C^-R^+\gamma^0\int d^4p(2\pi)^{-4}\, p^+e^{-ip\cdot(x-y)}/(p^2 + m^2 + i\varepsilon) \qquad (3.76)$$

Similarly the light-front Feynman propagator for the right-handed ψ_{TR}^{+} tachyon field is

$$iS^+_{TRF}(x,y) = \theta(x^+ - y^+)<0|\psi_{TR}{}^+(x)\psi_{TR}{}^{+\dagger}(y)\gamma^0|0> -$$
$$- \theta(y^+ - x^+)<0|\psi_{TR}{}^{+\dagger}(y)\gamma^0\psi_{TR}{}^+(x)|0>$$

$$= -\tfrac{1}{2}C^+R^+\gamma^0\int d^4p(2\pi)^{-4}\, p^+e^{-ip\cdot(x-y)}/(p^2 + m^2 + i\varepsilon) \qquad (3.77)$$

where the relative minus sign between eqs. 3.76 and 3.77 is due to the relative minus signs of the Fouier component operator anti-commutation relations in eq. 3.58 and 3.63.
Thus we find *tachyon* pole terms in the tachyon propagator as one would expect.

3.6 Complex Lorentz Group L_C Spin ½ Fermion Equations and Wave Functions

In this section we consider L_C group boosts and use them to develop a wider set of dynamical equations for free spin ½ fermions.[8] In chapter 2 we defined L_C boosts with

$$\Lambda_C(\mathbf{v}_c) = \exp[i\omega\hat{\mathbf{w}}\cdot\mathbf{K}] \qquad (2.61)$$
$$\omega = (\omega_r^2 - \omega_i^2 + 2i\omega_r\omega_i\, \hat{\mathbf{u}}_r\cdot\hat{\mathbf{u}}_i)^{1/2} \qquad (2.62)$$
$$\hat{\mathbf{w}} = (\omega_r\hat{\mathbf{u}}_r + i\omega_i\hat{\mathbf{u}}_i)/\omega \qquad (2.63)$$
$$\hat{\mathbf{w}}\cdot\hat{\mathbf{w}} = \hat{\mathbf{u}}_r\cdot\hat{\mathbf{u}}_r = \hat{\mathbf{u}}_i\cdot\hat{\mathbf{u}}_i = 1 \qquad 2.64a)$$
$$\mathbf{v}_c = \hat{\mathbf{w}}\tanh(\omega) \qquad (2.64b)$$

3.7 L_C Spinor "Normal" Lorentz Boosts & More Spin ½ Particle Types

Spinor boost transformations of extended Lorentz groups were used in sections 3.2 and 3.3 to develop the dynamical equations for Dirac fields and tachyon fields. In this section we will use L_C group spinor boosts to generate additional particle field dynamical equations.

The form of the L_C spinor boost transformation corresponding to the coordinate transformation eq. 2.61 is:

$$S_C(\omega, \mathbf{v}_c) = \exp(-i\omega\sigma_{0k}\hat{w}_k/2) = \exp(-\omega\gamma^0\boldsymbol{\gamma}\cdot\hat{\mathbf{w}}/2)$$
$$= \cosh(\omega/2)I + \sinh(\omega/2)\gamma^0\boldsymbol{\gamma}\cdot\hat{\mathbf{w}} \qquad (3.78)$$

The inverse transformation is

$$S_C^{-1}(\omega, \mathbf{v}_c) = \gamma^2\gamma^0 K^{-1} S_C^\dagger K\gamma^0\gamma^2 = \gamma^2\gamma^0 S_C^T\gamma^0\gamma^2 = \exp(\omega\gamma^0\boldsymbol{\gamma}\cdot\hat{\mathbf{w}}/2)$$
$$= \cosh(\omega/2)I - \sinh(\omega/2)\gamma^0\boldsymbol{\gamma}\cdot\hat{\mathbf{w}} \qquad (3.79)$$

---

[8] The complexon theory that we develop and use for quark dynamics in the Standard Model is <u>not</u> required. Our Standard Model could use Dirac fermion dynamics for the up-type quarks and tachyon dynamics for down-type quarks. We choose to use complexon dynamics for quarks because they have an internal SU(3)-like structure suggestive of color SU(3). More importantly, their spin dynamics is different and thus may resolve the differences between theory and experiment for the deep inelastic parton spin-dependent structure functions.

where the superscript T denotes the transpose and K is the complex conjugation operator (that also appears in the time-reversal operator). Note that S_C is not unitary just as in previous cases considered in this chapter.

We now redo the development of spin ½ dynamical equations of motion of sections 3.2 and 3.3 in this more general case of complex ω and $\hat{\mathbf{w}}$. Again we apply a boost to a Dirac equation for a positive energy plane wave particle of mass m at rest:

$$0 = S_C(\omega, \mathbf{v}_c))(m\gamma^0 - m)e^{-imt}w(0)$$

$$= [mS_C\gamma^0S_C^{-1} - m]e^{-imt}S_Cw(0) \qquad (3.80)$$

where $S_C = S_C(\omega, \hat{\mathbf{w}})$. After some algebra

$$mS_C\gamma^0S_C^{-1} = m[\cosh(\omega)\gamma^0 - \sinh(\omega)\boldsymbol{\gamma}\cdot\hat{\mathbf{w}}] \qquad (3.81)$$

Case 1: Parallel Real and Imaginary Relative Vector

If the real and imaginary relative vectors parts of $\hat{\mathbf{w}}$, namely $\hat{\mathbf{u}}_r$ and $\hat{\mathbf{u}}_i$, are parallel, then $\hat{\mathbf{u}}_r\cdot\hat{\mathbf{u}}_i = 1$ and

$$\omega = \omega_r + i\omega_i \qquad (3.82)$$

Eq. 2.64b enables us to re-express eq. 3.81 as

$$mS_C\gamma^0S_C^{-1} = m[\cosh(\omega_r)\cos(\omega_i) + i\sinh(\omega_r)\sin(\omega_i)]\gamma^0 -$$

$$- m[\sinh(\omega_r)\cos(\omega_i) + i\cosh(\omega_r)\sin(\omega_i)]\boldsymbol{\gamma}\cdot\hat{\mathbf{u}}_r \qquad (3.83)$$

or

$$mS_C\gamma^0S_C^{-1} = \cos(\omega_i)\gamma\cdot p_r + i\sin(\omega_i)\gamma\cdot p_i \qquad (3.84)$$

where

$$p_r^0 = m\cosh(\omega_r) \qquad p_i^0 = m\sinh(\omega_r) \qquad (3.85)$$

and

$$\mathbf{p}_r = m\hat{\mathbf{u}}_r\sinh(\omega_r) \qquad \mathbf{p}_i = m\hat{\mathbf{u}}_r\cosh(\omega_r) \qquad (3.86)$$

If $\omega_i = 0$, then we recover the momentum space Dirac equation eq. 3.8. If $\omega_i = \pi/2$, then we obtain the left-handed momentum space tachyon equation eq. 3.16. Since the range of ω_i is [0, ∞> (due to the cut along the real ω-plane axis) eq. 3.84 corresponds to the results of the Left-Handed Extended Lorentz group development discussed earlier.

Case 2: Anti-Parallel Real and Imaginary Relative Vector

If the real and imaginary relative vectors parts of $\hat{\mathbf{w}}$, $\hat{\mathbf{u}}_r$ and $\hat{\mathbf{u}}_i$, are anti-parallel $\hat{\mathbf{u}}_r = -\hat{\mathbf{u}}_i$, then $\hat{\mathbf{u}}_r \cdot \hat{\mathbf{u}}_i = -1$ and

$$\omega = \omega_r - i\omega_i \tag{3.87}$$

We can then express eq. 3.81 as

$$mS_C\gamma^0 S_C^{-1} = m[\cosh(\omega_r)\cos(\omega_i) - i\sinh(\omega_r)\sin(\omega_i)]\gamma^0 -$$

$$- m[\sinh(\omega_r)\cos(\omega_i) - i\cosh(\omega_r)\sin(\omega_i)]\boldsymbol{\gamma}\cdot\hat{\mathbf{u}}_r \tag{3.88}$$

or

$$mS_C\gamma^0 S_C^{-1} = \cos(\omega_i)\gamma\cdot p_r - i\sin(\omega_i)\gamma\cdot p_i \tag{3.89}$$

where

$$p_r^0 = m\cosh(\omega_r) \qquad p_i^0 = m\sinh(\omega_r) \tag{3.90}$$

and

$$\mathbf{p}_r = m\hat{\mathbf{u}}_r \sinh(\omega_r) \qquad \mathbf{p}_i = m\hat{\mathbf{u}}_r \cosh(\omega_r) \tag{3.91}$$

If $\omega_i = 0$, then we again recover the momentum space Dirac equation eq. 3.8. If $\omega_i = \pi/2$, then we obtain the right-handed momentum space tachyon equation. (The range of ω_i is again $[0, \infty>$.)

Note: Since the matrix elements in the boost eq. 2.65 depend on $\gamma = (1 - \beta^2)^{-1/2}$ with a singularities at $\beta = \pm 1$, which in turn corresponds to $\omega = \pm\infty$, there is a branch cut along the ω axis in the complex ω-plane. Therefore we point out again the product of three Left-handed extended Lorentz transformations is *not* equivalent to a Right-handed extended Lorentz transformation.

Case 3: Complexons: A New Type of Particle with Perpendicular Real and Imaginary 3-Momenta

If the real and imaginary relative vectors parts of $\hat{\mathbf{w}}$, namely $\hat{\mathbf{u}}_r$ and $\hat{\mathbf{u}}_i$, are perpendicular, $\hat{\mathbf{u}}_r \cdot \hat{\mathbf{u}}_i = 0$, then by eq. 2.62

$$\omega = (\omega_r^2 - \omega_i^2)^{1/2} \tag{3.92}$$

Thus ω is either pure real ($\omega_r \geq \omega_i$) or pure imaginary ($\omega_r < \omega_i$).

The momentum space equation generated by the corresponding L_C spinor boost is

$$\{m\cosh(\omega)\gamma^0 - m\sinh(\omega)\boldsymbol{\gamma}\cdot(\omega_r\hat{\mathbf{u}}_r + i\omega_i\hat{\mathbf{u}}_i)/\omega - m\}e^{-ip\cdot x}w_c(p) = 0 \tag{3.93}$$

Defining the momentum 4-vector

$$p = (p^0, \mathbf{p}) \tag{3.94}$$

where

$$p^0 = m\cosh(\omega) \qquad \mathbf{p} = \mathbf{p}_r + i\mathbf{p}_i \tag{3.95}$$

with

$$\mathbf{p}_r = m\omega_r\hat{\mathbf{u}}_r\sinh(\omega)/\omega \qquad \mathbf{p}_i = m\omega_i\hat{\mathbf{u}}_i\sinh(\omega)/\omega \tag{3.96a}$$

$$\mathbf{p}_r\cdot\mathbf{p}_i = 0 \tag{3.96b}$$

then eq. 3.93 becomes a positive energy Dirac-like equation

$$[p\cdot\gamma - m]e^{-ip\cdot x}w_c(p) = 0$$

or, explicitly,

$$[p^0\gamma^0 - (\mathbf{p}_r + i\mathbf{p}_i)\cdot\gamma - m]e^{-ip\cdot x}w_c(p) = 0 \tag{3.97}$$

with a complex 3-momentum $\mathbf{p}$ and the 4-momentum mass shell condition:

$$p^2 = p^{0\,2} - \mathbf{p}_r\cdot\mathbf{p}_r + \mathbf{p}_i\cdot\mathbf{p}_i = m^2 \tag{3.98}$$

Note

$$|\mathbf{v}| = |\mathbf{p}|/p^0 = [(\mathbf{p}_r + i\mathbf{p}_i)\cdot(\mathbf{p}_r + i\mathbf{p}_i)]^{\frac{1}{2}}/p^0 = \tanh(\omega) \tag{3.99}$$

and so the Lorentz factor

$$\gamma = \cosh(\omega) \tag{3.100}$$

Eq. 3.97 is the momentum space equivalent of the wave equation

$$[i\gamma^0\partial/\partial t + i\gamma\cdot(\nabla_r + i\nabla_i) - m]\psi_C(t, \mathbf{x}_r, \mathbf{x}_i) = 0 \tag{3.101}$$

where $\mathbf{x} = \mathbf{x}_r - i\mathbf{x}_i$, and where the grad operators ∇_r and ∇_i are with respect to $\mathbf{x}_r$ and $\mathbf{x}_i$ respectively. Since $\hat{\mathbf{u}}_r\cdot\hat{\mathbf{u}}_i = 0$, which in turn implies eq. 3.96b, we see that there is a subsidiary condition on the wave function

$$\nabla_r\cdot\nabla_i\,\psi_C(t, \mathbf{x}_r, \mathbf{x}_i) = 0 \tag{3.102a}$$

We will call the particles satisfying eqs. 3.101 and 3.102a *complexons*. In addition eq. 3.96b implies the anti-commutation relation

$$\{\gamma\cdot\mathbf{p}_r, \gamma\cdot\mathbf{p}_i\} = 0 \tag{3.102b}$$

which in turn implies

$$\gamma\cdot\nabla_r\gamma\cdot\nabla_i\psi_C(t, \mathbf{x}_r, \mathbf{x}_i) = \gamma\cdot\nabla_i\gamma\cdot\nabla_r\psi_C(t, \mathbf{x}_r, \mathbf{x}_i) = 0 \tag{3.102c}$$

also holds. We note that eq. 3.101 is covariant under the real Lorentz group and eq. 3.102 can be easily put into covariant form since the difference of these 4-vectors squared (which is real Lorentz group invariant): $[\gamma^0\partial/\partial t + \gamma\cdot(\nabla_r + i\nabla_i)]^2 - [\gamma^0\partial/\partial t + i\gamma\cdot(\nabla_r - i\nabla_i)]^2 = 4\nabla_r\cdot\nabla_i$.

Before considering a lagrangian formulation and the Fourier operator representation of $\psi_C(t, \mathbf{x}_r, \mathbf{x}_i)$ we will define the spinors and associated real and imaginary spin operators.

The spinor generated from a spin up Dirac spinor at rest by the complex Lorentz boost eqs. 3.78 is

$$w_c(p) = S_C(p)w(0) = [\cosh(\omega/2)I + \sinh(\omega/2)\gamma^0\boldsymbol{\gamma}\cdot\hat{\mathbf{w}}]w(0) \qquad (3.103)$$

Following a procedure similar to Appendix 3-A (which the reader may wish to examine first) we define four spinors for Dirac particles at rest:

$$w^k(0) = \begin{bmatrix} \delta_{1k} \\ \delta_{2k} \\ \delta_{3k} \\ \delta_{4k} \end{bmatrix} \qquad (3\text{-A}.2)$$

where Kronecker deltas appear in the brackets. Then by applying eq. 3.103 to the spinors defined by eq. 3-A.2 we find the L_C spinors

$$S_C w^k(0) = w_{Cr}{}^k(p) + i w_{Ci}{}^k(p) \qquad (3.104)$$

where

$$S_{Cr} = \cosh(\omega/2)I + (\omega_r/\omega)\sinh(\omega/2)\gamma^0\boldsymbol{\gamma}\cdot\hat{\mathbf{u}}_r$$

$$= [(m + E)/(2m)]^{1/2}I + [m(m + E)]^{-1/2}\gamma^0\boldsymbol{\gamma}\cdot\mathbf{p}_r = aI + b\gamma^0\boldsymbol{\gamma}\cdot\mathbf{p}_r \qquad (3.105)$$

by eqs. 3.95 and 3.96. Thus the "real" spinors $w_{Cr}{}^k(p)$ are the columns of

$$S_{Cr} = \begin{bmatrix} \underline{w_{Cr}{}^1(p)} & \underline{w_{Cr}{}^2(p)} & \underline{w_{Cr}{}^3(p)} & \underline{w_{Cr}{}^4(p)} \\ a & 0 & bp_{r\,z} & bp_{r-} \\ 0 & a & bp_{r+} & -bp_{r\,z} \\ bp_{r\,z} & bp_{r-} & a & 0 \\ bp_{r+} & -bp_{r\,z} & 0 & a \end{bmatrix} \qquad (3.106)$$

where $p_{r\pm} = p_{r\,x} \pm ip_{r\,y}$. The "imaginary" spinors are the columns of

$$S_{Ci} = (\omega_i/\omega)\sinh(\omega/2)\gamma^0\boldsymbol{\gamma}\cdot\hat{\mathbf{u}}_i = [m(m + E)]^{-1/2}\gamma^0\boldsymbol{\gamma}\cdot\mathbf{p}_i = b\gamma^0\boldsymbol{\gamma}\cdot\mathbf{p}_i \qquad (3.107)$$

$$S_{Ci} = \begin{bmatrix} \underline{w_{Ci}^{1}(p)} & \underline{w_{Ci}^{2}(p)} & \underline{w_{Ci}^{3}(p)} & \underline{w_{Ci}^{4}(p)} \\ 0 & 0 & bp_{i\,z} & bp_{i-} \\ 0 & 0 & bp_{i\,+} & -bp_{i\,z} \\ bp_{i\,z} & bp_{i-} & 0 & 0 \\ bp_{i\,+} & -bp_{i\,z} & 0 & 0 \end{bmatrix} \quad (3.108)$$

where $p_{i\,\pm} = p_{i\,x} \pm ip_{i\,y}$.

Eqs. 3.101 through 3.108 imply that the wave function solution of eq. 3.101, subject to the subsidiary condition eq. 102a, is[9, 10]

$$\psi_C(x_r, x_i) = \sum_{\pm s}\int d^3p_r d^3p_i\, N_C(p)\delta(\mathbf{p_r}\cdot\mathbf{p_i}/m^2)[b_C(p,s)u_C(p, s)e^{-i(p\cdot x\,+\,p^*\cdot x^*)/2} + \\ + d_C^{\dagger}(p,s)v_C(p, s)e^{+i(p\cdot x\,+\,p^*\cdot x^*)/2}] \quad (3.109)$$

where $\mathbf{p} = \mathbf{p_r} + i\mathbf{p_i}$ (eq. 3.95), $\mathbf{x} = \mathbf{x_r} - i\mathbf{x_i}$, $p\cdot x = p^0x^0 - \mathbf{p}\cdot\mathbf{x}$, and where we use

$$(p\cdot x + p^*\cdot x^*)/2 = p^0x^0 - \mathbf{p_r}\cdot\mathbf{x_r} - \mathbf{p_i}\cdot\mathbf{x_i} \quad (3.109a)$$

in the exponentials in order to avoid divergences that would appear in the calculation of the equal-time commutator, the Feynman propagator and other quantities of interest after second quantization. Note that

$$(\nabla_r + i\nabla_i)e^{-i(p\cdot x\,+\,p^*\cdot x^*)/2} = i(\mathbf{p_r} + i\mathbf{p_i})e^{-i(p\cdot x\,+\,p^*\cdot x^*)/2} \quad (3.109b)$$

and

$$(\nabla_r + i\nabla_i)e^{-ip^*\cdot x^*} = 0 \quad (3.109c)$$

for all p.

The wave function's conjugate (the hermitean conjugate modified by letting $\mathbf{x_i} \rightarrow -\mathbf{x_i}$ in addition to hermitean conjugation) is

$$\psi_C^{\dagger}(x) = \psi_C^{\dagger}(x_r, -x_i) = \sum_{\pm s}\int d^3p_r d^3p_i\, \delta(\mathbf{p_r}\cdot\mathbf{p_i}/m^2)N_C(p^*)\cdot \\ \cdot[b_C^{\dagger}(p^*,s)u_C^{\dagger}(p^*,s)e^{+i(p\cdot x^*\,+\,p^*\cdot x)/2} + d_C(p^*,s)v_C^{\dagger}(p^*,s)e^{-i(p\cdot x^*\,+\,p^*\cdot x)/2}] \quad (3.110)$$

---

[9] Note that when $|\mathbf{p_i}| \geq |\mathbf{p_r}|$ (for imaginary $\omega = (\omega_r^2 - \omega_i^2)^{1/2}$) the 3-momentum becomes imaginary $\mathbf{p}\cdot\mathbf{p} < 0$. However, since we will be identifying confined quarks with this type of particle – much modified by a confining color quark interaction – the issue of an imaginary 3-momentum in the hypothetical free quark case becomes moot. We note the energy gap between positive and negative energy states disappears so E = 0 is possible. Thus real Lorentz transformations can mix positive and negative energy states. The solution is to do all calculations in the light-front frame as we do for tachyons. Then the mixing issue is resolved. In the present case we second quantize on the "time-front" for illustrative purposes.

[10] We scale $\mathbf{p_r}\cdot\mathbf{p_i}$ with m^2 in the delta function for convenience. In the case of a zero mass particle some other scale could be used.

where $\mathbf{p} = \mathbf{p_r} + i\mathbf{p_i}$ (eq. 3.95), $\mathbf{x} = \mathbf{x_r} - i\mathbf{x_i}$, $p\cdot x = p^0x^0 - \mathbf{p}\cdot\mathbf{x}$, and † indicates hermitean hermitean conjugation.

The spinors are

$$\begin{aligned} u_C(p, s) &= S_C(p)w^1(0) \\ u_C(p, -s) &= S_C(p)w^2(0) \\ v_C(p, s) &= S_C(p)w^3(0) \\ v_C(p, -s) &= S_C(p)w^4(0) \qquad (3.110a) \\ u_C^\dagger(p^*, s) &= w^{1T}(0)S_C^\dagger(p^*) = w^{1T}(0)S_C(p) \\ u_C^\dagger(p^*, -s) &= w^{2T}(0)S_C^\dagger(p^*) = w^{2T}(0)S_C(p) \\ v_C^\dagger(p^*, s) &= w^{3T}(0)S_C^\dagger(p^*) = w^{3T}(0)S_C(p) \\ v_C^\dagger(p^*, -s) &= w^{4T}(0)S_C^\dagger(p^*) = w^{4T}(0)S_C(p) \end{aligned}$$

with the superscript "T" indicating the transpose. Note that

$$S_C^\dagger(p^*) = [S_C(p^*)]^\dagger = S_C(p) \qquad (3.110b)$$

The normalization factor $N_C(p)$ is

$$N_C(p) = [2m/((2\pi)^6p^0)]^{1/2} \qquad (3.110c)$$

Since $\mathbf{p_r} = \mathbf{p_i} = 0$ in the particle rest frame prior to the complex group boost, the boosted particle spin 4-vector s^μ satisfies

$$s^\mu p_r^\mu = s^\mu p_i^\mu = 0 \qquad (3.111)$$

Note that s^μ is itself complex[11] and, if the spin points in the z-direction prior to the complex boost (eq. 2.57), then the boosted s^μ has the form

$$s^\mu = (-\sinh(\omega)\hat{w}_z, (0,0,1) + (\cosh(\omega) - 1)\hat{w}_z\hat{\mathbf{w}}) \qquad (3.112)$$

with $\hat{\mathbf{w}}$ defined by eq. 2.63: $\hat{\mathbf{w}} = (\omega_r\hat{\mathbf{u}}_r + i\omega_i\hat{\mathbf{u}}_i)/\omega = \mathbf{p}/(m\sinh(\omega))$ using eq. 3.96a.

A Global SU(3) Symmetry Revealed

Before proceding to consider the second quantization of this case, we will consider a global SU(3) symmetry implicit in eqs. 3.101, 3.102 and the solution 3.109. The defining property of the group SU(3) is that it preserves the invariance of inner products of complex 3-vectors of the form:

---

[11] This feature of partons, which is not present in ordinary Dirac particles, might be the source of the discrepancies between theory and experiment in deep inelstic parton spin physics which is based on conventional real parton spins.

$$u^{*}\cdot v = u^{1*}v^{1} + u^{2*}v^{2} + u^{3*}v^{3} \qquad (3.113)$$

If we examine the dynamical equation eq. 3.101 we see that the differential operator is invariant under an SU(3) transformation U (using $\nabla_c = (\nabla_c{}^*)^* = \mathbf{D}_c{}^*$)

$$[i\gamma^0\partial/\partial t + i\mathbf{D}_c{}^*\cdot\gamma - m] = [i\gamma^0\partial/\partial t + i\mathbf{D}_c{}'^*\cdot\gamma' - m] \qquad (3.114)$$

where

$$\mathbf{D}_c{}^* = \nabla_c = \nabla_r + i\nabla_i$$

and

$$\gamma'^a = U^{ab}\gamma'^b$$
$$D_c'^{*a} = D_c'^{*b}U^{\dagger ab}$$

where U is a global SU(3) transformation and $U^\dagger = U^{-1}$. By theorem[12] all 4×4 γ matrices such as γ' are equivalent up to a unitary transformation V. Thus $V^\dagger\gamma'V = \gamma$ and eq. 3.114 is equivalent to

$$[i\gamma^0\partial/\partial t + i\mathbf{D}_c{}^*\cdot\gamma - m] = [i\gamma^0\partial/\partial t + i\mathbf{D}_c{}'^*\cdot\gamma - m] \qquad (3.115)$$

$$= [i\gamma^0\partial/\partial t + i\nabla_c{}'\cdot\gamma - m]$$

where $\nabla_c{}'_a = U^{ab}\nabla_{cb}$ showing that eq. 3.101 is invariant under an SU(3) transformation if

$$\psi_C(t, \mathbf{x}) = \psi_C(t, U\mathbf{x}) = \psi_C{}'(t, \mathbf{x}') \qquad (3.116)$$

where $\psi_C(t, \mathbf{x}) \equiv \psi_C(t, \mathbf{x}_r, \mathbf{x}_i)$.

The subsidiary condition eq. 3.102a can be seen to transform as

$$\nabla_r\cdot\nabla_i\, \psi_C(t, \mathbf{x}) = \nabla_r{}^*\cdot\nabla_i\, \psi_C(t, \mathbf{x}) = \nabla_r{}'^*\cdot\nabla_i{}'\psi_C{}'(t, \mathbf{x}') = 0 \qquad (3.117)$$

under an SU(3) rotation. The invariance of the orthogonality condition is preserved.

The wave function (eq. 3.109) transforms in the following way under the SU(3) transformation U. If we define

$$q^{*\mu} = (q^0, \mathbf{q}^*) = (p^0, \mathbf{p}_r + i\mathbf{p}_i) = (p^0, \mathbf{p}) = p^\mu \qquad (3.118)$$

then eq. 3.109 can be rewritten in an almost[13] manifestly invariant form under a SU(3) transformation:

---

[12] R. H. Good, Rev. Mod. Phys., **27**, 187 (1955).

$$\psi_C(x) = \sum_{\pm s}\int d^3q_r d^3q_i\, N_C(p^0)\delta(\mathbf{q}_r{}^*\cdot\mathbf{q}_i/m^2)[b_C(q^*,s)u_C(q^*,s)e^{-i(q^*\cdot x + q\cdot x^*)/2} +$$
$$+ d_C^{\dagger}(q^*,s)v_C(q^*,s)e^{+i(q^*\cdot x + q\cdot x^*)/2}] \quad (3.119)$$

subject to an examination of the transformation properties of the fourier coefficients and spinors. Note both terms in each exponential are separately invariant under global SU(3). (Note also $\mathbf{q}_r{}^* = \mathbf{q}_r$ since $\mathbf{q}_r$ is real.)

From the form of S_C (eq. 3.78) it is clear that an argument similar to that for the dynamical equations (eqs. 3.114 – 3.115) shows S_C is invariant under an SU(3) transformation and thus the spinors defined by eqs. 3.110a are also invariant under SU(3) transformations. The fourier coefficients, if second quantized in a direct generalization of the usual way, have covariant anti-commutation relations under an SU(3) transformation. For example

$$\{b_C(q,s),\, b_C^{\dagger}(q'^*,s')\} = \delta_{ss'}\delta^3(q_r - q'_{r'})\delta^3(q_i - q'_{i'}) \quad (3.120)$$

Under an SU(3) transformation, z = Uq and z' = Uq', the right side of eq. 3.120 transforms to

$$\delta^3(q_r - q'_{r'})\delta^3(q_i - q'_{i'}) \rightarrow \delta^3(z_r - z'_{r'})\delta^3(z_i - z'_{i'})/|\partial(q)/\partial(z)|$$
$$= \delta^3(z_r - z'_{r'})\delta^3(z_i - z'_{i'}) \quad (3.121)$$

where

$$|\partial(q)/\partial(z)| = |\partial(q_r{}^1,q_r{}^2,q_r{}^3,q_i{}^1,\, q_i{}^2,\, q_i{}^3)/\partial(z_r{}^1,z_r{}^2,z_r{}^3,z_i{}^1,\, z_i{}^2,\, z_i{}^3)| = 1 \quad (3.122)$$

is the Jacobian of the transformation U. Thus the fourier coefficients transform trivially under SU(3). For example,

$$b_C(q^*,s) \rightarrow b_C(z^*,s) \quad (3.123)$$

Since the integrand transforms as

$$\int d^3q_r d^3q_i \;\rightarrow\; \int d^3z_r d^3z_i\, |\partial(q)/\partial(z)| = \int d^3z_r d^3z_i \quad (3.124)$$

---

[13] The δ-function in eq. 3.119, $\delta(\mathbf{q}_r{}^*\cdot\mathbf{q}_i/m^2)$, is not invariant under a global SU(3) transformation rather it is a constraint that breaks the SU(3) symmetry to O(3)×O(2) rather like a δ-function term in the path integral formalism of gauge theories "breaks" the symmetry and fixes the gauge. The O(3)×O(2) symmetry can be visualized by mapping (conceptually) the $\mathbf{q}_r$ and $\mathbf{q}_i$ to 3-space and considering the perpendicular vector $\mathbf{q}$ to the pair as rotating under O(3) transformations, and $\mathbf{q}_r$ and $\mathbf{q}_i$ rotating in the plane perpendicular to $\mathbf{q}$ under O(2) transformations.

the wave function $\psi_C(t, \mathbf{x})$ transforms as an SU(3) scalar (eq. 3.116) up to an inessential unitary transformation V of γ matrices: $\psi_C(t, \mathbf{x}) \rightarrow V\psi_C(t, \mathbf{x})$.[14]

Global SU(3) Spin ½ Complexon Fields

Having uncovered an SU(3) symmetry in the scalar field equations of Case 3A the generalization of the scalar field equations to the 3 representation of SU(3) is direct:

$$\psi_C{}^a(x) = \sum_{\pm s} \int d^3p_r d^3p_i\, N_C(p)\delta(\mathbf{p}_r\cdot\mathbf{p}_i/m^2)[b_C(p,a,s)u_C{}^a(p, s)e^{-i(p\cdot x + p^*\cdot x^*)/2} + $$
$$+ d_C{}^\dagger(p,a,s)v_C{}^a(p, s)e^{+i(p\cdot x + p^*\cdot x^*)/2}] \qquad (3.125a)$$

for a = 1,2, 3 with $u_C{}^a(p, s)$ and $v_C{}^a(p, s)$ being the product a spinor of type eq. 3.110a and a 3 element column vector c^a with b^{th} element

$$c^a(b) = \delta^{ab} \qquad (3.125b)$$

Under a global SU(3) transformation U the 3 complexon wave functions transform as

$$\psi_C{}'^a(x) = U^{ab}\psi_C{}^b(x) \qquad (3.126)$$

In a subsequent chapter we will extend the global SU(3) symmetry described in these subsections to color local SU(3) upon the introduction of the Yang-Mills color gluon interaction and detach it from the global SU(3) coordinate transformation feature found in this section.

Lagrangian Formulation and Second Quantization of Complexons

In this subsection we will outline the canonical quantization of SU(3) singlet complexons with the quantum field equation

$$[i\gamma^0\partial/\partial t + i\gamma\cdot(\nabla_r + i\nabla_i) - m]\psi_C(t, \mathbf{x}_r, \mathbf{x}_i) = 0 \qquad (3.101)$$

and subsidiary condition

$$\nabla_r\cdot\nabla_i\, \psi_C(t, \mathbf{x}_r, \mathbf{x}_i) = 0 \qquad (3.102a)$$

[14] The spinors $u_C(q^*,s)$ and $v_C(q^*,s)$ are unchanged up to a unitary transformation of the γ matrices ($V^\dagger\gamma' V = \gamma$). Thus the term $(U\mathbf{w})^*\cdot\gamma' = \mathbf{w}^*\cdot V\gamma V^\dagger \equiv \mathbf{w}^*\cdot\gamma$ in the expressions for the $u_C(q^*,s)$ and $v_C(q^*,s)$ spinors.

We begin with the Lagrangian density

$$\mathcal{L} = \bar{\psi}_C(i\gamma^\mu D_\mu - m)\psi_C(x) \tag{3.127}$$

where $\bar{\psi}_C = \psi_C^{\dagger}\gamma^0$:

$$\psi_C^{\dagger} = [\psi_C(\mathbf{x}_r, \mathbf{x}_i)]^\dagger\,|_{\mathbf{x}_i = -\mathbf{x}_i} \tag{3.128}$$

$$D_0 = \partial/\partial x^0$$
$$D_k = \partial/\partial x^k + i\,\partial/\partial x_i^{\,k} \tag{3.129}$$

with $x^k = x_r^{\,k}$ for k = 1,2, 3. The invariant action (under real Lorentz transformations) is

$$I = \int d^7x\,\mathcal{L} \tag{3.130}$$

It is easy to show that the action is real

$$I^* = I \tag{3.131}$$

in a manner similar to the case considered in Appendix 3-A due to the form of $\psi_C^{\dagger}$ in eq. 3.128. (One has to change the integration over $\mathbf{x}_i$ to $-\mathbf{x}_i$ after taking the complex conjugate of I and performing manipulations similar to those in Appendix 3-A.)

The conjugate momentum is

$$\pi_{Ca} = \partial\mathcal{L}/\partial\dot{\psi}_{Ca} \equiv \partial\mathcal{L}/\partial(\partial\psi_{Ca}/\partial x^0) = i\psi_{C\,a}^{\dagger} \tag{3.132}$$

where a is a spinor index. It yields the non-zero anti-commutation relation

$$\{\psi_{C\,a}^{\dagger}(x), \psi_{Cb}(y)\} = \delta_{ab}\,\delta^3(x_r - y_r)\delta^3(x_i - y_i) \tag{3.133a}$$

However we will see that the constraint eq. 3.102a is required. So the correct anti-commutator turns out to be

$$\{\psi_{C\,a}^{\dagger}(x), \psi_{Cb}(y)\} = -\delta_{ab}\delta'(\nabla_r\cdot\nabla_i/m^2)[\delta^3(x_r - y_r)\delta^3(x_i - y_i)] \tag{3.133b}$$

where all ∇_r and ∇_i are ∇ derivatives with respect to x, and where $\delta'(\nabla_r\cdot\nabla_i)$ is the derivative of a delta function with the argument being differential operators such as those in eq. 3.102a. The minus sign is due to the presence of a *derivative* of a delta-function and not an issue.

The hamiltonian density is

$$\mathcal{H} = \pi_C\dot{\psi}_C - \mathcal{L} = \psi_C^{\dagger}(-i\boldsymbol{\alpha}\cdot\mathbf{D} + \beta m)\psi_C \tag{3.134}$$

and the (unsymmetrized) energy-momentum tensor is

$$\mathcal{T}_{\mu\nu} = -\, g_{\mu\nu}\,\mathcal{L} + \partial\mathcal{L}/\partial(D^{\mu}\psi_C)D_{\nu}\psi_C \tag{3.135}$$

The conserved energy and momentum are

$$P^0 = H = \int d^3x_r d^3x_i\ \mathcal{T}^{00} = \int d^3x_r d^3x_i\ \mathcal{H} \tag{3.136}$$

and

$$P^i = \int d^3x_r d^3x_i\ \mathcal{T}^{0i} \tag{3.137}$$

We now proceed to establish the canonical anti-commutation relations eq. 3.133. First, the second quantization of the complexon field (eqs. 3.109 and 3.110) uses the fourier coefficient anti-commutation relations eq. 3.120 (suitably rewritten):

$$\begin{aligned}
\{b_C(p,s), b_C^{\dagger}(p'^*,s')\} &= \delta_{ss'}\delta^3(\mathbf{p_r} - \mathbf{p'_{r'}})\delta^3(\mathbf{p_i} + \mathbf{p'_{i'}})\\
\{d_C(p,s), d_C^{\dagger}(p'^*,s')\} &= \delta_{ss'}\delta^3(\mathbf{p_r} - \mathbf{p'_{r'}})\delta^3(\mathbf{p_i} + \mathbf{p'_{i'}})\\
\{b_C(p,s), b_C(p'^*,s')\} &= \{d_C(p,s), d_C(p'^*,s')\} = 0\\
\{b_C^{\dagger}(p,s), b_C^{\dagger}(p'^*,s')\} &= \{d_C^{\dagger}(p,s), d_C^{\dagger}(p'^*,s')\} = 0\\
\{b_C(p,s), d_C^{\dagger}(p'^*,s')\} &= \{d_C(p,s), b_C^{\dagger}(p'^*,s')\} = 0\\
\{b_C^{\dagger}(p,s), d_C^{\dagger}(p'^*,s')\} &= \{d_C(p,s), b_C(p'^*,s')\} = 0
\end{aligned} \tag{3.138}$$

The delta-function arguments $\delta^3(\mathbf{p_i} + \mathbf{p'_{i'}})$ above have a positive sign in order to obtain $\delta^3(\mathbf{x_i} - \mathbf{y_i})$ in the field anti-commutator eq. 3.133b.

The spinors, eq. 3.110a, satisfy

$$\begin{aligned}
\sum_{\pm s} u_{\alpha}(p, s)\bar{u}_{\beta}(p^*, s) &= (2m)^{-1}(\not{p} + m)_{\alpha\beta}\\
\sum_{\pm s} v_{\alpha}(p, s)\bar{v}_{\beta}(p^*, s) &= (2m)^{-1}(\not{p} - m)_{\alpha\beta}
\end{aligned} \tag{3.139}$$

by eqs. 3.78, 3.81, 3.85 and 3.103 remembering

$$\bar{u}_C(p^*,s) = w^{1T}(0)S_C(p)\gamma^0 = w^{1T}(0)[\cosh(\omega/2)I + \sinh(\omega/2)\gamma^0\boldsymbol{\gamma}\cdot\hat{\mathbf{w}}]\gamma^0 \tag{3.140}$$

by eqs. 3.110 since $\hat{\mathbf{w}}^{**} = \hat{\mathbf{w}}$.

We will now evaluate the equal-time anti-commutation relation using eqs. 3.109 and 3.110:

$$\{\psi_{C\,a}^{\dagger}(x), \psi_{Cb}(y)\} = \sum_{\pm s,\, s'} \int d^3p_r d^3p_i\, d^3p'_r d^3p'_i\, \delta(\mathbf{p}_r\cdot\mathbf{p}_i/m^2)\delta(\mathbf{p}'_r\cdot\mathbf{p}'_i/m^2)\, N_C(p')N_C(p)\cdot$$
$$\cdot[\{b_C^{\dagger}(p^*,s)u_{Ca}^{\dagger}(p^*,s)e^{+i(p\cdot x^* + p^*\cdot x)/2},\, b_C(p',s')u_{Cb}(p',s')e^{-i(p'\cdot y + p'^*\cdot y^*)/2}\} +$$
$$+ \{d_C(p^*,s)v_{Ca}^{\dagger}(p^*,s)e^{-i(p\cdot x^* + p^*\cdot x)/2},\, d_C^{\dagger}(p',s')v_{Cb}(p',s')e^{+i(p'\cdot y + p'^*\cdot y^*)/2}\}]$$

$$= \int d^3p_r d^3p_i\, N_C^2(p)[\delta(\mathbf{p}_r\cdot\mathbf{p}_i/m^2)]^2[((\not{p} + m)\gamma^0)_{ba}e^{+i(p\cdot x^* + p^*\cdot x)/2 - i(p^*\cdot y + p\cdot y^*)/2} +$$
$$+((\not{p} - m)\gamma^0)_{ba}e^{-i(p\cdot x^* + p^*\cdot x)/2 + i(p^*\cdot y + p\cdot y^*)/2}]/(2m)$$

Next we use eq. 3.110c and the identity

$$[\delta(x - y)]^2 = -\tfrac{1}{2}\,\delta'(x - y) \equiv -\tfrac{1}{2}\, d\delta(x - y)/dx \tag{3.141}$$

which can be derived from the step function identity $\theta(x - y) = [\theta(x - y)]^2$ to obtain

$$\{\psi_{C\,a}^{\dagger}(x),\psi_{Cb}(y)\} = -\tfrac{1}{2}\int d^3p_r d^3p_i N_C^2(p)\delta'(\mathbf{p}_r\cdot\mathbf{p}_i/m^2)[((\not{p}+m)\gamma^0)_{ba}e^{-ipr\cdot(xr - yr) + ipi\cdot(xi - yi)} +$$
$$+ ((\not{p} - m)\gamma^0)_{ba}e^{+ipr\cdot(xr - yr) - ipi\cdot(xi - yi)}]/(2m)$$
$$= -\tfrac{1}{2}\delta_{ba}\int d^3p_r d^3p_i N_C^2(p)\delta'(\mathbf{p}_r\cdot\mathbf{p}_i/m^2)p^0e^{-ipr\cdot(xr - yr) + ipi\cdot(xi - yi)}/m$$

$$= -\,\delta_{ab}\,\delta'(\nabla_r\cdot\nabla_i/m^2)[\delta^3(x_r - y_r)\delta^3(x_i - y_i)] \tag{3.142}$$

The grad operators, ∇_r and ∇_i, are derivatives are with respect to x of the Dirac delta functions. The factor[15] $\delta'(\nabla_r\cdot\nabla_i)$ expresses the orthogonality constraint in coordinate space on the momenta eq. 3.102a. It is analogous to the transversality constraint on the electromagnetic vector potential commutator:

$$[\pi_A{}^j(x), A_k(y)] = -i\,\delta^{tr}{}_{jk}(x - y) \tag{3.143}$$

$$\delta^{tr}{}_{jk}(x - y) = (\delta_{jk} - \partial_j\partial_k/\nabla^2)\,\delta^3(x - y) \tag{3.144}$$

where $\partial_k = \partial/\partial x_k$.

Complexon Feynman Propagator

The complexon Feynman propagator for ψ_C is[16]

---

[15] A derivative of a delta function containing grad operators.

[16] The reader, upon seeing the additional integrations $\int d^3p_i$ might suspect that they would ultimately lead to divergence issues in perturbation theory calculations. However the $\delta'(\mathbf{p}_r\cdot\mathbf{p}_i/m^2)$ term compensates in part for the additional integrations by four powers of momentum since $\delta'(\mathbf{p}_r\cdot\mathbf{p}_i/m^2) = (|\mathbf{p}_r||\mathbf{p}_i|/m^2)^{-2}\delta'(\cos\theta_{ri})$ where θ_{ri} is the angle between the momenta. As a result only 2 fermion and 3 fermion loop integrations would potentially have difficulties if one uses the conventional approach to perturbation theory. If one uses the approach of Blaha (2003) and (2005a) then there are no divergences.

$$iS_C(x, y) = \theta(x^0 - y^0)\langle 0|\psi_C(x)\psi_C^\dagger(y)\gamma^0|0\rangle -$$
$$- \theta(y^0 - x^0)\langle 0|\psi_C^\dagger(y)\gamma^0\psi_C(x)|0\rangle \qquad (3.145)$$
$$= \int d^3p_r d^3p_i N_C^2(p)[\delta(\mathbf{p}_r\cdot\mathbf{p}_i/m^2)]^2\{\theta(x^0 - y^0)(\not{p} + m)e^{-i(p^*\cdot(x-y) + p\cdot(x^*-y^*))/2} -$$
$$- \theta(y^0 - x^0)(\not{p} - m)e^{+i(p^*\cdot(x-y) + p\cdot(x^*-y^*))/2}\}/(2m)$$

$$= -(4\pi)^{-1}\int dp^0 d^3p_r d^3p_i (2\pi)^{-6}\delta'(\mathbf{p}_r\cdot\mathbf{p}_i/m^2)(\not{p}+m)e^{-i(p^*\cdot(x-y) + p\cdot(x^*-y^*))/2}/(p^2-m^2+i\varepsilon)$$

$$= -\tfrac{1}{2}\int dp^0 d^3p_r d^3p_i\, \delta'(\mathbf{p}_r\cdot\mathbf{p}_i/m^2)(\not{p} + m)(2\pi)^{-7}\cdot$$
$$\cdot\exp[-ip^0(x^0 - y^0) + i\mathbf{p}_r\cdot(\mathbf{x}_r - \mathbf{y}_r) - i\mathbf{p}_i\cdot(\mathbf{x}_i - \mathbf{y}_i)]/(p^2 - m^2 + i\varepsilon) \qquad (3.146)$$

The integral can be written in the form:

$$I = \int dp^0 d^3p_r d^3p_i \delta'(\mathbf{p}_r\cdot\mathbf{p}_i/m^2)(\not{p}+m)\exp[-ip^0(x^0-y^0)+i\mathbf{p}_r\cdot(\mathbf{x}_r-\mathbf{y}_r)-i\mathbf{p}_i\cdot(\mathbf{x}_i-\mathbf{y}_i)]/(p^2-m^2+i\varepsilon)$$
$$= \int d^4p_r dM^2\delta'(\nabla_r\cdot\nabla_i/m^2)(p^0\gamma^0-(\mathbf{p}_r-\nabla_i)\cdot\gamma+m)\exp[-ip^0(x^0-y^0)+i\mathbf{p}_r\cdot(\mathbf{x}_r-\mathbf{y}_r)]\cdot$$
$$\cdot J(\mathbf{x}_i - \mathbf{y}_i, M^2)/(p_r^2-M^2 + i\varepsilon) \qquad (3.147)$$

where $p_r^2 = p^{0\,2} - \mathbf{p}_r\cdot\mathbf{p}_r$ and

$$J(\mathbf{x}_i - \mathbf{y}_i, M^2) = (2\pi)^{-3}\int d^3p_i\,\delta(M^2 + \mathbf{p}_i^2 - m^2)\exp[-i\mathbf{p}_i\cdot(\mathbf{x}_i - \mathbf{y}_i)] \qquad (3.148)$$

$$= (2\pi)^{-2}|\mathbf{x}_i - \mathbf{y}_i|^{-1}\theta(m^2 - M^2)\sin((m^2 - M^2)^{1/2}|\mathbf{x}_i - \mathbf{y}_i|)$$

The complexon Feynman propagator eq. 3.146 can be rearranged into the form of a spectral integral:

$$iS_C(x, y) = -\int dM\,(i\gamma^0\partial/\partial x^0 - i(\nabla_r - i\nabla_i)\cdot\gamma + m)\delta'(\nabla_r\cdot\nabla_i/m^2)\cdot$$
$$\cdot J(\mathbf{x}_i - \mathbf{y}_i, M^2)\Delta_F(x - y, M) \qquad (3.149)$$

where

$$\Delta_F(x - y, M) = (2\pi)^{-4}\int d^4p_r\exp[-ip^0(x^0 - y^0) + i\mathbf{p}_r\cdot(\mathbf{x}_r - \mathbf{y}_r)]/(p_r^2 - M^2 + i\varepsilon)$$
$$(3.150)$$

Case 4: Left-handed Tachyon Complexons

In this case $\hat{\mathbf{u}}_r \cdot \hat{\mathbf{u}}_i = 0$ again. However we add an imaginary term to ω to obtain a manifest Left-handed L_C boost[17]

$$\Lambda_{CL}(\mathbf{v}_c) = \exp[i(\omega + i\pi/2)\hat{\mathbf{w}}\cdot\mathbf{K}] \tag{3.151}$$

where ω remains

$$\omega = (\omega_r^{\,2} - \omega_i^{\,2})^{1/2} \tag{2.62}$$

and

$$\hat{\mathbf{w}} = (\omega_r \hat{\mathbf{u}}_r + i\omega_i \hat{\mathbf{u}}_i)/\omega \tag{2.63}$$
$$\hat{\mathbf{w}}\cdot\hat{\mathbf{w}} = \hat{\mathbf{u}}_r\cdot\hat{\mathbf{u}}_r = \hat{\mathbf{u}}_i\cdot\hat{\mathbf{u}}_i = 1 \tag{2.64a}$$
$$\mathbf{v}_c = \hat{\mathbf{w}}\tanh(\omega + i\pi/2) = \hat{\mathbf{w}}\,\text{cotanh}(\omega) \tag{3.152}$$

Letting $\omega_L = \omega + i\pi/2$ we find, as before,

$$\cosh(\omega_L) = i\sinh(\omega) = -\gamma = i\,\gamma_s \tag{2.11}$$
$$\sinh(\omega_L) = i\cosh(\omega) = -\beta\gamma = i\beta\gamma_s$$

with, $\beta = v_c = |\mathbf{v}_c| > 1$, $\gamma_s = (\beta^2 - 1)^{-1/2}$, and

$$\sinh(\omega) = \gamma_s \tag{2.12}$$
$$\cosh(\omega) = \beta\gamma_s$$

Thus we denote $\Lambda_{CL}(\mathbf{v}_c)$ by

$$\Lambda_{CL}(\mathbf{v}_c) \equiv \Lambda_{CL}(\omega, \hat{\mathbf{w}}) \tag{3.153}$$

The corresponding spinor boost transformation is:

$$S_{CL}(\Lambda_{CL}(\omega, \hat{\mathbf{w}})) = \exp(-i\omega_L\sigma_{0i}\hat{w}_i/2) = \exp(-\omega_L\gamma^0\boldsymbol{\gamma}\cdot\hat{\mathbf{w}}/2)$$
$$= \cosh(\omega_L/2)I + \sinh(\omega_L/2)\gamma^0\boldsymbol{\gamma}\cdot\hat{\mathbf{w}} \tag{3.154}$$

The momentum space equation generated by $S_{CL}(\Lambda_{CL}(\omega, \hat{\mathbf{w}}))$ is

$$\{m\cosh(\omega_L)\gamma^0 - m\sinh(\omega_L)\boldsymbol{\gamma}\cdot(\omega_r\hat{\mathbf{u}}_r + i\omega_i\hat{\mathbf{u}}_i)/\omega - m\}e^{+ip\cdot x}w_{cL}(p) = 0 \tag{3.155}$$

or

$$\{im\sinh(\omega)\gamma^0 - im\cosh(\omega)\boldsymbol{\gamma}\cdot(\omega_r\hat{\mathbf{u}}_r + i\omega_i\hat{\mathbf{u}}_i)/\omega - m\}e^{+ip\cdot x}w_{cL}(p) = 0 \tag{3.156}$$

---

[17] The reader can readily verify the form is consistent with an L_C boost transformation.

where $p{\cdot}x = Et - \mathbf{p{\cdot}x}$ after performing a corresponding left-handed superluminal coordinate transformation in the exponential factor based on eq. 2.10d. Thus the positive energy wave is transformed into a negative energy wave by the transformation.

The momentum 4-vector is defined by

$$p = (p^0, \mathbf{p}) \tag{3.157}$$

where

$$p^0 = m\sinh(\omega) \qquad \mathbf{p} = \mathbf{p_r} + i\mathbf{p_i} \tag{3.158}$$

with

$$\mathbf{p_r} = m\omega_r\hat{\mathbf{u}}_r\cosh(\omega)/\omega \qquad \mathbf{p_i} = m\omega_i\hat{\mathbf{u}}_i\cosh(\omega)/\omega \tag{3.159}$$

and

$$\mathbf{p_r}{\cdot}\mathbf{p_i} = 0 \tag{3.160}$$

then eq. 3.156 becomes the complexon tachyon equation

$$[ip{\cdot}\gamma - m]e^{+ip.x}w_{cL}(p) = 0 \tag{3.161}$$

with a complex 3-momentum $\mathbf{p}$ and the tachyon 4-momentum mass shell condition:[18]

$$p^2 = p^{0\,2} - \mathbf{p_r}^2 + \mathbf{p_i}^2 = -m^2 \tag{3.162}$$

Eq. 3.161 is the momentum space equivalent of the wave equation

$$[\gamma^0\partial/\partial t + \boldsymbol{\gamma}{\cdot}(\nabla_r + i\nabla_i) - m]\psi_{CL}(t, \mathbf{x_r}, \mathbf{x_i}) = 0 \tag{3.163a}$$

or

$$[\gamma{\cdot}\nabla - m]\psi_{CL}(t, \mathbf{x_r}, \mathbf{x_i}) = 0 \tag{3.163b}$$

with the subsidiary condition on the wave function

$$\nabla_r{\cdot}\nabla_i\,\psi_{CL}(t, \mathbf{x_r}, \mathbf{x_i}) = 0 \tag{3.164}$$

also holds. We note that eq. 3.163 is covariant under the real Lorentz group and eq. 3.164 can be easily put into (real Lorentz group) covariant form.

Before considering a lagrangian formulation and the Fourier operator representation of $\psi_{CL}(t, \mathbf{x_r}, \mathbf{x_i})$ we will define the tachyon spinors, and its associated real and imaginary spin operators.

The spinor generated from a spin up Dirac spinor at rest by the L_C spinor boost eqs. 3.154 is

---

[18] Note that the presence of the $\mathbf{p_i}^2$ term does not change the tachyon requirement that $\mathbf{p_r}^2 \geq m^2$ as seen in the previous cases.

$$w_{cL}(p) = S_{CL}w(0) = [\cosh(\omega_L/2)I + \sinh(\omega_L/2)\gamma^0\boldsymbol{\gamma}\cdot\hat{\mathbf{w}}]w(0) \qquad (3.165)$$

Following a procedure similar to Appendix 3-A (which the reader may wish to examine first) we define four spinors for Dirac particles at rest with eq. 3-A.2. Then by applying a boost to these rest spinors we find the L_C tachyon spinors:

$$S_{CL}w^k(0) = w_{cL}{}^k(p) \qquad (3.166)$$

and from these tachyon spinors we generalize to tachyon spinors $u_{CL}(p, s)$ and $v_{CL}(p, s)$ in a manner similar to eqs. 3.110a of the previous case.

Eqs. 3.161 through 3.164 imply that the wave function solution of eq. 3.161, subject to the subsidiary condition eq. 3.164, has the form

$$\psi_{CL}(x) = \sum_{\substack{\pm s \\ \mathbf{p}_r^2 \geq m^2}} \int d^3p_r d^3p_i\, N_{CL}(p)\delta(\mathbf{p}_r\cdot\mathbf{p}_i/m^2)[b_{CL}(p,s)u_{CL}(p, s)e^{-i(p\cdot x + p^*\cdot x^*)/2} + $$

$$+ d_{CL}{}^{\dagger}(p,s)v_{CL}(p, s)e^{+i(p\cdot x + p^*\cdot x^*)/2}] \qquad (3.167)$$

where $\mathbf{p} = \mathbf{p}_r + i\mathbf{p}_i$, $\mathbf{x} = \mathbf{x}_r - i\mathbf{x}_i$, $p\cdot x = p^0x^0 - \mathbf{p}\cdot\mathbf{x}$, and $b_{CL}(p, s)$ and $d_{CL}(p,s)$ are tachyon fourier coefficients.

Global SU(3) Symmetry

We can show that there is also a global SU(3) symmetry present here as shown in the previous case. The demonstration is similar to that of eqs. 3.113 – 3.126.

Light-Front Quantization of Tachyonic Complexons

Because of the momentum constraint $\mathbf{p}_r^2 \geq m^2$ the set of solutions of the form of eq. 3.167 is incomplete and the result of second quantization will not be an equal time anti-commutator expression consisting of derivatives of delta functions (eq. 3.142) but rather an analogue to previous unsuccessful attempts to create a second quantized tachyon theory.[19]

Therefore we will use light-front coordinates, and left and right handed field operators (as previously) to obtain a successful second quantization of this new type of tachyon.

The "missing" factor of i in the first term of eq. 3.163b requires the lagrangian to be different from the conventional Dirac lagrangian in order for the lagrangian to be real. The simplest, physically acceptable, free spin ½ tachyon lagrangian density for ψ_{CL} is:

---

[19] Such as G. Feinberg, Phys. Rev. **159**, 1089 (1967).

$$\mathcal{L}_{CL} = \psi_{CL}{}^{C}(x)(\gamma\cdot\nabla - m)\psi_{CL}(x) \tag{3.168}$$

where

$$\psi_{CL}{}^{C}(x) = [\psi_{CL}(x)]^{\dagger}|_{\mathbf{x}_i = -\mathbf{x}_i}\, i\gamma^0\gamma^5 \tag{3.169}$$

similar to eq. 3.128. In words, eq. 3.169 states take the hermitean conjugate of $\psi_{CL}(x)$; change $\mathbf{x}_i$ to $-\mathbf{x}_i$; and then post-multiply by the indicated factors.

The free complexon invariant action (under real Lorentz transformations) is

$$I = \int d^7x\,\mathcal{L}_{CL} \tag{3.170}$$

The action can be shown to be real

$$I^* = I \tag{3.171}$$

in a manner similar to the case considered in Appendix 3-A. The tachyonic complexon's energy-momentum tensor is

$$\mathfrak{T}_{CL\mu\nu} = -\, g_{\mu\nu}\,\mathcal{L}_{CL} + \partial\mathcal{L}_{CL}/\partial(D^{\mu}\psi_{CL})\, D_{\nu}\psi_{CL} \tag{3.172}$$

$$= i\psi_{CL}{}^{C}\gamma^0\gamma^5\gamma_{\mu}D_{\nu}\psi_{CL}$$

where

$$D_0 = \partial/\partial x^0$$
$$D_k = \partial/\partial x^k + i\,\partial/\partial x_i{}^k \tag{3.129}$$

and thus the conserved energy and momentum are

$$P^0 = H = \int d^3x_r d^3x_i\, \mathfrak{T}_{CL}{}^{00} = i\int d^3x_r d^3x_i \psi_{CL}{}^{C}\gamma^5(\boldsymbol{\alpha}\cdot\mathbf{D} + \beta m)\psi_{CL} \tag{3.173}$$

and

$$P^k = \int d^3x_r d^3x_i\, \mathfrak{T}_{CL}{}^{0k} = -\, i\int d^3x_r d^3x_i\, \psi_{CL}{}^{C}\gamma^5 D^k\psi_{CL} \tag{3.174}$$

Having defined a suitable tachyon lagrangian we can now proceed to its canonical quantization. The conjugate momentum can be calculated from the lagrangian density eq. 3.168:

$$\pi_{CLa} = \partial\mathcal{L}_{CL}/\partial\dot{\psi}_{CLa} \equiv \partial\mathcal{L}_{CL}/\partial(\partial\psi_{CLa}/\partial t) = -i([\psi_{CL}(x)]^{\dagger}|_{\mathbf{x}_i = -\mathbf{x}_i}\gamma^5)_a \tag{3.175}$$

The resulting non-zero, canonical anti-commutation relations are presumably

$$\{\pi_{CLa}(x),\, \psi_{CLb}(y)\} = i\,\delta_{ab}\,\delta^3(x_r - y_r)\delta^3(x_i - y_i)$$

based on locality in both real and imaginary coordinates, or

$$\{\psi_{CL\ a}^{\dagger}(x)|_{\mathbf{X_i} = -\mathbf{X_i}'}, \psi_{Tb}(y)\} = -[\gamma^5]_{ab}\,\delta^3(x_r - y_r)\delta^3(x_i - y_i) \qquad (3.176)$$

At this point we might attempt to complete the canonical quantization procedure in the conventional manner by fourier expanding the field and specifying anti-commutation relations for the fourier component amplitudes. However the incompleteness of the set of plane waves, which are limited by the restriction $\mathbf{p_r}^2 \geq m^2$, causes the equal time anti-commutator of the fields *not* to yield a δ-functions.

Therefore we turn to the previous successful approach to tachyon quantization[20] and decompose the tachyonic complexon field into left-handed and right-handed parts and then second quantize in light-front coordinates.

Separation into Left-Handed and Right-Handed Fields

As before we will use a transformed set of Dirac matrices (eq. 3.32) to develop our left-handed and right-handed tachyon formulations. The γ^5 chirality operator's eigenvalues define handedness: +1 corresponds to right-handed; and –1 corresponds to left-handed:

$$\gamma^5\psi_{CLL} = -\psi_{CLL} \qquad\qquad \gamma^5\psi_{CLR} = \psi_{CLR} \qquad (3.177)$$

As before we define left-handed and right-handed tachyon fields with the projection operators:

$$\begin{aligned} C^{\pm} &= \tfrac{1}{2}(I \pm \gamma^5) \\ C^{+} + C^{-} &= I \\ C^{\pm\,2} &= C^{\pm} \\ C^{+}C^{-} &= 0 \end{aligned} \qquad (3.34)$$

with the result

$$\begin{aligned} \psi_{CLL} &= C^{-}\psi_{CL} \\ \psi_{CLR} &= C^{+}\psi_{CL} \end{aligned} \qquad (3.178)$$

We can calculate the commutation relations of the left-handed and right-handed tachyonic complexon fields from eq. 3.176 by pre-multiplying and post-multiplying by $\tfrac{1}{2}(1-\gamma^5)$ and $\tfrac{1}{2}(1+\gamma^5)$. The results are:

$$\{\psi_{CLLa}^{\dagger}(x)|_{\mathbf{X_i} = -\mathbf{X_i}'}, \psi_{CLLb}(y)\} = C^{-}_{ab}\,\delta^6(x - y) \qquad (3.179)$$

---

[20] Blaha (2006) discusses this case in detail.

$$\{\psi_{CLRa}{}^\dagger(x)|_{\mathbf{x_i} = -\mathbf{x_i}}, \psi_{CLRb}(y)\} = -C^+{}_{ab}\,\delta^6(x - y) \qquad (3.180)$$

$$\{\psi_{CLLa}{}^\dagger(x)|_{\mathbf{x_i} = -\mathbf{x_i}}, \psi_{CLRb}(y)\} = \{\psi_{CLRa}{}^\dagger(x)|_{\mathbf{x_i} = -\mathbf{x_i}}, \psi_{CLLb}(x')\} = 0 \qquad (3.181)$$

where

$$\delta^6(x - y) = \delta^3(x_r - y_r)\delta^3(x_i - y_i) \qquad (3.182)$$

The lagrangian density of eq. 3.168 decomposes into left-handed and right-handed parts: (The change $\mathbf{x_i}$ to $-\mathbf{x_i}$ will be understood in $\psi_{CLL}{}^\dagger(x)$ and $\psi_{CLR}{}^\dagger(x)$ in the following.)

$$\mathcal{L}_{CL} = \psi_{CLL}{}^\dagger\gamma^0 i\gamma^\mu\partial_\mu\psi_{CLL} - \psi_{CLR}{}^\dagger\gamma^0 i\gamma^\mu\partial_\mu\psi_{CLR} - im[\psi_{CLR}{}^\dagger\gamma^0\psi_{CLL} - \psi_{CLL}{}^\dagger\gamma^0\psi_{CLR}] \qquad (3.183)$$

Further Separation into + and – Light-Front Complexon Fields

As previously we now use light-front coordinates and quantization to obtain a successful second quantization of this form of tachyon field. Light-front variables, in the present case where we have to contend with complex 3-vectors, are defined by the real coordinates and derivatives:

$$x^\pm = (x^0 \pm x_r{}^3)/\sqrt{2}$$
$$\partial/\partial x^\pm \equiv \partial^\mp \equiv (\partial/\partial x^0 \pm \partial/\partial x_r{}^3)/\sqrt{2} \qquad (3.184)$$

with the "transverse" real coordinate variables, $x_r{}^1$ and $x_r{}^2$, and imaginary coordinate variables $x_i{}^1$, $x_i{}^2$, and $x_i{}^3$.

The inner product of two 4-vectors has the form

$$x\cdot y = x^+y^- + y^+x^- + i[y_i{}^3(x^+ - x^-) + x_i{}^3(y^+ - y^-)]/\sqrt{2} + x_i{}^3y_i{}^3 - (\mathbf{x_{r\perp}} - i\mathbf{x_{i\perp}})\cdot(\mathbf{y_{r\perp}} - i\mathbf{y_{i\perp}}) \qquad (3.185)$$

with

$$\mathbf{x_{r\perp}} = (x_r{}^1, x_r{}^2) \qquad \mathbf{x_{i\perp}} = (x_i{}^1, x_i{}^2) \qquad (3.186)$$
$$\mathbf{y_{r\perp}} = (y_r{}^1, y_r{}^2) \qquad \mathbf{y_{i\perp}} = (y_i{}^1, y_i{}^2)$$

where $x = (x^0, \mathbf{x} = \mathbf{x_r} - i\mathbf{x_i})$ and $y = (y^0, \mathbf{y} = \mathbf{y_r} - i\mathbf{y_i})$. Momenta are always defined as $p = (p^0, \mathbf{p} = \mathbf{p_r} + i\mathbf{p_i})$.

The light-front definition of Dirac matrices is:

$$\gamma^\pm = (\gamma^0 \pm \gamma^3)/\sqrt{2} \qquad (3.42)$$

with transverse matrices γ^1 and γ^2 defined as usual. Note:

$$\gamma^{\pm 2} = 0$$

We define "+" and "–" tachyon fields with the projection operators:

$$R^{\pm} = ½(I \pm \gamma^0\gamma^3) \qquad (3.43)$$

$$\text{Left-handed, } \pm \text{ light-front fields:} \quad \psi_{CLL}^{\pm} = R^{\pm}C^{-}\psi_{CL} \qquad (3.187)$$
$$\text{Right-handed, } \pm \text{ light-front fields:} \quad \psi_{CLR}^{\pm} = R^{\pm}C^{+}\psi_{CL}$$

Transforming to light-front variables and fields as above we obtain the light-front free tachyon lagrangian:

$$\begin{aligned}\mathcal{L}_{CL} = {} & 2^{½}\psi_{CLL}^{+\dagger}i\partial^{-}\psi_{CLL}^{+} + 2^{½}\psi_{CLL}^{-\dagger}i\partial^{+}\psi_{CLL}^{-} - \psi_{CLL}^{+\dagger}\gamma^0[i\gamma_\perp\cdot\nabla_{r\perp} - \gamma\cdot\nabla_i]\psi_{CLL}^{-} - \\ & - \psi_{CLL}^{-\dagger}\gamma^0[i\gamma_\perp\cdot\nabla_{r\perp} - \gamma\cdot\nabla_i]\psi_{CLL}^{+} - 2^{½}\psi_{CLR}^{+\dagger}i\partial^{-}\psi_{CLR}^{+} - 2^{½}\psi_{CLR}^{-\dagger}i\partial^{+}\psi_{CLR}^{-} + \\ & + \psi_{CLR}^{+\dagger}\gamma^0[i\gamma_\perp\cdot\nabla_{r\perp} - \gamma\cdot\nabla_i]\psi_{CLR}^{-} + \psi_{CLR}^{-\dagger}\gamma^0[i\gamma_\perp\cdot\nabla_{r\perp} - \gamma\cdot\nabla_i]\psi_{CLR}^{+} - \\ & - im[\psi_{CLR}^{+\dagger}\gamma^0\psi_{CLL}^{-} - \psi_{CLL}^{+\dagger}\gamma^0\psi_{CLR}^{-} + \psi_{CLR}^{-\dagger}\gamma^0\psi_{CLL}^{+} - \psi_{CLL}^{-\dagger}\gamma^0\psi_{CLR}^{+}]\end{aligned} \qquad (3.188)$$

(Note the similarity to eq. 3.45 in the previous tachyon case.) Again the difference in signs between the left-handed and right-handed terms will be a crucial factor in the derivation of the left-handed features of the Standard Model.

Eq. 3.188 generates the equations of motion:

$$\begin{aligned} 2^{½}i\partial^{-}\psi_{CLL}^{+} - \gamma^0[i\gamma_\perp\cdot\nabla_{r\perp} - \gamma\cdot\nabla_i]\psi_{CLL}^{-} + im\gamma^0\psi_{CLR}^{-} &= 0 \\ 2^{½}i\partial^{-}\psi_{CLR}^{+} - \gamma^0[i\gamma_\perp\cdot\nabla_{r\perp} - \gamma\cdot\nabla_i]\psi_{CLR}^{-} + im\gamma^0\psi_{CLL}^{-} &= 0 \\ 2^{½}i\partial^{+}\psi_{CLL}^{-} - \gamma^0[i\gamma_\perp\cdot\nabla_{r\perp} - \gamma\cdot\nabla_i]\psi_{CLL}^{+} + im\gamma^0\psi_{CLR}^{+} &= 0 \\ 2^{½}i\partial^{+}\psi_{CLR}^{-} - \gamma^0[i\gamma_\perp\cdot\nabla_{r\perp} - \gamma\cdot\nabla_i]\psi_{CLR}^{+} + im\gamma^0\psi_{CLL}^{+} &= 0 \end{aligned} \qquad (3.189)$$

Eqs. 3.189 show that ψ_{CLL}^{-} and ψ_{CLR}^{-} are dependent fields that are functions of ψ_{CLL}^{+} and ψ_{CLR}^{+} on the light-front where x^{+} equals a constant. They can be expressed in an integral form as well. (The independent fields ψ_{CLL}^{+} and ψ_{CLR}^{+} play a fundamental role in tachyonic complexon theory and are used to define "in" and "out" tachyon states in perturbation theory.)

The conjugate momenta implied by eq. 3.188 are

$$\begin{aligned}\pi_{CLL}^{+} &= \partial\mathcal{L}/\partial(\partial^{-}\psi_{CLL}^{+}) = 2^{½}i\psi_{CLL}^{+\dagger} \\ \pi_{CLL}^{-} &= \partial\mathcal{L}/\partial(\partial^{-}\psi_{CLL}^{-}) = 0\end{aligned} \qquad (3.190)$$

$$\pi_{CLR}{}^{+} = \partial\mathcal{L}/\partial(\partial^{-}\psi_{CLR}{}^{+}) = -2^{1/2}i\psi_{CLR}{}^{+\dagger} \qquad (3.191)$$
$$\pi_{CLR}{}^{-} = \partial\mathcal{L}/\partial(\partial^{-}\psi_{CLR}{}^{-}) = 0$$

x^{+} plays the role of the "time" variable in light-front quantized theories. So we define canonical equal x^{+} anti-commutation relations for spin ½ tachyonic complexons also.

The canonical equal-light-front ($x^{+} = y^{+}$) anti-commutation relations of the independent fields would normally be:

$$\{\psi_{CLL}{}^{+\dagger}{}_{a}(x),\psi_{CLL}{}^{+}{}_{b}(y)\} = 2^{-1}[C^{-}R^{+}]_{ab}\delta(x^{-}-y^{-})\delta^{2}(x_{r}-y_{r})\delta^{3}(x_{I}-y_{i}) \qquad (3.192)$$

$$\{\psi_{CLR}{}^{+\dagger}{}_{a}(x), \psi_{CLR}{}^{+}{}_{b}(y)\} = -2^{-1}[C^{+}R^{+}]_{ab}\,\delta(x^{-}-y^{-})\delta^{2}(x_{r}-y_{r})\delta^{3}(x_{I}-y_{i}) \quad (3.193)$$

$$\{\psi_{CLL}{}^{+}{}_{a}{}^{\dagger}(x), \psi_{CLR}{}^{+}{}_{b}(y)\} = \{\psi_{CLR}{}^{+}{}_{a}{}^{\dagger}(x), \psi_{CLL}{}^{+}{}_{b}(y)\} = 0 \qquad (3.194)$$
$$\{\psi_{CLL}{}^{+}{}_{a}(x), \psi_{CLR}{}^{+}{}_{b}(y)\} = \{\psi_{CLR}{}^{+}{}_{a}{}^{\dagger}(x), \psi_{CLL}{}^{+\dagger}{}_{b}(y)\} = 0 \qquad (3.195)$$

But as in the previous case they will be modified.

Again we see that the right-handed tachyon anti-commutation relation (eq. 3.193) has a minus sign relative to the corresponding conventional right-handed anti-commutation relation (eq. 3.55).

The sign differences between the left-handed and right-handed lagrangian terms ultimately lead to parity violating features in the Standard Model lagrangian.

Left-Handed Tachyonic Complexons

The free, "+" light-front, left-handed tachyonic complexon Fourier expansion is:

$$\psi_{CLL}{}^{+}(x_{r}, x_{i}) = \sum_{\pm s}\int d^{2}p_{r}dp^{+}d^{3}p_{i}\, N_{CLL}{}^{+}(p)\theta(p^{+})\delta((p_{i}{}^{3}(p^{+}-p^{-})/\sqrt{2} + \mathbf{p}_{r\perp}\cdot\mathbf{p}_{i\perp})/m^{2})\cdot$$
$$\cdot[b_{CLL}{}^{+}(p, s)u_{CLL}{}^{+}(p, s)e^{-i(p\cdot x + p^{*}\cdot x^{*})/2} + d_{CLL}{}^{+\dagger}(p, s)v_{CLL}{}^{+}(p, s)e^{+i(p\cdot x + p^{*}\cdot x^{*})/2}] \qquad (3.196)$$

(Compare to eq. 3.109.) Its hermitean conjugate is

$$\psi_{CLL}{}^{+\dagger}(x_{r}, x_{i}) = \sum_{\pm s}\int d^{2}p_{r}dp^{+}d^{3}p_{i}\, N_{CLL}{}^{+}(p)\theta(p^{+})\delta((p_{i}{}^{3}(p^{+}-p^{-})/\sqrt{2} + \mathbf{p}_{r\perp}\cdot\mathbf{p}_{i\perp})/m^{2})\cdot$$
$$\cdot[b_{CLL}{}^{\dagger}(p^{*},s)u_{CLL}{}^{\dagger}(p^{*},s)e^{+i(p^{*}\cdot x + p\cdot x^{*})/2} + d_{CLL}(p^{*},s)v_{CLL}{}^{\dagger}(p^{*},s)e^{-i(p^{*}\cdot x + p\cdot x^{*})/2}] \qquad (3.197)$$

where $\mathbf{p} = \mathbf{p_r} + i\mathbf{p_i}$ (eq. 3.95), $\mathbf{x} = \mathbf{x_r} - i\mathbf{x_i}$, $p\cdot x = p^0x^0 - \mathbf{p}\cdot\mathbf{x}$, and † indicates hermitean conjugate. (Compare to eq. 3.110.) The spinors are

$$\begin{aligned}
u_{CLL}{}^{+}(p, s) &= C^- R^+ S_{CL}w^1(0)\\
u_{CLL}{}^{+}(p, -s) &= C^- R^+ S_{CL}w^2(0)\\
v_{CLL}{}^{+}(p, s) &= C^- R^+ S_{CL}w^3(0)\\
v_{CLL}{}^{+}(p, -s) &= C^- R^+ S_{CL}w^4(0)\\
u_{CLL}{}^{+\dagger}(p^*, s) &= w^{1T}(0)S_{CL}R^+C^-\\
u_{CLL}{}^{+\dagger}(p^*, -s) &= w^{2T}(0)S_{CL}R^+C^-\\
v_{CLL}{}^{+\dagger}(p^*, s) &= w^{3T}(0)S_{CL}R^+C^-\\
v_{CLL}{}^{+\dagger}(p^*, -s) &= w^{4T}(0)S_{CL}R^+C^-
\end{aligned} \tag{3.198}$$

using eq. 3.166 where the superscript "T" indicates the transpose (These spinors are described in Appendix 3-A.) and

$$N_{CLL}{}^{+}(p) = (2\pi)^{-3}(2m/p^+)^{1/2} \tag{3.199}$$

The anti-commutation relations of the Fourier coefficient operators are

$$\begin{aligned}
\{b_{CLL}(p,s), b_{CLL}{}^{\dagger}(p'^*,s')\} &= 2^{-1/2}\delta_{ss'}\delta(p^+ - p'^+)\delta^2(\mathbf{p_r} - \mathbf{p'_{r'}})\delta^3(\mathbf{p_i} + \mathbf{p'_{i'}})\\
\{d_{CLL}(p,s), d_{CLL}{}^{\dagger}(p'^*,s')\} &= 2^{-1/2}\delta_{ss'}\delta(p^+ - p'^+)\delta^2(\mathbf{p_r} - \mathbf{p'_{r'}})\delta^3(\mathbf{p_i} + \mathbf{p'_{i'}})\\
\{b_{CLL}(p,s), b_{CLL}(p'^*,s')\} &= \{d_{CLL}(p,s), d_{CLL}(p'^*,s')\} = 0\\
\{b_{CLL}{}^{\dagger}(p,s), b_{CLL}{}^{\dagger}(p'^*,s')\} &= \{d_{CLL}{}^{\dagger}(p,s), d_{CLL}{}^{\dagger}(p'^*,s')\} = 0\\
\{b_{CLL}(p,s), d_{CLL}{}^{\dagger}(p'^*,s')\} &= \{d_{CLL}(p,s), b_{CLL}{}^{\dagger}(p'^*,s')\} = 0\\
\{b_{CLL}{}^{\dagger}(p,s), d_{CLL}{}^{\dagger}(p'^*,s')\} &= \{d_{CLL}(p,s), b_{CLL}(p'^*,s')\} = 0
\end{aligned} \tag{3.200}$$

The delta-function arguments $\delta^3(\mathbf{p_i} + \mathbf{p'_{i'}})$ above have a positive sign as in eq. 3.138 in order to obtain $\delta^3(\mathbf{x_i} - \mathbf{y_i})$ in the field anti-commutators.

The spinors, eq. 3.198, satisfy

$$\begin{aligned}
\sum_{\pm s} u_{CLL}{}^{+}{}_{\alpha}(p, s)\bar{u}_{CLL}{}^{+}{}_{\beta}(p^*, s) &= (2m)^{-1}[C^-R^+(i\not{p} + m)R^-C^+]_{\alpha\beta}\\
\sum_{\pm s} v_{CLL}{}^{+}{}_{\alpha}(p, s)\bar{v}_{CLL}{}^{+}{}_{\beta}(p^*, s) &= (2m)^{-1}[C^-R^+(i\not{p} - m)R^-C^+]_{\alpha\beta}
\end{aligned} \tag{3.201}$$

where $\bar{u}_{CLL}{}^{+} = u_{CLL}{}^{+\dagger}\gamma^0$ and $\bar{v}_{CLL}{}^{+} = v_{CLL}{}^{+\dagger}\gamma^0$.

We now evaluate the canonical left-handed, light-front anti-commutation relation (eq. 3.192):

$$\{\psi_{CLL}{}^{+}{}_{a}(x), \psi_{CLL}{}^{+\dagger}{}_{b}(y)\} = \sum_{\pm s,s'} \int d^3p_i d^2p dp^+ \int d^3p_i' d^2p' dp'^+ N_{CLL}{}^{+}(p)\, N_{CLL}{}^{+}(p')\cdot$$

$$\cdot\theta(p^+)\theta(p'^+)\delta((p_i{}^3(p^+ - p^-)/\surd 2 + \mathbf{p}_{r\perp}\cdot\mathbf{p}_{i\perp})/m^2)\,\delta((p_i'^3(p'^+ - p'^-)/\surd 2 + \mathbf{p'}_{r\perp}\cdot\mathbf{p'}_{i\perp})/m^2)\cdot$$

$$\cdot[\{b_{CLL}{}^{+\dagger}(p'^*,s'), b_{CLL}{}^{+}(p,s)\}u_{CLL}{}^{+}{}_{a}(p,s)u_{CLL}{}^{+\dagger}{}_{b}(p'^*,s')e^{+i(p'^*\cdot y + p'\cdot y^*)2 - i(p\cdot x + p^*\cdot x^*)/2} +$$

$$+\{d_{CLL}{}^{+}(p'^*,s'), d_{CLL}{}^{+\dagger}(p,s)\}v_{CLL}{}^{+}{}_{a}(p,s)v_{CLL}{}^{+\dagger}{}_{b}(p'^*,s')e^{-i(p'^*\cdot y + p'\cdot y^*)/2 + i(p\cdot x + p^*\cdot x^*)/2}]$$

$$= 2^{-\frac{1}{2}}\sum_{\pm s}\int d^3p_i d^2p_r dp^+ [N_{CLL}{}^{+}(p)]^2\theta(p^+)[\delta((p_i{}^3(p^+ - p^-)/\surd 2 + \mathbf{p}_{r\perp}\cdot\mathbf{p}_{i\perp})/m^2)]^2$$

$$[u_{CLL}{}^{+}{}_{a}(p,s)u_{CLL}{}^{+\dagger}{}_{b}(p^*,s)e^{+i(p^*\cdot(y - x) + p\cdot(y^* - x^*))/2} +$$

$$+ v_{CLL}{}^{+}{}_{a}(p,s)v_{CLL}{}^{+\dagger}{}_{b}(p^*,s)e^{-i(p^*\cdot(y - x) + p\cdot(y^* - x^*))/2}]$$

and, using eq. 3.141,

$$= -2^{-3/2}\int d^3p_i d^2p dp^+\theta(p^+)[N_{CLL}{}^{+}(p)]^2\delta'((p_i{}^3(p^+ - p^-)/\surd 2 + \mathbf{p}_{r\perp}\cdot\mathbf{p}_{i\perp})/m^2)(2m)^{-1}$$

$$\{[\,C^-R^+(i\not{p} + m)\gamma^0R^+C^-]_{ab}e^{+i(p^*\cdot(y - x) + p\cdot(y^* - x^*))/2} +$$

$$+ [C^-R^+(i\not{p} - m)\gamma^0R^+C^-]_{ab}e^{-i(p^*\cdot(y - x) + p\cdot(y^* - x^*))/2}\}$$

$$= -(1/2)C^-R^+\delta_{ab}\int d^3p_i\, d^2p_\perp \int_0^\infty dp^+\,\delta'((p_i{}^3(p^+ - p^-)/\surd 2 + \mathbf{p}_\perp\cdot\mathbf{p}_{i\perp})/m^2)(2\pi)^{-6}\cdot$$

$$\cdot\{\, e^{+i\{p^+(y^- - x^-) - \mathbf{p}_{r\perp}\cdot(\mathbf{y}_{r\perp} - \mathbf{x}_{r\perp}) + \mathbf{p}_i\cdot(\mathbf{y}_i - \mathbf{x}_i)\}} + e^{-i\{p^+(y^- - x^-) - \mathbf{p}_{r\perp}\cdot(\mathbf{y}_{r\perp} - \mathbf{x}_{r\perp}) + \mathbf{p}_i\cdot(\mathbf{y}_i - \mathbf{x}_i)\}}\}$$

$$= -C^-R^+\delta_{ab}(4\pi)^{-1}\int_0^\infty dp^+\delta'(\nabla_r\cdot\nabla_i/m^2)\delta^3(y_i - x_i)\,\delta^2(y_r - x_r)\{e^{+ip^+(y^- - x^-)} + e^{-ip^+(y^- - x^-)}\}$$

whereupon we revert back to the original form of the constraint: $\delta(\nabla_r\cdot\nabla_i/m^2)$

$$\{\psi_{CLL}{}^{+}{}_{a}(x), \psi_{CLL}{}^{+\dagger}{}_{b}(y)\} = -(1/2)C^-R^+\delta_{ab}\,\delta'(\nabla_r\cdot\nabla_i/m^2)\delta(y^- - x^-)\delta^2(y_r - x_r)\delta^3(y_i - x_i) \qquad (3.202)$$

The result is the left-handed, light-front equivalent of the earlier non-tachyon result eq. 3.142. Again the constraint is apparent in the anti-commutator. (The factor of 2 difference is due to light-front coordinate definitions.)

Therefore we have left-handed, light-front quantized tachyonic complexons with the equivalent of canonical anti-commutation relations, and with localized tachyonic complexons. As a result we have a canonical tachyonic complexon Quantum Field Theory

Left-handed Case 4 Tachyonic Complexon Feynman Propagator

The light-front Feynman propagator for the left-handed $\psi_{CLL}{}^{+}$ *tachyonic* complexon field is

$$iS^{+}{}_{CLLF}(x,y) = \theta(x^{+} - y^{+})\langle 0|\psi_{CLL}{}^{+}(x)\psi_{CLL}{}^{++}(y)\gamma^{0}|0\rangle - \\ - \theta(y^{+} - x^{+})\langle 0|\psi_{CLL}{}^{++}(y)\gamma^{0}\psi_{CLL}{}^{+}(x)|0\rangle \tag{3.203}$$

$$= -\tfrac{1}{2}\int d^3p_i d^2p_r dp^{+}\theta(p^{+})N_{CLL}{}^{+2}\delta'((p_i{}^{3}(p^{+} - p^{-})/\surd 2 + \mathbf{p}_{r\perp}\cdot\mathbf{p}_{i\perp})/m^2)(2m)^{-1}C^{-}R^{+} \\ \{\theta(x^{+} - y^{+})[(i\not{p} + m)\gamma^{0}]e^{+i(p^{*}\cdot(y-x)+p\cdot(y^{*}-x^{*}))/2} + \\ + \theta(y^{+} - x^{+})[(i\not{p} - m)\gamma^{0}]e^{-i(p^{*}\cdot(y-x)+p\cdot(y^{*}-x^{*}))/2}\}R^{+}C^{-}\gamma^{0}$$

If we define the on-shell momentum variable $p_0{}^{-} = (p_{r0}{}^{1}p_{r0}{}^{1} + p_{r0}{}^{2}p_{r0}{}^{2} - \mathbf{p}_{i0}\cdot\mathbf{p}_{i0} - m^2)/(2p_0{}^{+})$, $p_0{}^{+} = p^{+}$, $p_{r0}{}^{j} = p_r{}^{j}$ (for j = 1, 2), $\mathbf{p}_{i0} = \mathbf{p}_i$, $p_{r\perp 0}{}^{2} = p_{r0}{}^{j}p_{r0}{}^{j}$ and $\not{p}_0 = p_0\cdot\gamma$ with $p_0 = (p^0, \mathbf{p}_{r0} + i\mathbf{p}_{r0})$ then the above equation can be rewritten as

$$= -\tfrac{1}{2}C^{-}R^{+}\int d^4p d^3p_i N_{CLL}{}^{+2}\delta'((p_{i0}{}^{3}(p_0{}^{+} - p_0{}^{-})/\surd 2 + \mathbf{p}_{r\perp 0}\cdot\mathbf{p}_{i\perp 0})/m^2)(4\pi m)^{-1}e^{+i(p^{*}\cdot(y-x)+p\cdot(y^{*}-x^{*}))/2}$$

$$\cdot\{\theta(p^{+})(i\not{p}_0 + m)\gamma^{0}]/[p^{-} - p_0{}^{-} + i\varepsilon] + \theta(-p^{+})(i\not{p}_0 - m)\gamma^{0}]/[p^{-} + p_0{}^{-} - i\varepsilon]\}R^{+}C\gamma^{0}$$

$$= -\tfrac{1}{2}\int d^4p_r d^3p_i\, N_{CLL}{}^{+2}\delta'((p_{i0}{}^{3}(p^{+} - p^{-})/\surd 2 + \mathbf{p}_{r\perp}\cdot\mathbf{p}_{i\perp})/m^2)(p^{+}/4\pi m)\, e^{+i(p^{*}\cdot(y-x)+p\cdot(y^{*}-x^{*}))/2}\,. \\ \cdot[C^{-}R^{+}(i\not{p} + m)\gamma^{0}R^{+}C^{-}\gamma^{0}][(p^2 + m^2 + i\varepsilon)]^{-1}$$

with $p_r = (p^0, \mathbf{p}_r)$ and $p = (p^0, \mathbf{p}_r + i\mathbf{p}_r)$. Substituting for N_{CLL} and using $x\delta'(x) = -\delta(x)$ we obtain

$$= -\tfrac{1}{2}\int d^4p_r d^3p_i(2\pi)^{-7}\delta'(\mathbf{p}_r\cdot\mathbf{p}_i/m^2)\exp[ip^0(y^0 - x^0) - i\mathbf{p}_r\cdot(\mathbf{y}_r - \mathbf{x}_r) + i\mathbf{p}_i\cdot(\mathbf{y}_i - \mathbf{x}_i)]\cdot \\ \cdot[C^{-}R^{+}(i\not{p} + m)R^{-}C^{+}]/(p^2 + m^2 + i\varepsilon)$$

since $C^{-}R^{+}(i\not{p} + m)\gamma^{0}R^{+}C^{-}\gamma^{0} = C^{-}R^{+}(i\not{p} + m)R^{-}C^{+}$. The integral can be written:

$$I = \int d^4p_r d^3p_i\delta'(\mathbf{p}_r\cdot\mathbf{p}_i/m^2)C^{-}R^{+}(i\not{p} + m)R^{-}C^{+}\cdot \\ \cdot\exp[-ip^0(x^0 - y^0) + i\mathbf{p}_r\cdot(\mathbf{x}_r - \mathbf{y}_r) - i\mathbf{p}_i\cdot(\mathbf{x}_i - \mathbf{y}_i)]/(p^2 + m^2 + i\varepsilon)$$

$$= \int d^4p_r dM^2\delta'(\nabla_r\cdot\nabla_i/m^2)C^{-}R^{+}(ip^{0}\gamma^{0} - (\nabla_r - i\nabla_i)\cdot\gamma + m)R^{-}C^{+}\cdot \\ \cdot\exp[-ip^0(x^0 - y^0) + i\mathbf{p}_r\cdot(\mathbf{x}_r - \mathbf{y}_r)]J_2(\mathbf{x}_i - \mathbf{y}_i, M^2)/(p_r{}^2 + M^2 + i\varepsilon)$$

where

$$J_2(\mathbf{x}_i - \mathbf{y}_i, M^2) = (2\pi)^{-3}\int d^3p_i\delta(M^2 - \mathbf{p}_i{}^2 - m^2)\exp[-i\mathbf{p}_i\cdot(\mathbf{x}_i - \mathbf{y}_i)] \tag{3.204}$$

$$= (2\pi)^{-2}|\mathbf{x_i} - \mathbf{y_i}|^{-1}\theta(M^2 - m^2)\sin((M^2 - m^2)^{1/2}|\mathbf{x_i} - \mathbf{y_i}|)$$

This tachyonic complexon Feynman propagator can be rearranged into the form of a spectral integral:

$$iS^+_{CLLF}(x, y) = -\int dM\, C^-R^+(\gamma^0\partial/\partial x^0 + (\nabla_r - i\nabla_i)\cdot\gamma - m)R^-C^+\delta'(\nabla_r\cdot\nabla_i/m^2)\cdot$$
$$\cdot J_2(\mathbf{x_i} - \mathbf{y_i}, M^2)\Delta_{FT}(x - y, M) \qquad (3.205)$$

with ∇_r and ∇_i derivatives with respect to $\mathbf{x_r}$ and $\mathbf{x_i}$ and where

$$\Delta_{FT}(x - y, M) = (2\pi)^{-4}\int d^4p_r \exp[-ip^0(x^0 - y^0) + i\mathbf{p_r}\cdot(\mathbf{x_r} - \mathbf{y_r})]/(p_r^2 + M^2 + i\varepsilon) \qquad (3.206)$$

Right-Handed Tachyonic Complexons

The case of right-handed tachyonic complexons is similar to left-handed complexons with only one difference: a minus sign in the canonical right-handed equal-time commutation relations resulting in a minus sign in the creation and annihilation operator anti-commutation relations. The right-handed tachyonic complexon wave function (eq. 3.187) light-front Fourier expansion is:

$$\psi_{CLR}^+(x_r, x_i) = \sum_{\pm s}\int d^2p_r dp^+ d^3p_i\, N_{CLR}^+(p)\theta(p^+)\delta((p_i^3(p^+ - p^-)/\surd 2 + \mathbf{p_{r\perp}}\cdot\mathbf{p_{i\perp}})/m^2)\cdot$$
$$\cdot[b_{CLR}^+(p, s)u_{CLR}^+(p, s)e^{-i(p\cdot x + p^*\cdot x^*)/2} + d_{CLR}^{+\dagger}(p, s)v_{CLR}^+(p, s)e^{+i(p\cdot x + p^*\cdot x^*)/2}] \qquad (3.207)$$

where

$$N_{CLR}^+(p) = (2\pi)^{-3}(2m/p^+)^{1/2} \qquad (3.208)$$

Its hermitean conjugate is

$$\psi_{CLR}^{+\dagger}(x_r, x_i) = \sum_{\pm s}\int d^2p_r dp^+ d^3p_i\, N_{CLR}^+(p)\theta(p^+)\delta((p_i^3(p^+ - p^-)/\surd 2 + \mathbf{p_{r\perp}}\cdot\mathbf{p_{i\perp}})/m^2)\cdot$$
$$\cdot[b_{CLR}^\dagger(p^*,s)u_{CLR}^\dagger(p^*,s)e^{+i(p^*\cdot x + p\cdot x^*)/2} + d_{CLR}(p^*,s)v_{CLR}^\dagger(p^*,s)e^{-i(p^*\cdot x + p\cdot x^*)/2}] \qquad (3.209)$$

where $\mathbf{p} = \mathbf{p_r} + i\mathbf{p_i}$ (eq. 3.95), $\mathbf{x} = \mathbf{x_r} - i\mathbf{x_i}$, $p\cdot x = p^0x^0 - \mathbf{p}\cdot\mathbf{x}$, and † indicates hermitean conjugate. (Compare to eq. 3.110.) The right-handed spinors are

$$u_{CLR}^+(p, s) = C^+ R^+ S_{CR}w^1(0)$$
$$u_{CLR}^+(p, -s) = C^+ R^+ S_{CR}w^2(0)$$
$$v_{CLR}^+(p, s) = C^+ R^+ S_{CR}w^3(0)$$
$$v_{CLR}^+(p, -s) = C^+ R^+ S_{CR}w^4(0) \qquad (3.210)$$

$$u_{CLR}^{+\dagger}(p^*, s) = w^{1T}(0)S_{CR}R^+C^+$$
$$u_{CLR}^{+\dagger}(p^*, -s) = w^{2T}(0)S_{CR}R^+C^+$$
$$v_{CLR}^{+\dagger}(p^*, s) = w^{3T}(0)S_{CR}R^+C^+$$
$$v_{CLR}^{+\dagger}(p^*, -s) = w^{4T}(0)S_{CR}R^+C^+$$

where the superscript "T" indicates the transpose. The anti-commutation relations of the Fourier coefficient operators are

$$\begin{aligned}
\{b_{CLR}(p,s), b_{CLR}^\dagger(p'^*,s')\} &= -2^{-\frac{1}{2}}\delta_{ss'}\delta(p^+ - p'^+)\delta^2(\mathbf{p}_r - \mathbf{p'}_{r'})\delta^3(\mathbf{p}_i + \mathbf{p'}_{i'}) \\
\{d_{CLR}(p,s), d_{CLR}^\dagger(p'^*,s')\} &= -2^{-\frac{1}{2}}\delta_{ss'}\delta(p^+ - p'^+)\delta^2(\mathbf{p}_r - \mathbf{p'}_{r'})\delta^3(\mathbf{p}_i + \mathbf{p'}_{i'}) \\
\{b_{CLR}(p,s), b_{CLR}(p'^*,s')\} &= \{d_{CLR}(p,s), d_{CLR}(p'^*,s')\} = 0 \\
\{b_{CLR}^\dagger(p,s), b_{CLR}^\dagger(p'^*,s')\} &= \{d_{CLR}^\dagger(p,s), d_{CLR}^\dagger(p'^*,s')\} = 0 \\
\{b_{CLR}(p,s), d_{CLR}^\dagger(p'^*,s')\} &= \{d_{CLR}(p,s), b_{CLR}^\dagger(p'^*,s')\} = 0 \\
\{b_{CLR}^\dagger(p,s), d_{CLR}^\dagger(p'^*,s')\} &= \{d_{CLR}(p,s), b_{CRR}(p'^*,s')\} = 0
\end{aligned} \quad (3.211)$$

The spinors satisfy

$$\sum_{\pm s} u_{CLR\ \alpha}^{+}(p, s)\bar{u}_{CLR\ \beta}^{+}(p^*, s) = (2m)^{-1}[C^+R^+(-i\not{p} + m)R^-C^-]_{\alpha\beta} \quad (3.212)$$

$$\sum_{\pm s} v_{CLR\ \alpha}^{+}(p, s)\bar{v}_{CLR\ \beta}^{+}(p^*, s) = (2m)^{-1}[C^+R^+(-i\not{p} - m)R^-C^-]_{\alpha\beta}$$

where $\bar{u}_{CLR}^{+} = u_{CLR}^{+\dagger}\gamma^0$ and $\bar{v}_{CLR}^{+} = v_{CLR}^{+\dagger}\gamma^0$.

The right-handed anti-commutation relation with a minus sign follows in particular because of the minus signs in eqs. 3.211.

Case 4 Right-handed Tachyonic Complexon Feynman Propagator

The Feynman propagator for right-handed tachyonic complexons can be obtained from eqs. 3.205 and 3.206 by changing the parity projection operator and some numerator signs in the integral (basically $p \rightarrow -p$) resulting in

$$iS^+_{CLRF}(x, y) = \int dM\, C^+R^+(\gamma^0\partial/\partial x^0 + (\nabla_r - i\nabla_i)\cdot\gamma - m)R^-C^-\delta'(\nabla_r\cdot\nabla_i/m^2)\cdot \cdot J_2(\mathbf{x}_i - \mathbf{y}_i, M^2)\Delta_{FT}(x - y,M) \quad (3.213)$$

with $\nabla_r + i\nabla_i$ derivatives with respect to $\mathbf{x}_r$ and $\mathbf{x}_i$ and where

$$\Delta_{FT}(x - y, M) = (2\pi)^{-4}\int d^4p_r \exp[-ip^0(x^0 - y^0) + i\mathbf{p}_r\cdot(\mathbf{x}_r - \mathbf{y}_r)]/(p_r^2 + M^2 + i\varepsilon) \quad (3.214)$$

Other Cases? No

The four cases considered above are the only cases having symmetry under the real Lorentz group L and a single real energy (with a corresponding single real time parameter) that is independent of the direction of the boost thus preserving (real) spatial rotation invariance. The realness of the time variable survives the breakdown to conventional Lorentz invariance.

One might think that using the other type of spinor boost operator (Compare to eq. 3.154.)

$$S_{CR}(\Lambda_{CR}(\omega, \hat{\mathbf{w}})) = \exp(-i\omega_R\sigma_{0i}w_i/2) = \exp(-\omega_R\gamma^0\boldsymbol{\gamma}\cdot\hat{\mathbf{w}}/2) \quad (3.215)$$

$$= \cosh(\omega_R/2)I + \sinh(\omega_R/2)\gamma^0\boldsymbol{\gamma}\cdot\hat{\mathbf{w}}$$

where $\omega_R = \omega - i\pi/2$ might lead to more possible forms of spin ½ wave equations and particles. In fact it merely leads to the same particle types but with the role of the left-handed and right-handed fields reversed. The result would be a "right-handed" Standard Model contrary to experiment.

3.8 L_C Spinor Lorentz Boosts Generate 4 Types of Particles that suggest they are Analogues of Leptons and Color Quarks

In this chapter we have found four types of fermions using a complexified form of Lorentz boosts, L_C boosts, that correspond in a natural way with the four general types of known fermions: charged leptons, neutrinos, up-type color quarks and down-type color quarks.[21]

Charged lepton fermions

The conventional Dirac equation and solutions.

Neutrinos

Simple tachyons with real energy and 3-momentum. Their free field equation is:

$$(\gamma^\mu\partial/\partial x^\mu - m)\psi_T(x) = 0 \quad (3.18)$$

and their left-handed $\psi_{TL}^{\;+}$ Feynman propagator is:

$$iS^+_{TLF}(x, y) = ½C^-R^+\gamma^0\int d^4p(2\pi)^{-4}\, p^+e^{-ip\cdot(x-y)}/(p^2 + m^2 + i\varepsilon) \quad (3.76)$$

---

[21] We call each type of fermion a *species*. Each species has three known generations.

Similarly the light-front Feynman propagator for the right-handed ψ_{TR}^{+} tachyon field is

$$iS^{+}_{TRF}(x,y) = -\tfrac{1}{2}C^{+}R^{+}\gamma^{0}\int d^{4}p(2\pi)^{-4}\, p^{+}e^{-ip\cdot(x-y)}/(p^{2} + m^{2} + i\varepsilon) \qquad (3.77)$$

Up-type Color Quarks

Up-type quarks are assumed[22] to be fermions with complex 3-momenta - complexons, and an internal color SU(3) symmetry, that are "normal" with $p^2 = m^2$. Their field equation with a color SU(3) index, denoted a, inserted is

$$[i\gamma^{0}\partial/\partial t + i\boldsymbol{\gamma}\cdot(\nabla_{r} + i\nabla_{i}) - m]\psi_{C}^{a}(t, \mathbf{x}_{r}, \mathbf{x}_{i}) = 0 \qquad (3.101)$$

with the subsidiary condition

$$\nabla_{r}\cdot\nabla_{i}\, \psi_{C}^{a}(t, \mathbf{x}_{r}, \mathbf{x}_{i}) = 0 \qquad (3.102a)$$

The free field solution is:

$$\psi_{C}^{a}(x) = \sum_{\pm s}\int d^{3}p_{r}d^{3}p_{i}\, N_{C}(p)\delta(\mathbf{p}_{r}\cdot\mathbf{p}_{i}/m^{2})[b_{C}(p,a,s)u_{C}^{a}(p, s)e^{-i(p\cdot x + p^{*}\cdot x^{*})/2} +$$
$$+ d_{C}^{\dagger}(p,a,s)v_{C}^{a}(p, s)e^{+i(p\cdot x + p^{*}\cdot x^{*})/2}] \qquad (3.125a)$$

The free Feynman propagator arranged into the form of a spectral integral is

$$iS_{C}^{ab}(x,y) = -\delta^{ab}\int dM\, (i\gamma^{0}\partial/\partial x^{0} - i(\nabla_{r} - i\nabla_{i})\cdot\boldsymbol{\gamma} + m)\delta'(\nabla_{r}\cdot\nabla_{i}/m^{2})\cdot$$
$$\cdot J(\mathbf{x}_{i} - \mathbf{y}_{i}, M^{2})\Delta_{F}(x - y, M) \qquad (3.149)$$

where

$$\Delta_{F}(x - y, M) = (2\pi)^{-4}\int d^{4}p_{r}\, \exp[-ip^{0}(x^{0} - y^{0}) + i\mathbf{p}_{r}\cdot(\mathbf{x}_{r} - \mathbf{y}_{r})]/(p_{r}^{2} - M^{2} + i\varepsilon)$$
$$(3.150)$$

and

$$J(\mathbf{x}_{i}, M^{2}) = (2\pi)^{-3}\int d^{3}p_{i}\,\delta(M^{2} + \mathbf{p}_{i}^{2} - m^{2})\exp[-i\mathbf{p}_{i}\cdot(\mathbf{x}_{i} - \mathbf{y}_{i})] \qquad (3.148)$$

$$= (2\pi)^{-2}|\mathbf{x}_{i} - \mathbf{y}_{i}|^{-1}\theta(m^{2} - M^{2})\sin((m^{2} - M^{2})^{1/2}|\mathbf{x}_{i} - \mathbf{y}_{i}|)$$

---

[22] The complexon theory that we develop and use for quark dynamics in the Standard Model is not required. Our Standard Model could use Dirac fermion dynamics for the up-type quarks and tachyon dynamics for down-type quarks. Then the (broken) Left-handed Extended Lorentz group would be the basic space-time group rather than L_C. We choose to use complexon dynamics for quarks because they have an internal SU(3)-like structure suggestive of color SU(3). More importantly, their spin dynamics is different and thus may resolve the differences between theory and experiment for the deep inelastic parton spin-dependent structure functions.

Down-type Color Quarks

Tachyonic complexons with complex 3-momenta, and an internal global SU(3) symmetry, that have mass shell condition $p^2 = -m^2$. Their field equation with a color SU(3) index, denoted a, inserted is

$$[\gamma^0\partial/\partial t + \boldsymbol{\gamma}\cdot(\nabla_r + i\nabla_i) - m]\psi_{CL}{}^a(t, \mathbf{x}_r, \mathbf{x}_i) = 0 \qquad (3.163a)$$

with the subsidiary condition on the wave function

$$\nabla_r\cdot\nabla_i\, \psi_{CL}{}^a(t, \mathbf{x}_r, \mathbf{x}_i) = 0 \qquad (3.164)$$

Its free field left-handed solution is:

$$\psi_{CLL}{}^{+a}(x_r, x_i) = \sum_{\pm s}\int d^2p_r dp^+ d^3p_i\, N_{CLL}{}^+(p)\theta(p^+)\delta((p_i{}^3(p^+ - p^-)/\surd 2 + \mathbf{p}_{r\perp}\cdot\mathbf{p}_{i\perp})/m^2)\cdot$$
$$\cdot[b_{CLL}{}^+(p,a,s)u_{CLL}{}^a(p,a,s)e^{-i(p\cdot x + p^*\cdot x^*)/2} + d_{CLL}{}^{+\dagger}(p,a,s)v_{CLL}{}^{+a}(p,a,s)e^{+i(p\cdot x + p^*\cdot x^*)/2}] \qquad (3.196)$$

and its right-handed solution is

$$\psi_{CLR}{}^{+a}(x_r, x_i) = \sum_{\pm s}\int d^2p_r dp^+ d^3p_i\, N_{CLR}{}^+(p)\theta(p^+)\delta((p_i{}^3(p^+ - p^-)/\surd 2 + \mathbf{p}_{r\perp}\cdot\mathbf{p}_{i\perp})/m^2)\cdot$$
$$\cdot[b_{CLR}{}^+(p,a,s)u_{CLR}{}^{+a}(p,a,s)e^{-i(p\cdot x+p^*\cdot x^*)/2} + d_{CLR}{}^{+\dagger}(p,a,s)v_{CLR}{}^{+a}(p,a,s)e^{+i(p\cdot x+p^*\cdot x^*)/2}] \qquad (3.207)$$

The free left-handed Feynman propagator arranged into the form of a spectral integral is

$$iS^+{}_{CLLF}{}^{ab}(x,y) = -\delta^{ab}\int dM\, C^- R^+(\gamma^0\partial/\partial x^0 + (\nabla_r - i\nabla_i)\cdot\boldsymbol{\gamma} - m)R^- C^+\,\delta'(\nabla_r\cdot\nabla_i/m^2)\cdot$$
$$\cdot J_2(\mathbf{x}_i - \mathbf{y}_i, M^2)\triangle_{FT}(x - y, M) \qquad (3.205)$$

with ∇_r and ∇_i derivatives with respect to $\mathbf{x}_r$ and $\mathbf{x}_i$ and where

$$\triangle_{FT}(x - y, M) = (2\pi)^{-4}\int d^4p_r \exp[-ip^0(x^0 - y^0) + i\mathbf{p}_r\cdot(\mathbf{x}_r - \mathbf{y}_r)]/(p_r{}^2 + M^2 + i\varepsilon) \qquad (3.206)$$

and

$$J_2(\mathbf{x}_i, M^2) = (2\pi)^{-3}\int d^3p_i\,\delta(M^2 - \mathbf{p}_i{}^2 - m^2)\exp[-i\mathbf{p}_i\cdot(\mathbf{x}_i - \mathbf{y}_i)] \qquad (3.204)$$

$$= (2\pi)^{-2}|\mathbf{x}_i - \mathbf{y}_i|^{-1}\theta(M^2 - m^2)\sin((M^2 - m^2)^{1/2}|\mathbf{x}_i - \mathbf{y}_i|)$$

The free right-handed Feynman propagator arranged into the form of a spectral integral is

$$iS^{+}{}_{CLRF}{}^{ab}(x, y) = \delta^{ab}\int dM\, C^{+}R^{+}(\gamma^{0}\partial/\partial x^{0} + (\nabla_{r} - i\nabla_{i})\cdot\gamma - m)R^{-}C^{-}\delta'(\nabla_{r}\cdot\nabla_{i}/m^{2})\cdot$$
$$\cdot J_{2}(\mathbf{x}_{i} - \mathbf{y}_{i}, M^{2})\triangle_{FT}(x - y, M) \qquad (3.213)$$

with ∇_r and ∇_i derivatives with respect to $\mathbf{x}_r$ and $\mathbf{x}_i$, and where

$$\triangle_{FT}(x - y, M) = (2\pi)^{-4}\int d^{4}p_{r}\, \exp[-ip^{0}(x^{0} - y^{0}) + i\mathbf{p}_{r}\cdot(\mathbf{x}_{r} - \mathbf{y}_{r})]/(p_{r}^{2} + M^{2} + i\varepsilon)$$
$$(3.214)$$

3.9 First Step towards a One Generation Standard Model

Thus we have found a set of four fermion species that corresponds to the known fermions of one fermion generation. In subsequent chapters we will derive the one generation model in detail. Then we will introduce three generations with mixing to complete the derivation of the form of the Standard Model. Then the only remaining major issue will be the values of the coupling constants and other numerical parameters.

The overall pattern that begins to emerge from the developments in this chapter divides particles and interactions into two categories (as seen in Nature):

Particles with real 4-Momenta	Complexons (Complex 3-Momenta)
Leptons	color quarks
SU(2)⊗U(1) Vector Bosons	Color SU(3) gluons
Higgs Particles	Possibly Higgs Particles

We will explore these issues in detail in the following chapters. But basically the leptons, SU(2)⊗U(1) Vector Bosons and a set of Higgs particles appear to be based on the Left-handed Extended Lorentz group. These particles have real energies and momenta although some are "normal" and some are tachyons.

The other category of particles, complexons, emerges from our study of L_C. These particles have real energies and complex 3-momenta. In perturbation theory the loop integrations of loops of these particles would consist of a 7-fold integration over energy and complex 3-momenta with corresponding 7-fold delta functions to enforce energy-momentum conservation. As pointed out earlier the complex 3-momenta of these types of fermions has an SU(3) symmetry that it is natural to generalize to local color SU(3). (The other category of fermions lacks this global SU(3) symmetry just as leptons lack color SU(3).) Thus we see the beginnings of the structure of the Standard Model in this chapter on spin ½ particles. The following chapters lead to a detailed derivation of the form of the Standard Model.

Appendix 3-A. Leptonic Tachyon Spinors

The general form of the solutions of the free tachyon Dirac equation eq. 3.18 can be written

$$\psi_T^{\,r}(x) = e^{-i\chi_r p\cdot x} w^r(p) \qquad (3\text{-A.1})$$

where $\chi_r = +1$ for r = 1, 2 and $\chi_r = -1$ for r = 3, 4. Denoting the spinors $w^r(p) = w^r(0)$ for a particle is at rest in a frame (E = m) we see they can take the form

$$w^r(0) = \begin{bmatrix} \delta_{1r} \\ \delta_{2r} \\ \delta_{3r} \\ \delta_{4r} \end{bmatrix} \qquad (3\text{-A.2})$$

where Kronecker deltas appear in the brackets. From eq. 3.15 we find

$$S_L(\Lambda_L(\omega, \mathbf{u}))w^r(0) = w_T^{\,r}(p) \qquad (3\text{-A.3})$$

Using eq. 3.11 for $S_L(\Lambda_L(\omega, \mathbf{u}))$ and

$$\mathbf{p} = m\mathbf{v}\gamma_s \qquad E = m\gamma_s \qquad (3\text{-A.4})$$

we see that eq. 3-A.3 implies the columns of the resulting $S_L(\Lambda_L(\omega, \mathbf{u}))$ matrix are

$$S_L(\Lambda_L(\omega, \mathbf{u})) = \begin{array}{c} \begin{array}{cccc} \underline{w_T^{\,3}(p)} & \underline{w_T^{\,4}(p)} & \underline{w_T^{\,1}(p)} & \underline{w_T^{\,2}(p)} \end{array} \\ \begin{bmatrix} \cosh(\omega_L/2) & 0 & \sinh(\omega_L/2)p_z/p & \sinh(\omega_L/2)p_-/p \\ 0 & \cosh(\omega_L/2) & \sinh(\omega_L/2)p_+/p & -\sinh(\omega_L/2)p_z/p \\ \sinh(\omega_L/2)p_z/p & \sinh(\omega_L/2)p_-/p & \cosh(\omega_L/2) & 0 \\ \sinh(\omega_L/2)p_+/p & -\sinh(\omega_L/2)p_z/p & 0 & \cosh(\omega_L/2) \end{bmatrix} \end{array}$$

(3-A.5)

based on the superluminal transformation of positive energy states to negative energy states (eqs. 3.15 and 3.16) with $p_\pm = p_x \pm ip_y$ and where $p = |\mathbf{p}|$. It is easy to verify

$$(i\not{p} - \chi_r m)w_T{}^r(p) = 0 \tag{3-A.6}$$

where $\chi_r = -1$ for r = 1, 2 and $\chi_r = +1$ for r = 3, 4.

The spinors that we defined in eq. 2.10 can be generalized in a manner similar to Dirac spinors. We will use a similar notation to the Dirac spinor notation:

$$\begin{aligned} u_T(p, s) &= w_T{}^1(p) \\ u_T(p, -s) &= w_T{}^2(p) \\ v_T(p, s) &= w_T{}^3(p) \\ v_T(p, -s) &= w_T{}^4(p) \end{aligned} \tag{3-A.7}$$

We define "double dagger" spinors:

$$\begin{aligned} u_T{}^\ddagger(p, s) &= u_T{}^\dagger(p, s)i\gamma\cdot\mathbf{p}/|\mathbf{p}| \\ u_T{}^\ddagger(p, -s) &= u_T{}^\dagger(p, -s)i\gamma\cdot\mathbf{p}/|\mathbf{p}| \\ v_T{}^\ddagger(p, s) &= v_T{}^\dagger(p, s)i\gamma\cdot\mathbf{p}/|\mathbf{p}| \\ v_T{}^\ddagger(p, -s) &= v_T{}^\dagger(p, -s)i\gamma\cdot\mathbf{p}/|\mathbf{p}| \end{aligned} \tag{3-A.8}$$

.

where † indicates hermitean conjugate, which appear in important spinor "completeness" sums:

$$\sum_{\pm s} u_{T\alpha}(p, s)u_T{}^\ddagger{}_\beta(p, s) = (2m)^{-1}(i\not{p} - m)_{\alpha\beta} \tag{3-A.9}$$

$$\sum_{\pm s} v_{T\alpha}(p, s)v_T{}^\ddagger{}_\beta(p, s) = (2m)^{-1}(i\not{p} + m)_{\alpha\beta} \tag{3-A.10}$$

or

$$\sum_{\pm s} u_{T\alpha}(p, s)u_T{}^\dagger{}_\beta(p, s) = -i(2m)^{-1}[(i\not{p} - m)\gamma\cdot\mathbf{p}/|\mathbf{p}|]_{\alpha\beta} \tag{3-A.11}$$

$$\sum_{\pm s} v_{T\alpha}(p, s)v_T{}^\dagger{}_\beta(p, s) = -i(2m)^{-1}[(i\not{p} + m)\gamma\cdot\mathbf{p}/|\mathbf{p}|]_{\alpha\beta} \tag{3-A.12}$$

Lastly we define light-front, left-handed tachyon spinors by

$$\begin{aligned} u_{TL}{}^+(p, s) &= C^- R^+ S_L(\Lambda_L(\omega, \mathbf{u}))w^1(0) \\ u_{TL}{}^+(p, -s) &= C^- R^+ S_L(\Lambda_L(\omega, \mathbf{u}))w^2(0) \end{aligned} \tag{3-A.13}$$

$$v_{TL}^{\;+}(p, s) = C^- R^+ S_L(\Lambda_L(\omega, \mathbf{u}))w^3(0)$$
$$v_{TL}^{\;+}(p, -s) = C^- R^+ S_L(\Lambda_L(\omega, \mathbf{u}))w^4(0)$$

$$u_{TL}^{\;+\dagger}(p, s) = w^{1T}(0)\, S_L^{\dagger}(\Lambda_L(\omega, \mathbf{u}))\, R^+C^-$$
$$u_{TL}^{\;+\dagger}(p, -s) = w^{2T}(0)\, S_L^{\dagger}(\Lambda_L(\omega, \mathbf{u}))R^+C^- \qquad (3\text{-A}.14)$$
$$v_{TL}^{\;+\dagger}(p, s) = w^{3T}(0)\, S_L^{\dagger}(\Lambda_L(\omega, \mathbf{u}))R^+C^-$$
$$v_{TL}^{\;+\dagger}(p, -s) = w^{4T}(0)\, S_L^{\dagger}(\Lambda_L(\omega, \mathbf{u}))R^+C^-$$

where the superscript "T" indicates the transpose and † indicates hermitean conjugate.

Appendix 3-B. Proof of the Reality of the Leptonic Tachyon Action

The tachyon lagrangian density and action are

$$\mathcal{L}_T = \psi_T^{\ S}(\gamma^\mu \partial/\partial x^\mu - m)\psi_T(x) \tag{3.19}$$

$$I = \int d^4x \mathcal{L}_T \tag{3.21}$$

Where

$$\psi_T^{\ S} = \psi_T^{\ \dagger}\, i\gamma^0\gamma^5 \tag{3.20}$$

The complex conjugate of the tachyon lagrangian density is

$$\mathcal{L}_T{}^* = -\psi_T^{\ T}\, i\gamma^0\gamma^5(\gamma^{\mu *}\partial/\partial x^\mu - m)\psi_T{}^*(x) \tag{3-B.1}$$

where the superscript ^{T} indicates the transpose. Eq. 3-B.1 can be expressed as a transpose:

$$\mathcal{L}_T{}^* = -i[\psi_T^{\ \dagger}(\gamma^{\mu\dagger}\overleftarrow{\partial}/\partial x^\mu - m)\gamma^5\gamma^0\psi_T(x)]^T \tag{3-B.2}$$

$$= -i[\psi_T^{\ \dagger}\gamma^5\gamma^0(-\gamma^\mu\overleftarrow{\partial}/\partial x^\mu - m)\psi_T(x)]^T \tag{3-B.3}$$

$$= [\psi_T^{\ \dagger} i\gamma^0\gamma^5(-\gamma^\mu\overleftarrow{\partial}/\partial x^\mu - m)\psi_T(x)]^T \tag{3-B.4}$$

$$= \psi_T^{\ \dagger} i\gamma^0\gamma^5(-\gamma^\mu\overleftarrow{\partial}/\partial x^\mu - m)\psi_T(x) \tag{3-B.5}$$

since eq. 3-B.4 is the transpose of a 1 by 1 matrix. Upon performing a partial integration in the action we find

$$I^* = \int d^4x[\psi_T^{\ \dagger} i\gamma^0\gamma^5(\gamma^\mu\partial/\partial x^\mu - m)\psi_T(x)] = I \tag{3-B.6}$$

4. Integer Spin Tachyons

4.0 Introduction

In this chapter we will begin by developing the quantum field theory of bosonic tachyons with real energy and 3-momenta. Then in section 4.4 we will consider complexon bosons of two types: one type with positive m^2, and the other type tachyonic ($m^2 < 0$). Both types have real energies and complex 3-momenta. They are the analogues of the complexon fermions with real energies and complex 3-momenta that were derived in the preceding chapter using L_C boosts.

4.1 Massive, Scalar Tachyons

The case of massive scalar tachyons would normally be the starting point for the discussion of tachyons. But the importance of spin ½ tachyons in the Standard Model led us to consider spin ½ tachyons first. We now turn to free, neutral, spin 0 tachyons, which we anticipate satisfies the tachyon equivalent of the Klein-Gordon equation:

$$(\Box - m^2)\phi_T(x) = 0 \tag{4.1}$$

where

$$\Box = \partial/\partial x_\mu \partial/\partial x^\mu \tag{4.2}$$

(The charged scalar tachyon case is analogous.)

Eq. 4.1 can be derived using the canonical procedure from the lagrangian density and action

$$\mathcal{L}_T = ½[\partial\phi_T/\partial x^\mu \partial\phi_T/\partial x_\mu + m^2\phi_T^2] \tag{4.3}$$

$$I = \int d^4x \mathcal{L}_T$$

We can canonically second quantize this theory using light-front coordinates. The lagrangian density then becomes

$$\mathcal{L}_T = \partial\phi_T/\partial x^+ \partial\phi_T/\partial x^- - ½\partial\phi_T/\partial x^i \partial\phi_T/\partial x^i + ½m^2\phi_T^2 \tag{4.4}$$

The conjugate momentum is

$$\pi_T = \partial\mathcal{L}/\partial(\partial^- \phi_T) = \partial^+ \phi_T \equiv \partial\phi_T/\partial x^- \tag{4.5}$$

and the equal x^+ commutation relations[23] are

$$[\pi_T(x), \phi_T(y)] = -i2^{-½}\delta(x^- - y^-)\delta^2(\mathbf{x}_\perp - \mathbf{y}_\perp) \tag{4.6}$$

We provisionally define the Fourier expansion of ϕ_T as

$$\phi_T(x) = \int d^2p dp^+ N_T(p)\theta(p^+)[a_T(p)e^{-ip\cdot x} + a_T^\dagger(p)e^{+ip\cdot x}] \tag{4.7}$$

where $N_T(p)$ is

$$N_T(p) = [(2\pi)^3 p^+]^{-½} \tag{4.8}$$

and the Fourier component operator *commutation* relations are

$$[a_T(q), a_T^\dagger(p)] = 2^{-½}\delta^2(\mathbf{q} - \mathbf{p})\delta(q^+ - p^+) \tag{4.9}$$
$$[a_T(q), a_T(p)] = [a_T(q), a_T(p)] = 0$$

We now calculate

$$[\pi_T(x), \phi_T(y)] = [\partial\phi_T(x)/\partial x^-, \phi_T(y)]$$
$$= \int d^2p dp^+ \int d^2p' dp'^+ N_T(p)N_T(p')\theta(p^+)\theta(p'^+)\cdot$$
$$\cdot\{-ip^+[a_T(p), a_T^\dagger(p')]e^{+ip'\cdot y - ip\cdot x} + ip^+[a_T^\dagger(p), a_T(p')]e^{-ip'\cdot y + ip\cdot x}\}$$

---

[23] Feinberg (G. Feinberg, Phys. Rev. **159**, 1089 (1967)) and others have suggested that scalar tachyons obey anti-commutation relations because a Lorentz transformation can change a positive energy to a negative energy (and vice versa). However in light-front coordinates a Lorentz or Superluminal boost in the z direction does not change the sign of the equivalent of energy p^-. Boosts in other directions may change the sign of p^-. However the light-front is a particular choice of variables in a specific frame. Since perturbative and other calculations lead to covariant results we can do all calculations on the light-front, and then, after expressing the results in covariant form, transform to any other reference frame. Then tachyon scattering events seen in the new coordinate system should be in agreement with the corresponding covariant calculation of the event. Therefore scalar tachyon quantization using commutators is acceptable and has the advantage of conforming to the general quantization program for scalar particles.

$$= -i2^{-\frac{1}{2}}\int d^2p_\perp \int_0^\infty dp^+ N_T^{+2}(p)p^+\{e^{+ip^+(y^- - x^-) - ip_\perp\cdot(y_\perp - x_\perp)} + e^{-ip^+(y^- - x^-) + ip_\perp\cdot(y_\perp - x_\perp)}\}$$

$$= -i2^{-\frac{1}{2}}\int d^2p_\perp \int_{-\infty}^\infty dp^+ (2\pi)^{-3}\, e^{+ip^+(y^- - x^-) - i\mathbf{p}_\perp\cdot(\mathbf{y}_\perp - \mathbf{x}_\perp)}$$

$$= -i2^{-\frac{1}{2}}\delta(x^- - y^-)\delta^2(\mathbf{x}_\perp - \mathbf{y}_\perp) \tag{4.10}$$

verifying the equal x^+ commutation relation.

Scalar Tachyon Feynman Propagator

The scalar tachyon Feynman propagator is defined by

$$i\Delta_{TF}(x - y) = \theta(x^+ - y^+)<0|\phi_T(x)\,\phi_T(y)|0> + \theta(y^+ - x^+)<0|\phi_T(y)\phi_T(x)|0>$$
$$= \int d^2p dp^+ \int d^2p'dp'^+\, N_T(p)N_T(p')\theta(p^+)\theta(p'^+)\cdot$$
$$\cdot\{\theta(x^+ - y^+)<0|a_T(p)a_T^\dagger(p')|0>e^{+ip'\cdot y - ip\cdot x} +$$
$$+ \theta(y^+ - x^+)<0|a_T(p')a_T^\dagger(p)|0>e^{-ip'\cdot y + ip\cdot x}\}$$

$$= \int d^2p_\perp \int_0^\infty dp^+ N_T^{+2}(p)\{\theta(x^+ - y^+)e^{+ip^-(y^+ - x^+) + ip^+(y^- - x^-) - i\mathbf{p}_\perp\cdot(\mathbf{y}_\perp - \mathbf{x}_\perp)} +$$

$$+ \theta(y^+ - x^+)e^{-ip^-(y^+ - x^+) - ip^+(y^- - x^-) + i\mathbf{p}_\perp\cdot(\mathbf{y}_\perp - \mathbf{x}_\perp)}\}$$
$$= i\int d^4p(2\pi)^{-4}e^{-ip\cdot(x - y)}/(p^2 + m^2 + i\varepsilon) \tag{4.11}$$

with the expected tachyon pole term.

4.2 Massive Vector Tachyons

The case of massive vector tachyons is of some interest since massive vector bosons, W and Z bosons, have been found in nature. Therefore there is a possibility that, hitherto undiscovered, massive vector tachyons may exist in nature and might eventually be created by particle accelerators. In this section we will second quantize a massive tachyon vector boson in light-front coordinates.

We begin with a standard, neutral, free, massive vector boson lagrangian with the sign of the mass term changed to make it a tachyon vector boson lagrangian:

$$\mathcal{L}_{TVB} = -\tfrac{1}{4}\, F_T^{\mu\nu}(x)F_{T\mu\nu}(x) - \tfrac{1}{2}\, m^2V_T^\mu V_{T\mu} \tag{4.12}$$

where

$$F_{T\mu\nu} = (\partial V_{T\mu}/\partial x^\nu - \partial V_{T\nu}/\partial x^\mu) \tag{4.13}$$

The equations of motion are

$$\partial F_T^{\mu\nu}/\partial x^\nu - m^2 V_T^\mu = 0 \tag{4.14}$$

Eq. 4.14 implies the subsidiary condition

$$\partial V_T^\mu/\partial x^\mu = 0 \tag{4.15}$$

which, in turn, implies

$$(\Box - m^2)V_T^\mu = 0 \tag{4.16}$$

where

$$\Box = \partial/\partial x_\mu \partial/\partial x^\mu \tag{4.2}$$

as previously.

Eq. 4.16 is immediately recognizable as a tachyon equation for each component. We now transform the lagrangian to light-front coordinates and proceed to quantize. Using the previous definition of light-front variables we define the fields:

$$\begin{aligned}
F_T^{+-} &= \partial^+ V_T^- - \partial^- V_T^+ \\
F_T^{+j} &= \partial^+ V_T^j - \partial^j V_T^+ \\
F_T^{-j} &= \partial^- V_T^j - \partial^j V_T^- \\
F_T^{ij} &= \partial^i V_T^j - \partial^j V_T^i \\
V_T^- &= 2^{-1/2}(V_T^0 - V_T^3) \\
V_T^+ &= 2^{-1/2}(V_T^0 + V_T^3)
\end{aligned} \tag{4.17}$$

The light-front equivalent of the lagrangian (eq. 4.12) is:

$$\mathcal{L}_{TVB} = -\tfrac{1}{2}(\partial V_{T\mu}/\partial x^\nu \partial V_T^\mu/\partial x_\nu - \partial V_T^\mu/\partial x_\nu \partial V_{T\nu}/\partial x^\mu) - \tfrac{1}{2}m^2 V_T^\mu V_{T\mu}$$

After using the constraint eq. 4.15 and discarding a total divergence, we see

$$\mathcal{L}_{TVB} \equiv -\tfrac{1}{2}\partial V_{T\mu}/\partial x^\nu \partial V_T^\mu/\partial x_\nu - \tfrac{1}{2}m^2 V_T^\mu V_{T\mu} \tag{4.18}$$

which becomes

$$\mathcal{L}_{TVB} \equiv -\partial^+ V_T^- \partial^- V_T^+ - \partial^+ V_T^+ \partial^- V_T^- + \partial^+ V_T^i \partial^- V_T^i + \partial^i V_T^+ \partial^i V_T^- +$$

$$+ \tfrac{1}{2}\partial^i V_T^{\,j} \partial^i V_T^{\,j} - \tfrac{1}{2}m^2(2V_T^{\,+} V_T^{\,-} - V_T^{\,i} V_T^{\,i}) \qquad (4.19)$$

using light-front coordinates with implied sums over i and j. The resulting equations of motion are

$$(\square - m^2)V_T^{\,-} = 0 \qquad (4.20)$$
$$(\square - m^2)V_T^{\,+} = 0$$
$$(\square - m^2)V_T^{\,i} = 0$$

for i = 1, 2.

The conjugate spacelike-surface momenta are

$$\pi^\mu = \partial\mathcal{L}/\partial(\partial^0 V_T^{\,\mu}) = -\,\partial V_T^{\,\mu}/\partial x^0 \qquad (4.21)$$

and the conjugate light-front momenta are

$$\pi^{+} = \partial\mathcal{L}/\partial(\partial^{-} V_T^{\,+}) = -\,\partial^+ V_T^{\,-} \equiv -\,\partial V_T^{\,-}/\partial x^{-} \qquad (4.22)$$
$$\pi^{-} = \partial\mathcal{L}/\partial(\partial^{-} V_T^{\,-}) = -\,\partial^+ V_T^{\,+} \equiv -\,\partial V_T^{\,+}/\partial x^{-} \qquad (4.23)$$
$$\pi^{i} = \partial\mathcal{L}/\partial(\partial^{-} V_T^{\,i}) = \partial^+ V_T^{\,i} \equiv \partial V_T^{\,i}/\partial x^{-} \qquad (4.24)$$

The equal x^+ commutation relations are

$$[\pi_T^{\,a}(x), V_T^{\,b}(y)] = -i2^{-\frac{1}{2}}\delta^{3ab}(x - y) \qquad (4.25)$$

where

$$\delta^{3ab}(x - y) = \int d^2k\,dk^+ e^{i[k^+(x^- - y^-) - \mathbf{k}\cdot(\mathbf{x} - \mathbf{y})]}[g^{ab} + k^a k^b/m^2]/(2\pi)^3 \qquad (4.26)$$

where $\mathbf{k} = (k^1, k^2)$, and $g^{-+} = g^{+-} = 1 = -g^{11} = = -g^{22}$ with all other $g^{ab} = 0$. The equal x^+ commutation relations satisfy the constraint:

$$\partial_a[\pi_T^{\,a}(x), V_T^{\,b}(y)] = \partial_b[\pi_T^{\,a}(x), V_T^{\,b}(y)] = 0 \qquad (4.27)$$

implied by eq. 4.15.

Next we define the Fourier expansion of $V_T^{\,\mu}$ as

$$V_T^{\mu}(x) = \sum_s \int d^2k dk^+ N_{TV}(k)\theta(k^+)\varepsilon^{\mu}(k, s)[a_T(k, s)e^{-ik\cdot x} + a_T^{\dagger}(k, s)e^{+ik\cdot x}] \quad (4.28)$$

where $k^2 = 2k^+k^- - k^{i\,2} = -m^2$, and where $N_{TV}(k)$ is

$$N_{TV}(k) = [(2\pi)^3 k^+]^{-1/2} \quad (4.29)$$

There are three spin orientations: two transverse orientations and a longitudinal orientation, $s = \pm 1, 0$. The spin polarization vector satisfies

$$k_{\mu}\varepsilon^{\mu}(k, s) = 0 \quad (4.30)$$

It also satisfies the normalization condition

$$\sum_s \varepsilon^{\mu}(k, s)\varepsilon^{\upsilon}(k, s) = -(g^{\mu\nu} + k^{\mu}k^{\nu}/m^2) \quad (4.31)$$

The Fourier component operator commutation relations are

$$[a_T(q, s), a_T^{\dagger}(p, s')] = 2^{-1/2}\delta_{ss'}\delta^2(\mathbf{q} - \mathbf{p})\delta(q^+ - p^+) \quad (4.32)$$
$$[a_T(q, s), a_T(p, s')] = [a_T(q, s), a_T(p, s')] = 0$$

Eqs. 4.28 – 4.32 imply the commutation relations eqs. 4.25.

Tachyon Vector Boson Feynman Propagator

The tachyon vector boson Feynman propagator is defined by

$$i\Delta_{TF}(x - y)^{\mu\nu} = \theta(x^+ - y^+)\langle 0|V_T^{\mu}(x)V_T^{\nu}(y)|0\rangle + \theta(y^+ - x^+)\langle 0|V_T^{\nu}(y)V_T^{\mu}(x)|0\rangle \quad (4.33)$$

and is equal to

$$= -i\int \frac{d^4k\, e^{-ik\cdot(x-y)}\,(g^{\mu\nu} + k^{\mu}k^{\nu}/m^2)}{(2\pi)^4\,(k^2 + m^2 + i\varepsilon)}$$

The propagator displays the tachyon poles as expected.

4.3 Massive Spin 2 Tachyons – Massive Tachyon Gravitons

Gravitons – the quanta of the gravitation – are massless as far as we know. Massive gravitons have been a subject of a number of theoretical investigations.

While there is no evidence for massive gravitons there is evidence that the universe in the large has additional forces at play that affect the rotation of galaxies and seem to be producing an accelerating expansion of the universe. Therefore it is sensible to consider the possibility that massive spin 2 tachyons may exist that could play a role in the understanding of unusual features of the universe in the large. Since the effect of new forces seems to be seen only at distances comparable to the size of galaxies or greater, it is possible that massive spin 2 tachyons may have a small mass of the order of [1/L] where L is the galactic radius of galaxies such as our galaxy.

4.4 L_C Spin 0 Bosons with Complex 3-Momenta

In this section we will develop the quantum field theory of complexon bosons with real energies and complex 3-momenta that are the bosonic analogues of complexon fermions with real energies and complex 3-momenta that we discovered using L_C boosts in chapter 3.

Complexon Scalar Bosons with Complex 3-Momenta

The energy and momentum of these bosons satisfy the mass shell condition

$$p^2 = p^{0\,2} - (\mathbf{p}_r + i\mathbf{p}_i)\cdot(\mathbf{p}_r + i\mathbf{p}_i) = p^{0\,2} - \mathbf{p}_r\cdot\mathbf{p}_r + \mathbf{p}_i\cdot\mathbf{p}_i = m^2 \tag{3.98}$$

where the complex 3-momentum is $\mathbf{p} = \mathbf{p}_r + i\mathbf{p}_i$ and

$$\mathbf{p}_r\cdot\mathbf{p}_i = 0 \tag{4.34}$$

The wave equation for these bosons is

$$(\partial^2/\partial t^2 - (\nabla_r + i\nabla_i)\cdot(\nabla_r + i\nabla_i) + m^2)\phi_C(t, \mathbf{x}_r, \mathbf{x}_i) = 0 \tag{4.35}$$

with the subsidiary condition (corresponding to eq. 3.102a)

$$\nabla_r\cdot\nabla_i\phi_C(t, \mathbf{x}_r, \mathbf{x}_i) = 0 \tag{4.36}$$

The lagrangian with equation of motion eq. 4.35 is

$$\mathcal{L}_{\phi C} = \tfrac{1}{2}[D^\mu\phi_C D_\mu\phi_C - m^2\phi_C^{\,2}] \tag{4.37}$$

where

$$D_0 = \partial/\partial x^0$$
$$D_k = \partial/\partial x_r^{\,k} + i\,\partial/\partial x_i^{\,k} \tag{4.38}$$

and its action[24] is

$$I_{\phi C} = \int d^7x \mathcal{L}_{\phi C} \tag{4.39}$$

The conjugate momentum is

$$\pi_C = \partial\mathcal{L}/\partial(\partial_0\phi_C) = \partial\phi_C/\partial t \equiv \partial_0\phi_C \tag{4.40}$$

and the equal time commutation relations are

$$[\pi_C(x), \phi_C(y)] = \tfrac{1}{2}\, i\delta'(\nabla_r\cdot\nabla_i/m^2)\delta^3(\mathbf{x}_r - \mathbf{y}_r)\delta^3(\mathbf{x}_i - \mathbf{y}_i) \tag{4.41}$$

where the mass squared in the delta-function could be omitted in the case of a zero mass particle. The Fourier expansion of ϕ_C is

$$\phi_C(x) = \int d^3p_r d^3p_i N_C(p)\delta(\mathbf{p}_r\cdot\mathbf{p}_i/m^2)[a_C(p)e^{-i(p\cdot x^* + p^*\cdot x)/2} + a_C^\dagger(p)e^{+i(p\cdot x^* + p^*\cdot x)/2}] \tag{4.42}$$

based on the analogous fermion case. Eq. 4.42 can be expressed as

$$\phi_C(x) = \int d^3p_r d^3p_i N_C(p)\delta(\mathbf{p}_r\cdot\mathbf{p}_i/m^2)[a_C(p)e^{-i(p^0x^0 - \mathbf{p}_r\cdot\mathbf{x}_r + \mathbf{p}_i\cdot\mathbf{x}_i)} + a_C^\dagger(p)e^{+i(p^0x^0 - \mathbf{p}_r\cdot\mathbf{x}_r + \mathbf{p}_i\cdot\mathbf{x}_i)}] \tag{4.43}$$

where

$$N_C(p) = [1/((2\pi)^6 p^0)]^{-1/2} \tag{4.44}$$

The Fourier component operator *commutation* relations are

$$[a_C(q), a_C^\dagger(p)] = \delta^3(\mathbf{q}_r - \mathbf{p}_r)\delta^3(\mathbf{q}_i - \mathbf{p}_i) \tag{4.45}$$
$$[a_C(q), a_C(p)] = [a_C(q), a_C(p)] = 0$$

We now calculate the equal-time commutator

$$[\pi_C(x), \phi_C(y)] = [\partial\phi_C(x)/\partial x^0, \phi_C(y)]$$
$$= \int d^3p_r d^3p_i \int d^3p'_r d^3p'_i N_C(p)N_C(p')\delta(\mathbf{p}_r\cdot\mathbf{p}_i/m^2)\,\delta(\mathbf{p}'_r\cdot\mathbf{p}'_i/m^2)\cdot$$
$$\cdot\{-ip^0[a_C(p), a_C^\dagger(p')]e^{+i(\mathbf{p}_r\cdot\mathbf{x}_r - \mathbf{p}_i\cdot\mathbf{x}_i) - i(\mathbf{p}'_r\cdot\mathbf{y}_r - \mathbf{p}'_i\cdot\mathbf{y}_i)} +$$

[24] The action can be shown to be real by performing partial integrations on the imaginary terms and then using the subsidiary condition. The lagrangian density resulting from these manipulations is real.

$$+\ ip^0[a_C^\dagger(p),\ a_C(p')]e^{-i(\mathbf{p_r\cdot x_r} - \mathbf{p_i\cdot x_i}) + i(\mathbf{p'_r\cdot y_r} - \mathbf{p'_i\cdot y_i})}\}$$

$$= \tfrac{1}{2}i\int d^3p_r d^3p_i N_C^{\ 2}(p)\delta'(\mathbf{p_r\cdot p_i}/m^2)p^0\{e^{-i(\mathbf{p_r\cdot(y_r - x_r)} - \mathbf{p_i\cdot(y_i - x_i)})} + e^{+i(\mathbf{p_r\cdot(y_r - x_r)} - \mathbf{p_i\cdot(y_i - x_i)})}\}$$

$$= \tfrac{1}{2}\ i\delta'(\nabla_r\cdot\nabla_i/m^2)\delta^3(\mathbf{x_r} - \mathbf{y_r})\delta^3(\mathbf{x_i} - \mathbf{y_i})$$

verifying the equal time commutation relation eq. 4.41.

Complexon Scalar boson Feynman Propagator

The complexon scalar Feynman propagator is defined by

$$\begin{aligned} i\Delta_{CF}(x - y) &= \theta(x^0 - y^0)<0|\phi_C(x)\phi_C(y)|0> + \theta(y^0 - x^0)<0|\phi_C(y)\phi_C(x)|0> \\ &= \int d^3p_r d^3p_i \int d^3p'_r d^3p'_i N_C(p)N_C(p')\delta(\mathbf{p_r\cdot p_i}/m^2)\ \delta(\mathbf{p'_r\cdot p'_i}/m^2)\cdot \\ &\quad\cdot\{\theta(x^0 - y^0)<0|a_C(p)a_C^\dagger(p')|0>e^{-i(p\cdot x^* + p^*\cdot x)/2 + i(p'\cdot y^* + p'^*\cdot y)/2} + \\ &\quad + \theta(y^0 - x^0)<0|a_C(p')a_C^\dagger(p)|0>e^{+i(p\cdot x^* + p^*\cdot x)/2 - i(p'\cdot y^* + p'^*\cdot y)/2}\} \end{aligned}$$

$$\begin{aligned} &= \int d^3p_r d^3p_i N_C^{\ 2}(p)[\delta(\mathbf{p_r\cdot p_i}/m^2)]^2\{\theta(x^0 - y^0)e^{-i(p\cdot(x^* - y^*) + p^*\cdot(x - y))/2} + \\ &\quad + \theta(y^0 - x^0)e^{+i(p\cdot(x^* - y^*) - p^*\cdot(x - y))/2}\} \end{aligned}$$

$$\begin{aligned} &= -\tfrac{1}{2}\int d^3p_r d^3p_i N_C^{\ 2}(p)\delta'(\mathbf{p_r\cdot p_i}/m^2)\{\theta(x^0 - y^0)e^{-ip^0(x^0 - y^0) + i(\mathbf{p_r\cdot(y_r - x_r)} - \mathbf{p_i\cdot(y_i - x_i)})} + \\ &\quad + \theta(y^0 - x^0)e^{+ip^0(x^0 - y^0) - i(\mathbf{p_r\cdot(y_r - x_r)} - \mathbf{p_i\cdot(y_i - x_i)})}\} \\ &= -\tfrac{1}{2}\int d^3p_r d^3p_i\ (2\pi)^{-6}\delta'(\mathbf{p_r\cdot p_i}/m^2)\{\theta(x^0 - y^0)e^{-ip^0(x^0 - y^0) + i(\mathbf{p_r\cdot(y_r - x_r)} - \mathbf{p_i\cdot(y_i - x_i)})} + \\ &\quad + \theta(y^0 - x^0)e^{+ip^0(x^0 - y^0) - i(\mathbf{p_r\cdot(y_r - x_r)} - \mathbf{p_i\cdot(y_i - x_i)})}\}/p^0 \\ &= i\int d^4p_r d^3p_i\ (2\pi)^{-7}\delta'(\mathbf{p_r\cdot p_i}/m^2)e^{-ip^0(x^0 - y^0) + i(\mathbf{p_r\cdot(y_r - x_r)} - \mathbf{p_i\cdot(y_i - x_i)})}/(p^2\ - m^2 + i\varepsilon) \end{aligned} \tag{4.46}$$

Eq. 4.46 can be put in the form of a spectral integral similar to eq. 3.149 but without the γ-matrix denominator.

Tachyonic Complexon Scalar Bosons

The energy and momentum of these bosons satisfy the mass shell condition

$$p^2 = p^{0\,2} - (\mathbf{p_r} + i\mathbf{p_i})\cdot(\mathbf{p_r} + i\mathbf{p_i}) = p^{0\,2} - \mathbf{p_r\cdot p_r} + \mathbf{p_i\cdot p_i} = -m^2 \tag{4.47}$$

where the complex 3-momentum is $\mathbf{p} = \mathbf{p_r} + i\mathbf{p_i}$ and satisfies eq. 4.34.

The wave equation for these bosons is

$$(\partial^2/\partial t^2 - (\nabla_r + i\nabla_i)\cdot(\nabla_r + i\nabla_i) - m^2)\phi_{CT}(t, \mathbf{x_r}, \mathbf{x_i}) = 0 \tag{4.48}$$

with the subsidiary condition (corresponding to eq. 3.102a)

$$\nabla_r\cdot\nabla_i\phi_{CT}(t, \mathbf{x_r}, \mathbf{x_i}) = 0 \tag{4.49}$$

The lagrangian with equation of motion eq. 4.35 is

$$\mathcal{L}_{\phi CT} = \tfrac{1}{2}[D^\mu\phi_{CT}D_\mu\phi_{CT} + m^2\phi_{CT}^{\ 2}] \tag{4.50}$$

where

$$D_0 = \partial/\partial x^0$$
$$D_k = \partial/\partial x_r^{\ k} + i\,\partial/\partial x_i^{\ k} \tag{4.51}$$

and its action[25] is

$$I_{\phi CT} = \int d^7x\mathcal{L}_{\phi CT} \tag{4.52}$$

As in previous cases we must second quantize the theory using light-front coordinates to avoid the problems of earlier tachyon theories. Therefore we use the light-front variables of eqs. 3.184 – 3.186:

$$x^\pm = (x^0 \pm x_r^{\ 3})/\surd 2 \tag{3.184}$$
$$\partial/\partial x^\pm \equiv \partial^\mp \equiv (\partial/\partial x^0 \pm \partial/\partial x_r^{\ 3})/\surd 2$$

with the "transverse" real coordinate variables, $x_r^{\ 1}$ and $x_r^{\ 2}$, and imaginary coordinate variables $x_i^{\ 1}$, $x_i^{\ 2}$, and $x_i^{\ 3}$.

The inner product of two 4-vectors has the form

$$x\cdot y = x^+y^- + y^+x^- + i[y_i^{\ 3}(x^+ - x^-) + x_i^{\ 3}(y^+ - y^-)]/\surd 2 + x_i^{\ 3}y_i^{\ 3} - (\mathbf{x}_{r\perp} - i\mathbf{x}_{i\perp})\cdot(\mathbf{y}_{r\perp} - i\mathbf{y}_{i\perp}) \tag{3.185}$$

with

$$\mathbf{x}_{r\perp} = (x_r^{\ 1}, x_r^{\ 2}) \qquad \mathbf{x}_{i\perp} = (x_i^{\ 1}, x_i^{\ 2}) \tag{3.186}$$
$$\mathbf{y}_{r\perp} = (y_r^{\ 1}, y_r^{\ 2}) \qquad \mathbf{y}_{i\perp} = (y_i^{\ 1}, y_i^{\ 2})$$

where $x = (x^0, \mathbf{x} = \mathbf{x}_r - i\mathbf{x}_i)$ and $y = (y^0, \mathbf{y} = \mathbf{y}_r - i\mathbf{y}_i)$.

In light-front variables the lagrangian (eq. 4.50) becomes

$$\mathcal{L}_{\phi CT} = \partial\phi_{CT}/\partial x^+\partial\phi_{CT}/\partial x^- - \tfrac{1}{2}\nabla_{r\perp}\phi_{CT}\cdot\nabla_{r\perp}\phi_{CT} + \tfrac{1}{2}\nabla_i\phi_{CT}\cdot\nabla_i\phi_{CT} + \tfrac{1}{2}m^2\phi_{CT}^{\ 2} \tag{4.53}$$

---

[25] The action can be shown to be real by performing partial integrations on the imaginary terms and then using the subsidiary condition. The lagrangian density resulting from these manipulations is real.

with subsidiary condition eq. 4.49. The conjugate momentum is

$$\pi_{CT} = \partial\mathcal{L}/\partial(\partial^-\phi_{CT}) = \partial^+\phi_{CT} \equiv \partial\phi_{CT}/\partial x^- \tag{4.54}$$

and the equal x^+ commutation relations are

$$[\pi_{CT}(x), \phi_{CT}(y)] = i(\tfrac{1}{2})^{3/2}\delta'(\nabla_r\cdot\nabla_i/m^2)\delta(x^- - y^-)\delta^2(\mathbf{x}_\perp - \mathbf{y}_\perp)\delta^3(\mathbf{x}_i - \mathbf{y}_i) \tag{4.55}$$

The Fourier expansion of ϕ_{CT} is

$$\phi_{CT}(x) = \int d^2p dp^+ \int d^3p_i N_{CT}(p)\delta(\mathbf{p}_r\cdot\mathbf{p}_i/m^2)\theta(p^+)[a_{CT}(p)e^{-i(p\cdot x^* + p^*\cdot x)/2} + \\ + a_{CT}^\dagger(p)e^{+i(p\cdot x^* + p^*\cdot x)/2}] \tag{4.56}$$

where p^2 satisfies eq. 4.47 and where $N_{CT}(p)$ is

$$N_{CT}(p) = [(2\pi)^3p^+]^{-\frac{1}{2}} \tag{4.57}$$

and the Fourier component operator *commutation* relations are

$$[a_{CT}(q), a_{CT}^\dagger(p)] = 2^{-\frac{1}{2}}\delta^2(\mathbf{q} - \mathbf{p})\delta(q^+ - p^+) \tag{4.58}$$
$$[a_{CT}(q), a_{CT}(p)] = [a_{CT}(q), a_{CT}(p)] = 0$$

We now calculate

$$[\pi_{CT}(x), \phi_{CT}(y)] = [\partial\phi_{CT}(x)/\partial x^-, \phi_{CT}(y)]$$
$$= \int d^2p dp^+ \int d^2p' dp'^+ \int d^3p_i \int d^3p'_i N_{CT}(p)N_{CT}(p')\theta(p^+)\theta(p'^+)\delta(\mathbf{p}_r\cdot\mathbf{p}_i/m^2)\delta(\mathbf{p'}_r\cdot\mathbf{p'}_i/m^2)\cdot$$
$$\cdot\{-ip^+[a_{CT}(p), a_{CT}^\dagger(p')]e^{-i(p\cdot x^* + p^*\cdot x)/2 + i(p'\cdot y^* + p'^*\cdot y)/2} + \\ + ip^+[a_{CT}^\dagger(p), a_{CT}(p')]e^{+i(p\cdot x^* + p^*\cdot x)/2 - i(p'\cdot y^* + p'^*\cdot y)/2}\}$$

$$= -i2^{-\frac{1}{2}}\int d^2p_\perp \int_0^\infty dp^+ \int d^3p_i N_{CT}^{+2}(p)[\delta(\mathbf{p}_r\cdot\mathbf{p}_i/m^2)]^2p^+\{e^{+ip^+(y^- - x^-) - ip_\perp\cdot(y_\perp - x_\perp) + ip_i\cdot(y_i - x_i)} + \\ + e^{-ip^+(y^- - x^-) + ip_\perp\cdot(y_\perp - x_\perp) - ip_i\cdot(y_i - x_i)}\}$$

$$= -i2^{-\frac{1}{2}}\int d^2p_\perp \int_{-\infty}^\infty dp^+ \int d^3p_i (2\pi)^{-6}[\delta(\mathbf{p}_r\cdot\mathbf{p}_i/m^2)]^2e^{+ip^+(y^- - x^-) - ip_\perp\cdot(y_\perp - x_\perp) + ip_i\cdot(y_i - x_i)}$$

$$= (\tfrac{1}{2})^{3/2}\, i\delta'(\nabla_r\cdot\nabla_i/m^2)\delta(x^- - y^-)\delta^2(\mathbf{x}_\perp - \mathbf{y}_\perp)\delta^3(\mathbf{x}_i - \mathbf{y}_i) \tag{4.59}$$

verifying the equal x^+ commutation relation.

Scalar Tachyonic Complexon Feynman Propagator

The scalar tachyonic complexon Feynman propagator is defined by

$$i\Delta_{CTF}(x - y) = \theta(x^+ - y^+)<0|\phi_{CT}(x)\,\phi_{CT}(y)|0> + \theta(y^+ - x^+)<0|\phi_{CT}(y)\phi_{CT}(x)|0>$$
$$= i\int d^4p_r d^3p_i (2\pi)^{-7}\delta'(\mathbf{p_r}\cdot\mathbf{p_i}/m^2)e^{-ip^+(x^- - y^-) - ip^-(x^+ - y^+) + i\mathbf{p}_\perp\cdot(\mathbf{x}_\perp - \mathbf{y}_\perp) - i\mathbf{p_i}\cdot(\mathbf{x_i} - \mathbf{y_i})}/(p^2 + m^2 + i\varepsilon) \quad (4.60)$$

Eq. 4.60 can be placed in the form of a spectral integral.

4.5 Complexon Vector Bosons with Complex 3-Momenta

In this section we will second quantize massive complexon vector bosons (with complex 3-momenta) with positive m^2, and negative m^2 (tachyonic complexons).

Complexon Vector Bosons

We begin with a standard, neutral, free, massive vector boson lagrangian with one change the spatial derivatives are complex:

$$\mathcal{L}_{CVB} = -\tfrac{1}{4} F_C^{\mu\nu}(x)F_{C\mu\nu}(x) + \tfrac{1}{2} m^2 V_C^\mu V_{C\mu} \quad (4.61)$$

where

$$F_{C\mu\nu} = D_\nu V_{C\mu} - D_\mu V_{C\nu} \quad (4.62)$$

where D_μ is defined by eq. 4.51. The equations of motion are

$$D_\nu F_C^{\mu\nu} + m^2 V_C^\mu = 0 \quad (4.63)$$

Eq. 4.63 implies the subsidiary condition

$$D_\mu V_C^\mu = 0 \quad (4.64)$$

which, in turn, implies

$$(\Box + m^2)V_C^\mu = 0 \quad (4.64)$$

where

$$\Box = D_\mu D^\mu \quad (4.65)$$

In addition we have the subsidiary condition on the complex 3-momenta

$$\nabla_r\cdot\nabla_i V_C^\mu = 0 \quad (4.66)$$

The conjugate momenta are

$$\pi^{\mu} = \partial\mathcal{L}/\partial(\partial^0 V_C{}^{\mu}) = -\,\partial V_C{}^{\mu}/\partial x^0 \qquad (4.67)$$

The equal time commutation relations are

$$[\pi_C{}^{\mu}(x), V_C{}^{\nu}(y)] = \tfrac{1}{2}\, i\delta'(\nabla_r\cdot\nabla_i/m^2)\delta^{\mu\nu}(x - y) \qquad (4.68)$$

where

$$\delta^{\mu\nu}(x - y) = \int d^3k_r d^3k_i e^{+i\mathbf{k}\cdot(\mathbf{x} - \mathbf{y})}[g^{\mu\nu} - k^{\mu}k^{\nu}/m^2]/(2\pi)^6 \qquad (4.69)$$

where $k = (k^0, \mathbf{k} = \mathbf{k_r} + i\mathbf{k_i})$, $x = (x^0, \mathbf{x} = \mathbf{x_r} - i\mathbf{x_i})$ and $y = (y^0, \mathbf{y} = \mathbf{y_r} - i\mathbf{y_i})$.

The equal time commutation relations satisfy the constraint:

$$\partial_{\mu}[\pi_C{}^{\mu}(x), V_C{}^{\nu}(y)] = \partial_{\nu}[\pi_C{}^{\mu}(x), V_C{}^{\nu}(y)] = 0 \qquad (4.70)$$

implied by eq. 4.64.

The Fourier expansion of $V_C{}^{\mu}$ is

$$V_C{}^{\mu}(x) = \sum_s \int d^3k_r d^3k_i N_{CV}(k)\varepsilon^{\mu}(k, s)\delta(\mathbf{k_r}\cdot\mathbf{k_i}/m^2)[a_C(k, s)e^{-i(k\cdot x^* + k^*\cdot x)/2} + $$
$$+ a_C{}^{\dagger}(k, s)e^{+i(k\cdot x^* + k^*\cdot x)/2}] \qquad (4.71)$$

where $N_{CV}(k)$ is

$$N_{CV}(k) = [(2\pi)^3 k^0]^{-1/2} \qquad (4.72)$$

There are three spin orientations: two transverse orientations and a longitudinal orientation, $s = \pm 1, 0$. The spin polarization vector satisfies

$$k_{\mu}\varepsilon^{\mu}(k, s) = 0 \qquad (4.73)$$

It also satisfies the normalization condition

$$\sum_s \varepsilon^{\mu}(k, s)\varepsilon^{\upsilon}(k, s) = -(g^{\mu\nu} - k^{\mu}k^{\nu}/m^2) \qquad (4.74)$$

The Fourier component operator commutation relations are

$$[a_C(q, s), a_C{}^{\dagger}(p, s')] = \delta_{ss'}\delta^3(\mathbf{q}_r - \mathbf{p}_r)\delta^3(\mathbf{q}_i - \mathbf{p}_i) \qquad (4.75)$$

$$[a_C(q, s), a_C(p, s')] = [a_C(q, s), a_C(p, s')] = 0$$

Eqs. 4.71 – 4.75 imply the commutation relations eqs. 4.68.

Complexon Vector Boson Feynman Propagator

The complexon vector boson Feynman propagator is defined by

$$i\Delta_{CF}(x-y)^{\mu\nu} = \theta(x^0 - y^0)\langle 0|V_C^{\ \mu}(x)V_C^{\ \nu}(y)|0\rangle + \theta(y^0 - x^0)\langle 0|V_C^{\ \nu}(y)V_C^{\ \mu}(x)|0\rangle \tag{4.76}$$

and is equal to

$$= \tfrac{1}{2}\, i \int \frac{d^4k_r d^3k_i\, \delta'(\mathbf{k_r}\cdot\mathbf{k_i}/m^2) e^{-ik\cdot(x-y)}\, (g^{\mu\nu} - k^\mu k^\nu/m^2)}{(2\pi)^7\, (k^2 - m^2 + i\varepsilon)}$$

Tachyonic Complexon Vector Bosons

This case is similar to the previous case except for the change in the sign of the mass squared, which necessitates light-front quantization. The tachyonic complexon vector boson lagrangian is:

$$\mathcal{L}_{CTVB} = -\tfrac{1}{4} F_{CT}^{\ \mu\nu}(x)F_{CT\mu\nu}(x) - \tfrac{1}{2} m^2 V_{CT}^{\ \mu} V_{CT\mu} \tag{4.77}$$

where

$$F_{CT\mu\nu} = D_\nu V_{CT\mu} - D_\mu V_{CT\nu} \tag{4.78}$$

where D_μ is defined by eq. 4.51. The equations of motion are

$$D_\nu F_{CT}^{\ \mu\nu} - m^2 V_{CT}^{\ \mu} = 0 \tag{4.79}$$

Eq. 4.79 implies the subsidiary condition

$$D_\mu V_{CT}^{\ \mu} = 0 \tag{4.80}$$

which, in turn, implies

$$(\Box - m^2)V_{CT}^{\ \mu} = 0 \tag{4.81}$$

where $\Box = D_\mu D^\mu$ by eq. 4.65. In addition we have the subsidiary condition on the complex 3-momenta:

$$\nabla_r \cdot \nabla_i V_{CT}^{\ \mu} = 0 \tag{4.82}$$

Eq. 4.81 is immediately recognizable as a tachyon equation for each component. We now transform the lagrangian to light-front coordinates and proceed to quantize. Based on the previous definition of light-front variables we define the fields:

$$\begin{aligned} F_{CT}^{+-} &= \partial^+ V_{CT}^- - \partial^- V_{CT}^+ \\ F_{CT}^{+j} &= \partial^+ V_{CT}^j - \partial^j V_{CT}^+ \\ F_{CT}^{-j} &= \partial^- V_{CT}^j - \partial^j V_{CT}^- \\ F_{CT}^{ij} &= \partial^i V_{CT}^j - \partial^j V_{CT}^i \\ V_{CT}^- &= 2^{-\frac{1}{2}}(V_{CT}^0 - V_{CT}^3) \\ V_{CT}^+ &= 2^{-\frac{1}{2}}(V_{CT}^0 + V_{CT}^3) \end{aligned} \qquad (4.83)$$

The light-front equivalent of the lagrangian (eq. 4.77) is:

$$\mathcal{L}_{CTVB} = -\tfrac{1}{2}(D^\nu V_{CT\mu} D_\nu V_{CT}^\mu - D_\nu V_{CT}^\mu D_\mu V_{CT}^\nu - \tfrac{1}{2}m^2 V_{CT}^\mu V_{CT\mu} \qquad (4.84)$$

After using the constraint eq. 4.80 and discarding a total divergence, we see

$$\mathcal{L}_{CTVB} \equiv -\tfrac{1}{2}D^\nu V_{CT\mu} D_\nu V_{CT}^\mu - \tfrac{1}{2}m^2 V_{CT}^\mu V_{CT\mu} \qquad (4.85)$$

which becomes

$$\begin{aligned} \mathcal{L}_{CTVB} \equiv & -\partial^+ V_{CT}^- \partial^- V_{CT}^+ - \partial^+ V_{CT}^+ \partial^- V_{CT}^- + \partial^+ V_{CT}^k \partial^- V_{CT}^k + E^i V_{CT}^+ E^i V_{CT}^- + \\ & + \tfrac{1}{2}E^i V_{CT}^k E^i V_{CT}^k - \tfrac{1}{2}m^2(2V_{CT}^+ V_{CT}^- - V_{CT}^k V_{CT}^k) \end{aligned} \qquad (4.86)$$

using light-front coordinates with implied sums over i = 1, 2, 3 and k = 1, 2; where

$$E^1 = \partial/\partial x_r^1 + i\partial/\partial x_i^1 \qquad E^2 = \partial/\partial x_r^2 + i\partial/\partial x_i^2 \qquad E^3 = i\partial/\partial x_i^3$$

The resulting equations of motion are

$$\begin{aligned} (\Box - m^2)V_{CT}^- &= 0 \\ (\Box - m^2)V_{CT}^+ &= 0 \\ (\Box - m^2)V_{CT}^i &= 0 \end{aligned} \qquad (4.87)$$

for i = 1, 2 in agreement with eq. 4.81.

The conjugate spacelike-surface momenta are

$$\pi^\mu = \partial\mathcal{L}/\partial(\partial^0 V_{CT}{}^\mu) = -\,\partial V_{CT}{}^\mu/\partial x^0 \tag{4.88}$$

and the conjugate light-front momenta are

$$\pi^+ = \partial\mathcal{L}/\partial(\partial^- V_{CT}{}^+) = -\,\partial^+ V_{CT}{}^- \equiv -\,\partial V_{CT}{}^-/\partial x^- \tag{4.89}$$
$$\pi^- = \partial\mathcal{L}/\partial(\partial^- V_{CT}{}^-) = -\,\partial^+ V_{CT}{}^+ \equiv -\,\partial V_{CT}{}^+/\partial x^- \tag{4.90}$$
$$\pi^i = \partial\mathcal{L}/\partial(\partial^- V_{CT}{}^i) = \partial^+ V_{CT}{}^i \equiv \partial V_{CT}{}^i/\partial x^- \tag{4.91}$$

The equal x^+ commutation relations are

$$[\pi_{CT}{}^\mu(x), V_{CT}{}^\nu(y)] = +i2^{-3/2}\delta^{6\mu\nu}(x-y) \tag{4.92}$$

where

$$\delta^{6\mu\nu}(x-y) = \int d^2k_r dk_r{}^+ d^3k_i \delta'(\mathbf{k}_r\cdot\mathbf{k}_i/m^2) e^{i[k_r{}^+(x^- - y^-) - \mathbf{k}_{r\perp}\cdot(\mathbf{x}_{r\perp} - \mathbf{y}_{r\perp}) + \mathbf{k}_i\cdot(\mathbf{x}_i - \mathbf{y}_i)]}\cdot$$
$$\cdot[g^{\mu\nu} + k^\mu k^\nu/m^2]/(2\pi)^6 \tag{4.93}$$

where $\mathbf{k}_{r\perp} = (k_r{}^1, k_r{}^2)$, and $g^{-+} = g^{+-} = 1 = -g^{11} = = -g^{22}$ with all other $g^{ab} = 0$ with variables defined as in eqs. 3.184 – 3.186. (In normal coordinates the 3-vector $\mathbf{k} = \mathbf{k}_r + i\mathbf{k}_i$.) The equal x^+ commutation relations satisfy the constraint:

$$\partial_\mu[\pi_{CT}{}^\mu(x), V_{CT}{}^\nu(y)] = \partial_\nu[\pi_{CT}{}^\mu(x), V_{CT}{}^\nu(y)] = 0 \tag{4.94}$$

as implied by eq. 4.80.

Next we define the Fourier expansion of $V_{CT}{}^\mu$ as

$$V_{CT}{}^\mu(x) = \sum_s \int d^2k dk^+ d^3k_i N_{CTV}(k)\delta(\mathbf{k}_r\cdot\mathbf{k}_i/m^2)\theta(k^+)\varepsilon^\mu(k, s)\cdot$$
$$\cdot[a_{CT}(k, s)e^{-i(k\cdot x^* + k^*\cdot x)/2} + a_{CT}{}^\dagger(k, s)e^{+i(k\cdot x^* + k^*\cdot x)/2}] \tag{4.95}$$

where $x = (x^0, \mathbf{x} = \mathbf{x}_r - i\mathbf{x}_r)$, $k = (k^0, \mathbf{k} = \mathbf{k}_r + i\mathbf{k}_i)$, and where $k^2 = -m^2$, and where $N_{CTV}(k)$ is

$$N_{CTV}(k) = [(2\pi)^3 k^+]^{-1/2} \tag{4.96}$$

There are three spin orientations: two transverse orientations and a longitudinal orientation, $s = \pm 1, 0$. The spin polarization vector satisfies

$$k_\mu\varepsilon^\mu(k, s) = 0 \tag{4.97}$$

It also satisfies the normalization condition

$$\sum_s \varepsilon^{\mu}(k, s)\varepsilon^{\upsilon}(k, s) = -(g^{\mu\nu} + k^{\mu}k^{\nu}/m^2) \qquad (4.98)$$

The Fourier component operator commutation relations are

$$[a_{CT}(q, s), a_{CT}^{\dagger}(p, s')] = 2^{-1/2}\delta_{ss'}\delta^2(\mathbf{q} - \mathbf{p'})\delta(q^+ - p^+) \qquad (4.99)$$
$$[a_{CT}(q, s), a_{CT}(p, s')] = [a_{CT}(q, s), a_{CT}(p, s')] = 0$$

Eqs. 4.95 – 4.99 imply the commutation relations eqs. 4.92.

Tachyonic Complexon Vector Boson Feynman Propagator

The tachyonic complexon vector boson Feynman propagator is defined by

$$i\Delta_{CTF}(x - y)^{\mu\nu} = \theta(x^+ - y^+)\langle 0|V_{CT}^{\mu}(x)V_{CT}^{\nu}(y)|0\rangle + $$
$$+ \theta(y^+ - x^+)\langle 0|V_{CT}^{\nu}(y)V_{CT}^{\mu}(x)|0\rangle \qquad (4.100)$$

and is equal to

$$= \tfrac{1}{2} i \int \frac{d^4k_r d^3k_i\, \delta'(\mathbf{k_r}\cdot\mathbf{k_i}/m^2) e^{-ik\cdot(x - y)} (g^{\mu\nu} + k^{\mu}k^{\nu}/m^2)}{(2\pi)^7 (k^2 + m^2 + i\varepsilon)}$$

with $k = (k^0, \mathbf{k} = \mathbf{k_r} + i\mathbf{k_i})$.

5. Free Tachyon Discrete Symmetries: C, P, T, and CPT

5.0 Introduction

In the first part of this chapter we will consider the behavior of quantum tachyon theories (developed from Extended Lorentz group (and L_C) boosts) under the discrete transformations P, C and T. In sections 5.8 and 5.9 we will consider the behavior of quantum field theories of complexons under the discrete transformations.

5.1 Tachyons and the Discrete Symmetries: C, P, and T

The discrete (improper) transformations, parity, time reversal and charge conjugation, are of major importance in analyzing the structure of ElectroWeak Theory and the Standard Model. In this chapter we will examine these transformations with respect to tachyon particles.

First, bosonic tachyons (spins 0, 1, and 2) have discrete transformation properties similar to ordinary bosons and so will not be considered further. The interested reader should read standard texts on this topic and notice the sign of the squared mass does not introduce any distinctive differences between tachyon and normal bosons.

In the case of fermions (odd half integer spin particles) there is a difference between tachyon fermions and conventional fermions. We will consider the case of spin ½ tachyons. Fundamental tachyon fermions of higher spin (should any exist) would also have distinctively different P, C, and T transformation properties.

We will use the manifestly covariant lagrangian

$$\mathcal{L} = \psi_T^\dagger i\gamma^0\gamma^5(\gamma^\mu\partial/\partial x^\mu - m)\psi_T(x) \tag{5.1}$$

with equations of motion

$$(\gamma^\mu\partial/\partial x^\mu - m)\psi_T(x) = 0 \tag{5.2}$$

and anti-commutator

$$\{\psi_{T\,a}^{\dagger}(x), \psi_{Tb}(x')\} = -[\gamma^5]_{ab}\,\delta^3(x - x') \tag{5.3}$$

as the starting points of our discussion.

5.2 Parity

In defining the parity transformation for spin ½ tachyons we try to retain as much similarity as possible to the Dirac spin ½ fermion parity transformation. By definition the parity transformation changes $\mathbf{x} \rightarrow -\mathbf{x}$. In the case of a Dirac field if the transformation is defined as:

$$\mathcal{P}\psi(\mathbf{x}, t)\mathcal{P}^{-1} = \gamma^0\psi(-\mathbf{x}, t) \tag{5.4}$$

then the free Dirac field lagrangian, field equation and and anti-commutators are invariant under this transformation.

If we now consider a spin ½ tachyon field and assume the same general form for the transformation:

$$\mathcal{P}\psi_T(\mathbf{x}, t)\mathcal{P}^{-1} = \gamma^0\psi_T(-\mathbf{x}, t) \tag{5.5}$$

then we find

$$\mathcal{P}\mathcal{L}(\mathbf{x}, t)\mathcal{P}^{-1} = -\mathcal{L}(-\mathbf{x}, t) \tag{5.6}$$

$$(\gamma^\mu\partial/\partial x'^\mu - m)\psi_T(-\mathbf{x}, t) = 0 \tag{5.7}$$

$$\{\psi_{T\,a}^{\dagger}(x'), \psi_{Tb}(y')\} = [\gamma^5]_{ab}\,\delta^3(x' - y') \tag{5.8}$$

where $x' = (-\mathbf{x}, t)$ and $y' = (-\mathbf{y}, t)$. The lagrangian and the anti-commutation relations change sign under the parity transformation. Therefore the physics of tachyons is not invariant under parity. This fact is directly evidenced by eq. 3.45 where the expression of the lagrangian in terms of left-handed and right-handed fields shows the lagrangian changes sign under the interchange of left and right handed fields (an effect of the parity transformation). Thus spin ½ tachyon theory, like nature, violates parity.

Note that parity violation is intrinsic to tachyons – even free tachyons. The discussion of the Standard Model in the following chapters will associate tachyon parity violation with Standard Model parity violation.

5.3 Charge Conjugation

The charge conjugation transformation is connected to the interchange of particle and antiparticle. If we assume that a spin ½ tachyon has charge and is coupled to the electromagnetic field then (assuming the usual gauge coupling) the tachyon lagrangian becomes

$$\mathcal{L} = \psi_T^\dagger i\gamma^0\gamma^5[\gamma^\mu(\partial/\partial x^\mu + ieA_\mu) - m]\psi_T(x) \tag{5.9}$$

If the theory were charge conjugation invariant then a unitary operator $\mathcal{C}$ would exist that would change the sign of the electromagnetic current $j^\mu(x)$, and the electromagnetic field A(x, t), while leaving the lagrangian invariant:

$$\mathcal{C}j^\mu(x)\mathcal{C}^{-1} = -j^\mu(x) \qquad ???? \tag{5.10}$$

$$\mathcal{C}\mathbf{A}(\mathbf{x}, t)\mathcal{C}^{-1} = -\mathbf{A}(\mathbf{x}, t) \qquad ???? \tag{5.11}$$

$$\mathcal{C}\mathcal{L}(\mathbf{x}, t)\mathcal{C}^{-1} = \mathcal{L}(\mathbf{x}, t) \qquad ???? \tag{5.12}$$

The tachyon electromagnetic current implied by the lagrangian eq. 5.9 is

$$j_T^\mu(x) = \tfrac{1}{2} e[\psi_T^\dagger\gamma^0\gamma^5, \gamma^\mu\psi_T] \tag{5.13}$$

where we antisymmetrize as in the case of Dirac fermions.

We extend the standard charge conjugation transformation[26] *with one modification* from that of a conventional Dirac spin ½ field to the case of spin ½ tachyons:

$$\psi_{TC}(\mathbf{x}, t) = \mathcal{C}_T\psi_T(\mathbf{x}, t)\mathcal{C}_T^{-1} = C_T\bar{\psi}_T^{\,T}(\mathbf{x}, t) = C_T\gamma^0\psi_T^{\,*}(\mathbf{x}, t) \tag{5.14}$$

where $\psi_{TC}(\mathbf{x}, t)$ is the antitachyon field of opposite charge, where the <u>super</u>script T indicating the transpose, and where the tachyon charge conjugation matrix (which differs from the Dirac field analogue) is

$$C_T = i\gamma^2\gamma^5\gamma^0 \qquad \text{and} \qquad C_T^{-1} = i\gamma^0\gamma^5\gamma^2 = -C_T = C_T^\dagger \tag{5.15}$$

The hermitean conjugate of the antitachyon field is

$$\psi_{TC}^\dagger(\mathbf{x}, t) = \mathcal{C}_T\psi_T^\dagger(\mathbf{x}, t)\mathcal{C}_T^{-1} = \bar{\psi}_T^{\,*}(\mathbf{x}, t)C_T^{-1} = \psi_T^{\,T}(\mathbf{x}, t)\gamma^0C_T^{-1} \tag{5.16}$$

Under this transformation we find the tachyon lagrangian, field equation, and anti-commutator are invariant under charge conjugation:

[26] In the Dirac representation. See for example Bjorken (1965) p. 115, or Kaku (1993) p. 117.

$$[\gamma^\mu(\partial/\partial x^\mu + ieA_\mu) - m]\psi_{TC}(x) = 0 \tag{5.17}$$

using $\mathcal{C}_T A_\mu \mathcal{C}_T^{-1} = -A_\mu$, and for the lagrangian in eq. 5.9

$$\mathcal{C}_T \mathcal{L}(\mathbf{x}, t)\mathcal{C}_T^{-1} = \mathcal{L}(\mathbf{x}, t) \tag{5.18}$$

The charge conjugate anti-commutator is

$$\{\psi_{TC}{}^\dagger{}_a(x), \psi_{TCb}(y)\} = -[\gamma^5]_{ab}\,\delta^3(x - y) \tag{5.19}$$

The charge conjugated current

$$j_{TC}{}^\mu(x) = \tfrac{1}{2}\, e[\psi_{TC}{}^\dagger\gamma^0\gamma^5, \gamma^\mu\psi_{TC}] = -\, j_T{}^\mu(x) \tag{5.20}$$

so that

$$\mathcal{C}_T j_T{}^\mu(x)A_\mu\mathcal{C}_T^{-1} = j_T{}^\mu(x)A_\mu \tag{5.21}$$

resulting in the tachyon charge conjugation invariant lagrangian eq. 5.9. Thus tachyons and antitachyons can be expected to have the same charge and mass.

5.4 CP Transformation

The CP transformation has been of major theoretical and experimental interest for some time. Experimentally[27] CP violation has been found in certain sectors: K meson and B meson decays.

Since we have seen that tachyons violate parity and do not violate charge conjugation we can see that tachyons inherently violate CP invariance.

5.5 Time Reversal

Time reversal invariance is also a significant theoretical and experimental issue. The standard Dirac fermion time reversal transformation is:

$$\mathcal{T}\psi(\mathbf{x}, t)\mathcal{T}^{-1} = T\psi(\mathbf{x}, -t) \tag{5.22}$$

where $\mathcal{T} = \mathcal{U}K$ where $\mathcal{U}$ is a unitary operator and K is the operator that takes the complex conjugate of all c-numbers, and where

$$T = i\gamma^1\gamma^3 \tag{5.23}$$

---

[27] B. Aubert et al, BaBar-PUB-07/001, arXiv:hp-ex/0702046 (2007) and references therein.

Due to the form of $\mathcal{L}$ (eq. 5.9), which assumes an electromagnetic interaction for the purpose of illustration, we find that the *tachyon* time reversal transformation is

$$\mathfrak{T}_T \psi_T(\mathbf{x}, t)\mathfrak{T}_T^{-1} = T_T \psi_T(\mathbf{x}, -t) \tag{5.24}$$

where $\mathfrak{T}_T = \mathcal{U}_T K$ where $\mathcal{U}_T$ is a unitary operator and K is the operator that takes the complex conjugate of all c-numbers, and where

$$T_T = i\gamma^5\gamma^1\gamma^3 \tag{5.25}$$

The matrix T_T satisfies:

$$T_T^{-1} = -i\gamma^3\gamma^1\gamma^5 = T_T \tag{5.26}$$

$$T_T\gamma^\mu T_T^{-1} = -\gamma_\mu \tag{5.27}$$

Under time reversal the current satisfies

$$\mathfrak{T}_T j_{T\mu}(\mathbf{x}, t)\mathfrak{T}_T^{-1} = j_T^{\ \mu}(\mathbf{x}, -t) \tag{5.28}$$

If we assume the electromagnetic field satisfies

$$\mathfrak{T}\mathbf{A}(\mathbf{x}, t)\mathfrak{T}^{-1} = -\mathbf{A}(\mathbf{x}, -t)$$

under time reversal, then the tachyon lagrangian

$$\mathcal{L} = \psi_T^{\dagger} i\gamma^0\gamma^5[\gamma^\mu(\partial/\partial x^\mu + ieA_\mu) - m]\psi_T(x) \tag{5.29}$$

satisfies

$$\mathfrak{T}_T\mathcal{L}(\mathbf{x}, t)\mathfrak{T}_T^{-1} = \mathcal{L}(\mathbf{x}, -t) \tag{5.30}$$

under time reversal. Although the action changes by a translation in time, Poincaré translation invariance implies the action is invariant. The equation of motion derived from the lagrangian eq. 5.9 is also invariant under the tachyon time reversal transformation.

Thus we find the dynamics of the tachyon lagrangian theory (eq. 5.9) to be invariant under the tachyon time reversal transformation.

5.6 Tachyon CPT Non-Invariance

The question of CPT invariance has long been of theoretical and experimental interest. For conventional particle theories the CPT Theorem implies

CPT invariance under very general conditions. We will examine the case of CPT invariance of a model tachyon theory with lagrangian eq. 5.9. For *Dirac* fermions

$$\mathcal{CPT}\psi_a(\mathbf{x}, t)\mathcal{T}^{-1}\mathcal{P}^{-1}\mathcal{C}^{-1} = i[\psi^\dagger(-\mathbf{x}, -t)\gamma^5]_a = i[\gamma^5\psi^*(-\mathbf{x}, -t)]_a \tag{5.31}$$

For spin ½ *tachyons*

$$\mathcal{C}_T\mathcal{P}\mathcal{T}_T\psi_{Ta}(\mathbf{x}, t)\mathcal{T}_T^{-1}\mathcal{P}^{-1}\mathcal{C}_T^{-1} = -i[\psi^\dagger(-\mathbf{x}, -t)\gamma^5]_a \tag{5.32}$$

where a is a spinor index. More succinctly,

$$\mathcal{C}_T\mathcal{P}\mathcal{T}_T\psi_T(\mathbf{x}, t)\mathcal{T}_T^{-1}\mathcal{P}^{-1}\mathcal{C}_T^{-1} = -i\gamma^5\psi^*(-\mathbf{x}, -t) \tag{5.33}$$

Eq. 5.33 differs only by a phase from eq. 5.31.

Therefore one might think that bilinear combinations of ψ and $\psi^\dagger$ which of necessity must be factors in a lagrangian will result in the cancellation of the –1 factors upon CPT transformation.

However the free field lagrangian

$$\mathcal{L} = \psi_T^\dagger(x)i\gamma^0\gamma^5(\gamma^\mu\partial/\partial x^\mu - m)\psi_T(x) \tag{5.34}$$

changes to

$$\mathcal{L}_{CPT} = \psi_T^\dagger(-x)i\gamma^0\gamma^5(\gamma^\mu\partial/\partial x^\mu + m)\psi_T(-x) \tag{5.35}$$

The mass term violates CPT invariance. Thus massive spin ½ tachyons inherently violate CPT invariance. If all spin ½ particles start out massless and acquire masses through spontaneous symmetry breaking then the breaking of CPT invariance by massive, spin ½ tachyons is another consequence of spontaneous symmetry breaking.

5.7 Microcausality and Tachyons

Since the CPT Theorem does not hold for spin ½ tachyons it is of interest to consider Jost's Theorem: CPT invariance is equivalent to weak local commutativity, which is a weak form of microcausality. In the case of spin ½ tachyons *weak local commutativity* (or weak microcausality) is defined as

$$<0|\{\psi_T^\dagger(x), \psi_T(y)\}|0> = 0 \quad \text{for} \quad (x-y)^2 < 0$$

(spacelike $(x - y)^2$).

The absence of CPT invariance leads us to inquire if microcausality, and/or weak microcausality, still hold?

To answer this question we evaluate the left-handed field commutator $\{\psi_{TL}{}^{+\dagger}(x), \psi_{TL}{}^{+}(y)\}$ to see if the normal microcausality condition holds:

$$\{\psi_{TL}{}^{+\dagger}(x), \psi_{TL}{}^{+}(y)\} = 0 \quad \text{for} \quad (x-y)^2 < 0 \tag{5.36}$$

We insert the Fourier expansions:

$$\{\psi_{TL}{}^{+}{}_{a}(x), \psi_{TL}{}^{+\dagger}{}_{b}(y)\} = \sum_{\pm s,s'} \int d^2p dp^+ \int d^2p' dp'^+ \, N_{TL}{}^{+}(p) N_{TL}{}^{+}(p')\theta(p^+)\theta(p'^+)\cdot$$
$$\cdot[\{b_{TL}{}^{+\dagger}(p',s'), b_{TL}{}^{+}(p,s)\} u_{TL}{}^{+}{}_{a}(p,s) u_{TL}{}^{+\dagger}{}_{b}(p',s') e^{+ip'\cdot y - ip\cdot x} +$$
$$+ \{d_{TL}{}^{+}(p',s'), d_{TL}{}^{+\dagger}(p,s)\} v_{TL}{}^{+}{}_{a}(p,s) v_{TL}{}^{+\dagger}{}_{b}(p',s') e^{-ip'\cdot y + ip\cdot x}]$$

$$= \sum_{\pm s} \int d^2p dp^+ \, N_{TL}{}^{+2}(p)\theta(p^+)[u_{TL}{}^{+}{}_{a}(p,s) u_{TL}{}^{+\dagger}{}_{b}(p,s) e^{+ip\cdot(y-x)} +$$
$$+ v_{TL}{}^{+}{}_{a}(p,s) v_{TL}{}^{+\dagger}{}_{b}(p,s) e^{-ip\cdot(y-x)}]$$

$$= i\int d^2p dp^+ \theta(p^+) N_{TL}{}^{+2}(p)(2m|\mathbf{p}|)^{-1}\{[C^- R^+(-i\not{p} + m)\boldsymbol{\gamma}\cdot\mathbf{p} R^+ C^-]_{ab} e^{+ip\cdot(y-x)} +$$
$$+ [C^- R^+(-i\not{p} - m)\boldsymbol{\gamma}\cdot\mathbf{p} R^+ C^-]_{ab} e^{-ip\cdot(y-x)}\}$$

$$= \tfrac{1}{2}[C^- R^+]_{ab} \int d^2p_\perp \int_0^\infty dp^+ (2\pi)^{-3} (e^{+ip\cdot(y-x)} + e^{-ip\cdot(y-x)})$$

where $p\cdot(y-x) = p^-(y^+ - x^+) + p^+(y^- - x^-) - \mathbf{p}_\perp\cdot(\mathbf{y}_\perp - \mathbf{x}_\perp)$. Since $p^2 = -m^2$, the integral can be rewritten, after letting $p^\mu = -p^\mu$, as

$$= \tfrac{1}{2}[C^- R^+]_{ab} \int d^2p_\perp \int_{-\infty}^{\infty} dp^+ (2\pi)^{-3} e^{-ip\cdot(y-x)}$$

where $p^- = (p_\perp{}^2 + m^2)/(2p^+)$. For spacelike $(x-y)^2 < 0$ we can always choose a coordinate system where $y^+ - x^+ = 0$ with the result

$$\{\psi_{TL}{}^{+\dagger}(x), \psi_{TL}{}^{+}(y)\} = 2^{-1} C^- R^+ \delta(y^- - x^-)\delta^2(\mathbf{y} - \mathbf{x}) \quad \underline{\text{if}} \quad y^+ - x^+ = 0 \tag{5.37}$$

Therefore

$$\{\psi_{TL}{}^{+\dagger}(x), \psi_{TL}{}^{+}(y)\} = 0 \quad \text{for} \quad (x-y)^2 < 0 \tag{5.38}$$

Consequently, free left-handed (or right-handed) tachyons with light-front quantization satisfy the microcausality condition.

5.8 C, P and T for Spin ½ Complexons ($m^2 > 0$)

In this section we will examine the behavior under C, P, and T of the spin ½ complexon quantum fields[28] that we found in chapter 3. The lagrangian that we defined for this type of fermion was

$$\mathcal{L} = \bar{\psi}_C(i\gamma^\mu D_\mu - m)\psi_C(x) \tag{3.127}$$

where $\bar{\psi}_C = \psi_C^{\dagger}\gamma^0$ and

$$\psi_C^{\dagger} = [\psi_C(\mathbf{x}_r, \mathbf{x}_i)]^\dagger \,|_{\mathbf{x}_i = -\mathbf{x}_i} \tag{3.128}$$

$$D_0 = \partial/\partial x^0$$
$$D_k = \partial/\partial x^k + i\,\partial/\partial x_i^{\,k} \tag{3.129}$$

with $x^k = x_r^{\,k}$ for k = 1,2, 3. The invariant action (under real Lorentz transformations) is

$$I = \int d^7x\mathcal{L} \tag{3.130}$$

Parity

Under a parity transformation

$$\mathcal{P}\psi_C(\mathbf{x}, t)\mathcal{P}^{-1} = \gamma^0\psi_C(-\mathbf{x}, t) \tag{5.39}$$

we find

$$\mathcal{P}\mathcal{L}(\mathbf{x}, t)\mathcal{P}^{-1} = \mathcal{L}(-\mathbf{x}, t) \tag{5.40}$$

$$(i\gamma^\mu D'_\mu - m)\psi_C(-\mathbf{x}, t) = 0 \tag{5.41}$$

where $x = x_r - ix_i$. Parity is conserved.

Charge Conjugation

To examine the fermion lagrangian under charge conjugation we introduce an electromagnetic field of the kind

$$\mathcal{L} = \psi_C^{\dagger}\gamma^0[i\gamma^\mu(D_\mu - ieA_\mu) - m]\psi_C(x) \tag{5.42}$$

[28] We will not consider the case of integer spin quantum fields with complex 3-momenta since they have straight-forward transformation properties under C, P and T.

$$j_C^{\mu}(x) = ½ [\psi_C^{\dagger}\gamma^0, \gamma^{\mu}\psi_C] \tag{5.42}$$

$$\psi_{C\mathcal{C}}(\mathbf{x}, t) = \mathcal{C}\psi_C(\mathbf{x}, t)\mathcal{C}^{-1} = C\bar{\psi}_C^{\,T}(\mathbf{x}, t) = C\gamma^0\psi_C^{\,*}(\mathbf{x}, t) \tag{5.43}$$

where $\psi_{C\mathcal{C}}(\mathbf{x}, t)$ is the antiparticle field of opposite charge, where the <u>super</u>script T indicating the transpose, and where the charge conjugation matrix (which is the same as the Dirac field analogue) is

$$C = i\gamma^2\gamma^0 = -C^T \quad \text{and} \quad C^{-1} = i\gamma^0\gamma^2 = -C = C^{\dagger} \tag{5.44}$$

The hermitean conjugate of the antiparticle field is

$$\psi_{C\mathcal{C}}^{\dagger}(\mathbf{x}, t) = \mathcal{C}\psi_C^{\dagger}(\mathbf{x}, t)|_{\mathbf{x}_i=-\mathbf{x}_i}\mathcal{C}^{-1} = \bar{\psi}_C^{\,*}(\mathbf{x}, t)|_{\mathbf{x}_i=-\mathbf{x}_i}C^{-1} = \psi_C^{\,T}(\mathbf{x}, t)|_{\mathbf{x}_i=-\mathbf{x}_i}\gamma^0 C^{-1} \tag{5.45}$$

Applying the charge conjugation transformation we find

$$\mathcal{C}j^{\mu}(x)\mathcal{C}^{-1} = -\,j^{\mu}(x) \tag{5.46}$$

and if $\mathcal{C}\mathbf{A}(\mathbf{x}, t)\mathcal{C}^{-1} = -\mathbf{A}(\mathbf{x}, t)$ then the action

$$I = \int d^7x\,\mathcal{L}(\mathbf{x}, t) = \int d^7x\;\mathcal{C}\mathcal{L}(\mathbf{x}, t)\mathcal{C}^{-1} \tag{5.47}$$

is invariant under charge conjugation. Note we change the sign of the imaginary coordinates $\mathbf{x}_i \rightarrow -\mathbf{x}_i$ in the action integration to obtain eq. 5.47.

Time Reversal

We use a modification of the standard Dirac fermion time reversal transformation:

$$\mathcal{T}\psi_C(\mathbf{x}, t)\mathcal{T}^{-1} = T\psi_C(\mathbf{x}^*, -t) \tag{5.48}$$

where $\mathcal{T} = \mathcal{U}K$ where $\mathcal{U}$ is a unitary operator and K is the operator that takes the complex conjugate of all c-numbers, where $\mathbf{x} = \mathbf{x}_r - i\mathbf{x}_i$, and where

$$T = i\gamma^1\gamma^3 \tag{5.49}$$

Assuming the lagrangian $\mathcal{L}$ (eq. 5.42) we find the time reversal transformation gives

$$I = \int d^7x \mathfrak{I}_T \mathcal{L}(\mathbf{x}, t)\mathfrak{I}_T^{-1} = \int d^7x \mathcal{L}(\mathbf{x}, -t) \tag{5.50}$$

under time reversal. Although the action changes by a translation in time, Poincaré time translation invariance implies the action is invariant. Note we change the sign of the imaginary coordinates $\mathbf{x}_i \rightarrow -\mathbf{x}_i$ in the action integration to obtain eq. 5.50.

CP Invariance

Since we have seen that particles with complex 3-momenta do not violate parity and do not violate charge conjugation we can see that CP is not violated.

CPT Invariance

The combined impact of the above subsections shows CPT is not violated by particles with complex 3-momenta.

5.9 C, P and T for Complexon Tachyons

In this section we will examine the behavior under C, P, and T of the spin ½ tachyonic complexon quantum fields. The lagrangian that we defined for this type of fermion was

$$\mathcal{L} = \bar{\psi}_{CT}(\gamma^\mu D_\mu - m)\psi_{CT}(x) \tag{5.51}$$

where $\bar{\psi}_{CT} = \psi_{CT}^\dagger i\gamma^0\gamma^5$ and

$$\psi_{CT}^\dagger = [\psi_{CT}(\mathbf{x}_r, \mathbf{x}_i)]^\dagger \big|_{\mathbf{X}_i = -\mathbf{X}_i} \tag{3.128}$$

$$D_0 = \partial/\partial x^0$$
$$D_k = \partial/\partial x^k + i\, \partial/\partial x_i^k \tag{3.129}$$

with $x^k = x_r^k$ for $k = 1,2,3$. The invariant action (under real Lorentz transformations) is

$$I = \int d^7x \mathcal{L} \tag{3.130}$$

Parity

Under a parity transformation

$$\mathcal{P}\psi_{CT}(\mathbf{x}, t)\mathcal{P}^{-1} = \gamma^0\psi_{CT}(-\mathbf{x}, t) \tag{5.52}$$

we find

$$\mathcal{P}\mathcal{L}(\mathbf{x}, t)\mathcal{P}^{-1} = -\mathcal{L}(-\mathbf{x}, t) \tag{5.53}$$

Parity is *not* conserved.

Charge Conjugation

To examine the fermion lagrangian under charge conjugation we introduce an electromagnetic field of the kind

$$\mathcal{L} = \psi_{CT}^{\dagger}\, i\gamma^0\gamma^5[\gamma^\mu(D_\mu - ieA_\mu) - m]\psi_{CT}(x) \tag{5.54}$$

$$j_{CT}^{\mu}(x) = ½\, [\psi_{CT}^{\dagger}\gamma^0\gamma^5, \gamma^\mu\psi_{CT}] \tag{5.55}$$

$$\psi_{CT\mathcal{C}}(\mathbf{x}, t) = \mathcal{C}_T\psi_{CT}(\mathbf{x}, t)\mathcal{C}_T^{-1} = C_T\bar{\psi}_{CT}^{T}(\mathbf{x}, t) = C_T\gamma^0\psi_{CT}^{*}(\mathbf{x}, t) \tag{5.56}$$

where $\psi_{CT\mathcal{C}}(\mathbf{x}, t)$ is the antiparticle field of opposite charge, where the <u>super</u>script T indicating the transpose, and where the charge conjugation matrix (which is the same as the Dirac field analogue) is

$$C_T = i\gamma^2\gamma^5\gamma^0 \qquad \text{and} \qquad C_T^{-1} = i\gamma^0\gamma^5\gamma^2 = -C_T = C_T^{\dagger} \tag{5.15}$$

The hermitean conjugate of the antiparticle field is

$$\begin{aligned}\psi_{CT\mathcal{C}}^{\dagger}(\mathbf{x}, t) &= \mathcal{C}_T\psi_{CT}^{\dagger}(\mathbf{x}, t)|_{\mathbf{x_i}=-\mathbf{x_i}}\mathcal{C}_T^{-1} \\ &= \bar{\psi}_{CT}^{*}(\mathbf{x}, t)|_{\mathbf{x_i}=-\mathbf{x_i}}C_T^{-1} = \psi_{CT}^{T}(\mathbf{x}, t)|_{\mathbf{x_i}=-\mathbf{x_i}}\gamma^0C_T^{-1}\end{aligned} \tag{5.57}$$

Applying the charge conjugation transformation we find

$$\mathcal{C}_T j^\mu(x)\mathcal{C}_T^{-1} = -\, j^\mu(x) \tag{5.58}$$

and if $\mathcal{C}\mathbf{A}(\mathbf{x}, t)\mathcal{C}^{-1} = -\mathbf{A}(\mathbf{x}, t)$ then the action

$$I = \int d^7x\mathcal{L}(\mathbf{x}, t) = \int d^7x\mathcal{C}_T\mathcal{L}(\mathbf{x}, t)\mathcal{C}_T^{-1} \tag{5.59}$$

is *invariant* under charge conjugation. Note we change the sign of the imaginary coordinates $\mathbf{x_i} \rightarrow -\mathbf{x_i}$ in the action integration to obtain eq. 5.59.

Time Reversal

We use a modification of the standard Dirac fermion time reversal transformation:

$$\mathcal{T}_T\psi_{CT}(\mathbf{x}, t)\mathcal{T}_T^{-1} = T_T\psi_{CT}(\mathbf{x}^*, -t) \tag{5.60}$$

where $\mathfrak{T}_T = \mathcal{U}_T K$ where $\mathcal{U}_T$ is a unitary operator and K is the operator that takes the complex conjugate of all c-numbers, where $\mathbf{x} = \mathbf{x_r} - i\mathbf{x_i}$, and where

$$T_T = i\gamma^5\gamma^1\gamma^3 \qquad (5.25)$$

The matrix T_T satisfies:

$$T_T^{-1} = -i\gamma^3\gamma^1\gamma^5 = T_T \qquad (5.26)$$

$$T_T\gamma^\mu T_T^{-1} = -\gamma_\mu \qquad (5.27)$$

Assuming the lagrangian $\mathcal{L}$ (eq. 5.51) we find the time reversal transformation gives

$$I = \int d^7x \mathfrak{T}_T \mathcal{L}(\mathbf{x}, t)\mathfrak{T}_T^{-1} = \int d^7x \mathcal{L}(\mathbf{x}, -t) \qquad (5.61)$$

under time reversal. Although the action changes by a translation in time, Poincaré time translation invariance implies the action is *invariant*. Note we change the sign of the imaginary coordinates $\mathbf{x_i} \rightarrow -\mathbf{x_i}$ in the action integration to obtain eq. 5.50.

CP Invariance

Since we have seen that tachyons with complex 3-momenta violate parity and do not violate charge conjugation we can see that tachyons inherently violate CP invariance.

CPT Invariance

The combined impact of the above subsections shows CPT is violated by tachyons with complex 3-momenta.

5.10 Summary of C, P, T, CP and CPT for Spin ½ Particles

In this section we summarize the results of the preceding sections. I indicates invariant under the transformation; N indicates not invariant under the transformation.

	P	C	T	CP	CPT
$M^2 > 0$	I	I	I	I	I
Complexon ($M^2 > 0$)	I	I	I	I	I
Tachyons	N	I	I	N	N
Tachyonic Complexons	N	I	I	N	N

Table 5.1. Summary of the C, P and T properties of the four types of spin ½ fermions.

Appendix 5-A. Tachyons and the Problem of the Uniformity of the Universe

A major cosmological problem has been the large-scale uniformity of the universe. The standard argument begins with the observation that if we trace back the various regions of the universe to very early times, parts of the universe that are now quite similar could not be in interaction with each other in the distant past because interactions are limited by the speed of light. Thus it is difficult to understand how these regions could be so similar today without any interaction between them that would have established uniformity. One proposed resolution of this difficulty is the inflationary scenario of Guth and others.

However, another solution to the uniformity problem presents itself. If we consider the universe at extremely early times when quarks were not confined, then tachyonic quarks and neutrinos could be the mechanism to resolve the uniformity problem since tachyons can exceed the speed of light. A detailed model of this possibility remains to be created.

Part 5. Two-Tier Renormalization – ***Quantum Theory of the Third Kind***

Quantum Theory of the Third Kind

A New Type of Divergence-free Quantum Field Theory Supporting a Unified Standard Model of Elementary Particles and Quantum Gravity based on a New Method in the Calculus of Variations

Stephen Blaha, Ph.D.*

Published in 2005

Pingree-Hill Publishing

* sblaha777@yahoo.com

ISBN: 0-9746958-3-1

PREFACE TO THE SECOND EDITION

This is the second edition of *A Finite Unified Quantum Field Theory of the Elementary Particle Standard Model and Quantum Gravity: Based on New Quantum Dimensions™ & a New Paradigm in the Calculus of Variations.* It contains new material such as a proof that Two-Tier quantum field theories are relativistic and corrects some typos in the first edition. It also provides a deeper discussion of the basis and rationale of this new form of quantum theory.

PREFACE

The Standard Model of Elementary Particles enjoys a measure of success in accounting for experimental results that is somewhat amazing in view of its "hodge-podge" nature, and in view of the manner in which it evolved from the 1940's until the present. Some issues that remain to be explicated are CP violation; the origin, and rationale, of the internal symmetries; the existence and origin of Higgs particles; and the unification of the Standard Model with Quantum Gravity. Practically the Standard Model is the only viable theory for performing calculations to compare to experiment.

The quantum field theoretic aspects of the Standard Model have been hitherto considered as satisfactory because the divergences that appear within it have been brought under control through renormalization techniques so that meaningful calculations can be made and compared with experiment. However these infinities (divergences) remain a source of uneasiness—particularly since known renormalization techniques cannot handle the infinities that crop up in Quantum Gravity—thus precluding a unified theory of all interactions.

One major attempt to solve the divergence issue as well as create a unified theory is Superstring theory. This theoretic approach may eventually be successful but the absence of any experimental evidence for the plethora of particles that superstring theories predict raises major questions as to its physical reality. Nature seems to be simpler than Superstring theory would have us believe. The use of compactified dimensions to generate internal symmetries also seems to beg the question. Compacted dimensions are little more than an artifice to insert extra dimensions. The physics of the compactification is unclear.

The type of quantum field theory theory presented *in this book* assumes that the universe has additional quantum dimensions that serve to eliminate divergences in quantum field theory – every particle has a quantum fluctuating cloud of quanta generating quantum dimensions around it. The dimensions are implemented as quantum fields and thus have no need for compactification.

The new type of theory, Two-Tier quantum field theory, is based on a new paradigm for the Calculus of Variations described in Appendix A. It leads to a new approach to quantization that we call quantum theory of the third kind. First there was quantum mechanics. Then there was quantum field theory. The new kind of quantum field theory defines quantum coordinates as the variable of quantum fields. (In conventional quantum field theory the coordinates are c-numbers – not operators.)

To my Wife, Margaret

With Love

CONTENTS

LIST OF FIGURES

1. Quantum Dimensions vs. Classical Dimensions

All beginnings are obscure.
H. Weyl – Space, Time, Matter

Beyond 4-Dimensional Space-time

There have been countless attempts since the 1920's to use additional dimensions beyond the known four dimensions of space and time to explicate and unify the fundamental forces of nature. The most noteworthy *recent* attempts along these lines have been Superstring theories and Technicolor theories.

In the opinion of this author the efforts in these directions are not justified by the results. The physics that these theories attempt to describe is simpler than the formulation of the theories with much fewer particles and interactions. Assigning high masses to undiscovered particles and placing undiscovered forces in the high energy regime beyond the limits of accelerators does not seem to be a satisfactory approach. Extrapolations of theories in the past, without the guidance and confirmation of experiment, have usually not been successful.

Therefore it seems reasonable to develop a deeper, sounder formulation of the Standard Model *as it is now* since it was developed through a close interplay of theory and experiment. It remains the preeminent theory of elementary particles—actually the *only* experimentally acceptable theory of elementary particles.

This book attempts to establish a deeper framework for the Standard Model that enables it to be combined with quantum gravity to form a divergence-free unified quantum field theory of nature.

It is clear from the existence of internal symmetries in the Standard Model that something is "going on" inside particles which is outside the framework of normal space and time. Otherwise, there would be no internal symmetries.

Therefore it is reasonable to consider the possibility of extra dimensions beyond normal space-time. The open question is how these extra dimensions enter into physical theory. Superstring theory appears to go too far in terms of the numbers of dimensions and the particles that it requires – not to mention – the complexity of its mathematics, appears to preclude all but the simplest calculations. Technicolor and extended Technicolor theories also introduce substantial additional complexities in order to explicate the pattern of symmetries of the Standard Model. We face the question of whether the cure is worst than the disease in these approaches.

Thus we ask if a more tractable theory is possible. Our first requirement is that it would improve the Standard Model by taming the divergences in quantum field theory so that the Standard Model can be unified with Quantum Gravity. This author has suggested[1] an alternate form of quantum field theory, Two-Tier quantum field theory, that is ultra-violet divergence-free to all orders for theories of the type of the Standard Model, and for Quantum Gravity. This theory not only resolves divergence issues in the Standard Model and Quantum Gravity but also eliminates the singularities at the point of the Big Bang in Cosmology.[2]

This theory is *not* based on extra dimensions that become compactified as in Superstring theories. Rather it postulates that extra dimensions are directly generated by a free quantum field and thus constitute *quantum dimensions*™. Quantum dimensions are fluctuating quantum degrees of freedom that make each elementary particle a "fuzzy ball" that partly exists in imaginary space as well as real space.

Ideally, in this author's view, we would start with a concept of a pre-dimensional entity – an entity that is perhaps initially formless which evolves, perhaps through a form of self-organization, to develop dimensions, energy and quantum particles. Physics is far from developing such a grand scheme and must, at present, content itself with assuming the existence of dimensions, a Lagrangian of some sort, and quantum dynamical entities (particles). Thus we will make assumptions as to the number and nature of additional dimensions.

Quantum Dimensions vs. Classical Dimensions

We are all familiar with the concept of dimension. Euclid gave a geometrical definition of dimensions and a procedure for determining the number of dimensions: move in a straight line for a distance; then make a 90° turn; again move in a straight line in the new direction for a while; then make a 90° turn such that the new direction is perpendicular to the previous two directions of motion; continue this procedure until

[1] See Blaha (2003) – the first edition of this book.
[2] See Blaha (2004).

it is no longer possible to make a 90^0 turn to move in a new direction. This process establishes the direction of dimensions and their number in flat space.

From a Cartesian point of view the number of dimensions can initially be simply viewed as the number of independent coordinate axes. Each axis is broken into intervals in such a way as to allow us to specify a position in space with an ordered set of numbers. We can then define functions that depend on these ordered sets of numbers – the coordinates of a point. Such a function can then have a range of values as the coordinates change from point to point in space. We can denote this function as $f(x_1, x_2, x_3, \ldots x_n)$ if the space has n dimensions.

The range of the dimensions in a flat Cartesian space is usually from $-\infty$ to $+\infty$. One can also specify cyclic or compact dimensions that "form a circle" – a coordinate can range in value say from 0 to 2π. The point at 2π can then be made to coincide with the point at 0 so that one can view the dimension as a circle.

Another approach, that was first introduced by Blaha(2003), is to use a new type of coordinate – a quantum coordinate. To understand this concept we imagine a 3-dimensional space of the normal sort with coordinates: x, y, z and values ranging from $-\infty$ to $+\infty$. Now suppose, for example, we introduce a sine function:

$$G(x, y) = \sin(x + y) \qquad (1.1)$$

and require

$$z = G(x, y) \qquad (1.2)$$

Then a function of the three coordinates of space f(x, y, z) becomes

$$f(x, y, z) = f(x, y, G(x, y)) \qquad (1.3)$$

on the surface defined by eq. 1.2. We see f still has values at each point in 3-dimensional space. However, since eq. 1.2.2 defines a 2-dimensional surface within the 3-space the expression for f in eq. 1.3 is properly viewed as defined on that surface.

Now if we replace G(x, y) with a second quantized field Q(x, y) in eq. 1.3 then we obtain a qualitatively new entity:

$$f(x, y, Q(x, y)) \qquad (1.4)$$

an operator expression. (We will assume that infinities and other issues are not present or resolvable.) Eq. 1.4 in itself does not have a value. It obtains a value when evaluated for quantum states |q>

$$q(x, y) = \langle q | f(x, y, Q(x, y)) | q \rangle \tag{1.5}$$

Thus the replacement of the z coordinate in f(x, y, z) with a field operator gives us a qualitatively new entity with well-defined values only when evaluated between states.

In a sense f(x, y, Q(x, y)) is dependent on three unknown quantities – the values of x and y, and the expectation value of f(x, y, Q(x, y)) between (as yet unidentified) quantum states. Thus we can regard f(x, y, Q(x, y)) as depending on two coordinates x and y, and on a quantum coordinate™ whose value is an undetermined quantum fluctuation quantity that only becomes known upon taking an expectation value.

We thus have developed a new form of dimension that is quite properly called a quantum dimension with values that are not c-numbers but in fact are q-numbers (determined only by taking expectation values between quantum states).

Quantum Theory of the Third Kind

First there was quantum mechanics – developed essentially in the period from 1914 - 1926. Quantum Mechanics postulated that position and momentum were non-commuting operators and so the position and momentum of a particle could not be measured simultaneously to arbitrary accuracy. Instead they satisfied the Heisenberg Uncertainty Principle:

$$\Delta x \Delta p \geq \hbar \tag{1.6}$$

Then quantum field theory (second quantization) was developed roughly in the period from 1935 – 1960. Quantum theory postulated that any quantum field and its conjugate momentum operator satisfied an uncertainty principle:

$$\Delta\phi \Delta p_\phi \geq c_\phi \hbar \tag{1.7}$$

for any quantum field ϕ and its conjugate momentum p_ϕ where c_ϕ is a factor dependent on the nature of the quantum field ϕ. As a result the arguments of the field operators ϕ - the position coordinates – were treated as ordinary parameters – c-numbers. Eq. 1.6 and eq. 1.7 were shown[3] to both be required for a consistent quantum theory. The standard example considered the effect of using a quantum electromagnetic field to measure the position and location of a particle.

[3] Heitler (1954) pp. 79-86.

We will suggest that there is a further quantum formulation beyond first first and second quantization in which quantum fields are functions of quantum coordinates. This type of *quantum theory of the third kind* resolves the divergence problems that have plagued second quantization.

Loosely speaking by making the coordinate arguments of quantum fields "fuzzy" we avoid the infinities associated with evaluating products of quantum field operators at precisely the same point. We will proceed to consider quantum field operators that are functions of quantum coordinates in the remainder of this book.

Quantum Coordinates in 4-dimensional Space-time

Now that we see the nature of a quantum dimension we will turn to developing the form of space-time for a universe with both ordinary space-time dimensions and additional dimensions to resolve divergence problems. In Blaha (2003) and Blaha (2004) we were concerned with resolving the divergences of the Standard Model quantum field theory and the divergences of Quantum Gravity as well as the singularity issue at the point of the Big Bang in the Standard Model of Cosmology. We pointed out all these issues could be resolved by using complex quantum coordinates

$$X_\mu(y) = y_\mu + i\, Y_\mu(y)/M_c^{\,2} \tag{1.8}$$

where $Y^\mu(y)$ is a real quantum field with properties identical to the free electromagnetic quantum field of Quantum Electrodynamics, y^μ is a Minkowski space-time 4-vector, and M_c is a large mass that is presumably of the order of or equal to the Planck mass.

All particle (boson and fermion) quantum fields were defined as q-number functions of the quantum coordinates $Y^\mu(y)$ and not of y^μ directly. Using a new method in the Calculus of Variations (Appendix A) that we called the "composition of extrema" we developed quantum field theories – called *Two-Tier quantum field theories* – for particles of spin 0, ½, 1, and 2.

For example, the Two-Tier Lagrangian for a free scalar particle is

$$I = \int \mathscr{L}_S\, d^4y \tag{1.9}$$

with

$$\mathscr{L}_S = J\, \mathscr{L}_F(\phi(X), \partial\phi/\partial X^\mu) + \mathscr{L}_C(X^\mu(y), \partial X^\mu(y)/\partial y^\nu) \tag{1.10}$$

where ϕ is a scalar field, J is the Jacobian for the transformation from $X^\mu(y)$ coordinates to y^μ coordinates, and

$$\mathscr{L}_F = \tfrac{1}{2}\,[(\partial\phi/\partial X^\nu)^2 - m^2\phi^2] \tag{1.11}$$

If we define $X^\mu(y)$ using eq. 1.8 then

$$\mathscr{L}_C = -\tfrac{1}{4}\, F_Y^{\mu\nu} F_{Y\mu\nu} \tag{1.12}$$

where

$$F_{Y\mu\nu} = \partial Y_\mu/\partial y^\nu - \partial Y_\nu/\partial y^\mu \tag{1.13}$$

Upon variation in ϕ with y^μ (and thus $Y^\mu(y)$) held fixed (see Appendix A for a discussion of composition of extrema in the Calculus of Variations) we find

$$\partial\mathscr{L}/\partial\phi - \partial/\partial X^\mu\,[\partial\mathscr{L}/\partial(\partial\phi/\partial X^\mu)] = 0 \tag{1.14}$$

which gives us the Klein-Gordon field equation for ϕ

$$(\Box + m^2)\,\phi(X) = 0 \tag{1.15}$$

where

$$\Box = \partial/\partial X^\nu\,\partial/\partial X_\nu \tag{1.16}$$

A Fourier representation of the solution of eq. 1.15 is:

$$\phi(X) = \int dp\,\delta(p^2 - m^2)\theta(p^0)\,[a(p){:}e^{-ip\cdot X}{:} + a^\dagger(p){:}e^{ip\cdot X}{:}] \tag{1.17}$$

where a(p) is a function of p, † indicates hermitean conjugation, and : : indicate normal ordering of the q-number expression in $X^\mu(y)$.

The manifold defined by $X^\mu(y)$ is found by variation of the lagrangian with respect to $Y^\mu(y)$. In brief it yields field equations (and gauge invariance) that are identical to the case of the electromagnetic field. This scalar particle case and other cases are considered in detail in the following chapters. So we will defer further consideration of the scalar particle case until then.

The following points were proved in Blaha (2003) and Blaha (2004):

1. Free field theories created with the Two-Tier formulation (including quantum gravity) have propagators that are the same as those in the corresponding conventional quantum field theory except that the Fourier representation of each particle propagator contains a gaussian momentum factor. This gaussian factor eliminates all ultra-violet divergences in perturbative quantum field theories such as the Standard Model and Quantum Gravity. The gauge invariance of $Y^{\mu}(y)$ was required in order to have well-behaved gaussian factors throughout the momentum region. Blaha (2003).

2. The resultant theories were proven to satisfy unitarity. Blaha (2003).

3. The resultant theories were proven to be Lorentz invariant in any gauge of $Y^{\mu}(y)$. Blaha (2004).

Thus this new approach to quantum field theory (Two-Tier quantum field theory), when applied to the Standard Model and Quantum Gravity, results in divergence-free quantum field theories enabling us to create a unified, divergence-free quantum field theory of all the forces of nature. (The question of creating this type of quantum field theory in curved space-time was successfully addressed in Blaha (2004).)

The nature of the point of the Big Bang in Cosmology has been an ongoing issue. When Two-Tier quantum field theory is applied to the Big Bang we find that it suggests the universe is very dense region of finite size and temperature at the time of the Big Bang with an inhomogeneous generalized Robertson-Walker metric. The Einstein equations of this metric are separable: one equation is dominated by classical physics; the other equation is quantum in nature. The universe is shown to rapidly change into a standard Robertson-Walker universe with the usual scale factor. Thus the Standard Model of Cosmology emerges shortly after the Big Bang but the singularity of the Standard Cosmological Model at the Big Bang is eliminated Blaha (2004).

The Two-Tier quantum field theoretic approach eliminates the divergence problems of the Standard Model, Quantum Gravity and the Standard Cosmological Model. In addition it justifies treating the Big Bang epoch of the universe as having an ultradense, quasi-free energy density in the form of a perfect fluid.[4] This result follows from the factorization of the scale factor of the generalized Robertson-Walker metric

[4] A recent experiment in which gold nucei collided at high energy confirms that the collision region briefly contained a perfect fluid consisting of a quark-gluon plasma. This result is consistent with our two-tier Standard Model and the two-tier cosmological model developed in Blaha (2004).

into a factor that is dominated by the macroscopic energy density and is thus primarily classical in nature; and another factor embodying quantum effects that is independent of the energy density.

Features of Two-Tier Quantum field Theories

The currently known theories of fundamental interactions and elementary particles fall into two broad categories: conventional quantum field theories and string-based theories that are united with supersymmetry to form superstring theories. Many physicists believe that only theories of these types can meet the reasonable physical requirements of Lorentz invariance and unitarity (positive probabilities summing to unity) while accounting for spin and internal quantum numbers.

It appears that our new form of elementary particle theory is viable. In this type of theory quantum fields are functions of q-number coordinates with imaginary quantum dimensions. We will investigate a unified quantum field theory of this type. Then we will create a unified theory of the Standard Model and Quantum Gravity. The unified theory has the following features:

1. Consistency with the current Standard Model in all respects at current energies up to current maximal energies of several TeV.

2. Consistency with the classical Theory of Gravity. Classical gravity is the "large distance", "low energy" limit of the theory.

3. General relativistic covariance of the field equations, Lorentz invariance of the S matrix, and unitarity as required in physically acceptable theories.

4. The unified theory is divergence-free – it contains no infinities.

5. All interactions are modified by a short distance, high-energy, substructure that begins at some energy presumably much above currently accessible energies. The energy scale could be set by the Planck scale (10^{19} GeV) or could be a much lower energy such as 10^5 GeV.

6. It allows modifications of the Standard Model such as further unification through broken higher symmetries without losing features 1 – 5 above.

7. It predicts a number of new phenomena at extremely short distances such as ultra-relativistic dilepton resonances consisting of two leptons of the same charge. An example of this type of bound state is a bound resonance consisting of two electrons. These dilepton states could be created at ultrahigh energies by penetrating the repulsive Coulomb barrier. Inside the barrier near r = 0 the modified Coulomb

potential is a linear potential. A dilepton bound state is highly unstable with a large decay rate due to quantum tunneling through the Coulomb barrier. Other exotic resonances are also possible.

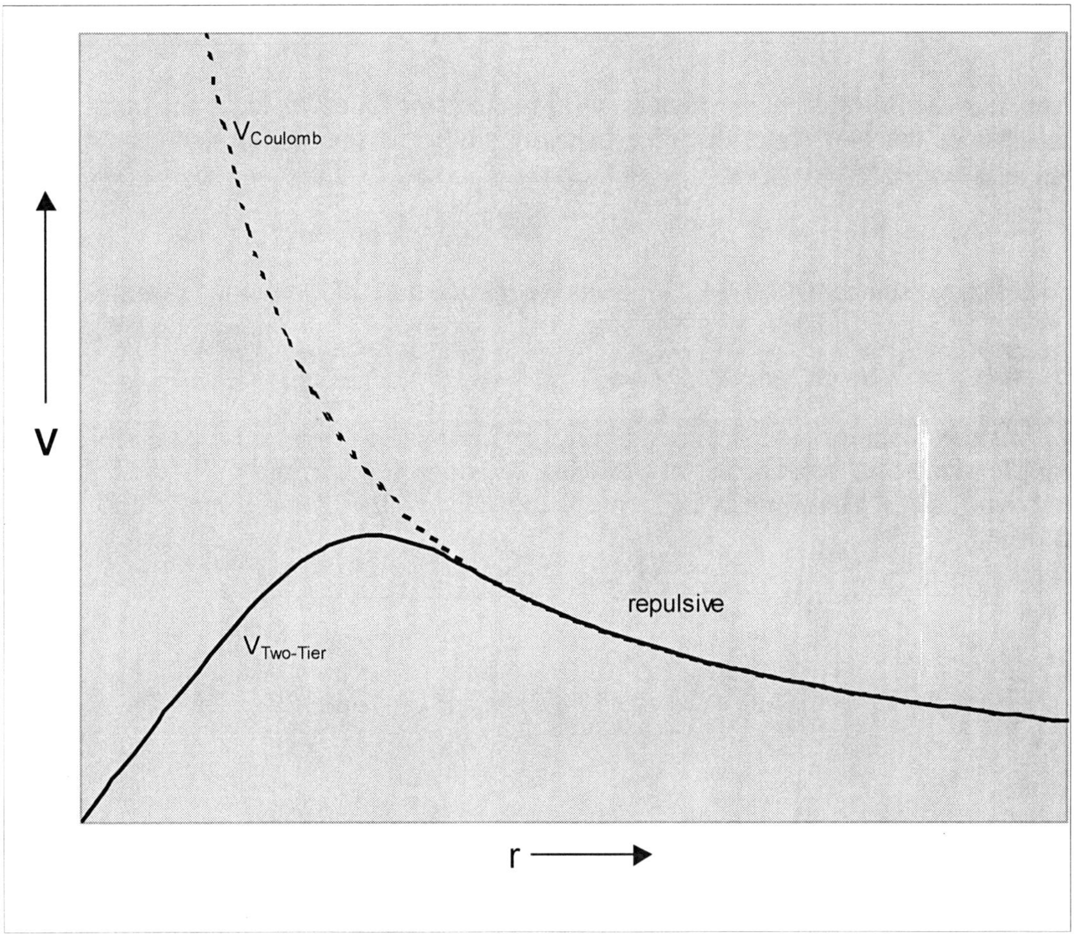

Figure 1.1. Modified electromagnetic Coulomb force between two particles with charges of the same sign (eg. two electrons) in Two-Tier QED and the Two-Tier Standard Model. The potential is repulsive at long distances and attractive at short distances. The potential becomes linear near the origin r = 0 opening up the possibility of unstable doubly charged resonances.

The curve in Fig. 1.1 is the Two-Tier potential between two particles of the same sign. The Two-Tier electromagnetic potential between two singly charged particles of the same sign is:

$$V_{new} = a\Phi(M_c^2\pi r^2)/r \tag{1.18}$$

where a is the fine structure constant, $\Phi(z)$ is the error function, M_c is the mass setting the scale of the new short distance behavior and r is the radial distance. At small distances ($\pi r^2 \ll M_c^{-2}$) we find

$$V_{new} \rightarrow a2\surd\pi\; M_c^2 r \tag{1.19}$$

and at large distances ($\pi r^2 \gg M_c^{-2}$) it becomes identical to the Coulomb potential:

$$V_{new} \rightarrow V_{Coul} = a/r \tag{1.20}$$

8. The short distance modifications of all four interactions open up the possibility of a more fundamental, smaller set of particles, of which the currently observed particles are bound states. Thus quarks and leptons might be bound states of more fundamental particles.

2. The Standard Model, Gravity and Superstring Theories

In this chapter we consider aspects of the Standard Model and Gravitation, and Superstring theories, that are relevant for the discussion of our new unified theory.

Standard Model

The Standard Model unites the Electroweak interactions and the Strong interactions in a successful theory that appears to be consistent with the known experimental data at all accessible energies. The theory satisfies the unitarity condition with positive probabilities adding to unity, Lorentz invariance, and renormalizability. There are several variants and proposed extensions of the Standard Model. But the major aspects of the theory are common to all variants.

The most important technical issue that faced the development of the Standard Model was the question of the renormalizability of the unification of the electromagnetic interactions and the weak interactions in the Electroweak Theory sector. The issue was resolved by the proof of renormalizability by 't Hooft and Veltmann in 1971.

At first glance Electroweak Theory, and its electromagnetic sector, Quantum Electrodynamics (QED), appear to have divergences when perturbative calculations are made of transition probabilities, or of physical quantities such as the anomalous magnetic moment of the electron or muon. However these divergences can be isolated into a finite number of infinite quantities (renormalization theory) which in turn can be absorbed into redefinitions of fundamental parameters of the theory (such as the "bare" electric charge or the "bare" mass of the electron.)

For example in QED the the "bare" charge e_0 is renormalized by infinite factors to give the "physical" charge e: $e = Z_1^{-1}Z_2\surd Z_3 e_0$ where Z_1, Z_2, and Z_3 are divergent renormalization constants. Numerous authors over the last fifty years or so have commented on the "unnaturalness" of renormalization which, after all, amounts to multiplying infinity by zero (e_0) to obtain a finite observable number, the electric charge e in this case.

The quantum field theory formalism that we shall develop resolves these issues by being divergence-free – no infinities appear because of a short distance modification that appears in all particle propagators. The result is a logically satisfactory theory that avoids divergences independent of the details of the interactions and symmetries, and, in fact, allows a wider range of interactions such as interactions with derivative couplings, which were previously totally unrenormalizable.

Quantization of Gravity

Classical gravity began with Newton's theory of gravity, which still remains an acceptable theory for most phenomena involving small masses and velocities. In the early twentieth century A. Einstein developed a new theory of gravitation based on geometrical concepts. This theory, the General Theory of Relativity, has been tested and found to be correct as far as we can determine at the classical level.

There have been many attempts to develop a quantum version of the General Theory of Relativity. While part of the overall framework of such a quantum theory is known, all attempts to develop it within the framework of conventional quantum field theory have failed due to the infinite number of divergences that appear when calculations are attempted in perturbation theory.[5] A theory with an infinite number of divergences cannot be handled with the renormalization procedures used to tame the divergences of QED, Yang-Mills theories, or the Electroweak Theory; and thus does not have predictive power. The gravitational sector of the unified theory that we will develop does not have these divergences and is in fact divergence-free.

Correspondence Principle

The development of Quantum Mechanics was guided by a principle developed by Bohr called the *Correspondence Principle.* Simply put, this principle states that quantum mechanical systems must approach the behavior of classical systems as the features of the quantum mechanical system become much larger than the quantum of action – Planck's constant h. Classical mechanics is thus the large scale limit of quantum mechanics.

Similarly, the Theory of Special Relativity also has a sort of correspondence principle – the predictions of the theory of Special Relativity for a system must approach the predictions of classical mechanics in the limit that all the velocities of the system become much smaller than the speed of light. Thus classical mechanics is a limiting case of Special Relativity.

Just as these twentieth century theories have a type of correspondence principle in which they must approximate an earlier theory in a certain limit, any new theory of elementary particles must approximate the Standard Model in the currently known range of energies. After all the Standard Model works! Therefore it must have an

---

[5] See for example B. S. DeWitt, Phys. Rev. **162**, 1239 (1967), and R. P. Feynman, Acta Physica Polonica **24**, 697 (1963).

element of physical truth and a more fundamental theory must embody that truth in the explored range of experimental energies.

Furthermore the extreme accuracy of QED calculations[6] must be matched by any theory purporting to account for elementary particle interactions. At the moment, and apparently for the foreseeable future, Superstring theories have difficulties when attempts are made to find the Standard Model, or something like it, as an approximation at current energies. Capturing the accuracy of QED predictions would seem to be in the distant future, if at all, in Superstring theories.

String and Superstring Theories

W. Pauli once remarked, "Man should not join together that which God has put asunder." In apparent contradiction (perhaps) to this dictum Superstring theory attempts to unite fermions (half-integer spin particles) and bosons (integer-spin particles) within a larger symmetry so that they can, in a sense, be rotated into one another. While support for this symmetry is currently lacking in elementary particle physics, there may be evidence for supersymmetry in nuclear physics.

However the price for supersymmetry in particle physics appears to be quite high: a large number of dimensions and a large number of particles. No evidence exists at the time of this writing for more than four space-time dimensions or for the supersymmetric partners of the known elementary particles. Thus, at best, supersymmetry is for the future when experimental energies reach energies where supersymmetric features are unequivocally seen (if in fact they exist.)

The theory of strings – two-dimensional substructures that constitute elementary particles – is grounded in phenomenology – more particularly, in the Veneziano-Suzuki formula for scattering amplitudes. Y. Nambu and T. Goto developed a vibrating string theory that accounted for the form of the Veneziano-Suzuki formula. So there appears to be some justification for a string model of elementary particles. In this model the elementary particles, which appear as structure-less and point-like at lower energies, can be seen to have a string-like structure at very high energies.

Thus a sub-structure for elementary particles may have some experimental justification.

Logic Does Not Necessarily Bring Progress in Physics

An often-expressed hope in the last two decades of the twentieth century was that some unique Superstring theory would emerge, and that this theory would prove to be the only logically reasonable theory. After thirty years of effort by a large group of extremely talented physicists (perhaps larger than any group of physics theorists devoted to one topic since the Manhattan project) this hope is yet to be realized.

Historically, the movement from level to deeper level to yet deeper level in Physics has not been the result of logical inquiry but rather have been the result of new

---

[6] T. Kinoshita, "The Fine Structure Constant", Cornell Univ. Preprint CLNS 96/1406 (1996).

experiments confronting existing theory wherein the existing theory is found to be wanting.[7] A rational physicist in the eighteenth century would not have conjectured the theory of special relativity as a logical possibility: "Why should mechanics change when the speed of an object becomes large?" A rational physicist in the nineteenth century would not have conjectured the theory of quantum mechanics as a logical possibility: "Why should it be impossible to measure the momentum and position of a particle with arbitrary precision?" These questions did not, and could not have arisen, in a rational, "down to earth" scientist. Furthermore, despite knowing that these theories are correct we do not know WHY they are implemented in Nature. We only know that they are.[8]

With this historical perspective in mind it appears that we should not expect to arrive at the form of the next deeper level of Physics through reason alone. Nature will most likely surprise us again (and again). Thus experiment is our teacher as we learn more of the nature of matter and space-time.

Goal: Unified Theory Without Renormalization Issues

There are a number of issues confronting elementary particle physics today. Issues such as CP violation, the nature of dark matter and of dark energy, the form of the symmetries of Nature, and the origin of the "numerous" particles and constants in the Standard Model. Practically speaking, perhaps the most important problem at this time is the development of a renormalizable unified theory of all the interactions. We have seen that we can cope with the incomplete unification in the Standard Model between the electroweak interactions and the strong interaction. The Standard Model is renormalizable and thus we can make perturbation theory calculations with confidence and compare the results with experiment. But we cannot make calculations in quantum gravity with confidence. As a result Planck scale physics is totally speculative and we cannot understand the nature of the Big Bang when the universe was contained within a region the size of the Planck length.

With these issues in mind we have developed a unified theory of all the known interactions that is divergence-free. Thus we can perform calculations to any order of perturbation theory in any sector and obtain finite results that can be compared with experiment. The theory has the Standard Model and classical General Relativity as "low energy" limits. At high energies a string-like sub-structure generates a smooth high-energy limit that eliminates the divergences in perturbation theory. Thus there is a correspondence principle for our theory – it has the right "low energy" limits.

The nature of the string-like sub-structure is not dependent on the details of the interactions and symmetries. The interactions in the Standard Model and quantum gravity can have any polynomial form (in the quantum fields). A variety of other interactions (such as those with derivative couplings) are allowed in this approach

[7] This historical process of Physics is described in some detail in the book Blaha(2002) by this author.

[8] A possible answer as to why Nature is quantum in nature has recently been proposed by Blaha (2005) based on Gödel's Theorem.

which do not affect the finiteness of the theory. Thus the range of possible extensions of the Standard Model is significantly widened.

3. Quantization of Coordinate Systems

Non-commuting Coordinates

Field theories with non-commuting coordinates are currently an active field of study.[9] Investigators are studying gauge theories, and in particular Quantum Electrodynamics, with non-commuting coordinates. Non-commuting coordinates are usually implemented quantum mechanically by positing non-zero commutators for coordinates:

$$[x^i, x^j] = i\theta^{ij} \tag{3.1}$$

New Approach to Non-Commuting Coordinates

In this book we will consider an alternative approach that postulates a q-number coordinate system X^μ with which all particle fields are defined. This coordinate system is realized as a mapping from a more fundamental c-number coordinate system y^ν, which we will call the subspace for want of a better term. We will treat X^μ as a vector of quantum fields, thus realizing a new type of non-commutative coordinates at unequal subspace times.

This approach is radically different from the non-commutative coordinate realizations hitherto discussed in the literature. It has a number of beneficial results to recommend it – the main result is the finiteness of quantum field theories that are defined within its framework. We will explore some of these results in the following chapters.

The X^μ coordinate system, as we define it, has a c-number real part and a q-number imaginary part. Thus particle fields which are normally defined on four-dimensional real space-time will now be defined on a complex four-dimensional space-time where four imaginary dimensions will appear as *Quantum Dimensions*™ embodied in a vector quantum field $Y^\mu(y)$.

---

[9] M. R. Douglas and N. A. Nekrasov, Rev. Mod. Phys. **73**, 977 (2002) and references therein; J. Harvey, hep-th/0102076; M. Hamanaka and K. Toda, hep-th/0211148; N. Seiberg and E. Witten, hep-th/9908142; R. J. Szabo, hep-th/0109162; G. Berrino, S. L. Cacciatori, A. Celi, L. Martucci, and A. Vicini, hep-th/0210171; S. Godfrey and A. Doncheski, DESY eprint 02-195; M. Caravati, A. Devoto, and W. W. Repko, hep-th/0211463; and references within these papers.

$$X_\mu(y) = y_\mu + i\, Y_\mu(y)/M_c^2$$

The $Y^\mu(y)$ field is a function of the subspace y coordinates. The real part of the space-time dimensions will be taken to be the subspace y coordinates.[10]

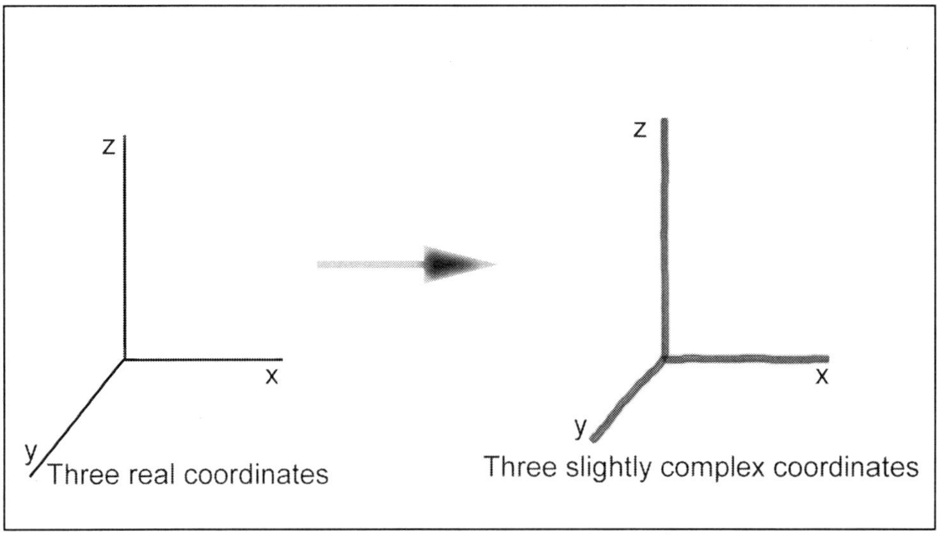

Figure 3.1. The change from purely real space to a slightly complex space with imaginary quantum fluctuations for each spatial axis in the Coulomb gauge of the Y field.

The imaginary part of space-time (which has not been experimentally seen) will simply be the quantum fluctuations of a massless vector quantum field that are suppressed further by a large mass scale – perhaps of the order of the Planck mass – that reduces the imaginary Quantum Dimensions™ to the infinitesimal. The effects of Quantum Dimensions™ only become appreciable in quantum field theory at energies of the order of M_c. At these energies the exponential Gaussian factor in each particle (and ghost) propagator that is generated by the Quantum Dimensions™ serves to make perturbation theory calculations ultra-violet finite – including calculations in Quantum Gravity.

The formalism that we will describe introduces a new form of interaction that does not have the form of the simple polynomial interactions that have hitherto

[10] In a deeper theory the real part might also be a quantum field that undergoes a condensation to generate c-number coordinates. We will not consider this possibility in this book.

dominated quantum field theories. This form of interaction takes place via the composition of quantum fields and can be called a *Dimensional Interaction*™ or an *Interdimensional Interaction*™ since it affects particle behavior through Quantum Dimensions™.

Quantization Using a C-Number X^μ

We will begin by considering the case of a scalar quantum field theory. We assume a real underlying y subspace. Since X^μ is a set of coordinates, we choose to define a scalar field ϕ as a function of X^μ, which in turn is a function of the y^ν coordinates. We will provisionally second quantize ϕ treating X^μ as c-number coordinates using a conventional approach.[11]

We assume a Lagrangian specified by eq. 1.10 that leads to the Klein-Gordon equation eq. 1.15. For that Lagrangian formulation, the momentum conjugate to ϕ is:

$$\pi_\phi = \partial L_F / \partial \phi' \equiv \partial L_F / \partial(\partial\phi/\partial X^0) \tag{3.2}$$

Following the canonical quantization procedure, π and ϕ become hermitian operators with equal time ($X^0 = X^{0\prime}$) commutation rules:

$$[\phi(X), \phi(X')] = [\pi_\phi(X), \pi_\phi(X')] = 0 \tag{3.3}$$

$$[\pi_\phi(X), \phi(X')] = -i\,\delta^3(\mathbf{X} - \mathbf{X}') \tag{3.4}$$

The hamiltonian is defined by eq. A.112. (Appendix A contains a detailed development of the formalism for the scalar particle case. It was placed there because there are many formal similarities to conventional quantum field and this approach allows us to proceed more quickly to the main points of difference between conventional quantum field theory and Two-Tier quantum field theory in the present chapter. Appendix A also describes a new type of method for the Calculus of Variations – the composition of extrema. *Equations numbered A.xxx are in Appendix A.*) We assume a metric $\eta_{\mu\nu}$ where $\eta_{00} = +1$, $\eta_{0i} = 0$, and $\eta_{ij} = -1$ for i, j = 1,2,3.

The standard Fourier expansion of the solution to the Klein-Gordon equation (eq. A.34) is:

---

[11] Some texts are: Bogoliubov, N. N., Shirkov, D. V., *Introduction to the Theory of Quantized Fields* (Wiley-Interscience Publishers Inc., New York, 1959); Bjorken, J. D., Drell, S. D., *Relativistic Quantum Fields* (McGraw-Hill, New York, 1965); Huang, K., *Quarks, Leptons & Gauge Fields Second Edition* (World Scientific, River Edge, NJ, 1992); Kaku, M., *Quantum Field Theory* (Oxford University Press, New York, 1993); Weinberg, S., *The Quantum Theory of Fields* (Cambridge University Press, New York, 1995).

$$\phi(X) = \int d^3p\ N_m(p)\ [a(p)\ e^{-ip\cdot X} + a^\dagger(p)\ e^{ip\cdot X}] \tag{3.5}$$

where

$$N_m(p) = [(2\pi)^3 2\omega_p]^{-1/2} \tag{3.6}$$

and

$$\omega_p = (\mathbf{p}^2 + m^2)^{1/2} \tag{3.7}$$

The commutation relations of the Fourier coefficient operators are:

$$[a(p), a^\dagger(p')] = \delta^3(\mathbf{p} - \mathbf{p}') \tag{3.8}$$

$$[a^\dagger(p), a^\dagger(p')] = [a(p), a(p')] = 0 \tag{3.9}$$

The reader will recognize the quantization procedure is formally identical to the standard canonical quantization procedure of a free scalar quantum field.

In the case of spin ½, spin 1 and spin 2 fields the standard quantization procedure *in terms of the X coordinate system* can also be followed in a way similar to the procedure in standard texts. We will see these quantization procedures in the following chapters. In the next section we will quantize the transformation from the y coordinate system to the X coordinate system.

The procedures developed in this section and the following sections may disturb some readers since we are combining operators with Dirac delta functions and using other unusual operator expressions. These concerns should be put at rest when we show that a path integral formulation presented later gives precisely the same results as the present development.

Coordinate Quantization

In this section we will quantize the coordinates X^μ as a vector field defined on a fundamental c-number coordinate system y^ν of the same dimensionality. We will assume the y^ν space is a "normal" flat Minkowski space with three spatial and one time dimensions. Generalizations to spaces with more dimensions are straightforward but will not be considered here.

Thus we will assume X^μ has three spatial dimensions and one time dimension. For reasons primarily of simplicity (primarily to avoid multiple time coordinates) we will assume the X^μ fields are similar to the free electromagnetic vector potential A^μ with the Lagrangian:

$$\mathcal{L}_C = +\tfrac{1}{4}\, M_c^{\,4} F^{\mu\nu} F_{\mu\nu} \tag{3.10}$$

$$F_{\mu\nu} = \partial X_\mu/\partial y^\nu - \partial X_\nu/\partial y^\mu \tag{3.11}$$

where M_c^4 is a mass scale to the fourth power that is required on dimensional grounds and serves to set the scale for new Physics as we will see later. *Note the sign in eq. 3.10 is not negative – superficially contrary to the conventional electromagnetic Lagrangian. The reason for this difference is that the field part of X^μ is imaginary.* Thus $\mathcal{L}_C$ winds up having the correct sign after taking account of the factors of i in the field strength $F_{\mu\nu}$.

We assume X^μ is complex[12] with the form:

$$X_\mu(y) = y_\mu + i\, Y_\mu(y)/M_c^2 \tag{3.12}$$

where $Y_\mu(y)$ is a quantum field, M_c is a mass scale, and the real part is the c-number 4-vector y_μ. If X^μ has this form, then

$$F_{\mu\nu} = i\,(\partial Y_\mu/\partial y^\nu - \partial Y_\nu/\partial y^\mu)/M_c^2 \tag{3.13}$$

Defining

$$F_{Y\mu\nu} = \partial Y_\mu/\partial y^\nu - \partial Y_\nu/\partial y^\mu \tag{3.14}$$

we see the Lagrangian assumes the form of the conventional electromagnetic Lagrangian:

$$\mathcal{L}_C = -\tfrac{1}{4}\, F_Y^{\mu\nu} F_{Y\mu\nu} \tag{3.15}$$

This Lagrangian can be used to develop field equations and a canonical quantization that is completely analogous to Quantum Electrodynamics.

Gauge Invariance

The gauge invariance of the Lagrangian allows us to choose a convenient gauge. The gauge invariance of the full Lagrangian

---

[12] Theories of quantum mechanics, and quantum fields, in complex and quaternion spaces have been considered by numerous authors. For example see C. M. Bender, D. C. Brody and H. F. Jones, "Complex Extension of Quantum Mechanics" Phys. Rev. Letters **89**, 270401-1 (2002) and references therein; S. L. Adler and A. C. Millard, "Generalized Quantum Dynamics as Pre-Quantum Mechanics", Princeton Univ. preprint arXiv:hep-th/9508076 (1995) and references therein. These theories are all very different from the theories presented herein.

$$\mathscr{L}_s = \mathscr{L}_F(\phi(X), \partial\phi/\partial X^\mu)\, J + \mathscr{L}_C(X^\mu(y), \partial X^\mu(y)/\partial y^\nu) \qquad (A.96)$$

is based on the standard gauge invariance of $\mathscr{L}_C$, and the gauge invariance of $J\mathscr{L}_F$ in the form of translational invariance

$$X^\mu(y) \rightarrow X^\mu(y) + \delta X^\mu(y) \qquad (A.97)$$

for the special case of a translation of X in the form of a gauge transformation:

$$\delta X^\mu(y) = \partial\Lambda(y)/\partial y_\mu$$

In this case eq. A.106 implies

$$\int d^4y\, \Lambda(y)\, \partial\, [\, J\, \partial/\partial X^\mu\, \mathscr{T}_{F\mu\nu}\,]/\partial y_\nu \;= 0$$

after a partial integration and so gives the differential conservation law:

$$\partial\, [\, J\, \partial/\partial X^\mu\, \mathscr{T}_{F\mu\nu}\,]/\partial y_\nu \;= 0 \qquad (3.16)$$

since $\Lambda(y)$ is arbitrary. This conservation law is trivially obeyed since, by eq. A.108:

$$\partial/\partial X^\mu\, \mathscr{T}_{F\mu\nu} \;= 0 \qquad (A.108)$$

Thus translational invariance in the $\mathscr{L}_F$ sector together with standard gauge invariance in the $\mathscr{L}_C$ sector automatically guarantees Y field gauge invariance of the total Lagrangian. Basically we use the separate invariance of each term of

$$L = \int d^4y\, [\mathscr{L}_F\, J + \mathscr{L}_C\,] = \int d^4X\, \mathscr{L}_F + \int d^4y\, \mathscr{L}_C = L_F + L_C$$

under a constant translation $X^\mu \rightarrow X^\mu + \delta X^\mu$ where δX^μ is constant to establish eq. A.108. Then we consider a position dependent translation/gauge transformation to derive eq. 3.16, which taken together with eq. A.108, establishes invariance under the position dependent translation/gauge transformation eq. A.97.

An alternate approach that leads to the same result is to start with the particle part of the Lagrangian L_F rewritten to be invariant under general coordinate transformations, as it must, when we generalize to include General Relativity. Since position dependent translations are a form of general coordinate transformation the full theory must be invariant under position dependent translations due to invariance under general coordinate transformations.

Having established invariance under gauge transformations we now choose to use the most convenient gauge – the Coulomb gauge[13]:

$$\partial Y^i/\partial y^i = 0 \tag{3.17a}$$

which, in the absence of external sources, allows us to set

$$Y^0 = 0 \tag{3.17b}$$

since Y^0 does not have a canonically conjugate momentum. A conventional treatment leads to the equal time commutation relations:

$$[Y^\mu(\mathbf{y}, y^0), Y^\nu(\mathbf{y}', y^0)] = [\pi^\mu(\mathbf{y}, y^0), \pi^\nu(\mathbf{y}', y^0)] = 0 \tag{3.18}$$

$$[\pi^j(\mathbf{y}, y^0), Y_k(\mathbf{y}', y^0)] = -i\,\delta^{tr}{}_{jk}(\mathbf{y} - \mathbf{y}') \tag{3.19}$$

(Note the locations of the j indexes in eq. 3.19 introduce a minus sign.) where

$$\pi^k = \partial \mathscr{L}_C \;/\; \partial Y_k{}' \tag{3.20}$$

$$\pi^0 = 0 \tag{3.21}$$

$$\delta^{tr}{}_{jk}(\mathbf{y} - \mathbf{y}') = \int d^3k\; e^{i\,\mathbf{k}\cdot(\mathbf{y} - \mathbf{y}')}(\delta_{jk} - k_j k_k/\mathbf{k}^2)/(2\pi)^3 \tag{3.22}$$

$$Y_k{}' = \partial Y_k/\partial y^0 \tag{3.23}$$

The Coulomb gauge reveals the two degrees of freedom that are present in the vector potential. The Fourier expansion of the vector potential is:

$$Y^i(y) = \int d^3k\; N_0(k) \sum_{\lambda=1}^{2} \varepsilon^i(k, \lambda)[a(k,\lambda)\, e^{-ik\cdot y} + a^\dagger(k,\lambda)\, e^{ik\cdot y}] \tag{3.24}$$

---

[13] It is also possible to quantize using an indefinite metric that preserves manifest Lorentz covariance as was done by Gupta and Bleuler for the electromagnetic field. We will use the Gupta-Bleuler approach later to establish covariance under special relativity later. Now we opt for manifest positivity and use the Coulomb gauge.

where

$$N_0(k) = [(2\pi)^3 2\omega_k]^{-½} \tag{3.25}$$

and (since m = 0)

$$\omega_k = (\mathbf{k}^2)^{½} = k^0 \tag{3.26}$$

with $\vec{\epsilon}(k, \lambda)$ being the polarization unit vectors for $\lambda = 1,2$ and $k^\mu k_\mu = 0$.

The commutation relations of the Fourier coefficient operators are:

$$[a(k,\lambda), a^\dagger(k',\lambda')] = \delta_{\lambda\lambda'}\delta^3(\mathbf{k} - \mathbf{k}') \tag{3.27}$$

$$[a^\dagger(k,\lambda), a^\dagger(k',\lambda')] = [a(k,\lambda), a(k',\lambda')] = 0 \tag{3.28}$$

and the polarization vectors satisfy

$$\sum_{\lambda=1}^{2} \varepsilon_i(k, \lambda)\varepsilon_j(k, \lambda) = (\delta_{ij} - k_i k_j/\mathbf{k}^2) \tag{3.29}$$

It will be convenient to divide the Y field into positive and negative frequency parts:

$$Y^+_i(y) = \int d^3k\, N_0(k) \sum_{\lambda=1}^{2} \varepsilon_i(k, \lambda)\, a(k,\lambda)\, e^{-ik\cdot y} \tag{3.30}$$

and

$$Y^-_i(y) = \int d^3k\, N_0(k) \sum_{\lambda=1}^{2} \varepsilon_i(k, \lambda)\, a^\dagger(k,\lambda)\, e^{ik\cdot y} \tag{3.31}$$

For later use we note the commutator between the positive and negative frequency parts is:

$$[\,Y^-_j(y_1), Y^+_k(y_2)] = -\int d^3k\, e^{ik\cdot(y_1 - y_2)}\, (\delta_{jk} - k_j k_k/\mathbf{k}^2)/[(2\pi)^3 2\omega_k] \tag{3.32}$$

Bare ϕ Particle States

We now turn to the ϕ particle states. The creation and annihilation operators can be used to define “bare” free particle states. Bare free particle states are states that are not dressed with coherent states of Y quanta. For example a bare one-particle state of momentum p is

$$|p> = a^{\dagger}(p)\,|0_{\phi}> \tag{3.33}$$

with corresponding bare bra state

$$<p| = <0_{\phi}|\,a(p) \tag{3.34}$$

where the vacuum is defined as usual:

$$a(p)\,|0_{\phi}> = 0 \tag{3.35}$$

$$<0_{\phi}|\,a^{\dagger}(p) = 0 \tag{3.36}$$

Multi-particle bare states can also be defined in the conventional way with products of creation and annihilation operators applied to the vacuum.

Y Fock Space Imaginary Coordinate States

States can also be defines for the quantized Y field. These states will be similar in form to electromagnetic photon states but play a different role in our approach since they are in fact coordinate excitation states for the imaginary part of X^{μ}. Thus the scalar field (and other particle fields) will exist in a real four-dimensional space with quantum excitations into imaginary Quantum Dimensions™. These excitations become significant at high energies. At the low energies, with which we are familiar, space-time appears real; at very high energies space-time becomes slightly complex.

There are two types of imaginary coordinate excitations: 1.) Quantum excitations into Fock states consisting of superpositions of states with a definite finite number of Y "particles" and 2.) Imaginary coordinate excitations into coherent Y states with an "infinite" number of particles. Coherent states can be viewed as representing "classical" fields.

In this section we will consider Y field states with a definite number of excitations ("particles"). The creation and annihilation operators of the Y field can be used to define free particle states. For example a one particle state can be defined by

$$|k, \lambda> = a^{\dagger}(k, \lambda)\,|0_{Y}> \tag{3.37}$$

with corresponding bra state

$$<k, \lambda| = <0_{Y}|\,a(k, \lambda) \tag{3.38}$$

where the "coordinate vacuum" is defined as usual:

$$a(k, \lambda)\,|\,0_Y> = 0 \tag{3.39}$$

$$<0_Y\,|\,a^\dagger(k, \lambda) = 0 \tag{3.40}$$

Multi-particle states can also be defined in the conventional way with products of the creation and annihilation operators applied to the vacuum. The set of all states, each containing a finite number of "particles", constitutes a Fock space.

A state with a finite number of Y "particles" represents a quantum fluctuation into imaginary Quantum Dimensions™. Such states do not appear in Two-Tier quantum field theory since the Y field is a free field and has no source. Thus they appear only as part of normal particles. A normal particle, such as a ϕ particle, has a coherent state of Y quanta associated with it, which play a role in interactions. The Y coherent state part of a normal particle can be viewed as boring an infinitesimal "hole" into an extra pair of imaginary dimensions in a neighborhood of the particle of a radial extent set by the length M_c^{-1}.

Y Coherent Imaginary Coordinate States

Coherent Y states bring us closer what we might consider to be "classical" imaginary dimensions – dimensions that we can, in principle, experience as we do normal dimensions. Let us define the coherent state[14]

$$|\,y, p> = e^{-\mathbf{p}\cdot\mathbf{Y}^-(y)/M_c^2}\,|\,0_Y> \tag{3.41}$$

This state is an eigenstate of the coordinate operator $Y^+(y')$:

$$Y^+_j(y_1)\,|\,y_2, p> = -[Y^+_j(y_1), \mathbf{p}\cdot\mathbf{Y}^-(y_2)]/M_c^2\,|\,y, p> \tag{3.42}$$

$$= -\int d^3k\,[N_0(k)]^2\,e^{ik\cdot(y_2 - y_1)}\,(p_j - k_j\mathbf{p}\cdot\mathbf{k}/\mathbf{k}^2)/M_c^2\,|\,y, p> \tag{3.43}$$

$$= p^i\Delta_{Tij}(y_1 - y_2)/M_c^2\,|\,y, p> \tag{3.44}$$

where $p^i\Delta_{Tij}(y_1 - y_2)/M_c^2$ is the eigenvalue of $Y^+_j(y_1)$. As we will see in the next chapter, the eigenvalue of Y^+ becomes large as $(y_1 - y_2)^2 \rightarrow 0$. Thus the imaginary Quantum Dimensions™ become significant at very short distances, and significantly modify the

[14] Coherent states are well known in the physics literature. See for example T. W. B. Kibble, J. Math. Phys. **9**, 315 (1968) and references therein; V. Chung, Phys. Rev. **140**, B1110 (1965); J. R. Klauder, J. McKenna, and E. J. Woods, J. Math. Phys. **7**, 822 (1966) and references therein.

high-energy behavior of quantum field theories. In particular, Quantum Dimensions™ have a significant effect when

$$(y_1 - y_2)^2 \lessapprox (4\pi^2 M_c^2)^{-1} \tag{3.45}$$

according to eq. 4.13 in the next chapter. We are assuming the mass scale M_c is very large – perhaps of the order of the Planck mass (1.221×10^{19} GeV/c²). Thus imaginary Quantum Dimensions™ are far from detectable in today's "low" energy experiments. Their effect is significant in the analysis of the first instants after the Big Bang.[15]

The Dynamical Generation of New Dimensions

Effectively, the imaginary dimensions that we have constructed raise the total number of real and Quantum Dimensions™ to 8 with 6 space dimensions and two time dimensions. As we will see later the requirement of gauge invariance for the quantized Y field reduces the number of time dimensions to one and constrains the six space dimensions to five degrees of freedom giving a 5+1 dimensional space. Since X is a function of y we can also view the four dimensional world that we live in as a four-dimensional surface in a 6-dimensional space-time.

Generation of Quantum Dimensions™ by the ϕ(X) field

The ϕ(X) field generates Quantum Dimensions™ via coherent states from the vacuum. From eq. 3.5 and 3.12 we see

$$\phi(X) = \int d^3p\, N_m(p)\, [a(p)\, e^{-ip\cdot(y + iY/M_c^2)} + a^\dagger(p)\, e^{ip\cdot(y + iY/M_c^2)}] \tag{3.46}$$

with the result

$$\phi(X)\,|0> = \int d^3p\, N_m(p)\, a^\dagger(p)\, e^{ip\cdot(y + iY/M_c^2)}\,|0> \tag{3.47}$$

is a superposition of coherent Y states plus one scalar particle. The vacuum state |0> is the product of the ϕ and Y vacuum states $|0> = |0_Y>|0_\phi>$. We will use |0> in most of the following discussions.

We can also define coherent Y states with total momentum q using the expression:

$$|q\, Y> = \int d^4y\, e^{iq\cdot X(y)}\,|0> = \int d^4y\, e^{iq\cdot(y + iY/M_c^2)}\,|0> \tag{3.48}$$

---

15 Blaha (2004).

Expanding the Y part of the exponential in eq. 3.48 gives

$$|q\, Y\rangle = \sum_{n=0}^{\infty}(-1)^n (n!)^{-1} \prod_{j=1}^{n}\left(\int d^3k_j N_0(k_j)\right)\delta^4\left(q - \sum_{s=1}^{n} k_s\right)\prod_{r=1}^{n}\sum_{\lambda_r=1}^{2} q\cdot \varepsilon(k_r, \lambda_r)\, a^\dagger(k_r,\lambda_r)\,|0\rangle \tag{3.49}$$

which indicates that the sum of the Y particle momenta for each term in the expansion is q.

Hamiltonian for Particle and Coordinate States

The hamiltonian for the separable (field hamiltonian term separate from the Y hamiltonian term – see Appendix A), coordinate quantized, scalar quantum field theory is:

$$H_s = \int d^3y\ \mathscr{H}_s \tag{A.79}$$

with

$$\mathscr{H}_s = J\mathscr{H}_F + \mathscr{H}_C \tag{A.82}$$

$$\mathscr{H}_F(\phi(X), \pi_\phi, \partial\phi/\partial X^i) = \pi_\phi\, \phi' - \mathscr{L}_F \tag{A.83}$$

$$\mathscr{H}_C(X^\mu(y), \pi_X{}^\mu, \partial X^\mu(y)/\partial y^j, y^\nu) = \pi_X{}^\mu\, X_\mu' - \mathscr{L}_C \tag{A.84}$$

$$\mathscr{L}_F = ½\,[\,(\partial\phi/\partial X^\nu)^2 - m^2\phi^2\,] \tag{A.33}$$

$$\mathscr{L}_C = -¼\, M_c{}^4 F_Y{}^{\mu\nu}F_{Y\mu\nu} \tag{3.15}$$

We note

$$\mathscr{H}_F = ½\,[\pi_\phi{}^2 + (\partial\phi/\partial X^i)^2 + m^2\phi^2] \tag{3.50}$$

is the conventional scalar particle hamiltonian when viewed as a function of the X coordinates. $\mathscr{H}_C$ has the same form as the conventional electromagnetic hamiltonian when eq. 3.12 is used to specify X in terms of the Y fields.

$$\mathscr{H}_C = ½\,(E_Y{}^2 + B_Y{}^2) \tag{3.51}$$

where

$$E_Y{}^i = -\partial Y^i/\partial y^0 \tag{3.52}$$

$$B_Y^{\ i} = \varepsilon^{ijk}\, \partial Y_j/\partial y^k \tag{3.53}$$

Using the fourier expansions of ϕ and X^μ (eqs. 3.5 and 3.24) we obtain the following expression for the normal-ordered hamiltonian H_S:

$$P_s^{\ 0} \equiv H_s = \int :\mathscr{H}_s: d^3y \tag{3.54}$$

$$H_s = \int d^3p\, (\mathbf{p}^2 + m^2)^{1/2} a^\dagger(p)a(p) + \int d^3k \sum_{\lambda=1}^{2} (\mathbf{k}^2)^{1/2} a^\dagger(k,\lambda)a(k,\lambda) \tag{3.55}$$

where : : indicates normal ordering and where we perform a functional integration over X (Note the Jacobian is present within $\mathscr{H}_S$.) for the particle part of the hamiltonian $\mathscr{H}_F$. The hamiltonian is manifestly positive definite.

The spatial momentum is specified by

$$P_s^{\ j} = -\int d^3X :\pi_\phi(X)\partial\phi(X)/\partial X_j: + \int d^3y :E_Y^{\ i}\partial Y^i/\partial y_j: \tag{3.56}$$

$$= \int d^3p\, p^j a^\dagger(p)a(p) + \int d^3k \sum_{\lambda=1}^{2} k^j a^\dagger(k,\lambda)a(k,\lambda) \tag{3.57}$$

where the first term in eq. 3.57 follows because of $\int d^3X$ in eq. 3.56. The momentum operator generates displacements in ϕ

$$[P_s^{\ \mu}, \phi(X)] = -i\partial\phi/\partial X_\mu \tag{3.58}$$

Second Quantized Coordinates

At this point we have developed a formalism for a scalar particle quantum field theory based on our non-commutative coordinates. In the following chapters we will proceed to use this formalism to develop a unified quantum field theory of the known forces of nature.

4. Scalar Two-Tier Quantum Field Theory

It appears then that there must be fundamental changes in our basic formulation of quantum field theory, so that unrenormalized masses and unrenormalized coupling constants can become finite.
T. D. Lee & G. C. Wick[16]

Introduction

In this chapter we will examine a new formulation of quantum field theory that we call *Two-Tier quantum field theory* in more detail for the case of a free scalar particle. This type of quantum field theory incorporates a structure similar to a string-like substructure within a quantum field theoretic framework. In the following chapters we will apply the approach to QED, Electroweak Theory, the Standard Model and lastly to a unified model for the known forces of nature.

"Two-Tier" Space

In the preceding chapter we developed quantized coordinates X^μ defined on an underlying c-number coordinate system y^ν with the equations:

$$X_\mu(y) = y_\mu + iY_\mu(y)/M_c^2 \qquad (3.12)$$

$$Y^i(y) = \int d^3k\, N_0(k) \sum_{\lambda=1}^{2} \varepsilon^i(k, \lambda)[a(k,\lambda)\, e^{-ik\cdot y} + a^\dagger(k,\lambda)\, e^{ik\cdot y}] \qquad (3.24)$$

We also developed a free scalar quantum field theory with the Fourier expansion:

$$\phi(X) = \int d^3p\, N_m(p)\, [a(p)\, e^{-ip\cdot X} + a^\dagger(p)\, e^{ip\cdot X}] \qquad (3.5)$$

[16] T. D. Lee and G. C. Wick, Phys. Rev. **D2**, 1033 (1970). Lee and Wick's model QED is totally unrelated to our approach.

We will now consider the implications of the separable Lagrangian:

$$\mathscr{L}_s = \mathscr{L}_F(\phi(X), \partial\phi/\partial X^\mu)\, J + \mathscr{L}_C(X^\mu(y), \partial X^\mu(y)/\partial y^\nu) \tag{A.96}$$

where

$$\mathscr{L}_F = ½\,[\,(\partial\phi/\partial X^\nu)^2 - m^2\phi^2\,] \tag{A.33}$$

and

$$\mathscr{L}_C = -¼\, M_c^4 F_Y^{\mu\nu} F_{Y\mu\nu} \tag{3.10}$$

with

$$F_{Y\mu\nu} = \partial Y_\mu/\partial y^\nu - \partial Y_\nu/\partial y^\mu \tag{3.14}$$

M_c is the mass that sets the scale at which the imaginary part of X^μ becomes significant.

This quantum field theory behaves as a conventional quantum field theory until energies reach the magnitude of M_c. At energies of the order of M_c, and above, the imaginary part of X^μ becomes significant and alters the high-energy behavior of the theory in a major way. This modification leads to the elimination of divergences that normally appear in perturbation theory when interactions are introduced. Yet the low energy behavior of the theory remains the same as conventional scalar quantum field theory. Thus the precise calculations of QED that have been verified to an amazing degree of accuracy remain valid when a Two-Tier formulation of QED is created (in chapter 5). And the "low energy" results found in other conventional quantum field theories such as Electroweak Theory and the Standard Model also are equal to their corresponding Two-Tier versions at currently accessible energies.

The straightforward use of the above equations[17] (and the canonical quantization described in the preceding chapters) leads to a scalar quantum field with the Fourier expansion:

$$\phi(X) = \int d^3p\, N_m(p)\, [a(p)e^{-ip\cdot(y + iY/M_c^2)} + a^\dagger(p)e^{ip\cdot(y + iY/M_c^2)}] \tag{4.1}$$

using eq. 3.5 above. We note the equal time commutation relations of ϕ and π_ϕ are the same as the conventional equal time commutation relations of a scalar field despite the fact that X^μ and Y^μ are themselves quantum fields since $[Y^\mu(\mathbf{y}, y^0), Y^\nu(\mathbf{y}', y^0)] = 0$ for $\mathbf{y} \neq \mathbf{y}'$. In addition, we note the ϕ and π_ϕ fields are not hermitean.

The Fourier expansion of ϕ does require one refinement – the exponential terms in X^μ must be *normal ordered* to avoid infinities in the unequal time commutation relations:

---

[17] The use of functionals in quantum field theory is, of course, far from new as one can see in texts such as Bogoliubov (1959) (see for example pp. 198-226).

$$\phi(X) = \int d^3p\ N_m(p)\ [a(p)\ :e^{-ip\cdot(y + iY/M_c^2)}: + a^\dagger(p)\ :e^{ip\cdot(y + iY/M_c^2)}:] \qquad (4.2)$$

Since the hamiltonian as well as other quantities are normal ordered in quantum field theory the additional requirement of normal ordering in the field operator is merely an extension of a standard procedure to a more complex situation and is not disturbing. The unequal time commutation relation of the normal ordered ϕ field is:

$$[\phi(X^\mu(y_1)), \phi(X^\mu(y_2))] = i\Delta(y_1 - y_2) + \mathcal{O}(1/M_c^2) \qquad (4.3)$$

where

$$\Delta(y_1 - y_2) = -i \int d^3k\ (e^{-ik\cdot(y_1 - y_2)} - e^{ik\cdot(y_1 - y_2)})/[(2\pi)^3 2\omega_k] \qquad (4.4)$$

is a familiar c-number invariant singular function. The additional terms in eq. 4.3 are q-number terms that become significant at very short distances of the order M_c^{-1}. Thus precise measurements of field strengths at larger distances are limited by standard quantum effects as indicated by the commutation relation.

The principle of *microscopic causality* is violated at extremely short distances of the order M_c^{-1} since the commutator (eq. 4.3) is non-zero, in general, for space-like distances of the order of M_c^{-1} due to the q-number terms. This violation is not experimentally measurable now – and for the foreseeable future – and reflects a type of non-locality at extremely short distances.

We will see that the short distance behavior of Two-Tier quantum field theory leads to the elimination of divergences resulting in finite interacting quantum field theories.

Vacuum Fluctuations

While the expectation value of a *conventional* free scalar field $\phi_{conv}(X)$ is zero in a conventional quantum field theory:

$$<0|\phi_{conv}(X)|0> = 0 \qquad (4.5)$$

the vacuum fluctuations of *conventional* scalar quantum field theory are quadratically divergent:

$$<0|\phi_{conv}(X)\phi_{conv}(X)|0> = \int d^3p/[(2\pi)^3 2\omega_p] \qquad (4.6)$$

In "Two-Tier" quantum field theory we find the vacuum expectation value of a free field is zero (like eq. 4.5) *and the expectation value of the square of the field is also zero:*

$$<0|\phi(X)\phi(X)|0> = \int d^3p\ e^{-p^ip^j\Delta_{Tij}(0)/M_c^4}/[(2\pi)^3 2\omega_p] = 0 \qquad (4.7)$$

since the exponential factor in the integral is $-\infty$. The exponent contains

$$\Delta_{Tij}(z) = \int d^3k\ e^{-ik\cdot z}\ (\delta_{ij} - k_ik_j/\mathbf{k}^2)/[(2\pi)^3 2\omega_k] \qquad (4.8)$$

where "T"is for "Two-Tier". Thus *vacuum fluctuations are zero in Two-Tier quantum field theory.* Correspondingly, we will see that renormalization constants are finite in the Two-Tier versions of QED, Electroweak Theory, the Standard Model and Quantum Gravity.

The Feynman Propagator

The Feynman propagator for a Two-Tier free scalar quantum field is:

$$i\Delta_F^{TT}(y_1 - y_2) = <0|T(\phi(X(y_1)),\phi(X(y_2)))|0> \qquad (4.9)$$

$$\equiv <0|\phi(X(y_1))\phi(X(y_2))|0>\ \theta(y_1^0 - y_2^0) +$$

$$+\ \phi(X(y_2))\phi(X(y_1))|0>\ \theta(y_2^0 - y_1^0) \qquad (4.10)$$

Since $X^0 = y^0$ in the Coulomb gauge of the X^μ field there is no ambiguity in the choice of the relevant time variable. A straightforward calculation shows:

$$i\Delta_F^{TT}(y_1 - y_2) = i\int d^4p\ e^{-ip\cdot(y_1 - y_2)}\ R(\mathbf{p}, y_1 - y_2)/[(2\pi)^4(p^2 - m^2 + i\varepsilon)] \qquad (4.11)$$

where

$$R(\mathbf{p}, y_1 - y_2) = \exp[-p^ip^j\Delta_{Tij}(y_1 - y_2)/M_c^4] \qquad (4.12)$$

$$= \exp\{-p^2[A(v) + B(v)\cos^2\theta]/[4\pi^2 M_c^4 z^2]\} \qquad (4.13)$$

with

$$z^\mu = y_1^\mu - y_2^\mu \qquad (4.14)$$

$$z = |\mathbf{z}| = |\mathbf{y}_1 - \mathbf{y}_2| \tag{4.15}$$

$$p = |\mathbf{p}| \tag{4.16}$$

$$v = |z^0|/z \tag{4.17}$$

$$A(v) = (1 - v^2)^{-1} + .5v \ln[(v - 1)/(v + 1)] \tag{4.18}$$

$$B(v) = v^2(1 - v^2)^{-1} - 1.5v \ln[(v - 1)/(v + 1)] \tag{4.19}$$

$$\mathbf{p}\cdot\mathbf{z} = pz \cos\theta \tag{4.20}$$

and $|\mathbf{p}|$ denoting the length of a spatial vector **p** while $|z^0|$ is the absolute value of z^0.

As eq. 4.11 indicates, the Gaussian damping factor R(p, z) for large momentum p is the same for both the positive and negative frequency parts of the Two-Tier Feynman propagator. It is also important to note that R(p, z) does not depend on p^0 (in the Y Coulomb gauge) and thus the integration over p^0 proceeds in the usual way to produce time-ordered positive and negative frequency parts.

Large Distance Behavior of Two-Tier Theories

The large distance behavior of the Two-Tier Feynman propagator approaches the behavior of the conventional Feynman propagator since

$$R(\mathbf{p}, y_1 - y_2) \to 1 \tag{4.21}$$

when $(y_1 - y_2)^2$ becomes much larger than M_c^{-2} as eq. 4.13 shows. Thus the behavior of a conventional quantum field theory naturally emerges at large distance. We will see that the conventional Standard Model is the large distance limit of the Two-Tier Standard Model thus *realizing a form of Correspondence Principle for Quantum Field Theory*. Some features of the conventional Standard Model that depend specifically on the existence of divergences, such as the axial anomaly, will be different in the Two-Tier Standard Model since it is a divergence-free theory.

Short Distance Behavior of Two-Tier Theories

At short distances the Gaussian factor dominates and radically changes the behavior of the Feynman propagator eliminating its short distance singular behavior, and thus paving the way to finite quantum field theories. Near the light cone, $M_c^{-2} \gg -(y_1 - y_2)^2 \to 0$, we can approximate eq. 4.11 with

$$i\Delta_F^{TT}(y_1 - y_2) \approx \int d^3p\, [N(p)]^2\, R(\mathbf{p}, y_1 - y_2) \qquad (4.22)$$

since $e^{-ip\cdot(y_1 - y_2)}$ is approximately unity for small $(y_1 - y_2)$. We assume the mass of the ϕ particle is zero or is negligible at high energies so we set $m = 0$ to study the high energy behavior of eq. 4.22. Upon performing the integrations in eq. 4.22 for space-like $(y_1 - y_2)^2$ (and analytically continuing to the time-like regions[18,19]) we find

$$i\Delta_F^{TT}(y_1 - y_2) \approx [z^2 M_c^4/(4i\surd A\surd B)]\ \ln[(\surd A + i\surd B)/(\surd A - i\surd B)] \qquad (4.23)$$

with A and B defined in eqs. 4.18 and 4.19. As $(y_1 - y_2)^2 \to 0$ from the space-like or time-like side of the light cone we find eq. 4.23 becomes:

$$i\Delta_F^{TT}(y_1 - y_2) \to \pi M_c^4\, |\,(y_1 - y_2)^\mu (y_1 - y_2)_\mu\,|\,/8 \qquad (4.24)$$

Eq. 4.24 has several noteworthy points:

1. The propagator is well behaved on the light cone and approaches zero smoothly from both space-like and time-like directions. In contrast, the conventional scalar Feynman propagator diverges as $[(y_1 - y_2)^\mu (y_1 - y_2)_\mu]^{-2}$. This good behavior near the light cone will be seen later for other particle propagators with the net result that the usual infinities found in conventional quantum field theory are absent in Two-Tier quantum field theories.

2. The quadratic form of the propagator in eq. 4.24 is suggestive of attempts to formulate a relativistic harmonic oscillator model of elementary particles[20] and more recent attempts to achieve quark confinement. The fact that the absolute value of the quadratic term appears in eq. 4.24 neatly avoids the common pitfall seen in fully relativistic harmonic oscillator attempts.

3. The quadratic behavior *in coordinate space* of the propagator at short distances is equivalent to a high-energy behavior of

$$p^{-6} \qquad (4.25)$$

---

[18] See S. Blaha, "Relativistic Bound State Models with Quasi-Free Constituent Motion", Phys. Rev. **D12**, 3921 (1975) and references therein.

[19] It should be noted that A and B in eq. 4.23 have the same sign for $0 \le v < 1.1243$ thus making for easy analytic continuation across the light cone (which corresponds to $v = 1$ in eqs. 4.18 and 4.19).

[20] H. Yukawa, H., Phys. Rev. **91**, 416 (1953); Y. S. Kim and M. E. Noz, Phys. Rev. **D8**, 3521 (1973) and references therein.

in momentum space. Thus we get the equivalent *of a higher derivative theory* in Two-Tier quantum field theory at high energies while retaining a positive definite energy spectrum. The problems of negative metric states that have plagued conventional higher derivative quantum field theories are avoided.[21]

String-like Substructure of the Theory

Imaginary Quantum Dimensions™ endow a particle with an extended structure that resembles to some extent the extended structure seen in bosonic string and Superstring theories. For example, Bailin (1994) use the operator[22]

$$V_\Lambda(k) = \int d^2\sigma \sqrt{-h}\, W_\Lambda(\tau, \sigma)\, e^{-ik\cdot X} \qquad (4.26)$$

where X^μ is a quantized fourier expansion of the string fields (see eq. 7.22 of Bailin (1994)).

We note our X^μ coordinate-field has two transverse degrees of freedom due to gauge invariance, which also invites comparison to the bosonic string. A point of difference is that we will create a well-defined quantum field theoretic formulation in conventional space-time that has the Standard Model as its "large distance" behavior thus introducing a note of reality that is not (yet?) very apparent in Superstring theories. We see that the interacting quantum field theories based on this approach also have good, finite, short distance behavior just as string theories.

The scalar, and other particles', Feynman propagators can be viewed as describing the propagation of a particle cloaked (accompanied) by a cloud of Y particles (represented by the $R(\mathbf{p}, y_1 - y_2)$ factor in the propagator of eq. 4.11). If we examine the fourier transform of R(p, z) we see:

$$(2\pi)^4 R(\mathbf{p}, q) = \int d^4z\, e^{iq\cdot z}\, R(\mathbf{p}, z) = \int d^4z\, e^{iq\cdot z} \exp[-p^i p^j \Delta_{Tij}(z)/M_c^4] \qquad (4.27)$$

and we find

$$R(\mathbf{p},q) = \sum_{n=0}^{\infty} [i(2\pi M_c)^4]^{-n} (n!)^{-1} \prod_{j=1}^{n} [\int d^4k_j\, \theta(k_j^0)(\mathbf{p}^2 - (\mathbf{p}\cdot\mathbf{k}_j)^2/\mathbf{k}_j^2)/(k_j^2 + i\varepsilon)]\, \delta^4(q - \sum_r k_r) \qquad (4.28)$$

---

[21] S. Blaha, Phys.Rev. **D10**, 4268 (1974); S. Blaha, Phys.Rev. **D11**, 2921 (1975); S. Blaha, Nuovo Cim. **A49**, :113 (1979); S. Blaha, "Generalization of Weyl's Unified Theory to Encompass a Non-Abelian Internal Symmetry Group" SLAC-PUB-1799, Aug 1976; S. Blaha, "Quantum Gravity and Quark Confinement" Lett. Nuovo Cim. **18**, 60 (1977); Nakanishi, N., Suppl. Prog. Theo. Phys. **51**, 1 (1972); and references therein.

[22] D. Bailin and A. Love, *Supersymmetric Gauge Field Theory and String Theory* (Institute of Physics Publishing, Philadelphia, PA, 1994) page 272.

which can be interpreted as a "cloud" of Y particles dressing the "bare" particle propagator. (The manifest divergences in eq. 4.28 for R(p, q) are an artifact of the expansion and the subsequent fourier transformation. They are not present in the R(**p**, $y_1 - y_2$) factor in the propagator of eq. 4.11.) See Fig. 4.1 for the Feynman diagram of the Two-Tier cloaked propagator as compared to the normal scalar particle Feynman propagator. The Two-Tier scalar Feynman propagator is basically a conventional scalar propagator that is modified by coherent Y particle emission.[23]

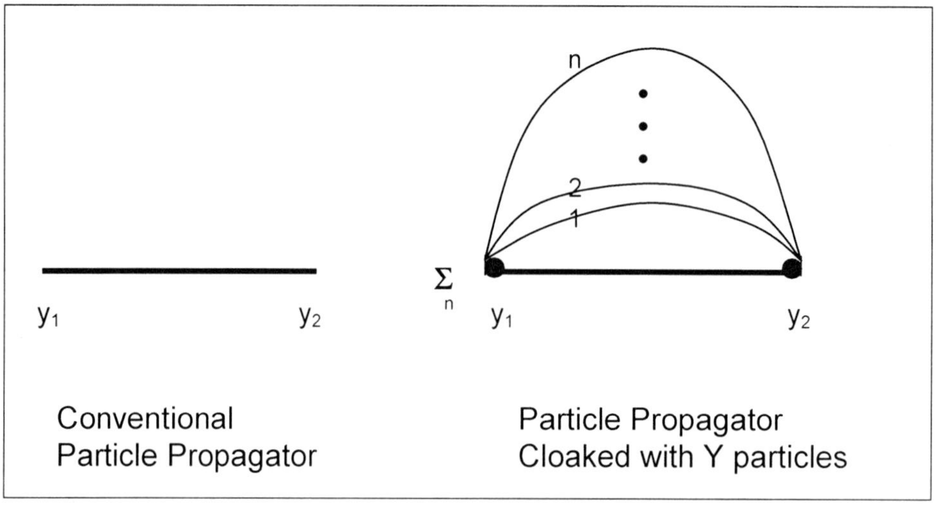

Figure 4.1. Feynman diagram for conventional and cloaked Two-Tier propagators.

We note that R(p, q) satisfies the convolution theorem:

$$\int d^4k\ R(\mathbf{p}, k)\ R(\mathbf{p}, q - k) = [R(\mathbf{p}, q)]^2 \tag{4.29a}$$

or

$$(2\pi)^4 \int d^4z\ e^{iq\cdot z}\ R(\mathbf{p}, z)\ R(\mathbf{p}, z) = [\ \int d^4z\ e^{iq\cdot z}\ R(\mathbf{p}, z)\ ]^2 \tag{4.29b}$$

The proof follows from eq. 4.28 and the Binomial theorem.

---

[23] T. W. B. Kibble, Phys. Rev. **173**, 1527 (1968) and references therein. In particular see p. 1532 of Kibble's paper.

Parity

Parity can appear in two guises within the framework of Two-Tier quantum field theory. One can consider a parity operation where the space parts of X^{μ} are reversed while y^{μ} is unchanged. Or one can consider a second type of parity where the space parts of y^{μ} are reversed.

X Parity

Under this form of parity operation y^{μ} is unchanged while the arguments of ϕ *appear* to change by

$$X^i(y) \rightarrow -X^i(y) \tag{4.30}$$

$$X^0(y) \rightarrow X^0(y) \tag{4.31}$$

We will denote the parity operator of this type $\mathscr{P}_X$. Under $\mathscr{P}_X$ the arguments of the scalar quantum field operator ϕ change according to eqs. 4.30-1 so that ϕ transforms as

$$\mathscr{P}_X \phi(\mathbf{X}(y), X^0(y)) \mathscr{P}_X^{-1} = \phi(-\mathbf{X}(y), X^0(y)) \tag{4.32}$$

From the form of ϕ in eq. 4.2 we see we can implement eq. 4.32 by requiring:

$$\mathscr{P}_X a(\mathbf{p}, p^0) \mathscr{P}_X^{-1} = a(-\mathbf{p}, p^0) \tag{4.33}$$

$$\mathscr{P}_X a^{\dagger}(\mathbf{p}, p^0) \mathscr{P}_X^{-1} = a^{\dagger}(-\mathbf{p}, p^0) \tag{4.34}$$

$$\mathscr{P}_X X^0(y) \mathscr{P}_X^{-1} = X^0(y) \tag{4.35}$$

$$\mathscr{P}_X X^i(y) \mathscr{P}_X^{-1} = X^i(y) \tag{4.36}$$

$$\mathscr{P}_X Y^i(y) \mathscr{P}_X^{-1} = Y^i(y) \tag{4.37}$$

where i = 1,2,3.

This parity transformation is analogous to the standard form of parity transformation in conventional quantum field theory. The separable Lagrangian in eq. A.96 (and listed at the beginning of this chapter) is invariant under this parity transformation.

y Parity

This form of parity transformation in which $y^i \rightarrow -y^i$ has significant differences from the normal parity transformation. We specify this parity transformation for a scalar quantum field by:

$$\mathscr{P}_y \phi(\mathbf{X}(\mathbf{y}, y^0), X^0(\mathbf{y}, y^0)) \mathscr{P}_y^{-1} = \phi(\mathbf{X}(-\mathbf{y}, y^0), X^0(-\mathbf{y}, y^0)) \quad (4.38)$$

This transformation can be implemented through the following set of transformations:

$$\mathscr{P}_y a(\mathbf{p}, p^0) \mathscr{P}_y^{-1} = a(-\mathbf{p}, p^0) \quad (4.39)$$

$$\mathscr{P}_y a^\dagger(\mathbf{p}, p^0) \mathscr{P}_y^{-1} = a^\dagger(-\mathbf{p}, p^0) \quad (4.40)$$

$$\mathscr{P}_y X^0(\mathbf{y}, y^0) \mathscr{P}_y^{-1} = X^0(-\mathbf{y}, y^0) \quad (4.41)$$

$$\mathscr{P}_y Y^i(\mathbf{y}, y^0) \mathscr{P}_y^{-1} = -Y^i(-\mathbf{y}, y^0) \quad (4.42a)$$

$$\mathscr{P}_y a(\mathbf{k}, k^0, 1) \mathscr{P}_y^{-1} = a(-\mathbf{k}, k^0, 1) \quad (4.42b)$$

$$\mathscr{P}_y a(\mathbf{k}, k^0, 2) \mathscr{P}_y^{-1} = -a(-\mathbf{k}, k^0, 2) \quad (4.42c)$$

where i = 1,2,3 and assuming: $\boldsymbol{\varepsilon}(\mathbf{k},k^0,1) = -\boldsymbol{\varepsilon}(-\mathbf{k},k^0,1)$ and $\boldsymbol{\varepsilon}(\mathbf{k},k^0,2) = +\boldsymbol{\varepsilon}(-\mathbf{k},k^0,2)$.

Forms of the Parity Transformations

The parity transformations for a scalar particle are

$$\mathscr{P}_X = \exp\{-i\pi \int d^3p\, [a^\dagger(\mathbf{p}, p^0) a(\mathbf{p}, p^0) - a^\dagger(\mathbf{p}, p^0) a(-\mathbf{p}, p^0)]/2\} \quad (4.43a)$$

$$\mathscr{P}_y = \mathscr{P}_X \exp\{-i\pi \int d^3k\, [\sum_{\lambda=1}^{2} a^\dagger(\mathbf{k},k^0,\lambda) a(\mathbf{k},k^0,\lambda) - a^\dagger(\mathbf{k}, k^0, 1) a(-\mathbf{k},k^0,1) + $$

$$+ a^\dagger(\mathbf{k}, k^0, 2) a(-\mathbf{k},k^0,2)]/2\} \quad (4.43b)$$

The separable Lagrangian of eq. A.96 is invariant under these parity transformations.

Charge Conjugation

Charge conjugation is implemented in a way similar to that of conventional quantum field theory. In particular

$$\mathscr{C} X^{\mu}(\mathbf{y}, y^0)\mathscr{C}^{-1} = X^{\mu}(\mathbf{y}, y^0) \tag{4.44}$$

Time Reversal

Since $X^0 = y^0$ in the Y Coulomb gauge in Two-Tier quantum theory the only non-trivial form of time reversal transformation $\mathscr{T}$ is based on $y^0 = -y^0$. This time reversal transformation is similar in part to to the conventional time reversal transformation in conventional quantum field theory. Therefore we will define $\mathscr{T}$ as the product of the operation of taking the complex conjugate of all c-numbers times a unitary operator $\mathscr{U}_Y$. Under $\mathscr{T}$ a scalar quantum field operator ϕ transforms as

$$\mathscr{T}\phi(\mathbf{X}(\mathbf{y}, y^0), X^0(\mathbf{y}, y^0))\mathscr{T}^{-1} = \phi(\mathbf{X}(\mathbf{y}, -y^0), X^0(\mathbf{y}, -y^0)) \tag{4.45}$$

From the form of in ϕ eq. 4.2 we see that

$$\mathscr{T} a(\mathbf{p}, p^0)\mathscr{T}^{-1} = a(-\mathbf{p}, p^0) \tag{4.46}$$

$$\mathscr{T} a^{\dagger}(\mathbf{p}, p^0)\mathscr{T}^{-1} = a^{\dagger}(-\mathbf{p}, p^0) \tag{4.47}$$

$$\mathscr{T} X^i(\mathbf{y}, y^0)\mathscr{T}^{-1} = X^i(\mathbf{y}, -y^0) \tag{4.48}$$

$$\mathscr{T} Y^i(\mathbf{y}, y^0)\mathscr{T}^{-1} = -Y^i(\mathbf{y}, -y^0) \tag{4.49a}$$

$$\mathscr{T} a(\mathbf{k}, k^0,1)\mathscr{T}^{-1} = a(-\mathbf{k}, k^0,1) \tag{4.49b}$$

$$\mathscr{T} a(\mathbf{k}, k^0,2)\mathscr{T}^{-1} = -a(-\mathbf{k}, k^0,2) \tag{4.49c}$$

where i = 1,2,3 and assuming: $\varepsilon(\mathbf{k},k^0,1) = -\varepsilon(-\mathbf{k},k^0,1)$ and $\varepsilon(\mathbf{k},k^0,2) = +\varepsilon(-\mathbf{k},k^0,2)$.

The unitary operator $\mathscr{U}_Y$ is given by

$$\mathscr{U}_X = \exp\{-i\pi\int d^3p\ [a^{\dagger}(\mathbf{p}, p^0)a(\mathbf{p}, p^0) - a^{\dagger}(\mathbf{p}, p^0)a(-\mathbf{p}, p^0)]/2\} \tag{4.50a}$$

and

$$\mathscr{U}_y = \mathscr{U}_X \exp\{-i\pi \int d^3k\, [\sum_{\lambda=1}^{2} a^\dagger(\mathbf{k},k^0,\lambda)a(\mathbf{k},k^0,\lambda) - a^\dagger(\mathbf{k}, k^0, 1)a(-\mathbf{k},k^0,1) +$$

$$+ a^\dagger(\mathbf{k}, k^0, 2)a(-\mathbf{k},k^0,2)]/2\} \tag{4.50b}$$

The separable Klein-Gordon Lagrangian (eq. A.96) is invariant under our definition of time reversal.

We note

$$\mathscr{U}_y = \mathscr{P}_y \tag{4.50c}$$

Although the present theory is somewhat more complicated than conventional quantum field theory the overall nature of the $\mathscr{P}$, $\mathscr{C}$, and $\mathscr{T}$ transformations is the same.

5. Interacting Quantum Field Theory – Perturbation Theory

Introduction

The form of quantum field theory that we have developed in chapters 3 and 4 can be used as the basis of new formulations of QED, Electroweak Theory, the Standard Model and a divergence-free, unified theory of all the known interactions. The development of these theories requires a number of topics be addressed. This chapter covers perturbation theory. As much as possible, we attempt to retain the features of the standard approach so that the reader will more readily follow the discussion and more readily accept this new formalism. In physics originality is secondary to reality. The perturbation theory that we will develop will be shown to be identical to the perturbation theory that we develop later using a path integral formalism.

Two-Tier theory will be shown to satisfy unitarity in chapter 6 and invariance under special relativity in Appendix B.

An Auxiliary Asymptotic Field

The definition of the asymptotic "free" in and out states is an issue in Two-Tier quantum field theory because the "free particle field" of the theory $\phi(X(y))$ is a "dressed" particle, ab initio, since it is cloaked in a cloud of Y particles as discussed in the passage following eq. 4.27.

While one could use $\phi(X(y))$ directly to define in and out asymptotic states it is more convenient initially to introduce a "fictitious" auxiliary asymptotic quantum field $\Phi(y)$ that will represent the equally fictitious "bare ϕ particle" in and out states.

We will consider the case of a scalar field. We define a free, scalar Klein-Gordon particle field with the physical mass m of the physical $\phi(X(y))$ particle.

$$\Phi(y) = \int d^3p\, N_m(p)\, [a(p)\, e^{-ip\cdot y} + a^\dagger(p)\, e^{ip\cdot y}] \tag{5.1}$$

using the creation and annihilation operators of $\phi(X(y))$ (in eq. 4.2). The set of particle states of $\Phi(y)$ has the familiar Fock space form

$$| p_1, p_2, \ldots p_n> = a^\dagger(p_1)a^\dagger(p_2) \ldots a^\dagger(p_n) | 0> \tag{5.2}$$

with powers of creation operators allowed since Φ particles are bosons. The set of particle states constitutes a complete orthonormal set of states. The corresponding bra states are defined by hermitean conjugation:

$$<p_1, p_2, \ldots p_n| = (| p_1, p_2, \ldots p_n>)^\dagger \tag{5.3}$$

We note that the energy spectrum of these states is positive definite with the hamiltonian

$$H_\Phi = P_\Phi^{\ 0} = \int d^3y \ \tfrac{1}{2} \ [\pi_\Phi^{\ 2} + (\partial\Phi/\partial X^i)^2 + m^2\Phi^2] \tag{5.4a}$$

$$= \int d^3p \ (\mathbf{p}^2 + m^2)^{1/2} a^\dagger(p)a(p) \tag{5.4b}$$

and momentum vector:

$$\mathbf{P}_\Phi = \int d^3p \ \mathbf{p} \ a^\dagger(p)a(p) \tag{5.5}$$

We will use this set of energy-momentum eigenstates to define asymptotic "in" and "out" states in perturbation theory.

Transformation Between Φ(y) and ϕ(X(y))

For later use in the definition of the perturbation theory expansion, we will determine the transformations between the in and out Φ(y) fields, and the in and out $\phi(X(y))$ fields. Let us define a transformation $W_a(y)$ that transforms in and out Φ(y) fields to in and out $\phi(X(y))$ fields respectively:

$$\phi_a(X(y)) = :W_a(y)\Phi_a(y)W_a^{-1}(y): \tag{5.6}$$

where the label a = "in" or a = "out", where : ... : signifies normal ordering, and where

$$\Phi_{in}(y) = \int d^3p \ N_m(p) \ [a_{in}(p) \ e^{-ip\cdot y} + a_{in}^\dagger(p) \ e^{ip\cdot y}] \tag{5.7}$$

$$\Phi_{out}(y) = \int d^3p \ N_m(p) \ [a_{out}(p) \ e^{-ip\cdot y} + a_{out}^\dagger(p) \ e^{ip\cdot y}] \tag{5.8}$$

$$\phi_{in}(X) = \int d^3p\, N_m(p)\, [a_{in}(p) :e^{-ip\cdot(y + iY/M_c^2)}: + a_{in}^{\dagger}(p) :e^{ip\cdot(y + iY/M_c^2)}:] \quad (5.9)$$

$$\phi_{out}(X) = \int d^3p\, N_m(p)\, [a_{out}(p) :e^{-ip\cdot(y + iY/M_c^2)}: + a_{out}^{\dagger}(p) :e^{ip\cdot(y + iY/M_c^2)}:] \quad (5.10)$$

Note that the transformation eq. 5.6 includes normal ordering. While this transformation may seem strange it is no stranger than the time reversal operator, in which the complex conjugate of all c-number terms is taken in addition to applying a unitary transformation.

In the Coulomb gauge of Y it is easy to show that

$$W_a(y) = \exp(-\mathbf{Y}(y)\cdot\mathbf{P}_{\Phi a}/M_c^2) \quad (5.11)$$

and

$$W_a^{-1}(y) = \exp(\mathbf{Y}(y)\cdot\mathbf{P}_{\Phi a}/M_c^2) \quad (5.12)$$

where the label a = "in" or a = "out", where the inner products in the exponentials are the usual spatial vector inner product, and where

$$\mathbf{P}_{\Phi a} = -\int d^3y\, \partial\Phi_a(y)/\partial y^0\, \nabla\Phi_a(y) = \int d^3p\, \mathbf{p}\, a_a^{\dagger}(p) a_a(p) \quad (5.12a)$$

is a spatial vector (the Φ spatial momentum operator) that is written solely in terms of $\Phi_a(y)$'s creation and annihilation operators.

In addition to performing the transformation in eq. 5.6 $W_a(y)$ also performs a "translation" in Y^{μ}:

$$W_a(y) Y^i(y') W_a^{-1}(y) = Y^i(y') + i\Delta^{trij}(y' - y) P_{\Phi a}^{\ j}/M_c^2 \quad (5.13a)$$

where

$$i\Delta^{trij}(y' - y) = \int d^3k\, (e^{-ik\cdot(y' - y)} - e^{ik\cdot(y' - y)})(\delta_{jk} - k_j k_k/\mathbf{k}^2)/[(2\pi)^3 2\omega_k] \quad (5.13b)$$

We note that $W_a(y)$ is not a unitary operator but it is pseudo-unitary:

$$W_a^{-1}(y) = V W_a^{\dagger}(y) V^{-1} = V W_a(y) V^{-1} \quad (5.14)$$

where

$$V = \exp(-i\pi \sum_{\lambda = 1}^{2} \int d^3k\, a^{\dagger}(k, \lambda) a(k, \lambda)) \quad (5.15)$$

is a unitary operator with the property

$$VY^j(y)V^{-1} = -Y^j(y) \tag{5.16}$$

for j = 1,2,3. We note

$$V^\dagger = V^{-1} = V \tag{5.17}$$

and thus

$$V^2 = I \tag{5.18}$$

V will be shown to be a metric operator in the following discussion.[24] We note that the Y "particle" (hermitean) number operator appears in eq. 5.9 in the expression for V:

$$N_Y = \sum_{\lambda=1}^{2} \int d^3k\, a^\dagger(k, \lambda)a(k, \lambda) \tag{5.19}$$

Thus states with an even number of Y "particles" have a V eigenvalue of one, and states with an odd number of Y "particles" have a V eigenvalue of minus one.

Model Lagrangian with ϕ^4 Interaction

We will develop our perturbation theory using a scalar Lagrangian model with a ϕ^4 interaction term:

$$\mathscr{L}_S = J\mathscr{L}_F + \mathscr{L}_C \tag{5.20}$$

with

$$\mathscr{L}_F = ½ [(\partial\phi/\partial X^\nu)^2 - m^2\phi^2] + \mathscr{L}_{Fint} \tag{5.21}$$

and

$$\mathscr{L}_C = -¼ F_Y^{\mu\nu}F_{Y\mu\nu} \tag{5.22}$$

with

$$F_{Y\mu\nu} = \partial Y_\mu/\partial y^\nu - \partial Y_\nu/\partial y^\mu \tag{5.23}$$

and

$$\mathscr{L}_{Fint} = {}^1\!/_{4!}\, \chi_0\, \phi(X(y))^4 + ½ (m^2 - m_0^2)\phi^2 \tag{5.24}$$

where J is the Jacobian (as in Appendix A), χ_0 is the bare coupling constant, and m_0 is the bare mass.

---

[24] P. A. M. Dirac, Proc. R. Soc. London A **180**, 1 (1942); T. D. Lee and G. C. Wick, Nucl. Phys. **B9**, 209 (1969); C. M. Bender, D. C. Brody and H. F. Jones, "Complex Extension of Quantum Mechanics" Phys. Rev. Letters **89**, 270401-1 (2002) and references therein.

The conserved momentum operator is:

$$P_{F\beta} = \int d^3X\, \mathscr{T}_{F0\beta} \tag{5.25}$$

where

$$\mathscr{T}_{F\mu\nu} = -\, g_{\mu\nu}\mathscr{L}_F + \partial\mathscr{L}_F \,/\, \partial(\partial\phi/\partial X_\mu)\; \partial\phi/\partial X^\nu \tag{5.26}$$

is the ϕ field energy-momentum tensor with conservation law (eq. A.110):

$$\partial P_{F\beta}/\partial X^0 = 0 \tag{5.27}$$

due to eq. A.108.

The hamiltonian density (eq. A.83) is

$$\mathscr{H}_F = \mathscr{T}_{F0\beta} = \mathscr{H}_{F0} + \mathscr{H}_{Fint} \tag{5.28}$$

with

$$\mathscr{H}_{F0} = \tfrac{1}{2}\, [\pi_\phi^{\,2} + (\partial\phi/\partial X^i)^2 + m^2\phi^2] \tag{5.29}$$

$$\mathscr{H}_{Fint} = -\, \tfrac{1}{4!}\, \chi_0\, \phi(X(y))^4 + \tfrac{1}{2}\, (m^2 - m_0^{\,2})\phi(X(y))^2 \tag{5.30}$$

In-states and Out-States

In this section we will develop properties of in-fields and out-fields. We will use a somewhat more complicated procedure to set up the perturbation theory for the S matrix due to the introduction of imaginary coordinates. The procedure can be schematized as:

$$\Phi_{in}(y) \Rightarrow \phi_{in}(X(y)) \Rightarrow \phi(X(y)) \Rightarrow \phi_{out}(X(y)) \Rightarrow \Phi_{out}(y) \tag{5.31}$$

In-states are constructed using the auxiliary field Φ_{in} which are then effectively transformed into $\phi_{in}(X(y))$ expressions in order to make contact with our Lagrangian formalism. Then $\phi_{in}(X(y))$ is related to the interacting field $\phi(X(y))$ as a limit ($y^0 \to -\infty$). Similarly out-states are constructed using the auxiliary field Φ_{out} which are then expressed in terms of $\phi_{out}(X(y))$. Then $\phi_{out}(X(y))$ is related to the interacting field $\phi(X(y))$ using the LSZ limiting process ($y^0 \to +\infty$).

Since much of the development differs only trivially from the standard treatment in textbooks we will simply "list" relevant equations and let the reader pursue them further in quantum field theory textbooks.

ϕ In-Field

In order to define a perturbation theory for particle scattering we will next specify features of the in-field $\phi_{in}(X(y))$ and in-field states – the field and states representing physical particles as $X^0 = y^0 \rightarrow -\infty$.

A. The in-field $\phi_{in}(X(y))$ satisfies the Klein-Gordon equation in the X variable:

$$(\Box_X + m^2)\, \phi_{in}(X) = 0 \tag{5.32}$$

where

$$\Box_X = (\partial/\partial X^\nu)(\partial/\partial X_\nu)$$

B. Under coordinate displacements and Lorentz transformations $\Phi_{in}(y)$, $\phi_{in}(X(y))$, and $\phi(X(y))$ transform in the same way:

$$[P^\mu, \Phi_{in}(y)] = -\,i\partial\Phi_{in}/\partial y_\mu \tag{5.33a}$$

$$[P^\mu, \phi_{in}(X)] = -\,i\partial\phi_{in}/\partial y_\mu \tag{5.33b}$$

$$[P^\mu, \phi(X)] = -\,i\partial\phi/\partial y_\mu \tag{5.34}$$

with the energy-momentum vector P^μ specified by eq. A.57.

C. We can relate the asymptotic in-field $\phi_{in}(X(y))$ to the interacting field $\phi(X(y))$ using the equation of motion of $\phi(X(y))$

$$(\Box_X + m^2)\, \phi(X) = j(X) \tag{5.35}$$

where $j(X)$ embodies the interaction. Using the physical mass m we find

$$(\Box_X + m^2)\, \phi(X) = j(X) + (m^2 - m_0^2)\phi(X) = j_{tot}(X) \tag{5.36}$$

If the current is taken to be the source of the scattered waves we may write

$$\sqrt{Z}\, \phi_{in}(X(y)) = \phi(X(y)) - \int d^4X(y')\, \Delta_{ret}(y - y')\, j_{tot}(X(y')) \tag{5.37}$$

$$= \phi(X(y)) - \int d^4y' \, J \, \Delta_{ret}(y - y') \, j_{tot}(X(y')) \quad (5.38)$$

where Z is a wave function renormalization constant, J is the Jacobian, and Δ_{ret} is a retarded Green's function.

D. We can define Φ_{in} in-field states with expressions like

$$| \, p_1, p_2, \ldots \, p_n \text{ in}> = a_{in}^{\dagger}(p_1) a_{in}^{\dagger}(p_2) \ldots a_{in}^{\dagger}(p_n) \, | \, 0> \quad (5.39)$$

with powers of creation operators allowed since Φ_{in} is a boson field. The set of all particle states constitutes a complete orthonormal set of states. The corresponding bra states are defined by hermitean conjugation:

$$<p_1, p_2, \ldots \, p_n \text{ in} \, | = (| \, p_1, p_2, \ldots \, p_n \text{ in}>)^{\dagger} \quad (5.40)$$

ϕ Out-Field

In order to define a perturbation theory for particle scattering we begin by listing aspects of the out-field $\phi_{out}(X(y))$ and out-field states – the field and states representing physical particles as $X^0 = y^0 \to -\infty$.

A. The out-field $\phi_{out}(X(y))$ satisfies the Klein-Gordon equation in the X variable:

$$(\Box_X + m^2) \, \phi_{out}(X) = 0 \quad (5.41)$$

where

$$\Box_X = (\partial/\partial X^{\nu})(\partial/\partial X_{\nu})$$

B. Under coordinate displacements and Lorentz transformations $\Phi_{out}(y)$, $\phi_{out}(X(y))$, and $\phi(X(y))$ transform in the same way:

$$[P^{\mu}, \Phi_{out}(y)] = - i\partial\Phi_{out}/\partial y_{\mu} \quad (5.42a)$$

$$[P^{\mu}, \phi_{out}(X)] = - i\partial\phi_{out}/\partial y_{\mu} \quad (5.42b)$$

$$[P^{\mu}, \phi(X)] = - i\partial\phi/\partial y_{\mu} \quad (5.43)$$

with the energy-momentum vector P^{μ} specified by eq. A.57.

C. We can relate the asymptotic out-field $\phi_{out}(X(y))$ to the interacting field $\phi(X(y))$ using the equation of motion of $\phi(X(y))$ specified by eq. 5.36:

$$\sqrt{Z}\, \phi_{out}(X(y)) = \phi(X(y)) - \int d^4X(y')\, \Delta_{adv}(y - y')\, j_{tot}(X(y')) \tag{5.44}$$

$$= \phi(X(y)) - \int d^4y'\, J\, \Delta_{adv}(y - y')\, j_{tot}(X(y')) \tag{5.45}$$

where Z is a wave function renormalization constant, J is the Jacobian, and Δ_{adv} is an advanced Green's function.

D. We can define Φ_{out} out-field states with expressions like

$$|\, p_1, p_2, \ldots\, p_n \text{ out}> = a_{out}^\dagger(p_1)a\Phi_{out}^\dagger(p_2)\, \ldots\, a\Phi_{out}^\dagger(p_n)\,|0> \tag{5.46}$$

with powers of creation operators allowed since Φ_{out} is a boson field. The set of all particle states constitutes a complete orthonormal set of states. The corresponding bra states are defined by hermitean conjugation:

$$<p_1, p_2, \ldots\, p_n \text{ out}| = (|\, p_1, p_2, \ldots\, p_n \text{ out}>)^\dagger \tag{5.47}$$

The Y Field

The Y field in the present model Lagrangian (eq. 5.20) is a free field and thus:

$$Y_{in}(y) = Y_{out}(y) = Y(y) \tag{5.48}$$

The states of the Y field have two general forms: 1) States in a Fock space consisting of particle states that are eigenstates of the Y particle number operator (eq. 5.19); and 2) Coherent states in a non-Fock space of generalized coherent states in an infinite tensor product space.[25]

The coherent ket states that arise in Two-Tier quantum field theory have the general form (eq. 3.41):

$$|y,\ p> = e^{-\mathbf{p}\cdot\mathbf{Y}^-(y)/M_c^2}\,|0> \tag{3.41}$$

as can be seen from an examination of $\phi_{in}(X(y))$. The corresponding bra state is:

$$<y, p| = (V|\ y, p>)^\dagger = <0|\,e^{+\mathbf{p}\cdot\mathbf{Y}^+(y)/M_c^2} \tag{5.49}$$

[25] See Kibble and other references on coherent states.

with V, the metric operator, reversing the sign of Y in the exponential. The inner product of coherent states is:

$$<y_1, p_1 | y_2, p_2> = \exp[-p_1{}^i p_2{}^j \Delta_{Tij}(y_1 - y_2)/M_c{}^4] \tag{5.50}$$

showing the set of coherent states is not orthonormal and, in fact, is overcomplete. Comparing eq. 5.50 to eq. 4.12 gives

$$<y_1, p | y_2, p> = R(p, y_1 - y_2) \tag{5.50a}$$

The completeness of the set of states for each time y^0 can be verified by examining the projection operator:

$$\mathscr{R}_Y(y^0) = \because \exp[-i \int d^3y\, Y^-_i(y) | 0><0 | \pi^{+i}(y)] \because \tag{5.51}$$

where

$$\pi^{+i}(y) = -\partial Y^{+i}(y)/\partial y^0 \tag{5.52}$$

and where $\because$ represents an extended normal ordering operator:

$$\because \ldots \because$$

which is defined as placing creation operators to the left, projection operators in the center, and annihilation operators to the right. Thus eq 5.51 can be written

$$\mathscr{R}_Y = \sum_n (-i/n!)^n \int d^3y_1 \ldots \int d^3y_n Y^{-i_1}(y_1) Y^{-i_2}(y_2) \ldots Y^{-i_n}(y_n) | 0><0 | \pi^+_{i_1}(y_1) \pi^+_{i_2}(y_2) \ldots \pi^+_{i_n}(y_n) \tag{5.53}$$

where we have used the fact that $|0><0|$ is a projection operator, and reduced $|0><0|$ $|0><0|$ ... $|0><0|$ to $|0><0|$ in eq. 5.53. The vacuum state is the product of the Y and ϕ vacuum states:

$$|0> = |0_Y> |0_\phi> \tag{5.53a}$$

We note

$$\mathscr{R}_Y (y^0) | \mathbf{y}, y^0\, p> = | \mathbf{y}, y^0\, p> \tag{5.54}$$

using eq. 3.22 and $\int d^3y_2\, p^i\, \Delta^{tr}_{ij}(y_1 - y_2)Y^{+j}(y_2) = \mathbf{p}\cdot\mathbf{Y}^+(y_1)$. Also

$$\mathscr{R}_Y(y^0)\,|n\rangle = |n\rangle \tag{5.55}$$

where |n> is any Y particle Fock state of finite particle number. In view of eqs. 5.54 and 5.55, we see that $\mathscr{R}_Y$ is the identity operator in the Fock space and in the space of generalized coherent Y field states. Thus the set of Y coherent states forms an overcomplete set of states. We will define the S matrix for any combination of Φ Fock space states and coherent Y states. The $\mathscr{R}_Y$ operator can be generalized to include Φ Fock space states:

$$\mathscr{R}_{\Phi Y}(y^0) = \because \exp[-i \int d^3y\; Y^-_j(y)\mathscr{R}_\Phi \pi^{+j}(y)] \because \tag{5.56}$$

with

$$\mathscr{R}_\Phi = \sum_n |n\rangle\langle n| \tag{5.57}$$

being a sum over all Φ Fock space states with vacuum state given by eq. 5.53a. Since $\mathscr{R}_\Phi$ is a projection:

$$[\mathscr{R}_\Phi]^N = \mathscr{R}_\Phi$$

for any power N, we find:

$$\mathscr{R}_{\Phi Y}(y^0) = \sum_n (-i)^n \int d^3y_1 \ldots \int d^3y_n Y^{-j_1}(y_1)Y^{-j_2}(y_2)\ldots Y^{-j_n}(y_n)\mathscr{R}_\Phi \pi^+_{j_1}(y_1)\pi^+_{j_2}(y_2)\ldots\pi^+_{j_n}(y_n) \tag{5.58}$$

As a result we have

$$\mathscr{R}_{\Phi Y}(y^0)\,|y, p; n_\Phi\rangle = |y, p; n_\Phi\rangle \tag{5.59}$$

for any combination of Y coherent states and Φ Fock space states n_Φ. Also

$$\mathscr{R}_{\Phi Y}(y^0)\,|n_\Phi\rangle = |n_\Phi\rangle \tag{5.60}$$

Thus $\mathscr{R}_{\Phi Y}$ is the identity operator on this space – the (over) complete space of in and out states which we will use to define the S matrix of the scalar field theory specified by the Lagrangian eq. 5.20.

S Matrix

Following the standard definition of the S matrix we have:

$$S_{\alpha\beta} = <\alpha \text{ out}|\beta \text{ in}> \tag{5.61}$$

$$= <\alpha \text{ in}|S|\beta \text{ in}> \tag{5.62}$$

$$|0> = |0 \text{ in}> = |0 \text{ out}> = S|0 \text{ in}> \tag{5.63}$$

$$\Phi_{\text{in}}(y) = S\Phi_{\text{out}}(y)S^{-1} \tag{5.64}$$

and the other standard properties of the S matrix with the sole exception being the form of the unitarity relation (discussed later).

LSZ Reduction for Scalar Fields

In this section we will determine the reduction formula for the S matrix for scalar ϕ fields. Consider the S matrix element corresponding to an in state of particles β plus one ϕ particle of momentum p, and an out state α:

$$S_{\alpha\beta p} = <\alpha \text{ out}|\beta p \text{ in}> \tag{5.65}$$

After standard manipulations we have:

$$S_{\alpha\beta p} = <\alpha - p \text{ out}|\beta \text{ in}> - i<\alpha \text{ out}|\int d^3y\, f_p(y)\, \overleftrightarrow{\partial_0}\, [\Phi_{\text{in}}(y) - \Phi_{\text{out}}(y)]\ |\beta \text{ in}> \tag{5.66}$$

where $<\alpha - \text{p out}|$ is an out state with a particle of momentum p removed (if present) and where

$$f(y^0)\, \overleftrightarrow{\partial_0}\, g(y^0) = f(y^0)\, \partial g(y^0)/\partial y^0 - \partial f(y^0)/\partial y^0\, g(y^0) \tag{5.67}$$

and

$$f_p(y) = N_m(p)e^{-ip\cdot y} \tag{5.68}$$

with $N_m(p)$ specified by eq. 3.6.
We now express

$$S_{\alpha\beta p} = S_{\alpha-p\beta} - i<\alpha\ \text{out}|\int d^3y\ f_p(y)\ \overleftrightarrow{\partial}_0 W^{-1}[\phi_{in}(X(y)) - \phi_{out}(X(y))]W|\beta\ \text{in}> \tag{5.69}$$

using $W(y) = W_{in}(y)$ with

$$\Phi_a(y) = W_a^{-1}(y)\phi_a(X(y))W_a(y) \tag{5.70}$$

where the label a = "in" or a = "out", and where

$$W_a(y) = \exp(-\mathbf{Y}(y)\cdot\mathbf{P}_{\Phi a}/M_c^2) \tag{5.71}$$

and

$$W_a^{-1}(y) = \exp(\mathbf{Y}(y)\cdot\mathbf{P}_{\Phi a}/M_c^2) \tag{5.72}$$

in the Coulomb gauge of Y with $\mathbf{P}_{\Phi a}$ the momentum spatial vector defined by eq. 5.12a.

We note that the interacting $\phi(X(y))$ approaches the in and out fields $\phi_{in}(X(y))$ and $\phi_{out}(X(y))$ in the limit that $y^0 \rightarrow -\infty$ and $y^0 \rightarrow +\infty$ respectively in the sense of Lehmann, Symanzik and Zimmermann[26] which we *symbolize* as:

$$\phi(X(y)) \rightarrow \sqrt{Z}\ \phi_{in}(X(y)) \quad \text{as} \quad y^0 \rightarrow -\infty \tag{5.73}$$

$$\phi(X(y)) \rightarrow \sqrt{Z}\ \phi_{out}(X(y)) \quad \text{as} \quad y^0 \rightarrow +\infty \tag{5.74}$$

with $\sqrt{Z}$ defined in eqs. 5.37 and 5.44. Thus we can rewrite eq. 5.69 as

$$S_{\alpha\beta p} = S_{\alpha-p\beta} + iZ^{-1/2}\left(\lim_{y_0\rightarrow+\infty} - \lim_{y_0\rightarrow-\infty}\right)<\alpha\ \text{out}|\int d^3y\ f_p(y)\ \overleftrightarrow{\partial}_0 W^{-1}\phi(X(y))W|\beta\ \text{in}> \tag{5.75}$$

which after standard manipulations becomes

[26] H. Lehmann, K. Symanzik and W. Zimmermann, Nuov. Cim., **1**, 1425 (1955); W. Zimmermann, Nuov. Cim., **10**, 567 (1958); O. W. Greenberg, Doctoral Dissertation, Princeton University 1956.

$$S_{\alpha\beta p} = S_{\alpha-p\beta} + iZ^{-1/2}\int d^4y\, f_p(y)(\Box_y + m^2)<\alpha\ \text{out}|\, W(y)^{-1}\phi(X(y))W(y)\,|\beta\ \text{in}> \tag{5.76}$$

Eq. 5.76 is similar to the usual LSZ reduction formula except for the appearance of the W(y) operator and its inverse. We note that $W(y) = W_{in}(y)$ still because $\mathbf{P}_{\Phi in}$ is independent of y^0.

Similarly an out ϕ particle can be reduced from an S matrix part. For example,

$$\begin{aligned}<\alpha\ \text{out}|W^{-1}(y)\phi(X(y))W(y)\,|\beta\ \text{in}> &= <\alpha-p'\ \text{out}|W^{-1}(y)\phi(X(y))W(y)\,|\beta-p'\ \text{in}> \\ &- i<\alpha-p'\ \text{out}|\int d^3y'\,[W^{-1}(y')\phi_{in}(X(y'))W(y')W^{-1}(y)\phi(X(y))W(y) - \\ &- W^{-1}(y)\phi(X(y))W(y)W^{-1}(y')\phi_{out}(X(y'))W(y')]\,|\beta\ \text{in}>\overleftrightarrow{\partial}_0 f_{p'}^{*}(y')\end{aligned} \tag{5.77}$$

which becomes

$$\begin{aligned}<\alpha\ \text{out}|W^{-1}(y)\phi(X(y))W(y)\,|\beta\ \text{in}> &= <\alpha-p'\ \text{out}|\varphi(y)\,|\beta-p'\ \text{in}> + \\ &+ iZ^{-1/2}\int d^4y' <\alpha-p'\ \text{out}|T(\varphi(y')\varphi(y))\,|\beta\ \text{in}>(\overleftarrow{\Box}_{y'} + m^2)\, f_{p'}^{*}(y')\end{aligned} \tag{5.78}$$

where the time ordered product is defined with respect to ordering with y^0 and where

$$\varphi(y) = W^{-1}(y)\phi(X(y))W(y) \tag{5.79}$$

These results directly generalize to multi-particle in and out states:

$$<p_1, p_2, \ldots p_n\ \text{out}|\ q_1, q_2, \ldots q_m\ \text{in}> = \ldots <0|T(\varphi(y'_1)\ \ldots\ \varphi(y'_n)\varphi(y_1)\ \ldots\ \varphi(y_m))\,|0> \ldots \tag{5.80}$$

thus reducing the development of the perturbation theory of the S matrix to the evaluation of time ordered products such as

$$<0|T(\varphi(y_1)\ \ldots\ \varphi(y_n))\,|0> \tag{5.81}$$

6. Perturbation Theory II

The U Matrix

The U matrix for a Two-Tier theory is developed in a way similar to conventional field theory starting from the defining relations:

$$\phi(X(y)) = U^{-1}\phi_{in}(X(y))U \qquad (6.1)$$

$$\pi_\phi(X(y)) = U^{-1}\pi_{\phi in}(X(y))U \qquad (6.2)$$

From eq. 5.29 we define the free field hamiltonian

$$H_{F0in}(\phi_{in}, \pi_{\phi in}) = \int d^3X\, \mathscr{H}_{F0}(\phi_{in}, \pi_{\phi in}) \qquad (6.3)$$

Noting $X^0 = y^0$ in the Y Coulomb gauge we find

$$\partial\phi_{in}/\partial y^0 = i[H_{F0in}, \phi_{in}(X)] \qquad (6.4)$$

$$\partial\pi_{\phi in}/\partial y^0 = i[H_{F0in}, \pi_{\phi in}(X)] \qquad (6.5)$$

For the entire hamiltonian (eq. 5.28) we have

$$\partial\phi/\partial y^0 = i[H_F, \phi(X)] \qquad (6.6)$$

$$\partial\pi_\phi/\partial y^0 = i[H_F, \pi_\phi(X)] \qquad (6.7)$$

with

$$H_F(\phi, \pi_\phi) = :\int d^3X\, \mathscr{H}_F(\phi, \pi_\phi): \qquad (6.8)$$

(Note the *entire* interaction term is normal ordered since d^3X is a q-number. Combining the above equations in the standard way yields a familiar differential equation for the U matrix:

$$i\partial U(y^0)/\partial y^0 = (H_{Fint} + E_0(t))U(y^0) \tag{6.9}$$

where $E_0(t)$ is a c-number function of y^0 that we can set equal to 0 (as it would be cancelled later in any case), and where

$$H_{Fint}(\phi_{in}, \pi_{\phi in}) = :\int d^3X\, \mathscr{H}_{Fint}(\phi_{in}, \pi_{\phi in}): \tag{6.10}$$

with $\mathscr{H}_{Fint}$ given by eq. 5.30. Solving for U gives the familiar time ordered exponential:

$$U(y^0) = T\left(\exp[-i\int_{-\infty}^{t} dy^0\ H_{Fint}]\right) \tag{6.11a}$$

which is a symbolic notation for:

$$U(y^0) = 1 + \sum_{n=1}^{\infty} (-i)^{-n}(n!)^{-1} \int_{-\infty}^{y^0} dy_1^{\,0} \ldots \int_{-\infty}^{y^0} dy_n^{\,0}\, T(H_{Fint}(y_1^{\,0}) \ldots H_{Fint}(y_n^{\,0}))$$

$$\tag{6.11b}$$

We note for later use that the hermiticity of H_{Fint} is not used in the derivation of eq. 6.11. Thus eq. 6.11 would still hold if H_{Fint} were not hermitean.

Reduction of Time Ordered φ Products

In the previous chapter we reduced the calculation of the S matrix to the evaluation of time ordered products of the form

$$\tau(y_1, \ldots, y_n) = \langle 0|T(\varphi(y_1) \ldots \varphi(y_n))|0\rangle \tag{6.12}$$

where φ(y) is specified by eq. 5.79. Expanding the terms within eq. 6.12 using eq. 5.79 we find

$$\varphi(y_1) \ldots \varphi(y_n) = W^{-1}(y_1)\phi(X(y_1))W(y_1)W^{-1}(y_2)\phi(X(y_2))W(y_2) \ldots W^{-1}(y_n)\phi(X(y_n))W(y_n) \tag{6.13}$$

which can be re-expressed as

$$W^{-1}(y_1)U^{-1}(y_1^{\,0})\phi_{in}(X(y_1))U(y_1^{\,0})W(y_1)W^{-1}(y_2)U^{-1}(y_2^{\,0})\phi_{in}(X(y_2))U(y_2^{\,0})W(y_2) \ldots \tag{6.14}$$

using eq. 6.1 and denoting $W_{in}(y)$ as $W(y)$. Defining

$$\mathscr{U}(y_1, y_2) = U(y_1^0)W(y_1)W^{-1}(y_2)U^{-1}(y_2^0) \tag{6.15}$$

we see eq. 6.14 can be rewritten as

$$W^{-1}(y_1)U^{-1}(y_1^0)\phi_{in}(X(y_1))\mathscr{U}(y_1, y_2)\phi_{in}(X(y_2))\,\mathscr{U}(y_2, y_3)\phi_{in}(X(y_3))\ldots\phi_{in}(X(y_n))U(y_n^0)W(y_n) \tag{6.16}$$

From eqs. 5.71 and 5.72

$$\mathscr{U}(y_1, y_2) = U(y_1^0)\exp((\mathbf{Y}(y_2) - \mathbf{Y}(y_1))\cdot\mathbf{P}_{\Phi a}/M_c^2)U^{-1}(y_2^0) \tag{6.17}$$

Defining

$$W(y_1, y_2) = \exp((\mathbf{Y}(y_2) - \mathbf{Y}(y_1))\cdot\mathbf{P}_{\Phi a}/M_c^2) \tag{6.18}$$

and looking ahead to the Wick expansion of the time ordered product of eq. 6.12 we note that the only time ordered products involving $W(y_1, y_2)$ that would appear in the expansion are

$$<0|T(\phi_{in}(X(y))W(y_1, y_2))|0> = 0 \tag{6.19a}$$

$$<0|T(Y(y)W(y_1, y_2))|0> = 0 \tag{6.19b}$$

$$<0|T(\partial Y(y)/\partial y^\mu\, W(y_1, y_2))|0> = 0 \tag{6.19c}$$

$$<0|T(\partial Y(y)/\partial y^\mu\, \phi_{in}(X(y)))|0> = 0 \tag{6.19d}$$

$$<0|T(W(y_1, y_2)W(y_3, y_4))|0> = 1 \tag{6.19e}$$

due to the factor of $\mathbf{P}_{\Phi a}$ that appears in $W(y_1, y_2)$. Also

$$<0|T(\phi_{in}(X(y))Y(y_1))|0> = 0 \tag{6.20}$$

due to the $a_{in}(p)$ and $a_{in}^\dagger(p)$ factors appearing in $\phi_{in}(X(y))$.

Thus the $W(y_1, y_2)$ factor in eq. 6.17 may be set to the value one with the result

$$\mathscr{U}(y_1, y_2) \equiv U(y_1^{\,0})U^{-1}(y_2^{\,0}) = U(y_1^{\,0}, y_2^{\,0}) \tag{6.21}$$

where $U(y_1^{\,0}, y_2^{\,0})$ is the conventionally defined U matrix satisfying

$$i\partial\, U(y_1^{\,0}, y_2^{\,0})/\partial y_1^{\,0} = iH_{Fint}\, U(y_1^{\,0}, y_2^{\,0}) \tag{6.22}$$

with the boundary condition

$$U(y^0, y^0) = 1 \tag{6.23}$$

This result would still be true if the $W(y_1, y_2)$ exponentials were expanded in their "power series" form.

Then, paralleling the standard approach we find an expression for the U matrix:

$$U(y_1^{\,0}, y_2^{\,0}) = T\Big(\exp[-i \int_{y_2^{\,0}}^{y_1^{\,0}} dy'^0{:}d^3X(y')\, \mathscr{H}_{Fint}(\phi_{in}(X(y')), \pi_{\phi in}(X(y'))){:}]\Big) \tag{6.24}$$

The $U(y_1^{\,0}, y_2^{\,0})$ matrix satisfies the conventional multiplication rule:

$$U(y_1^{\,0}, y_3^{\,0}) = U(y_1^{\,0}, y_2^{\,0})U(y_2^{\,0}, y_3^{\,0}) \tag{6.25}$$

The inverse of $U(y_1, y_2)$ is

$$U^{-1}(y_1^{\,0}, y_2^{\,0}) = U(y_2^{\,0}, y_1^{\,0}) \tag{6.26}$$

We now return to eq. 6.16, which can now be written in the form:

$$U^{-1}(y^0)U(y^0, y_1^{\,0})\phi_{in}(X(y_1))U(y_1^{\,0}, y_2^{\,0})\phi_{in}(X(y_2))U(y_2^{\,0}, y_3^{\,0}) \ldots \phi_{in}(X(y_n))U(y_n^{\,0}, -y^0)U(-y^0) \tag{6.27}$$

where y^0 is a reference time that is later than all other times, and $-y^0$ is earlier than all the other times, in the time-ordered product. As a result the time-ordered product in eq. 5.80 can be expressed in a symbolic notation as:

$$<0|U^{-1}(y^0)T(\phi_{in}(X(y_1))\phi_{in}(X(y_2)) \ldots \phi_{in}(X(y_n))U(y^0, -y^0))U(-y^0)|0> \tag{6.28}$$

The analysis of eq. 6.28 as $y^0 \to \infty$ follows the standard path, which begins by noting

$$U(-y)\,|0> = \lambda_-|0> \qquad \text{when } y^0 \to \infty \tag{6.29a}$$

$$U(y)\,|0> = \lambda_+|0> \qquad \text{when } y^0 \to \infty \tag{6.29b}$$

following a standard textbook proof, which, in turn, leads to:

$$\lambda_-\lambda_+{}^* = <0|T\Big(\exp[\,+i\int_{-\infty}^{\infty} dy'^0{:}d^3X(y')\mathscr{H}_{Fint}(\phi_{in}(X(y')),\,\pi_{\phi in}(X(y'))){:}]\Big)\,|0> \tag{6.30}$$

$$= \Big[<0|T\Big(\exp\,[\,-i\int_{-\infty}^{\infty} dy'^0 d^3X(y')\;\mathscr{H}_{Fint}(\phi_{in}(X(y')),\,\pi_{\phi in}(X(y')))]\Big)\,|0>\Big]^{-1} \tag{6.31}$$

Thus the time ordered product of eq. 6.12, which appears in the evaluation of the S matrix element in eq. 5.80, can be symbolically written as:

$$\tau(\,y_1, \ldots, y_n) = \frac{<0|T(\phi_{in}(X(y_1))\;\ldots\;\phi_{in}(X(y_n))U(\infty,-\infty))\,|0>}{<0|T\Big(\exp\,[-i\int dy'^0{:}d^3X(y')\mathscr{H}_{Fint}(\phi_{in}(X(y')),\pi_{\phi in}(X(y'))){:}]\Big)\,|0>} \tag{6.32}$$

in the limit $y^0 \to \infty$.

The $\int d^3X$ Integration

The integration over the X space coordinates presents the difficulty of a functional integration of a q-number that needs to be properly defined. Since

$$X^\mu(y) = y^\mu + \; i\,Y^\mu(y)/M_c^{\,2} \tag{3.12}$$

by definition and since, in the Y Coulomb gauge we have $X^0(y) = y^0$ due to $Y^0 = 0$, the classical Jacobian for the transformation from y to X coordinates is the absolute value:

$$J = \Big|\,\varepsilon^{ijk}\Big(\delta^{1i} + \frac{i}{M_C{}^2}\frac{\partial Y^1}{\partial y^i}\Big)\Big(\delta^{2j} + \frac{i}{M_C{}^2}\frac{\partial Y^2}{\partial y^j}\Big)\Big(\delta^{3k} + \frac{i}{M_C{}^2}\frac{\partial Y^3}{\partial y^k}\Big)\Big| \tag{6.33}$$

The Jacobian appears in a change of integration variables:

$$\int d^3X = \int d^3y\ J \tag{6.34}$$

and

$$\int d^4X = \int d^4y\ J \tag{6.35}$$

in the Y Coulomb gauge.

A change of variables for c-number coordinate transformations is well known. The situation changes when one set of coordinates are in fact q-numbers. The second quantization of the Y field requires the definition of J to be clarified since the product of fields at the same position is normally undefined. The normal ordering of the interaction hamiltonian term in eqs. 6.34 and 6.32 resolves the issue. Therefore eq. 6.33 must be considered as inserted within a normal ordered expression.

While normal ordering eliminates the infinities that would otherwise be present, J still presents a problem because it is still effectively part of the interaction term. This situation appears to be unsatisfactory in the present, scalar quantum field theory in which Y is not intended to play a direct dynamical role but rather a passive role as a coordinate. The normal ϕ field is supposed to be the only in, out, and interacting field.

The problem of J is resolved by eqs. 6.19c and 6.19d, which reduces the effect of the derivative terms in eq. 6.33 to zero in the Wick expansion of the time ordered product in eq. 6.32 if no Y quanta appear in or out S matrix states. Thus

$$J \equiv 1 \tag{6.36}$$

As a result the time ordered product (eq. 6.32) becomes:

$$\tau(y_1,\ldots, y_n) = \frac{<0|T(\phi_{in}(X(y_1))\ldots\phi_{in}(X(y_n))\exp[-i\int d^4y'\,\mathscr{H}_{Fint}(\phi_{in}(X(y')))])\ |0>}{<0|T\big(\exp\ [-i\int d^4y'\,\mathscr{H}_{Fint}(\phi_{in}(X(y')))]\big)\ |0>} \tag{6.37}$$

Y In and out states

The Y fields have no interactions and are thus free fields in the model Lagrangian under consideration and in the Two-Tier quantum field theories that we will construct later. Therefore “in” Y quanta are the same as “out” Y quanta.

Since the Lagrangians that we consider do not have interaction terms explicitly containing Y field factors, the S matrix is “block diagonal” in the sense that if an in-state does not contain Y quanta, (or Y coherent states) then out-states will not contain Y quanta (or coherent Y states). The proof is based on the expansion of S matrix elements using Wick's theorem in products of time ordered products of pairs of in field operators. Eqs. 6.19, 6.20 and 6.36, and in particular,

$$<0|\ T(\phi_{in}(X(y_1))Y^j(y_2))\,|0> = 0 \tag{6.39}$$

and

$$<0|T(\phi_{in}(X(y_1))e^{-q\cdot Y^-(y)/M_c^2})\,|0> = <0|T(\phi_{in}(X(y_1))e^{+q\cdot Y^+(y)/M_c^2})\,|0> = 0 \tag{6.40}$$

prove S matrix elements with no incoming Y quanta or coherent states will have zero matrix elements to produce outgoing Y quanta or coherent states. In addition any non-zero S matrix element with n incoming Y quanta must have n outgoing Y quanta. For example an incoming state with 5 Y quanta and 2 ϕ particles can only become an outgoing state with 5 Y quanta and two or more ϕ particles. Therefore we have proved the general result:

Theorem 6.I: ***Any non-zero S matrix element has the same number of incoming Y quanta and outgoing Y quanta.***

This theorem is true in any Two-Tier quantum field theory. In order to have a tractable theory we will require all in-states and out-states <u>*not*</u> *to contain Y quanta or coherent states. All normal in-state and out-state particles will contain factors of* $:e^{\pm p\cdot Y/M_c^2}:$ *in the fourier expansions of their corresponding fields.*

Unitarity

For many years it has been evident that modified field theories[17, 27] might offer some hope of avoiding the divergences of conventional quantum field theory. Usually these theories suffer from unitarity problems: negative norms and negative probabilities. In the absence of a physically acceptable interpretation of negative probabilities, these theories have been thought to be unsatisfactory.

The Two-Tier type of quantum field theory *superficially* also appears to have a unitarity problem due to the non-hermitean nature of Two-Tier hamiltonians. The lack of hermiticity is due entirely to the appearance of iY^μ in the X^μ field coordinates. *In fact*

[27] S. Blaha, Phys.Rev. **D10**, 4268 (1974); S. Blaha, Phys.Rev. **D11**, 2921 (1975); S. Blaha, Nuovo Cim. **A49**, :113 (1979); S. Blaha, "Generalization of Weyl's Unified Theory to Encompass a Non-Abelian Internal Symmetry Group" SLAC-PUB-1799, Aug 1976; S. Blaha, "Quantum Gravity and Quark Confinement" Lett. Nuovo Cim. **18**, 60 (1977); S. Blaha, "The Local Definition of Asymptotic Particle States" Nuovo Cim. **A49**, 35 (1979) and references therein.

Two-Tier quantum field theories satisfy unitarity for physical states. Physical states are defined to consist of any number of normal Two-Tier particles and NO Y quanta.

Two-Tier interaction hamiltonians, such as the one in eq. 6.37, are not hermitean. For example,

$$H_{Fint} = \int d^3y' \mathscr{H}_{Fint}(\phi_{in}(y' + iY(y')/M_c^2)) \tag{6.41}$$

and

$$H_{Fint}^{\dagger} = \int d^3y' \mathscr{H}_{Fint}(\phi_{in}(y' - iY(y')/M_c^2)) \neq H_{Fint} \tag{6.42}$$

The relation between H_{Fint} and its hermitean conjugate is

$$H_{Fint} = VH_{Fint}^{\dagger}V \tag{6.43}$$

where $V^2 = I$ is the metric operator defined in eqs. 5.15 – 5.18. Thus the S matrix is not unitary; the S matrix is *pseudo-unitary*:

$$S^{-1} = VS^{\dagger}V \tag{6.44}$$

and so

$$VS^{\dagger}VS = I \tag{6.45}$$

We will now show that the S matrix is *unitary between physical states.* To prove this point, consider eq. 6.45 between physical states |i> and <f| – each consisting of a number of ϕ particles and no Y quanta.

$$\delta_{fi} = \langle f\,|I|\,i\rangle = \langle f\,|VS^{\dagger}VS|\,i\rangle$$

$$= \sum_{n,\,m,\,p} \langle f\,|V|\,p\rangle\langle p\,|S^{\dagger}|\,n\rangle\langle n\,|V|\,m\rangle\langle m\,|S|\,i\rangle$$

$$= \sum_{n,\,m,\,p} \langle f\,|S^{\dagger}|\,m\rangle\langle m\,|S|\,i\rangle \tag{6.46}$$

since V has the eigenvalue 1 between states consisting of no Y quanta. Due to eqs. 6.19a – 6.19e and 6.20 since there are no incoming Y quanta there are no outgoing Y quanta.

The block diagonality of S (and the diagonality of V) limits the intermediate states |n> and |m> to states containing ϕ particles and no Y quanta – although normalization factors R(**p**, z) will appear (described later) due to the presence of $:e^{\pm p\cdot Y/M_c^2}:$ factors within quantum field fourier expansions that embody Y coherent state effects. Thus

$$S_{phys}^{\dagger} S_{phys} = I \tag{6.47}$$

and

$$S_{phys}^{\dagger} = S_{phys}^{-1} \tag{6.48}$$

proving unitarity between physical states – states consisting of ϕ particles and no Y quanta that are properly normalized. A detailed example is presented starting on page 67.

Finite Renormalization of External Legs

In the previous section we showed the theory satisfies unitarity for states that are properly normalized. However the use of the non-unitary operator W(y) (eq. 5.6) to transform $\Phi_{in}(y)$ fields into $\phi_{in}(X(y))$ fields in the LSZ procedure in eq. 5.69, and related equations, does not preserve the norm of input and output ϕ particle legs. Thus a finite renormalization is needed for each external particle leg in order to have a unitary S-matrix.

We define this renormalization of external legs within the framework of a perturbation theory example in the section beginning on page 67.

Perturbation Expansion

Perturbation theory in Two-Tier quantum field theory is very similar to conventional perturbation theory. The difference is in the form of the propagators, which have a high energy damping factor R(**p**, z) that eliminates infinities that normally appear at high energy in conventional quantum field theories.

In order to develop a feeling for Two-Tier perturbation theory we will calculate a few low order diagrams in the perturbation theory of the model scalar ϕ^4 theory that we have been using as an example in this chapter.

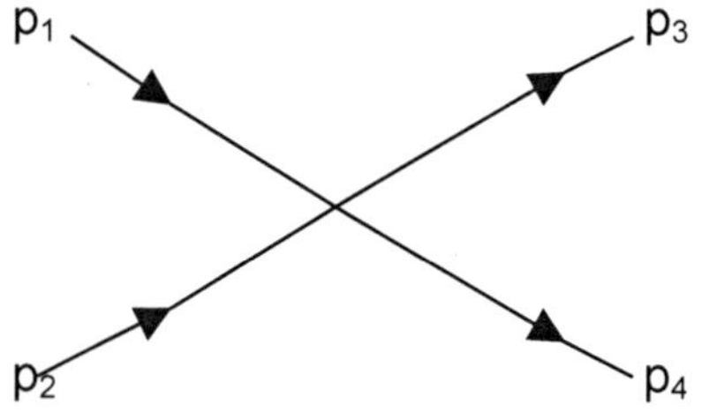

Figure 6.1. Lowest order quartic interaction diagram.

Fig. 6.1 contains the lowest order diagram for the scattering of two ϕ particles into a two ϕ particle out-state. The S matrix element for this diagram is

$$S_1 = i^4(\tfrac{1}{4!}\, i\varkappa_0) \prod_{j=1}^{4} \int d^4y_j\, d^4y\, f_{Zp_1}(y_1) f_{Zp_2}(y_2) f_{Zp_3}^{\,*}(y_3) f_{Zp_4}^{\,*}(y_4)(\Box_{y_1} + m^2)\cdot$$
$$\cdot(\Box_{y_2}+m^2)(\Box_{y_3}+m^2)(\Box_{y_4}+m^2)<0|T(\phi_{in}(X(y_1))\ldots\phi_{in}(X(y_4)):(\phi_{in}(X(y))^4:)|0> \tag{6.49}$$

with $f_{Zp}(y)$ specified by

$$f_{Zp}(y) = [(2\pi)^3 2p^0 Z_p]^{-1/2}\, e^{-ip\cdot y} \tag{6.49a}$$

where Z_p is a normalization factor that will be specified later.

Expanding the time ordered product, and realizing there are 4! ways of combining the four field factors in the interaction hamiltonian, we find:

$$S_1 = i^4(i\varkappa_0) \prod_{j=1}^{4} \int d^4y_j\, d^4y\, f_{Zp_1}(y_1) f_{Zp_2}(y_2) f_{Zp_3}^{\,*}(y_3) f_{Zp_4}^{\,*}(y_4)(\Box_{y_1} + m^2)\cdot$$
$$\cdot(\Box_{y_2}+m^2)(\Box_{y_3}+m^2)(\Box_{y_4}+m^2) i\Delta_F^{TT}(y_1-y) i\Delta_F^{TT}(y_2-y) i\Delta_F^{TT}(y_3-y) i\Delta_F^{TT}(y_4-y) \tag{6.50}$$

where

$$i\Delta_F^{TT}(y_1 - y_2) = <0|T(\phi(X(y_1)),\phi(X(y_2)))|0> \tag{6.51}$$

$$= i \int \frac{d^4p\, e^{-ip\cdot(y_1 - y_2)}\, R(\mathbf{p}, y_1 - y_2)}{(2\pi)^4\, (p^2 - m^2 + i\varepsilon)} \tag{6.52}$$

with

$$R(\mathbf{p}, y_1 - y_2) = \exp[-p^i p^j \Delta_{Tij}(y_1 - y_2)/M_c^4] \tag{6.53}$$

(summations are over space indices only in the Y Coulomb gauge) and

$$\Delta_{Tij}(z) = \int d^3k\, e^{-ik\cdot z}(\delta_{ij} - k_i k_j/\mathbf{k}^2)/[(2\pi)^3 2\omega_k] \tag{6.54}$$

From chapter 4 we have:

$$R(\mathbf{p}, y_1 - y_2) = \exp\{-p^2[A(v) + B(v)\cos^2\theta] / [4\pi^2 M_c^{\,4} z^2]\} \tag{6.55}$$

with

$$z^\mu = y_1^{\ \mu} - y_2^{\ \mu} \tag{6.56}$$

$$z = |\mathbf{z}| = |\mathbf{y}_1 - \mathbf{y}_2| \tag{6.57}$$

$$p = |\mathbf{p}| \tag{6.58}$$

$$v = |z^0| / z \tag{6.59}$$

$$A(v) = (1 - v^2)^{-1} + .5v\ \ln[(v-1)/(v+1)] \tag{6.60}$$

$$B(v) = v^2(1 - v^2)^{-1} - 1.5v\ \ln[(v-1)/(v+1)] \tag{6.61}$$

$$\mathbf{p}\cdot\mathbf{z} = pz\cos\theta \tag{6.62}$$

and with $|\mathbf{p}|$ denoting the length of the spatial vector $\mathbf{p}$, while $|z^0|$ is the absolute value of z^0.

We note

$$R(\mathbf{p}, y) = R(\mathbf{p}, -y) \tag{6.62a}$$

for later use.

Letting $y_i = w_i + y$ yields

$$S_1 = i^4(i\chi_0)(2\pi)^4\,\delta^4(p_3 + p_4 - p_1 - p_2)\mathcal{N}^+(p_4)\mathcal{N}^+(p_3)\mathcal{N}(p_2)\mathcal{N}(p_1) \tag{6.63}$$

where

$$\mathcal{N}(p) = iZ_p^{-1/2}\int d^4w\ f_p(w)(\Box + m^2)\Delta_F^{TT}(w) \tag{6.64}$$

$$\mathcal{N}^+(p) = iZ_p^{-1/2}\int d^4w\ f_p{}^*(w)(\Box + m^2)\Delta_F^{TT}(w) \tag{6.65}$$

are "normalizations" of the "external legs" – the in and out states due to the Y field cloud around each particle with $Z^{-1/2}$ a renormalization factor to be determined later. In the limit of low momentum ($p \ll M_C$):

$$\mathcal{N}(p) = \mathcal{N}^{+}(p) \rightarrow -iZ_p^{-1/2}[(2\pi)^3\, 2p^0\,]^{-1/2} \tag{6.66}$$

which the reader will note is the standard normalization factor for external scalar field legs in conventional quantum field theory modulo the $Z_p^{-1/2}$ factor. The factor $Z_p^{-1/2}$ performs the finite renormalization of external legs discussed in the preceding unitarity discussion.

Higher Order Diagram Containing a Loop

We will now consider the simplest one loop scattering diagrams in the scalar ϕ^4 theory.

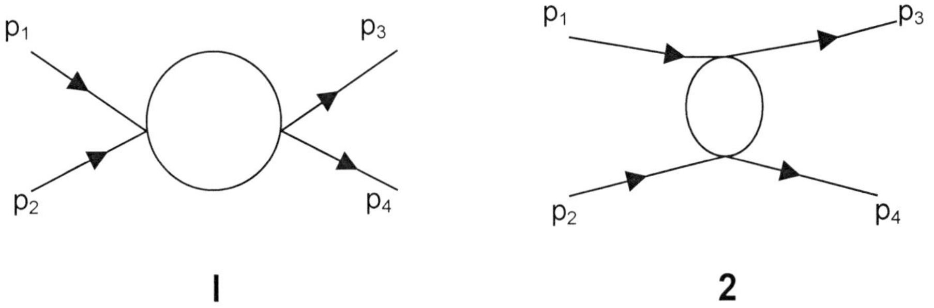

Figure 6.2. Low order loop scattering diagrams.

The S matrix element for these diagrams (and some other disconnected diagrams) is contained in

$$S_2 = i^4(1/4!\, i\chi_0)^2 \prod_{j=1}^{4} \int d^4y_j\, d^4y'_1\, d^4y'_2\, f_{Zp_1}(y_1) f_{Zp_2}(y_2) f^*_{Zp_3}(y_3) f^*_{Zp_4}(y_4)(\Box_{y_1} + m^2)\cdot$$
$$\cdot(\Box_{y_2}+m^2)(\Box_{y_3}+m^2)(\Box_{y_4}+m^2)<0|\,T(\phi_{in}(X(y_1))\ldots\phi_{in}(X(y_4))$$
$$:(\phi_{in}(X(y'_1))^4::(\phi_{in}(X(y'_2))^4:)\,|\,0>/2! \tag{6.67}$$

together with contributions from some other disconnected diagrams.

Expanding the time ordered product and keeping only the terms corresponding to Fig. 6.2 gives:

$$S_2 = i^4(i\chi_0)^2/2 \prod_{j=1}^{4} \int d^4y_j\, d^4y_1'\, d^4y_2'\, f_{Zp_1}(y_1) f_{Zp_2}(y_2) f_{Zp_3}^{\;*}(y_3) f_{Zp_4}^{\;*}(y_4)\cdot$$

$$\cdot(\Box_{y_1} + m^2)\,(\Box_{y_2} + m^2)(\Box_{y_3} + m^2)(\Box_{y_4} + m^2)\cdot$$

$$\cdot\{i\Delta_F^{TT}(y_1 - y_1')i\Delta_F^{TT}(y_2 - y_1')i\Delta_F^{TT}(y_3 - y_2')i\Delta_F^{TT}(y_4 - y_2') +$$

$$+ i\Delta_F^{TT}(y_1 - y_1')i\Delta_F^{TT}(y_2 - y_2')i\Delta_F^{TT}(y_3 - y_1')i\Delta_F^{TT}(y_4 - y_2')\}\cdot$$

$$\cdot i\Delta_F^{TT}(y_1' - y_2')i\Delta_F^{TT}(y_1' - y_2') \tag{6.68}$$

Following a similar procedure to the previous calculation yields

$$S_2 = i^4[(i\chi_0)^2/2](2\pi)^4\delta^4(p_3 + p_4 - p_1 - p_2)\mathcal{N}^+(p_4)\mathcal{N}^+(p_3)\mathcal{N}(p_2)\mathcal{N}(p_1)\cdot$$

$$\cdot \int d^4z\, [e^{-i(p_1 + p_2)\cdot z} + e^{-i(p_1 - p_3)\cdot z}]\, [i\Delta_F^{TT}(z)]^2 \tag{6.69}$$

revealing a similar normalization of the external legs to that found in eq. 6.63, and a momentum conserving delta function as in eq. 6.63. The loop integrals have the form:

$$I(q) = \int d^4z\, e^{-iq\cdot z}\, [i\Delta_F^{TT}(z)]^2 \tag{6.70}$$

The behavior of the Two-Tier Feynman propagator $\Delta_F^{TT}(z)$ was studied at long and short distances in eqs. 4.21-4.24. The large distance behavior of the Two-Tier Feynman propagator $\Delta_F^{TT}(z)$ approaches the behavior of the conventional Feynman propagator since

$$R(\mathbf{p}, z) \rightarrow 1 \tag{6.71}$$

as $z^2 = z^\mu z_\mu$ becomes much larger than M_c^{-2} ($z^2 \gg M_c^{-2}$) (eq. 6.55). Thus I(q) approaches the standard one loop expression of conventional field theory at large distance (or small momentum). Again we see that Two-Tier *quantum field theory realizes a form of Correspondence Principle by approaching conventional quantum field theory at large distances.*

At short distances the Gaussian factor $R(\mathbf{p}, z)$ dominates. The Two-Tier Feynman propagator $\Delta_F^{TT}(z)$ is radically different from the conventional Feynman propagator at very short distances (very high momentum). The singular behavior of the conventional Feynman propagator is replaced with a well-behaved, high-energy (short distance) behavior. Near the light cone $M_c^{-2} \gg z^2 \to 0$ (or $p^2 \gg M_c^2$) we can approximate eq. 6.52 with

$$i\Delta_F^{TT}(z) \approx \int d^3p\, [N(p)]^2\, R(\mathbf{p}, z) \tag{6.72}$$

since $e^{-ip \cdot z}$ is approximately unity for small z. We assume the mass of the ϕ particle is negligible on this scale. Upon performing the integrations (see eq. 4.23 for the exact result) we find eq. 6.72 approaches:

$$i\Delta_F^{TT}(z) \to \pi\, M_c^4 |z^2|/8$$

as $z^2 = z^\mu z_\mu \to 0$ from the space-like or time-like side of the light cone where $|\ |$ represents the absolute value.

Therefore $I(q)$ is finite and well-behaved. At high energy ($q^2 \gg M_c^2$)

$$I(q) \sim q^{-8}$$

since the fourier transform of $\Delta_F^{TT}(z)$ (momentum space) is

$$\Delta_F^{TT}(p) = \int d^4z\, e^{-ip\cdot z}\, \Delta_F^{TT}(z) \ \sim p^{-6}$$

for large p ($p^2 \gg M_c^2$). (Compare the preceding high energy behavior of I(q) with the conventional logarithmically divergent one loop result $I(q) \sim \ln(q^2/\Lambda^2)$ with Λ a cutoff.)

Thus Two-Tier quantum provides the benefits of a higher derivative theory without its drawbacks.

Finite Renormalization of External Particle Legs & Unitarity Example

The renormalization factor $Z_p^{-1/2}$ appearing in eqs. 6.64 and 6.65 that is due to the use of the non-unitary operator W(y) (eq. 5.6) to transform $\Phi_{in}(y)$ fields into $\phi_{in}(X(y))$ fields in the LSZ procedure in eq. 5.69, and related equations, does not preserve the norm of input and output ϕ particle legs. $Z_p^{-1/2}$ performs a finite renormalization for each external particle leg to compensate for the effects of W(y).

The required renormalization is nicely illustrated by considering the unitarity sum in the imaginary part of the preceding example.

The transition matrix T_{fi} is defined in terms of the S matrix by

$$S_{fi} = \delta_{fi} - i\,(2\pi)^4\,\delta^4(P_f - P_i)T^{(+)}{}_{fi}$$

The unitarity condition is

$$T^{(+)}{}_{fi} - T^{(-)}{}_{fi} = -i\sum_n (2\pi)^4\,\delta^4(P_n - P_i)\,T^{(-)}{}_{fn}\,T^{(+)}{}_{ni} \qquad (6.73)$$

Therefore we see that the first term on the right side of eq. 6.69 gives a transition matrix term:

$$T^{(+)}{}_{2a} = -\,i[\chi_0{}^2/2]N^+(p_4)N^+(p_3)N(p_2)N(p_1)\int d^4z\; e^{-iP\cdot z}[i\Delta_F{}^{TT}(z)]^2 \qquad (6.69a)$$

where $P = p_1 + p_2$. Substituting for $i\Delta_F{}^{TT}$ (using eq. 4.11) we find that the imaginary part of $T^{(+)}{}_{2a}$ is given by (Note $R(\mathbf{p}, z)$ is real.)

$$T^{(+)}{}_{2a} - T^{(-)}{}_{2a} = -i[\chi_0{}^2/2]N^+(p_4)N^+(p_3)N(p_2)N(p_1)\int d^4z\; e^{-iP\cdot z}\cdot$$
$$\cdot\left[i\int d^4p\,\theta(p_0)\,\delta(p^2 - m^2)e^{-ip\cdot z}\,R(\mathbf{p}, z)/(2\pi)^3\right]^2$$

If we express the $R(\mathbf{p}, z)$ factors in terms of their fourier transforms (see eq. 4.27):

$$R(\mathbf{p}, z) = \int d^4q\; e^{-iq\cdot z}\,R(\mathbf{p}, q)$$

Then we find

$$T^{(+)}{}_{2a} - T^{(-)}{}_{2a} = -i[\chi_0{}^2/2]N^+(p_4)N^+(p_3)N(p_2)N(p_1)\int d^4z\; e^{-iP\cdot z}\cdot$$
$$\cdot\left[i\int d^4k_1\, d^4q_1\theta(k_1{}^0)\,\delta(k_1{}^2 - m^2)e^{-ik_1\cdot z}\,e^{-iq_1\cdot z}\,R(\mathbf{k_1}, q_1)/(2\pi)^3\right]\cdot$$
$$\cdot\left[i\int d^4k_2\, d^4q_2\theta(k_2{}^0)\,\delta(k_2{}^2 - m^2)e^{-ik_2\cdot z}\,e^{-iq_2\cdot z}\,R(\mathbf{k_2}, q_2)/(2\pi)^3\right]$$

Performing the integral over z gives

$$T^{(+)}{}_{2a} - T^{(-)}{}_{2a} = +i[\chi_0{}^2/2]N^+(p_4)N^+(p_3)N(p_2)N(p_1)(2\pi)^4\cdot$$

$$\cdot \int d^4k_1 d^4q_1 d^4k_2 d^4q_2 \theta(k_1{}^0)\, \delta(k_1{}^2 - m^2)\, \theta(k_2{}^0)\, \delta(k_2{}^2 - m^2)\cdot$$
$$\cdot R(\mathbf{k_1}, q_1)R(\mathbf{k_2}, q_2)\, \delta^4(P + k_1 + q_1 + k_2 + q_2)/(2\pi)^6$$

Introducing delta functions enables us to re-express this equation as

$$T^{(+)}{}_{2a} - T^{(-)}{}_{2a} = +i[\chi_0{}^2/2]N^+(p_4)N^+(p_3)N(p_2)N(p_1) \int d^4r_1 d^4r_2(2\pi)^4 \delta^4(P - r_1 - r_2)\cdot$$

$$\cdot \int d^4k_1 d^4q_1\, \delta^4(r_1 + k_1 + q_1)\theta(k_1{}^0)\, \delta(k_1{}^2 - m^2)\, R(\mathbf{k_1}, q_1)\cdot$$

$$\cdot \int d^4k_2 d^4q_2 \theta(k_2{}^0)\, \delta(k_2{}^2 - m^2)\, \delta^4(r_2 + k_2 + q_2)R(\mathbf{k_2}, q_2)/(2\pi)^6$$

which becomes

$$T^{(+)}{}_{2a} - T^{(-)}{}_{2a} = +i[\chi_0{}^2/2]N^+(p_4)N^+(p_3)N(p_2)N(p_1) \int d^4r_1 d^4r_2(2\pi)^4 \delta^4(P - r_1 - r_2)\cdot$$

$$\cdot \int d^4k_1 \theta(k_1{}^0)\, \delta(k_1{}^2 - m^2)\, R(\mathbf{k_1}, -k_1 - r_1)\cdot$$

$$\cdot \int d^4k_2 \theta(k_2{}^0)\, \delta(k_2{}^2 - m^2) R(\mathbf{k_2}, -k_2 - r_2)/(2\pi)^6$$

$R(\mathbf{k_2}, -k_2 - r_2)$ can be expressed in terms of its fourier transform $R(\mathbf{k_2}, z)$ using eq. 4.27.
We can now rewrite the above expression in terms of intermediate states:

$$T^{(+)}{}_{2a} - T^{(-)}{}_{2a} = -i \int d^4r_1 d^4r_2\, (2\pi)^4 \delta^4(P - r_1 - r_2)\cdot$$

$$\cdot\, i\chi_0 N^+(p_4)N^+(p_3) \int d^4k_1 \theta(k_1{}^0)\, \delta(k_1{}^2 - m^2)\, [R(\mathbf{k_1}, -k_1 - r_1)/(2\pi)^3]\cdot$$

$$\cdot \int d^4k_2 \theta(k_2{}^0)\, \delta(k_2{}^2 - m^2)[R(\mathbf{k_2}, -k_2 - r_2)/(2\pi)^3] i\chi_0 N(p_2)N(p_1)/2$$

which has the form:

$$T^{(+)}{}_{2a} - T^{(-)}{}_{2a} = -i\int d^4r_1 d^4r_2 (2\pi)^4 \delta^4(P - r_1 - r_2)\left[\int d^4k_1 \theta(k_1^{\,0})\, \delta(k_1^{\,2} - m^2)\, R(\mathbf{k_1}, -k_1 - r_1)/(2\pi)^3\right]\left[\int d^4k_2 \theta(k_2^{\,0})\, \delta(k_2^{\,2} - m^2) R(\mathbf{k_2}, -k_2 - r_2)/(2\pi)^3\right] T^{(-)}{}_{fn} T^{(+)}{}_{ni}/2!$$

where

$$T^{(-)}{}_{fn} = \chi_0 \mathcal{N}^+(p_4)\mathcal{N}^+(p_3)\mathcal{N}(r_2)\mathcal{N}(r_1)$$

$$T^{(+)}{}_{ni} = \chi_0 \mathcal{N}^+(r_2)\mathcal{N}^+(r_1)\mathcal{N}(p_2)\mathcal{N}(p_1)$$

if

$$\mathcal{N}^+(p) = \mathcal{N}(p) = 1 \qquad (6.74)$$

Eq. 6.74 implies the (finite) external leg renormalization must be

$$Z_p = -\left[\int d^4w\, f_p(w)(\Box + m^2)\Delta_F^{\,TT}(w)\right]^2 \qquad (6.74a)$$

by 6.64 and 6.65. Thus all external legs must be "lopped off."

The result is a theory that satisfies the unitarity condition (eq. 6.73) as shown in the above detailed discussion.

If we define

$$\mathfrak{N}(r) = \int d^4k\; \theta(k^0)\delta(k^2 - m^2)R(\mathbf{k}, -k - r) \qquad (6.75a)$$

$$= (2\pi)^{-4}\int d^4k\; d^4z\; \theta(k^0)\delta(k^2 - m^2)\; e^{-i(k + r)\cdot z}\; R(\mathbf{k}, z) \qquad (6.75b)$$

then the Two-Tier completeness expression becomes:

$$S_{fi} = \sum_n (2\pi)^{-3n}(n!)^{-1}\int\left(\prod_{j=1}^{n} d^4r_j \mathfrak{N}(r_j)\right)S_{fn}S_{ni}^{\dagger}\,\delta^4(P_n - \sum_{k=1}^{n} r_k) \qquad (6.75c)$$

This expression reflects the fact that ϕ particles are surrounded by a "cloud" of Y quanta. Thus we have achieved unitarity! For small momenta $r_j \ll M_c$, we find $\mathfrak{N}(r_j) \simeq \theta(r_j^{\,0})\delta(r_j^{\,2} - m^2)$ (eq. 6.75b with $R(k, q) \simeq 1$). $\theta(r_j^{\,0})\delta(r_j^{\,2} - m^2)$ is the form seen in conventional quantum field theory. For large momenta $\mathfrak{N}(r_j)$ is very different.

General Form of Propagators

In this chapter we have considered a scalar Two-Tier quantum field theory. We have seen that the Two-Tier Feynman propagator is well behaved near the light cone resulting in a finite ϕ^4 theory. This finite ϕ^4 theory approximates the results of conventional ϕ^4 theory at low energy thus implementing a correspondence principle: *At low energy, results in Two-Tier quantum field theory approach the corresponding results of the corresponding conventional quantum field theory.*

The observations on Two-Tier field theory made in this chapter generally apply to Two-Tier versions of Quantum Electrodynamics, ElectroWeak Theory and the Standard Model as well as Two-Tier Quantum Gravity:

1. At low energy ($p^2 \ll M_c^2$ or large distances $z^2 \gg M_c^{-2}$) the Two-Tier quantum field theory is the same as the corresponding conventional quantum field theory to good approximation. (Correspondence Principle)
2. At high energy ($p^2 \gg M_c^2$ or short distances: $z^2 \ll M_c^{-2}$) Two-Tier quantum field theories (of physical interest) are well-behaved and finite.
3. Two-Tier quantum field theories (of physical interest) satisfy unitarity and Lorentz invariance (and in the case of quantum gravity their dynamical equations satisfy the requirements of general relativity).

The generality of these results is based on:

1. The expansion of the S matrix in time ordered products of field operators.
2. Wick's Theorem
3. The general form of all particle propagators in Two-Tier quantum field theories. All particle Feynman propagators have the form:

$$iG_F^{TT}{}_{\ldots}(y_1 - y_2) = \langle 0|T(\chi_{\ldots}(X(y_1)),\chi_{\ldots}(X(y_2)))|0\rangle \qquad (6.76)$$

$$= \int d^4p\, iG_{F\ldots}(p)e^{-ip\cdot(y_1 - y_2)} R(\mathbf{p}, y_1 - y_2) \qquad (6.77)$$

where $iG_{F\ldots}(p)$ is the conventional momentum space $\chi_{\ldots}$ particle propagator, and where ... represents the relevant tensor and matrix indices. $R(\mathbf{p}, y_1 - y_2)$ introduces a damping factor in each particle propagator that eliminates divergences.

Scalar Particle Propagator

The Two-Tier propagator for the case of a free scalar particle is:

$$i\Delta_F^{TT}(y_1 - y_2) = \langle 0|T(\phi(X(y_1)),\phi(X(y_2)))|0\rangle \qquad (6.51)$$

$$= i \int \frac{d^4p\, e^{-ip\cdot(y_1 - y_2)}\, R(\mathbf{p}, y_1 - y_2)}{(2\pi)^4\,(p^2 - m^2 + i\varepsilon)} \tag{6.52}$$

Since the mass m is not relevant at high energy we set $m = 0$. This enables us to obtain a more tractable expression for the propagator. After some manipulation the massless scalar propagator can be represented as:

$$i\Delta_F^{TT}(z) = -\beta[16\pi^3(AB)^{1/2}]^{-1} \int_{-\infty}^{\infty} dy_1 \int_{-\infty}^{\infty} dy_2 \cdot$$

$$\cdot\{\theta(z_0)\exp[-\beta((y_1 - z_0)^2B + (y_2 + z)^2A)/(4AB)] +$$

$$+ \theta(-z_0)\exp[-\beta((y_1 + z_0)^2B + (y_2 - z)^2A)/(4AB)]\}/(y_1^2 - y_2^2) \tag{6.78}$$

with $\beta = 4\pi^2M_c^4\mathbf{z}^2$. Using

$$(y_1^2 - y_2^2)^{-1} = -0.5 \int_0^{\infty} dq_1 \int_{-\infty}^{\infty} dq_2\, \theta(q_1^2 - q_2^2)\exp[iq_1y_1 - iq_2y_2] \tag{6.79}$$

we obtain the representation

$$i\Delta_F^{TT}(z^\mu) = (8\pi^2)^{-1} \int_0^{\infty} dq_1 \int_{-\infty}^{\infty} dq_2\, \theta(q_1^2 - q_2^2) \cdot$$

$$\cdot \exp\{iq_1|z_0| + iq_2z - [A'q_1^2 + B'q_2^2]/[\beta'(\mathbf{z}^2 - z_0^2)]\} \tag{6.80}$$

where $|z_0|$ is the absolute value of z_0, $z^2 - z_0^2 = -z^\mu z_\mu$ and

$$A = A'/(1 - v^2) \tag{6.81}$$

$$B = B'/(1 - v^2) \tag{6.82}$$

$$\beta = 4\pi^2M_c^4\mathbf{z}^2 = \beta' z^2 \tag{6.83}$$

with $\mathbf{z} = |\vec{\mathbf{z}}|$ – the magnitude of the spatial vector $\vec{\mathbf{z}}$, and A and B given by eqs. 6.60 – 6.61.

The representation of $i\Delta_F^{TT}$ in eq. 6.80 is particularly useful in determining its low energy ($\ll M_c$), and its high energy ($\gg M_c$) behavior. The low energy behavior is governed by the linear terms in the exponential in eq. 6.80 since $\beta'(\mathbf{z}^2 - z_0^2)$ is very large in this limit:

$$i\Delta_F^{TT}(z^\mu)_{low} \simeq (8\pi^2)^{-1} \int_0^\infty dq_1 \int_{-\infty}^\infty dq_2\, \theta(q_1^2 - q_2^2)\exp\{\, iq_1|z_0| + iq_2 z\} \tag{6.84}$$

$$= [4\pi^2(\mathbf{z}^2 - z_0^2)]^{-1} \tag{6.85}$$

$$= i\Delta_F(z^\mu)$$

equaling the exact massless, free, spin 0 Feynman propagator of conventional quantum field theory.

In the high energy limit when $\beta'(\mathbf{z}^2 - z_0^2)$ is small since $\mathbf{z}^2 \approx z_0^2$ (i.e. near the light cone), the quadratic terms in the exponential in eq. 6.80 dominate and $A' \simeq B'$. We then find

$$i\Delta_F^{TT}(z^\mu)_{high} \simeq (8\pi^2)^{-1} \int_0^\infty dq_1 \int_{-\infty}^\infty dq_2\, \theta(q_1^2 - q_2^2)\exp\{A'(q_1^2 + q_2^2)/[\beta'(\mathbf{z}^2 - z_0^2)]\} \tag{6.86}$$

$$= \pi M_c^4 |(\mathbf{z}^2 - z_0^2)|/8 \tag{6.87}$$

as in eq. 4.24. As pointed out earlier, eq. 6.87 corresponds to k^{-6} behavior in momentum space:

$$i\Delta_F^{TT}(k)_{high} \backsim k^{-6} \tag{6.87a}$$

Spin ½ Particle Propagator

For the case of a free, spin ½ particle the propagator is:

$$iS_F^{TT}(y_1 - y_2) = \langle 0|T(\bar{\psi}(X(y_1))\,\psi(X(y_2)))|0\rangle \tag{6.88}$$

$$= i \int \frac{d^4p\, e^{-ip\cdot(y_1 - y_2)} (\not{p} + m)\, R(\mathbf{p}, y_1 - y_2)}{(2\pi)^4 (p^2 - m^2 + i\varepsilon)}$$

Again setting m = 0 we find a convenient representation in the form:

$$S_F^{TT}(z^\mu) = i(8\pi^2)^{-1} \int_0^\infty dq_1 \int_{-\infty}^\infty dq_2\, \theta(q_1^2 - q_2^2)(\epsilon(z_0)q_1\gamma_0 - q_2\vec{\mathbf{z}}\cdot\vec{\gamma}/z)\cdot$$

$$\cdot \exp\{iq_1|z_0| + iq_2 z - [A'q_1^2 + B'q_2^2]/[\beta'(\mathbf{z}^2 - z_0^2)]\} \quad (6.89)$$

using the same symbols and notation as eq. 6.80, and with $\epsilon(z_0) = +1$ if $z_0 \geq 0$ and -1 otherwise.

The representation of S_F^{TT} in eq. 6.89 is useful in determining its low energy ($\ll M_c$), and high energy ($\gg M_c$) behavior. The low energy behavior is governed by the linear terms in the exponential in eq. 6.89 since $\beta'(\mathbf{z}^2 - z_0^2)$ is large in this limit:

$$S_F^{TT}(z^\mu)_{low} \simeq (8\pi^2)^{-1} \int_0^\infty dq_1 \int_{-\infty}^\infty dq_2\, \theta(q_1^2 - q_2^2)(\epsilon(z_0)q_1\gamma_0 - q_2\vec{\mathbf{z}}\cdot\vec{\gamma}/z)\cdot$$

$$\cdot \exp\{iq_1|z_0| + iq_2 z\} \quad (6.90)$$

$$= \not{z}[2\pi^2(\mathbf{z}^2 - z_0^2)^2]^{-1} \quad (6.91)$$
$$= S_F(z^\mu)$$

equaling the exact massless, spin ½ Feynman propagator of conventional quantum field theory. If we had not set m = 0 initially, we would have obtained the usual massive, spin ½ Feynman propagator.

In the high energy limit when $\beta'(\mathbf{z}^2 - z_0^2)$ is small since $\mathbf{z}^2 \approx z_0^2$ (i.e. near the light cone), the quadratic terms in the exponential in eq. 6.89 dominate and $A' \simeq B'$. We then find

$$S_F^{TT}(z^\mu)_{high} \simeq (8\pi^2)^{-1} \int_0^\infty dq_1 \int_{-\infty}^\infty dq_2\, \theta(q_1^2 - q_2^2)(\epsilon(z_0)q_1\gamma_0 - q_2\vec{\mathbf{z}}\cdot\vec{\gamma}/z)\cdot$$

$$\cdot\ \exp\{A'(q_1^{\ 2} + q_2^{\ 2})/[\beta'(\mathbf{z}^2 - z_0^{\ 2})]\} \qquad (6.92)$$

$$= i(8\pi^2)^{-1}\{\mathbf{z}^{-1}(\mathbf{z}^2 - z_0^{\ 2})^{3/2}2^{3/2}\pi^{7/2}M_c^{\ 6}z_0\gamma_0 -$$
$$- 4i(\mathbf{z}^2 - z_0^{\ 2})^2\pi^5 M_c^{\ 8}\vec{\mathbf{z}}\cdot\vec{\gamma})\} \qquad (6.93)$$

The leading momentum dependence of the fourier transform of $S_F^{\ TT}(z^\mu)_{high}$ is

$$S_F^{\ TT}(p)_{high} \backsim M_c^{\ 6}p^{-7}\gamma_0 \qquad (6.94)$$

Massless Spin 1 Particle Propagator

The Two-Tier Feynman propagator for the case of a free, massless, spin 1, gauge field particle (coupled to a conserved current) such as a photon is:

$$iD_F^{\ TT}(z)_{\mu\nu} = -i\int\frac{d^4p\ e^{-ip\cdot z}\ g_{\mu\nu}R(\mathbf{p}, y_1 - y_2)}{(2\pi)^4\ (p^2 + i\varepsilon)} \qquad (6.95)$$

The form of eq. 6.95 is the same as the scalar particle propagator multiplied by $-g_{\mu\nu}$. As a result we have the representation:

$$iD_F^{\ TT}(z)_{\mu\nu} = -(8\pi^2)^{-1}\int_0^\infty dq_1\int_{-\infty}^\infty dq_2\ \theta(q_1^{\ 2} - q_2^{\ 2})\ g_{\mu\nu}\ \cdot$$
$$\cdot\ \exp\{iq_1|z_0| + iq_2 z - [A'q_1^{\ 2} + B'q_2^{\ 2}]/[\beta'(\mathbf{z}^2 - z_0^{\ 2})]\} \qquad (6.96)$$

As before in the scalar particle case, the low energy behavior is governed by the linear terms in the exponential in eq. 6.96 since $\beta'(\mathbf{z}^2 - z_0^{\ 2})$ is very large in this limit:

$$iD_F^{\ TT}(z)_{\mu\nu low} \simeq -g_{\mu\nu}(8\pi^2)^{-1}\int_0^\infty dq_1\int_{-\infty}^\infty dq_2\ \theta(q_1^{\ 2} - q_2^{\ 2})\exp\{iq_1|z_0| + iq_2 z\} \qquad (6.97)$$

$$= -g_{\mu\nu}[4\pi^2(\mathbf{z}^2 - z_0^{\ 2})]^{-1} \qquad (6.98)$$

$$= -ig_{\mu\nu}\Delta_F(z)$$

equaling the exact free, massless, spin 1 Feynman gauge field propagator of conventional quantum field theory.

In the high energy limit when $\beta'(\mathbf{z}^2 - z_0^2)$ is small since $\mathbf{z}^2 \approx z_0^2$ (i.e. near the light cone), the quadratic terms in the exponential in eq. 6.96 dominate, and $A' \simeq B'$. We then find

$$iD_F^{TT}(z)_{\mu\nu high} \simeq -(8\pi^2)^{-1}\int_0^\infty dq_1 \int_{-\infty}^\infty dq_2 \theta(q_1^2 - q_2^2) g_{\mu\nu} \exp\{ A'(q_1^2 + q_2^2)/[\beta'(\mathbf{z}^2 - z_0^2)]\} \tag{6.99}$$

$$= -g_{\mu\nu}\pi M_c^4 \,|(\mathbf{z}^2 - z_0^2)|/8 \tag{6.100}$$

Eq. 6.100 corresponds to k^{-6} behavior in momentum space:

$$iD_F^{TT}(k)_{\mu\nu high} \backsim g_{\mu\nu} M_c^4 k^{-6} \tag{6.101}$$

Spin 2 Particle Propagator

The Two-Tier propagator for the case of a free, massless, spin 2 particle such as a graviton is:

$$i\Delta_{F2}^{TT}(z)_{\mu\nu\rho\sigma} = i\int \frac{d^4p\, e^{-ip\cdot z}\, b_{\mu\nu\rho\sigma}(p) R(\mathbf{p}, y_1 - y_2)}{(2\pi)^4 (p^2 + i\varepsilon)} \tag{6.102}$$

in an appropriate gauge where $b_{\mu\nu\rho\sigma}(p)$ is a tensor that is independent of the coordinates. We can express eq. 6.102 in the form:

$$i\Delta_{F2}^{TT}(z)_{\mu\nu\rho\sigma} = (8\pi^2)^{-1}\int_0^\infty dq_1 \int_{-\infty}^\infty dq_2\, \theta(q_1^2 - q_2^2)\, \tilde{b}(z_0, z, q_1, q_2)_{\mu\nu\rho\sigma} \cdot$$

$$\cdot \exp\{ iq_1|z_0| + iq_2 z - [A'q_1^2 + B'q_2^2]/[\beta'(\mathbf{z}^2 - z_0^2)]\} \tag{6.103}$$

where $\tilde{b}(z_0, z, q_1, q_2)_{\mu\nu\rho\sigma}$ is a tensor generated from the $b_{\mu\nu\rho\sigma}(p)$ tensor.

As before in the scalar particle case, the low energy behavior is governed by the linear terms in the exponential in eq. 6.103 since $\beta'(\mathbf{z}^2 - z_0^2)$ is very large in this limit and we find that the covariant piece[28] behaves like:

$$i\Delta_{F2}^{TT}(z)_{\mu\nu\rho\sigma\text{lowCov}} \simeq \tilde{\tilde{b}}_{\mu\nu\rho\sigma}(8\pi^2)^{-1}\int_0^\infty dq_1 \int_{-\infty}^\infty dq_2\, \theta(q_1^2 - q_2^2)\exp\{ iq_1|z_0| + iq_2 z\} \quad (6.104)$$

$$= \tilde{\tilde{b}}_{\mu\nu\rho\sigma}[4\pi^2(\mathbf{z}^2 - z_0^2)]^{-1} \quad (6.105)$$

$$= i\Delta_F(z^\mu)\,\tilde{\tilde{b}}_{\mu\nu\rho\sigma}$$

where

$$\tilde{\tilde{b}}_{\mu\nu\rho\sigma} = \tfrac{1}{2}\,[\eta_{\mu\rho}\eta_{\nu\sigma} + \eta_{\mu\sigma}\eta_{\nu\rho} - \eta_{\mu\nu}\eta_{\rho\sigma}] \quad (6.106)$$

so that the expression in eq. 6.105 equals the corresponding covariant piece of the exact free, massless, spin 2 Feynman propagator of conventional quantum field theory.

In the high energy limit when $\beta'(\mathbf{z}^2 - z_0^2)$ is small since $\mathbf{z}^2 \approx z_0^2$ (i.e. near the light cone), the quadratic terms in the exponential in eq. 6.103 dominate, and $A' \simeq B'$. We then find

$$i\Delta_{F2}^{TT}(z)_{\mu\nu\rho\sigma\text{high}} \simeq (8\pi^2)^{-1}\int_0^\infty dq_1 \int_{-\infty}^\infty dq_2\,\theta(q_1^2 - q_2^2)\ \tilde{b}(z_0, z, q_1, q_2)_{\mu\nu\rho\sigma}\cdot$$

$$\cdot\exp\{A'(q_1^2 + q_2^2)/[\beta'(\mathbf{z}^2 - z_0^2)]\} \quad (6.107)$$

and the covariant piece behaves like

$$i\Delta_{F2}^{TT}(z)_{\mu\nu\rho\sigma\text{highCov}} \simeq \tilde{\tilde{b}}_{\mu\nu\rho\sigma}\pi M_c^4|(\mathbf{z}^2 - z_0^2)|/8 \quad (6.108)$$

The coordinate space behavior of eq. 6.108 corresponds to k^{-6} behavior in momentum space:

$$i\Delta_{F2}^{TT}(k)_{\mu\nu\rho\sigma\text{highCov}} \backsim \tilde{\tilde{b}}_{\mu\nu\rho\sigma}\, k^{-6} \quad (6.109)$$

[28] S. Weinberg, Phys. Rev. **135**, B1049 (1964); Phys. Rev. **138**, B988 (1965).

The high-energy behavior of the spin 2 propagator in momentum space results in a Two-Tier theory of quantum gravity that has no high-energy divergences and is thus finite. See chapter 9 for a detailed discussion.

7. Two-Tier Quantum Electrodynamics

Formulation

There have been numerous attempts to develop a finite theory of Quantum Electrodynamics (QED). Among the noteworthy attempts are the Lee-Wick[29] formulation of QED and the Johnson-Baker-Willey model.[30] The unification of the electromagnetic interaction with the weak interaction in Electroweak Theory, and the proof that it is renormalizable, has switched the focus of interest away from QED. However the extremely precise experimental tests of QED, which are among the most accurate measurements made by science, and the impressive agreement[31] with the theoretical predictions of QED, make QED of interest in its own right.

In this chapter we will describe the formulation of Two-Tier QED. We will see that it is finite, and yet it is in complete agreement with the highly accurate calculations of QED if M_c is sufficiently large – such as of the order of the Planck mass. We will also see a modification of the Coulomb potential in the Two-Tier model that makes possible the existence of (unstable) bound states of particles with the same charge such as a two electron bound state. The Two-Tier QED Coulomb potential is linear at ultra-short distances and zero at $r = 0$.

A New Quantum Electrodynamics with Non-Commuting Coordinates

Two-Tier QED is formulated in a way that is similar to conventional QED and captures the excellent results of QED while making the theory finite. We will consider the case of QED for electrons. The results apply directly to any charged spin ½ field and with a few changes to charged particles of other spin. The Two-Tier QED Lagrangian that we will investigate is:

---

[29] T. D. Lee and G. C. Wick, Phys. Rev. **D2**, 1033 (1970); T. D. Lee and G. C. Wick, Nucl. Phys. **B9**, 209 (1969) and references therein.
[30] S. Blaha, "An Approximate Calculation of the Eigenvalue Function in Massless Quantum Electrodynamics", Phys.Rev. **D9**, 2246 (1974) and references therein.
[31] T. Kinoshita, "The Fine Structure Constant", Cornell University preprint CLNS 96/1406 (1996); V. W. Hughes and T. Kinoshita, Rev. Mod. Phys. **71**, S133 (1999).

$$\mathscr{L} = J\mathscr{L}_F + \mathscr{L}_C(X^\mu(y), \partial X^\mu(y)/\partial y^\nu, y) \tag{7.1}$$

with J the Jacobian and

$$\mathscr{L}_F = \bar{\psi}(X(y))((i\nabla\!\!\!\!/_X - e_0 A\!\!\!/(X(y)) - m_0)\psi(X(y)) - {}^1\!/\!_4\, F^{\mu\nu}(X(y))F_{\mu\nu}(X(y)) \tag{7.2}$$

with

$$F_{\mu\nu}(X(y)) = \partial A_\mu(X(y))/\partial X^\nu - \partial A_\nu(X(y))\,/\partial X^\mu \tag{7.3}$$

and

$$\mathscr{L}_C(X^\mu(y), \partial X^\mu(y)/\partial y^\nu, y) = -{}^1\!/\!_4\, F_Y{}^{\mu\nu}F_{Y\mu\nu} \tag{7.4}$$

$$F_{Y\mu\nu} = \partial Y_\mu/\partial y^\nu - \partial Y_\nu\,/\partial y^\mu \tag{7.5}$$

We note the Lagrangian $\mathscr{L}_F$ has the form of the conventional electromagnetic Lagrangian[32] except for the functional dependence on X(y).

Since the Lagrangian in eq. 7.2 is separable we will follow the same procedure as we did for the scalar field theory in the development of eqs. A.74 – A.112. Thus we obtain the hamiltonian (as in eq. A.112 for the scalar case):

$$H_F = :\!\int d^3X\,(\mathscr{H}_{F0} + \mathscr{H}_{Fint})\!: \tag{7.6}$$

where

$$\mathscr{H}_{F0} = \bar{\psi}(X(y))(i\nabla\!\!\!\!/_X - m_0)\psi(X(y)) + {}^1\!/\!_2(\mathbf{E}^2 + \mathbf{B}^2) \tag{7.7}$$

$$\mathscr{H}_{Fint} = e_0\bar{\psi}(X(y))A\!\!\!/(X(y))\psi(X(y)) \tag{7.8}$$

with **E** being the electric field and **B** being the magnetic field:

$$E^i = -\partial A^i/\partial y^0$$

$$B^i = \varepsilon^{ijk}\,\partial A_j/\partial y^k$$

The field equations are:

---

32 We follow the conventions of Kaku (1993), and Bjorken (1995). It will be evident that the proof that two-tier QED is finite will not require detailed knowledge of specific conventions.

$$(i\nabla\!\!\!/_X - e_0 A\!\!\!/(X(y)) - m_0)\psi(X(y)) = 0 \tag{7.9}$$

and

$$\partial(F^{\mu\nu}(X(y)))/\partial X^\nu = e_0\bar{\psi}(X(y))\gamma^\mu\psi(X(y)) \tag{7.10}$$

The Y Field

The X^μ coordinate field is related to the Y^μ field via:

$$X_\mu(y) = y_\mu + i\,Y_\mu(y)/M_c^2 \tag{3.12}$$

The Y field has the free Lagrangian eq. 7.4 (based on eqs. 3.10 – 3.14), and the hamiltonian

$$\mathscr{H}_C = \tfrac{1}{2}\,(E_Y^{\,2} + B_Y^{\,2}) \tag{7.11}$$

where

$$E_Y^{\,i} = -\partial Y^i/\partial y^0 \tag{7.12}$$

$$B_Y^{\,i} = \varepsilon^{ijk}\,\partial Y_j/\partial y^k \tag{7.13}$$

The Y field equations are

$$\partial F_Y^{\mu\nu}(y)/\partial y^\nu = 0 \tag{7.14}$$

The quantization of the Y field in the Coulomb gauge is described in chapter 3. We will use the Y Coulomb gauge throughout our discussions of various Two-Tier quantum field theories.

Quantization of the Free Dirac Field

The quantization procedure is formally identical to that of conventional QED. The standard equal time anti-commutation relations for the spin ½ field are:

$$\{\psi_\alpha(X), \psi_\beta(X')\} = \{\pi_{\psi\alpha}(X), \pi_{\psi\beta}(X')\} = 0 \tag{7.15}$$

$$\{\pi_{\psi\alpha}(X), \psi_\beta(X')\} = i\,\delta_{\alpha\beta}\,\delta^3(\mathbf{X} - \mathbf{X}') \tag{7.16}$$

where α and β are the spinor indices and where

$$\pi_{\psi a}(X) = i\,\psi_a^{\dagger}(X) \tag{7.17}$$

The spin ½ field can be expanded in a fourier series:

$$\psi(X(y)) = \sum_{\pm s}\int d^3p\ N^d{}_m(p)\ [b(p,s)u(p,s)\ :e^{-ip\cdot(y\,+\,iY/M_c^{\,2})}: + \\ + d^{\dagger}(p,s)v(p,s)\ :e^{ip\cdot(y\,+\,iY/M_c^{\,2})}:] \tag{7.18}$$

$$\psi^{\dagger}(X(y)) = \sum_{\pm s}\int d^3p\ N^d{}_m(p)\ [b^{\dagger}(p,s)\bar{u}(p,s)\gamma^0\ :e^{+ip\cdot(y\,+\,iY/M_c^{\,2})}: + \\ + d(p,s)\bar{v}(p,s)\gamma^0\ :e^{-ip\cdot(y\,+\,iY/M_c^{\,2})}:] \tag{7.19}$$

where

$$N^d{}_m(p) = [m/((2\pi)^3E_p)]^{1/2} \tag{7.20}$$

and

$$E_p = (\mathbf{p}^2 + m^2)^{1/2} \tag{7.21}$$

The commutation relations of the Fourier coefficient operators are:

$$\{b(p,s),\, b^{\dagger}(p',s')\} = \delta_{ss'}\delta^3(\mathbf{p} - \mathbf{p'}) \tag{7.22}$$

$$\{d(p,s),\, d^{\dagger}(p',s')\} = \delta_{ss'}\delta^3(\mathbf{p} - \mathbf{p'}) \tag{7.23}$$

$$\{b(p,s),\, b(p',s')\} = \{d(p,s),\, d(p',s')\} = 0 \tag{7.24}$$

$$\{b^{\dagger}(p,s),\, b^{\dagger}(p',s')\} = \{d^{\dagger}(p,s),\, d^{\dagger}(p',s')\} = 0 \tag{7.25}$$

$$\{b(p,s),\, d^{\dagger}(p',s')\} = \{d(p,s),\, b^{\dagger}(p',s')\} = 0 \tag{7.26}$$

$$\{b^{\dagger}(p,s),\, d^{\dagger}(p',s')\} = \{d(p,s),\, b(p',s')\} = 0 \tag{7.27}$$

The spinors u(p,s) and v(p,s) are defined in the conventional way (as in Kaku (1993), and in Bjorken (1965)).

Quantization of the Electromagnetic Field

The gauge invariance of Two-Tier QED can be seen by examining the field equation

$$(i\not{\nabla}_X - e_0\not{A}(X(y)) - m_0)\psi(X(y)) = 0 \qquad (7.9)$$

and considering the effect of a gauge gauge transformation:

$$A^\mu(X(y)) \rightarrow A^\mu(X(y)) - \partial\Lambda(X(y))/\partial X_\mu \qquad (7.28)$$

The field equation eq. 7.10 remains unchanged and the change in eq. 7.9 can be accommodated by a change of phase of the Dirac field:

$$\psi(X(y)) \rightarrow \exp(ie_0\Lambda(X(y))\ \psi(X(y)) \qquad (7.29)$$

The only novelty is that $\Lambda(X(y))$ in general becomes a complex q-number quantity at extremely short distances since X is a complex q-number.

The gauge invariance of the Lagrangian eq. 7.2 allows us to choose a convenient gauge. It appears that the most convenient gauge is the Coulomb gauge[33]:

$$\partial A^i/\partial X^i = 0 \qquad (7.30)$$

where the sum is over spatial components labeled with i. We also set

$$A^0 = 0 \qquad (7.31)$$

in the absence of an external source.
A conventional treatment leads to the equal time commutation relations:

$$[A^\mu(X(y)), A^\nu(X(y'))] = [\pi_A{}^\mu(X(y)), \pi_A{}^\nu(X(y'))] = 0 \qquad (7.32)$$

$$[\pi_A{}^j(X(y)), A_k(X(y'))] = -i\,\delta^{tr}{}_{jk}(\mathbf{X}(y) - \mathbf{X}(y')) \qquad (7.33)$$

(note the locations of the j component label in eq. 7.33 introduces a minus sign) where

[33] It is also possible to quantize in two-tier QED using an indefinite metric that preserves manifest Lorentz covariance as was done by Gupta and Bleuler for conventional QED. See Heitler (1954) or Bogoliubov (1959).

$$\pi_A{}^k = \partial\mathscr{L}_F/\partial\dot{A}_k \tag{7.34}$$

$$\pi_A{}^0 = 0 \tag{7.35}$$

$$\delta^{tr}{}_{jk}(X(y) - X(y')) = \int d^3k\, e^{i\,\mathbf{k}\cdot(\mathbf{X}(y) - \mathbf{X}(y))}(\delta_{jk} - k_j k_k/\mathbf{k}^2)/(2\pi)^3 \tag{7.36}$$

$$\dot{A}_k = \partial A_k/\partial X^0 \tag{7.37}$$

The Coulomb gauge reveals the two transverse degrees of freedom that are present in the vector potential. The Fourier expansion of the vector potential is:

$$A^i(X(y)) = \int d^3k\, N_0(k) \sum_{\lambda=1}^{2} \varepsilon^i(k, \lambda)[a(k,\lambda)\, e^{-ik\cdot X(y)} + a^\dagger(k,\lambda)\, e^{ik\cdot X(y)}] \tag{7.38}$$

where

$$N_0(k) = [(2\pi)^3 2\omega_k]^{-1/2} \tag{7.39}$$

(m = 0) and

$$\omega_k = (\mathbf{k}^2)^{1/2} = k^0 \tag{7.40}$$

with $\epsilon(k, \lambda)$ being the polarization unit vectors for $\lambda = 1, 2$.

The commutation relations of the Fourier coefficient operators are:

$$[a(k,\lambda), a^\dagger(k',\lambda')] = \delta_{\lambda\lambda'}\delta^3(\mathbf{k} - \mathbf{k}') \tag{7.41}$$

$$[a^\dagger(k,\lambda), a^\dagger(k',\lambda')] = [a(k,\lambda), a(k',\lambda')] = 0 \tag{7.42}$$

and the polarization vectors satisfy

$$\sum_{\lambda=1}^{2} \varepsilon_i(k, \lambda)\varepsilon_j(k, \lambda) = (\delta_{ij} - k_i k_j/\mathbf{k}^2) \tag{7.43}$$

Particle propagators

The electron and photon Feynman propagators differ from the conventional QED propagators by having the Gaussian factor $R(\mathbf{p}, z)$ in their fourier expansions:

$$iS_F^{TT}(y_1 - y_2) = <0|T(\bar{\psi}(X(y_1))\psi(X(y_2)))|0> \tag{7.44}$$

where the time ordering is with respect to y_1^0 and y_2^0. Expanding the free fields leads to the fourier representation:

$$iS_F^{TT}(y_1 - y_2) = i\int \frac{d^4p\, e^{-ip\cdot(y_1 - y_2)}\, (\not{p} + m)\, R(\mathbf{p}, y_1 - y_2)}{(2\pi)^4\, (p^2 - m^2 + i\varepsilon)} \tag{7.45}$$

with the Gaussian factor $R(\mathbf{p}, z)$ specified in eq. 6.53. The photon propagator is

$$iD_F^{trTT}(y_1 - y_2)_{\mu\nu} = <0|T(A_\mu(X(y_1))A_\nu(X(y_2)))|0> \tag{7.46}$$

$$= -ig_{\mu\nu}\int \frac{d^4k\, e^{-ik\cdot(y_1 - y_2)}\, R(\mathbf{k}, y_1 - y_2)}{(2\pi)^4\, (k^2 + i\varepsilon)} \tag{7.47}$$

plus gauge terms and minus the Coulomb term. The presence of the Gaussian factor $R(\mathbf{p}, z)$ results in a theory of QED that has no divergences and thus is finite. See eqns. 6.94 and 6.101 for their large momentum behavior.

Coulomb Interaction

The Coulomb interaction in Two-Tier QED is different at short distances from the conventional Coulomb interaction. The Coulomb potential between singly charged (same sign) particles in Two-Tier QED is:

$$V_{TT}(y_1 - y_2) = e^2\int \frac{d^4p\, e^{-ip\cdot(y_1 - y_2)}\, R(\mathbf{p}, y_1 - y_2)}{(2\pi)^4\, \mathbf{k}^2} \tag{7.48}$$

$$= \alpha\, \Phi(M_c^2\pi r^2)\, \delta(y_1^0 - y_2^0)/r \tag{7.49}$$

where α is the fine structure constant, $\Phi(z)$ is the error function, M_c is the mass that sets the scale of the short distance behavior, and $r = \surd(y_1 - y_2)^2$ is the radial distance. At small distances ($\pi r^2 \ll M_c^{-2}$) the Two-Tier potential becomes linear in r:

$$V_{TT} \rightarrow 2\alpha r\surd\pi\, M_c^{\,2}\delta(y_1^{\,0} - y_2^{\,0}) \qquad (7.50)$$

and at large distances ($\pi r^2 \gg M_c^{-2}$) the Two-Tier potential approaches the conventional Coulomb potential:

$$V_{TT} \rightarrow V_{Coul} = \alpha\, \delta(y_1^{\,0} - y_2^{\,0})/r \qquad (7.51)$$

using the error function normalization $\Phi(\infty) = 1$. The modified Coulomb potential V_{TT} of eq. 7.49 (modulo the delta function in time) is plotted in Fig. 7.5 using $M_c = 200$ GeV/c^2 and Fig 7.6 using M_c = Planck mass. At large distances the Coulomb potential (which has been verified experimentally with great precision) can be approximated arbitrarily closely by Two-Tier QED by simply letting M_c become larger. Conceivably M_c can be as large as the Planck mass (1.221×10^{19} GeV/c^2) or even larger. Thus conventional QED is the “large" distance limit of Two-Tier QED.

The short distance behavior of the Two-Tier Coulomb potential opens the possibility of quasi-bound states of particles of the same sign such as a two electron bound state. The normally repulsive potential has a linear behavior near $r = 0$, and a potential barrier, before becoming like the conventional Coulomb potential at larger distances. A pair of electrons, if localized within the linear region of the potential, would be bound but the “bound state” would quickly decay via electron tunneling through the barrier. States of this type conceivably might have existed in the first instants after the Big Bang and influenced the earliest evolution of the universe. Creating these dilepton states does not appear to be feasible if M_c is extremely large.

Asymptotic States

The development of perturbation theory in chapter 6 applies to the Two-Tier theory of QED with only superficial changes.

First we note that the form of the photon propagator has exactly the form of eq. 7.47 since terms proportional to k_μ or k_ν that would appear in the evaluation of eq. 7.46 do not contribute due to current conservation. In addition the instantaneous Coulomb interaction cancels the remaining Coulomb-like term appearing in the evaluation of eq. 7.46.

Thus we are left with the electron propagator (eq. 7.45) and the effective photon propagator (eq. 7.47). The formalism and role of the Y field is the same as in the scalar Two-Tier quantum field theory considered earlier.

In-states and Out-states

The dependence of the Lagrangian, and the particle fields in particular, on X^μ rather than directly on the coordinates y^μ leads to a "fuzziness" of the definition of asymptotic particle states that we have chosen to resolve with the construction of an asymptotic free field for each particle species of the "normal" sort. In the scalar case we defined an auxiliary field $\Phi_{in}(y)$ using the creation and annihilation operators of the free scalar field $\phi_{in}(X(y))$. In actuality $\Phi_{in}(y) \equiv \phi_{in}(y)$. The change from the argument y^μ to $X^\mu(y)$ is a form of translation that can be implemented using the (non-unitary) exponentiated momentum operator as we did in eq. 5.70:

$$\phi_a(y) \equiv \Phi_a(y) = W_a^{-1}(y)\phi_a(X(y))W_a(y)$$

for a = "in" or "out." *The benefit from this approach is a clean simple definition of asymptotic particle states of definite momentum (and spin etc.).* We will follow the same strategy in Two-Tier QED.

Fermion In-states and Out-States

In this section we will develop properties of fermion in-fields and out-fields. The LSZ procedure can be schematized as:

$$\psi_{in}(y) \Rightarrow \psi_{in}(X(y)) \Rightarrow \psi(X(y)) \Rightarrow \psi_{out}(X(y)) \Rightarrow \psi_{out}(y) \qquad (7.52)$$

In-states are constructed using $\psi_{in}(y)$ which is then transformed into $\psi_{in}(X(y))$ in order to make contact with our Lagrangian formalism. The interacting field $\psi(X(y))$ is related to $\psi_{in}(X(y))$ using the standard LSZ limiting ($y^0 \to -\infty$) process. Similarly out-states are constructed using $\psi_{out}(y)$ which is transformed into $\psi_{out}(X(y))$. Again the interacting field $\psi(X(y))$ is related to $\psi_{out}(X(y))$ as part of the familiar LSZ limiting ($y^0 \to +\infty$) process.

Since much of the development differs only trivially from the standard treatment in textbooks we will simply list relevant equations and let the reader pursue them further in quantum field theory introductory textbooks.

ψ In-Field

In order to define a perturbation theory for particle scattering we will use a free Dirac field $\psi_{in}(y)$ that satisfies

$$(i\nabla\!\!\!/_y - m)\psi_{in}(y) = 0 \qquad (7.53)$$

where

$$\nabla\!\!\!/_y = \gamma^\nu \partial/\partial y^\nu$$

Defining the fourier expansion for the "bare" and "cloaked" fermion fields:

$$\psi_{in}(y) = \sum_{\pm s} \int d^3p\ N^d{}_m(p)\ [b_{in}(p,s)u(p,s)\ e^{-ip\cdot y} + d_{in}{}^\dagger(p,s)v(p,s)\ e^{ip\cdot y}] \quad (7.54a)$$

$$\psi_{in}(X(y)) = \sum_{\pm s} \int d^3p\ N^d{}_m(p)\ [b_{in}(p,s)u(p,s){:}e^{-ip\cdot X(y)}{:} + d_{in}{}^\dagger(p,s)v(p,s){:}e^{ip\cdot X(y)}{:}] \quad (7.54b)$$

we can define ψ_{in} in-field states with expressions like

$$|(p_n s_n), \ldots, (p_1 s_1); (\bar{p}_m \bar{s}_m), \ldots (\bar{p}_1 \bar{s}_1)\ in\rangle = b_{in}{}^\dagger(p_n s_n) \ldots b_{in}{}^\dagger(p_1 s_1)\ d_{in}{}^\dagger(\bar{p}_m \bar{s}_m) \ldots d_{in}{}^\dagger(\bar{p}_1 \bar{s}_1)|0\rangle \quad (7.55)$$

for n electrons and m positrons. The development parallels the conventional development of fermion in-states.

ψ Out-Field

Similarly we can define fermion out-states for the free field $\psi_{out}(y)$. Again we will use a free Dirac field $\psi_{out}(y)$ that satisfies

$$(i\nabla\!\!\!/_y - m)\psi_{out}(y) = 0 \quad (7.56)$$

Defining the fourier expansions for the "bare" and "cloaked" fermion fields:

$$\psi_{out}(y) = \sum_{\pm s} \int d^3p\ N^d{}_m(p)\ [b_{out}(p,s)u(p,s)\ e^{-ip\cdot y} + d_{out}{}^\dagger(p,s)v(p,s)\ e^{ip\cdot y}] \quad (7.57a)$$

$$\psi_{out}(X(y)) = \sum_{\pm s} \int d^3p\ N^d{}_m(p)\ [b_{out}(p,s)u(p,s){:}e^{-ip\cdot X(y)}{:} + d_{out}{}^\dagger(p,s)v(p,s){:}e^{ip\cdot X(y)}{:}] \quad (7.57b)$$

we can define bare ψ_{out} out-field states with expressions like

$$|(p_n s_n), \ldots, (p_1 s_1); (\bar{p}_m \bar{s}_m), \ldots (\bar{p}_1 \bar{s}_1)\ out\rangle = b_{out}{}^\dagger(p_n s_n) \ldots b_{out}{}^\dagger(p_1 s_1) \cdot \cdot d_{out}{}^\dagger(\bar{p}_m \bar{s}_m) \ldots d_{out}{}^\dagger(\bar{p}_1 \bar{s}_1)|0\rangle \quad (7.58)$$

for n electrons and m positrons. The development again parallels the conventional development of fermion out-states.

Photon In and Out States

In this section we will develop properties of photon in-fields and out-fields. The LSZ procedure can be schematized as:

$$A_{in}^{\mu}(y) \Rightarrow A_{in}^{\mu}(X(y)) \Rightarrow A^{\mu}(X(y)) \Rightarrow A_{out}^{\mu}(X(y)) \Rightarrow A_{out}^{\mu}(y) \qquad (7.59)$$

In-states are constructed using a "bare" field $A_{in}^{\mu}(y)$ which are then transformed into $A_{in}^{\mu}(X(y))$ in order to make contact with our Lagrangian formalism. The interacting field $A^{\mu}(X(y))$ is related to $A_{in}^{\mu}(X(y))$ using the standard LSZ limiting ($y^0 \rightarrow -\infty$) process. Similarly out-states are constructed using $A_{out}^{\mu}(y)$ which are transformed into $A_{out}^{\mu}(X(y))$. Again the interacting field $A^{\mu}(X(y))$ is related to $A_{out}^{\mu}(X(y))$ as part of the familiar LSZ limiting ($y^0 \rightarrow +\infty$) process.

The fourier expansions of the free "bare" and "cloaked" photon in and out fields are

$$A_a^{\,i}(y) = \int d^3k\, N_0(k) \sum_{\lambda=1}^{2} \varepsilon^i(k, \lambda)[a_a(k,\lambda)\, e^{-ik\cdot y} + a_a^{\dagger}(k,\lambda)\, e^{ik\cdot y}] \qquad (7.60a)$$

and

$$A_a^{\,i}(X(y)) = \int d^3k\, N_0(k) \sum_{\lambda=1}^{2} \varepsilon^i(k, \lambda)[a_a(k,\lambda){:}e^{-ik\cdot X(y)}{:} + a_a^{\dagger}(k,\lambda){:}e^{ik\cdot X(y)}{:}] \qquad (7.60b)$$

where a = "in" or 'out."

We can define bare photon in-field states with expressions like

$$|(k_n\lambda_n), \ldots, (k_1\lambda_1)\ in\rangle = a_{in}^{\dagger}(k_n\lambda_n) \ldots a_{in}^{\dagger}(k_1\lambda_1)|0\rangle \qquad (7.61)$$

for n photons. The development parallels the conventional development of photon in-states as does the definition of photon out-states.

S Matrix

The S matrix is defined in a familiar way by

$$\psi_{in}(y) = S\psi_{out}(y)S^{-1} \qquad (7.62)$$

$$A^{\mu}{}_{in}(y) = SA^{\mu}{}_{out}(y)S^{-1} \qquad (7.63)$$

and the other standard properties of the S matrix with the sole exception being the form of the unitarity relation (which was discussed in the previous chapter).

LSZ Reduction

In this section we will determine the reduction formula for fermions and photons for the S matrix in Two-Tier QED.

Dirac Fields

Consider a charged Dirac particle such as an electron. The S matrix element corresponding to an in-state: β plus one Dirac particle of momentum p and spin s, and an out state α can be represented by

$$S_{\alpha\beta_{ps}} = \langle\alpha \text{ out}|\beta \text{ (ps) in}\rangle \tag{7.64}$$

which becomes

$$S_{\alpha\beta_{ps}} = \langle\alpha - \text{(ps) out}|\beta \text{ in}\rangle + \langle\alpha \text{ out}|\int d^3y\, U_{ps}(y)[\psi_{in}^{\dagger}(y) - \psi_{out}^{\dagger}(y)]|\beta \text{ in}\rangle \tag{7.65}$$

through standard manipulations where $\langle\alpha - \text{(ps) out}|$ is an out state with a particle of momentum p and spin s removed (if present) and where

$$U_{ps}(y) = \{m/[(2\pi)^3E_p]\}^{1/2}u(p,s)e^{-ip\cdot y} \tag{7.66a}$$

and for later use

$$V_{ps}(y) = \{m/[(2\pi)^3E_p]\}^{1/2}v(p,s)e^{ip\cdot y} \tag{7.66b}$$

Eq. 7.65 can be reexpressed as

$$S_{\alpha\beta ps} = S_{\alpha-ps\beta} + \langle\alpha \text{ out}|\int d^3y U_{ps}(y)W_{QED}^{-1}[\psi_{in}^{\dagger}(X(y)) - \psi_{out}^{\dagger}(X(y))]W_{QED}|\beta \text{ in}\rangle \tag{7.67}$$

using $W_{QED}(y) = W_{QEDin}(y)$ with

$$\psi_a(y) = W_{QEDa}^{-1}(y)\psi_a(X(y))W_{QEDa}(y) \tag{7.68a}$$

$$\psi_a^{\dagger}(y) = VW_{QEDa}^{-1}(y)\psi_a^{\dagger}(X(y))W_{QEDa}(y)V \tag{7.68b}$$

where the label a = “in” or a = “out”, and where

$$W_{QEDa}(y) = \exp(-\mathbf{Y}(y)\cdot\mathbf{P}_{QEDa}/M_c^2) \tag{7.69}$$

and

$$W_{QEDa}^{-1}(y) = \exp(\mathbf{Y}(y)\cdot\mathbf{P}_{QEDa}/M_c^2) \tag{7.70}$$

in the Coulomb gauge of Y with $\mathbf{P}_{QEDa}$ being the spatial momentum vector for the free fermion and photon fields defined by

$$\mathbf{P}_{QEDa} = \sum_{\pm s} \int d^3p\, \mathbf{p}\, [b_a^\dagger(p,s)b_a(p,s) + d_a^\dagger(p,s)d_a(p,s)] + \sum_{\lambda=1}^{2} \int d^3k\, \mathbf{k}\, a_a^\dagger(k,\lambda)a_a(k,\lambda) \tag{7.71}$$

using the free fermion and photon creation and annihilation operators.

We note that $W_{QEDa}(y)$ is not a unitary operator – a situation similar to that of the scalar particle quantum field theory – but it is pseudo-unitary:

$$W_{QEDa}^{-1}(y) = V\, W_{QEDa}^{\dagger}(y)\, V^{-1} \tag{7.72}$$

where (letting $a_Y^\dagger(k, \lambda)$ and $a_Y(k, \lambda)$ represent the creation and annihilation operators of the Y field) V is given by

$$V = \exp(-i\pi \sum_{\lambda=1}^{2} \int d^3k\, a_Y^\dagger(k, \lambda)a_Y(k, \lambda)) \tag{7.73}$$

V is a unitary operator with the property

$$VY^j(y)V^{-1} = -Y^j(y) \tag{7.74}$$

for j = 1,2,3. We note (as in the scalar field discussion)

$$V^\dagger = V^{-1} = V \tag{7.75}$$

and thus

$$V^2 = I \tag{7.76}$$

V is a metric operator in the sense of Dirac as discussed earlier. We also note:

$$X^\mu(y) = V[X^\mu(y)]^\dagger V^{-1} \tag{7.77}$$

$$\psi_a^\dagger(X(y)) = V[\psi_a(X(y))]^\dagger V^{-1} \tag{7.78}$$

$$A^{\mu}{}_{a}(X(y)) = V[A^{\mu}{}_{a}(X(y))]^{\dagger} V^{-1} \tag{7.79}$$

for a = "in" or "out." These properties are required for eqs. 7.68a and 7.68b to hold.

The interacting $\psi(X(y))$ field approaches the in and out fields $\psi_{in}(X(y))$ and $\psi_{out}(X(y))$ in the limit that $y^0 \to -\infty$ and $y^0 \to +\infty$ respectively in the sense of Lehmann, Symanzik and Zimmermann which we *symbolize* as:

$$\psi(X(y)) \to \sqrt{Z_2}\, \psi_{in}(X(y)) \quad \text{as} \quad y^0 \to -\infty \tag{7.72}$$

$$\psi(X(y)) \to \sqrt{Z_2}\, \psi_{out}(X(y)) \quad \text{as} \quad y^0 \to +\infty \tag{7.73}$$

with $\sqrt{Z_2}$ the wave function renormalization constant. Using eqs. 7.72 and 7.73 and following the standard LSZ reduction procedure leads to:

$$S_{\alpha\beta ps} = S_{\alpha-ps\beta} - iZ_2^{-1/2}\int d^4y \langle \alpha\ \text{out}|W_{QED}^{-1}\bar{\psi}(X(y))W_{QED}|\beta\ \text{in}\rangle(-i\overleftarrow{\nabla\!\!\!\!/}_y - m)U_{ps}(y) \tag{7.74}$$

Eq. 7.74 is similar to the usual LSZ reduction formula for a fermion extracted from an in-state except for the appearance of the W(y) operator and its inverse. We note that $W(y) = W_{in}(y)$ still because $\mathbf{P}_{QEDin}$ is independent of y^0.

The expressions for the other possible reductions of a fermion and its anti-particle are:

1. Reduction of an anti-particle from an in-state

$$iZ_2^{-1/2}\int d^4y \bar{V}_{\bar{p}\bar{s}}(y)(i\nabla\!\!\!\!/_y - m)\langle \alpha\ \text{out}|W_{QED}^{-1}\psi(X(y))W_{QED}|\beta\ \text{in}\rangle \tag{7.75}$$

2. Reduction of a particle from an out-state

$$-iZ_2^{-1/2}\int d^4y \bar{U}_{p's'}(y)(i\nabla\!\!\!\!/_y - m)\langle \alpha\ \text{out}|W_{QED}^{-1}\psi(X(y))W_{QED}|\beta\ \text{in}\rangle \tag{7.76}$$

3. Reduction of an anti-particle from an out-state

$$iZ_2^{-1/2}\int d^4y \langle \alpha\ \text{out}|W_{QED}^{-1}\bar{\psi}(X(y))W_{QED}|\beta\ \text{in}\rangle(-i\overleftarrow{\nabla\!\!\!\!/}_y - m)V_{\bar{p}'\bar{s}'}(y) \tag{7.77}$$

where

$$V_{ps}(y) = \{m/[(2\pi)^3 E_p]\}^{1/2} v(p,s)e^{ip\cdot y} \tag{7.78}$$

Electromagnetic Field

The LSZ reduction of a photon from an S matrix element:

$$S_{\alpha\beta k\lambda} = \langle\alpha \text{ out}|\beta\ \gamma(k\lambda) \text{ in}\rangle \tag{7.79}$$

begins with

$$S_{\alpha\beta k\lambda} = \langle\alpha - \gamma(k\lambda) \text{ out}|\beta \text{ in}\rangle - iZ_3^{-1/2}\int d^3y\ A_{k\lambda}^{\mu *}(y)\langle\alpha \text{ out}|\,[A_{in}^{\mu}(y) - A_{out}^{\mu}(y)]\,|\beta \text{ in}\rangle \tag{7.80}$$

where Z_3 is a normalization constant and

$$A_{k\lambda}^{\mu}(y) = [(2\pi)^3\omega_k]^{-1/2} e^{-ik\cdot y}\, \varepsilon^{\mu}(k, \lambda) \tag{7.81}$$

Using the LSZ symbolic notation we see

$$A^{\mu}(X(y)) \rightarrow \sqrt{Z_3}\, A_{in}^{\mu}(X(y)) \quad \text{as } y^0 \rightarrow -\infty \tag{7.82}$$

$$A^{\mu}(X(y)) \rightarrow \sqrt{Z_3}\, A_{out}^{\mu}(X(y)) \quad \text{as } y^0 \rightarrow \infty \tag{7.83}$$

and

$$A_a^{\mu}(y) = W_{QEDa}^{-1}(y) A_a^{\mu}(X(y)) W_{QEDa}(y) \tag{7.84}$$

where a = "in" or "out". Next we arrive at the reduction expression following steps that parallel the scalar field reduction:

$$S_{\alpha\beta k\lambda} = \langle\alpha - \gamma(k\lambda) \text{ out}|\beta \text{ in}\rangle - iZ_3^{-1/2}\int d^4y\ A_{k\lambda}^{\mu *}(y)\Box_y\langle\alpha \text{ out}|\,W_{QED}^{-1} A^{\mu}(X(y)) W_{QED}\,|\beta \text{ in}\rangle \tag{7.85}$$

Time Ordered Products and Perturbation Theory

By repeated application of the LSZ procedure outlined above, an S matrix element is reduced to the vacuum expectation value of time ordered products of fermion and photon fields.

$\langle\alpha \text{ out}|\beta \text{ in}\rangle = \ldots \langle 0|T(\ldots W^{-1}(y_m)U^{-1}(y_m^{0})\psi_{in}(X(y_m))U(y_m^{0})W(y_m) \ldots$

$$W^{-1}(y_n)U^{-1}(y_n{}^0)\bar{\psi}_{in}(X(y_n))U(y_n{}^0)W(y_n)\ \ldots$$
$$W^{-1}(y_p)U^{-1}(y_p{}^0)A_{in}{}^{\mu}(X(y_p))U(y_p{}^0)W(y_p)\ \ldots\ )\,|\,0> \ldots$$
$$(7.86)$$

Following the same development of the U matrix as described in chapter 6 with minor changes in details leads to the time ordered product for S matrix elements:

$$\tau(y_1,\ldots,y_n) = \frac{<0\,|\,T(\ldots\psi_{in}(X(y_m))\ldots\bar{\psi}_{in}(X(y_n))\ldots A_{in}{}^{\mu}(X(y_p))\ldots\exp[-i\int d^4y'\,\mathscr{H}_{FintQED}])\,|\,0>}{<0\,|\,T(\exp\,[-i\int d^4y'\,\mathscr{H}_{FintQED}])\,|\,0>} \qquad (7.87)$$

where the QED interaction hamiltonian is

$$\mathscr{H}_{FintQED} = e_0{:}\bar{\psi}_{in}(X(y))\not{A}_{in}(X(y))\,\psi_{in}(X(y)){:} \qquad (7.88)$$

plus mass counter terms.

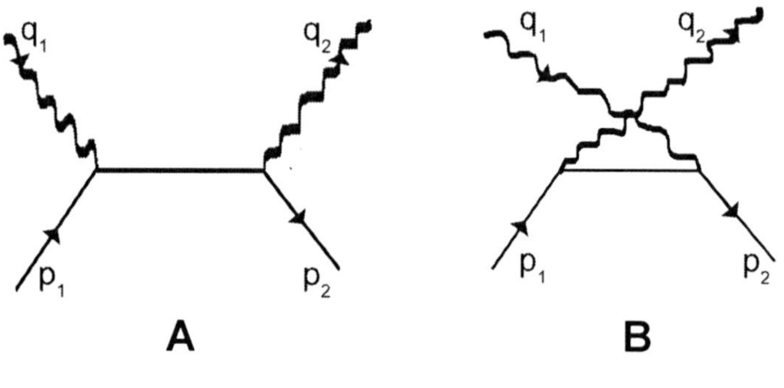

Figure 7.1. Lowest order elastic photon-electron scattering diagrams.

Example - Photon-Electron Elastic Scattering

In order to illustrate perturbative calculations in Two-Tier QED we will calculate the lowest order photon-electron elastic scattering S matrix element (Fig. 7.1).

We will see that at large distances (where the momenta are $\ll M_c$) the result is the same as the conventional QED calculation. At short distances (where the momenta are $\gg M_c$) the result differs markedly due to the effects of the Y field. The S matrix element containing the contribution of these diagrams is

$$S_{\gamma e} = (Z_2 Z_3)^{-1} e_0^{\,2} \prod_{j=1}^{6} \int d^4 y_j\, \bar{U}_{p_2 s_2}(y_4) A_{k_1 \lambda_1}^{\;\mu_1}(y_2) A_{k_2 \lambda_2}^{\;\mu_2 *}(y_3) \Box_{y_2} \Box_{y_3} \cdot$$

$$\cdot (i\nabla\!\!\!\!/_{y_4} - m) <0| T(\bar{\psi}_{in}(X(y_1))\, \psi_{in}(X(y_4)) A_{in}^{\;\mu_1}(X(y_2)) A_{in}^{\;\mu_2}(X(y_3)) \cdot$$

$$\cdot :\bar{\psi}_{in}(X(y_5)) A\!\!\!/_{in}(X(y_5))\, \psi_{in}(X(y_5)) :: \bar{\psi}_{in}(X(y_6)) A\!\!\!/_{in}(X(y_6))\, \psi_{in}(X(y_6)):$$

$$)\,|0> \overleftarrow{(-i\nabla\!\!\!\!/_{y_1}} - m) U_{p_1 s_1}(y_1) \tag{7.89}$$

Diagram A in Fig. 7.1 corresponds to the Wick expansion:

$$S_{\gamma e A} = (Z_2 Z_3)^{-1} e_0^{\,2} \prod_{j=1}^{6} \int d^4 y_j\, A_{k_1 \lambda_1}^{\;\mu_1}(y_2) A_{k_2 \lambda_2}^{\;\mu_2 *}(y_3) \bar{U}_{p_2 s_2}(y_4) \{$$

$$(i\nabla\!\!\!\!/_{y_4} - m) i S_F^{\,TT}(y_4 - y_6) \gamma^{\nu} i S_F^{\,TT}(y_6 - y_5) \gamma^{\mu} i S_F^{\,TT}(y_5 - y_1) \overleftarrow{(-i\nabla\!\!\!\!/_{y_1}} - m)\}$$

$$U_{p_1 s_1}(y_1)\, \Box_{y_2} \Box_{y_3}\, i D_F^{\,trTT}(y_5 - y_2)_{\mu_1 \mu}\, i D_F^{\,trTT}(y_3 - y_6)_{\mu_2 \nu} \tag{7.90}$$

After some manipulation eq. 7.90 can be placed in the form:

$$S_{\gamma e A} = (Z_2 Z_3)^{-1} e_0^{\,2} (2\pi)^4 \delta^4(p_2 + k_2 - p_1 - k_1) \bar{u}(p_2, s_2) \mathscr{S}_L^{\,TT}(p_2)\, \not{\epsilon}(k_2, \lambda_2) \cdot$$

$$\cdot \mathscr{S}^{\,TT}(p_1 + k_1)\, \not{\epsilon}(k_1, \lambda_1) \mathscr{S}_R^{\,TT}(p_1) u(p_1, s_1) \mathscr{D}^{\,TT}(k_1) \mathscr{D}^{\,TT}(k_2) \tag{7.91}$$

where

$$\mathscr{S}_L^{\,TT}(p) = i N_{fp} \int d^4 y\, e^{ip\cdot y} (i\nabla\!\!\!\!/_y - m)\, S_F^{\,TT}(y) \tag{7.92}$$

$$\mathscr{S}^{\,TT}(p) = i \int d^4 y\, e^{ip\cdot y}\, S_F^{\,TT}(y) \tag{7.93}$$

$$\mathscr{S}_R^{\,TT}(p) = i N_{fp} \int d^4 y\, e^{ip\cdot y}\, S_F^{\,TT}(y)\, \overleftarrow{(i\nabla\!\!\!\!/_y} - m) \tag{7.94}$$

$$\mathscr{D}^{TT}(k) = iN_{\gamma k} \int d^4y\, e^{ik\cdot y}\, \Box_y\, D_F^{trTT}(y)_{00} \tag{7.95}$$

with

$$N_{fp} = \{m/[(2\pi)^3 E_p]\}^{1/2} \tag{7.96}$$

and

$$N_{\gamma k} = [(2\pi)^3 \omega_k]^{-1/2} \tag{7.97}$$

The factors $\mathscr{S}_L^{TT}(p_2)$, $\mathscr{S}_R^{TT}(p_1)$, and $\mathscr{D}^{TT}(k_1)$ and $\mathscr{D}^{TT}(k_2)$ serve to normalize the in and out particle legs. They are a consequence of the dressing of the legs by Y particle "clouds." At long distance (low momentum) they approach the corresponding values of conventional QED:

$$\mathscr{S}_L^{TT}(p) \to iN_{fp} \tag{7.98}$$

$$\mathscr{S}_R^{TT}(p) \to iN_{fp} \tag{7.99}$$

$$\mathscr{D}^{TT}(k) \to iN_{\gamma k} \tag{7.100}$$

if $p, k \ll M_c$.

Here again we must "lop off" external legs as in eqs. 6.74-6.75 to achieve a theory satisfying unitarity. We leave this as an exercise for the reader.

Deep e-p Inelastic Scattering Partons?

The form of eq. 7.93 shows the fermion propagator factor $\mathscr{S}^{TT}(p)$ is not the simple form of a free fermion propagator. Rather it consists of a fermion traveling within a "stream" of free Y quanta. This picture is reminiscent of Feynman's picture of deep inelastic e-p scattering in which the proton is viewed as a stream of partons. This question will be addressed in a future publication.

External Leg "Normalizations"

The external leg factors $\mathscr{S}_L^{TT}(p)$, $\mathscr{S}_R^{TT}(p)$ and $\mathscr{D}^{TT}(k)$ change the normalization of the external legs due to the Y particle "cloud" surrounding each particle. We *must "lop off" external legs as in eqs. 6.74-6.75 to achieve a theory satisfying unitarity. In this section we examine external leg factors to see the effect of the Y quanta cloud around particles.*

In order to find their form for large momenta we will first evaluate the large momentum limit of the fermion propagator as $z^2 \to 0$ (the light cone). Starting from eq. 7.45 one can show that (space-like limit)

$$iS_F^{TT}(z) \to \gamma^0 \epsilon(z^0) M_c^3 \, [-M_c^2 \pi z^2/2]^{3/2} + \mathscr{O}(z^5) \tag{7.101}$$

where $z^2 = z_0^2 - \mathbf{z}^2$. Therefore on dimensional grounds we see that

$$\mathscr{S}^{TT}(p) \backsim \gamma^0 M_c^6 \, p^{-7} + \mathscr{O}(p^{-9}) \tag{7.102}$$

as p gets very large ($p \gg M_c$). At low momenta ($p \ll M_c$) the standard momentum space form of the fermion propagator is found:

$$\mathscr{S}^{TT}(p) \backsim (\not{p} - m + i\epsilon)^{-1} \tag{7.103}$$

Similarly we find

$$\mathscr{S}_L^{TT}(p) \backsim M_c^6 \, p^{-6} + \mathscr{O}(p^{-8}) \tag{7.104}$$

$$\mathscr{S}_R^{TT}(p) \backsim M_c^6 \, p^{-6} + \mathscr{O}(p^{-8}) \tag{7.105}$$

as p gets very large ($p \gg M_c$), and

$$\mathscr{D}^{TT}(k) \backsim M_c^4 \, k^{-4} \tag{7.106}$$

from eqs. 7.47 and 4.24 as k gets very large ($k \gg M_c$).

For small momenta compared to M_c we find the usual normalization:

$$\mathscr{S}_L^{TT}(p) = \mathscr{S}_R^{TT}(p) = iN_{fp} \tag{7.107}$$

$$\mathscr{D}^{TT}(k) = iN_{\gamma k} \tag{7.108}$$

Thus for low momenta (p, k $\ll M_c$) $S_{\gamma_{eA}}$ yields the standard result of QED while at large momenta (p, k $\gg M_c$) we see a high power of inverse momentum showing the well behaved nature of the theory at short distances.

Renormalization of Two-Tier QED

Two-Tier QED is a finite quantum field theory satisfying the unitarity condition. The degree of divergence of a Feynman diagram term in *conventional QED* is

$$D = 4k - 2b - f \tag{7.109}$$

where

k = the number of internal momentum integrations
b = the number of internal photon lines
f = the number of internal electron lines

Many diagrams are thus divergent in conventional QED and a renormalization program must be followed to achieve a theory with all divergences formally absorbed into renormalizations of the fundamental parameters of the theory. Despite the success of this approach and the excellent agreement of QED with experiment the presence of divergences in QED is logically unsatisfactory and suggests QED, and its successors Electroweak Theory and the Standard Model, are at best interim theories. Numerous attempts have been made to modify QED in order to eliminate its divergences. Some noteworthy attempts include the Lee-Wick formulation of QED and the Johnson-Baker-Willey model of QED. None of these attempts have succeeded for one reason or another.[34]

The degree of divergence formula is different in Two-Tier QED. It demonstrates there are no divergences in Two-Tier QED. Thus it would be more aptly named the "degree of convergence." The formula is:

$$D^{TT} = 4k - 6b - 7f \tag{7.110}$$

with k, b and f as above. The coefficient of b is 6 in Two-Tier QED because the Two-Tier photon propagator behaves as k^{-6} at high momentum (See eq. 7.47 for the photon propagator. Eq. 4.25 shows its high momentum behavior). The coefficient of f is 7 in Two-Tier QED because the Two-Tier fermion propagator behaves as k^{-7} at high momentum (see eq. 7.102).

34 Lee, T. D. and Wick, G. C., Phys. Rev. **D2**, 1033 (1970); T. D. Lee and G. C. Wick, Nucl. Phys. **B9**, 209 (1969) and references therein; M. Baker and K. Johnson, Phys. Rev. **D3**, 2516 (1971); M. Baker and K. Johnson, Phys. Rev. **D8**, 1110 (1973); K. Johnson, M. Baker, and R. Willey, Phys. Rev. **136**, B1111 (1964); K. Johnson, R. Willey, and M. Baker, Phys. Rev. **163**, 1699 (1967); S. Blaha,, Phys. Rev. **D9**, 2246 (1974) and references therein; S. Adler, Phys. Rev. **D5**, 3021 (1972).

We note the degree of Divergence in Two-Tier QED is always negative – indicating a finite theory!

$$D^{TT} < 0 \qquad (7.111)$$

For example the degree of divergence in Two-Tier QED of the lowest order in α (see Fig. 7.2): i) vacuum polarization diagram is $D^{TT} = -10$, ii) fermion self-energy is $D^{TT} = -9$

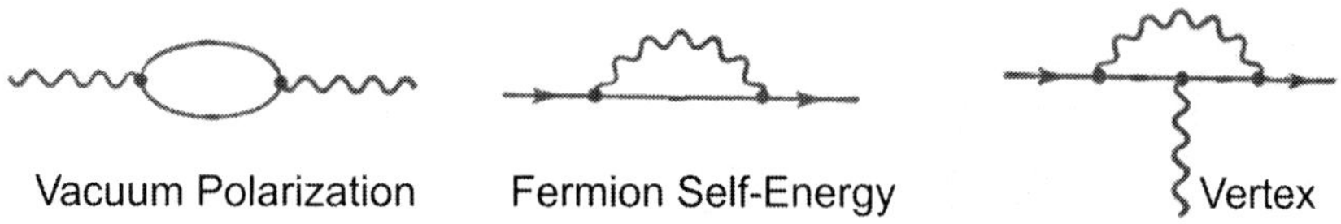

and iii) electromagnetic vertex correction is $D^{TT} = -16$.

Figure 7.2 Some low order (normally divergent) diagrams for the vacuum polarization, fermion self-energy and electromagnetic vertex correction.

It is easy to see that all Two-Tier QED Feynman diagrams are ultra-violet finite.

Unitarity of Two-Tier QED

The remaining major issue is unitarity. Two-Tier QED satisfies the unitarity condition between physical states. The argument demonstrating unitarity parallels the discussion of unitarity for scalar ϕ^4 quantum field theory in chapter 6.

Two-Tier QED *superficially* appears to have a unitarity problem due to the non-hermitean nature of its hamiltonian. The lack of hermiticity is entirely due to the appearance of iY^μ in the X^μ field coordinates. Thus the interaction hamiltonian in eq. 7.88 is not hermitean:

$$H_{FintQED} = \int d^3y' \mathscr{H}_{FintQED}(\bar{\psi}_{in}(y' + iY(y')/M_c^2),\ A_{in}{}^\mu(y' + iY(y')/M_c^2),\ \psi_{in}(y' + iY(y')/M_c^2)) \qquad (7.112)$$

and

$$H_{FintQED} \neq H_{FintQED}{}^\dagger = \int d^3y' \mathscr{H}_{FintQED}(\bar{\psi}_{in}(y' - iY(y')/M_c^2),\ A_{in}{}^\mu(y' - iY(y')/M_c^2),\ \psi_{in}(y' - iY(y')/M_c^2)) \qquad (7.113)$$

The metric operator (eq. 7.73) establishes the relation between $H_{FintQED}$ and its hermitean conjugate is

$$H_{FintQED} = V\, H_{FintQED}{}^\dagger\, V \qquad (7.114)$$

Thus the Two-Tier QED S matrix is not unitary – S is pseudo-unitary in general:

$$S^{-1} = V\, S^{\dagger}\, V \tag{7.115}$$

which implies

$$VS^{\dagger} VS = I \tag{7.116}$$

Two-Tier *QED S matrix satisfies unitarity between physical asymptotic states* which in Two-Tier QED are states consisting of charged fermions, and photons. The proof is identical to eqs. 6.46 – 6.48.

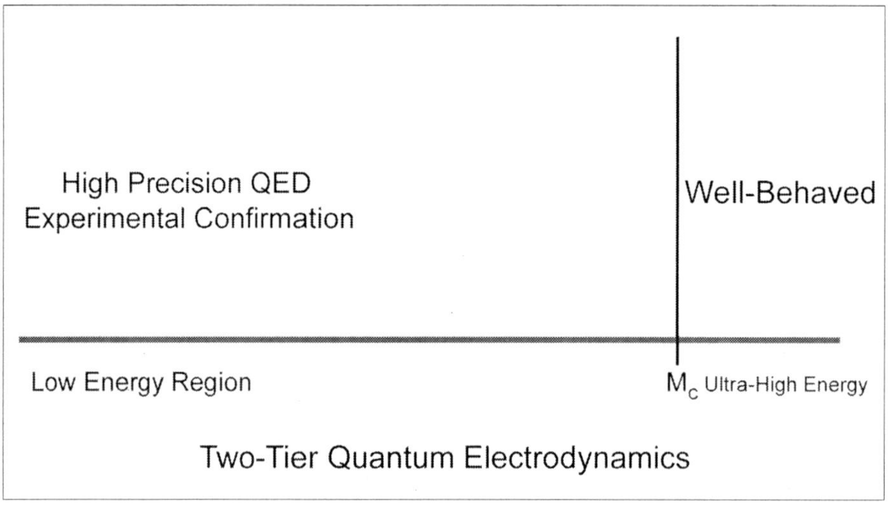

Figure 7.3. A Schematic View of Two-Tier QED.

Correspondence Principle for Two-Tier QED

The Two-Tier QED theory that we have developed has the behavior of conventional QED at “low energies.” Since M_c sets the scale at which deviations from the normal QED results become significant we can set M_c to an extremely high value such as the Planck mass and obtain as close an agreement with conventional QED as we wish. Thus Two-Tier QED implements a type of Correspondence Principle with conventional QED as its “low energy” limit.

Consistency with Precision QED Experiments

Since we can obtain results as close to conventional QED with a sufficiently high choice of M_c, Two-Tier QED has the same excellent agreement[35] with experiment as conventional QED.

Calculational Rules for Feynman Diagrams

This section describes the procedure for the calculation of S matrix terms corresponding Feynman diagrams in Two-Tier QED and other quantum field theories. By design the approach parallels the perturbative approaches of conventional QED and other quantum field theories. The examples considered earlier in this chapter, and the preceding chapter, illustrate Two-Tier perturbation theory.

Procedure: Follow the conventional procedure to form the expression for each Feynman diagram in the perturbative calculation. Then replace each propagator in each expression with the corresponding Two-Tier propagator as specified below. Then evaluate the expression.

Feynman Propagators and their Two-Tier Propagator Equivalents

Two-Tier propagators are denoted with the superscript "TT".

Spin 0 Propagator Case

$$\Delta_F(p) = (p^2 - m^2 + i\varepsilon)^{-1} \quad \rightarrow \quad \int d^4z\, e^{+ip\cdot z}\, \Delta_F^{TT}(z) \tag{7.120}$$

where $\Delta_F^{TT}(z)$ is given by eq. 4.9.

Spin 1 Photon Propagator Case

$$D_F(p)_{\mu\nu} = -g_{\mu\nu}(p^2 + i\varepsilon)^{-1} \quad \rightarrow \quad \int d^4z\, e^{+ip\cdot z}\, D_F^{TT}(z)_{\mu\nu} \tag{7.121}$$

where $D_F^{TT}(z)_{\mu\nu}$ is given by eq. 7.47.

Spin ½ Fermion Propagator Case

$$S_F(p) = (\not{p} - m + i\varepsilon)^{-1} \quad \rightarrow \quad \int d^4z\, e^{+ip\cdot z}\, S_F^{TT}(z) \tag{7.122}$$

where $S_F^{TT}(z)$ is given by eq. 6.88.

---

35 See, for example, T. Kinoshita, "The Fine Structure Constant", Cornell University preprint CLNS 96/1406 (1996); V. W. Hughes and T. Kinoshita, Rev. Mod. Phys. **71**, S133 (1999) and references therein.

Spin 2 Massless Boson (graviton) Propagator Case

$$\Delta_{F2}(p)_{\mu\nu\rho\sigma} = b_{\mu\nu\rho\sigma}(p^2 - m^2 + i\varepsilon)^{-1} \quad \rightarrow \quad \int d^4z\, e^{+ip\cdot z}\, \Delta_{F2}^{TT}(z)_{\mu\nu\rho\sigma} \qquad (7.123)$$

where $\Delta_{F2}^{TT}(z)_{\mu\nu\rho\sigma}$ the graviton propagator.

Two-Tier Coulomb Potential vs. Conventional Coulomb Potential

The familiar Coulomb potential is (for two particles of opposite unit electric charge):

$$V_{Coulomb} = -a/|\mathbf{r}| \qquad (7.124)$$

The Two-Tier QED Coulomb potential (eq. 7.49) is:

$$V_{Two\text{-}TierCoul} = -a\Phi(M_c^2\pi|\mathbf{r}|^2)/|\mathbf{r}| \qquad (7.125)$$

where $\Phi(x)$ is the error function.[36] At small distances ($\pi r^2 \ll M_c^{-2}$)

$$V_{Two\text{-}TierCoul} \rightarrow -2a\sqrt{\pi}\, M_c^2|\mathbf{r}| \qquad (7.126)$$

a linear potential, and at large distances ($\pi r^2 \gg M_c^{-2}$)

$$V_{Two\text{-}TierCoul} \rightarrow V_{Coulomb} \qquad (7.127)$$

The Two-Tier Coulomb potential has a minimum at

$$M_c^2\pi|\mathbf{r}|^2 = 1 \qquad (7.128)$$

---

36 W. Magnus and F. Oberhettinger, *Formulas and Theorems for the Special Functions of Mathematical Physics* (Chelsea Publishing Co., New York, 1949) page 96.

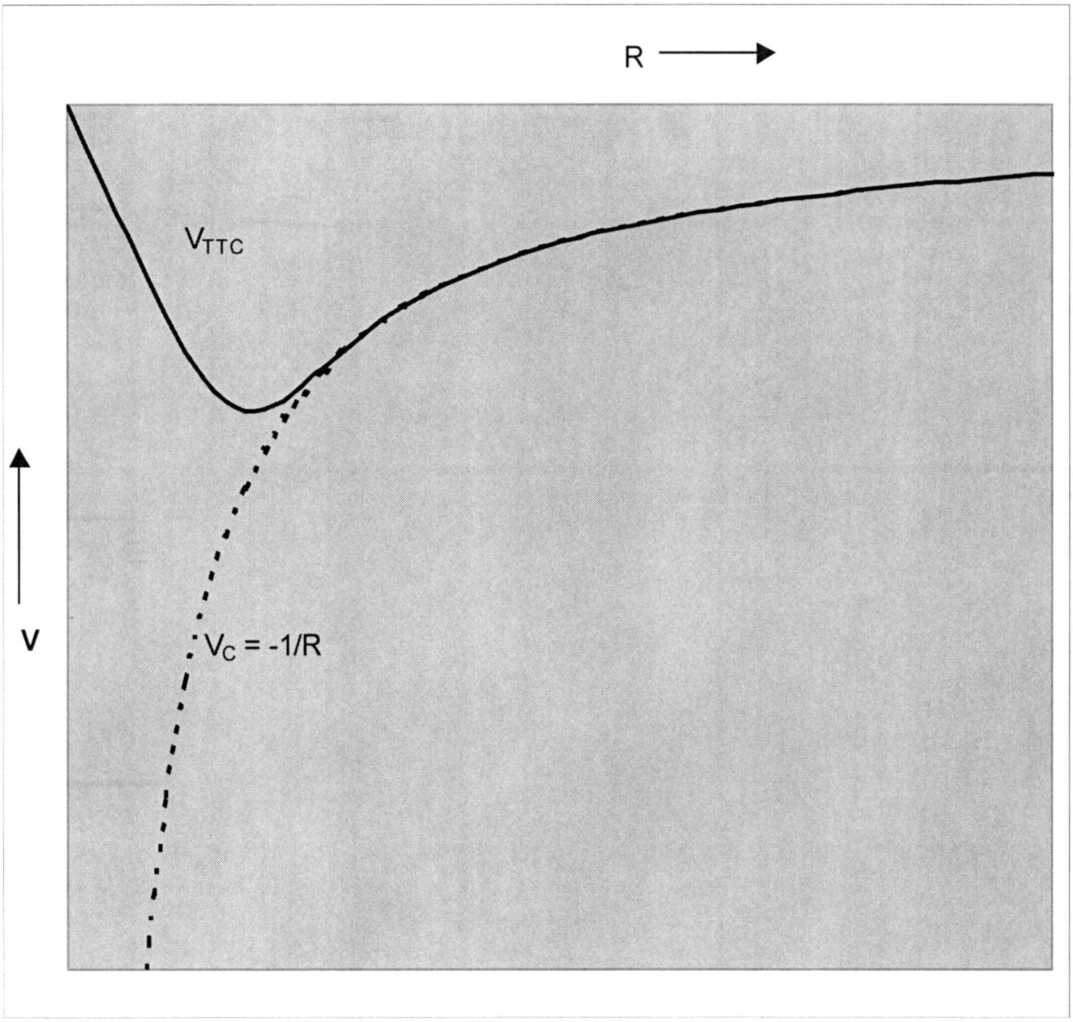

Figure 7.4. Plot of the form of the Two-Tier Coulomb "attractive" potential between particles of opposite unit charge divided by αM_c: $V_{TTC} = V_{Two\text{-}TierCoul}/(\alpha M_c)$. V_{TTC} is dimensionless. The dotted line is the conventional Coulomb attractive potential divided by αM_c: $V_c = V_{Coulomb}/(\alpha M_c) = 1/R$. Note the Two-Tier Coulomb force between particles of opposite charge is repulsive at short distance.

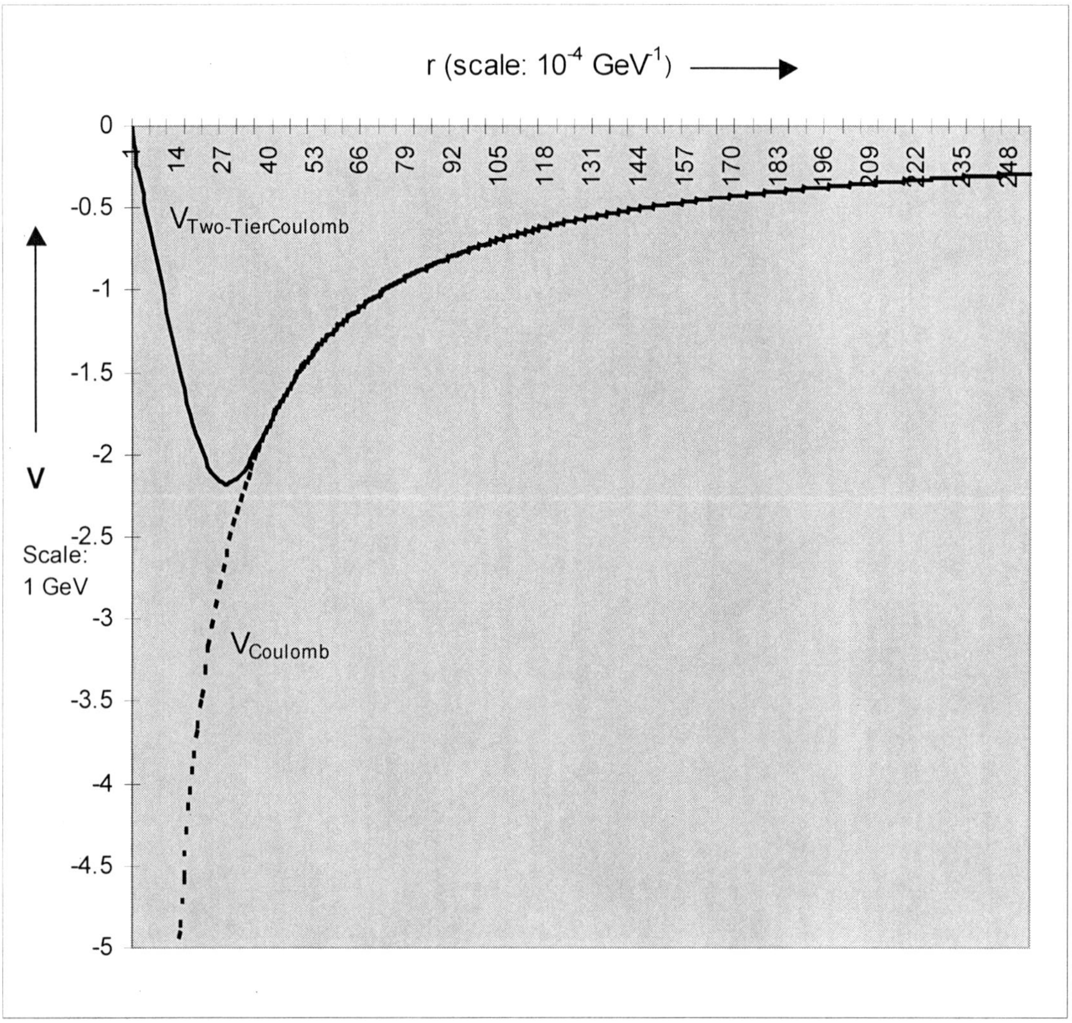

Figure 7.5. Two-Tier Coulomb Potential compared to conventional Coulomb potential for M_c = 200 GeV/c^2. Radial distance is measured in units of 10^{-4} GeV^{-1}. The potential energy for two opposite unit charges is measured in GeV units.

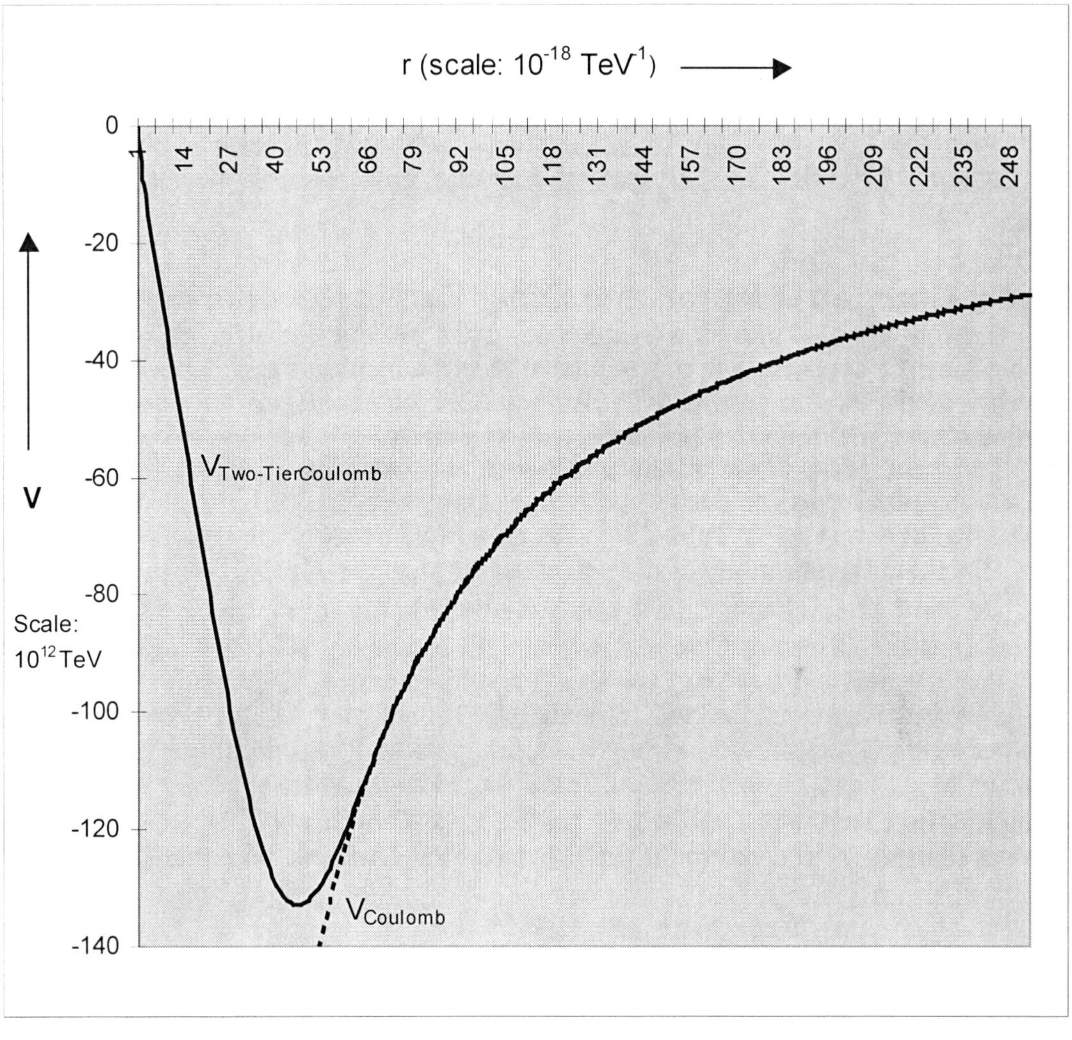

Figure 7.6. Two-Tier Coulomb Potential compared to conventional Coulomb potential for $M_c = M_{planck} = 1.22 \times 10^{19}$ GeV/c^2. Radial distance is in units of 10^{-18} TeV^{-1}. The potential energy for two opposite unit charges is measured in units of 10^{12} TeV.

At the minimum $V_{Two\text{-}TierCoul}$ has the value:

$$V_{Two\text{-}TierCoulMIN} = -.8427\alpha\sqrt{\pi}\, M_c \tag{7.129}$$

If we define distance in terms of the mass scale

$$|\mathbf{r}| = R/M_c \qquad (7.130)$$

then

$$V_{Two\text{-}TierCoul} = -\,\alpha M_c \Phi(\pi R^2)/R \qquad (7.131)$$

Fig. 7.4 displays a plot of $V_{TTC} = V_{Two\text{-}TierCoul}/(\alpha M_c)$ showing the general form of the Two-Tier Coulomb potential for particles of opposite unit charge. We also plot $V_C = V_{Coulomb}/(\alpha M_c) = 1/R$.

Doubly-Charged Dilepton and Other Exotic Resonances

Doubly charged dilepton resonances such as e^-e^- or $\mu^-\mu^-$ resonances are possible since it appears that the repulsive electromagnetic force between similarly charged leptons goes to zero as the distance between them goes to zero creating a potential barrier that might, in principle, lead to binding.

However because the distance between the particles at which the repulsive Coulomb potential starts to decline (Fig. 7.5) is very small, a very broad resonance is the best that one can expect. Tunneling will cause rapid decay of the bound state.

Note that the minimum of the potential, for $M_c = 200$ GeV/c^2, is at $r = 2.8 \times 10^{-5}$ GeV^{-1} $\equiv 5.5 \times 10^{-19}$ cm (by eq. 7.128) while the electron Compton wavelength is 3.861×10^{-13} m. Thus the electron's Compton wavelength is roughly 10^6 times larger than the distance of the maximum of the Two-Tier Coulomb potential. As a result the electrons in a dilepton resonance would immediately tunnel out of the binding region (*zitterbewegung*!) despite the strength of the potential at minimum: 21.7 TeV/c^2 according to eq. 7.129. Even if one could create a resonance state of two τ^- leptons with τ Compton wavelengths of 1.135×10^{-14} cm the τ wavelength would be 10^5 times larger than the distance of the minimum of the Two-Tier Coulomb potential if $M_c = 200$ GeV/c^2.

If M_c is much larger than 200 GeV/c^2 then the location of the peak of the potential is even smaller as shown in Fig. 7.6 where M_c equals the Planck mass. Thus a dilepton resonance can be expected to be very broad and very short-lived if indeed it exists at all.

Other exotic resonances might be possible involving the strong force which also interchanges repulsion and attraction at very short distances.

8. Two-Tier Standard Model Theory

Introduction

The Standard Model is a unified model of the electromagnetic, weak and strong interactions. It is formed from Electroweak Theory and an SU(3) color gauge theory of the strong interaction called Quantum Chromodynamics (QCD). Each part is separately renormalizable, as is the combined theory. This ad hoc theory of all known interactions except gravitation has been remarkably successful in accounting for the experimental data at energies currently accessible at accelerators. There are a number of phenomena that are not understood within the framework of the Standard Model such as dark matter and dark energy, some CP violation phenomena, and the spin dependence observed in certain experiments. There is also an appreciation of the ad hoc nature of the unification of the strong and electroweak sectors. They are "just glued together". An overall symmetry encompassing both sectors (broken as it would be) would be more reassuring and more elegant. However the theory works, and appears to work well, for most particle physics phenomena.

Superstring theory has become a seriously considered alternative to the Standard Model. It appears to offer a finite theoretical framework without divergences. It offers the hope of a rational grand unification. It is mathematically interesting and elegant. However, it has almost no experimental support. It predicts many particles that remain to be found. It requires additional dimensions that are not evidenced. It does not have a demonstrable, satisfactory "low energy limit" that resembles the Standard Model although ad hoc Superstring-like Standard Model variants have been studied. It has the problem of too many possibilities. After thirty years of effort one cannot point to a Superstring theory and say it is THE Superstring theory, and here is the method of symmetry breaking that results in the Standard Model. Considering that far more theoretical effort has been expended on Superstring theory by more theorists then on any other theory of physics, its physical justification, its nature, and its contact with experiment is still not well understood.

In the hope of developing a sound theory that clearly makes contact with known experimental data, that has a direct connection to the Standard Model, that is finite (no divergences), and that can be directly extended to encompass gravitation, we have developed a Two-Tier formulation of the Standard Model and its variants. It is admittedly far more modest in its scope than Superstring theory. Yet it allows us to have a finite theory – a major goal of Superstring theory – and to address cosmological issues (Blaha(2004)) by its easy unification with Two-Tier quantum gravity (next chapter) – also a finite theory without divergences.

Thus we have a theory that accounts for all known interactions and yet incrementally extends known successful theories – in a significant way. With (perhaps) the exception of General Relativity, the growth of particle physics theory in the twentieth-century has been largely incremental.

The nature of the mechanism that Two-Tier quantum field theory uses to avoid divergences has the important feature that it does not depend on "magic cancellations", does not depend on details of the symmetries, and does not depend on details of the form of the theories. Thus all the variations and extensions of the Standard Model have an equivalent Two-Tier version that is finite. This feature gives the theorist the flexibility to modify and extend the Two-Tier Standard Model without worrying about the introduction of new divergences.

The other major advantage is that the theory is very much like the world we see at current energies – four dimensions and no large number of new particles. It leaves open the door for new interactions and new particles IF they are found at higher energies. Thus it is not a showstopper. Nor is it a prediction of a vast desert extending to the highest energies. It gives a finite, unified theory with an "Open Door" policy for the future.

Formulation of Two-Tier Standard Models

There are many excellent texts[37] on the Standard Model. Therefore this author sees no reason to recapitulate the features of the Standard Model. We will focus on the form and features of a generic Two-Tier version of the Standard Model.

We will assume that the form of the Standard Model Lagrangian is:

$$\mathscr{L}_{\mathrm{SM}} = \mathrm{J}\mathscr{L}_{\mathrm{F}}^{\mathrm{SM}}(\mathrm{X}^{\mu}) + \mathscr{L}_{\mathrm{C}}(\mathrm{X}^{\mu}(\mathrm{y}), \partial\mathrm{X}^{\mu}(\mathrm{y})/\partial\mathrm{y}^{\nu}) \qquad (8.1)$$

where $\mathscr{L}_{\mathrm{F}}^{\mathrm{SM}}$ is the complete "normal" quantum field theory Lagrangian for the Standard Model variant under consideration and J is a Jacobian (eq. A.21). *All particle fields in* $\mathscr{L}_{\mathrm{F}}^{SM}$ *are assumed to be functions of the* X^{μ} *coordinate only. The dependence of the particle fields on the "underlying" coordinates* y^{μ} *is assumed to be solely through* X^{μ}. The Lagrangian $\mathscr{L}_{\mathrm{SM}}$ is a separable Lagrangian of the type of eq. A.26 embodying the composition of extrema described in Appendix A. Scalar particle examples of separable

---

[37] Some of the many excellent texts: P. Ramond, *Journeys Beyond the Standard Model* (Westview Press, New York, 1999); W. N. Cottingham and D. A. Greenwood, *An Introduction to the Standard Model of Particle Physics* (Cambridge University Press, New York, 1999); K. Huang, *Quarks, Leptons and Gauge Fields* (World Scientific, River Edge, NJ 1992); T. P, Cheng and L. F. Li, *Gauge Theories of the Elementary Particles* (Oxford University Press, New York, 1988); J. Donoghue, E. Golowich and B. Holstein, *Dynamics of the Standard Model* (Cambridge University Press, New York, 1994).

Lagrangians appear in eqs. A.96, 3.10 and 3.15. A scalar particle ϕ^4 theory with a separable Lagrangian is specified by eqs. 5.20 – 5.24 and described in detail in chapters 5 and 6. The separable Lagrangian Two-Tier version of QED is specified by eqs. 7.1 – 7.5 and described in some detail in chapter 7.

In all of these cases the coordinate part of the Lagrangian $\mathscr{L}_C$ as

$$\mathscr{L}_C = -\tfrac{1}{4}\, F_Y^{\mu\nu} F_{Y\mu\nu} \tag{3.15}$$

with

$$F_{Y\mu\nu} = \partial Y_\mu/\partial y^\nu - \partial Y_\nu/\partial y^\mu \tag{3.14}$$

The field equations of the theory are found using the conventional approach. The Hamiltonian, energy-momentum tensor and conserved quantities are also found using a conventional approach as illustrated in chapters 2, 3 and 4. The canonical quantization procedure is also followed. Thus the application of these procedures to the Standard Model will yield the standard results with the sole difference being that all fields are functions of X^μ and all derivatives of fields are derivatives with respect to X^μ. A perturbation theory along the lines of chapters 5 and 6 can then be directly developed.

The net result is that the free field Feynman propagators in a Feynman diagram approach to perturbative calculations are each replaced with the corresponding Two-Tier propagator. All coordinate space integrations are over the y coordinates.

Rule 1: *If the conventional coordinate space Feynman propagator for a free particle has the form:*

$$G_{\ldots}(z) = \int d^4p\, e^{-ip\cdot z}\, G_{\ldots}(p)/(2\pi)^4 \tag{8.2}$$

where ... represents any space-time indices, spin indices and internal symmetry indices, then the equivalent coordinate space (y – coordinates) Two-Tier propagator is

$$G_{\ldots}^{TT}(z) = \int d^4p\, e^{-ip\cdot z}\, G_{\ldots}(p) R(p, z)/(2\pi)^4 \tag{8.3}$$

where R(p, z) is specified in eqs. 4.12 – 4.20.

The vertex expressions remain identically the same as in the conventional Standard Model. All integrations are done in the y-coordinate space as shown in the perturbation discussion leading to eq. 6.37 and the perturbation theory calculations at the end of chapter 6 and in chapter 7.

Procedure to Construct the Two-Tier Equivalent of a Conventional Feynman Diagram Expression

After we form the expression for a Feynman diagram in a conventional Standard Model variant, we can construct the Two-Tier equivalent by taking *every* particle or ghost ghost propagator factor of the form:

$$G\ldots(p)$$

in momentum space, and replacing it with

$$G\ldots^{TT}(p) = \int d^4z\, e^{ip\cdot z}\, G\ldots^{TT}(z)$$

$$= \int d^4z\, e^{ip\cdot z} \int d^4p'\, e^{-ip'\cdot z}\, G\ldots(p')R(p', z)/(2\pi)^4$$

The resulting expression is the Two-Tier expression for the Feynman diagram.

"Low Energy" Behavior of the Two-Tier Standard Model

Since $R(p, z) \cong 1$ for $p \ll M_c$ the Two-Tier Standard Model is virtually identical to the conventional, corresponding Standard Model variant for energies much less than M_c. Since M_c can be very large – perhaps equal to the Planck mass or larger, the Two-Tier Standard Model predictions at current energies can be made arbitrarily close to the "low energy" results of the corresponding Standard Model. The Standard Model is a limiting case of the Two-Tier Standard Model – thus implementing a *Correspondence Principle.*

"High Energy" Behavior of the Two-Tier Standard Model

At high energies (short distances) where p is of the order of M_c or larger, R(p, q) provides a Gaussian damping factor that makes all perturbation theory calculations ultra-violet finite. This has been shown in exhaustive detail for fermions and bosons in earlier chapters.

Massive Vector Bosons – No Divergence Problems

The only new case is the case of massive vector bosons. The propagator has the general form:

$$D_{VM}(k)_{\mu\nu} = -(g_{\mu\nu} - k_\mu k_\nu/m^2)(k^2 - m^2 + i\varepsilon)^{-1} \tag{8.4}$$

in momentum space in conventional quantum field theory modulo internal symmetry indices.

The Two-Tier quantum field theory equivalent is:

$$D_{VM}{}^{TT}(z)_{\mu\nu} = \int d^4k\, e^{-ip\cdot z}\, D_{VM}(k)_{\mu\nu} R(k, z)/(2\pi)^4 \quad (8.5)$$

in y-coordinate space. The leading momentum space behavior at high-energy of $D_{VM}{}^{TT}$ is generated by the $g_{\mu\nu}$ term (unlike the situation in conventional quantum field theory):

$$D_{VM}{}^{TT}(k)_{\mu\nu} \sim g_{\mu\nu} k^{-6} \quad (8.6)$$

as in eq. 6.101 for the massless vector boson case. Thus a Higgs mechanism is not needed to give vector bosons mass while maintaining renormalizability in the Electroweak sector.

Parenthetically we note that the Two-Tier version of the ghost propagators appearing in the gauge theory sectors is also handled by Rule 1. These propagators have the same high energy leading momentum behavior as massless scalar bosons, k^{-6}.

Thus the leading high-energy behavior of all Two-Tier propagators of Standard Model particles and ghosts is a high negative power of momentum. As a result all perturbative calculations are ultra-violet finite.

Path Integral Formulation

Until now we have been viewing quantum field theory calculations in terms of conventional, Feynman diagram-based, perturbation theory. The appearance of a number of internal symmetries in the Standard Model (usually SU(2)⊗SU(1)⊗SU(3)) that are implemented with Yang-Mills gauge fields often makes it convenient to use a path integral formalism. Since there are many excellent introductions to the path integral formalism in relation to Yang-Mills fields and the Standard Model we will confine our discussion to aspects specifically related to the Two-Tier formalism for quantum field theory that we have been developing.

We will consider the case of a Two-Tier scalar particle quantum field theory initially for the sake of simplicity, and then consider Two-Tier Yang-Mills gauge field theories. The primary quantity in the path integral formulation is

$$Z(J) = \langle 0^+ | 0^- \rangle \quad (8.8)$$

$$= N \int D^4Y D\phi \exp\{ i\int d^4y\, [\mathscr{L} + j_\mu(y)Y^\mu(y) + J(y)\phi(X)]\} \quad (8.9)$$

$$= \langle 0 | T\big(\exp\{i\int d^4y\, [\mathscr{L} + j_\mu(y)Y^\mu(y) + J(y)\phi(X)]\}\big) | 0 \rangle \quad (8.10)$$

with X defined by eq. 3.12. Eq. 8.9 expresses the path integral as a product of integrals over classical field points, and eq. 8.10 provides an operator formulation of the path integral. N is a normalization factor. (As earlier we use units where $\hbar = 1$.) We note that $<0^+|0^->$ is the probability amplitude that the vacuum state at $y^0 = -\infty$ transitions to the vacuum state at $y^0 = +\infty$. The external sources $j_\mu(y)$ and $J(y)$ are arbitrary c-number functions of the respective variables with the restriction: as $y^0 \equiv Y^0 \to \pm\infty$ (in the Y Coulomb gauge that we are using) both external sources approach zero:

$$\lim_{y0\to\pm\infty} j_\mu(y) = \lim_{y0\to\pm\infty} J(y) = 0 \qquad (8.11)$$

The functional derivatives satisfy

$$\delta J(y_1)/\delta J(y_2) = \delta^4(y_1 - y_2) \qquad (8.12)$$

and

$$\delta j_\mu(y_1)/\delta j^\nu(y_2) = g_{\mu\nu}\delta^4(y_1 - y_2) \qquad (8.13)$$

General Procedure

In the general case we will assume that the Lagrangian (with external sources added) has the form:

$$\mathscr{L} = \mathcal{J}(Y, y)[\mathscr{L}_{Fint} + \sum_i \mathscr{L}_{F0i}] + \mathscr{L}_C \qquad (8.14)$$

where $\mathscr{L}_{Fint}$ is the "interaction Lagrangian" that contains all non-quadratic field operator product terms, $\mathscr{L}_{F0i}$ is the "free field" Lagrangian for particle species i, $\mathscr{L}_C$ is the (free) Lagrangian for the Y^μ abelian gauge field described in detail in earlier chapters, and $\mathcal{J}(Y, y)$ is the Jacobian for the transformation of y^μ coordinates to X^μ coordinates. (We now use script $\mathcal{J}$ for the Jacobian because the sources use the symbol J.) $\mathscr{L}_{Fint}$ and $\mathscr{L}_{F0i}$ are functions of the fields corresponding to the physical particles of the theory. Each field is solely a function of X^μ and all derivatives are with respect to X^μ as developed in the Two-Tier models of preceding chapters:

$$X_\mu(y) = y_\mu + i\, Y_\mu(y)/M_c^2 \qquad (3.12)$$

In the present discussion we will deal directly with the Y^μ field and use eq. 3.12 as a notational convenience. We begin with eqs. 8.9 and 8.14 for the case of n scalar Klein-Gordon particles, which we write as:

$$Z(J) = N\,\{\exp\{\, i\int d^4y\, \mathcal{J}(-i\delta/\delta j^\nu(y)\,,\, y)\mathcal{L}_{Fint}(-i\,\delta/\delta J_1(y),\, \ldots,\, -i\,\delta/\delta J_n(y))\}\cdot$$

$$\cdot\prod_k [\,\int D\phi_k \exp\{\, i\int d^4y\, [\mathcal{J}(-i\delta/\delta j^\nu(y),\, y)\mathcal{L}_{F0k} + J_k(y)\phi_k(X)]\}\,\Big|_{j_\mu = 0}\,]\cdot$$

$$\cdot\int D^4Y \exp\{\, i\int d^4y[\mathcal{L}_C + j_\mu(y)Y^\mu(y)]\}\}\,\Big|_{j_\mu = 0} \tag{8.15}$$

using functional derivatives with respect to the sources. We note that we set $j^\mu = 0$ *after* evaluating each of the n free field factors in eq. 8.15. The physical justification for this procedure is that we wish the free field propagators to be truly independent free field propagators and not depend on external sources, and not be convoluted together via the Y^μ field. The result will be a simpler, physically reasonable, perturbation theory in which each ϕ particle is truly a free field (independent of the other ϕ fields) that emits Y quanta and then absorbs all its emitted Y quanta. (An *alternative, different*, theory would allow a ϕ particle to emit a Y quantum that would then be absorbed by a different ϕ particle – thus ϕ particles could interact via the exchange of Y quanta. The exchange of Y quanta would make these seemingly free fields actually interacting fields – contrary to our starting asumption that the $\mathcal{L}_{F0i}$ terms each define a free field. This type of theory would also be a calculational nightmare since a ϕ particle emits an infinite superposition of Y quanta. Thus the lowest order diagram in ϕ^4 perturbation would have an infinite number of terms. We therefore will not consider this possibility in the hope that Nature opted for our Two-Tier theory.)

We will now evaluate eq. 8.15 in stages since a number of novel features appear in its evaluation although the path integral evaluations themselves are conventional. We start by performing the Y integration. The Y field is a free abelian gauge field. We choose the same Coulomb gauge as in previous chapters $\nabla\cdot \mathbf{Y} = 0$ and $Y^0 = 0$. Consider

$$Z_Y(j_\mu) = \int D^4Y \exp\{\, i\int d^4y[\mathcal{L}_C + j_\mu(y)Y^\mu(y)]\} \tag{8.16}$$

with $\mathcal{L}_C$ given by eq. 3.15. After some manipulations we find

$$Z_Y(j_\mu) = \exp\{\, -i\int d^4y_1 d^4y_2\, j^i(y_1)D_{Fij}(y_1 - y_2)j^j(y_2)/2\} \tag{8.17}$$

with

$$D_{Fij}(y_1 - y_2) = \int d^4k\, e^{-ik\cdot(y_1 - y_2)}\, (\delta_{ij} - k_i k_j/\mathbf{k}^2)/[(2\pi)^4(k^2 + i\varepsilon)] \tag{8.18}$$

the spatial components of the massless vector boson Feynman propagator in the Coulomb gauge.

We now evaluate one of the factors in the exponentiated n free boson Lagrangian:

$$Z_{\phi_k}(J_k) = [\int D\phi_k \exp\{\, i\int d^4y\, [\mathcal{J}(-i\delta/\delta j^\nu(y), y)\mathcal{L}_{F0k} + J_k(y)\phi_k(X)]\}\, Z_Y(j_\mu)]\,|_{j_\mu = 0} \tag{8.19}$$

The free Klein-Gordon Lagrangian is:

$$\mathcal{L}_{F0k} = \tfrac{1}{2}\,[\,(\partial\phi_k/\partial X^\nu)^2 - m_k^{\,2}\phi_k^{\,2}] \tag{8.20}$$

As a result the classical action can be written as

$$S_{0k}[\phi_k] = \int d^4y\, \mathcal{J}(-i\delta/\delta j^\nu(y), y)\mathcal{L}_{F0k} \tag{8.21}$$

$$= -\tfrac{1}{2}\int d^4X_{op}\; \phi_k(X_{op})(\Box_X + m_k^{\,2})\phi_k(X_{op}) \tag{8.22}$$

under the usual assumptions of good behavior at infinity and utilizing the Jacobian factor with

$$X_{op\mu}(y) = y_\mu + M_c^{-2}\,\delta/\delta j^\mu(y) \tag{8.23}$$

and with

$$\Box_X = (\partial/\partial X_{op}^{\;\nu})(\partial/\partial X_{op\nu}) \tag{8.24}$$

Inserting eq. 8.21 in eq. 8.19 we obtain:

$$Z_{\phi_k}(J_k) = [\int D\phi_k \exp\{i\int d^4X_{op}[-\tfrac{1}{2}\phi_k(X_{op})(\Box_X + m_k^{\,2})\phi_k(X_{op}) +$$

$$+ J_k(y)\phi_k(y)/\mathcal{J}(-i\delta/\delta j^\nu(y), y)]\}\, Z_Y(j_\mu)]\,|_{j_\mu = 0} \tag{8.25}$$

The (Gaussian in ϕ_k) path integral in eq. 8.25 can be performed (see below) using standard techniques to yield:

$$Z_{\phi_k}(J_k) = \Big[\exp\{-i\int d^4X_{op}(y_1)d^4X_{op}(y_2)[J_k(y_1)/\mathcal{J}(-i\delta/\delta j^\nu(y_1), y_1)]\cdot$$

$$\cdot\Delta_{Fk}^{TT}(y_1 - y_2)[J_k(y_2)/\mathcal{J}(-i\delta/\delta j^\nu(y_2), y_2)]/2\}\, Z_Y(j_\mu)\Big]\Big|_{j_\mu = 0}$$

$$= \exp\{-i\int d^4y_1 d^4y_2 J_k(y_1)\Delta_{Fk}^{TT}(y_1 - y_2)J_k(y_2)/2\} \tag{8.26}$$

after changing the integration variables in the exponential using the Jacobian with $\Delta_{Fk}^{TT}(y_1 - y_2)$ given by eqs. 6.51-52. The added index k serves to identify the propagator $\Delta_{Fk}^{TT}(y_1 - y_2)$ as having the mass m_k.

The derivation of eq. 8.25 uses functional techniques in a straightforward way. We can derive the form of $\Delta_{Fk}^{TT}(y_1 - y_2)$ best in momentum space. The operator $(\Box_X + m_k^2)^{-1}$ can be represented via

$$(\Box_X + m_k^2)^{-1}g(X_{op}(y_1))Z_Y(j_\mu) = (\Box_X + m_k^2)^{-1}\int d^4p\{e^{-ip\cdot X_{op}(y_1)}g(p)/(2\pi)^4\}Z_Y(j_\mu)$$

$$= -\int d^4p\{e^{-ip\cdot X_{op}(y_1)}g(p)/[(2\pi)^4(p^2 - m_k^2 + i\varepsilon)]\}Z_Y(j_\mu)$$

$$= -\int d^4X_{op}(y_2)\Delta_{Fkop}^{TT}(X_{op}(y_1) - X_{op}(y_2))g(X_{op}(y_2))Z_Y(j_\mu)$$

$$= -\int d^4X_{op}(y_2)\Delta_{Fkop}^{TT}(X_{op}(y_1) - X_{op}(y_2))Z_Y(j_\mu)g(X(y_2))$$

We now calculate

$$\Delta_{Fkop}^{TT}(X_{op}(y_1) - X_{op}(y_2))Z_Y(j_\mu) = \int d^4p\ \{e^{-ip\cdot(y_1 - y_2)}/[(2\pi)^4(p^2 - m_k^2 + i\varepsilon)]\}\cdot$$

$$\cdot\exp(-ip^\nu[\delta/\delta j^\nu(y_1) - \delta/\delta j^\nu(y_2)]/M_c^2)Z_Y(j_\mu)$$

$$= Z_Y(j_\mu)\int d^4p\ \{e^{-ip\cdot(y_1 - y_2)}/[(2\pi)^4\,(p^2 - m_k^2 + i\varepsilon)]\}\cdot$$

$$\cdot \exp(-ip^ip^j[D_{Fij}(y_1 - y_2) + D_{Fij}(y_2 - y_1)]/(2M_c^4)) \tag{8.27}$$

where $D_{ij}(y_1 - y_2)$ is given by eq. 8.18. The last factor in eq. 8.27 is the Gaussian damping factor that appears in Two-Tier propagators as repeatedly seen earlier – for example in eq. 6.53 and 6.54:

$$R(\mathbf{p}, y_1 - y_2) = \exp(-ip^ip^j[D_{Fij}(y_1 - y_2) + D_{Fij}(y_2 - y_1)]/(2M_c^4)) \tag{8.28}$$

Eq. 8.26 then follows with the Two-Tier scalar Klein-Gordon Feynman propagator.

Therefore the n scalar Klein-Gordon particle path integral of eq. 8.15 simplifies to

$$Z(J) = N\,\{\exp\{\,i\int d^4y\,\mathcal{J}(-i\delta/\delta j^\nu(y), y)\,\mathscr{L}_{Fint}(-i\delta/\delta J_1(y), \ldots, -i\delta/\delta J_n(y))\}\cdot$$

$$\cdot\exp\{-\sum_k[\,i\int d^4y_1d^4y_2J_k(y_1)\Delta_{Fk}^{TT}(y_1 - y_2)J_k(y_2)/2]\}\,Z_Y(j_\mu)\,\}\Big|_{j_\mu = 0} \tag{8.29}$$

The only dependence on functional derivatives with respect to j^μ in eq. 8.29 is in the Jacobian of the interaction Lagrangian term. It is readily seen that the Jacobian effectively reduces to one if there is no external physical source for Y particles (as we have assumed.) Consider

$$\begin{aligned}\mathcal{J}(-i\delta/\delta j^\nu(y),y)Z_Y(j_\mu) = \{&\varepsilon^{ijk}(\delta_{1i} + M_C^{-2}(\partial/\partial y^i)\delta/\delta j^1(y))\cdot\\ &\cdot(\delta_{2j} + M_C^{-2}(\partial/\partial y^j)\,\delta/\delta j^2(y))\cdot\\ &\cdot(\delta_{3k} + M_C^{-2}(\partial/\partial y^k)\,\delta/\delta j^3(y))Z_Y(j_\mu)\}\end{aligned} \tag{8.30a}$$

Since the interaction Hamiltonian – including the q-number Jacobian – is normal-ordered (eq. 6.8) the functional derivatives in eq. 8.30a can only apply to $Z_Y(j_\mu)$. Therefore

$$\begin{aligned}\mathcal{J}(-i\delta/\delta j^\nu(y),y)Z_Y(j_\mu) = Z_Y(j_\mu)\varepsilon^{ijk}\big(&\delta_{1i} + M_C^{-2}(\partial/\partial y^i)(-i/2)[\int d^4y_2D_{F1a}(y - y_2)j^a(y_2) +\\ &+ \int d^4y_1\,j^a(y_1)D_{Fa1}(y_1 - y)]\big)\cdot\big(\delta_{2j} + M_C^{-2}(\partial/\partial y^j)\,(-i/2)[\int d^4y_2\,D_{F2a}(y - y_2)j^a(y_2) +\\ &+ \int d^4y_1\,j^a(y_1)D_{Fa2}(y_1 - y)]\big)\cdot\big(\delta_{3k} + M_C^{-2}(\partial/\partial y^k)\,(-i/2)[\int d^4y_2\,D_{F3a}(y - y_2)j^a(y_2) +\end{aligned}$$

$$+ \int d^4y_1\, j^a(y_1) D_{Fa3}(y_1 - y)]\Big) \qquad (8.30a)$$

After setting $j_\mu = 0$ *for the case of no incoming or outgoing Y quanta* we see

$$\mathcal{J}(-i\delta/\delta j^\nu(y), y) \equiv 1 \qquad (8.30b)$$

This result is consistent with our physical notion that the integration, in itself, does not generate dynamical effects. *Otherwise* a c-number interaction Lagrangian term would generate a non-trivial interacting quantum field theory effects – contrary to our physical expectations. Eq. 8.30b eliminates this physically unreasonable possibility.

Thus eq. 8.29 becomes

$$Z(J) = N \exp\{ i\int d^4y\, \mathscr{L}_{Fint}(-i\delta/\delta J_1(y), \ldots, -i\delta/\delta J_n(y))\}\cdot$$

$$\cdot\exp\{-i\int d^4y_1 d^4y_2 \sum_k J_k(y_1)\Delta_{Fk}{}^{TT}(y_1 - y_2)J_k(y_2)/2]\} \qquad (8.31)$$

which gives the same perturbation theory found earlier using canonical quantum field theory that is built on the U matrix expansion.

Derivative Coupling Case

Eq. 8.31 is based on the assumption that derivative couplings do not appear in the interaction Lagrangian. (There are no derivative couplings in the Standard Model but there are derivative couplings in quantum gravity so we must deal with this issue if we wish to create a unified theory.) If the interaction Lagrangian does contain derivatives of scalar fields with respect to the X variable then the interaction Lagrangian has a superficial dependence on the functional derivative with respect to j^μ, which we can symbolize in the modified path integral expression:

$$Z(J) = N \Big\{\exp\{ i\int d^4y\, \mathscr{L}_{Fint}(\partial/\partial(y^\nu + M_c^{-2}\delta/\delta j_\nu(y)), -i\delta/\delta J_1(y), \ldots, -i\delta/\delta J_n(y))\}\cdot$$

$$\cdot\exp\{-\sum_k [\, i\int d^4y_1 d^4y_2 J_k(y_1)\Delta_{Fk}{}^{TT}(y_1 - y_2)J_k(y_2)/2]\}\ Z_Y(j_\mu)\Big\}\Big|_{j_\mu = 0} \qquad (8.32)$$

obtained from eq. 8.31.

The j^μ dependence now appears only in the interaction Lagrangian. For good reason we will now show the j^μ dependence of the interaction Lagrangian in eq. 8.32 can be eliminated.

Our approach will be to separate the coordinate dependence in the propagator into two parts: the coordinate dependence in the Gaussian factor, and the coordinate dependence in the $e^{ip\cdot x}$ factor. We can then express the derivatives of fields in the interaction lagrangian as derivatives with respect to the coordinates in the $e^{ip\cdot x}$ factor appearing in integral representations of the Two-Tier propagator when a perturbative expansion of the path integral solution is made.

We begin by noting that

$$i\Delta_{Fk}^{TT}(y_1 - y_2) = <0|T(\phi(X(y_1)),\phi(X(y_2)))|0> \tag{6.51}$$

$$= i \int \frac{d^4p\, e^{-ip\cdot(y_1 - y_2)} R(\mathbf{p}, y_1 - y_2)}{(2\pi)^4 (p^2 - m_k^2 + i\varepsilon)} \tag{6.52}$$

with

$$R(\mathbf{p}, y_1 - y_2) = \exp[-p^i p^j \Delta_{Tij}(y_1 - y_2)/M_c^4] \tag{6.53}$$

(summations are over space indices only in the Y Coulomb gauge) and

$$\Delta_{Tij}(z) = \int d^3k\, e^{-ik\cdot z} (\delta_{ij} - k_i k_j/\mathbf{k}^2)/[(2\pi)^3 2\omega_k] \tag{6.54}$$

We now define *a more general* Two-Tier propagator by introducing a distinction between the spatial dependence of the gaussian and exponential terms:

$$i\Delta_{Fk}^{TT}(y_1 - y_2, z) = i \int \frac{d^4p\, e^{-ip\cdot(y_1 - y_2)} R(\mathbf{p}, z)}{(2\pi)^4 (p^2 - m_k^2 + i\varepsilon)} \tag{8.33}$$

We then note that

$$\partial i\Delta_{Fk}^{TT}(y_1 - y_2)/\partial X^\mu = \partial i\Delta_{Fk}^{TT}(y_1 - y_2, z)/\partial y_1^\mu |_{z=y_1-y_2} \tag{8.34}$$

comparing eqs. 6.51 and 6.52 with eq. 8.33. As a result we can write eq. 8.32 symbolically as:

$$Z(J) = N\{\exp\{ i\int d^4y\, \mathscr{L}_{Fint}(\partial/\partial y^\nu, -i\delta/\delta J_1(y), \ldots, -i\delta/\delta J_n(y))\}\cdot$$

$$\cdot\exp\{-\sum_k [\, i\int d^4y_1 d^4y_2 J_k(y_1)\Delta_{Fk}^{TT}(y_1 - y_2, z)J_k(y_2)/2]\}\}\big|_{z=y_1-y_2} \tag{8.35}$$

Eq. 8.35 is interpreted as the following:

1. For a given process take appropriate functional derivatives of $Z(J)$ with respect to $J_1(y), \ldots, J_n(y)$.

2. Then expand the exponential factors in a perturbation series applying any derivatives with respect to y in $\mathscr{L}_{Fint}$. Do not perform any of the $\int d^4y_1 d^4y_2$ integrals.

3. Then set $z = y_1 - y_2$ in each $\Delta_{Fk}^{TT}(y_1 - y_2, z)$ propagator.

4. Lastly perform all $\int d^4y_1 d^4y_2$ integrals.

Thus we thus achieve a path integral formulation that is very similar to the corresponding expression in conventional field theory – the only difference is the form of the free field propagators, which now contain a Gaussian factor.

ϕ^4 Theory Path Integral Formulation Example

We will now consider the case of Two-Tier scalar field theory using the specific example of the ϕ^4 Lagrangian of eqs. 5.20 – 5.24:

$$\mathscr{L}_{Fint}(X^\mu) = \tfrac{1}{4!}\, \chi_0\, \phi(X)^4 + \tfrac{1}{2}\,(m^2 - m_0^2)\phi(X)^2 \tag{8.36}$$

with

$$\mathscr{L}_{F0} = \tfrac{1}{2}\,[\,(\partial\phi/\partial X^\nu)^2 - m^2\phi^2] \tag{8.37}$$

In this theory eq. 8.31 can be written as:

$$Z(J) = N\exp\{i\int d^4y\, \mathscr{L}_{Fint}(-i\delta/\delta J(y))\}\, \exp\{-i\int d^4y_1 d^4y_2 J(y_1)\Delta_{Fk}^{TT}(y_1 - y_2)J(y_2)/2\} \tag{8.38}$$

The perturbation theory generated from eq. 8.38 is the same as conventional perturbation theory except for the differing free field propagators.

Two-Tier Yang-Mills Gauge Fields

Two-Tier Yang-Mills gauge field theories have many similarities to conventional Yang-Mills theories.[38] We assume the reader in familiar with internal symmetries and the conventional Yang-Mills formulation.

General Rule: All gauge fields and derivatives of gauge fields, as well as group properties and other features such as the Faddeev-Popov method, are expressed solely in terms of the X coordinate system (which in turn is a function of the y coordinates).

We note all matter fields are assumed to be functions of X coordinates only as done in previous chapters. The general rule is implemented by defining:

1. The covariant derivative of any matter field $\Psi(X)$ by

$$D^{\mu}\Psi(X) = [\partial/\partial X_{\mu} + igA^{\mu}(X)]\Psi(X) \tag{8.39}$$

with $A^{\mu}(X)$ being an element of a Lie algebra:

$$A^{\mu}(X) = A_a^{\ \mu}(X)L_a \tag{8.40}$$

where L_a is a generator of a Lie algebra (a and b are internal symmetry indexes) with commutation relations:

$$[L_a, L_b] = ic_{ab}^{\ \ c}L_c \tag{8.41}$$

with $c_{ab}^{\ \ c}$ being real numbers called the *structure constants* of the Lie algebra.

2. The field strengths are defined as the commutator of covariant derivatives:

$$F^{\mu\nu} = [D^{\mu},D^{\nu}] = \partial A^{\nu}(X)/\partial X_{\mu} - \partial A^{\mu}(X)/\partial X_{\nu} + ig[A^{\mu}(X), A^{\nu}(X)] \tag{8.42}$$

3. The Lagrangian for a Yang-Mills gauge field interacting with a matter field $\psi(X)$ has the form:

$$\mathscr{L}_{YM} = [\mathscr{L}_F^{\ YM}(X^{\mu}) + \mathscr{L}_F^{\ Matter}(X^{\mu})]\, J + \mathscr{L}_C(X^{\mu}(y), \partial X^{\mu}(y)/\partial y^{\nu}) \tag{8.43}$$

where J is the Jacobian and

[38] C. N. Yang and R. L. Mills, Phys. Rev. **96**, 191 (1954).

$$\mathcal{L}_F^{YM}(X^\mu) = -\tfrac{1}{4}\, F_a^{\mu\nu}\, F_{a\mu\nu} \tag{8.44}$$

with $F^{\mu\nu} = F_a^{\mu\nu} L_a$ and

$$\mathcal{L}_F^{Matter}(X^\mu) = \mathcal{L}_F^{Matter}(\psi(X), D^\mu\psi(X)) \tag{8.45}$$

and $\mathcal{L}_C$ is defined as previously and specifies the Y field evolution.

The generalization to multiple gauge fields interacting with multiple matter fields is direct. The overall form of the Standard Model Lagrangian is:

$$\begin{aligned} L = \int d^4y\ \{J[\mathcal{L}_F^{Matter}(X^\mu) + \mathcal{L}_F^{GaugeFields}(X^\mu) + \mathcal{L}_F^{Higgs}(X^\mu)] + \\ + \mathcal{L}_C(X^\mu(y), \partial X^\mu(y)/\partial y^\nu)\} \end{aligned} \tag{8.46}$$

where J is the Jacobian for the transformation from y to X coordinate integration. Eq. 8.46 can be rewritten as

$$\begin{aligned} L = \int d^4X[\mathcal{L}_F^{Matter}(X^\mu) + \mathcal{L}_F^{GaugeFields}(X^\mu) + \mathcal{L}_F^{Higgs}(X^\mu)] + \\ + \int d^4y\,\mathcal{L}_C(X^\mu(y), \partial X^\mu(y)/\partial y^\nu) \end{aligned} \tag{8.47}$$

It is clear from eq. 8.47 that the conventional Standard Model equations of motion and canonical quantization procedure emerge in the Two-Tier formulation if all expressions are written as functions of the X coordinates as shown previously in our discussions of separable Lagrangians. The second quantization of X as a function of the y coordinates leads to the gaussian factor in all free particle propagators (except the Y propagator).

Thus the gauge field and matter field parts of the Lagrangian, the gauge field transformation laws and other related operations solely depend on the X coordinate system as stated in the general rule above. The quantization and field equations are the same as the conventional case – except that they are specified solely in terms of the X coordinate system.

Path Integral Formulation and Faddeev-Popov Method

We now turn to the Two-Tier path integral formulation of Yang-Mills gauge theories and in particular to the Two-Tier version of the Faddeev-Popov method. The Two-Tier path integral for a gauge field can be written symbolically as:

$$Z(J^\mu) = N \int DADY\Delta_{FP}(A)\delta(F(A))\ \exp\{i\int d^4y\ [\mathcal{L} +$$

$$+\ j_\mu(y)Y^\mu(y)+J^\mu(y)A_\mu(X)]\}\,\Big|_{j_\mu=0} \tag{8.48}$$

where $\delta(F(A))$ specifies the gauge and $\Delta_{FP}(A)$ is the Faddeev-Popov determinant. The Lagrangian is

$$\mathscr{L} = J\mathscr{L}_F^{YM}(X^\mu) + \mathscr{L}_C(X^\mu(y), \partial X^\mu(y)/\partial y^\nu) \tag{8.49}$$

with $\mathscr{L}_F^{YM}$ specified by eq. 8.44 and J the Jacobian for the transformation from X coordinates to y coordinates. The Faddeev-Popov determinant may be calculated in the standard way. First we note that the delta function fixing the gauge can be written as a delta function in the gauge times a determinant:

$$\delta(F(A^\omega)) = \delta(\omega - \omega_0)\,\Big|\det \delta F(A_\mu{}^\omega(X))/\delta\omega(X)\,\Big|^{-1}\,\Big|_{F(A)=0} \tag{8.50}$$

where ω_0 is a reference gauge, where

$$A_\mu{}^\omega(X) = A_\mu(X) + \partial\omega(X)/\partial X^\mu \tag{8.51}$$

and where

$$\Delta_{FP}(A) = \Big|\det \delta F(A_\mu{}^\omega(X))/\delta\omega(X)\,\Big|\,\Big|_{F(A)=0} \tag{8.52}$$

We will choose the Lorentz gauge to evaluate the Faddeev-Popov determinant:

$$F_a(A) = \partial A_a{}^\mu(X)/\partial X^\mu = 0 \tag{8.53}$$

Under an infinitesimal gauge transformation:

$$A_{a\mu}{}^\omega(X) = A_{a\mu}(X) + g^{-1}\partial\omega_a/\partial X^\mu + c_{ab}{}^c\,\omega_b(X)A_c{}^\mu(X) \tag{8.54}$$

we find

$$F_a(A_\mu{}^\omega(X)) = \partial(A_{a\mu}(X) + g^{-1}\partial\omega_a(X)/\partial X^\mu + c_{ab}{}^c\,\omega_b(X)A_c{}^\mu(X))/\partial X^\mu$$

$$= g^{-1}\,\Box_X\,\omega_a(X) + c_{ab}{}^c\,\partial\omega_b(X)/\partial X^\mu\ A_c{}^\mu(X) \tag{8.55}$$

Thus

$$\delta F_a(A_\mu{}^\omega(X))/\delta\omega_b(X) = g^{-1}\,\delta_{ab}\Box_X + c_{ab}{}^c A_c{}^\mu(X)\partial/\partial X^\mu \qquad (8.56)$$

and

$$\Delta_{FP}(A) = \left|\det\,(g^{-1}\,\delta_{ab}\Box_X + c_{ab}{}^c A_c{}^\mu(X)\partial/\partial X^\mu)\right|\Big|_{F(A)=0} \qquad (8.57)$$

where | ... | represent absolute value.

We note the Two-Tier Faddeev-Popov determinant is solely a function of the X coordinates. Thus we can follow the standard procedure and rewrite the determinant as a path integral over anti-commuting c-number fields with a ghost Lagrangian that is solely a function of X – just like the gauge particles and other particles in Two-Tier theories:

$$\Delta_{FP}(A) = \int D\chi^* D\chi \exp[\,i\int d^4X\, \mathscr{L}^{ghost}(X^\mu)] \qquad (8.58)$$

where

$$\mathscr{L}^{ghost}(X^\mu) = \chi_a{}^*(X)[\delta_{ab}\Box_X + g\, c_{ab}{}^c A_c{}^\mu(X)\partial/\partial X^\mu]\chi_b(X) \qquad (8.59)$$

Thus the complete path integral is

$$Z(J^\mu) = N\int DAD\chi^* D\chi DY \delta(F(A)) \exp\{i\int d^4y\, [\mathscr{L} +$$

$$+\, j_\mu(y)Y^\mu(y) + J^\mu(y)A_\mu(X)]\}\,|_{j_\mu=0} \qquad (8.60)$$

where $\delta(F(A))$ specifies the gauge and

$$\mathscr{L} = J\mathscr{L}_F{}^{YM}(X^\mu) + J\mathscr{L}^{ghost}(X^\mu) + \mathscr{L}_C(X^\mu(y), \partial X^\mu(y)/\partial y^\nu) \qquad (8.61)$$

with $\mathscr{L}_F^{YM}$ specified by eq. 8.44.

At this point it is obvious that we can follow almost identical steps as we did in the scalar particle case starting from eq. 8.9 and obtain an expression similar to eq. 8.38. The result is a perturbation theory for the Yang-Mills gauge field that is identical to the usual theory except that the free propagators for the gauge fields, *and the ghost fields*, acquire the gaussian factor R(p,z) as stated earlier in the discussions of eqs. 6.52, 6.88, 6.95, 6.102, and 7.120-123.

Two-Tier Massive Vector Fields

Massive vector fields have been a problem for perturbation theory due to the $k_\mu k_\nu/m^2$ term appearing in the free field propagator. This term makes conventional interacting quantum field theories of massive vector bosons non-renormalizable. The Higgs mechanism is used in Electroweak theory to evade the non-renormalizability of massive gauge fields. It gives mass to the vector bosons mediating the weak force while maintaining the renormalizability of the theory. It also implements symmetry breaking.

In Two-Tier quantum field theory a massive vector boson does not create renormalization issues. For example the Two-Tier version of the Weinberg-Salam model of the electromagnetic and weak forces (with massive vector bosons) is finite! Thus the need for spontaneous symmetry breaking to give mass to vector bosons in Electroweak theory is not present. Higgs particles are not needed (although they may exist. Their existence is an experimental question – not a theoretical necessity for a Two-Tier Electroweak theory with massive vector bosons.)

Massive Vector Boson Propagator

The massive free vector particle propagator in <u>*conventional*</u> quantum field theory has the representation:

$$i\Delta_{FV}(y_1 - y_2)_{\mu\nu} = -i \int \frac{d^4k\, e^{-ik\cdot(y_1 - y_2)}\,(g_{\mu\nu} - k_\mu k_\nu/m^2)}{(2\pi)^4\,(k^2 - m^2 + i\varepsilon)} \qquad (8.62)$$

A Two-Tier massive vector boson theory can be constructed in a straightforward way from the Lagrangian:

$$\mathscr{L} = J[-\tfrac{1}{4} F_V^{\mu\nu}(X(y))F_{V\mu\nu}(X(y)) - \tfrac{1}{2} m^2 V^\mu V_\mu] + \mathscr{L}_C(X^\mu(y), \partial X^\mu(y)/\partial y^\nu, y) \qquad (8.63)$$

with J the Jacobian and with

$$F_{V\mu\nu}(X(y)) = \partial V_\mu(X(y))/\partial X^\nu - \partial V_\nu(X(y))\,/\partial X^\mu \qquad (8.64)$$

and the usual Two-Tier Y Lagrangian terms

$$\mathscr{L}_C(X^\mu(y), \partial X^\mu(y)/\partial y^\nu, y) = -\tfrac{1}{4} F_Y^{\mu\nu}F_{Y\mu\nu} \qquad (7.4)$$

$$F_{Y\mu\nu} = (\partial Y_\mu/\partial y^\nu - \partial Y_\nu\,/\partial y^\mu) \qquad (7.5)$$

Following steps similar to the previously considered scalar particle, fermion and massless vector particle (photon) cases we find the Two-Tier massive vector boson Feynman propagator to be:

$$i\Delta_{FV}^{TT}(y_1 - y_2)_{\mu\nu} = -i \int \frac{d^4k\, e^{-ik\cdot(y_1 - y_2)}\, (g_{\mu\nu} - k_\mu k_\nu/m^2)\, R(\mathbf{k}, y_1 - y_2)}{(2\pi)^4\, (k^2 - m^2 + i\varepsilon)} \qquad (8.65)$$

No Need for Higgs Mechanism in Electroweak Theory

The leading coordinate space dependence at high energy (short distance) of the fourier transform of Δ_{FV}^{TT} is the same as Δ_F^{TT} since it comes from the $g_{\mu\nu}$ term in Δ_F^{TT}

$$\Delta_{FV}^{TT}(y_1 - y_2)_{\mu\nu} \sim g_{\mu\nu}\, (y_1 - y_2)^2 \qquad (8.66)$$

which in momentum space is equivalent to

$$\Delta_{FV}^{TT}(p)_{\mu\nu} \sim g_{\mu\nu}\, p^{-6} \qquad (8.67)$$

The $k_\mu k_\nu/m^2$ term appearing in the free vector field Two-Tier propagator actually is of higher order in the large momentum limit (short distance):

$$\int \frac{d^4k\, e^{-ik\cdot(y_1 - y_2)} k_\mu k_\nu/m^2)\, R(\mathbf{k}, y_1 - y_2)}{(2\pi)^4\, (k^2 - m^2 + i\varepsilon)} \sim (y_1 - y_2)^4 \qquad (8.68)$$

which corresponds to momentum space behavior of

$$p^{-8} \qquad (8.69)$$

Two-Tier propagators such as the propagator in eq. 8.65 have the feature that the gaussian factor "inverts" the high-energy behavior of the terms in the numerator of the integrand: terms with higher powers of momentum are less significant at short distances. The term with the lowest power of momentum generates the leading behavior at high energy (short distances).

Thus Two-Tier massive vector boson theories are ultra-violet convergent and do not constitute a problem as they do in conventional quantum field theory. *Therefore there is no need for the Higgs mechanism in the Electroweak sector of the Standard*

Model in order to obtain a renormalizable theory. Ordinary massive vector bosons can be used and the resulting Two-Tier theory is finite!

General Short Distance (High Momentum) Behavior of Two-Tier Propagator

The higher the power of the momentum in the numerator of the integrand of a Two-Tier propagator, the more convergent the large momentum behavior of the fourier transform of the Two-Tier propagator.

The short distance behavior of a term with n factors of momentum in a Two-Tier propagator has the leading short distance coordinate space behavior:

$$\int \frac{d^4k\, e^{-ik\cdot(y_1 - y_2)}\, k_{\mu_1} k_{\mu_2} \ldots k_{\mu_n} R(\mathbf{k}, y_1 - y_2)}{(2\pi)^4 (k^2 - m^2 + i\varepsilon)} \sim (y_1 - y_2)^{2+n} \tag{8.70}$$

which corresponds to the high energy behavior:

$$p^{-6-n} \tag{8.71}$$

In contrast to conventional quantum field theory the more powers of momentum in the numerator of a Two-Tier propagator, the better the short distance behavior!

Higgs Particles

A previous section shows that the Higgs Mechanism is not needed in the Two-Tier Electroweak sector of the Standard Model. Massive vector bosons such as **W**'s would not make the Electroweak theory non-renormalizable. *As we have seen a Two-Tier Electroweak theory with massive vector bosons is finite.*

Nevertheless we would like to point out a Two-Tier version of the Higgs particle sector of the Standard Model can be defined that largely parallels the conventional treatment of the Higgs sector. Higgs particles continue to be of interest since they may play a role in the origin of particle masses and symmetry breaking.

The Two-Tier scalar Higgs field Lagrangian terms (plus $\mathscr{L}_C$) can be written:

$$\mathscr{L}_{\text{Higgs}} = J[D_\mu\phi^\dagger D^\mu\phi - V(\phi(X))] + \mathscr{L}_C \tag{8.72}$$

$$V(\phi(X)) = m_0^{\,2}\phi^\dagger(X(y))\phi(X(y)) + \lambda[\phi^\dagger(X(y))\phi(X(y))]^2 +$$

$$+ G_c[\bar{L}(X)\phi(X)R(X) + \bar{R}(X)\phi^\dagger(X)R(X)] \tag{8.73}$$

with the covariant derivative defined with the usual B_μ and $W_\mu^{\ a}$ gauge fields of the Standard Model, and L representing a left-handed fermion isodoublet, and R representing a right-handed fermion isosinglet. We note all items in eqs. 8.72 and 8.73 are written solely as functions of the X coordinates with the exception of $\mathscr{L}_C$, which is the Lagrangian term for the Y field.

The conventional effective potential method can be followed to implement the Higgs mechanism. In particular we may write

$$\phi(X(y)) = <\phi> + \eta(X) \tag{8.74}$$

with $<\phi>$ the vacuum expectation value, which is a constant by translational invariance. Then vector bosons can acquire mass via the Higgs mechanism. The quantum part of the Higgs field has a Two-Tier propagator with p^{-6} behavior for large momentum. Thus all sectors of the Standard Model wind up with Two-Tier propagators and all perturbative calculations are finite.

General Form of the Two-Tier Standard Model Path Integral

The general form of the path integral for the Two-Tier version of the Standard Model is:

$$Z(J) = N\,\{\exp\{\, i\int d^4y\ \mathscr{L}_{Fint}(\partial/\partial y^\nu, -i\delta/\delta J_1(y), \ldots, -i\delta/\delta J_n(y))\}\cdot$$

$$\cdot \exp\{-\sum_k [\, i\int d^4y_1 d^4y_2 J_k(y_1)\Delta_{Fk}^{TT}(y_1 - y_2, z)J_k(y_2)/2]\}\}\,|_{z=y1-y2} \tag{8.75}$$

where the sum over k is a sum over all matter fields, gauge fields, Higgs fields, and ghost fields. The index k, and indices on the functional derivatives, represents all space-time indices and internal symmetry indices that are relevant for each particle. We assume the total number of particle and ghost fields is n. Notice all dependence on the Y field has been "integrated away."

Also any derivatives with respect to y appearing in the interaction Lagrangian are applied to Two-Tier propagators using the method represented by eqs. 8.33 and 8.34. This procedure results in the same momentum polynomials for Feynman diagrams in a Two-Tier version as appear in the corresponding conventional theory. Thus we have the same algebraic structure (both in the momentum polynomials and internal symmetries) in the Two-Tier version of a conventional theory.

We note that the large distance behavior of the Two-Tier theory is the same as the Standard Model in all respects including gauge symmetries. Deviations from the conventional Standard Model results only appear at extremely high energies of the order of M_c.

Renormalization - Finite

Since all particle (and ghost) propagators are Two-Tier propagators in the Two-Tier Standard Model, the Two-Tier Standard Model yields finite results to all orders in perturbation theory. We note the large momentum behavior of the various types of particle propagators in the Two-Tier Standard Model is:

Fermion Propagators

$$p^{-7}$$

Vector Boson (Gauge Field) Propagators

$$p^{-6}$$

Ghost Propagators

$$p^{-6}$$

Higgs Particle Propagators

$$p^{-6}$$

Thus all Feynman diagrams are highly ultra-violet convergent and the Two-Tier Standard Model is finite. This result is independent of the details of the internal symmetries, particle spectrum, and particle masses.

Unitarity

The Two-Tier Standard Model *superficially* appears to have a unitarity problem due to the non-hermitean nature of its hamiltonian. The lack of hermiticity is due entirely to the appearance of iY^μ in the X^μ field coordinates.

Thus the Two-Tier Standard Model interaction hamiltonian is not hermitean:

$$H_{\mathrm{Fint}} = \int d^3y' \mathscr{H}_{\mathrm{Fint}}(y' + iY(y')/M_c^2) \tag{8.76}$$

and

$$H_{\mathrm{Fint}}^{\dagger} = \int d^3y' \mathscr{H}_{\mathrm{Fint}}(y' - iY(y')/M_c^2) \neq H_{\mathrm{Fint}} \tag{8.77}$$

The relation between H_{Fint} and its hermitean conjugate is

$$H_{Fint} = V\, H_{Fint}^{\dagger}\, V \qquad (8.78)$$

where $V^2 = I$ is the metric operator defined in eqs. 5.16 – 5.18. Eq. 8.78 implies that the Two-Tier Standard Model S matrix is not unitary. The Two-Tier Standard Model S matrix is pseudo-unitary:

$$S^{-1} = V\, S^{\dagger}\, V \qquad (8.79)$$

Therefore

$$S^{\dagger} V S = V \qquad (8.80)$$

The Two-Tier Standard Model S matrix satisfies unitarity between physical asymptotic states – states consisting of only physical particles: leptons, quarks, photons, W and Z particles, gluons, and Higgs particles (if they exist). Put another way: physical states can consist of any set of particles in the Two-Tier Standard Model except ghosts and Y particles. The proof is identical to eqs. 6.46 – 6.48.

Anomalies

The axial anomaly (Adler-Bell-Jackiw anomaly) follows from the linear divergence of a fermion triangle graph (Fig. 8.1) in the conventional Standard Model. All higher order terms are divergence-free. These terms do not contribute to the axial anomaly. Thus the axial anomaly can properly be regarded as an artifact of the regularization of the divergence of the fermion triangle diagram.

In Two-Tier theory the axial anomaly is not present. Fermion triangle diagrams in Two-Tier quantum field theories are finite. Thus the source of the anomaly in conventional theories is absent in Two-Tier theories.

A massless Dirac field theory is formally invariant under a chiral transformation implying a conserved axial-vector current. The Two-Tier axial-vector current is

$$j_5^{\mu}(X(y)) = \bar{\psi}(X(y))\gamma^{\mu}\gamma_5\psi(X(y)) \qquad (8.84)$$

with formal conservation law:

$$\partial j_5^{\mu}(X(y))/\partial X^{\mu} = 2m\ j_5(X(y)) = 2m\ \bar{\psi}(X(y))\gamma_5\psi(X(y)) \qquad (8.85)$$

Eq. 8.85 implies

$$\partial j_5^{\mu}(X(y))/\partial X^{\mu} = 0 \qquad (8.86)$$

in the limit m → 0. The question we now address is whether eq. 8.86 holds in Two-Tier perturbation theory – perhaps in the same form as the conventional axial anomaly:

$$\partial j_5^{\mu}(X(y))/\partial X^{\mu} = 2m\, j_5(X(y)) + a_0(4\pi)^{-1}\varepsilon^{\mu\nu\alpha\beta}F_{\alpha\beta}F_{\mu\nu} \quad ? \tag{8.87}$$

where a_0 is the unrenormalized fine structure constant.

The simplest manifestation of the axial anomaly in conventional field theory is the fermion triangle diagram, which we will now examine in Two-Tier quantum field theory. As stated earlier, the Two-Tier triangle diagram is finite and zero unlike the conventional quantum field theory result. *Thus the axial anomaly does not exist in Two-Tier quantum field theory. The axial anomaly is a result of the divergence of the triangle diagram in conventional quantum field theory.*

The absence of the anomaly reflects the absence of divergences in Two-Tier quantum field theory, which preserves chiral invariance. Unlike Pauli-Villars regularization, for example, the finiteness of Two-Tier theory follows from the Gaussian factors. Unlike the dimensional regularization approach (where there is no equivalent to γ_5), Two-Tier theory can use the normal γ_5 matrix.

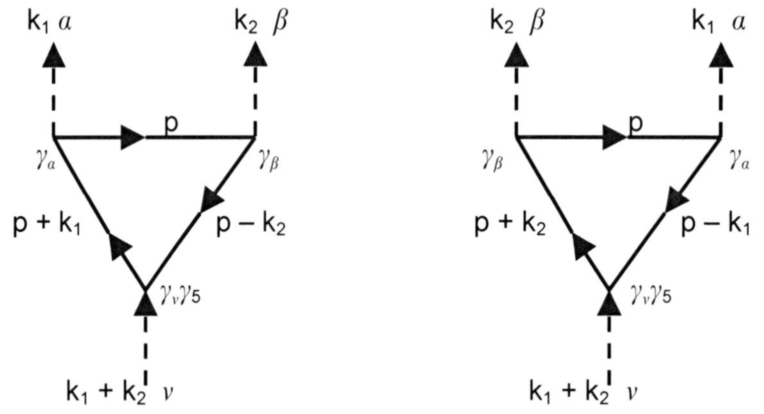

Figure 8.1. The V-V-A triangle diagrams.

The expression for the Two-Tier triangle diagrams is:

$$T_{\alpha\beta\nu}(k_1, k_2) = S_{\alpha\beta\nu}(k_1, k_2) + S_{\beta\alpha\nu}(k_2, k_1) \tag{8.88}$$

where

$$S_{\alpha\beta\nu}(k_1, k_2)\delta^4(k_1 + k_2 - q) = -iN\int d^4y_1 d^4y_2 d^4y_3\, e^{ik_1\cdot y_1 + ik_2\cdot y_2 - iq\cdot y_3} \cdot$$
$$\cdot\, Tr\{S_F^{TT}(y_1 - y_3)\gamma_\alpha S_F^{TT}(y_2 - y_1)\gamma_\beta S_F^{TT}(y_3 - y_2)\gamma_\nu\gamma_5\} / (2\pi)^4 \quad (8.89)$$

where N is a constant, and where $S_F^{TT}(z)$ is specified by eq. 7.45. We now define the fourier transform:

$$S_F^{TT}(z) = -i\int d^4p\, e^{-ip\cdot z}\, \mathscr{S}^{TT}(p) / (2\pi)^4 \quad (8.90)$$

where $\mathscr{S}^{TT}(p)$ defined by eq. 7.93. We then substitute the fourier transform in eq. 8.89 and perform the coordinate integrations to obtain:

$$S_{\alpha\beta\nu}(k_1, k_2) = N\int d^4p\, Tr\{\mathscr{S}^{TT}(p + k_1)\gamma_\alpha \mathscr{S}^{TT}(p)\gamma_\beta \mathscr{S}^{TT}(p - k_2)\gamma_\nu\gamma_5\} / (2\pi)^4$$

We note that

$$k_1^{\alpha}T_{\alpha\beta\nu}(k_1, k_2) \neq 0 \quad (8.91)$$

$$k_2^{\beta}T_{\alpha\beta\nu}(k_1, k_2) \neq 0 \quad (8.92)$$

$$(k_1 + k_2)^{\nu}T_{\alpha\beta\nu}(k_1, k_2) \neq 0 \quad (8.93)$$

in Two-Tier quantum field theory because the conservation laws are expressed with respect to the X coordinates – not the y coordinates. Thus since k_1^{α} corresponds $\partial/\partial y_\alpha$, and not $\partial/\partial X_\alpha$ there is no reason for eqs. 8.91-93 to be zero. However at "large distances" relative to M_c^{-1} we see

$$k_1^{\alpha}T_{\alpha\beta\nu}(k_1, k_2) \cong 0 \quad (8.94)$$

$$k_2^{\beta}T_{\alpha\beta\nu}(k_1, k_2) \cong 0 \quad (8.95)$$

$$(k_1 + k_2)^{\nu}T_{\alpha\beta\nu}(k_1, k_2) \cong 0 \quad (8.96)$$

to very good approximation since the gaussian damping factor in the fermion propagators is approximately unity and thus the Two-Tier expression becomes essentially the same as the conventional field theory expression.

On the other hand at very short distances the anomaly is absent since Two-Tier theory is very well behaved at high energy with

$$\mathscr{S}^{TT}(p) \backsim \gamma^0 M_c^{\,6}\, p^{-7} + \mathscr{O}(p^{-9}) \tag{7.102}$$

As a result we see

$$k_1^{\,\alpha} T_{\alpha\beta\nu}(k_1, k_2) \sim p^{4-21} \sim p^{-17} \tag{8.97}$$

as $p \rightarrow \infty$ is highly convergent. Thus there is no high energy divergence unlike conventional field theory where the integral is linearly divergent. And so no anomaly is generated.

Asymptotic Freedom and Quark Confinement

Two-Tier quantum field theory is totally consistent with the Standard Model at currently accessible energies. Thus if the Standard Model Quantum Chromodynamics (QCD) sector is asymptotically free, and also gives quark confinement at large distance (compared to M_c^{-1}), then similar statements would also be true in the Two-Tier version of QCD.

We assume M_c is extremely large – much beyond current energies – and possibly of the order of the Planck mass. Fig. 8.2 depicts the various regions in Two-Tier QCD assuming a very large M_c.

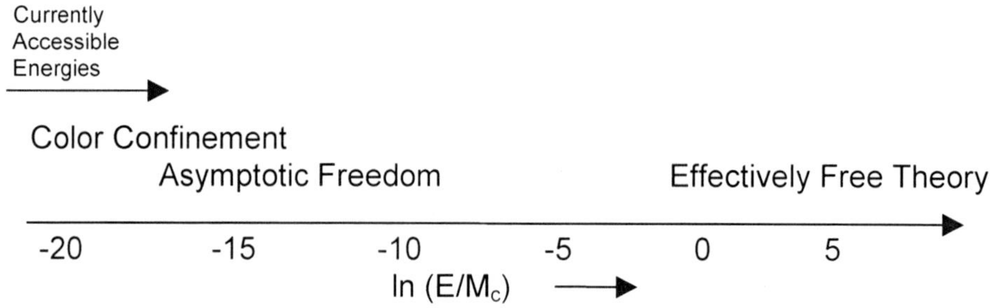

Figure 8.2. A depiction of the Two-Tier QCD regions as a function of the logarithm of the energy assuming M_c is of the order of the Planck mass.

Two-Tier Coordinates

We have developed a physical picture of Two-Tier coordinate systems in which we define two sets of related coordinates. This picture views ordinary real 4-dimensional space as a low energy approximation to a complex 4-dimensional space that only becomes apparent at ultra-high energies with the scale set by M_c. The imaginary part of the coordinates is based on the excitations of a quantum field Y^μ. The real part of the coordinates is the familiar 4-dimensional c-number coordinate system.

Another way to view the Two-Tier X and y coordinate systems is to view them as defining a four-dimensional hyperplane in a six (or eight) dimensional real space-time. See Blaha (2004) for a detailed discussion.

It would also be interesting to consider an extension of the theory to the case where both the real and imaginary parts of the coordinate system are quantum fields. In this case the c-number coordinates of daily experience would emerge as a "condensation", or spontaneous symmetry breaking, phenomena.

One could also generalize the Y quantum field to a non-abelian gauge field, which, properly handled, could be the origin of internal symmetries such as QCD SU(3) using a Kaluza-Klein approach within a general relativistic *vierbein* framework. This approach is an alternative to the compactification of dimensions that is a cornerstone of SuperString theories.

9. Two-Tier Quantum Gravity: Finite!

Introduction

There are numerous excellent books and monographs on classical gravity and a large literature on quantum gravity.[39] Therefore our discussion will assume the reader is familiar with classical General Relativity and aware of attempts to create quantum theories of gravity.

We will begin by establishing the general form of Two-Tier classical General Relativity and then proceed to define a quantization procedure. We will work in Minkowski space with three space and one time dimension. The flat-space metric $\eta_{\alpha\beta}$ is defined as diagonal with $\eta_{00} = 1$ and $\eta_{ij} = -\delta_{ij}$ for i, j = 1, 2, 3.

Two-Tier General Relativity

In developing Quantum Gravity we will make the same ansatz that we have made throughout our development of Two-Tier quantum field theories: all field expressions are functions of the X coordinate field system, which in turn are functions of the "ordinary" y space-time coordinate system. The Two-Tier Theory of Quantum Gravity is invariant under special relativistic transformations. The dynamical field equations, which are strictly functions of the X coordinates, are covariant under general relativistic transformations. The rationale for these assumptions is described in detail in chapter 10.

---

[39] H. Weyl, *Space, Time, Matter* (Dover, New York, 1950); L. D. Landau and E. M. Lifshitz, *The Classical Theory of Fields*, (Addison-Wesley, New York, 1962); S. Weinberg, *Gravitation and Cosmology*, (John Wiley & Sons, New York, 1972); C. W. Misner, K. S. Thorne and J. A. Wheeler, *Gravitation*, (W. H. Freeman, San Francisco, 1973); B. S. DeWitt, Phys. Rev. **162**, 1239 (1967), **162**, B1195 (1967); R. P. Feynman, Acta Physica Polonica **24**, 697 (1963); S. Deser and P. van Nieuwenhuizen, Phys. Rev. Letters **32**, 245 (1974); S. Deser, H.-S. Tsao and P. van Nieuwenhuizen, "One Loop Divergences of the Einstein-Yang-Mills System", Brandeis Univ. preprint (1974); S. Weinberg, Phys. Rev. **138**, B988 (1965); L. Smolin, *Three Roads to Quantum Gravity*, (Basic Books, New York, 2001); L. Smolin, "How Far are We From the Quantum Theory of Gravity", (Univ. Waterloo preprint (2003) and references therein; T. Thiemann, "Lectures on Loop Quantum Gravity", Preprint AEI-2002-087, Albert Einstein Insitute, Golm, Germany (2002) and references therein; A. Pais and G. E. Uhlenbeck, Phys. Rev. **79**, 145 (1950); G. E. Uhlenbeck, "Lecture Notes on General Relativity", The Rockefeller University (1967), unpublished; S. Blaha, "Generalization of Weyl's Unified Theory to Encompass a Non-Abelian Internal Symmetry Group" SLAC-PUB-1799, Aug 1976; S. Blaha, "Quantum Gravity and Quark Confinement" Lett. Nuovo Cim. **18**, 60 (1977); R. Utiyama, Phys. Rev. **101**, 1597 (1956); T. W. B. Kibble, J. Math. Phys. **2**, 212 (1961); R. Arnowitt, S. Deser, and C. W. Misner, Phys. Rev. **117**, 1595 (1960); and references therein.

We define the proper time differential $d\tau$ as

$$d\tau^2 = g_{\mu\nu}(X(y))dX^\mu dX^\nu \tag{9.1}$$

where, as usual,

$$X^\mu(y) = y^\mu + i\, Y^\mu(y)/M_c^{\;2} \tag{3.12}$$

Thus eq. 9.1 could be written:

$$d\tau^2 = g_{\mu\nu}(X(y))(\eta^\mu{}_\alpha + iM_c^{-2}\partial Y^\mu/\partial y^\alpha)(\eta^\nu{}_\beta + iM_c^{-2}\partial Y^\nu/\partial y^\beta)dy^\alpha dy^\beta \tag{9.2}$$

The inverse of $g_{\mu\nu}$, denoted $g^{\nu\lambda}$, satisfies

$$g_{\mu\nu}(X(y))g^{\nu\lambda}(X(y)) = \delta_\mu^{\;\lambda} \tag{9.3}$$

Since the algebraic manipulation of the tensor indices is the same as that of the conventional theory of gravitation the Two-Tier affine connection is:

$${}_X\Gamma^\sigma{}_{\lambda\mu} = ½\, g^{\nu\sigma}\{\partial g_{\mu\nu}/\partial X^\lambda + \partial g_{\lambda\nu}/\partial X^\mu - \partial g_{\lambda\mu}/\partial X^\nu\} \tag{9.4}$$

The Two-Tier Riemann-Christoffel curvature tensor is:

$${}_XR^\lambda{}_{\mu\nu\kappa} \equiv \partial_X\Gamma^\lambda{}_{\mu\nu}/\partial X^\kappa - \partial_X\Gamma^\lambda{}_{\mu\kappa}/\partial X^\nu + {}_X\Gamma^\alpha{}_{\mu\nu}\,{}_X\Gamma^\lambda{}_{\kappa\alpha} - {}_X\Gamma^\alpha{}_{\mu\kappa}\,{}_X\Gamma^\lambda{}_{\nu\alpha} \tag{9.5}$$

and the Two-Tier Ricci tensor is

$${}_XR_{\mu\nu} = {}_XR^\alpha{}_{\mu\alpha\nu} \tag{9.6}$$

The Two-Tier curvature scalar is

$${}_XR = g^{\mu\nu}\,{}_XR_{\mu\nu} \tag{9.7}$$

We also define

$${}_XR_{\lambda\mu\nu\kappa} = g_{\lambda\alpha}\,{}_XR^\alpha{}_{\mu\nu\kappa} \tag{9.8}$$

with the result

$${}_XR_{\lambda\mu\nu\kappa} = ½[\partial^2 g_{\lambda\nu}/\partial X^\kappa\partial X^\mu - \partial^2 g_{\mu\nu}/\partial X^\kappa\partial X^\lambda - \partial^2 g_{\lambda\kappa}/\partial X^\nu\partial X^\mu + \partial^2 g_{\mu\kappa}/\partial X^\nu\partial X^\lambda] +$$

$$+\ g_{\alpha\beta}[{}_X\Gamma^{\alpha}{}_{\nu\lambda}\ {}_X\Gamma^{\beta}{}_{\mu\kappa}\ -\ {}_X\Gamma^{\alpha}{}_{\kappa\lambda}\ {}_X\Gamma^{\beta}{}_{\mu\nu}] \tag{9.9}$$

We denote the fact that all quantities in eqs. 9.4 – 9.8 are only functions of X by placing a left subscript X on each quantity.

The algebraic properties, and the Bianchi identities, satisfied by ${}_X R_{\lambda\mu\nu\kappa}$ in the Two-Tier theory of gravitation are identical to those of the conventional theory with all derivatives being with respect to X.

The Two-Tier version of Einstein's field equations is:

$${}_X R_{\mu\nu} - \tfrac{1}{2}\, g_{\mu\nu}\, {}_X R = -8\pi G\, T_{\mu\nu} \tag{9.10}$$

where G is Newton's gravitational constant ($6.674 \times 10^{-11}\ m^3 kg^{-1} s^{-2}$) and $T_{\mu\nu}$ is the energy-momentum tensor – also strictly a function of X. It is convenient to define the coupling constant

$$\kappa = \sqrt{4\pi G} \tag{9.11}$$

Lagrangian Formulation

We will now formulate a Two-Tier Quantum Gravity theory following the same ansatz that we have used throughout this book.

Unified Standard Model and Quantum Gravity Lagrangian

We define the Lagrangian, and action, for the unified quantum field theory of gravitation and the Standard Model as

$$L_{Unified} = \int d^4y\, \mathcal{L}_{Unified} \tag{9.12}$$

$$\mathcal{L}_{Unified} = J\sqrt{g(X)}\, (\mathcal{L}_F^{Grav}(X^\mu) + \mathcal{L}_F^{SM}(X^\mu)) + \mathcal{L}_C \tag{9.13}$$

with

$$\mathcal{L}_F^{Grav}(X^\mu) = (2\kappa^2)^{-1}\, {}_X R \tag{9.14}$$

where $\mathcal{L}_F^{SM}$ is the complete "normal" Quantum Field theory Lagrangian for the Standard Model version under consideration written in a general covariant form, g(X) is the absolute value of the determinant of $g_{\mu\nu}$, and J is the Jacobian of eq. A.21. *All particle fields in $\mathcal{L}_F^{SM}$ are assumed to be functions of the X^μ coordinate only. The dependence of the particle fields on the "underlying" coordinates y^μ is assumed to be* *solely through X^{μ}*. The Lagrangian $\mathscr{L}_{\text{Unified}}$ is a separable Lagrangian of the type of eq. A.26 embodying the composition of extrema described in Appendix A.

As in all of cases that we have considered, we have specified the coordinate part of the Lagrangian $\mathscr{L}_C$ as

$$\mathscr{L}_C = -\tfrac{1}{4}\, F_Y^{\mu\nu} F_{Y\mu\nu} \tag{3.15}$$

with

$$F_{Y\mu\nu} = \partial Y_\mu/\partial y^\nu - \partial Y_\nu/\partial y^\mu \tag{3.14}$$

and

$$F_Y^{\mu\nu} = \eta^{\mu\alpha}\eta^{\nu\beta}F_{Y\alpha\beta} \tag{9.15}$$

Why Are the Y Field Dynamics Independent of the Gravitational Field?

It is evident from eqs. 9.12-3 and 9.15 that the Y field is truly free and, in particular, does not depend on the gravitational field as represented by $\sqrt{g}$ and $g_{\mu\nu}$. Our rationale for this formulation is described in detail in chapter 10. For the moment it suffices to make the following remarks. The Y field is a quantum field at each point in space-time including regions with ultra-strong gravitational fields such as the neighborhoods of black holes. If Y were to depend on the gravitational field then the Y field could be appreciable in such regions and might even be a "classical" field. In this case we would have new dimensions, allbeit imaginary, for which no evidence exists.

Furthermore, the Y part of the Lagrangian establishes a functional relation between the imaginary Y coordinates and the real space-time y coordinates. The Principle of Equivalence applies only to real coordinates and has not been shown to apply to imaginary Quantum Dimensions™. Thus there is no reason to require the Y part of the action to be invariant under general coordinate transformations.

Lastly, the non-invariance of the Y part of the action under general coordinate transformations effectively creates an "absolute" coordinate system – actually a class of "absolute coordinate systems" – namely the class of inertial reference frames that are related to each other by special relativistic transformations. This feature does not conflict with our knowledge of the universe. The universe appears to be almost flat. The large-scale distribution of masses is responsible for this flatness. The flatness, or flattened space if it is slightly curved, together with Mach's Principle (inertial forces are absent in the reference frame determined by the distribution of masses in the universe) selects a preferred class of local reference frames – local inertial reference frames. Since space is almost flat, or flat, these local reference frames occupy a large volume (if we exclude regions with intense gravitational fields.) We can define the Y field within this class of local inertial reference frames in each locale and establish a satisfactory quantum field theory. Thus we have a dynamics defined in the variable X, which we require to be covariant under general coordinate transformations, and a local "ground state" that "breaks" general coordinate invariance down to special relativistic invariance. See chapter 10 for a more complete discussion.

No "Space-time Foam"

The fact that our unified theory of the known forces of Nature *self-consistently* has a weak gravitational field at high energies (the graviton sector is finite to all orders in perturbation theory) supports the formulation of eq. 9.12-3. Gravity becomes weaker at ultra-short distances. Therefore space-time is not quantum foam at ultra-short distances but rather smooth and flat a là special relativity – consistent with our formulation.

Quantum Gravity – Scalar Particle Model Lagrangian

While the application of the Two-Tier approach to the unified theory is a straightforward extension of the concepts and approaches described in the preceding chapters, it is useful to consider a simplified model that minimizes the tensorial verbiage so that the concepts and features might better stand out. The procedure differs only in detail from the case of gauge fields.

The introduction of spinor fields requires the use of a Two-Tier vierbein formalism, which is straightforward to develop. A Two-Tier vierbein field $e^{\mu}{}_{a}$ is a function of X, $e^{\mu}{}_{a}(X)$, with $g_{\mu\nu} = e_{\mu a}(X)e_{\nu}{}^{a}(X)$ where the index a is an index of a flat tangent space defined at each space-time point. The Two-Tier formulation of a vierbein theory is similar to the other Two-Tier formulations that we have considered and will not be developed here.

Thus we will consider the Lagrangian model for a scalar particle field interacting with the $g_{\mu\nu}$ gravitational field:

$$L_{GS} = \int d^4y \sqrt{g(X)}\, \mathscr{L}_{GS} + \mathscr{L}_{C} \tag{9.16}$$

$$\mathscr{L}_{GS} = J\, \mathscr{L}_{F}^{Grav}(X^{\mu}) + J\, \mathscr{L}_{F\phi}(X^{\mu}) \tag{9.17}$$

with covariant versions of eqs. 5.21 and 5.24:

$$\mathscr{L}_{F\phi} = \tfrac{1}{2}\, [\, g^{\mu\nu}\partial\phi/\partial X^{\mu}\, \partial\phi/\partial X^{\nu} - m^2\phi^2] + \mathscr{L}_{F\phi int} \tag{9.18}$$

$$\mathscr{L}_{F\phi int} = \tfrac{1}{4!}\, \chi_0\, \phi(X(y))^4 + \tfrac{1}{2}\,(m^2 - m_0^{\,2})\phi^2 \tag{9.19}$$

A Justifiable Weak Field Approximation for Quantum Gravity

Many discussions of quantizing conventional gravity make a weak field approximation for the gravity sector which, in view of divergences in the resulting quantum field theory, are impossible to justify:

$$g_{\mu\nu} = \eta_{\mu\nu} + \kappa h_{\mu\nu} \tag{9.20}$$

where $\eta_{\mu\nu}$ is the flat space-time metric and $h_{\mu\nu}$ is a "small" deviation ($\langle h_{\mu\nu}\rangle \ll 1$) from the flat space-time metric.

The Two-Tier formulation of quantum gravity is finite and the effective field becomes increasingly weaker at short distances. Thus the weak field approximation becomes *more accurate* at short distances:

$$g_{\mu\nu}(X(y)) \simeq \eta_{\mu\nu} + \kappa h_{\mu\nu}(X(y)) \tag{9.21a}$$

At short distances space-time can be considered approximately flat (except possibly in the neighborhood of singularities) with quantum fluctuations embodied in $h_{\mu\nu}$. Thus eq. 9.21a is reasonable within the context of Two-Tier Quantum Gravity.

To first order in $h_{\mu\nu}$ the square root of the absolute value of the determinant of the metric tensor is:

$$\sqrt{g(X)} \simeq 1 + ½\, \kappa h^{\sigma}{}_{\sigma}(X(y)) \tag{9.21b}$$

Quantization of Quantum Gravity – Scalar Particle Model

We now proceed to quantize gravity based on the linearization of the gravitational field equations in the weak field approximation. Assuming eq. 9.21a and keeping terms to first order in $h_{\mu\nu}$ gives the affine connection:

$${}_X\Gamma^{\sigma}{}_{\mu\nu} = ½\, \kappa \eta^{\sigma\alpha}[\partial h_{\alpha\nu}/\partial X^{\mu} + \partial h_{\alpha\mu}/\partial X^{\nu} - \partial h_{\mu\nu}/\partial X^{\alpha}] + \mathcal{O}(h^2) \tag{9.22}$$

and the Ricci tensor:

$${}_X R_{\mu\nu} = \partial_X \Gamma^{\lambda}{}_{\lambda\mu}/\partial X^{\nu} - \partial_X \Gamma^{\lambda}{}_{\mu\nu}/\partial X^{\lambda} + \mathcal{O}(h^2) \tag{9.23}$$

Thus the linearized gravitation lagarangian terms are

$$L^{Grav} = \int d^4y\, \sqrt{g(X)}\, J \mathscr{L}_F^{Grav}(X^{\mu}) \rightarrow L^{Grav}{}_{linear} = \int d^4y\, J \mathscr{L}^{Grav}{}_{linear}(X^{\mu}) \tag{9.24}$$

The scalar particle Lagrangian terms become

$$L^{\phi} = \int d^4y \sqrt{g(X)} J \mathscr{L}_{F\phi} \rightarrow \int d^4y J \{[½(\eta^{\mu\nu}\partial_{\mu}\phi\partial_{\nu}\phi - m^2\phi^2) + \mathscr{L}_{F\phi int}] +$$

$$+ \tfrac{1}{2}\kappa h^{\mu\nu}\partial_\mu\phi\partial_\nu\phi + \tfrac{1}{4}\,\kappa h(\eta^{\mu\nu}\partial_\mu\phi\partial_\nu\phi - m^2\phi^2) +$$

$$+ \tfrac{1}{2}\,\kappa h\mathscr{L}_{F\phi int}\} \tag{9.25}$$

with the notation $h = h^\sigma{}_\sigma$ and using

$$\partial_\mu \equiv \partial/\partial X^\mu \tag{9.26}$$

$\eta^{\mu\nu}$ and $\eta_{\mu\nu}$ are used to raise and lower indices in keeping with the linearized, weak field approximation.

The Y terms in the Lagrangian are (as previously):

$$L^Y = \int d^4y\mathscr{L}_C = -\tfrac{1}{4}\int d^4y\eta^{\mu\nu}\eta^{\alpha\beta}F_{Y\mu\alpha}F_{Y\nu\beta} \tag{9.27}$$

We will lump the higher order terms (in h) in the gravity part of the Lagrangian, and the scalar particle part of the Lagrangian, into

$$L_{Higher} = \int d^4y\, J\mathscr{L}_{Higher}(h,\, \phi) \tag{9.28}$$

Thus the complete lagragian for a scalar particle interacting with gravitons is

$$L_{GS} = L^{Grav}{}_{linear} + L^{\phi}{}_{linear} + L^Y + L_{Higher} \tag{9.29}$$

The linearized gravitational Lagrangian term $L^{Grav}{}_{linear}$ generates the field equations:

$$\Box_X h_{\mu\nu} + \partial_\nu\partial_\mu h - \partial_\alpha\partial_\nu h^\alpha{}_\mu - \partial_\alpha\partial_\mu h^\alpha{}_\nu = \kappa S_{\mu\nu} \tag{9.30}$$

where

$$\partial_\mu S^\mu{}_\nu = \tfrac{1}{2}\,\partial_\nu S^\sigma{}_\sigma \tag{9.31}$$

to 0th order in h, and where

$$\Box_X = (\partial/\partial X^\nu)(\partial/\partial X_\nu) \tag{9.32}$$

The most general coordinate transformation that maintains the weakness of the gravitational field has the form:

$$y^a \rightarrow \; y'^a = y^a + \chi^a(X(y)) \tag{9.33}$$

This transformation induces a gauge transformation in $h_{\mu\nu}$ to:

$$h'_{\mu\nu} = h_{\mu\nu} - \partial_\mu \chi_\nu - \partial_\nu \chi_\mu \tag{9.34}$$

It is easy to verify that eq. 9.30 is satisfied by $h'_{\mu\nu}$ if it is satisfied by $h_{\mu\nu}$.

Let us assume that we perform a gauge transformation making $h_{\mu\nu}$ traceless:

$$h^\sigma{}_\sigma = 0 \tag{9.35}$$

and choose the gauge

$$\partial^\mu h_{\mu\nu} = 0 \tag{9.36}$$

then eq. 9.30 becomes the wave equation:

$$\Box_X \, h_{\mu\nu} = \kappa S_{\mu\nu} \tag{9.37}$$

Another gauge transformation of the free field $h_{\mu\nu}$ (if $S_{\mu\nu} = 0$) makes

$$h_{\mu 0} = h_{0\mu} = 0 \tag{9.38}$$

while retaining

$$h_{\mu\nu} = h_{\nu\mu} \tag{9.39}$$

The general solution[40] for the free field $h_{\mu\nu}$ (with $S_{\mu\nu} = 0$ in eq. 9.37) can be expressed as a fourier expansion:

$$h_{\mu\nu}(X(y)) = \int d^3k\, N_0(k) \sum_{\lambda=1}^{2} \varepsilon_{\mu\nu}(k, \lambda)[a(k,\lambda)\, e^{-ik\cdot X} + a^\dagger(k,\lambda)\, e^{ik\cdot X}] \tag{9.40}$$

where $\lambda = 1,2$ labels the ± 2 helicity states, and where $N_0(k)$ is specified by eq. 3.25. The equal time ($y'^0 = y^0$) commutation relations are:

$$[h_{\mu\nu}(X(y)), h_{\alpha\beta}(X(y'))] = [\pi_{\mu\nu}(X(y)), \pi_{\alpha\beta}(X(y'))] = 0 \tag{9.41}$$

---

[40] S. Weinberg, Phys. Rev. **135**, B1049 (1964); Phys. Rev. **138**, B988 (1965)

$$[h_{\alpha\beta}(X(y')), \pi_{\mu\nu}(X(y))] = i\, \mathscr{D}_{\alpha\beta\mu\nu}(\mathbf{X}(y) - \mathbf{X}(y')) \tag{9.42}$$

for μ, $\nu = 1, 2, 3$ where

$$\pi_{\mu\nu}(X(y)) = \partial h_{\mu\nu}(X(y))/\partial y^0 \tag{9.43}$$

in the Y Coulomb gauge, and where $X^0 = y^0$. $\mathscr{D}_{\alpha\beta\mu\nu}$ is specified by:

$$\mathscr{D}_{\alpha\beta\mu\nu}(X(y) - X(y')) = \int d^3k\, e^{i\,\mathbf{k}\cdot(\mathbf{X}(y) - \mathbf{X}(y'))}\, \Pi_{\alpha\beta\mu\nu}(\mathbf{k})/(2\pi)^3 \tag{9.44}$$

$$\Pi_{\alpha\beta\mu\nu}(\mathbf{k}) = ½\, [(\delta_{\alpha\mu} - k_\alpha k_\mu/\mathbf{k}^2)(\delta_{\beta\nu} - k_\beta k_\nu/\mathbf{k}^2) + (\delta_{\alpha\nu} - k_\alpha k_\nu/\mathbf{k}^2)(\delta_{\beta\mu} - k_\beta k_\mu/\mathbf{k}^2) -$$

$$- (\delta_{\alpha\beta} - k_\alpha k_\beta/\mathbf{k}^2)(\delta_{\mu\nu} - k_\mu k_\nu/\mathbf{k}^2)] \tag{9.45}$$

where α, β, μ, $\nu = 1, 2, 3$.

The "transverse" graviton propagator can be represented as a time-ordered product of field operators:

$$i\Delta_{F2}{}^{TT}(y_1 - y_2)_{\lambda\tau\rho\sigma} = \langle 0|\, T(h_{\lambda\tau}(X(y_1)), h_{\rho\sigma}(X(y_2)))\, |0\rangle \tag{9.46}$$

$$= -i \int \frac{d^4k\, e^{-ik\cdot(y_1 - y_2)}\, b_{\lambda\tau\rho\sigma}(k) R(\mathbf{k}, y_1 - y_2)}{(2\pi)^4\, (k^2 + i\varepsilon)}$$

where $R(\mathbf{k}, y_1 - y_2)$ is the gaussian factor appearing in propagators throughout Two-Tier theories, $\mathbf{k}$ is a spatial 3-vector, and where $b_{\mu\nu\rho\sigma}(k)$ is a function of k only:

$$b_{\alpha\beta\mu\nu}(k) = ½[(\eta_{\alpha\mu} - k_\alpha k_\mu/\mathbf{k}^2)(\eta_{\beta\nu} - k_\beta k_\nu/\mathbf{k}^2) + (\eta_{\alpha\nu} - k_\alpha k_\nu/\mathbf{k}^2)(\eta_{\beta\mu} - k_\beta k_\mu/\mathbf{k}^2) -$$

$$- (\eta_{\alpha\beta} - k_\alpha k_\beta/\mathbf{k}^2)(\eta_{\mu\nu} - k_\mu k_\nu/\mathbf{k}^2)] \tag{9.47}$$

where α, β, μ, $\nu = 0, 1, 2, 3$.

The quantum gravitational interaction also has an "instantaneous" part (similar to the instantaneous Coulomb interaction of QED) in addition to the transverse interaction embodied in eq. 9.46. This "instantaneous" interaction contains the Newtonian potential (described later) as its large distance limit. The sum of the instantaneous interaction and the transverse interaction gives the total gravitational interaction.

The above graviton propagator has the form given in eq. 6.102. The caculation of the leading behavior is the same as that of the Two-Tier scalar boson propagator except for the presence of factors such as $\eta_{\rho\sigma}$. The leading momentum dependence of the graviton propagator in momentum space is

$$i\Delta_{F2}{}^{TT}(p)_{\lambda\tau\rho\sigma} \backsim p^{-6} \tag{9.48}$$

The graviton vertices in Two-Tier Quantum Gravity will be described within the framework of the path integral formulation.

Quantum Gravity–Scalar Particle Model Path Integral

A path integral formalism can be developed for Two-Tier Quantum Gravity interacting with matter fields. In this section we will consider the case of a matter field consisting of massive scalar bosons with a quartic interaction. The path integral formalism that we develop is similar to that of Yang-Mills theories in the previous chapter.

The Two-Tier path integral for a Quantum Gravity–Scalar Particle Theory can be written as:

$$Z(J, J^{\mu\nu}) = N \int D\phi Dh DY \Delta_{FPG}(h)\delta(F(h)) \exp\{i\int d^4y[\mathscr{J}(\mathscr{L}^{Grav}{}_{linear}(X^\mu) +$$

$$+ \mathscr{L}^{\phi}{}_{linear}(X^\mu) + \mathscr{L}_{Higher}(h, \phi)) + \mathscr{L}_C(X, y) +$$

$$+ j_\mu(y)Y^\mu(y) + J(y)\phi(X) + J^{\mu\nu}(y)h_{\mu\nu}(X)]\}\Big|_{j_\mu = 0} \tag{9.49}$$

where $\delta(F(h))$ specifies the gauge as a functional delta function, and $\Delta_{FPG}(h)$ is the corresponding Faddeev-Popov determinant. $\mathscr{J}$ is the Jacobian for the transformation from y coordinates to X coordinates. The Faddeev-Popov determinant $\Delta_{FPG}(h)$ can be calculated in the standard way. First we note

$$\delta(F(h^\chi)) = \delta(\chi - \chi_0)\ \left|\det \delta F(h_{\mu\nu}{}^\chi(X))/\delta\chi(X)\right|^{-1}\Big|_{F(h)=0} \tag{9.50}$$

where

$$h_{\mu\nu}{}^{\chi} = h_{\mu\nu} - \partial_\mu \chi_\nu - \partial_\nu \chi_\mu \qquad (9.34)$$

Then

$$\Delta_{FPG}(h) = \left| \det \delta F(h^{\chi}(X))/\delta\chi(X) \right| \Big|_{F(h)=0} \qquad (9.51)$$

We will choose the gauge of eq. 9.36 to evaluate the Faddeev-Popov determinant. Under an infinitesimal gauge transformation of the form:

$$h_{\mu\nu}{}^{\chi}(X) = h_{\mu\nu}(X) - \partial_\mu \chi_\nu - \partial_\nu \chi_\mu \qquad (9.52)$$

which preserves the weak field nature of $h_{\mu\nu}$, we find

$$F_\nu(h^{\chi}) = \partial^\mu \left(h_{\mu\nu}(X) - \partial_\mu \chi_\nu - \partial_\nu \chi_\mu\right)$$

$$= -\Box_X \chi_\nu(X) - \partial_\nu\partial^\mu \chi_\mu \qquad (9.53)$$

Thus

$$\delta F_\mu(h^{\chi}(X))/\delta\chi^\nu(X) = -\eta_{\mu\nu}\Box_X - \partial_\mu\partial_\nu \qquad (9.54)$$

and

$$\Delta_{FP}(A) = \left| \det\left(-\eta_{\mu\nu}\Box_X - \partial_\mu\partial_\nu\right) \right| \Big|_{F(h)=0} \qquad (9.55)$$

We note the Two-Tier Faddeev-Popov determinant is solely a function of the X coordinates. The determinant only introduces an overall multiplicative constant that can be absorbed into the normalization constant N. This fact becomes evident if we follow the standard procedure and rewrite the determinant as a path integral over anti-commuting c-number fields with a ghost Lagrangian. Then we see that the ghost does not interact with the other fields and thus only generates an overall multiplicative constant that can be absorbed in N:

$$\Delta_{FPG}(h) = \int Dc^* Dc \exp[\, i\int d^4X\, \mathscr{L}^{ghost}(X^\mu)] \qquad (9.56)$$

where

$$\mathscr{L}^{ghost}(X^\mu) = c^{\mu *}(X)[\eta_{\mu\nu}\Box_X + \partial_\mu\partial_\nu]c^\nu(X) \qquad (9.57)$$

We now go through the same analysis as we did in the ϕ^4 theory path integral example and the Yang-Mills path integral example (with some superficial differences). First we integrate the linear part of the Y field Lagrangian as we did previously. Then we integrate the linear part of the ϕ field Lagrangian as done previously. Lastly we integrate the linear part of the gravitation Lagrangian to obtain the path integral for the perturbative expansion with the result:

$$Z(J, J^{\mu\nu}) = N\ \{\exp[i\int d^4y \mathscr{L}_{Higher}(\partial/\partial y^\nu, -i\delta/\delta J^{\mu\nu}(y), -i\delta/\delta J(y))]\cdot$$

$$\cdot \exp[-½\ i\int d^4y_1 d^4y_2 J^{\mu\nu}(y_1)\Delta_{F2}{}^{TT}(y_1 - y_2, z)_{\mu\nu\rho\sigma} J^{\rho\sigma}(y_2)]\cdot$$

$$\cdot \exp[-½\ i\int d^4y_1 d^4y_2 J(y_1)\Delta_F{}^{TT}(y_1 - y_2, z) J(y_2)]\} \Big|_{z=y_1-y_2} \qquad (9.58)$$

There are two issues that arise in the development of eq. 9.58:

1.) The integral over y in $\int d^4y \mathscr{L}_{Higher}$ which began as the integral $\int d^4y \mathscr{J} \mathscr{L}_{Higher}$ = $\int d^4X \mathscr{L}_{Higher}$ in eq. 9.49; and

2.) The handling of derivatives with respect to X in $\mathscr{L}_{Higher}$.

These are resolved by the following respective observations:

1.) See the discussions following eqs. 6.34 and 8.29 that apply here as well without change.

2.) See the discussion following eq. 8.32, which applies here with only superficial changes. In particular we note that the derivative with respect to X of the graviton propagator (eq. 9.46-7) is specified by the following:

$$\partial i\Delta_{F2}{}^{TT}(y_1 - y_2)_{\lambda\tau\rho\sigma}/\partial X^\mu(y_1) = \partial[i\Delta_{F2}{}^{TT}(y_1 - y_2, z)_{\lambda\tau\rho\sigma}]/\partial y_1{}^\mu \Big|_{z = y_1 - y_2} \qquad (9.59)$$

where

$$i\Delta_{F2}{}^{TT}(y_1 - y_2, z)_{\lambda\tau\rho\sigma} = -\,i\int \frac{d^4k\; e^{-ik\cdot(y_1 - y_2)}\, b_{\lambda\tau\rho\sigma}(k)R(\mathbf{k}, z)}{(2\pi)^4\,(k^2 + i\varepsilon)} \tag{9.60}$$

Thus

$$\frac{\partial\; i\Delta_{F2}{}^{TT}(y_1 - y_2)_{\lambda\tau\rho\sigma}}{\partial X^\mu(y_1)} = -\,i\int \frac{d^4k\; e^{-ik\cdot(y_1 - y_2)}\,(-\,ik_\mu)b_{\lambda\tau\rho\sigma}(k)R(\mathbf{k}, y_1 - y_2)}{(2\pi)^4\,(k^2 + i\varepsilon)} \tag{9.61}$$

Therefore derivatives with respect to X in the interaction Lagrangian terms can be replaced by derivatives with respect to y if the graviton propagator is generalized to eq. 9.60. After taking all derivatives with respect to y, we set z equal to the respective $y_1 - y_2$ (actually the difference of the appropriate variables) in each propagator with results similar to eq. 9.61.

$$\begin{aligned} Z(J, J^{\mu\nu}) = N\,\{&\exp\left[i\textstyle\int d^4y\,\mathscr{L}_{Higher}(\partial/\partial y^\nu, -i\delta/\delta J^{\mu\nu}(y), -i\delta/\delta J(y))\right]\cdot \\ &\cdot \exp[-\tfrac{1}{2}\, i\textstyle\int d^4y_1 d^4y_2 J^{\mu\nu}(y_1)\Delta_{F2}{}^{TT}(y_1 - y_2, z)_{\mu\nu\rho\sigma} J^{\rho\sigma}(y_2)]\cdot \\ &\cdot \exp[-\tfrac{1}{2}\, i\textstyle\int d^4y_1 d^4y_2 J(y_1)\Delta_F{}^{TT}(y_1 - y_2, z)J(y_2)]\}\,\Big|_{z=y_1-y_2} \end{aligned} \tag{9.58a}$$

To be precise eq. 9.58a is interpreted as executing the following steps:

1. For a given process take appropriate functional derivatives of Z(J) with respect to J and $J^{\mu\nu}$.

2. Then expand the exponential factors in a perturbation series applying any derivatives with respect to y in $\mathscr{L}_{Higher}$. Do not perform any of the $\int d^4y_1 d^4y_2$ integrals.

3. Then set $z = y_1 - y_2$ in each $\Delta_{Fk}{}^{TT}(y_1 - y_2, z)$ and $\Delta_{F2}{}^{TT}(y_1 - y_2, z)_{\mu\nu\rho\sigma}$ propagator.

4. Lastly perform all $\int d^4y_1 d^4y_2$ integrals.

Thus we achieve a path integral formulation that is very similar to the corresponding expression in conventional field theory – the only difference is in the form of the free field propagators, which each now contain a Gaussian factor. The net consequence is that graviton vertices result in exactly the same polynomials in momenta as the conventional theory.

Thus Two-Tier gravity generates a perturbative expansion identical to conventional quantum gravity except that each graviton propagator has a gaussian damping factor $R(\mathbf{k},y_1-y_2)$. At low energies the tree diagrams of conventional gravity theory emerge to good approximation in Two-Tier gravity. All diagrams with loops converge. Thus Two-Tier gravity is finite.

Finiteness of Quantum Gravity–Scalar Particle Model

Two-Tier Quantum Gravity perturbation theory is finite. Calculations are highly convergent at large momentum ($\gtrsim M_c$). At low momentum the Two-Tier theory is similar to conventional gravity – particularly for tree diagrams and other convergent diagrams in conventional quantum gravity.

For pure *conventional* Quantum Gravity DeWitt[41] finds the superficial degree of divergence of a diagram to be:

$$D = -2L_i + 2\sum_n V_n + 4K \tag{9.62}$$

where L_i is the number of internal lines, V_n is the number of n-pronged vertices, and K is the number of independent momentum integrations. DeWitt further points out

$$K = L_i - \sum_n V_n + 1 \tag{9.63}$$

Thus the superficial degree of divergence of a <u>*conventional*</u> Quantum Gravity diagram is:

$$D = 2(K + 1) \tag{9.64}$$

for $K \geq 1$, displaying an ever increasing degree of divergence as the order of the diagram increases.

In the case of *Two-Tier Quantum Gravity* the superficial degree of divergence of a diagram is:

---

[41] B. S. DeWitt, Phys. Rev. **162**, 1239 (1967).

$$D_{TT} = -6L_i + 2\sum_n V_n + 4K \tag{9.65}$$

(from eq. 9.48) with the result (taking account of eq. 9.63):

$$D_{TT} = -2L_i - 2\sum_n V_n + 2 \tag{9.66}$$

Since any diagram with a loop has $L_i \geq 1$ and $\sum_n V_n \geq 1$ we see that $D \leq -2$. Thus *all* diagrams are convergent and *the Two-Tier formulation of Quantum Gravity theory is finite. The addition of arbitrary species of other Two-Tier fields – matter and gauge fields – does not introduce divergences in the combined Two-Tier theory.*

Unitarity of Quantum Gravity–Scalar Particle Model

The Two-Tier Quantum Gravity – Scalar Particle Model *superficially* appears to have a unitarity problem due to the non-hermitean nature of its hamiltonian. The lack of hermiticity is due entirely to the appearance of iY^μ in the X^μ field coordinates.

The interaction Lagrangian is not hermitean:

$$L_{Higher} = \int d^3y' \mathscr{L}_{Higher}(y' + iY(y')/M_c^2) \tag{9.67}$$

and

$$L_{Higher}^\dagger = \int d^3y' \mathscr{L}_{Higher}(y' - iY(y')/M_c^2) \neq L_{Higher} \tag{9.68}$$

The relation between L_{Higher} and its hermitean conjugate is

$$L_{Higher} = V\, L_{Higher}^\dagger\, V \tag{9.69}$$

where $V^2 = I$ is the metric operator defined in eqs. 5.16 – 5.18. By eq. 6.37 we see as a result that the Two-Tier S matrix is not unitary – it is pseudo-unitary:

$$S^{-1} = V\, S^\dagger\, V \tag{9.70}$$

Therefore

$$S^\dagger VS = V \tag{9.71}$$

The S matrix satisfies the unitarity condition between physical asymptotic states – states consisting of only scalar ϕ particles and gravitons. The proof is identical in form to eqs. 6.46 – 6.48. The S matrix of the unified theory of the Standard Model and Quantum Gravity can be similarly shown to satisfy the unitarity condition.

The Mass Scale M_c

The mass scale of Two-Tier theories is set by M_c. This mass scale cannot be ascertained with any degree of certainty at current, experimentally accessible, accelerator energies. Cosmic ray data also does not seem to give any clues as to the value of M_c. It appears that M_c is probably above 10^3 GeV/c^2 and may be of the order of (or equal to) the Planck mass:

$$M_{planck} = \sqrt{\hbar c/G} = 1.22\times 10^{19}\ \mathrm{GeV}/c^2 \tag{9.75}$$

If M_c is of the 1,000 GeV/c^2 or larger the differences between its predictions at current accelerator energies and the predictions of conventional renormalized perturbation theory will be negligible. Actually a much lower value of M_c would still be consistent with the current stringent QED theoretical predictions as well as other predictions of conventional renormalized perturbation theory.

Planck Scale Physics

A finite theory of Quantum Gravity can provide information on the issues that have been of concern for many years – including the short distance behavior of the gravitational metric and ultra-small black holes.

Quantum Foam

Some theorists have conjectured that the classical view of smooth, almost flat space-time does not hold in the quantum regime at energies of the order of the Planck mass. Suggestions that space-time dissolves into quantum foam have appeared.

The finite Two-Tier formulation of Quantum Gravity is well-behaved at short distances and suggests that the quantum behavior of gravity and space-time in the short distance limit does not have limitless quantum fluctuations that result in a foam-like space-time picture.

Measurement of the Quantum Gravity Field

A number of conceptual problems have been raised about the effects of quantized General Relativity. Two-Tier Quantum Gravity seems to resolve these issues.

Measurement of Time Intervals

Wigner[42] has studied the measurement of time intervals in General Relativity and sees a problem in the measurement of extremely short intervals. According to Wigner: the measurement of a time inteval in a region of space requires the measurement of the length of time required for an event to happen. The measurement requires an accurate clock. But the accuracy of the clock is limited by the energy-time uncertainty relation:

$$\Delta E \Delta t \geq \hbar \tag{9.76}$$

Thus the uncertainty in the clock's time measurement is related to the uncertainty in the clock's energy which is, in turn, related to the uncertainty in the clock's mass:

$$\Delta E = (\Delta m)c^2 \tag{9.77}$$

To obtain "infinite" accuracy the uncertainty (fluctuations) in the clock's mass must be infinite and thus the clock's mass must be infinite. Infinite fluctuations in the clock's mass will produce corresponding infinite fluctuations in the gravitational field.

$$\Delta h \propto \Delta E \qquad \text{(in conventional General Relativity)} \tag{9.78}$$

As a result the notion of space-time and time intervals (which depend on the geometry through General Relativity) become uncertain. Thus, according to Wigner and others, the concept of time intervals and space-time points becomes questionable.

The Two-Tier version of Quantum Gravity offers a potential way out of this dilemma. The gravitational force becomes stronger as one goes to shorter distances (higher energies) down to a distance (up to an energy) whose scale is set by M_c. At shorter distances (higher energies) the gravitational force becomes weaker and declines to zero at zero distance. Thus at very high energy the gravitational field fluctuations (Δh) are at worst inversely proportional to the energy (and probably decline by a higher power of inverse energy.) (The same considerations would apply if one chooses to consider fluctuations in the Riemann-Christoffel symbols.)

$$\Delta h < c_1/E < c_1/(\Delta E) \qquad \text{(in Two-Tier Quantum Gravity)} \tag{9.79}$$

Thus Wigner's conclusion does not hold in the Two-Tier version of Quantum Gravity as gravitational fluctuations actually become smaller at energies above a critical energy whose scale is set by M_c.

In fact, combining eqs. 9.79 and 9.76 we see

[42] E. P. Wigner, Rev. Mod. Phys. **29**, 255 (1957); J. Math. Phys. **2**, 207 (1961).

$$c_1\Delta t/\Delta h \geq \hbar \tag{9.80}$$

at sufficiently high energy. Therefore the time uncertainty Δt, and the gravitational field fluctuations Δh, can both decrease while maintaining the energy-time uncertainty relation. *Thus the notion of a space-time point "is saved" in Two-Tier quantum gravity.*

Vacuum Fluctuations in the Gravitation Fields

While the expectation value of the free graviton field $h_{\mu\nu\text{conv}}(X)$ is zero in a conventional quantum field theoric approach:

$$<0|\, h_{\mu\nu\text{conv}}(X)\, |0> = 0 \tag{9.81}$$

the vacuum fluctuations of the *conventional* quantum graviton field is quadratically divergent since

$$<0|\, h_{\mu\nu\text{conv}}(X) h_{\alpha\beta\text{conv}}(X)\, |0> = \int d^3p\; b'_{\mu\nu\alpha\beta}(p)/[(2\pi)^3\; 2\omega_p] = \infty \tag{9.82}$$

where $b'_{\mu\nu\alpha\beta}(p)$ is a rational function of the momentum p.

In "Two-Tier" quantum field theory we find

$$<0|\, h_{\mu\nu}(X) h_{\alpha\beta}(X)\, |0> = \int d^3p\; b'_{\mu\nu\alpha\beta}(p)\; e^{-\,p^ip^j\Delta_{\text{T}ij}(0)}/[(2\pi)^3 2\omega_p] = 0 \tag{9.83}$$

since the exponential factor in the integrand is $-\infty$. The exponent contains

$$\Delta_{\text{T}ij}(z) = \int d^3k\; e^{-ik\cdot z}(\delta_{ij} - k_ik_j/\mathbf{k}^2)/[(2\pi)^3 2\omega_k] \tag{4.8}$$

Thus the vacuum fluctuations of $h_{\mu\nu}$ are zero in "Two-Tier" quantum field theory.

The Two-Tier Gravitational Potential vs. Newton's Gravitational Potential

The familiar gravitational potential of Newton is:

$$V_{\text{Newton}} = -G/|\mathbf{r}| \tag{9.84}$$

The Two-Tier gravitational potential is:

$$V_{\text{Two-Tier}} = -\,G\Phi(M_c^{\,2}\pi\,|\mathbf{r}|^2)/\,|\mathbf{r}| \tag{9.85}$$

where $\Phi(y)$ is the error function.[43] It can be calculated in Two-Tier Quantum Gravity from Two-Tier Quantum Gravity propagator terms similar to corresponding terms in the Two-Tier photon propagator that led to the Two-Tier Coulomb potential (eqs. 7.48 – 7.51). At small distances ($\pi r^2 \ll M_c^{-2}$)

$$V_{\text{Two-Tier}} \rightarrow -\,G2\surd\pi\; M_c^{\,2}|\mathbf{r}| \tag{9.86}$$

a linear potential, and at large distances ($\pi r^2 \gg M_c^{-2}$)

$$V_{\text{Two-Tier}} \rightarrow V_{\text{Newton}} = -\,G/\,|\mathbf{r}| \tag{9.87}$$

the Newtonian potential.

The Two-Tier gravitational potential has a minimum at

$$M_c^{\,2}\pi r_{\text{MIN}}^{\;2} = 1 \tag{9.88}$$

At the minimum $V_{\text{Two-Tier}}$ has the value:

$$V_{\text{Two-TierMIN}} = -\,.8427G\surd\pi\; M_c \tag{9.89}$$

Figs.9.1 – 9.2 display plots of $V_{\text{Two-Tier}}$ for M_c = 1 TeV/c^2, and M_c = 1.22 10^{19} GeV/c^2 = $G^{-1/2}$ – the Planck mass.

[43] W. Magnus and F. Oberhettinger, *Formulas and Theorems for the Special Functions of Mathematical Physics* (Chelsea Publishing Co., New York, 1949) page 96.

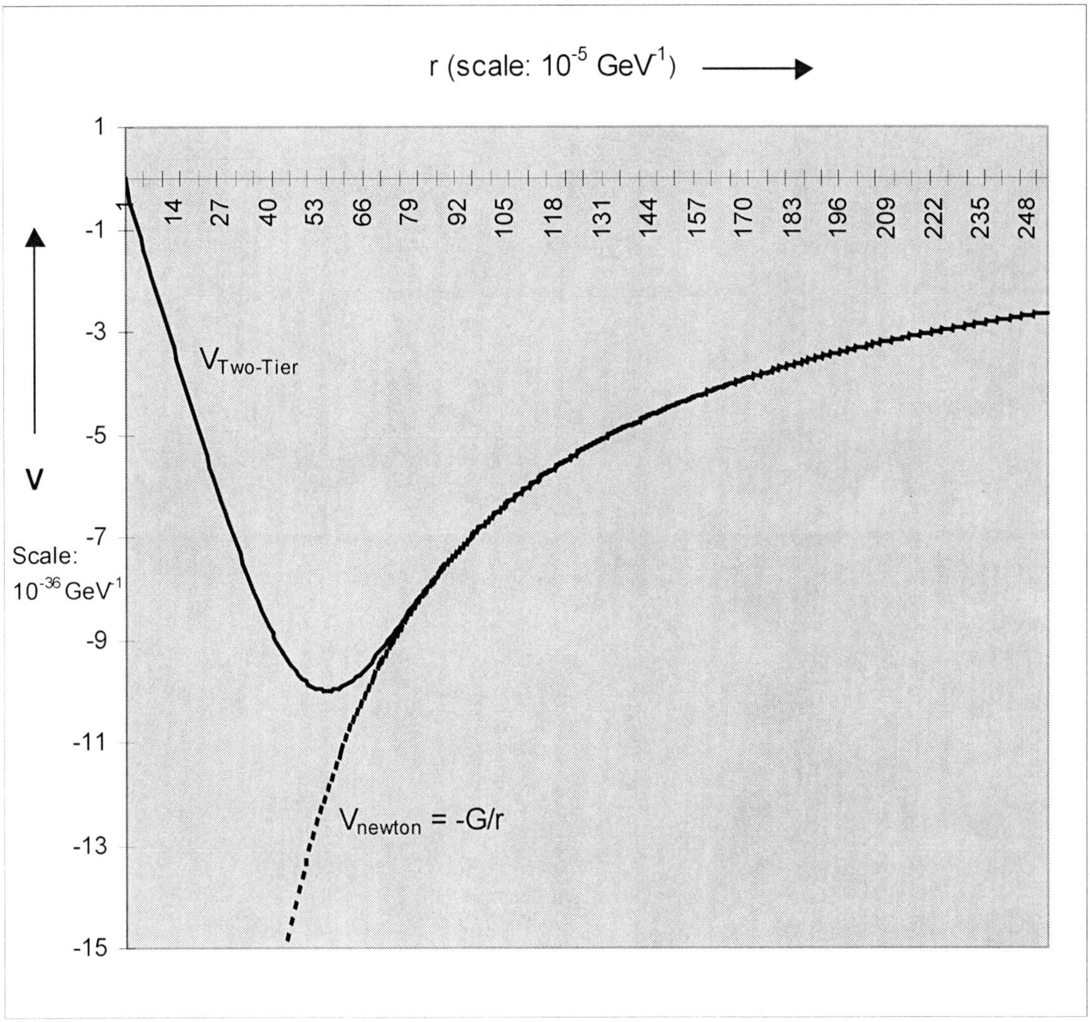

Figure 9.1. Plot of Two-Tier gravitational potential for M_c = 1 TeV/c^2 and Newton's gravitational potential. The potentials are measured in units of 10^{-36} GeV^{-1}. The radial distance is measured in units of 10^{-5} GeV^{-1}. The plot of the Two-Tier potential shows the force of gravity is repulsive for small $r < 5.7 \times 10^{-4}$ GeV^{-1}.

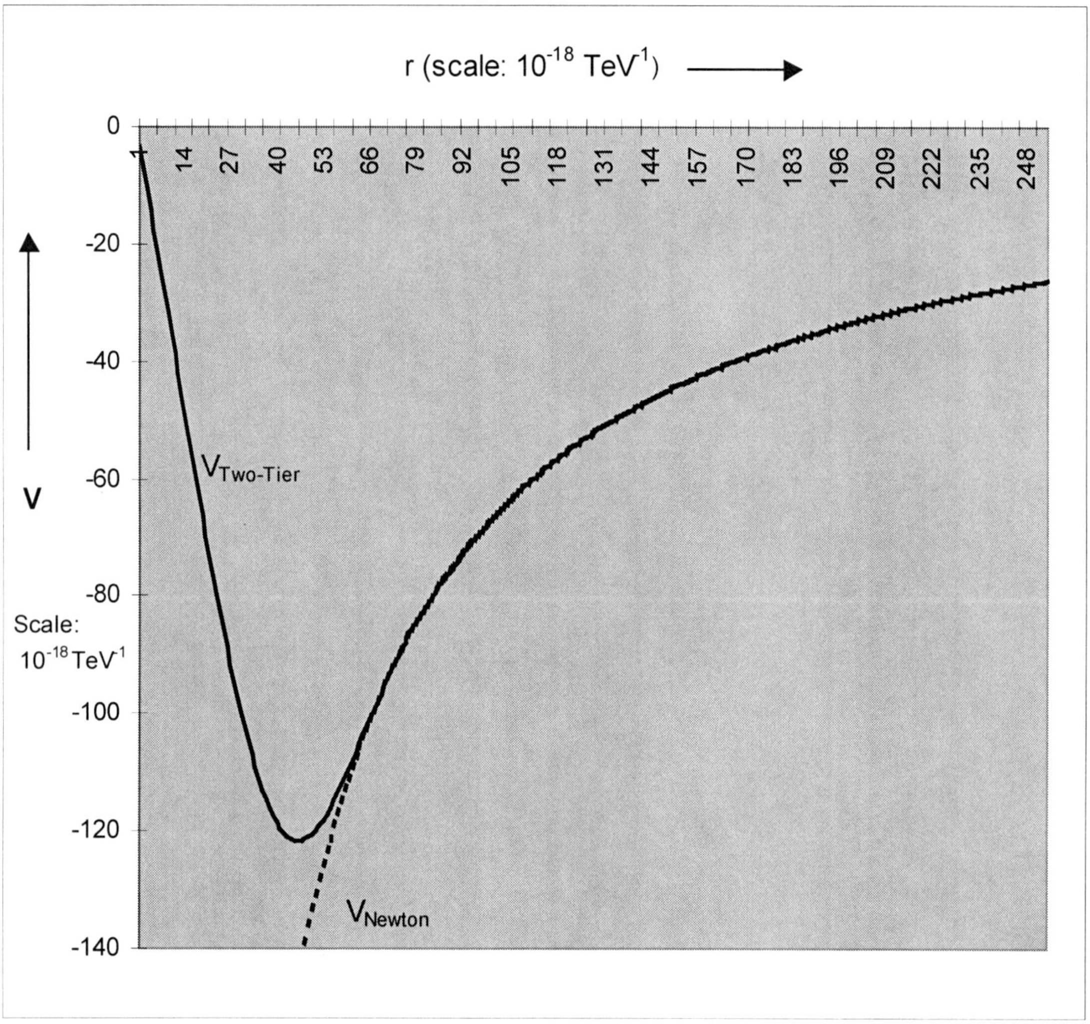

Figure 9.2. Plot of Two-Tier gravitational potential for $M_c = 1.22 \times 10^{19}$ GeV/c^2 (the Planck mass) and Newton's gravitational potential. Potentials are measured in units of 10^{-18} TeV^{-1}. The radial distance is measured in units of 10^{-18} TeV^{-1}.

Black Holes

The existence of microscopic black holes has been the subject of much speculation. It appears that arbitrarily small black holes can exist in classical General Relativity. The divergences associated with the short distance behavior of its conventional quantization raise the possibility of additional singular behavior at short distances as well.

On the other hand, in Two-Tier Quantum Gravity, at short distances, when the distance scale becomes less than M_c^{-1} (and thus the energy scale becomes greater than M_c), the Two-Tier gravitational force grows smaller and become zero in the limit of zero distance or infinite energy. The preceding figures (Figs. 9.1 – 9.2) show the Two-Tier gravitational potential linearly approaches zero at short distances unlike the Newtonian gravitational potential which approaches $-\infty$ as r approches zero. (The transverse gravitational propagator also approaches zero at short distances.) Thus the short distance behavior of Two-Tier gravity suggests that black holes of ultra-small size may not exist in Two-Tier Quanum Gravity.

If we examine the Two-Tier gravitational potential we note that it is similar to the Newtonian potential until the separation distance approaches the minimum of the potential. Thus we might expect that conventional classical General Relativity would be approximately valid down to distances of the order of the location of the minimum of the Two-Tier potential. Based on this assumption and on the assumption that M_c equals the Planck mass:

Assumption: $$M_c = M_{Planck} = G^{-1/2} \tag{9.90}$$

we can calculate the mass of a black hole whose radius equals the minimum of the Two-Tier potential. From eq. 9.88 we obtain

$$r_{MIN} = (G/\pi)^{1/2} = r_{BlackHole} = 2GM_{BlackHoleMIN} \tag{9.91}$$

with the result

$$M_{BlackHoleMIN} = (4\pi G)^{-1/2} = \kappa^{-1} \tag{9.92}$$

by eq. 9.11 and

$$M_{BlackHoleMIN} \cong .282\ M_{Planck} \tag{9.93}$$

or 6.15×10^{-6} grams. This lower limit on black hole mass is substantially greater than the collision energy than can be achieved in any current particle accelerator. Thus the production of ultra-small black holes in particle accelerators is unlikely.

Since corrections to conventional quantum gravity are at most of the order of M_c^{-2} it appears that the value of $M_{BlackHoleMIN}$ is consistent with the approximate validity of classical expression for a black hole radius. We note

$$(M_{BlackHoleMIN}/M_c)^2 \cong .0795 \tag{9.94}$$

and so corrections to eq. 9.93 would be very small.

10. Curved Space-time Generalization of Two-Tier Quantum Gravity

Inertial Reference Frames & Absolute Space-time

The concept of a flat, absolute space-time can be defined in two ways:

1. There exists a specific reference frame, an *absolute reference frame*, with space-time coordinates that we will denote as y^μ for $\mu = 0, 1, 2, 3$. Any reference frame whose space-time coordinates y'^μ are related to the y^μ coordinates by equations of the form:

$$y'^j = R^j_{\ i} y^i + v^j y^0 + c^j \tag{10.1}$$

$$y'^0 = y^0 + c^0 \tag{10.2}$$

where $R^j_{\ i}$ is a constant, real, orthogonal matrix, and where v^j and c^μ are constants for $j = 1, 2, 3$ and $\mu = 0, 1, 2, 3$ is an equivalent inertial reference frame. The set of these reference frames is called the set of *inertial reference frames*. The form of the equations of motion for a set of point-like particles is the same in any inertial reference frame. Thus these reference frames are physically equivalent.

2. There is a class of reference frames called inertial reference frames whose coordinates are related by equations of the form of eqs. 10.1 and 10.2. The form of the equations of motion for a set of point-like particles is the same in any inertial reference frame and for slowly moving particles have the form of Newton's equations without inertial forces. No inertial reference frame has any special significance.

Current cosmological data suggest that space is almost flat, or flat. Thus we can establish either local inertial reference frames (curved space case) or global inertial reference frames (flat space case) as experiment eventually will indicate.

Since the form of the equations of motion is the same in all local inertial frames it would appear that there is no way to physically distinguish between definitions 1 and 2. However, the observed characteristics of cosmic background radiation (CBR), of the redshift – distance relationship, and of the cosmological X-ray background effectively define a preferred local reference frame in each spatial locale. Effectively the CBR plays the role of an *aether selecting a preferred local inertial reference frame in each spatial locale*. The set of all such preferred inertial reference frames for the universe effectively defines an Absolute Reference Frame. IF space is truly flat, then the Absolute Reference Frame consists of one inertial reference frame.

The views of physicists on the question of an absolute reference frame have oscillated over time. Newton chose the first definition and talked of "absolute space" in the famous quote, "Absolute space in its own nature, and with regard to anything external, always remains similar and immovable." Mach challenged Newton's view and enunciated what became known as Mach's Principle postulating that there existed a class of inertial frames (second type of definition) that were defined by the mass distribution, and its movement, of the universe. Einstein established an "intermediate" position between Newton and Mach in his General Theory of Relativity. General Relativity implicitly defines the equivalent of an absolute space and shows that the presence of masses is not required since inertial frames exist in the empty space solutions of the equations of general relativity.

However the dynamical equations of General Relativity are covariant under any general relativistic transformation. *One can view General Relativity as a theory embodying invariance under general relativistic transformations with the invariance "broken" by a class of equivalent local "ground states" called local inertial reference frames that are determined by Mach's Principle, and eqs. 10.1 and 10.2 – an analogue of spontaneous broken symmetry.*

In this chapter we shall examine the Newtonian, Machian, and Einsteinian views in more detail and discuss them in relation to Two-Tier quantum field theory.

Newtonian Mechanics Embodies an Absolute Space-time

Newton developed his formulation of mechanics with equations of motion for groups of point-like particles. These equations had the same form in all inertial reference frames. He then asserted that the class of inertial reference frames was selected because they were either at rest or in a state of constant velocity with respect to a particular reference frame that corresponded to *absolute space*. He postulated the existence of a *physical* absolute space partly for theological reasons.

Mach's Principle Embodies an Absolute Space-time

Ernst Mach disagreed with Newton's concept that the basis for the special properties of inertial reference frames was their relation to the reference frame of an absolute space. Following Leibniz and others he proposed another view that is now called *Mach's Principle*.

An example[44] that illustrates Mach's thinking is:

> Consider a universe consisting of two identical spheres, not necessarily in close proximity, upon each of which an ant is stationary on the sphere's equator. Assume the spheres are rotating with respect to each other with parallel axes of rotation. The ant on each sphere sees the ant on the other sphere rotating with the sphere. Questions: which ant experiences an "upward" centrifugal force? Which ant experiences a Coriolis force (which is proportional to the angular velocity of rotation)?

These questions do not have answers in the universe of two spheres subject only to Newton's laws according to Ernst Mach. Either sphere could be considered to be the non-rotating sphere and a corresponding valid coordinate system defined in which the other sphere would be rotating.

Mach resolved this issue by noting the universe is very large and populated with large masses in all directions at great distance. In his view the distribution and motion of all the masses in the universe defined a preferred reference frame, although it is not usually portrayed in that manner. Mach would then have resolved the two spheres issue by saying the rotation of each sphere must be determined relative to the distribution and motion of the rest of the matter in the *real* universe.

As R. H. Dicke has remarked,[45] "If one were to remove this matter [at great distances], then according to Mach, the inertial force would disappear. If one were to reduce the matter to negligible proportions, there would be striking changes in local inertial effects. To summarize, according to Mach's point of view, we should interpret inertial effects as a consequence of interactions of matter at great distances in the universe with accelerated bodies in the laboratory."

Although many physicists, including Einstein, were strongly influenced by Mach's arguments many physicists were also uneasy about Mach's Principle. As Eddington[46] remarked, on the use of matter at infinity to define inertial frames, "the main feeling seems to be that it is unsatisfactory to have certain conditions prevailing in the world, which can be traced away to infinity and so have, as it were, their source at infinity; and there is a desire to find some explanation of the inertial frame as built up through conditions at a finite distance."

Thus Mach's Principle was viewed with mixed feelings both before, and long after, Einstein's development of his general theory of relativity.

---

44 Mach (1991).
45 R. H. Dicke, lecture entitled *The Many Faces of Mach* (1963).
46 Eddington (1995) p. 157.

General Relativity Embodies an Absolute Space-time

Einstein, himself, was strongly influenced by Mach's arguments. Mach's Principle was certainly on his mind when he was formulating his equivalence principle and the general theory of relativity. To some extent he felt his theory embodied Mach's Principle. However in the view of almost all observers General Relativity only partly implements Mach's Principle. General Relativity *also* embodies aspects of an absolute space-time.

R. H. Dicke,[47] points out:

> "Einstein's theory is not relativistic in the Machian sense. In his theory, space has physical properties and constitutes a physical structure even in the absence of all matter. ... general relativity does not appear to describe Mach's principle properly. This can be seen by noting that, in the absence of all matter, the metric tensor describes a flat space and this flat space possesses inertial properties. Even Schwarzschild's famous solution is unsatisfactory, from the point of view of Mach. As one moves to infinity, and the mass source (the source of inertial forces according to Mach) disappears in the distance, the space becomes flat and continues to possess inertial properties in contradiction with the expectations of Mach. ...
>
> We have ... the return to the idea that we are dealing with an absolute space-time. From the viewpoint of Synge, general relativity describes the geometry of an absolute space. According to him, certain things are measurable about this space in an absolute way. There exist curvature invariants that characterize this space, and one can, in principle, measure these invariants. Bergmann has pointed out that the mapping of these invariants throughout space is, in a sense, a labeling of the points of this space with invariant labels (independent of coordinate system). These are concepts of an absolute space, and we have here a return to the old notions of an absolute space."

The Case for Absolute Space-time

With the very strong case for absolute space-time made by Synge, Dicke and Bergmann, three of the great general relativists, we now ask whether a return to absolute space-time is in order in the sense of definition 2 above.

The arguments for an absolute space-time are:

1. It is embodied in the two successful theories of mechanics: Newtonian mechanics and general relativity.

2. It supports a local definition of physics consistent with the spirit of Riemannian geometry, general relativity and quantum field theory.

---

[47] R. H. Dicke, lecture entitled *The Many Faces of Mach* (1963).

3. **Experimental data is consistent with the existence of absolute rotation. As Eddington[48] notes, "The great stumbling block for a philosophy which denies absolute space is the experimental detection of absolute rotation." Most interestingly, current cosmological experimental data suggests space is very close to flat if not flat. Thus we are living in an absolute reference frame if the universe is considered in the large (with local masses averaged over cosmological distances.)**

4. It appears to be impossible to construct a theory of classical mechanics that is consistent with experiment that does not explicitly, or implicitly, embody absolute space-time in the form of definitions 1 or 2.

Experimental Determination of an Absolute Reference Frame

Experimentally we have found that space is close to flat although it appears to have enough curvature to form a closed space. It may be flat.

Experimentally we have also found that the observed characteristics of cosmic background radiation (CBR), of the redshift – distance relationship, and of the X-ray background effectively define a preferred local reference frame. The near flatness of space on large distance scales suggests this preferred local reference frame is almost an Absolute Minkowski reference frame. Thus current experiment has found an absolute reference frame – the only question is whether it is local or global. *As Peebles (1993) points out*[49]

"Blackbody radiation can appear isotropic only in one frame of motion. An observer moving relative to this frame finds that the Doppler shift makes the radiation hotter than average in the direction of motion, cooler in the backward direction. That means CBR acts as an aether, giving a local definition for preferred motion. ... In the standard interpretation, the same preferred comoving rest frame is defined by the CBR, the redshift-distance relation for galaxies, and the X-ray background. ... The evidence[50] *is that the frames are consistent to perhaps 300 km s^{-1}."*

Thus we have an experimental definition of an almost flat, or flat, absolute reference frame. We conclude with Synge:[51] "The Principle of Equivalence performed the essential office of midwife at the birth of general relativity, but, as Einstein remarked, the infant would never have gotten beyond its longclothes had it not been for

---

[48] Eddington (1995) p. 152.
[49] Eddington (1995) p. 151-2.
[50] M. Aaronson et al, Astrophysical Journal **302**, 536 (1986); R. A. Shafer and A. C. Fabian, in *Early Evolution of the Universe and its Present Structure*, ed. G. O. Abell and G. Chincarini, p. 333 (1983); Rubin (19878).
[51] Synge (1960) pp. ix-x.

Minkowski's concept. I suggest that the midwife be now buried with appropriate honours and the facts of an absolute space-time faced."

In the preceding chapters we have defined a unified quantum field theory that embodies the notion of an absolute inertial reference frame (or a set of local preferred inertial reference frames that apply to large locales) in a more direct way than classical general relativity. We defined X coordinates and a Y field in the preferred inertial reference frame of a locale, and then defined Two-Tier theories that are invariant under special relativistic transformation to other inertial reference frames.

Curved Space-time Generalization of Two-Tier Quantum Gravity

Thus the preceding chapters developed a divergence-free theory of scalar particles and quantum gravity in a flat space-time. In this section we show that a curved space-time version of Two-Tier quantum field theories including quantum gravity can be developed along the lines pioneered by DeWitt and collaborators. Two-Tier curved space-time quantum field theory is based on a mapping from a flat space-time parametrized by y coordinates to a curved space-time parametrized by X coordinates.

The physical picture of the mapping can be visualised using the simple example of a sphere of radius one in three-dimensional space with a coordinate system on the sphere and two planes – one above the sphere and one below it – each with its own flat space coordinate system. Both planes are assumed to be parallel to the disk defined by the crossection of the sphere bounded by the equator of the sphere. A minimum of two coordinate patches are needed to cover a sphere in three dimensions since it necessarily has coordinate singularities.

Let us place a rectangular coordinate system on the top plane. Points on this plane can be mapped onto its northern hemisphere of the sphere in a simple one-to-one fashion. Similarly a rectangular coordinate system can be placed on the bottom plane which can be mapped in a one to one fashion onto the southern hemisphere of the sphere. The top and bottom planes each have a two-dimensional coordinate system that we can choose to be a Cartesian coordinate system in both cases. We will label the coordinates on the top plane x_t^1 and x_t^2, and the points on the bottom plane as x_b^1 and x_b^2. Each plane has a flat space metric $g_{tij} = g_{bij} = \delta_{ij}$ for i, j = 1,2 with δ_{ij} the Kronecker delta.

In addition, just for concreteness, we will place the origin of the top plane coordinate system vertically above the north pole of the sphere, and the origin of the bottom plane coordinate system vertically below the south pole of the sphere.

If we place the sphere at the center of a three dimensional, coordinate system then the points on the sphere (x,y,z) all satisfy:

$$x^2 + y^2 + z^2 = 1 \qquad (10.3)$$

We can defined coordinates u^1 and u^2 for each hemisphere on the surface of the sphere with equations of the form:

$$x_n = f_{1n}(u_n^1, u_n^2) \tag{10.4}$$
$$y_n = f_{2n}(u_n^1, u_n^2) \tag{10.5}$$
$$z_n = f_{3n}(u_n^1, u_n^2) \tag{10.6}$$

for the northern hemisphere, and

$$x_s = f_{1s}(u_s^1, u_s^2) \tag{10.7}$$
$$y_s = f_{2s}(u_s^1, u_s^2) \tag{10.8}$$
$$z_s = f_{3s}(u_s^1, u_s^2) \tag{10.9}$$

for the southern hemisphere.

In addition, we choose $u_n^1 = u_n^2 = 0$ at the north pole and $u_s^1 = u_s^2 = 0$ at the south pole. The surface of the sphere is curved and each (u^1, u^2) coordinate system has a metric, g_{nij} and g_{sij} for i, j = 1,2 respectively, and a non-zero curvature tensor R_{nijkl} and R_{sijkl}.

Now we are allowed to define a simple map of points on the northern hemisphere of the sphere to points on the top plane such as:

$$x_t^1 = u_n^1 \tag{10.10}$$
$$x_t^2 = u_n^2 \tag{10.11}$$

and of points on the southern hemisphere of the sphere to points on the bottom plane:

$$x_b^1 = u_s^1 \tag{10.12}$$
$$x_b^2 = u_s^2 \tag{10.13}$$

Thus we can specify the location of events on the sphere on our planes. Note that eqs. 10.4 – 10.9 are *not* a coordinate transformation of the (u^1, u^2) coordinate systems on the sphere and thus the plane can have a different (flat) metric from the sphere.

The preceding example can be simplified by using a cylinder enclosing the sphere instead of two planes. The cylinder, which is a flat surface technically, is aligned so that its axis is parallel to, and cenetered on, the north-south axis of the sphere. Then a map can be made from points on the sphere to points on the cylinder that is similar to a Mercator projection, or from points on the sphere to the cylinder that maps the poles to the ends of the cylinder at + and – infinity.

The preceding discussion shows a clear analogy to our map from the y Minkowski space-time to the curved X space-time using

$$X^\mu = y^\mu + i\,Y^\mu(y)/M_c^2 \qquad (10.14)$$

modulo the imaginary term. The y Minkowski space-time has a flat space-time in which we are allowed to choose the Minkowski metric $\eta_{\mu\nu}$. The curved X space-time has an appropriate metric $g_{\mu\nu}(X)$ that can only be transformed to locally inertial coordinates with perhaps a Minkowski metric in the neighborhood of a point. The additional imaginary term does not alter this picture except that the curved X space-time is now a slightly complex manifold in complex space-time.

Therefore we conclude that our Two-Tier quantum field theoretic formalism that is erected on eq. 10.14, where the real part of the X space-time was flat, can be extended to curved space-time while maintaining eq. 10.14 if the y space-time consists of coordinate patches analogous to the two planes (or the cylinder) in the example of the sphere. The difference is that we now use a curved space-time background metric $g_{\mu\nu}(X)$ instead of $\eta_{\mu\nu}$ throughout the lagrangian with the exception of L^Y (eq. 9.27).

In L^Y we continue to use $\eta_{\mu\nu}$ as the metric. As a result L^Y breaks the invariance of the complete lagrangian under general coordinate transformations. Thus an implicit absolute space-time is implied – as it is implicitly in classical General Relativity and in cosmological experiments. This consequence is not disturbing and is physically acceptable for the following reasons:

1. As Bergmann and Synge point out classical general relativity implicitly embodies an absolute space-time.
2. Experiment shows that space in the large (of the order of the Hubble length) is nearly flat although space does appear to be closed. CBR, and other, experimental data suggests that an absolute reference frame exists.

Thus our universe does appear to be in a state of broken general coordinate transformation invariance. Two-Tier quantum field theory in curved space-time is not in contradiction with our previous classical general relativistic theories or with our experimental knowledge of the universe. *The full lagrangian theory L is invariant under special relativity. $L - L^Y$ is formally invariant under general coordinate transformations in the X coordinates.*

Why Are the Y Field Dynamics Independent of the Gravitational Field?

It is evident that the Y^a field is a truly free field in our formulation. In particular, it does not depend on, or interact directly with, the gravitational field as represented by $\sqrt{g}$ and $g_{\mu\nu}$ factors. On the other hand, these quantities depend on the Y^a field through their dependence on the variable X^μ.

Thus the role of Y^a is strictly that of a coordinate, and of a field that is parametrized by a set of inertial frame coordinates y^μ. The arguments of Mach supplemented by the arguments of Bergmann and Synge show that a de facto absolute reference frame exists (actually it is the set of inertial reference frames). Therefore we can chose to formulate our theory in an inertial reference frame and require that the theory only be invariant under Lorentz transformations to other inertial reference frames.

In this context it is allowed to have one or more fields like Y^a whose dynamics are not invariant under general coordinate transformations. It it is reasonable to require the particle and gravitational dynamical equations be covariant under general coordinate transformations in X. *Thus a part of the dynamics is invariant under Lorentz transformations – the Y^a sector – but this part of the dynamics is not directly observable; and a part of the dynamics – the observable part – is invariant under general coordinate transformations.*

Some reasons for having a free Y^a field are:

1. It is required to avoid divergences that would appear in perturbation theory if the Y^a were allowed to interact with gravitons. For example an hhYY interaction term causes a divergence to appear by generating a Y particle loop in graviton-graviton scattering.
2. If the Y^a particle interacted with gravity then measurable, classical Y^a fields could be generated in regions with ultra-strong gravitational fields such as the neighborhoods of black holes. In this case we would have new dimensions, allbeit imaginary, for which no experimental evidence currently exists.
3. The Principle of Equivalence has only been shown to apply on the classsical level for real coordinates. Any quantization that uses Minkowskian coordinates, or quasi-Minkowskian coordinates, causes general coordinate transformation invariance to be abandoned ab initio in the quantum regime.

11. A Unified Quantum Field Theory of the Known Forces of Nature

Formulation of Unified Theory

The unification of QED and weak interactions in Electroweak Theory interrelated the theories within the framework of an overall SU(2)⊗U(1) symmetry and thus was significantly more then merely "glueing" theories together.

The unification of Electroweak theory with Quantum Chromodynamics (QCD) into the Standard Model was a direct combination of these theories in which the symmetries of each respective theory were directly combined: SU(2)⊗U(1) symmetry from Electroweak Theory and SU(3) from QCD to produce the Standard Model with SU(2)⊗U(1)⊗SU(3) symmetry. Electroweak Theory was "glued together" with QCD to produce the Standard Model without an underlying rationale. Nevertheless, the Standard Model is a renormalizable theory that accounts for the vast majority of experimental data.

The present work has two goals: 1.) To make QED, Electroweak Theory, QCD and Quantum Gravity finite and thus remove a major long term defect in Quantum Field Theory, and 2.) To create a unified theory of the known forces of Nature from these pieces. Item 1 has been achieved in the preceding chapters using Two-Tier Quantum Field Theory. This chapter discusses item 2.

In this chapter we propose a finite unified theory of the Standard Model (and any of its variants) and Quantum Gravity within the framework of Two-Tier Quantum Field Theory. As we saw in the case of the Standard Model our unified theory amounts to glueing together the Two-Tier version of the Standard Model with the Two-Tier version of Quantum Gravity. Therefore we regard the theory as provisional in the sense that a deeper unification remains to be formulated.

Nevertheless the Two-Tier Unified Theory may be of some importance beyond the satisfaction of having a finite theory of Nature. It might be a starting point for a deeper, more unified theory. It can be used to address cosmological questions such as the state of the universe immediately after the Big Bang when the size of the universe was of the order of the Planck length or smaller – see Blaha (2004). The interactions in the unified theory become weaker at short distances and thus low order perturbation theory becomes a better approximation to the exact results.

Unified Two-Tier Standard Model and Quantum Gravity Lagrangian

We define the Lagrangian, and action, for the Two-Tier unified quantum field theory of gravitation and the Standard Model as

$$L_{Unified} = \int d^4y\, \mathscr{L}_{Unified} \tag{11.1}$$

$$\mathscr{L}_{Unified} = J\, \sqrt{g(X)}\, \left(\mathscr{L}_F^{Grav}(X) + \mathscr{L}_F^{SM}(X)\right) + \mathscr{L}_C \tag{11.2}$$

with

$$\mathscr{L}_F^{Grav}(X) = (2\kappa^2)^{-1}{}_X R(X) \tag{11.3}$$

where $\mathscr{L}_F^{SM}$ is the complete "normal" Quantum Field theory Lagrangian for the Standard Model variant under consideration written in a general coordinate covariant form, g(X) is the absolute value of the determinant of $g_{\mu\nu}$, and J is the Jacobian of eq. A.21. *All particle fields in* $\mathscr{L}_F^{SM}$ *are assumed to be functions of the X coordinate only. The dependence of the particle fields on the "underlying" coordinates* y^μ *is assumed to be solely through* X^μ. The Lagrangian $\mathscr{L}_{Unified}$ is a separable Lagrangian of the type of eq. A.26 embodying the composition of extrema described in Appendix A.

As in all cases considered we define the coordinate part of the Lagrangian $\mathscr{L}_C$ as

$$\mathscr{L}_C = -\tfrac{1}{4}\, F_Y^{\mu\nu} F_{Y\mu\nu} \tag{11.4}$$

with

$$F_{Y\mu\nu} = \partial Y_\mu/\partial y^\nu - \partial Y_\nu/\partial y^\mu \tag{11.5}$$

and

$$F_Y^{\mu\nu} = \eta^{\mu\alpha}\eta^{\nu\beta} F_{Y\alpha\beta} \tag{11.6}$$

The development of the physics embodied in the Lagrangian proceeds along the lines described in the previous chapters, except that the gravitational sector must be in the form of a vierbein theory since the full theory contains spin 1/2 particles.

"Low Energy" Behavior

The low energy sector of the unified theory is defined as the sector with momenta whose values are much less than M_c. It is clear from the discussions of the previous chapters that the low energy sector of the unified theory is effectively identical to that of the corresponding conventional quantum field theory if M_c is sufficiently large.

In addition the low energy behavior of the Two-Tier unified theory in the QED sector closely approximates the results of QED calculations which have been found to agree well with experiment to an extremely high degree of accuracy.

Thus the Two-Tier unified theory satisfies a Correspondence Principle in the Standard Model sector: The low energy behavior of the Two-Tier Standard Model sector is the same as the behavior of the conventional Standard Model to a high degree of approximation. In addition the Two-Tier vierbein Quantum Gravity sector tree diagrams are the same as the conventional vierbein Quantum Gravity tree diagrams to a high degree of accuracy.

Negative Degree of Divergence – A Finite Unified Theory

The previous discussions of the perturbation theory of matter fields, gauge fields and gravitons show that the theory is finite.

Unitarity

The unitarity discussions of the various sectors of the unified theory in previous chapters show the unified theory satisfies unitarity. As long as Y excitations are not allowed in in-states they will not appear in out-states. The S matrix is block diagonal and unitary within the physical asymptotic states sector.

Appendix A. Composition of Extrema in the Calculus of Variations

A New Paradigm in the Calculus of Variations

The Calculus of Variations has a long and venerable history in Physics and Mathematics. Many problems in Physics and Mathematics have been treated with approaches based on techniques in the Calculus of Variations (see the references at the end of the book). In this book we have developed a unified quantum field theory of the known forces of nature based on a new type of problem, or paradigm, in the Calculus of Variations. One way of viewing the spectrum of problems in the calculus of variations is the following progression.

A Classification of Variational Problems

1. Variational problems in a Euclidean, or Minkowski, flat space such as the minimal distance between two points or the extrema of a field theory Lagrangian.

2. Variational problems seeking extrema on a curved surface such as the shortest distance between points on the surface of a sphere.

The development in this book suggests a third and fourth, possibility, that to the author's knowledge, has not been addressed in the literature:

3. Variational problems where the extrema are determined on a surface that is itself defined as an extremum. The discussions in this book exemplify this pardigm.

4. Variational problems where the extrema are determined on a surface that is itself defined as an extremum that depends on the extrema on the surface. More simply put the extrema, and the surface upon which they are defined, are jointly determined and are interrelated. Fortunately, our unified theory does not use this paradigm. A future theory might.

In the unified theory that we will develop all particle fields including the graviton field are defined as a mapping of a Minkowski space-time y to a "particle" space-time X with the mapping determined as an extremum of a variation of a fundamental field (a type 3 variational problem in the above classification). Our theory could be generalized to include a back-reaction of the particle fields on the fundamental field (a type 4 variational problem in the above classification). We will not discuss this possibility in this book.

Simple Physical Example – Strings On Springs

In this section we will describe a simple physical example that illustrates a variational problem of type 3 in the Calculus of Variations. We view it as a composition of extrema. (This problem can be addressed using other calculus of Variations techniques.) The approach used in the solution of this problem is similar to the approach used in Two-Tier quantum field theory.

A Strings on Springs Mechanics Problem

Consider a long string or bar that can oscillate (undulate) in a direction perpendicular to its length. Further assume that one end of this bar or string is attached to a spring that cause the entire bar or string to oscillate back and forth in a direction parallel to its long side. This configuration is illustrated in Fig. A.1.

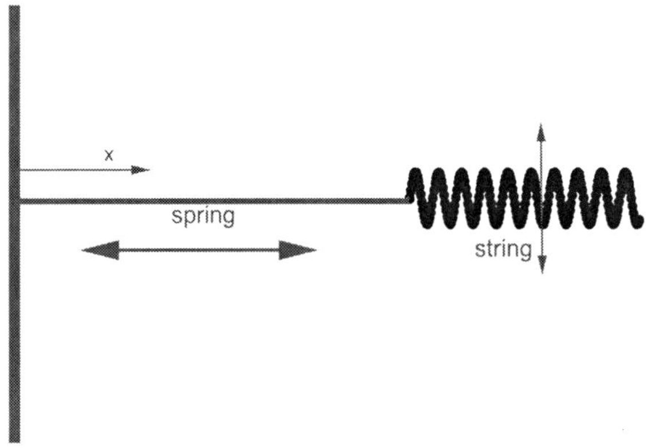

Figure A.1. An oscillating string attached to a spring.

Let x denote the distance to a point on the string when the spring is at equilibrium. If 2π times the frequency of the spring is ω_1, then the location of this point when the spring is oscillating is

$$X(t) = x + A \sin(\omega_1 t + \phi_1) \qquad (A.1)$$

where ϕ_1 is a phase, and A is the amplitude of the spring oscillation. Then the vertical displacement of a traveling wave on the *string* can take the form

$$\psi(t) = B \sin(\omega_2 t - k_2(x + A \sin(\omega_1 t + \phi_1)) + \phi_2) \quad (A.2)$$

where B is the amplitude of the string wave, and k_2, ω_2 and ϕ_2 are the parameters of the string wave. These simple mechanical formulae are well known. But they lead to an interesting new application of the ideas of the Calculus of Variations.

Suppose we treat X as an independent variable with X given by eq. (A.1), and with eq. (A.2) written as:

$$\psi(t) = B \sin(\omega_2 t - k_2 X + \phi_2) \quad (A.3)$$

Defining

$$\psi = \psi(X(t), t) \quad (A.4)$$

we can specify the dynamics of the above motion by finding the extrema of

$$I = \int \mathscr{L}_\psi \, dX(t) + \int \mathscr{L}_X \, dt \quad (A.5)$$

where the Lagrangian terms are

$$\mathscr{L}_\psi = ½ \{ \mu \, (\partial\psi/\partial t)^2 - Y \, (\partial\psi/\partial X)^2 \} \quad (A.6)$$

with μ and Y being constants, and

$$\mathscr{L}_X = ½ \{ m(\partial X/\partial t)^2 - k(X - x)^2 \} \quad (A.7)$$

where m and k are constants, and where x is a parameter. Applying Hamilton's Principle, and performing independent variations of X and ψ yields the Lagrangian equations:

$$\frac{\partial \mathscr{L}_\psi}{\partial \psi} - \frac{\partial}{\partial X}\frac{\partial \mathscr{L}_\psi}{\partial(\partial\psi/\partial X)} - \frac{\partial}{\partial t}\frac{\partial \mathscr{L}_\psi}{\partial(\partial\psi/\partial t)} = 0 \quad (A.8)$$

and

$$\frac{\partial \mathscr{L}_X}{\partial X} - \frac{\partial}{\partial t}\frac{\partial \mathscr{L}_X}{\partial(\partial X/\partial t)} = 0 \quad (A.9)$$

The resulting equations of motion are:

$$\mu\, \partial^2\psi/\partial t^2 - Y\, \partial^2\psi/\partial X^2 = 0 \quad (A.10)$$

and

$$m\, \partial^2 X/\partial t^2 + k(X - x) = 0 \quad (A.11)$$

with the solutions given in eqs. A.1 and A.2.

The procedure that we use to obtain these results may look a bit strange but they illustrate a type 3 problem in the Calculus of Variations involving the composition of extrema—the composition of an extremum that specifies a manifold in a space (possibly including all of space in a $R^n \rightarrow R^n$ mapping) with an extremum determining a function on that manifold. The procedure is described in detail in the next section.

The Composition of Extrema – Lagrangian Formulation

In this section we will explore the general case of the composition of extrema for fields. We will discuss the case of a scalar field ϕ that is a function of a vector field X^μ in a D-dimensional space with coordinate variables that we will denote as y^μ. (The discussion for other types of fields is a straightforward extension of this discussion.) Thus

$$\phi = \phi(X) \quad (A.12)$$

and

$$X^\mu = X^\mu(y) \quad (A.13)$$

We assume that the dynamics can be described by a Lagrangian formulation using an extension of Hamilton's principle:

$$I = \int \mathscr{L}\, d^4y \quad (A.14)$$

with

$$\mathscr{L} = \mathscr{L}(\phi(X), \partial\phi/\partial X^\nu, X^\mu(y), \partial X^\mu(y)/\partial y^\nu, y) \quad (A.15)$$

If we perform a standard variation[52] in ϕ for fixed y (and thus fixed X) we find

$$\delta I = \int [\delta\phi\, \partial\mathscr{L}/\partial\phi + \delta(\partial\phi/\partial X^\nu)\, \partial\mathscr{L}/\partial(\partial\phi/\partial X^\nu)]\, d^4y \quad (A.16)$$

---

52 Bogoliubov, N. N., & Shirkov, D. V., Volkoff, G. M. (tr), *Introduction to the Theory of Quantized Fields* (Wiley-Interscience, New York, 1959); Goldstein H., *Classical Mechanics* (Addison-Wesley, Reading, MA 1965).

We can rewrite the variation in the derivative of ϕ as

$$\delta(\partial\phi/\partial X^{\nu}) = \partial(\delta\phi)/\partial X^{\nu} \tag{A.17}$$

$$= \partial y^{\mu}/\partial X^{\nu}\, \partial(\delta\phi)/\partial y^{\mu} \tag{A.18}$$

with an implied summation over repeated indices. After substituting eq. A.18 in eq. A.16, and performing an integration by parts (and discarding the surface term which is assumed to yield zero in the standard fashion) we obtain:

$$\delta I = \int \delta\phi\, \{\partial\mathscr{L}/\partial\phi \;-\partial/\partial y^{\mu}\, [\partial\mathscr{L}/\partial(\partial\phi/\partial X^{\nu})\, \partial y^{\mu}/\partial X^{\nu})\,]\}\; d^4y$$

Since the variation of $\delta\phi$ is arbitrary we conclude

$$\partial\mathscr{L}/\partial\phi \;-\partial/\partial y^{\mu}\, [\partial\mathscr{L}/\partial(\partial\phi/\partial X^{\nu})\, \partial y^{\mu}/\partial X^{\nu})] = 0 \tag{A.19a}$$

The second term in eq. A.19a shows the effect of the dependence of ϕ on the field X, $\phi = \phi(X)$, rather than directly on the coordinate system y.

Similarly we can perform a variation in X^{μ} and obtain

$$\partial\mathscr{L}/\partial X^{\mu} \;-\partial/\partial y^{\nu}\, [\partial\mathscr{L}/\partial(\partial X^{\mu}/\partial y^{\nu})] = 0 \tag{A.19b}$$

The X field defines a "manifold" or, more properly, specifies a transformation from $R^n \rightarrow R^n$. If we make standard assumptions about the mapping: that it is continuous and piece-wise invertible, then we can establish the following lemmas:

Lemma 1: *If the transformation $X^{\mu} = X^{\mu}(y)$ is a transformation from $R^n \rightarrow R^n$ that is of class C′ and piece-wise invertible, then*

$$\frac{\partial}{\partial y^{\nu}}\,\frac{\partial y^{\nu}}{\partial X^{\mu}} \;=\; -\,\frac{\partial \ln J}{\partial X^{\mu}} \tag{A.20}$$

where

$$J = |\partial(X)/\partial(y)| \tag{A.21}$$

is the absolute value of the Jacobian of the transformation.

Proof:
Consider two equivalent forms of an integral:

$$I = \int \mathscr{L} J \, d^4y = \int \mathscr{L} \, d^4X$$

where $\mathscr{L}$ is specified as in eq. A.15. Then the first expression for I leads to eq. A.19a which can be written in the form

$$\partial\mathscr{L}/\partial\phi - \partial/\partial X^\mu [\partial\mathscr{L}/\partial(\partial\phi/\partial X^\mu)] - \partial\mathscr{L}/\partial(\partial\phi/\partial X^\mu)\{\partial[J\partial y^\nu/\partial X^\mu]/\partial y^\nu\} = 0$$

Using the second expression for I above we obtain the following equation by variation in ϕ:

$$\partial\mathscr{L}/\partial\phi - \partial/\partial X^\mu [\partial\mathscr{L}/\partial(\partial\phi/\partial X^\mu)] = 0$$

Comparing these two expressions and realizing that $\partial[J\partial y^\nu/\partial X^\mu]/\partial y^\nu$ is totally independent of ϕ and its derivatives leads us to conclude

$$\partial[J\partial y^\nu/\partial X^\mu]/\partial y^\nu = 0 \qquad (A.22)$$

It is a general relationship for a transformation between X and y based on continuity and piece-wise invertibility. After a few elementary manipulations eq. A.22 can be rewritten in the form of eq. A.20. ■

Lemma 2: *If the transformation $X^\mu = X^\mu(y)$ is a transformation from $R^n \rightarrow R^n$ that is of class C′ and piece-wise invertible and* $\mathscr{L} = \mathscr{L}(\phi(X), \partial\phi/\partial X^\nu, X^\mu(y), \partial X^\mu(y)/\partial y^\nu, y)$, *then*

$$\partial\mathscr{L}/\partial(\partial\phi/\partial X^\nu)\, \partial y^\mu/\partial X^\nu = \partial\mathscr{L}/\partial(\partial\phi/\partial y^\mu) \qquad (A.23)$$

Proof:
Let us express $\mathscr{L}$ as a power series in derivatives of ϕ:

$$\mathscr{L} = \sum_{n=0} a_{n\mu_1\mu_2\ldots\mu_n}(\phi(X), X^\mu(y), \partial X^\mu(y)/\partial y^\nu, y) \prod_{j=1}^{n} \partial\phi/\partial X^{\mu_j}$$

which can rewritten using piece-wise invertibility as

$$\mathscr{L} = \sum_{n=0} a_{n\mu_1\mu_2\cdots\mu_n}(\phi(X), X^\mu(y), \partial X^\mu(y)/\partial y^\nu, y) \prod_{j=1}^{n} \partial\phi/\partial y^{\nu_j}\, \partial y^{\nu_j}/\partial X^{\mu_j}$$

Taking the derivative of this equation with respect to $\partial\phi/\partial y^\mu$ immediately yields the result. ■

Eq. A.23 enables us to rewrite eq. A.19a as:

$$\partial\mathscr{L}/\partial\phi \;-\partial/\partial y^\mu\, [\partial\mathscr{L}/\partial(\partial\phi/\partial y^\mu)] = 0 \qquad \text{(A.24)}$$

which is as one would expect.

In order to get a feeling for the effect of eq. A.19a we will look at a simple example where we specify the relation of the X and y variables directly. Then we will look at the composition of extrema where the transformation between X and y is itself determined as an extremum solution.

Example: a hyperplane

We assume eq. A.19b yields the transformation:

$$X^i = ay^i \qquad \text{for } i = 1,2,3$$

$$X^0 = 0$$

Then eq. A.19a becomes

$$\partial\mathscr{L}/\partial\phi \;-\partial/\partial y^i\, [\partial\mathscr{L}/\partial(\partial\phi/\partial y^i)] = 0 \qquad \text{(A.25)}$$

with the time derivative disappearing. Effectively the variation of ϕ on the hyperplane $X^0 = 0$ is determined by the differential equation generated by A.25. On this hyperplane the transformation between the X and y variables is invertible.

Coordinate Transformation Determined as an Extremum Solution

We now develop a formalism that determines a mapping from space onto itself as the solution of an extremum problem and also determines the dynamics of one or more fields as a function of this mapping. To this author's knowledge this area in the Calculus of Variations – the determination of an extremum on a manifold where the manifold itself is determined by an extremum – has not been previously explored. We will also develop a hamiltonian formulation. Then we will proceed to quantize the theory.

Separable Lagrangian Case

Although there are many forms that the composition of extrema could take, one fairly general form that is directly useful in quantum field theory applications is based on a Lagrangian that can be split into two parts which we will call a *separable Lagrangian*:

$$\mathscr{L} = \mathscr{L}_F J + \mathscr{L}_C(X^\mu(y), \partial X^\mu(y)/\partial y^\nu, y) \qquad (A.26)$$

where J is defined in eq. A.21, where $\mathscr{L}_F$ contains all the dynamics of the fields and their interactions, and where $\mathscr{L}_C$ defines the coordinate mapping as an extremum solution. The procedure to determine the differential equations that specify the mapping, and the field equations that specify field interactions and evolution, is to vary in the coordinates X^μ and in the fields independently, using Hamilton's Principle. The extrema are to be determined for

$$I = \int \mathscr{L}\, d^4y \qquad (A.27)$$

We will begin by considering the case of one scalar field:

$$\mathscr{L}_F = \mathscr{L}_F(\phi(X), \partial\phi/\partial X^\nu) \qquad (A.28)$$

and

$$\mathscr{L}_C = \mathscr{L}_C(X^\mu(y), \partial X^\mu(y)/\partial y^\nu, y) \qquad (A.29)$$

Eq. A.27 can be written in the form:

$$I = \int \mathscr{L}_F(\phi(X), \partial\phi/\partial X^\nu)\, dX + \int \mathscr{L}_C(X^\mu(y), \partial X^\mu(y)/\partial y^\nu, y)\, d^4y \qquad (A.30)$$

using the Jacobian to transform to an integral over dX in the first term. A standard variation of ϕ and the application of Hamilton's Principle yields

$$\partial\mathscr{L}_F/\partial\phi - \partial/\partial X^\mu\, [\partial\mathscr{L}_F/\partial(\partial\phi/\partial X^\mu)] = 0 \qquad (A.31)$$

reflecting the fact that ϕ is a function of X^μ only, with X^μ a function of the y coordinates.

Next we perform a variation of X^μ determining the mapping from $y \to X$ as an extremum of the integral in eq. A.27. We note the piece-wise invertibility of the coordinate mapping $X^\mu(y)$ allows us to write the Jacobian J as a function of y^μ only. A standard variation of X^μ and the application of Hamilton's Principle yields

$$\partial\mathscr{L}_C/\partial X^\mu - \partial/\partial y^\nu\,[\partial\mathscr{L}_C/\partial(\partial X^\mu/\partial y^\nu)] = 0 \qquad (A.32)$$

Klein-Gordon Example

The Klein-Gordon scalar field theory furnishes us with a simple example of the application of the preceding development. The Lagrangian is

$$\mathscr{L}_F = \tfrac{1}{2}\,[\,(\partial\phi/\partial X^\nu)^2 - m^2\phi^2\,] \qquad (A.33)$$

From eq. A.31 we obtain the field equation:

$$(\Box + m^2)\phi(X) = 0 \qquad (A.34)$$

where

$$\Box = \partial/\partial X^\nu\,\partial/\partial X_\nu \qquad (A.34a)$$

A fourier representation of the solution of eq. A.34 is:

$$\phi(X) = \int dp\,\delta(p^2 - m^2)\theta(p^0)\,[A(p)\,e^{-ip\cdot X} + A(p)^*\,e^{ip\cdot X}] \qquad (A.35)$$

where A(k) is a function of k and * indicates complex conjugation.

The determination of $X^\mu(y)$ depends on the Lagrangian $\mathscr{L}_C$ and the solutions of eq. A.3A. If we chose

$$\mathscr{L}_C = -\tfrac{1}{2}\,(\partial X^\mu/\partial y^\nu)^2 \qquad (A.36)$$

Then we obtain the equation

$$\Box\,X^\mu = 0 \qquad (A.37)$$

with the solution

$$X^\mu = \int dk\,\delta(k^2)\theta(k^0)\,[a^\mu(k)\,e^{-ik\cdot y} + a^\mu(k)^*\,e^{ik\cdot y}] \qquad (A.38)$$

where $a^\mu(k)$ are complex vector functions of k in general. (We ignore positivity issues for the moment.) Substitution of eq. A.38 in eq. A.35 yields an expression with a form reminiscent of bosonic string expressions.[53] We will take up this point later in subsequent chapters.

The Composition of Extrema – Hamiltonian Formulation

The previous section established a Lagrangian formulation of dynamics based on the composition of extrema. In this section we will develop an equivalent hamiltonian formulation. We will assume a Minkowskian space-time with X^0 and y^0 playing the role of the time coordinates in the respective coordinate systems.

Initially, we will assume a scalar field ϕ with a Lagrangian of the form in eq. A.15 and define canonical momenta with

$$\Pi_\phi = \partial\mathscr{L}/\partial\dot{\phi} \equiv \partial\mathscr{L}/\partial(\partial\phi/\partial X^\mu)\,\partial y^0/\partial X^\mu \tag{A.39}$$

$$\Pi_X{}^\mu = \partial\mathscr{L}/\partial\dot{X}_\mu \tag{A.40}$$

where

$$\dot{\phi} = \partial\phi/\partial y^0 \equiv \partial\phi/\partial X^\mu\,\partial X^\mu/\partial y^0 \tag{A.41}$$

$$\dot{X}^\mu = \partial X^\mu/\partial y^0 \tag{A.42}$$

Then we define the hamiltonian density as

$$\mathscr{H} = \Pi_\phi\,\dot{\phi} + \Pi_X{}^\mu\,\dot{X}_\mu - \mathscr{L}(\phi(X), \partial\phi/\partial X^\nu, X^\mu(y), \partial X^\mu(y)/\partial y^\nu, y) \tag{A.43}$$

and the hamiltonian

$$H = \int \mathscr{H}\, d^3y \tag{A.44}$$

The hamiltonian density has the general form

$$\mathscr{H} = \mathscr{H}(\phi(X), \partial\phi/\partial X^i, \Pi_\phi, X^\mu(y), \partial X^\mu(y)/\partial y^j, \Pi_X{}^\mu, y^\nu) \tag{A.45}$$

for the case of one scalar field where the indices i and j represent space coordinates; time coordinates are assigned index value 0.

If we calculate the differential change in H using eq. A.45 we obtain

[53] See for example Polchinski (1998) and Bailin (1994).

$$dH = \int \{ \partial\mathcal{H}/\partial\phi \, d\phi + \partial\mathcal{H}/\partial\Pi_\phi \, d\Pi_\phi - \partial/\partial y^\nu [\partial\mathcal{H}/\partial(\partial\phi/\partial X^j)\partial y^\nu/\partial X^j] d\phi + $$
$$+ \partial\mathcal{H}/\partial X^\mu \, dX^\mu + \partial\mathcal{H}/\partial\Pi_X{}^\mu \, d\Pi_X{}^\mu - \partial/\partial y^j [\partial\mathcal{H}/\partial(\partial X^\mu/\partial y^j)] \, dX^\mu \} \, d^3y \tag{A.46}$$

after some partial integrations. (Repeated indices indicate summations. Indices labeled i and j indicate space coordinates. Greek indices include all space-time components of a variable.)

Expressing the differential in H using eq. A.43 we obtain

$$dH = \int dy \, \{ \Pi_\phi \, d\dot{\phi} + \dot{\phi} \, d\Pi_\phi - \partial\mathcal{L}/\partial\phi \, d\phi - \partial\mathcal{L}/\partial(\partial\phi/\partial X^\mu) d(\partial\phi/\partial X^\mu) + $$
$$+ \Pi_X{}^\mu \, d\dot{X}^\mu + \dot{X}^\mu \, d\Pi_X{}^\mu - \partial\mathcal{L}/\partial X^\mu \, dX^\mu - \partial\mathcal{L}/\partial(\partial X^\mu/\partial y^j) d(\partial X^\mu/\partial y^j) \} \tag{A.47a}$$

After some manipulations we find

$$dH = \int \{ \dot{\phi} \, d\Pi_\phi + \dot{X}_\mu d\Pi_X{}^\mu - \partial/\partial y^0 \, \Pi_\phi \, d\phi - \partial/\partial y^0 \, \Pi_X{}^\mu \, dX_\mu \} \, dy \tag{A.47b}$$

using the equations of motion eqs. A.19a and A.19b.

Comparing eqs A.46 and A.47 we obtain Hamilton's equations in the case of the composition of extrema:

$$\dot{\phi} = \partial\mathcal{H}/\partial\Pi_\phi \tag{A.48a}$$

$$\dot{\Pi}_\phi = -\partial\mathcal{H}/\partial\phi + \partial/\partial y^\nu \, [\partial\mathcal{H}/\partial(\partial\phi/\partial X^j) \, \partial y^\nu/\partial X^j] \tag{A.48b}$$

$$\dot{X}_\mu = \partial\mathcal{H}/\partial\Pi_X{}^\mu \tag{A.48c}$$

$$\dot{\Pi}_X{}^\mu = -\partial\mathcal{H}/\partial X^\mu + \partial/\partial y^j \, [\partial\mathcal{H}/\partial(\partial X^\mu/\partial y^j)] \tag{A.48d}$$

where

$$\dot{\Pi}_\phi = \partial \, \Pi_\phi/\partial y^0$$
$$\dot{\Pi}_X{}^\mu = \partial\Pi_X{}^\mu/\partial y^0 \tag{A.49}$$

Translational Invariance

If the Lagrangian of a field theory has no explicit dependence on the coordinates then one expects translational invariance accompanied by a conservation law for an energy-momentum stress tensor. We will show this is the case for Lagrangians implementing the composition of extrema. We assume a Lagrangian without an explicit dependence on the coordinates y^ν:

$$\mathscr{L} = \mathscr{L}(\phi(X), \partial\phi/\partial X^\nu, X^\mu(y), \partial X^\mu(y)/\partial y^\nu) \qquad (A.50)$$

Under an infinitesimal displacement,

$$y'^\nu = y^\nu + \epsilon^\nu \qquad (A.51a)$$

$$\delta\phi = \phi(X(y + \epsilon)) - \phi(X(y))$$

$$= \epsilon^\alpha\, \partial\phi/\partial y^\alpha \qquad (A.51b)$$

$$\delta X^\mu = \epsilon^\alpha\, \partial X^\mu/\partial y^\alpha \qquad (A.51c)$$

$$\delta(\partial\phi/\partial X^\mu) = \epsilon^\alpha\, \partial(\partial\phi/\partial y^\alpha)/\partial X^\mu \qquad (A.51d)$$

$$\delta(\partial X^\mu/\partial y^\nu) = \epsilon^\alpha\, \partial(\partial X^\mu/\partial y^\alpha)/\partial y^\nu \qquad (A.51e)$$

the Lagrangian changes by

$$\delta\mathscr{L} = \epsilon^\alpha\, \partial\mathscr{L}/\partial y^\alpha \qquad (A.52)$$

The change can also be expressed in terms of the changes in the fields, their derivatives and the mapping X^μ:

$$\delta\mathscr{L} = \partial\mathscr{L}/\partial\phi\, \delta\phi + \partial\mathscr{L}/\partial(\partial\phi/\partial X^\mu)\, \delta(\partial\phi/\partial X^\mu) + \partial\mathscr{L}/\partial X^\mu\, \delta X^\mu +$$
$$+ \partial\mathscr{L}/\partial(\partial X^\mu/\partial y^\nu)\, \delta(\partial X^\mu/\partial y^\nu) \qquad (A.53)$$

Combining eqs. A.51, A.52 and A.53 we obtain (after some manipulations):

$$\in^{\nu} \partial/\partial y_\mu \mathscr{T}_{\mu\nu} = 0 \tag{A.54}$$

where

$$\mathscr{T}_{\mu\nu} = -g_{\mu\nu}\mathscr{L} + \partial\mathscr{L}/\partial(\partial\phi/\partial X^\delta)\ \partial y_\mu/\partial X^\delta\ \partial\phi/\partial y^\nu + \partial\mathscr{L}/\partial(\partial X^\delta/\partial y_\mu)\partial X^\delta/\partial y^\nu \tag{A.55a}$$

or, alternately using Lemma 2,

$$\mathscr{T}_{\mu\nu} = -\ g_{\mu\nu}\mathscr{L} + \partial\mathscr{L}/\partial(\partial\phi/\partial y_\mu)\ \partial\phi/\partial y^\nu + \partial\mathscr{L}/\partial(\partial X^\delta/\partial y_\mu)\ \partial X^\delta/\partial y^\nu \tag{A.55b}$$

Since $\in^a$ is an arbitrary displacement we obtain the conservation law:

$$\partial/\partial y_\mu \mathscr{T}_{\mu\nu} = 0 \tag{A.56}$$

Eq. A.56 implies the energy-momentum vector

$$P_\beta = \int d^3y\ \mathscr{T}_{0\beta} \tag{A.57}$$

is conserved. We note

$$\partial/\partial y^0\, P_\beta = 0 \tag{A.58}$$

since eq. A.56 and A.57 can be used to obtain the integral of a divergence, which results in zero.

The hamiltonian (eqs. A.43-44) is

$$H = P_0 \tag{A.59}$$

We note for later use that the total energy, H, which is conserved, contains a term that represents the energy in the X^μ mapping. Thus energy can be exchanged in principle between the ϕ field sector and the X^μ sector.

Lorentz Invariance and Angular Momentum Conservation

We can also verify Lorentz invariance and obtain the form of the conserved angular momentum by considering the effect of an infinitesimal Lorentz transformation. We will consider the case of a scalar field ϕ.

Under an infinitesimal Lorentz transformation ($\in_{\mu\nu} = -\in_{\nu\mu}$):

$$y'_\mu = y_\mu + \delta y_\mu = y_\mu + \epsilon_{\mu\nu} y^\nu \qquad \text{(A.60a)}$$

$$\delta\phi = \phi(X(y')) - \phi(X(y))$$

$$= \epsilon^{\mu\nu}\, y_\nu\, \partial\phi/\partial X^\alpha\, \partial X^\alpha/\partial y^\mu \qquad \text{(A.60b)}$$

$$\delta X^\mu = S^\mu{}_\alpha X^\alpha(y') - X^\mu(y) \qquad \text{(A.60c)}$$

$$= \epsilon^\mu{}_\alpha X^\alpha(y) + \partial X^\mu/\partial y^\beta\, \delta y^\beta \qquad \text{(A.60d)}$$

where $S^\mu{}_\alpha$ is the matrix for the Lorentz transformation of a vector. (If X^μ were a gauge field then an additional operator gauge term would have to be added to eq. A.60d.)

The Lagrangian changes by

$$\delta\mathscr{L} = \epsilon^{\mu\nu}\, y_\nu\, \partial\mathscr{L}/\partial y^\mu \qquad \text{(A.61)}$$

under the infinitesimal Lorentz transformation. The change in the Lagrangian can also be expressed as:

$$\delta\mathscr{L} = \partial\mathscr{L}/\partial\phi\,\delta\phi + \partial\mathscr{L}/\partial(\partial\phi/\partial X^\mu)\,\delta(\partial\phi/\partial X^\mu) + \partial\mathscr{L}/\partial X^\mu\,\delta X^\mu + $$
$$+ \partial\mathscr{L}/\partial(\partial X^\mu/\partial y^\nu)\,\delta(\partial X^\mu/\partial y^\nu) \qquad \text{(A.62)}$$

Combining eqs. A.61 and A.62, and substituting and simplifying terms leads to:

$$\epsilon_{\mu\nu}\,\partial/\partial y^\sigma\,\mathscr{M}^{\sigma\mu\nu} = 0 \qquad \text{(A.63)}$$

where

$$\mathscr{M}^{\sigma\mu\nu} = (g^{\mu\sigma}y^\nu - g^{\nu\sigma}y^\mu)\mathscr{L} + \partial\mathscr{L}/\partial(\partial\phi/\partial X^\alpha)\,\partial y^\sigma/\partial X^\alpha\,(y^\mu\partial\phi/\partial y_\nu - y^\nu\partial\phi/\partial y_\mu) + $$
$$+ \partial\mathscr{L}/\partial(\partial X^\delta/\partial y^\sigma)\,(g^{\delta\nu}X^\mu - g^{\delta\mu}X^\nu + y^\mu\,\partial X^\delta/\partial y_\nu - y^\nu\,\partial X^\delta/\partial y_\mu) \qquad \text{(A.64)}$$

The conserved angular momentum is:

$$M^{\mu\nu} = \int d^3y\,\mathscr{M}^{0\mu\nu} \qquad \text{(A.65)}$$

with

$$\partial M^{\mu\nu}/\partial y^0 = 0 \tag{A.66}$$

The angular momentum density can be written in the familiar form:

$$\mathcal{M}^{\sigma\mu\nu} = y^\mu\ \mathcal{T}^{\sigma\nu} - y^\nu\ \mathcal{T}^{\sigma\mu} + \partial\mathcal{L}/\partial(\partial X^\delta/\partial y^\sigma)\ (g^{\delta\nu}X^\mu - g^{\delta\mu}X^\nu) \tag{A.67}$$

taking account of the vector nature of X^μ. The spatial part of $M^{\mu\nu}$ is the angular momentum.

Internal Symmetries

We will now consider the case of a set of scalar fields ϕ_r in a Lagrangian with an internal symmetry. Under a local transformation

$$\phi_r(X) \rightarrow \phi_r(X) - i\epsilon\lambda_{rs}\ \phi_s(X) \tag{A.68}$$

If the Lagrangian is invariant under this transformation, then

$$\delta\mathcal{L} = 0 = \partial\mathcal{L}/\partial\phi_r\delta\phi_r + \partial\mathcal{L}/\partial(\partial\phi_r/\partial X^\alpha)\ \delta(\partial\phi_r/\partial X^\alpha) \tag{A.69}$$

Using the equation of motion eq. A.19a satisfied by all the components ϕ_r we obtain a conserved current:

$$\mathcal{J}^\nu = -i\ \partial\mathcal{L}/\partial(\partial\phi_r/\partial X^\delta)\ \partial y^\nu/\partial X^\delta\ \lambda_{rs}\ \phi_s \tag{A.70}$$

which satisfies

$$\partial\mathcal{J}^\nu/\partial y^\nu = 0 \tag{A.71}$$

The conserved charge is

$$Q = \int d^3y\,\mathcal{J}^0 \tag{A.72}$$

$$\partial Q/\partial y^0 = 0 \tag{A.73}$$

Separable Lagrangians

We now consider the case of a separable Lagrangian such as in eq. A.26. Adopting the definitions:

$$\phi' = \partial\phi/\partial X^0 \tag{A.74}$$

$$X_\mu' = \partial X_\mu/\partial y^0 \tag{A.75}$$

we define canonical momenta as

$$\pi_\phi = \partial\mathscr{L}/\partial\phi' \equiv \partial\mathscr{L}/\partial(\partial\phi/\partial X^0) \tag{A.76}$$

$$\pi_X{}^\mu = \partial\mathscr{L}/\partial X_\mu' \equiv \partial\mathscr{L}/\partial(\partial X_\mu/\partial y^0) \tag{A.77}$$

We now define the separable hamiltonian density as

$$\mathscr{H}_s = J\pi_\phi\,\phi' + \pi_X{}^\mu\, X_\mu' - \mathscr{L}_s \tag{A.78}$$

where J is the Jacobian (eq. A.21) and

$$H_s = \int \mathscr{H}_s\, d^3y \tag{A.79}$$

The separable Lagrangian (from eq. A.26) is:

$$\mathscr{L}_s = \mathscr{L}_F(\phi(X), \partial\phi/\partial X^\mu)\, J + \mathscr{L}_C(X^\mu(y), \partial X^\mu(y)/\partial y^\nu, y) \tag{A.80}$$

In the case of one scalar field the separable hamiltonian density has the general form

$$\mathscr{H}_s = \mathscr{H}_s(\phi(X), \pi_\phi, \partial\phi/\partial X^i, X^\mu(y), \pi_X{}^\mu, \partial X^\mu(y)/\partial y^j, y^\nu) \tag{A.81}$$

where the indices i and j indicate spatial components. In particular, the terms in the separable hamiltonian are:

$$\mathscr{H}_s = \mathscr{H}_F\, J + \mathscr{H}_C \tag{A.82}$$

with

$$\mathscr{H}_F(\phi(X), \pi_\phi, \partial\phi/\partial X^i) = \pi_\phi\,\phi' - \mathscr{L}_F \tag{A.83}$$

$$\mathscr{H}_C(X^\mu(y), \pi_X{}^\mu, \partial X^\mu(y)/\partial y^j, y^\nu) = \pi_X{}^\mu\, X_\mu' - \mathscr{L}_C \tag{A.84}$$

where J is the absolute value of the Jacobian defined in A.21.

We now define the time integral of H as we did in eq. A.14 when considering the Lagrangian formulation:

$$G = \int dy^0 \, H_s \tag{A.85}$$

Thus G is an integral over all space-time coordinates. Using G we can develop a hamiltonian formulation. First we calculate the differential change in G. Using eqs. A.81-2 and A.85 we obtain

$$\begin{aligned} dG = \int \{ & J\, \partial\mathscr{H}_F / \partial\phi \, d\phi + J\, \partial\mathscr{H}_F / \partial\pi_\phi \, d\pi_\phi + \\ & + J\, \partial\mathscr{H}_F / \partial(\partial\phi/\partial X^j) \, d(\partial\phi/\partial X^j) + \partial\mathscr{H}_C / \partial X^\mu \, dX^\mu + \\ & + \partial\mathscr{H}_C / \partial\pi_X{}^\mu \, d\pi_X{}^\mu + \partial\mathscr{H}_C / \partial(\partial X^\mu/\partial y^j) \, d(\partial X^\mu/\partial y^j) \} \, d^4y \end{aligned} \tag{A.86}$$

with summations implied by repeated indices. (Index labels i and j label spatial coordinates only; Greek indices label space-time coordinates.) Rewriting dG as two integrals and performing partial integrations yields:

$$\begin{aligned} dG = & \int d^4X \{ \partial\mathscr{H}_F / \partial\phi \, d\phi + \partial\mathscr{H}_F / \partial\pi_\phi \, d\pi_\phi - \partial/\partial X^j [\partial\mathscr{H}_F / \partial(\partial\phi/\partial X^j)] \, d\phi \} + \\ & + \int d^4y \{ \partial\mathscr{H}_C / \partial X^\mu \, dX^\mu + \partial\mathscr{H}_C / \partial\pi_X{}^\mu \, d\pi_X{}^\mu - \partial/\partial y^j [\partial\mathscr{H}_C / \partial(\partial X^\mu/\partial y^j)] \, dX^\mu \} \end{aligned} \tag{A.87}$$

Alternately, expressing the differential in G using eqs. A.82-4 we obtain

$$\begin{aligned} dG = & \int d^4X \{ \pi_\phi \, d\phi' + \phi' d\pi_\phi - \partial\mathscr{L}_F / \partial\phi \, d\phi - \partial\mathscr{L}_F / \partial(\partial\phi/\partial X^\mu) d(\partial\phi/\partial X^\mu) \} + \\ & + \int d^4y \{ \pi_{X\mu} \, dX^{\mu\prime} + X^{\mu\prime} \, d\pi_{X\mu} - \partial\mathscr{L}_C / \partial X^\mu \, dX^\mu - \partial\mathscr{L}_C / \partial(\partial X^\mu/\partial y^j) d(\partial X^\mu/\partial y^j) \} \end{aligned} \tag{A.88}$$

which becomes

$$dG = \int d^4X \{ -\pi_\phi{}' \, d\phi + \phi' \, d\pi_\phi \} + \int d^4y \{ -\pi_{X\mu}{}' \, dX^\mu + X^{\mu\prime} \, d\pi_{X\mu} \} \tag{A.89}$$

using the equations of motion eqs. A.31-2.

Comparing eqs A.87 and A.89 we obtain Hamilton's equations for the case of the composition of extrema for a separable Lagrangian:

$$\phi' = \partial \mathscr{H}_F / \partial \pi_\phi \tag{A.90}$$

$$\pi_\phi{}' = - \partial \mathscr{H}_F / \partial \phi + \partial / \partial X^i \, [\partial \mathscr{H}_F / \partial(\partial \phi / \partial X^i)] \tag{A.91}$$

$$X_\mu{}' = \partial \mathscr{H}_C / \partial \pi_X{}^\mu \tag{A.92}$$

$$\pi_{X\mu}{}' = - \partial \mathscr{H}_C / \partial X^\mu + \partial / \partial y^i \, [\partial \mathscr{H}_C / \partial(\partial X^\mu / \partial y^i)] \tag{A.93}$$

where

$$\pi_\phi{}' = \partial \, \pi_\phi / \partial X^0 \tag{A.94}$$

$$\pi_{X\mu}{}' = \partial \, \pi_{X\mu} / \partial X^0 \tag{A.95}$$

Notice that $\mathscr{L}_F$, $\mathscr{H}_F$ and π_ϕ have precisely the same form, as a function of X^μ, as one sees in a conventional field theory formalism. Yet X^μ is a mapping/function of the coordinates y. In reality, it can be viewed as a field as we shall see.

Separable Lagrangians and Translational Invariance

The general rule for conventional Lagrangians is: if a Lagrangian has no explicit dependence on the coordinates then translational invariance follows accompanied by a conservation law for an energy-momentum tensor. We will show that this rule needs modification for separable Lagrangians that implement the composition of extrema.

Consider the Lagrangian:

$$\mathscr{L}_s = J \, \mathscr{L}_F(\phi(X), \partial\phi / \partial X^\mu) + \mathscr{L}_C(X^\mu(y), \partial X^\mu(y) / \partial y^\nu) \tag{A.96}$$

in which the X^μ play a dual role as both fields and coordinates. Let us consider a variation in X^μ:

$$X^\mu(y) \rightarrow X^\mu(y) + \delta X^\mu(y) \tag{A.97}$$

where $\delta X^\mu(y)$ is an arbitrary function of y that vanishes at the endpoints of the integration region of the integral. The action is:

$$I = \int \mathscr{L}_s d^4y \tag{A.98}$$

We will show that a variation in $X^\mu(y)$ leads to a conserved energy-momentum tensor. But we will use integrals of the Lagrangian density since it provides a simpler derivation of the result. Under the variation of eq. A.97 we find

$$\delta\phi = \phi(X(y) + \delta X^\mu(y)) - \phi(X(y))$$

$$= \delta X^\mu \, \partial\phi/\partial X^\mu \tag{A.99a}$$

$$\delta(\partial\phi/\partial X^\nu) = \delta X^\mu \, \partial(\partial\phi/\partial X^\mu)/\partial X^\nu \tag{A.99b}$$

$$\delta(\partial X^\mu/\partial y^\nu) = \partial(\delta X^\mu)/\partial y^\nu \tag{A.99c}$$

The integral in eq. A.98 changes by

$$\delta I = \int d^4y \, \delta\mathscr{L}_s = \int d^4y \, [\, \delta(J\mathscr{L}_F) + \delta\mathscr{L}_C \,] \tag{A.100a}$$

which becomes:

$$\delta I = \int d^4y \, [\delta X^\mu \, \partial(J\mathscr{L}_F)/\partial X^\mu + \partial(\delta X^\mu \partial\mathscr{L}_C \,/\, \partial(\partial X^\mu/\partial y^\nu))/\partial y^\nu] \tag{A.100b}$$

due to the equations of motion of X^μ (eq. A.19b) in X^μ's role. Since the second term is a total divergence its contribution to δI is zero. Thus we can express eq. A.100b as:

$$\delta I = \int d^4y \, [\, J\, \delta\mathscr{L}_F + \mathscr{L}_F \, \delta J \,] \tag{A.101}$$

realizing that the Jacobian J depends on y and thus X:

$$\delta J = \delta X^\mu \, \partial J/\partial X^\mu \tag{A.102}$$

A partial integration gives

$$\mathscr{L}_F \, \delta J = \delta X^\mu \, \partial(J\mathscr{L}_F)/\partial X^\mu - \delta X^\mu J \, \partial\mathscr{L}_F/\partial X^\mu \tag{A.103}$$

Evaluating $\delta\mathscr{L}_F$ we find:

$$\delta \mathscr{L}_F = \partial \mathscr{L}_F / \partial \phi \, \delta \phi + \partial \mathscr{L}_F / \partial(\partial \phi / \partial X^\mu) \, \delta(\partial \phi / \partial X^\mu) \quad \text{(A.104)}$$

which gives

$$\delta \mathscr{L}_F = \delta X^\nu \, \partial / \partial X^\mu \, [\partial \mathscr{L}_F / \partial(\partial \phi / \partial X^\mu) \, \partial \phi / \partial X^\nu] \quad \text{(A.105)}$$

using the equations of motion eq. A.31, and using eq. A.99b. Combining eqs. A.100, A.101, A.103 and A.105 we obtain:

$$\int d^4y \, J \, \delta X^\nu \, \partial / \partial X_\mu \, \mathscr{T}_{F\mu\nu} = \int d^4X \, \delta X^\nu \, \partial / \partial X_\mu \, \mathscr{T}_{F\mu\nu} = 0 \quad \text{(A.106)}$$

where

$$\mathscr{T}_{F\mu\nu} = - g_{\mu\nu} \mathscr{L}_F + \partial \mathscr{L}_F / \partial(\partial \phi / \partial X_\mu) \, \partial \phi / \partial X^\nu \quad \text{(A.107)}$$

after some manipulations. Since δX^ν is an arbitrary function of y the differential conservation law follows:

$$\partial / \partial X_\mu \, \mathscr{T}_{F\mu\nu} = 0 \quad \text{(A.108)}$$

Eq. A.108 implies the energy-momentum vector

$$P_{F\beta} = \int d^3X \, \mathscr{T}_{F0\beta} \quad \text{(A.109)}$$

is conserved:

$$\partial / \partial X^0 \, P_{F\beta} = 0 \quad \text{(A.110)}$$

The hamiltonian density (eq. A.83) is

$$\mathscr{H}_F = \mathscr{T}_{F0\beta} \quad \text{(A.111)}$$

Thus the field energy

$$H_F = P_{F0} = \int d^3X \, \mathscr{T}_{F00} \quad \text{(A.112)}$$

is conserved with respect to the "time" X^0. Later we will see that H_F is trivially conserved in the Coulomb gauge of X_μ. (We will also establish an electromagnetic-like quantum field theory for X_μ with gauge invariance.) In other gauges the conservation of H_F is not trivial.

Separable Lagrangians and Angular Momentum Conservation

We can also verify Lorentz invariance and obtain the form of the conserved angular momentum for a separable Lagrangian by considering the effect of an infinitesimal Lorentz transformation. We will consider the case of a scalar field ϕ.

Under an infinitesimal Lorentz transformation as specified by eqs. A.60a – A.60d the separable Lagrangian changes by

$$\delta \mathscr{L}_s = \epsilon^{\mu\nu}\, y_\nu\, \partial \mathscr{L}_s / \partial y^\mu \tag{A.113}$$

which can also be expressed as

$$\delta \mathscr{L}_s = \partial \mathscr{L}_s / \partial \phi\, \delta\phi + \partial \mathscr{L}_s / \partial(\partial\phi/\partial X^\mu)\, \delta(\partial\phi/\partial X^\mu) + \partial \mathscr{L}_s / \partial X^\mu\, \delta X^\mu + + [\partial \mathscr{L}_s / \partial(\partial X^\mu/\partial y^\nu)]\, \delta(\partial X^\mu/\partial y^\nu) \tag{A.114}$$

Combining eqs. A.113 and A.114 leads to:

$$\epsilon_{\mu\nu}\, \partial/\partial y^\sigma \mathscr{M}_s^{\ \sigma\mu\nu} = 0 \tag{A.115}$$

where

$$\mathscr{M}_s^{\ \sigma\mu\nu} = J\, \mathscr{M}_F^{\ \sigma\mu\nu} + \mathscr{M}_C^{\ \sigma\mu\nu} + \mathscr{M}_M^{\ \sigma\mu\nu} \tag{A.116}$$

$$\mathscr{M}_F^{\ \sigma\mu\nu} = (g^{\mu\sigma}y^\nu - g^{\nu\sigma}y^\mu)\mathscr{L}_F + \partial\mathscr{L}_F / \partial(\partial\phi/\partial y_\sigma)\, (y^\mu \partial\phi/\partial y_\nu - y^\nu \partial\phi/\partial y_\mu) \tag{A.117}$$

$$\mathscr{M}_C^{\ \sigma\mu\nu} = (g^{\mu\sigma}y^\nu - g^{\nu\sigma}y^\mu)\mathscr{L}_C + + \partial\mathscr{L}_C / \partial(\partial X^\delta/\partial y^\sigma)(g^{\delta\nu}X^\mu - g^{\delta\mu}X^\nu + y^\mu\, \partial X^\delta/\partial y_\nu - y^\nu\, \partial X^\delta/\partial y_\mu) \tag{A.118}$$

$$\mathscr{M}_M^{\ \sigma\mu\nu} = \mathscr{L}_F\, \partial J / \partial(\partial X^\delta/\partial y^\sigma)(g^{\delta\nu}X^\mu - g^{\delta\mu}X^\nu + y^\mu\, \partial X^\delta/\partial y_\nu - y^\nu\, \partial X^\delta/\partial y_\mu) \tag{A.119}$$

where the third term originates in the dependence of J on derivatives of X^μ. Eq. A.117 was obtained in part by using the identity:

$$\partial\mathscr{L}/\partial(\partial\phi/\partial y^{\sigma}) = \partial\mathscr{L}/\partial(\partial\phi/\partial X^{a})\ \partial y^{\sigma}/\partial X^{a} \tag{A.120}$$

where $\mathscr{L}$ and ϕ have the form specified in eq. A.15.

The conserved angular momentum is:

$$M_s^{\mu\nu} = \int dy\ \mathscr{M}_s^{0\mu\nu} \tag{A.121}$$

with

$$\partial M_s^{\mu\nu}/\partial y^0 = 0 \tag{A.122}$$

Angular Momentum and $\mathscr{L}_F$

An alternate conserved angular momentum can be obtained by considering the "field" part of the Lagrangian $\mathscr{L}_F$ under an infinitesimal Lorentz transformation ($\epsilon_{\mu\nu} = -\epsilon_{\nu\mu}$):

$$X'_{\mu} = X_{\mu} + \delta X_{\mu} \tag{A.123a}$$

$$\delta\phi = \phi(X'(y)) - \phi(X(y))$$

$$= \delta X^{\mu}\ \partial\phi/\partial X^{\mu} \tag{A.123b}$$

$$\delta X^{\mu} = S^{\mu}{}_{a}X^{a}(y) - X^{\mu}(y) \tag{A.123c}$$

$$= \epsilon^{\mu}{}_{a}X^{a}(y) \tag{A.123d}$$

where $S^{\mu}{}_{a}$ is the Lorentz transformation matrix for a vector. (If X^{μ} is a gauge field then an additional operator gauge term would have to be added to eq. A.123d.)

The Lagrangian changes by

$$\delta\mathscr{L}_F = \epsilon^{\mu\nu}X_{\nu}\ \partial\mathscr{L}_F/\partial X^{\mu} \tag{A.124}$$

under an infinitesimal Lorentz transformation. The change can also be expressed as:

$$\delta\mathscr{L}_F = \partial\mathscr{L}_F/\partial\phi\ \delta\phi + \partial\mathscr{L}_F/\partial(\partial\phi/\partial X^{\mu})\ \delta(\partial\phi/\partial X^{\mu}) \tag{A.125}$$

Combining eqs. A.124 and A.125 leads to:

$$\in_{\mu\nu}\partial/\partial X^{\sigma}\ \mathscr{M}_{FX}{}^{\sigma\mu\nu} = 0 \qquad \text{(A.126)}$$

where

$$\mathscr{M}_{FX}{}^{\sigma\mu\nu} = (g^{\mu\sigma}X^{\nu} - g^{\nu\sigma}X^{\mu})\mathscr{L}_F + \partial\mathscr{L}_F/\partial(\partial\phi/\partial X^{\sigma})\,(X^{\mu}\partial\phi/\partial X_{\nu} - X^{\nu}\partial\phi/\partial X_{\mu}) \qquad \text{(A.127)}$$

The conserved angular momentum associated with the X coordinates is:

$$M_{FX}{}^{\mu\nu} = \int d^3X\ \mathscr{M}_{FX}{}^{0\mu\nu} \qquad \text{(A.128)}$$

with

$$\partial M_{FX}{}^{\mu\nu}/\partial X^0 = 0 \qquad \text{(A.129)}$$

The angular momentum density can be written in the familiar form:

$$\mathscr{M}_{FX}{}^{\sigma\mu\nu} = X^{\mu}\ \mathscr{T}_F^{\sigma\nu} - X^{\nu}\ \mathscr{T}_F^{\sigma\mu} \qquad \text{(A.130)}$$

using eq. A.107.

Separable Lagrangians and Internal Symmetries

We will now consider the case of a set of scalar fields ϕ_r in a separable Lagrangian with an internal symmetry under a local transformation

$$\phi_r(X) \rightarrow \phi_r(X) - i\in\lambda_{rs}\ \phi_s(X) \qquad \text{(A.131)}$$

If the Lagrangian is invariant under this transformation, then

$$\delta\mathscr{L}_S \equiv \delta\mathscr{L}_F = 0 = \partial\mathscr{L}_F/\partial\phi_r\,\delta\phi_r + \partial\mathscr{L}_F/\partial(\partial\phi_r/\partial X^{\alpha})\ \delta(\partial\phi_r/\partial X^{\alpha}) \qquad \text{(A.132)}$$

Using the equation of motion eq. A.31, which is satisfied by all components ϕ_r, we obtain a conserved current:

$$\mathscr{J}^{\nu} = -i\ \partial\mathscr{L}_F/\partial(\partial\phi_r/\partial X^{\nu})\ \lambda_{rs}\ \phi_s \qquad \text{(A.133)}$$

satisfying

$$\partial \mathscr{J}^{\nu}/\partial X^{\nu} = 0 \tag{A.134}$$

The conserved charge is

$$Q = \int d^3X \, \mathscr{J}^{0} \tag{A.135}$$

$$\partial Q/\partial X^0 = 0 \tag{A.136}$$

We note eq. A.71 provides a corresponding conservation law for the y coordinate system.

Appendix B. Invariance of Two-Tier Quantum Field Theory under Special Relativity

Invariance of Two-Tier QFT under Special Relativistic Transformations

Turning now to the issue of invariance under special relativistic transformations we begin by noting that the transverse gauge of the Y^a field is not manifestly relativistic. In addition Two-Tier Feynman propagators calculated in the transverse gauge are also not manifestly relativistic.

The situation is similar to the case of the electromagnetic field yet differs because the Y^a field plays the role of coordinates in the lagrangian. We will now show that the manifestly Lorentz invariant Two-Tier *Lorentz gauge formulation* is equivalent to the transverse gauge formulation. The classical Lorentz gauge condition:

$$\partial Y^\mu/\partial y^\mu = 0 \tag{B.1}$$

is too stringent to make into an operator relation. We will therefore define the Lorentz gauge formulation of the Y^a quantization by implementing a condition on the space of physical states.

We begin with the covariant equal time commutation relations:

$$[Y^\mu(\mathbf{y}, y^0), Y^\nu(\mathbf{y}', y^0)] = [\pi^\mu(\mathbf{y}, y^0), \pi^\nu(\mathbf{y}', y^0)] = 0 \tag{B.2}$$

$$[\pi^\mu(\mathbf{y}, y^0), Y^\nu(\mathbf{y}', y^0)] = i\eta^{\mu\nu}\delta^3(\mathbf{y} - \mathbf{y}') \tag{B.3}$$

where

$$\pi^\mu = \partial\mathscr{L}_C/\partial Y_\mu{}' \tag{B.4}$$

$$Y^{\mu\prime} = \partial Y^\mu/\partial y^0 \tag{B.5}$$

The Fourier expansion of Y^μ is:

$$Y^\mu(y) = \int d^3k\ N_0(k)[\ a^\mu(k)\ e^{-ik\cdot y} + a^{\mu\dagger}(k)\ e^{ik\cdot y}] \tag{B.6}$$

where

$$N_0(k) = [(2\pi)^3 2\omega_k]^{-1/2} \tag{B.7}$$

and

$$\omega_k = (\mathbf{k}^2)^{1/2} = k^0 \tag{B.8}$$

with $k^\mu k_\mu = 0$.

The commutation relations of the Fourier coefficient operators are:

$$[a^\mu(k), a^{\nu\dagger}(k')] = -\eta^{\mu\nu}\delta^3(\mathbf{k} - \mathbf{k}') \tag{B.9}$$

$$[a^{\mu\dagger}(k), a^{\nu\dagger}(k')] = [a^\mu(k), a^\nu(k')] = 0 \tag{B.10}$$

It will be convenient to divide the Y field into positive and negative frequency parts:

$$Y^{\mu+}(y) = \int d^3k\, N_0(k) a^\mu(k)\, e^{-ik\cdot y} \tag{B.11}$$

and

$$Y^{\mu-}(y) = \int d^3k\, N_0(k)\, a^{\mu\dagger}(k)\, e^{ik\cdot y} \tag{B.12}$$

We define

$$a^4(k) = i\, a^0(k) \tag{B.13}$$

and

$$a^{4\dagger}(k) = i\, a^{0\dagger}(k) \tag{B.14}$$

with the resulting commutation relations:

$$[a^\mu(k), a^{\nu\dagger}(k')] = \delta^{\mu\nu}\delta^3(\mathbf{k} - \mathbf{k}') \tag{B.15}$$

for $\mu,\nu = 1, 2, 3, 4$.

Having redefined the operators we then follow the familiar Gupta-Bleuler procedure and introduce an indefinite Dirac metric η with $\eta^2 = 1$ and $\eta^\dagger = \eta$ that will enable us to avoid negative probabilities when inner products are calculated. In particular we note,

$$\eta a^4(k) = -a^4(k)\eta \tag{B.16}$$

$$\eta a^i(k) = a^i(k)\eta \tag{B.17}$$

for $i = 1, 2, 3$. Thus

$$\eta = (-1)^{n_4} \tag{B.18}$$

where n_4 is the number of time-like Y^4 "particles". Let us now define a particle state Φ_{n_4} with n_4 time-like Y^4 "particles." Then the time-like raising and lowering operators change the number of particles in a state:

$$a^4(k)\ \Phi_{n_4+1} = (n_4 + 1)^{1/2}\ \Phi_{n_4} \qquad \text{(B.19a)}$$

and

$$a^{4\dagger}(k)\ \Phi_{n_4} = (n_4 + 1)^{1/2}\ \Phi_{n_4+1} \qquad \text{(B.19b)}$$

We now chose the coordinate system so the z direction is the direction of "propagation." (A general specification of the coordinate system does not change the result.) Consequently, with the **k** direction fixed, we can write

$$\partial Y^1/\partial y^1 + \partial Y^2/\partial y^2 = 0 \qquad \text{(B.20)}$$

and the Lorentz gauge condition, which becomes a condition on the physical states,[54] denoted Φ_L, can be written as:

$$(a^3(k) + i\ a^4(k))\Phi_L = 0 \qquad \text{(B.21)}$$

for all k.

Any physical state Φ_L can be written as a superposition of states with sharp Y^3 and Y^4 particle numbers $\Phi_{n,m}$:

$$\Phi_{Ln} = \sum^{n} c_m\ \Phi_{n,n-m} \qquad \text{(B.22)}$$

where the number of transverse Y "particles" is fixed. In fact, due to our unitarity requirement the number of transverse Y "particles" is always fixed to zero in physical states.

Defining the Lorentz operator

$$L = \partial Y^\mu/\partial y^\mu = L_+ + L_- \qquad \text{(B.23)}$$

where we separate L into positive and negative frequency parts as in eqs. B.11 and B.12 we see that the Lorentz gauge condition (eq. B.21) becomes

$$L_+\Phi_L = 0 \qquad \text{(B.24)}$$

54 Similar approaches to defining the set of physical states appear in string and superstring theories. See, for example, Bailin and Love (1994) pp. 160 –167 and pp. 186 – 190.

with hermitean conjugate (We reserve the † superscript for later use: $\Phi_L^{\dagger} = \Phi_L^{*}\eta$.)

$$\Phi_L{}^{*}\, L_+{}^{\dagger} = \Phi_L{}^{*}\, \eta L_- = 0 \qquad \text{(B.25)}$$

which is obtained by multiplying the hermitean conjugate of eq. B.24 on the right by η and using $L_+{}^{\dagger} = \eta L_- \eta$ and $\eta^2 = 1$. Therefore the inner product equals zero

$$(\Phi_L{}^{*}\, \eta L \Phi_L) = 0 \qquad \text{(B.26)}$$

as does

$$(\Phi_L{}^{*}\, \eta L^2 \Phi_L) = 0 \qquad \text{(B.27)}$$

which shows the square of the Lorentz condition also vanishes between physical states.

Now, the consideration of the relation between matrix elements in the covariant Lorentz gauge and the transverse gauge requires us to inquire more deeply as to the nature of the physical states defined above. We note that it is easy to show

$$(\Phi_{Ln}{}^{*}\, \eta \Phi_{Lm}) = 0 \qquad \text{(B.28)}$$

if n or m is not equal to zero. In the case n = m = 0 the normalization can be chosen such that

$$(\Phi_{L0}{}^{*}\, \eta \Phi_{L0}) = 1 \qquad \text{(B.29)}$$

Thus the states Φ_{Lm} are zero norm states for $m \neq 0$ and orthogonal to the vacuum state Φ_{L0} which happens to be the vacuum state $|0\rangle$ used in our discussions in the earlier chapters. The state $|0\rangle$ contains no transverse, longitudinal or time-like Y "particles."

The zero norm states containing superpositions of longitudinal and time-like Y "particles" are needed due to their role in supporting the gauge invariance of the theory.

Demonstration that the Expectation Values of Y Field Products in the Lorentz Gauge Equal their Value in the Transverse Gauge

We will now show that

$$(\Phi_L{}^{*}\, \eta B \Phi_L) = (\Phi_{tr}{}^{*}\, \eta B \Phi_{tr}) = (\Phi_{tr}{}^{*}\, B_{tr} \Phi_{tr}) \qquad \text{(B.30)}$$

$$(\Phi_{L0}{}^{*}\, \eta B \Phi_{L0}) = (\Phi_{tr0}{}^{*}\, \eta B \Phi_{tr0}) = (\Phi_{tr0}{}^{*}\, B_{tr} \Phi_{tr0}) \equiv \langle 0|B_{tr}|0\rangle \qquad \text{(B.31)}$$

where $|0>$ and $<0|$ are the vacua used in previous sections, where Φ_{L0} and $\Phi_{tr0} = |0>$ are the Lorentz gauge and transverse gauge vacuum states respectively, and where

$$B = \sum_{\lambda} C(\partial/\partial y_1^{\mu}, \partial/\partial y_2^{\nu}, \ldots, \partial/\partial y_m^{\varrho}) Y_{\lambda_1}(y_1)\ Y_{\lambda_2}(y_2) \ldots Y_{\lambda_n}(y_n) \tag{B.32}$$

and

$$B_{tr} = \sum_{\lambda} C(\partial/\partial y_1^{\mu}, \partial/\partial y_2^{\nu}, \ldots, \partial/\partial y_m^{\varrho}) Y_{\lambda_1}{}^{tr}(y_1)\ Y_{\lambda_2}{}^{tr}(y_2) \ldots Y_{\lambda_n}{}^{tr}(y_n) \tag{B.33}$$

with C a polynomial.

We now write eqs. B.11 and B.12 in the form

$$Y^{\mu+}(y) = \int d^3k\ N_0(k) e^{-ik\cdot y} \{\sum_{\lambda=1}^{2} \varepsilon^{\mu}(k, \lambda)\ a(k,\lambda) + (k^{\mu}/|\mathbf{k}| - \varepsilon_0{}^{\mu}) a_3(k) + $$
$$+ \varepsilon_0{}^{\mu}[a_3(k) - a_0(k)]\} \tag{B.34}$$

and $Y^{\mu-}(y)$ in a corresponding form. Then

$$Y^{\mu+}(y)\Phi_L = \{Y^{\mu+tr}(y) + \partial\Lambda/\partial y_{\mu} + \eta^{\mu 0}\ D^{+}(y)\}\Phi_L \tag{B.35}$$

where $\partial\Lambda/\partial y_{\mu}$ can be eliminated with a gauge transformation and where $D^{+}(y)\Phi_L = 0$. As a result

$$Y^{\mu+}(y)\Phi_L = Y^{\mu+tr}(y)\Phi_L \qquad \text{and} \qquad \Phi_L{}^{\dagger}Y^{\mu-}(y) = \Phi_L{}^{\dagger}Y^{\mu-tr}(y) \tag{B.36}$$

where $\Phi_L{}^{\dagger} = \Phi_L{}^{*}\eta$. Consequently

$$(\Phi_L{}^{\dagger}\ Y^{\mu}(y)\ \Phi_L) = (\Phi_L{}^{\dagger}\ Y^{\mu\ tr}(y)\ \Phi_L) \tag{B.37}$$

for all physical Lorentz gauge states Φ_L. More generally

$$(\Phi_L{}^{\dagger}\ B\ \Phi_L) = (\Phi_{tr}{}^{\dagger}\ B\ \Phi_{tr}) = (\Phi_{tr}{}^{\dagger}\ B_{tr}\ \Phi_{tr}) \equiv <0|B_{tr}|0> \tag{B.38}$$

with B and B_{tr} given by eqs. B.32 and B.33, thus proving eqs. B.30 and B.31. The proof of eq. B.38 is directly based on the representation of $Y^{\mu\pm}(y)$ in eq. B.34, and

$$[D^{+}(k), Y^{\mu\pm}(y)] = 0 \tag{B.39}$$

for all k and y with $D^+(k) = a_3(k) - a_0(k)$, and

$$[D^+(k), D^-(k')] = 0 \tag{B.40}$$

for all k and k' with $D^-(k) = a_3^\dagger(k) - a_0^\dagger(k)$, and

$$D^-(k)\Phi_{L0} = 0 \quad \text{and} \quad \Phi_{L0}^\dagger D^+(k) = 0 \tag{B.41}$$

where Φ_{L0} is the Lorentz gauge vacuum state. As a result of these relations the only surviving terms in B in eqs. B.30 and B.31 are the commutators of $Y^{\mu-tr}$ and $Y^{\nu+tr}$ with the consequence

$$(\Phi_L{}^* \eta B\Phi_L) = (\Phi_{tr}{}^* \eta B\Phi_{tr}) = (\Phi_{tr}{}^* B_{tr}\Phi_{tr}) \tag{B.30}$$

Therefore we have shown the expectation value of a product of Y fields in the Lorentz covariant Lorentz gauge is equal to the expectation value of the corresponding product of transverse gauge Y fields, thus establishing the Lorentz invariance of Two-Tier quantum field theories.

Lorentz Invariant S Matrix Elements

Having established the Lorentz invariance of Two-Tier quantum field theories we now describe a calculational procedure that transforms seemingly non-covariant amplitudes into manifestly covariant amplitudes.

Consider an interaction with a certain number of incoming particles and (possibly) a different number of outgoing particles. Assume a calculation of an S matrix element is performed according to the rules in Blaha (2003), and the present work, in the Y transverse gauge and in the center of mass of the incoming particles. The total momentum in the center of mass is $P^\mu = (P^0, \mathbf{0})$. The S matrix amplitude thus calculated will have a non-manifestly covariant form in the center of mass frame:

$$S_{ab} = S_{ab}(p^0{}_1, p^0{}_2, \ldots, p^0{}_n, |\mathbf{p}_1|, |\mathbf{p}_2|, \ldots |\mathbf{p}_n|, \mathbf{p}_1\cdot\mathbf{p}_2, \ldots \mathbf{p}_i\cdot\mathbf{p}_j, \ldots) \tag{B.42}$$

It can easily be rewritten in covariant form using the total momentum P^μ:

$$p^0{}_i = P^\mu p_{i\mu} / (P^\nu P_\nu)^{1/2} \tag{B.43}$$

$$|\mathbf{p}_i| = [\,(P^\mu p_{i\mu})^2/(P^\nu P_\nu) - p_i^\mu p_{i\mu}\,]^{1/2} \tag{B.44}$$

$$\mathbf{p}_i \cdot \mathbf{p}_j = (P^\mu p_{i\mu} P^\alpha p_{j\alpha})/(P^\nu P_\nu) - p_i^\mu p_{j\mu} \tag{B.45}$$

for i, j = 1, 2, …, n. After substitutions are made we obtain a completely covariant form:

$$S_{ab} = S_{ab}(P^\nu P_\nu, P^\alpha p_{1\alpha}, P^\beta p_{2\beta}, \ldots, P^\kappa p_{n\kappa}, p_1^\rho p_{2\rho}, \ldots, p_i^\sigma p_{j\sigma}, \ldots) \tag{B.46}$$

Lastly we note that a gauge transformation can always be made after a Lorentz transformation in the y coordinates that restores the transverse gauge of the Y field.

Dirac Metric Operator

The metric operator η introduced earlier

$$\eta = (-1)^{n_4} \tag{B.47}$$

can be combined with the metric operator V (eq. 5.15) introduced in earlier chapters for the transverse gauge

$$V = \exp[-i\pi \sum_{\lambda=1}^{2} \int d^3k\, a^\dagger(k, \lambda)a(k, \lambda)] \tag{B.48}$$

with the property

$$V\, Y^j(y)\, V^{-1} = -Y^j(y) \tag{B.49}$$

for j = 1,2,3.

The result is the invariant metric operator for the Lorentz gauge:

$$V_L = \exp[-i\pi \int d^3k\, a^{\mu\dagger}(k)a_\mu(k)] \tag{B.50}$$

which could be used in inner products such as

$$(\Phi_{L0}{}^* V_L B\Phi_{L0}) = (\Phi_{tr0}{}^* V_L B\Phi_{tr0}) = (\Phi_{tr0}{}^* VB_{tr}\Phi_{tr0}) \equiv \langle 0|VB_{tr}|0\rangle \tag{B.51}$$

that occur in the evaluation of S matrix elements.

We note the following properties of V_L:

$$V_L^\dagger = V_L^{-1} = V_L \tag{B.52}$$

$$V_L^2 = I \tag{B.53}$$

REFERENCES

Akhiezer, N. I., Frink, A. H. (tr), 1962, *The Calculus of Variations* (Blaisdell Publishing, New York, 1962).

Bailin, D. & Love, A., 1994, *Supersymmetric Gauge Field Theory and String Theory* (Institute of Physics Publishing, Philadelphia, PA, 1994).

Bergmann, P. G., 1942, *Introduction to the Theory of General Relativity* (Prentice-Hall, Englewood Cliffs, NJ, 1942).

Bjorken, J. D., Drell, S. D., 1965, *Relativistic Quantum Fields* (McGraw-Hill, New York, 1965).

Blaha, S., 2002, *Cosmos and Consciousness Second Edition* (Pingree-Hill Publishing, Auburn, NH, 2002).

Blaha, S., 2003, *A Finite Unified Quantum Field Theory of the Elementary Particle Standard Model and Quantum Gravity Based on New Quantum Dimensions™ and a New Paradigm in the Calculus of Variations* (Pingree-Hill Publishing, Auburn, NH, 2003).

Blaha, S., 2004, *Quantum Big Bang Cosmology: Complex Space-time General Relativity, Quantum Coordinates, Dodecahedral Universe, Inflation, and New Spin 0, ½, 1 & 2 Tachyons & Imagyons* (Pingree-Hill Publishing, Auburn, NH, 2004).

Blaha, S., 2005, *The Equivalence of Elementary Particle Theories and Computer Languages: Quantum Computers, Turing Machines, Standard Model, Superstring Theory, and a Proof that Gödel's Theorem Implies Nature Must Be Quantum* (Pingree-Hill Publishing, Auburn, NH, 2005).

Bogoliubov N. N., & Shirkov, D. V., Volkoff, G. M. (tr), 1959, *Introduction to the Theory of Quantized Fields* (Wiley-Interscience, New York, 1959).

Buck, R. C., (1956), *Advanced Calculus* (McGraw-Hill Publishing, New York, 1956).

Cottingham, W. N. and Greenwood, D. A., 1998, *An Introduction to the Standard Model of Particle Physics* (Cambridge University press, Cambridge, UK, 1998).

Dicke, R. H., 1970, *Gravitation and the Universe* (American Philosophical Society, Philadelphia, 1970).

Dodelson, S., 2003, *Modern Cosmology* (Academic Press, London, 2003).

Donoghue, J. F., Golowich, E. and Holstein, B. R., 1992, *Dynamics of the Standard Model* (Cambridge University Press, Cambridge, UK, 1992).

Eddington, A. S., 1924, *The Mathematical Theory of Relativity* (Cambridge University Press, Cambridge, UK, 1924).

Eddington, A. S., 1995, *Space, Time & Gravitation* (Cambridge University Press, Cambridge, UK, 1995).

Gelfand, I. M., Fomin, S. V., Silverman, R. A. (tr), 2000, *Calculus of Variations* (Dover Publications, Mineola, NY, 2000).

Giaquinta, M., Modica, G., Souchek, J., 1998, *Cartesian Coordinates in the Calculus of Variations* Volumes I and II (Springer-Verlag, New York, 1998).

Giaquinta, M., Hildebrandt, S., 1996, *Calculus of Variations* Volumes I and II (Springer-Verlag, New York, 1996).

Goertzel, G., Tralli, N., 1960, *Some Mathematical Methods of Physics* (McGraw-Hill, New York, 1960).

Goldstein H., 1965, *Classical Mechanics* (Addison-Wesley, Reading, MA 1965).

Guth, A. H., 1997, *The Inflationary Universe* (Perseus Books, Cambridge, MA, 1997).

Hamermesh, M., 1962, *Group Theory* (Addison-Wesley Publishing, New York, 1962).

Heitler, W., 1954, *The Quantum Theory of Radiation* (Oxford University Press, London, 1954).

Huang, K., 1992, *Quarks, Leptons & Gauge Fields Second Edition* (World Scientific, River Edge, NJ, 1992).

Huang, K., 1998, *Quantum Field Theory* (John Wiley, New York, 1998).

Hübsch, T., 1994, *Calabi-Yau Manifolds* (World Scientific, London, 1994).

Jost, J., Li-Jost, X., 1998, *Calculus of Variations* (Cambridge University Press, New York, 1998).

Kaku, M., 1999, *Introduction to Superstrings and M-Theory Second Edition* (Springer-Verlag, New York, 1999).

Kaku, M., 1993, *Quantum Field Theory* (Oxford University Press, New York, 1993).

Kreyszig, E., 1991, *Differential Geometry* (Dover Publications, New York, 1991).

Landau, L. D. and Lifshitz, E. M., 1962, *The Classical Theory of Fields* (Addison-Wesley, New York, 1962).

Lang, S., 1987, *Introduction to Complex Hyperbolic Spaces* (Springer-Verlag, New York, 1987).

Mach, E., 1991, *Die Mechanik in ihrer Entwicklung* (Wissenschaftliche Buchgesellschaft, Berlin, 1991).

Magnus, W. and Oberhettinger, F., 1949, *Formulas and Theorems for the Special Functions of Mathematical Physics* (Chelsea Publishing Co., New York, 1949).

Misner, C. W., Thorne, K. S., Wheeler, J. A., 1973, *Gravitation* (W. H. Freeman, New York, 1973).

Pauli, W., 1958, *The Theory of Relativity* (Pergamon Press, London, 1958).

Polchinski, J., 1998, *String Theory* (Cambridge University Press, New York, 1998).

Quigg, C., 1997, *Gauge Theories of the Strong, Weak and Electromagnetic Interactions* (Westwood Press, 1997).

Rubin, V. C. and Coyne, G. V., *Large-Scale Motions in the Universe* (Princeton University Press, Princeton, 1988).

Sagan, H., 1993, *Introduction to the Calculus of Variations* (Dover Publications, Mineola, NY, 1993).

Schutz, B. F., 2002, *A First Course in General Relativity* (Cambridge University Press, Cambridge, UK, 2002).

Schwerdtfeger, H., 1979, *Geometry of Complex Numbers: Circle Geometry, Moebius transformations, Non-Euclidean Geometry* (Dover Publications, New York, 1979).

Smolin, L., 2001, *Three Roads to Quantum Gravity* (Basic Books, New York, 2001).

Streater, R. F. and Wightman, 2000, A. S., *PCT, Spin and Statistics, and All That* (Princeton University Press, Princeton, NJ, 2000).

Synge, J. L., 1960, *Relativity, The General Theory* (North-Holland, Amsterdam, 1960).

Weinberg, S., 1995, *The Quantum Theory of Fields Volume I* (Cambridge University Press, New York, 1995).

Weinberg, S., 1996, *The Quantum Theory of Fields Volume II* (Cambridge University Press, New York, 1996).

Weyl, H., 1950, *Space, Time, Matter* (Dover, New York, 1950).

INDEX - ... THIRD KIND

ABOUT THE AUTHOR

Stephen Blaha is an internationally known physicist with extensive interests in Science, the Arts, and Technology. He received his Ph.D. in Theoretical Physics from Rockefeller University (NY). He has written a highly regarded book on physics, consciousness and philosophy – *Cosmos and Consciousness*, a book on Science and Religion entitled *The Reluctant Prophets*, a book applying physics concepts to the history of civilizations, and books on Java and C++ programming. He developed a mathematical theory of civilizations that is described in *The Life Cycle of Civilizations*. Recently he completed a major new study of Cosmology: *Quantum Big Bang Cosmology: Complex Space-time General Relativity, Quantum Coordinates, Dodecahedral Universe, Inflation, and New Spin 0, ½, 1 & 2 Tachyons & Imagyons*. He has served on the faculties of several major universities. He was an Associate of the Harvard Physics Faculty for twenty years (1983-2003). He was also a Member of the Technical Staff at Bell Laboratories, a member of management at the Boston Globe Newspaper, a Director at Wang Laboratories, President of Blaha Software Inc and Janus Associates Inc. (NH), and 2008 Program Chair of the International Society for the Comparative Study of Civilizations (ISCSC) as well as publisher of its journal The Comparative Civilizations Review, and Conference Site Selection Chair.

Among other achievements he was a co-discoverer of the "r potential" for heavy quark binding developing the first (and still the only demonstrable) non-abelian gauge theory with an "r" potential; first suggested the existence of topological structures in superfluid He-3; first proposed Yang-Mills theories would appear in condensed matter phenomena with non-scalar order parameters; first developed a grammar-based formalism for quantum computers and applied it to elementary particle theories; first developed a new form of quantum field theory without divergences (thus solving a major 60 year old problem that enabled a unified theory of the Standard Model and Quantum Gravity without divergences to be developed); first developed a formulation of complex General Relativity based on analytic continuation from real space-time; first developed a generalized non-homogeneous Robertson-Walker metric that enabled a quantum theory of the Big Bang to be developed without singularities at $t = 0$; first generalized Cauchy's theorem and Gauss' theorem to complex curved multi-dimensional spaces; first developed a physically acceptable theory of faster-than-light particles – tachyons – of any spin; first showed a universe with three complex spatial dimensions has an icosahedral symmetry; first developed the form of the composition of extrema in the Calculus of Variations; first quantitatively suggested that inflationary periods in the history of the universe were not needed; first proved Gödel's Theorem implies Nature must be quantum, first derived the form of the Standard Model, first showed how to resolve logical paradoxes including Gödel's Undecidability Theorem by developing Operator Logic and Quantum operator Logic, first developed a quantitative harmonic oscillator-like model of the life cycle, and interactions, of civilizations, and first developed an axiomatic derivation of the forms of The Standard Model with WIMPs from geometry – space-time properties.

Blaha was also a pioneer in the development of UNIX for financial and scientific applications, in financial modelling software, in database benchmarking, in networking (1982), in the development of Desktop Publishing (1980's), and in the development a hybrid shell programming technique (1982) that was a precursor to the PERL programming language. He received Honorable Mention in the Gravity Research Foundation Essay Competition in 1978, and was nominated for three "Awards for Technical Excellence" in 1987 by PC Magazine for PC software products that he designed and developed.

CPSIA information can be obtained at www.ICGtesting.com
Printed in the USA
269387BV00003B/1/P

9 780984 553037